Applied Statistics (Continued)

HALD · Statistical Theory with Engineering Applications
HANSEN, HURWITZ, and MADOW · Sample Survey Methods and Theory, Volume I
HOEL · Elementary Statistics
KEMPTHORNE · An Introduction to Genetic Statistics
MEYER · Symposium on Monte Carlo Methods
MUDGETT · Index Numbers
RICE · Control Charts
ROMIG · 50–100 Binomial Tables
SARHAN and GREENBERG · Contributions to Order Statistics
TIPPETT · Technological Applications of Statistics
WILLIAMS · Regression Analysis
WOLD and JURÉEN · Demand Analysis
YOUDEN · Statistical Methods for Chemists

Books of Related Interest

ALLEN and ELY · International Trade Statistics
ARLEY and BUCH · Introduction to the Theory of Probability and Statistics
CHERNOFF and MOSES · Elementary Decision Theory
HAUSER and LEONARD · Government Statistics for Business Use, *Second Edition*
STEPHAN and McCARTHY · Sampling Opinions—An Analysis of Survey Procedures

Mathematical Statistics

A WILEY PUBLICATION IN MATHEMATICAL STATISTICS

Mathematical Statistics

SAMUEL S. WILKS
Professor of Mathematical Statistics
Princeton University

John Wiley & Sons, Inc.
New York · London

Second Printing with Corrections, 1963

LIBRARY OF CONGRESS CATALOG CARD NUMBER: 61-17365

PRINTED IN THE UNITED STATES OF AMERICA

to Gena

Preface

The purpose of this book is to introduce mathematical statistics to readers with good undergraduate backgrounds in mathematics. No previous knowledge of probability or statistics is assumed on the part of the reader, although having had one or more good undergraduate courses in these subjects certainly will be found useful. The book has been prepared mainly from material which has been successively revised and presented to graduate students at Princeton University since World War II. An early version of some of the material was issued in 1943 in lithoprinted form by the Princeton University Press under the title: *Mathematical Statistics*.

The field of mathematical statistics and its applications has been growing at a spectacular rate for more than a quarter of a century—a rate which now results in a flow of new material with which no single individual can keep pace. Although most of the research results in the field during this period have appeared in some half dozen specialized journals, substantial numbers of important papers have been published and continue to be published in many scattered scientific journals.

No attempt has been made here to write a comprehensive treatment of the main results in this body of literature. Instead, I have made a selection of basic material in mathematical statistics in accordance with my own preferences and prejudices, with inclinations toward trying to make a unified and systematic presentation of classical results of mathematical statistics, together with some of the more important contemporary results, in a framework of modern probability theory, without going into too many ramifications. It is therefore inevitable that some topics will be considered to have been slighted or inexcusably omitted, and understandably so, by enthusiasts and specialists on those topics. On the other hand, the reader who may become interested in topics lightly covered or barely mentioned will find references for further reading. Indeed, it is my hope that any mathematically qualified reader without previous knowledge of mathematical statistics who studies this book and becomes interested in further systematic study of the subject, or parts of it, will be able to continue such study in the literature with the guidance, initially at least, of the references cited throughout the book and listed at the end.

More than four hundred problems are included at the ends of the various chapters of the book to provide the reader with opportunities to increase his understanding and facility with the substance of mathematical statistics. Many of these problems also serve as vehicles for introducing additional topics and interesting results in brief form which could not be discussed in detail in the book on account of space limitations.

Some readers will wonder why more discussion was not interwoven between the mathematical results in the book and the statistical methodology which rests upon these results. Discussion of this kind has been kept deliberately at a minimal level. Consideration of statistical methodology and its applications is no less important than the treatment of the underlying mathematical theory of this methodology. Experience indicates, however, that both aspects of statistics, that is, the mathematical theory and the statistical methodology based on this theory, are most effectively combined in research papers, monographs, and books restricted to specific topics. In a fairly comprehensive book on mathematical statistics such as this it is my conviction that it would be most unwise to attempt to deal with both aspects of each topic with equal emphasis. A careful presentation of basic mathematical statistics and the underlying mathematical theory of a wide variety of topics in statistics, with just enough discussion and examples to clarify the basic concepts, such as that attempted here, is a much more feasible undertaking. I believe this approach to be prerequisite to a fuller understanding of statistical methodology, not to mention the one I find most satisfying.

Modern mathematical statistics depends heavily on the theory of probability. An attempt has been made therefore to set this entire treatment onto an adequate foundation in modern probability theory without actually constructing the foundation. That would be a task beyond the scope of this book and also unnecessary, since several excellent books already exist on the subject which are referred to at appropriate points throughout the book in case the reader may wish to do some further reading in probability theory.

In a book of this kind which covers such a span of topics, the problem of devising uniform notation and terminology throughout the book has not been easy. Some of the notation and terms will look unfamiliar to mathematical statisticians, but this is due partly to the price which has to be paid for consistency and partly to the sheer need for introducing new terms and notation in connection with topics which have never been treated systematically in the literature of mathematical statistics.

First drafts of most parts of this book concerning order statistics and nonparametric statistical inference were written when I was a Fulbright research scholar at Cambridge University in the spring and summer of 1951, at which time I planned to write a monograph on that subject. The opportunity for research and writing provided by that appointment is gratefully acknowledged. It was later decided, however, to incorporate that material into the present more comprehensive book on mathematical statistics. This more ambitious project could not have been undertaken without support such as that which has been provided by the Office of Naval Research. I take this occasion to express my deep appreciation for this support.

I have had the benefit of valuable advice and criticism of colleagues, former research associates, and graduate students at Princeton throughout the preparation of this book. A wide range of comments and advice generously given by J. W. Tukey and discussions with F. J. Anscombe have been especially useful. Criticism and suggestions by V. S. Varadarajan concerning some basic points in probability theory were very helpful. The author is indebted to D. M. Brown, D. A. Freedman, I. Guttman, A. T. James, and F. M. Sand for reading major portions of the manuscript and suggesting improvements. Thanks also go to D. R. Brillinger, who worked through more than three hundred and fifty problems contained in the manuscript before the final fifty-odd problems were added, and to J. A. Hartigan who read proof and weeded out numerous errors and mistakes and otherwise improved the final product. The advice of these colleagues and associates and of many other friends was always highly valued even though some of it was not accepted. I alone am responsible for errors and inaccuracies which remain and I shall appreciate having them called to my attention by readers who discover them. Finally, I express my appreciation to Mrs. Rebecca Werkman and to Mrs. Emily Sorenson for their diligence in typing and for their patience in photostating the manuscript.

SAMUEL S. WILKS

November, 1961

Remarks Concerning Second Printing

An attempt has been made in this second printing to correct various errors and inaccuracies in the first printing which have been called to my attention by several readers. I would like to especially acknowledge my thanks to D. R. Cox, P. C. Fishburn, I. Guttman, E. J. Hannan, W. Hoeffding, S. Kullback, M. Kupperman and G. P. Patil for the errors they pointed out.

SAMUEL S. WILKS

April, 1963

Contents

CHAPTER		PAGE
1	PRELIMINARIES	1
	1.1 Sample spaces and events	1
	1.2 Definitions and rules for combining and decomposing events	2
	1.3 Fields of sets	8
	1.4 Probability measure	10
	1.5 Extension of a probability measure	15
	1.6 Statistical independence	16
	1.7 Random variables	19
	1.8 Integration of random variables	21
	1.9 Conditional probability	24
	1.10 Conditional random variables	25
	Problems	26
2	DISTRIBUTION FUNCTIONS	30
	2.1 Preliminary remarks	30
	2.2 Distribution functions of one-dimensional random variables	31
	2.3 Common types of one-dimensional random variables	34
	2.4 Distribution functions of two-dimensional random variables	39
	2.5 Common types of two-dimensional random variables	43
	2.6 Distribution functions of k-dimensional random variables	49
	2.7 Common types of k-dimensional random variables	51
	2.8 Functions of random variables	53
	2.9 Conditional distribution functions	59
	2.10 Finite stochastic processes	68
	Problems	69

3 MEAN VALUES AND MOMENTS OF RANDOM VARIABLES 72

3.1 Introduction 72
3.2 Mean value of a random variable 73
3.3 Moments of one-dimensional random variables 74
3.4 Moments of two-dimensional random variables 77
3.5 Moments of k-dimensional random variables 79
3.6 Means, variances, and covariances of linear functions of random variables 82
3.7 Mean values of conditional random variables 83
3.8 Least squares linear regression 87
Problems 92

4 SEQUENCES OF RANDOM VARIABLES 96

4.1 Definition of a stochastic process 96
4.2 Probability measure for a stochastic process 96
4.3 Convergence in probability 99
4.4 Almost certain convergence 106
4.5 Kolmogorov's inequality 107
4.6 The strong law of large numbers 108
Problems 110

5 CHARACTERISTIC FUNCTIONS AND GENERATING FUNCTIONS 113

5.1 Case of a one-dimensional random variable 113
5.2 Case of a k-dimensional random variable 119
5.3 Characteristic functions of independent random variables . . 120
5.4 Characteristic functions of a sequence of random variables . 122
5.5 Determination of distribution functions from moments . . 125
Problems 129

6 SOME SPECIAL DISCRETE DISTRIBUTIONS 133

6.1 The hypergeometric distribution 133
6.2 The binomial distribution 136
6.3 The multinomial distribution 138
6.4 The Poisson distribution 140
6.5 Discrete waiting-time distributions 141
6.6 Distributions in the theory of runs 144
Problems 150

7 SOME SPECIAL CONTINUOUS DISTRIBUTIONS 155

7.1 The rectangular distribution 155
7.2 The normal distribution 156
7.3 The bivariate normal distribution 158
7.4 The k-variate normal distribution 163
7.5 The gamma distribution 170
7.6 The beta distribution 173
7.7 The Dirichlet distribution 177
7.8 Distributions involved in the analysis of variance 183
Problems 187

8 SAMPLING THEORY 195

8.1 Definition of a random sample 195
8.2 Means and variances of mean, variance, and other symmetric functions of a sample 198
8.3 Sampling theory of sample sums and means 203
8.4 Sampling theory of certain quadratic forms in samples from a normal distribution 208
8.5 Sampling from a finite population. 214
8.6 Matrix sampling 222
8.7 Sampling theory of order statistics 234
8.8 Order statistics in samples from finite populations 243
Problems 245

9 ASYMPTOTIC SAMPLING THEORY FOR LARGE SAMPLES 254

9.1 Convergence of sample mean in probability 254
9.2 Limiting distribution of sample sums and means 256
9.3 Asymptotic distribution of functions of sample means . . . 259
9.4 Asymptotic expansion of distribution of sample sum . . . 262
9.5 Limiting distributions of linear functions in large samples from large finite populations 266
9.6 Asymptotic distributions concerning order statistics . . . 268
Problems 274

10 LINEAR STATISTICAL ESTIMATION 277

10.1 Introductory comments 277
10.2 Minimum variance estimators for the mean and variance of a population from random samples 279
10.3 Estimators for parameters in linear regression analysis . . . 283

10.4 Interval and ellipsoidal estimators for the parameters in normal regression theory 289
10.5 Simultaneous confidence intervals: multiple comparisons . . 290
10.6 Normal linear regression analysis in experimental designs . . 297
10.7 Estimation of variance components from linear combinations of random variables 305
10.8 Estimators for variance components in experimental designs . 308
10.9 Linear estimators for means of stratified populations . . . 313
10.10 Linear estimator for mean of stratified populations in two-stage sampling 318
Problems 323

11 NONPARAMETRIC STATISTICAL ESTIMATION 329

11.1 Introductory remarks 329
11.2 Confidence intervals for quantiles 329
11.3 Confidence intervals for quantile intervals 332
11.4 Confidence intervals for quantiles in finite populations . . 333
11.5 Tolerance limits 334
11.6 One-sided confidence contours for a continuous distribution function 336
11.7 Confidence bands for a continuous distribution function . . 339
Problems 342

12 PARAMETRIC STATISTICAL ESTIMATION 344

12.1 Differentiation of parametric distribution functions . . . 345
12.2 Point estimation 350
12.3 Point estimation from large samples 358
12.4 Interval estimation 365
12.5 Interval estimation from large samples 371
12.6 Multidimensional point estimation 376
12.7 Multidimensional point estimation from large samples . . . 379
12.8 Multidimensional confidence regions 381
12.9 Asymptotically smallest confidence regions from large samples 384
Problems 389

13 TESTING PARAMETRIC STATISTICAL HYPOTHESES 394

13.1 Introductory remarks and definitions 394
13.2 Test of a simple hypothesis. 398
13.3 The likelihood ratio test 402
13.4 Asymptotic distribution of likelihood ratio in large samples . 408

13.5 Consistency of likelihood ratio test 411
13.6 Asymptotic power of likelihood ratio test 413
13.7 The likelihood ratio test of a simple hypothesis 417
13.8 The likelihood ratio test of a composite hypothesis 419
Problems 422

14 TESTING NONPARAMETRIC STATISTICAL HYPOTHESES 428

14.1 The quantile test 428
14.2 The nonparametric simple statistical hypothesis 430
14.3 The problem of two samples from continuous distributions . 441
14.4 The method of randomization 462
Problems 468

15 SEQUENTIAL STATISTICAL ANALYSIS 472

15.1 Introductory remarks 472
15.2 The basic structure of a sequential test 474
15.3 Cartesian sequential tests 479
15.4 The probability ratio sequential test 482
15.5 Application of probability ratio sequential test to binomial distribution 494
15.6 Sequential estimation 496
Problems 498

16 STATISTICAL DECISION FUNCTIONS 502

16.1 General remarks 502
16.2 Definitions and terminology 502
16.3 Minimax solution of the decision problem 504
16.4 Bayes solutions of the statistical decision problem 508
16.5 Remarks on extensions and generalizations 511
Problems 512

17 TIME SERIES 514

17.1 Introductory remarks 514
17.2 Stationary time series 515
17.3 The spectral function of a stationary time series. 517
17.4 Estimation of mean and covariance function of a stationary time series 522
17.5 Estimation of spectral distribution 523

17.6 Statistical tests for parametric time series 526
17.7 Testing a normal noise for whiteness 533
17.8 Linear prediction in time series 535
Problems 537

18 MULTIVARIATE STATISTICAL THEORY 540

18.1 Multidimensional statistical scatter 540
18.2 The Wishart distribution 547
18.3 Independence of means and internal scatter matrix in samples from k-dimensional normal distributions 555
18.4 Hotelling's generalized Student distribution 556
18.5 The multidimensional Model I analysis of variance test . . 561
18.6 Principal components 564
18.7 Discriminant analysis 573
18.8 Distribution of eigenvalues in discriminant analysis . . . 581
18.9 Canonical correlation 587
Problems 592

REFERENCES AND AUTHOR INDEX 603

SUBJECT INDEX 623

CHAPTER 1

Preliminaries

1.1 SAMPLE SPACES AND EVENTS

Mathematical statistics is founded on the theory of probability, which, in turn, depends on the theory of measure and integration for its precise description and treatment. The measure-theoretic description of probability, which will be used in this book, was formulated by Kolmogorov (1933a). It will be sufficient for our purposes to present here an introduction covering only the basic definitions, concepts, and machinery of probability theory in a form useful for mathematical statistics. The reader interested in a more detailed and comprehensive account of the theory of probability should refer to books by Doob (1953), Gnedenko and Kolmogorov (1954), Feller (1957), Kolmogorov (1933a), Lévy (1925, 1937), and Loève (1955). For similar information concerning measure theory he is referred to Halmos (1950) and Munroe (1953).

In presenting the essentials of probability theory needed in this book, we must first deal with the notions of sample space and event, description of events, and combination and decomposition of events. The theory of sets furnishes the machinery for handling these concepts.

We shall denote by R a set of elements e which will be called *sample points* or *event points* or more briefly *points*. The number of sample points may be finite or infinite. R is called the *sample space* or *outcome space*. We shall use the former of these two terms. A sample point may be thought of as a possible outcome of a *trial*, *experiment*, or *operation*, performed under a given *set of conditions*, although we shall make no attempt to define these terms formally. The sample space R is simply the set of *all possible outcomes* which could be realized when an operation is performed under the given set of conditions.

Examples. Some illustrative examples may help fix the ideas. In tossing a coin once where e represents an arbitrary face turning up, the sample space R

consists of two sample points: *Head*, *Tail*. In tossing a coin twice where e represents an arbitrary combination of faces turning up, the sample space R consists of four sample points which may be abbreviated as HH, HT, TH, TT. Thus, if a coin is actually tossed twice, one of these four sample points is *realized*.

In dealing a single hand of 13 bridge cards where e represents an arbitrary bridge hand, the sample space R consists of $\binom{52}{13}$ sample points, each sample point being one of the possible hands. Thus, if a hand of 13 cards is actually dealt, one of the $\binom{52}{13}$ sample points in R is *realized*.

If a light bulb is allowed to burn continuously until it "expires" and e represents the possible length of time the bulb burns, then e is a positive number and the sample space R (idealized) consists of all positive numbers. If k light bulbs $B_1, B_2, \ldots, B_k$ are allowed to burn continuously until they all "expire," then e may be taken as a set of k positive numbers $(x_1, x_2, \ldots, x_k)$ representing the possible burning lives of $B_1, B_2, \ldots, B_k$, and the sample space R (idealized) consists of the points in k-dimensional Euclidean space whose coordinates are all positive.

In all of these examples it should be noted that an operation is performed and the result of the operation is described by means of a sample point e which in turn belongs to a sample space R.

A sample point e is sometimes called an *elementary event*, and a *set* E of sample points, which is a *subset* of the points in R, is called an *event*. An event, of course, may consist of only one point. When we say that an event E *occurs*, we mean that the sample point e (representing the outcome of the experiment, trial, or operation) is contained in E. In general, there is more interest in events and classes of events than there is in individual sample points as such. Or to state the matter a little more precisely: We are usually more interested in probabilities associated with events than probabilities associated with individual sample points. The only events we are interested in are *measurable events*, that is, sets having associated probabilities. The notion of measurability will be discussed in Section 1.4.

In set theory terminology, we shall be interested in certain classes of subsets of R and probabilities (set functions) to be defined on the sets belonging to such classes. A set of points in the sample space R, however, is also called an event. Actually, *it will be convenient to use the terms "set" and "event" interchangeably*.

Before proceeding further in the discussion of these classes we shall introduce the basic principles of the algebra of sets.

1.2 DEFINITIONS AND RULES FOR COMBINING AND DECOMPOSING EVENTS

In this section, each of the sets $E, E_1, E_2, \ldots$ referred to will be events consisting of sample points in a sample space R. Such sets are sometimes referred to as e sets. A finite sequence of n sets will be written as $E_1, \ldots, E_n$, and a countably infinite sequence as $E_1, E_2, \ldots$.

If e is an element of E, we denote this by writing

$$e \in E. \tag{1.2.1}$$

We also say that *e belongs to E.*

If E_1 and E_2 are two sets such that *every point in E_1 is contained in* E_2, then E_1 is called a subset of E_2 and we write

$$E_1 \subset E_2 \quad \text{or} \quad E_2 \supset E_1. \tag{1.2.2}$$

Note that if E_1 consists of exactly one point, say e, then $E_1 \subset E_2$ is equivalent to the statement $e \in E_2$.

Two sets, E_1 and E_2, are *equal* if $E_1 \subset E_2$ and $E_1 \supset E_2$, in which case we write

$$E_1 = E_2. \tag{1.2.3}$$

The *empty set* (or *null set*) is the set which contains no points. If E contains no points, we write

$$E = \phi. \tag{1.2.4}$$

The null set ϕ is a subset of every set $E \subset R$.

The set of all points contained in *both* E_1 and E_2 is called the *intersection* or *product* of E_1 and E_2 and will be denoted by

$$E_1 \cap E_2, \tag{1.2.5}$$

sometimes read "E_1 cap E_2."

The sets E_1 and E_2 are *disjoint*, that is, contain no common points, if

$$E_1 \cap E_2 = \phi. \tag{1.2.6}$$

The intersection of a sequence of sets $E_1, E_2, \ldots$ is denoted by

$$\bigcap_{\alpha=1}^{n} E_\alpha \tag{1.2.7}$$

if there are n of these sets, and by

$$\bigcap_{\alpha=1}^{\infty} E_\alpha \tag{1.2.8}$$

if the sequence is denumerably infinite. More generally, for an arbitrary collection of sets $\{E_\alpha : \alpha \in T\}$ the intersection of the sets is denoted by $\bigcap_{\alpha \in T} E_\alpha$, or more briefly by $\bigcap_{\alpha} E_\alpha$ if the *range* T of values of α is clear from the text. T can be, of course, not only a set of integers, but elements in a more general space. In nearly all instances with which we shall be concerned in this book T will be a sequence (a finite or denumerable set of integers). If $E_\alpha \cap E_\beta = \phi$, $\alpha \neq \beta$, $E_1, \ldots, E_n$ are *pairwise* or *mutually disjoint*.

The set of all points contained in *at least one* of the sets E_1 and E_2 is called the *union* or *sum* of E_1 and E_2 and will be denoted by*

(1.2.9) $$E_1 \cup E_2,$$

sometimes read "E_1 cup E_2." Note that if E_1 is any *e* set and if $E_2 = \phi$, then $E_1 \cup E_2 = E_1$.

The union of a finite sequence of sets $E_1, \ldots, E_n$ is denoted by

(1.2.10) $$\bigcup_{\alpha=1}^{n} E_\alpha,$$

and if n is denumerably infinite, we write

(1.2.11) $$\bigcup_{\alpha=1}^{\infty} E_\alpha,$$

or for an arbitrary collection of sets $\{E_\alpha : \alpha \in T\}$ we may write $\bigcup_\alpha E_\alpha$.

The set consisting of all points of E_1 not contained in E_2 is called the *difference* between E_1 and E_2 and is written as

(1.2.12) $$E_1 - E_2.$$

If $E_1 \subset E_2$, then clearly $E_1 - E_2$ is the empty set. If $E_1 \cap E_2 = \phi$, then $E_1 - E_2 = E_1$. If $E_2 \subset E_1$, $E_1 - E_2$ is called the *proper difference* of E_1 and E_2. In this case it is sometimes convenient to say that $E_1 - E_2$ is obtained by *subtracting E_2 from E_1*.

The proper difference $E_1 - E_2$ is sometimes called the *complement of E_2 relative to E_1* and we write

(1.2.13) $$E_1 - E_2 = \bar{E}_2(E_1).$$

In the special but important case where E_1 is the entire sample space R then $R - E_2$ is called the *complement* of E_2 and is written

(1.2.14) $$R - E_2 = \bar{E}_2.$$

More generally, we have

(1.2.15) $$E_1 - E_2 = E_1 \cap \bar{E}_2.$$

To illustrate schematically the notions of the union, intersection, and difference of sets, we use a *Venn diagram*. Let the sample space R be

* The union (sum) $E_1 \cup E_2$ is sometimes written as $E_1 + E_2$ and the intersection (product) $E_1 \cap E_2$ as $E_1 \cdot E_2$.

represented by the set of points inside the rectangle in Fig. 1.1 and let E_1 and E_2 be the sets of points inside the large and small circles respectively. Then the *union* $E_1 \cup E_2$ is the set of points enclosed by at least one of the circles; the *intersection* $E_1 \cap E_2$ is the set of points inside both circles; the *difference* $E_1 - E_2$ $(= E_1 \cap \bar{E}_2)$ is the set of points inside the larger circle but not the smaller; $E_2 - E_1 (= \bar{E}_1 \cap E_2)$ has a similar interpretation; $\bar{E}_1 \cap \bar{E}_2 = R - (E_1 \cup E_2)$ is the set of points inside the rectangle but not contained within either circle.

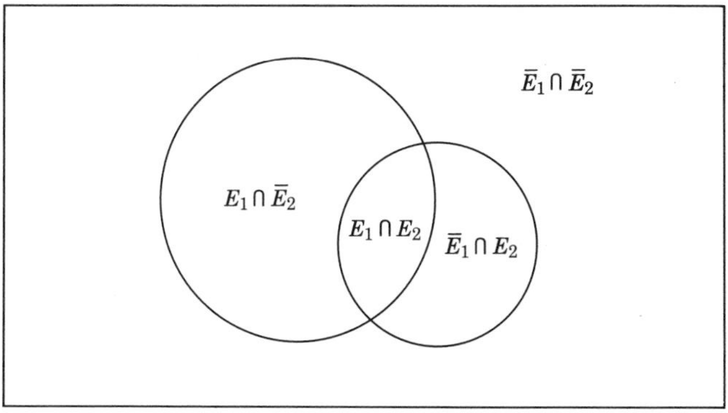

FIG. 1.1 Venn diagram illustrating the four basic disjoint sets $E_1 \cap E_2$, $E_1 \cap \bar{E}_2$, $\bar{E}_1 \cap E_2$, and $\bar{E}_1 \cap \bar{E}_2$ generated by E_1 and E_2.

These set theory notions have the following interpretations in the language of events:*

(i) $E_1 \subset E_2$, means that the occurrence of E_1 implies the occurrence of E_2, that is, a sample point e cannot occur in E_1 without occurring in E_2.

(ii) $E = \phi$, the empty set, means that event E cannot occur.

(iii) $E = R$, the entire sample space, means that event E must occur.

(iv) The intersection $E_1 \cap E_2$ is the event consisting of the joint occurrence of events E_1 and E_2.

(v) The union $E_1 \cup E_2$ is the event consisting of the occurrence of at least one of the events E_1 and E_2.

(vi) The difference $E_1 - E_2$ is the event denoting occurrence of E_1 but not E_2.

* It should be noted that all the definitions and operations (1.2.1) through (1.2.15) apply not only to sample points and sets of sample points (events) but also to elements and sets of elements of any kind. The elements may be point sets, people, automobiles, etc. For instance, if $\mathscr{F}$ is a collection of sets containing the set E, we write $E \in \mathscr{F}$; if every set in the collection $\mathscr{F}$ is contained in the collection $\mathscr{G}$, we write $\mathscr{F} \subset \mathscr{G}$ and so on.

The reader can easily verify that the operations of taking the union and product of sets are *commutative*, that is,

(1.2.16) $$E_1 \cup E_2 = E_2 \cup E_1, \qquad E_1 \cap E_2 = E_2 \cap E_1$$

are *associative*, that is,

(1.2.17) $$\begin{aligned}(E_1 \cup E_2) \cup E_3 = E_1 \cup (E_2 \cup E_3)\\ (E_1 \cap E_2) \cap E_3 = E_1 \cap (E_2 \cap E_3)\end{aligned}$$

and *distributive*, that is,

(1.2.18) $$E_1 \cap (E_2 \cup E_3) = (E_1 \cap E_2) \cup (E_1 \cap E_3).$$

It is also evident that E_1 and the proper difference $E_2 - (E_1 \cap E_2)$ are disjoint and that

(1.2.19) $$E_1 \cup E_2 = E_1 \cup [E_2 - (E_1 \cap E_2)].$$

Similarly, $E_1 \cap E_2$ and the proper differences $E_1 - (E_1 \cap E_2)$ and $E_2 - (E_1 \cap E_2)$ are disjoint such that

$$E_1 \cup E_2 = (E_1 \cap E_2) \cup [E_1 - (E_1 \cap E_2)] \cup [E_2 - (E_1 \cap E_2)].$$

It should be observed that the intersection of any two subsets E_1 and E_2 of R can be expressed in terms of unions and differences of sets by the following formula:

(1.2.20) $$E_1 \cap E_2 = R - (\bar{E}_1 \cup \bar{E}_2).$$

Similarly, we have for the union

(1.2.21) $$E_1 \cup E_2 = R - (\bar{E}_1 \cap \bar{E}_2).$$

We also have the following relationships:

(1.2.22) $$(E_1 \cap E_2) \subset E_\alpha, \qquad \alpha = 1, 2,$$

and

(1.2.23) $$E_\alpha \subset (E_1 \cup E_2), \qquad \alpha = 1, 2.$$

It will be useful for later sections to state extensions of several of the preceding results to finite or infinite collections of sets. The proofs are straightforward and are left to the reader.

1.2.1 *For an arbitrary collection of sets* $\{E_\alpha : \alpha \in T\}$,

(1.2.20*a*) $$\bigcap_\alpha E_\alpha = R - \bigcup_\alpha \bar{E}_\alpha$$

(1.2.21*a*) $$\bigcup_\alpha E_\alpha = R - \bigcap_\alpha \bar{E}_\alpha$$

An extension of (1.2.19) may be stated as follows:

1.2.2 *If* $E_1, E_2, \ldots,$ *is a sequence (finite or infinite) of sets, then* E_1, *and the proper differences* $E_2 - (E_2 \cap E_1)$, $E_3 - [E_3 \cap (E_2 \cup E_1)]$, $\ldots, E_n - [E_n \cap (E_{n-1} \cup \cdots \cup E_1)], \ldots,$ *are disjoint sets whose union is* $\bigcup_\alpha E_\alpha$.

1.2.2 provides a simple rule for decomposing the union of a sequence of sets into a disjoint sequence of sets. It should also be noted that E_1, and the differences $E_2 - E_1, E_3 - E_1 \cup E_2, \ldots, E_n - E_1 \cup \cdots \cup E_{n-1}, \ldots,$ are disjoint sets whose union is $\bigcup_\alpha E_\alpha$.

If $E_1, E_2, \ldots$ is a countably infinite sequence of sets in R, the set E^* consisting of all points which belong to infinitely many of the sets in this sequence is called the *superior limit* of the sequence and will be written as

$$E^* = \lim\sup_\alpha E_\alpha. \tag{1.2.24}$$

Similarly, the set E_* consisting of all points which belong to all but a finite number of sets in the sequence is called the *inferior limit* of the sequence and is written as

$$E_* = \lim\inf_\alpha E_\alpha. \tag{1.2.25}$$

If $E^* = E_*$, we say that *the limit of the sequence* $E_1, E_2, \ldots$ exists and we denote it by

$$\lim_{\alpha \to \infty} E_\alpha, \quad \text{or} \quad \lim_\alpha E_\alpha. \tag{1.2.26}$$

For an *expanding* or *increasing* sequence, $E_1 \subset E_2 \subset \cdots$, it is evident that the limit of the sequence exists and is equal to the union of all sets $\{E_\alpha\}$ that is,

$$\lim_\alpha E_\alpha = \bigcup_\alpha E_\alpha. \tag{1.2.27}$$

For a *contracting* or *decreasing* sequence, that is, one for which $E_1 \supset E_2 \supset \cdots$, it is seen that the limit exists and is equal to the intersection of all sets, and we write

$$\lim_\alpha E_\alpha = \bigcap_\alpha E_\alpha. \tag{1.2.28}$$

In the case of an arbitrary sequence of sets $E_1, E_2, \ldots$ in R, we have

$$\lim\sup E_\alpha = \bigcap_{n=1}^{\infty} \bigcup_{\alpha=n}^{\infty} E_\alpha, \qquad \lim\inf E_\alpha = \bigcup_{n=1}^{\infty} \bigcap_{\alpha=n}^{\infty} E_\alpha.$$

Expanding and contracting sequences are, by far, the most important sequences in probability and statistics. If $E_1, E_2, \ldots$ is either an expanding or a contracting sequence of sets, it is convenient to call it a *monotone sequence*. Thus, the limit of a monotone sequence always exists. If the limit is denoted by E, then $E = \bigcup_\alpha E_\alpha$ or $E = \bigcap_\alpha E_\alpha$, depending on whether the sequence is expanding or contracting. We shall find in later sections and chapters that monotone sequences of sets play a fundamental role in probability theory.

1.3 FIELDS OF SETS

As indicated earlier, we shall be more interested in certain classes of subsets (events) of the sample space R than in the individual sample points in R. In particular, we shall be concerned with *fields* of sets, that is, classes of sets, satisfying certain rules to be stated below.

A nonempty class $\mathscr{F}$ of sets in R is called a *Boolean field* of sets if it satisfies the following properties:

A1 *If* $E \in \mathscr{F}$, *then* $\bar{E} \in \mathscr{F}$.

A2 *If* $E_1 \in \mathscr{F}$ *and* $E_2 \in \mathscr{F}$, *then* $E_1 \cup E_2 \in \mathscr{F}$.

A class $\mathscr{F}$ of sets is called a *Borel field** of sets if the following additional property is satisfied:

A3 *If* $E_1, E_2, \ldots$ *is a countably infinite sequence of sets belonging to* $\mathscr{F}$, *then* $\bigcup_\alpha E_\alpha \in \mathscr{F}$.

Actually, **A2** is superfluous in defining a Borel field of sets since it is implied by **A3**. It will be seen that by choosing $E_2 = \bar{E}_1$ in **A2**, $E_1 \cup \bar{E}_1 = R$, and hence $R \in \mathscr{F}$. Also, it will be noted that if R is chosen for E in **A1**, then $\bar{E}$ is the empty set and hence the empty set is always contained in a field of sets, whether Boolean or Borel.

It follows from successive applications of **A2** that the union of any finite number of sets belonging to a Boolean field $\mathscr{F}$ also belong to $\mathscr{F}$. It can be verified from **A1** and **A2** and (1.2.20) that the intersection of a finite or countably infinite sequence of sets in a Borel field $\mathscr{F}$ is also in $\mathscr{F}$.

If R contains a finite number N of (distinct) sample points, then the class of all possible events is finite and this class of events is clearly a Boolean field. Actually, the Boolean field $\mathscr{F}$ in this case consists of the following 2^N sets (events): the empty set; all one-point sets; all two-point sets, . . . , and finally R itself (the N-point set). If we denote by $\mathscr{F}_0$ the class consisting of all N one-point sets, it is clearly possible to obtain any one of the finite sets belonging to the Boolean field $\mathscr{F}$ by a finite number of applications of **A1** and **A2**. Thus, the Boolean field $\mathscr{F}$ *can be generated from the initial*

* A Boolean field of sets is also called a *Boolean algebra* or a *finitely additive class* of sets; a Borel field of sets is also called a *σ-algebra* or a *completely additive class* of sets.

class $\mathscr{F}_0$. It should be emphasized that there are many different ways in which an initial class $\mathscr{F}_0$ may be selected from which the Boolean field $\mathscr{F}$ of sets can be generated; the class of one-point sets already mentioned is perhaps the simplest.

The Boolean field generated from a finite class of sets will, of course, contain only a finite number of events. But if R contains an infinite number of sample points—even a countably infinite number—a Boolean field provides us with an inadequate class of events for treating many problems of interest.

A3 is introduced to make sure that machinery can be developed for dealing with wider classes of events when R contains infinitely many sample points.

Now suppose R is a sample space and $\mathscr{F}_0$ is a non-empty class of events in R. Suppose $\mathscr{B}$ is any Borel field of sets which contains all the sets in $\mathscr{F}_0$. That at least one such Borel field exists is evident, since $\mathscr{B}$ may be taken as the class of all possible subsets of R, which, of course, contains the class $\mathscr{F}_0$. Suppose $\mathscr{B}(\mathscr{F}_0)$ is the class consisting of all sets which belong to every Borel field containing $\mathscr{F}_0$, that is, $\mathscr{B}(\mathscr{F}_0)$ is the intersection of all Borel fields containing the class $\mathscr{F}_0$. The sets in $\mathscr{B}(\mathscr{F}_0)$ satisfy **A1**, **A2**, and **A3**, and hence $\mathscr{B}(\mathscr{F}_0)$ is a Borel field.

Furthermore, $\mathscr{B}(\mathscr{F}_0)$, by definition, is contained in every Borel field containing $\mathscr{F}_0$. Consequently, it is the unique smallest Borel field containing $\mathscr{F}_0$, and is called the *Borel field generated* by the class $\mathscr{F}_0$.

We may summarize as follows:

1.3.1 *Let $\mathscr{F}_0$ be a non-empty class of sets in R. Then, there exists a unique Borel field $\mathscr{B}(\mathscr{F}_0)$ such that if $\mathscr{B}$ is any Borel field containing $\mathscr{F}_0$, $\mathscr{B}(\mathscr{F}_0) \subset \mathscr{B}$.*

In the special but important case where the sample space R is the axis of real numbers R_1 (the sample points e being the real numbers) and where $\mathscr{F}_0$, which may be referred to as the *initial class*, is taken as the class of all half-open intervals* $(a, b]$, the sets of the Borel field $\mathscr{B}(\mathscr{F}_0)$ are called *Borel sets of the real line*, and their class $\mathscr{B}(\mathscr{F}_0)$ will be denoted by $\mathscr{B}_1$. We obtain the same Borel field if $\mathscr{F}_0$ is chosen as the class of intervals of type (a, b), or $[a, b)$, or $[a, b]$, or more generally as the class of all open sets of R_1 or the class of all closed sets of R_1. The Borel sets of R_1 constitute a class which *includes* among its sets all sets which can be obtained

* We shall use the usual convention of letting $(a, b]$ denote the interval $a < x \leqslant b$, with corresponding meanings for (a, b), $[a, b)$, and $[a, b]$. If x is a k-dimensional real vector, that is, if $x_1, \ldots, x_k$ is a point in k-dimensional Euclidean space, we shall let $(a_1, \ldots, a_k; b_1, \ldots, b_k]$ or more briefly $(a; b]_k$ denote the k-dimensional interval $a_i < x_i \leqslant b_i$, $i = 1, \ldots, k$, with corresponding meanings for $(a; b)_k$, $[a; b)_k$, and $[a; b]_k$.

by performing finite or countable unions, intersections, complements, and differences, starting with a finite or countably infinite number of intervals.

More generally, if the sample space R is the k-dimensional Euclidean space R_k, and if the initial class $\mathscr{F}_0$ is taken as the class of all k-dimensional half-open intervals $(a; b]_k$, the Borel field $\mathscr{B}(\mathscr{F}_0)$ thus generated is called the *Borel sets of* R_k and will be denoted by $\mathscr{B}_k$. There are, of course, other similar choices of the initial class $\mathscr{F}_0$ which generate the Borel sets of R_k, such as the class of intervals of type $(a; b]_k$, $[a; b)_k$, or $[a; b]_k$. It should be particularly noted that the class of Borel sets in R_k *includes* any set which can be formed by a finite or countably infinite number of unions, intersections, differences, or complements, starting with intervals of any of the types mentioned above.

1.4 PROBABILITY MEASURE

In the preceding sections we have been concerned with the description and manipulation of events. We now consider the problem of assigning *probabilities* to these events and of setting up rules for manipulating probabilities. A common notion of the *probability of an event* is that it is an abstraction of the idea of the *relative frequency* with which an event occurs in a sequence of trials of an experiment under "a given set of conditions." Thus, suppose E is an event in the basic sample space R and that each time the "basic" experiment is performed the outcome corresponds to some sample point e in R. If the experiment is performed m times, let m_E be the number of times the resulting event point e belongs to E, that is, the frequency with which E occurred. The relative frequency of E in the m trials is m_E/m. This ratio clearly lies on the interval $[0, 1]$; it necessarily has the value 1 if $E = R$, and 0 if E is the empty set. If E_1 and E_2 are disjoint events, the frequency with which $E_1 \cup E_2$ occurs is $m_{E_1} + m_{E_2}$, and the relative frequency of $E_1 \cup E_2$ is $(m_{E_1} + m_{E_2})/m = m_{E_1}/m + m_{E_2}/m$. This shows that the relative frequency of the union of two disjoint events is equal to the sum of the relative frequencies of the two events. A similar expression relates the relative frequencies of any finite number of disjoint events to the relative frequency of their union.

If we think of the relative frequency of an event E in an (obviously unperformed!) indefinitely long series of trials of our "basic" experiment, we are not far from a reasonable interpretation of the *probability of an event E*. However, certain cumbersome difficulties, which need not be discussed here, arise if we try to establish a theory of probability by strictly formalizing the idea of relative frequency, on the assumption that convergence properties in an indefinitely long series of trials hold. This approach has been considered by von Mises (1931). A simpler and perhaps more fundamental formulation proposed originally by Kolmogorov (1933a) consists essentially of assuming that in dealing with any event of interest in a probability problem, numbers on the interval [0, 1] can be assigned as probabilities of events in some initial class of relatively simple events, and that these initial probabilities, together with rules for determining probabilities of more complex events, will enable one to determine the probability of any event in a class of events broad enough to include that of interest in the original probability problem. In the actual assignment of probabilities to events, one is usually guided by an hypothesis based on what one expects the relative frequencies of the events to be in a large series of trials of the "basic" experiment. Under the Kolmogorov formulation, the important point is that the mathematical theory begins *after* the assignment of the probabilities. We can, of course, question whether the probabilities are correctly assigned in any given situation. A formal treatment of this question is a problem in testing statistical hypotheses which will be considered in later chapters.

We formalize these ideas by defining *probability measure* or a *probability distribution* for a class of events.

A *set function* P defined for all sets in a Boolean field $\mathscr{F}$ and having the following three properties will be referred to as a *probability measure on the Boolean field* $\mathscr{F}$:

B1 *For every event E in $\mathscr{F}$ there is associated a real non-negative number $P(E)$, called the probability of event E. $P(E)$ is sometimes written as $P(e \in E)$.*

B2 *If $E_1, E_2, \ldots$ is a countably infinite sequence of mutually disjoint sets in $\mathscr{F}$ whose union is in $\mathscr{F}$, then*

$$P\left(\bigcup_{\alpha=1}^{\infty} E_\alpha\right) = \sum_{\alpha=1}^{\infty} P(E_\alpha).$$

B3 $P(R) = 1$.

If P is a set function defined for all sets in a Borel field $\mathscr{F}$ and satisfying **B1**, **B2**, and **B3**, P is called a *probability measure on the Borel field* $\mathscr{F}$. In this case, of course, $\bigcup_{\alpha=1}^{\infty} E_\alpha$ belongs to $\mathscr{F}$ by definition of a Borel field.

The triple $(R, \mathscr{F}, P)$ is called a *probability space*. It should be noted that **B2** holds for all finite collections of disjoint sets, say $E_1, \ldots, E_n$, since $E_{n+1}, E_{n+2}, \ldots$ can all be taken as the empty set ϕ, and $P(\phi) = 0$.

A set function P such that $P(E)$ is finite for every $E \in \mathscr{F}$ which satisfies **B1** (without the restriction of non-negativeness) and **B2** is sometimes called a *completely additive* set function.

We are usually interested in the *minimal* Borel field $\mathscr{B}(\mathscr{F})$ generated by some Boolean field $\mathscr{F}$, in which case, of course, $\mathscr{B}(\mathscr{F})$ contains $\mathscr{F}$. Thus, we may state several simple, but useful theorems concerning probabilities of events in $\mathscr{B}(\mathscr{F})$ which, of course, automatically hold for events in $\mathscr{F}$ since a probability measure on $\mathscr{B}(\mathscr{F})$ also provides probabilities for all sets in $\mathscr{F}$.

1.4.1 *If E is any event in $\mathscr{B}(\mathscr{F})$ then $P(E) + P(\bar{E}) = 1$.*

It follows at once from **1.4.1** and **B3** that $P(\phi) = 0$. A non-null event E for which $P(E) = 0$ is sometimes referred to as an *event* (or set) of *zero probability*.

1.4.2 *If E_1 and E_2 are events in $\mathscr{B}(\mathscr{F})$ such that $E_1 \supset E_2$, then $0 \leqslant P(E_1 - E_2) = P(E_1) - P(E_2)$, and $P(E_1) \geqslant P(E_2)$.*

1.4.3 *If E_1 and E_2 are events in $\mathscr{B}(\mathscr{F})$, then $P(E_1 \cup E_2) = P(E_1) + P(E_2 - E_1 \cap E_2)$. Also $P(E_1 \cup E_2) = P(E_1) + P(E_2) - P(E_1 \cap E_2)$.*

1.4.4 *More generally, if $E_1, \ldots, E_n$ are n sets belonging to $\mathscr{B}(\mathscr{F})$ then we have*

$$(1.4.1)\quad P\left(\bigcup_{\alpha=1}^{n} E_\alpha\right) = P(E_1) + P(E_2 - E_2 \cap E_1) + \cdots + P(E_n - E_n \cap [E_1 \cup \cdots \cup E_{n-1}]),$$

also

$$(1.4.1a)\quad P\left(\bigcup_{\alpha=1}^{n} E_\alpha\right) = \sum_{\alpha=1}^{n} P(E_\alpha) - \sum_{\beta>\alpha=1}^{n} P(E_\alpha \cap E_\beta) + \cdots + (-1)^{n-1} P(E_1 \cap \cdots \cap E_n).$$

Proofs of **1.4.1**, **1.4.2**, **1.4.3**, and **1.4.4** are left as exercises for the reader. If we consider an infinite sequence of sets $E_1, E_2, \ldots,$ in $\mathscr{B}(\mathscr{F})$, then (1.4.1) and (1.4.1*a*) become, respectively,

(1.4.1)′

$$P\left(\bigcup_{\alpha=1}^{\infty} E_\alpha\right) = P(E_1) + P(E_2 - E_2 \cap E_1) + P(E_3 - E_3 \cap [E_1 \cup E_2]) + \cdots$$

and (assuming convergence of the sums on the right),

$$(1.4.1a)'\quad P\left(\bigcup_{\alpha=1}^{\infty} E_\alpha\right) = \sum_{\alpha=1}^{\infty} P(E_\alpha) - \sum_{\beta>\alpha=1}^{\infty} P(E_\alpha \cap E_\beta) + \sum_{\gamma>\beta>\alpha=1}^{\infty} P(E_\alpha \cap E_\beta \cap E_\gamma) - \cdots.$$

Note that (1.4.1) and (1.4.1*a*) are special cases of (1.4.1)′ and (1.4.1*a*)′ respectively, obtained by taking $E_{n+1}, E_{n+2}, \ldots$ as empty sets.

The following theorem is of basic importance in Chapter 2:

1.4.5 *If $E_1, E_2, \ldots$ is a monotone sequence of sets in the Borel field $\mathscr{B}(\mathscr{F})$ then*

$$\lim_{\alpha\to\infty} P(E_\alpha) = P\left(\lim_{\alpha\to\infty} E_\alpha\right).$$

To prove **1.4.5**, we consider first the case of an increasing sequence $E_1 \subset E_2 \subset \cdots$. Then $E_1, E_2 - E_1, \ldots, E_\alpha - E_{\alpha-1}$, $\alpha = 2, 3, \ldots$ are mutually disjoint, such that

$$E_\alpha = E_1 \cup (E_2 - E_1) \cup \cdots \cup (E_\alpha - E_{\alpha-1}). \tag{1.4.2}$$

Taking limits as $\alpha \to \infty$, we have

$$\lim_{\alpha\to\infty} E_\alpha = E_1 \cup (E_2 - E_1) \cup \cdots \tag{1.4.3}$$

Applying **B2**, we have

$$P\left(\lim_{\alpha\to\infty} E_\alpha\right) = P(E_1) + P(E_2 - E_1) + \cdots \tag{1.4.4}$$

But

$$\begin{aligned} P(E_1) + P(E_2 - E_1) + \cdots &= \lim_{\alpha\to\infty} [P(E_1) + P(E_2 - E_1) + \cdots + P(E_\alpha - E_{\alpha-1})] \\ &= \lim_{\alpha\to\infty} P(E_1 \cup (E_2 - E_1) \cup \cdots \cup (E_\alpha - E_{\alpha-1})). \end{aligned} \tag{1.4.5}$$

Making use of (1.4.2) we have

$$\lim_{\alpha\to\infty} P(E_1 \cup (E_2 - E_1) \cup \cdots \cup (E_\alpha - E_{\alpha-1})) = \lim_{\alpha\to\infty} P(E_\alpha).$$

Therefore, we conclude that

$$P\left(\lim_{\alpha\to\infty} E_\alpha\right) = \lim_{\alpha\to\infty} P(E_\alpha), \tag{1.4.6}$$

thus completing the argument for **1.4.5** for the case of an expanding sequence of sets.

For a contracting sequence of events $E_1 \supset E_2 \supset \cdots$ the complements $\bar{E}_1, \bar{E}_2, \ldots$ form an expanding sequence of events and it follows from the argument just given that

$$P\left(\lim_{\alpha\to\infty} \bar{E}_\alpha\right) = \lim_{\alpha\to\infty} P(\bar{E}_\alpha), \tag{1.4.7}$$

that is,

$$P\left(\lim_{\alpha\to\infty}(R - E_\alpha)\right) = \lim_{\alpha\to\infty} P(R - E_\alpha), \tag{1.4.8}$$

which may be written as

$$P\left(R - \lim_{\alpha\to\infty} E_\alpha\right) = \lim_{\alpha\to\infty} P(R - E_\alpha). \tag{1.4.9}$$

Using **1.4.2** this reduces to (1.4.6), where, of course, in the contracting case

$$\lim_{\alpha\to\infty} E_\alpha = \bigcap_{\alpha=1}^{\infty} E_\alpha.$$

The following *covering theorem* will be useful in later sections:

1.4.6 *If* $E, E_1, E_2, \ldots$ *are events in* $\mathscr{B}(\mathscr{F})$ *such that* $E \subset \bigcup_{\alpha=1}^{\infty} E_\alpha$, *then*

$$P(E) \leqslant \sum_{\alpha=1}^{\infty} P(E_\alpha).$$

Whenever $E, E_1, E_2, \ldots$ is a finite or infinite sequence of sets such that $E \subset \bigcup_{\alpha=1}^{\infty} E_\alpha$ we shall say that $\bigcup_{\alpha=1}^{\infty} E_\alpha$ is a *covering* for E.

To establish **1.4.6**, it is sufficient to note that since $E \subset \bigcup_{\alpha=1}^{\infty} E_\alpha$ we can write

$$E = E \cap \left(\bigcup_{\alpha=1}^{\infty} E_\alpha\right). \tag{1.4.10}$$

Applying **1.2.2** we find that

$$E = [E \cap E_1] \cup [E \cap (E_2 - E_2 \cap E_1)] \cup \cdots. \tag{1.4.11}$$

The sets in [ ] are disjoint and belong to $\mathscr{B}(\mathscr{F})$. Hence

$$P(E) = P(E \cap E_1) + P(E \cap (E_2 - E_2 \cap E_1)) + \cdots. \tag{1.4.12}$$

But

$$[E \cap E_1] \subset E_1, \quad [E \cap (E_2 - E_2 \cap E_1)] \subset E_2, \ldots.$$

Therefore, applying **1.4.2**, we find

$$P(E) \leqslant P(E_1) + P(E_2) + \cdots \tag{1.4.13}$$

which concludes the argument for **1.4.6**.

By taking $E_{n+1} = E_{n+2} = \cdots = \phi$, it will be seen that the formula in **1.4.6** reduces to

$$P(E) \leqslant \sum_{\alpha=1}^{n} P(E_\alpha),$$

which would, of course, also hold if $E, E_1, \ldots, E_n$ are sets in $\mathscr{F}$.

Remark. If the sample space R contains only a finite number of sample points it is evident that if one takes as an initial class of events $\mathscr{F}_0$ all one-point events and assigns a probability to every element in $\mathscr{F}_0$ then this assignment of probabilities determines a Boolean probability measure. This means that we can determine the probability of *any* event in the Boolean field $\mathscr{F}$ generated by $\mathscr{F}_0$ from the probabilities assigned to the sets in $\mathscr{F}_0$.

1.5 EXTENSION OF A PROBABILITY MEASURE

(a) Uniqueness of Extension of Probability Measure

It will be recalled from Section 1.3 that if one starts with any initial Boolean field $\mathscr{F}$, there is a minimal Borel field $\mathscr{B}(\mathscr{F})$ containing $\mathscr{F}$. Now suppose we have a probability measure on the Boolean field $\mathscr{F}$; the question arises whether we can perform certain operations on the probabilities defined for sets in $\mathscr{F}$ so as to obtain a probability measure defined on $\mathscr{B}(\mathscr{F})$, without changing any of the probabilities already assigned to sets in $\mathscr{F}$. This can be done. It can be regarded as extension (or generation) of a probability measure on $\mathscr{B}(\mathscr{F})$ from a probability measure on $\mathscr{F}$.

Before considering whether there exists a method of extending a probability measure on $\mathscr{F}$ to one on $\mathscr{B}(\mathscr{F})$ we state the following theorem on the uniqueness of an extension:

1.5.1 *Let $\mathscr{F}$ be a Boolean field of sets and let P_1 and P_2 be probability measures defined on the Borel field $\mathscr{B}(\mathscr{F})$. If $P_1(E) = P_2(E)$ for every $E \in \mathscr{F}$ then $P_1(E) = P_2(E)$ for every $E \in \mathscr{B}(\mathscr{F})$.*

If we let $\mathscr{H}$ be the class of sets in $\mathscr{B}(\mathscr{F})$ such that for any $E \in \mathscr{H}$, $P_1(E) = P_2(E)$, it can be verified that $\mathscr{H}$ is a Borel field, and, of course, contains the (minimal) Borel field $\mathscr{B}(\mathscr{F})$. But since $\mathscr{H} \subset \mathscr{B}(\mathscr{F})$, we conclude that $\mathscr{H} = \mathscr{B}(\mathscr{F})$ and hence $P_1(E) = P_2(E)$ for any E in $\mathscr{B}(\mathscr{F})$.

Theorem **1.5.1** essentially states that there cannot be two probability measures defined on the Borel field $\mathscr{B}(\mathscr{F})$ generated by a Boolean field $\mathscr{F}$ which are equal on sets in $\mathscr{F}$ but not equal on sets in $\mathscr{B}(\mathscr{F}) - \mathscr{F}$. Thus, if we can find a way of starting with a probability measure on a Boolean field $\mathscr{F}$ and *extending* it to the Borel field $\mathscr{B}(\mathscr{F})$, then it follows from **1.5.1** that the probability measure will be unique.

(b) Extension of Probability Measure from $\mathscr{F}$ to $\mathscr{B}(\mathscr{F})$

Such an extension as discussed above can be achieved by using Carathéodory's (1927) theory of *outer measure*. [Also see Halmos (1950).] In defining outer measure for this problem we begin with a probability measure P for sets in a Boolean field $\mathscr{F}$. We then take any set E in $\mathscr{B}(\mathscr{F})$

and let $E_1, E_2, \ldots$ be a sequence of sets in $\mathscr{F}$ such that $E \subset \bigcup_{\alpha=1}^{\infty} E_\alpha$, that is, $\bigcup_{\alpha=1}^{\infty} E_\alpha$ is a covering for E. The *outer measure* of E, say $P^*(E)$, is defined as the greatest lower bound of $\sum_{\alpha=1}^{\infty} P(E_\alpha)$ for all possible coverings constructed from sets in $\mathscr{F}$, that is,

$$P^*(E) = \inf \sum_{\alpha=1}^{\infty} P(E_\alpha). \tag{1.5.1}$$

Since for any Boolean field $\mathscr{F}$ we can take $E_1 = R, E_2 = E_3 = \cdots = \phi$ as a covering for *any* set E in R, it is evident that $P^*(E)$ exists for every set in R. Actually, however, we shall be interested only in $P^*(E)$ for sets in $\mathscr{B}(\mathscr{F})$.

The following properties of P^* can be verified by the reader:

$$P^*(\phi) = 0. \tag{1.5.2}$$

$$P^*(R) = 1. \tag{1.5.3}$$

$$P^*(E) = P(E), \qquad \text{if } E \in \mathscr{F}. \tag{1.5.4}$$

$$P^*(E) \leqslant P(F), \qquad \text{if } E \subset F. \tag{1.5.5}$$

$$P^*(E) \leqslant \sum_{\alpha} P^*(E_\alpha), \qquad \text{if } E \subset \bigcup_{\alpha} E_\alpha. \tag{1.5.6}$$

The basic theorem for the extension of the probability measure P on a Boolean field $\mathscr{F}$ to a probability measure on the Borel field $\mathscr{B}(\mathscr{F})$ can be stated as follows:

1.5.2 *Let P be a probability measure on a Boolean field $\mathscr{F}$. Then $P^*(E)$ as defined in* (1.5.1) *is a probability measure on $\mathscr{B}(\mathscr{F})$ such that $P^*(E) = P(E)$ for every $E \in \mathscr{F}$.*

For proof, the reader is referred to Halmos (1950).

1.6 STATISTICAL INDEPENDENCE

(a) Cartesian Products of Sample Spaces

The *Cartesian product* R of the sample spaces $R^{(1)}$ and $R^{(2)}$ is the set of all ordered pairs $(e^{(1)}, e^{(2)})$, where $e^{(1)} \in R^{(1)}$ and $e^{(2)} \in R^{(2)}$. We write

$$R = R^{(1)} \times R^{(2)}. \tag{1.6.1}$$

Similarly, if $E^{(1)}$ and $E^{(2)}$ are events in $R^{(1)}$ and $R^{(2)}$ respectively, the Cartesian product of $E^{(1)}$ and $E^{(2)}$ is the event E whose sample points $e = (e^{(1)}, e^{(2)})$ have the property that $e^{(1)} \in E^{(1)}$ and $e^{(2)} \in E^{(2)}$. We write

$$E = E^{(1)} \times E^{(2)}. \tag{1.6.2}$$

E is an empty set in R if and only if $E^{(1)}$ or $E^{(2)}$ is an empty set. Also it is evident that if $E^{(1)}$ and $F^{(1)}$ are events in $R^{(1)}$, and $E^{(2)}$ and $F^{(2)}$ are events in $R^{(2)}$, then $(E^{(1)} \times E^{(2)}) \subset (F^{(1)} \times F^{(2)})$ if and only if $E^{(1)} \subset F^{(1)}$ and $E^{(2)} \subset F^{(2)}$, and furthermore that $(E^{(1)} \times E^{(2)}) = (F^{(1)} \times F^{(2)})$ if and only if $E^{(1)} = F^{(1)}$ and $E^{(2)} = F^{(2)}$.

We sometimes say that E defined by (1.6.2) is the *joint occurrence* of $E^{(1)}$ and $E^{(2)}$. Also, it is convenient to say that $E^{(1)}$ is the *projection* of E onto $R^{(1)}$, a similar statement holding for $E^{(2)}$. $R^{(1)}$ and $R^{(2)}$ are called *component* or *marginal* sample spaces of R. If we take all points $(e^{(1)}, e^{(2)})$ in R for which $e^{(1)} \in E^{(1)}$ we obtain a *cylinder set* in R which is, in fact, the Cartesian product $E^{(1)} \times R^{(2)}$. Similarly, $R^{(1)} \times E^{(2)}$ is the cylinder set for which $e^{(2)} \in E^{(2)}$. Thus, $E^{(1)} \times E^{(2)}$, $E^{(1)} \times R^{(2)}$, $R^{(1)} \times E^{(2)}$ are events in the sample space $R = R^{(1)} \times R^{(2)}$ such that the Cartesian product $E^{(1)} \times E^{(2)}$ is the intersection of the cylinder sets $E^{(1)} \times R^{(2)}$ and $R^{(1)} \times E^{(2)}$, that is,

$$E^{(1)} \times E^{(2)} = (E^{(1)} \times R^{(2)}) \cap (R^{(1)} \times E^{(2)}). \tag{1.6.3}$$

We remark with special emphasis that events E in R of the Cartesian product type form a special class of events which play an important role in probability theory.

The notion of Cartesian products extends in a straightforward manner to any finite or countably infinite number of events $E^{(1)}, E^{(2)}, \ldots$ in sample spaces $R^{(1)}, R^{(2)}, \ldots$, respectively.

Examples. As an illustration of Cartesian products of sample spaces and events, consider the special case where $R^{(1)}$, $R^{(2)}$, $E^{(1)}$, and $E^{(2)}$ are sets of points on the axis of real numbers R_1. Let $R^{(1)}$ be the closed interval $[a_1, b_1]$ on the $e^{(1)}$-axis; $R^{(2)}$ the closed interval $[a_2, b_2]$ on the $e^{(2)}$-axis; $E^{(1)}$ a closed set in $R^{(1)}$ and $E^{(2)}$ a closed set in $R^{(2)}$ as shown in Fig. 1.2. The Cartesian product space $R = R^{(1)} \times R^{(2)}$ consists of all points in the large rectangle $ABCD$ including its boundary. The cylinder set $E^{(1)} \times R^{(2)}$ consists of the set of points contained in the cross-hatched vertical strips, and the cylinder set $R^{(1)} \times E^{(2)}$ consists of the set of points contained in the cross-hatched horizontal strips. The Cartesian product $E = E^{(1)} \times E^{(2)}$ consists of the four black rectangular sets of points. Thus, E is the intersection of the two cylinder sets referred to in R, that is, $E = (E^{(1)} \times R^{(2)}) \cap (R^{(1)} \times E^{(2)})$.

It must be emphasized that the operation of taking Cartesian products of events can be applied to events which are more general than those represented by real numbers, that is, those represented by sets of points in a line, plane or in a multidimensional Euclidean space. For example, in playing two hands of bridge suppose $e^{(1)}$ is the hand player A obtains on the first deal, and $e^{(2)}$ is the hand he obtains on the second deal. Then the number of sample points $e^{(1)}$ and $e^{(2)}$ comprising event spaces $R^{(1)}$ and $R^{(2)}$, respectively, is $\binom{52}{13}$ in each case. Furthermore, the $\binom{52}{13}^2$ possible pairs of hands $(e^{(1)}, e^{(2)})$ A could obtain on the two deals correspond to the $\binom{52}{13}^2$ sample points of the Cartesian product space $R = R^{(1)} \times R^{(2)}$. Thus, if $E^{(1)}$ denotes the subset of $R^{(1)}$ consisting of all hands with exactly 2 aces and $E^{(2)}$ the subset of $R^{(2)}$ consisting of all hands of exactly 3 kings, then $E = E^{(1)} \times E^{(2)}$ will be the event in R containing all pairs of hands $(e^{(1)}, e^{(2)})$ such that the first hand $e^{(1)}$ contains exactly 2 aces and the second hand $e^{(2)}$ exactly 3 kings. There are $\left[\binom{4}{2}\binom{48}{11}\right]\left[\binom{4}{3}\binom{48}{10}\right]$ such pairs of hands, that is, sample points in E.

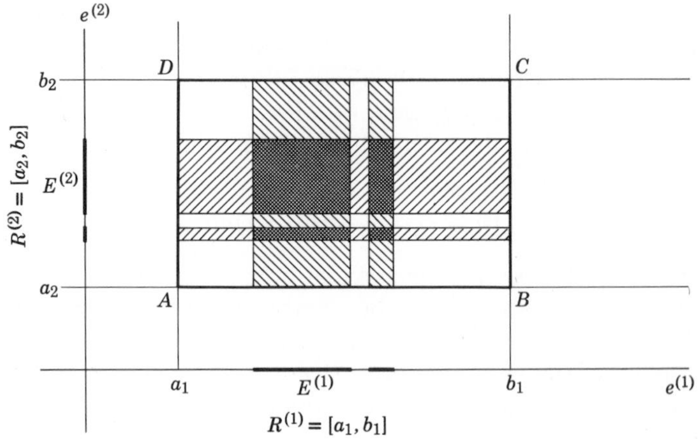

FIG. 1.2 Figure illustrating Cartesian product of events and sample spaces.

Suppose $R^{(1)}$ and $R^{(2)}$ are two sample spaces and that $\mathscr{B}(\mathscr{F}^{(1)})$ and $\mathscr{B}(\mathscr{F}^{(2)})$ are the Borel fields generated by Boolean fields of sets $\mathscr{F}^{(1)}$ and $\mathscr{F}^{(2)}$ in these two sample spaces, respectively. Now consider the class of sets $\mathscr{F}_0$ in $R = R^{(1)} \times R^{(2)}$ of type $E = E^{(1)} \times E^{(2)}$, where $E^{(1)} \in \mathscr{B}(\mathscr{F}^{(1)})$ and $E^{(2)} \in \mathscr{B}(\mathscr{F}^{(2)})$. Starting with $\mathscr{F}_0$ as the initial class, it can be used to generate a minimal Borel field of sets in R which will be designated by

$$\mathscr{B}(\mathscr{F}_0). \tag{1.6.4}$$

As a matter of fact it can be shown that this Borel field can also be generated by taking as the initial class $\mathscr{F}_0$ the sets of type $E^{(1)} \times E^{(2)}$ where $E^{(1)} \in \mathscr{F}^{(1)}$ and $E^{(2)} \in \mathscr{F}^{(2)}$.

(b) Products of Probability Measures

Now suppose $P^{(1)}$ and $P^{(2)}$ are probability measures on $\mathscr{B}(\mathscr{F}^{(1)})$ and $\mathscr{B}(\mathscr{F}^{(2)})$, respectively. For any set $E = E^{(1)} \times E^{(2)}$ in $\mathscr{F}_0$, where

$E^{(1)} \in \mathscr{B}(\mathscr{F}^{(1)})$ and $E^{(2)} \in \mathscr{B}(\mathscr{F}^{(2)})$, let us assign probability to E by the formula

$$P(E) = P^{(1)}(E^{(1)}) \cdot P^{(2)}(E^{(2)}). \tag{1.6.5}$$

It can be shown that P is a Boolean probability measure on $\mathscr{F}_0$ which can be uniquely extended to $\mathscr{B}(\mathscr{F}_0)$.

Whenever a probability measure P on the Borel field $\mathscr{B}(\mathscr{F}_0)$ satisfies (1.6.5) for every set $E = E^{(1)} \times E^{(2)}$ where $E^{(1)} \in \mathscr{B}(\mathscr{F}^{(1)})$ and $E^{(2)} \in \mathscr{B}(\mathscr{F}^{(2)})$ we say that the *component* probability spaces $(R^{(1)}, \mathscr{B}(\mathscr{F}^{(1)}), P^{(1)})$ and $(R^{(2)}, \mathscr{B}(\mathscr{F}^{(2)}), P^{(2)})$ are *(statistically) independent*. The probability space

$$(R, \mathscr{B}(\mathscr{F}_0), P) \tag{1.6.6}$$

where P is a probability measure defined by (1.6.5) is called the *Cartesian product* of the two component probability spaces

$$(R^{(1)}, \mathscr{B}(\mathscr{F}^{(1)}), P^{(1)}) \quad \text{and} \quad (R^{(2)}, \mathscr{B}(\mathscr{F}^{(2)}), P^{(2)}).$$

The notion of statistical independence extends, of course, to more than two probability spaces.

1.7 RANDOM VARIABLES

Let $(R, \mathscr{B}, P)$ be a probability space. Let $x(e)$ be a real, single-valued function defined at every sample point e in R such that for each real number b, the event E_b in R, for which $x(e) \leqslant b$, belongs to $\mathscr{B}$. The function $x(e)$ is called a *random variable* relative to $\mathscr{B}$. We sometimes say that $x(e)$ is $\mathscr{B}$-measurable. Note that if $\{E_b\}$ denotes the class of events in R corresponding to all real numbers b, then $\mathscr{B}(\{E_b\})$ is the Borel field generated by this class. It is convenient to call the set of numbers $x(e)$ can take on for all $e \in R$ the *sample space* of $x(e)$.

Thus, if $\mathscr{B}_1$ is the class of Borel sets on the real line R_1, the random variable $x(e)$ maps the sample points e in R into sample points x in R_1 in such a way that for every Borel set $E' \in \mathscr{B}_1$, there is an event $E \in \mathscr{B}$ consisting of all sample points in R for which $x(e) \in E'$. It is convenient to denote E by $x^{-1}(E')$. By setting $P'(E') = P(x^{-1}(E'))$, the probability space $(R, \mathscr{B}, P)$ associated with the basic sample space R can be used to define the probability space $(R_1, \mathscr{B}_1, P')$ associated with the real line R_1. In these circumstances we may say that $(R_1, \mathscr{B}_1, P')$ is *induced* from $(R, \mathscr{B}, P)$ by the random variable $x(e)$. Verification of the fact that $(R_1, \mathscr{B}_1, P')$ is a probability space is left as an exercise for the reader.

If, for the given probability space $(R, \mathscr{B}, P)$, $x_1(e), \ldots, x_k(e)$ are k random variables, the event $E_{b_1, \ldots, b_k}$ in R of form

$$\{e : x_i(e) \leqslant b_i, \quad i = 1, \ldots, k\},$$

where $b_1, \ldots, b_k$ are arbitrary real numbers, also belongs to $\mathscr{B}$, and $P(E_{b_1,\ldots,b_k}) = P(x_1(e) \leqslant b_1, \ldots, x_k(e) \leqslant b_k)$. Furthermore, if E' is any set in $\mathscr{B}_k$ (the Borel sets of Euclidean k-dimensional space R_k) the set of e points, say E, for which $(x_1(e), \ldots, x_k(e)) \in E'$ is also contained in $\mathscr{B}$, so that by making the assignment $P'(E') = P(E)$, we obtain a probability space $(R_k, \mathscr{B}_k, P')$ induced from $(R, \mathscr{B}, P)$ by $(x_1(e), \ldots, x_k(e))$. It is convenient to refer to $(x_1(e), \ldots, x_k(e))$ as a *k-dimensional* or *vector random variable* relative to $\mathscr{B}$ or as a $\mathscr{B}$-measurable k-dimensional random variable.

If for each sample point e in R we have $|x(e)| < M$ where $M < +\infty$ then $x(e)$ is called a *bounded random variable.* A k-dimensional random variable is bounded if each of its components is bounded.

Examples. The notion of a random variable can, perhaps, be further clarified by a few simple illustrations.

In considering all possible hands of 13 cards which can be dealt from a pack of playing cards, the set of sample points e in the basic sample space R consists of the $\binom{52}{13}$ possible hands of 13 cards. If $x(e)$ is a random variable which has a value equal to the number of aces contained in e, then $x(e)$ is defined at every sample point in R and has the value 0, 1, 2, 3, or 4 on each point in R. Thus, $x(e)$ maps every sample point e in R into one of the points 0, 1, 2, 3, 4 on the real line R_1. Since the number of sample points in R is finite, the class of all possible events in R (including R and the null set) forms a Boolean field $\mathscr{F}$. The event in R for which $x(e) \leqslant b$, for each real number b, is clearly one of the events in the class $\mathscr{F}$. If probability is assigned to each sample point in R $\left(\text{in the case of “thorough” shuffling they are all assigned the value } 1\Big/\binom{52}{13}\right)$ then we have a Boolean probability space $(R, \mathscr{F}, P)$ which is adequate for obtaining the probability for the event $E_b = \{e : x(e) \leqslant b\}$ for any real number b. Thus, if E' is any (Borel) set in R_1, its probability is provided by the probability space $(R_1, \mathscr{B}_1, P')$ induced from $(R, \mathscr{F}, P)$ by $x(e)$. The only sets in $\mathscr{B}_1$ of interest here, of course, are the 5-point set $\{0, 1, 2, 3, 4\}$ together with its subsets, that is, the sample space of $x(e)$ together with its subsets.

If $(x_1(e), x_2(e))$ is a two-dimensional random variable where $x_1(e)$ is the number of aces, and $x_2(e)$ is the number of spades contained in sample point e, we have a two-dimensional random variable such that the probability $P(x_1(e) \leqslant b_1, x_2(e) \leqslant b_2)$ can be computed from the Boolean probability space $(R, \mathscr{F}, P)$. By assigning the above probability to the interval $(-\infty, -\infty; b_1, b_2]$ for all pairs of real numbers (b_1, b_2), a probability space $(R_2, \mathscr{B}_2, P')$ is thus induced by the vector random variable $(x_1(e), x_2(e))$ which provides the probability of any event E in R_2. Again we point out that the only events E of R_2 of practical interest are those in the 62-point set $\{(x_1, x_2)\colon x_1 = 0, 1, 2, 3, 4;\ x_2 = 0, 1, \ldots, 13, \text{with } 0 \leqslant x_1 + x_2 \leqslant 14$ excluding (4, 0) and (0, 13)$\}$, that is, the sample space of $(x_1(e), x_2(e))$, together with its subsets.

Consider another example. Suppose two light bulbs B_1 and B_2 are set to burn continuously and their burning lives are t_1 and t_2, respectively. The basic sample space R is the first quadrant of the t_1t_2-plane and the sample points e are points in this quadrant. The event, say E_b, that neither bulb burns more than b units of time longer than the other consists of all points in R for which $|t_1 - t_2| \leq b$. Thus, we have a random variable $x(e)$ defined as $|t_1 - t_2|$ at each point $e = (t_1, t_2)$ in R. The probability that $x(e) \leq b$ can be computed once a probability space $(R, \mathscr{B}, P)$ is set up where $\mathscr{B}$ is a Borel field containing the events $|t_1 - t_2| \leq b$ for every real b, and P, of course, is a probability measure on $\mathscr{B}$. The random variable $|t_1 - t_2|$ maps the points in R into the real line R_1, so that $P'(E_b') = P(E_b)$ where E_b' is the interval $(-\infty, b]$ in R_1 and E_b is the set in R for which $|t_1 - t_2| \leq b$, and b is any real number. Thus, we have an initial class of events $\{E_b : b \text{ real}\}$ which, together with probabilities $\{P(E_b)\}$, can be used to generate a minimal Borel field $\mathscr{B}(\{E_b\})$ and probability space $(R, \mathscr{B}(\{E_b\}), P)$ from which a probability space $(R_1, \mathscr{B}_1, P')$ is induced by $x(e)$ and which provides the probability that $|t_1 - t_2|$ belongs to any set E' in $\mathscr{B}_1$.

If we let $x_1(e) = t_1 + t_2$ and $x_2(e) = t_1 - t_2$, we have two random variables defined at every point in R. If a probability space $(R, \mathscr{B}(\{E_{b_1b_2}\}), P)$ is generated from the initial class of events $\{E_{b_1b_2} : b_1, b_2 \text{ real}\}$ and their probabilities $\{P(E_{b_1b_2})\}$ where $E_{b_1b_2} = \{e : x_1(e) \leq b_1, x_2(e) \leq b_2\}$, then $(x_1(e), x_2(e))$ induces a probability space $(R_2, \mathscr{B}_2, P')$ from $(R, \mathscr{B}(\{E_{b_1b_2}\}, P)$ which provides the probability that $(t_1 + t_2, t_1 - t_2)$ belongs to any set E' in $\mathscr{B}_2$.

1.8 INTEGRATION OF RANDOM VARIABLES

In this section we restate in terms of probability terminology and without proof, some basic Lebesgue-Stieltjes integration theory. Proofs can be found in books such as those by Halmos (1950), Loève (1955), McShane and Botts (1959), and Saks (1937).

Suppose $(R, \mathscr{B}, P)$ is a probability space and $x(e)$ is a random variable relative to $\mathscr{B}$. If $x(e)$ takes on only a finite number of different values $x_1, \ldots, x_k$ such that for $x(e) = x_i$ for all $e \in E_i$, $i = 1, \ldots, k$, then $x(e)$ is called a *simple* random variable. In this case let us write

$$\int_R x(e)\, dP(e) = \sum_{i=1}^{k} x_i P(E_i). \tag{1.8.1}$$

More generally, if E is any set in $\mathscr{B}$, we write

$$\int_E x(e)\, dP(e) = \sum_{i=1}^{k} x_i P(E \cap E_i). \tag{1.8.2}$$

Now consider the case where $x(e)$ is bounded and can take on infinitely many values. It can be shown that there exists a sequence of *simple* random variables $x_1(e), x_2(e), \ldots$ such that

$$\lim_{\alpha \to \infty} x_\alpha(e) = x(e) \tag{1.8.3}$$

uniformly for all $e \in R$, and furthermore that for each $E \in \mathscr{B}$

$$\lim_{\alpha\to\infty} \int_E x_\alpha(e)\, dP(e) \tag{1.8.4}$$

exists and this limit is independent of the particular sequence that satisfies (1.8.3). This limit which we denote by

$$\int_E x(e)\, dP(e) \tag{1.8.5}$$

is called the *Lebesgue-Stieltjes integral* of $x(e)$, with respect to $(R, \mathscr{B}, P)$, over E. In the case of a simple random variable $x(e)$, the Lebesgue-Stieltjes integral of $x(e)$ over E is given by (1.8.2).

In the more general case where $x(e)$ is not necessarily bounded, but non-negative, it is *integrable* if there exists a sequence of bounded random variables $x_1(e) \leqslant x_2(e) \leqslant \cdots$ for all $e \in R$, such that

$$\lim_{\alpha\to\infty} x_\alpha(e) = x(e), \quad \text{all } e \in R \tag{1.8.6}$$

and

$$\lim_{\alpha\to\infty} \int_E x_\alpha(e)\, dP(e) < +\infty. \tag{1.8.7}$$

This limit is the integral of $x(e)$ with respect to $(R, \mathscr{B}, P)$ over E and is denoted by

$$\int_E x(e)\, dP(e).$$

Under these conditions it can be shown that for any $E \in \mathscr{B}$

$$\lim_{\alpha\to\infty} \int_E x_\alpha(e)\, dP(e) \tag{1.8.8}$$

exists and does not depend on the particular sequence $x_1(e) \leqslant x_2(e) \leqslant \cdots$ chosen which satisfies (1.8.6) and (1.8.7).

Finally, if $x(e)$ is arbitrary, it is said to be integrable if it can be written as a difference $x'(e) - x''(e)$ where $x'(e)$ and $x''(e)$ are non-negative integrable random variables, in which case we define

$$\int_E x(e)\, dP(e) = \int_E x'(e)\, dP(e) - \int_E x''(e)\, dP(e). \tag{1.8.9}$$

It can be shown that the value of the left-hand side of (1.8.9) is independent of the particular choice of $x'(e)$ and $x''(e)$.

The Lebesgue-Stieltjes integral has the following important properties, expressed in terms of random variables, where E is any set in $\mathscr{B}$.

1.8.1 *If $x(e) = k$, a constant, for all $e \in R$,*

$$\int_E x(e)\,dP(e) = kP(E).$$

1.8.2 *If $a \leqslant x(e) \leqslant b$ for all $e \in R$, where a and b are constants, then*

$$aP(E) \leqslant \int_E x(e)\,dP(e) \leqslant bP(E).$$

1.8.3 *If $x_1(e)$ and $x_2(e)$ are integrable and if a and b are any constants, $ax_1(e) + bx_2(e)$ is integrable and*

$$\int_E (ax_1(e) + bx_2(e))\,dP(e) = a\int_E x_1(e)\,dP(e) + b\int_E x_2(e)\,dP(e).$$

1.8.4 *If $x_1(e) \leqslant x(e) \leqslant x_2(e)$ for all $e \in R$ where $x_1(e)$, $x(e)$, and $x_2(e)$ are integrable, then*

$$\int_E x_1(e)\,dP(e) \leqslant \int_E x(e)\,dP(e) \leqslant \int_E x_2(e)\,dP(e).$$

1.8.5 *$x(e)$ is integrable if and only if $|x(e)|$ is integrable, and furthermore,*

$$\left|\int_E x(e)\,dP(e)\right| \leqslant \int_E |x(e)|\,dP(e).$$

1.8.6 *If $x_1(e), x_2(e), \ldots$ is a sequence of random variables such that, for all $e \in R$,*

$$\lim_{\alpha\to\infty} x_\alpha(e) = x(e), \qquad \textit{and} \qquad |x_\alpha(e)| \leqslant y(e),$$

where $y(e)$ is integrable, then $x(e)$ is integrable and

$$\lim_{\alpha\to\infty} \int_E |x_\alpha(e) - x(e)|\,dP(e) = 0.$$

In particular

$$\lim_{\alpha\to\infty} \int_E x_\alpha(e)\,dP(e) = \int_E x(e)\,dP(e)$$

uniformly for all $E \in \mathscr{B}$.

1.8.7 *If $(R_k, \mathscr{B}_k, P')$ is the probability space induced from the probability space $(R, \mathscr{B}, P)$ by the vector random variable $(x_1(e), \ldots, x_k(e))$, if $g(x_1, \ldots, x_k)$ is measurable with respect to $\mathscr{B}_k$ (and thus $g(x_1(e), \ldots, x_k(e))$ is measurable with respect to $\mathscr{B}$), and if E is the set in R for which $(x_1(e), \ldots, x_k(e)) \in E_k'$, where $E_k' \in \mathscr{B}_k$, then we have*

$$\int_E g(x_1(e), \ldots, x_k(e))\,dP(e) = \int_{E_k'} g(x_1, \ldots, x_k)\,dP'(x_1, \ldots, x_k)$$

in the sense that if either integral is finite so is the other and the two are equal.

Finally, we remark that if, in **1.8.1** through **1.8.6**, we replace the phrase "all $e \in R$" by "all $e \in R$ except possibly for a set F for which $P(F) = 0$," the conclusions remain unchanged.

1.9 CONDITIONAL PROBABILITY

Suppose $(R, \mathscr{B}, P)$ is a probability space and let E_1 and E_2 be events in $\mathscr{B}$ such that $P(E_1) > 0$. Let us write

$$P(E_2 \mid E_1) = \frac{P(E_1 \cap E_2)}{P(E_1)}. \tag{1.9.1}$$

This ratio is called the *conditional probability* of event E_2, given that E_1 occurs.

It can be verified that for any fixed E_1 in $\mathscr{B}$ such that $P(E_1) > 0$, $(R, \mathscr{B}, P(\cdot \mid E_1))$ is a probability space, where $P(\cdot \mid E_1)$ is a measure which takes the value $P(E_2 \mid E_1)$ on $E_2 \in \mathscr{B}$.

Note that we can rewrite (1.9.1) as follows:

$$P(E_1 \cap E_2) = P(E_1) \cdot P(E_2 \mid E_1). \tag{1.9.2}$$

If $P(E_2 \mid E_1) = P(E_2)$, then (1.9.2) reduces to the case of *independent events* E_1 and E_2, and we have

$$P(E_1 \cap E_2) = P(E_1)P(E_2). \tag{1.9.2a}$$

More generally, it can be shown that:

1.9.1 *If $E_1, E_2, \ldots$ is a finite or countably infinite sequence of events in $\mathscr{B}$ having probability measure P such that $P(E_1) > 0$, $P(E_1 \cap E_2) > 0$, $P(E_1 \cap E_2 \cap E_3) > 0, \ldots$ then*

$$P(E_1 \cap E_2 \cap \cdots) = P(E_1) \cdot P(E_2 \mid E_1) \cdot P(E_3 \mid E_1 \cap E_2) \cdots \cdot P(E_n \mid E_1 \cap E_2 \cap \cdots \cap E_{n-1}) \cdots. \tag{1.9.3}$$

In the case of *mutual independence* of $E_1, E_2, \ldots$ (1.9.3) reduces to

$$P(E_1 \cap E_2 \cap E_3 \cap \cdots) = P(E_1)P(E_2)P(E_3) \cdots. \tag{1.9.3a}$$

Another useful result is the following:

1.9.2 *If $E_1, E_2, \cdots$ is a finite or countably infinite sequence of disjoint events in $\mathscr{B}$ having nonzero probabilities, where $\bigcup_\alpha E_\alpha = R$, and if E is any set in $\mathscr{B}$, then*

$$P(E) = P(E_1)P(E \mid E_1) + P(E_2)P(E \mid E_2) + \cdots. \tag{1.9.4}$$

Verification of this statement is left to the reader.

1.10 CONDITIONAL RANDOM VARIABLES

For a given probability space $(R, \mathscr{B}, P)$, suppose $(x_1(e), x_2(e))$ is a two-dimensional random variable relative to $\mathscr{B}$. If sets E_1 and E_2 in (1.9.1) are chosen, respectively, as the sets for which $x_1(e) \in E_1'$ and $x_2(e) \in E_2'$ where E_1' and E_2' are sets in $\mathscr{B}_1$, then (1.9.1) becomes a conditional probability formula concerning random variables. It gives the conditional probability of $x_2(e) \in E_2'$ given that $x_1(e) \in E_1'$. In particular, suppose E_1' is a single point, say x_1. If $P(x_1(e) = x_1) > 0$ there is no difficulty with (1.9.1). But if $P(x_1(e) = x_1) = 0$ the question arises: is there some sense in which we can give meaning to the conditional probability $P(x_2(e) \in E_2' \mid x_1(e) = x_1)$?

In most cases of interest in mathematical statistics we shall show in Section 2.9 that we can give meaning to this conditional probability by fairly elementary considerations. But under more general conditions, an answer to the question is provided by the Radon-Nikodym theorem which may be stated as follows:

1.10.1 *If $(R, \mathscr{B}, P)$ is a probability space, and if Q is a (finite) completely additive set function on $\mathscr{B}$ such that $Q(E) = 0$ for every set $E \in \mathscr{B}$ for which $P(E) = 0$, then there exists a random variable $g(e)$ such that*

$$Q(E) = \int_E g(e)\, dP \tag{1.10.1}$$

for every $E \in \mathscr{B}$. Furthermore, if $g(e)$ and $h(e)$ are two such random variables, then $P(g(e) \neq h(e)) = 0$.

We shall omit the proof of this theorem, referring the reader to Halmos (1950) for a slightly more general formulation and proof. It should be particularly noted that the values of Q are not restricted to be non-negative.

If two completely additive set functions P and Q defined on $\mathscr{B}$ are such that $Q(E) = 0$ for every set E for which $P(E) = 0$, then Q is said to be *absolutely continuous* with respect to P.

Suppose $(R, \mathscr{B}, P)$ is a probability space and $x(e)$ is a random variable. Suppose, for a given $E \in \mathscr{B}$, and a fixed number x_1, we want to find the conditional probability $P(E \mid x(e) = x_1)$. As we have seen, definition (1.9.1) becomes meaningless if $P(x(e) = x_1) = 0$.

Intuitively, however, we would like to define this conditional probability as a sort of "limit" of $P(E \mid x(e) \in N_\alpha(x_1))$ as $\alpha \to \infty$, where $N_1(x_1)$, $N_2(x_1), \ldots$ is a sequence of neighborhoods of x_1 converging to the point x_1. In general, this limit will not exist everywhere. But it is intuitively plausible that if it does exist in a well-behaved domain containing F, where $F \in \mathscr{B}_1$, one should be able to obtain $P(E \mid x(e) \in F)$ by suitably averaging $P(E \mid x(e) = x_1)$ over all possible $x_1 \in F$.

In general, it is difficult to establish such a limit as that mentioned above except under special conditions. However, Kolmogorov (1933a) has suggested the following approach.

For any $E \in \mathscr{B}$ and $F \in \mathscr{B}_1$, let

$$Q(F) = P(E \cap x^{-1}(F)). \tag{1.10.2}$$

If $P(x^{-1}(F)) = 0$, $Q(F) = 0$, and it is clear that Q is a completely additive set function on $\mathscr{B}_1$. By **1.10.1** there is a real-valued $\mathscr{B}_1$-measurable function $f(x)$ such that

$$Q(F) = \int_F f(x)\, dP'(x) \tag{1.10.3}$$

where $P'(F) = P(x^{-1}(F))$ for all sets $F \in \mathscr{B}_1$. Therefore, if we write $g(e) = f(x(e))$ we have

$$P(E \cap x^{-1}(F)) = \int_{x^{-1}(F)} g(e)\, dP(e) \tag{1.10.4}$$

and $g(e)$, that is, $f(x(e))$, is "the conditional probability of E given $x(e)$". It should be noted that $f(x(e))$ is unique in the sense of **1.10.1**. From this definition of "conditional probability of E given $x(e)$" it can be shown that, except possibly for a set in R_1 of probability zero, we have

$$f(x_1) = \lim_{\alpha \to \infty} P(E \mid x(e) \in N_\alpha(x_1)) \tag{1.10.5}$$

where $\{N_\alpha(x_1),\ \alpha = 1, 2, \ldots\}$ is a sequence of measurable neighborhoods of x_1 which converge to x_1.

The approach discussed above extends in a straightforward manner to the case of a vector random variable $(x_1(e), \ldots, x_k(e))$.

PROBLEMS

1.1 Two names are picked "at random" from a list of N different names and alphabetized. Describe the sample space generated by this operation and state how many sample points it contains. Generalize to the case where n names are drawn "at random" from the list of N names.

1.2 The birthdays (month and day of month) of two persons A and B picked "at random" from *Who's Who* are recorded. Ignoring leap years, describe the sample space generated by this operation and state the number of sample points in it. Give the numbers of sample points in the following events:

(a) "A and B have identical birthdays."

(b) "A's and B's birthdays are not more than r days apart"?

(c) "A's and B's birthdays are in different months."

1.3 A store opens at 9 A.M. and closes at 5 P.M. A shopper taken "at random" walks into this store at time x and out at time y (both x and y being measured in hours on the time axis with 9 A.M. as origin). Describe the sample space of (x, y). Describe, in terms of x and y, the following events:

(a) "The shopper is in the store less than one hour."

(b) "The shopper is in the store at time z."

(c) "The shopper went into the store before time u and out after time v."

1.4 A box of N light bulbs has r bulbs ($r < N$) with broken filaments and a person tests them one by one until a defective bulb (that is, one with a broken filament) is found, observing only whether a bulb lights up or not when tested. Describe the sample space generated by this operation. How many points are in the sample space? Generalize to the case where bulbs are tested one by one until exactly s defectives are found.

1.5 Let R be the set of all students at University A. Let E_1 denote the set of all students in A who subscribe to magazine M_1, E_2 the set who subscribe to magazine M_2, and E_3 the set who subscribe to magazine M_3.

(a) Describe, in words, the following sets:

$$E_1 \cup E_2 \cup E_3; \qquad E_1 \cap E_2 \cap E_3; \qquad R - (E_1 \cup E_2);$$
$$\bar{E}_1 \cup \bar{E}_2 \cup \bar{E}_3; \qquad E_1 \cap E_2 \cap \bar{E}_3; \qquad \bar{E}_1 \cap \bar{E}_2 \cap \bar{E}_3.$$

(b) Express in terms of operations on R, E_1, E_2, E_3 the following sets:

(i) The students who subscribe to two or more of the three magazines.

(ii) The students who subscribe to not more than one of the three magazines.

1.6 Let R be the set of all possible hands of 13 cards in an ordinary pack of playing cards. Let E_1, E_2, E_3, E_4 be the sets of different hands of 13 cards containing, respectively, the ace of spades, ace of hearts, ace of diamonds, and ace of clubs. Describe in words the following sets and state how many hands there are in each set:

$$E_1 \cap E_2; \qquad E_1 - (E_2 \cap E_3); \qquad (E_1 \cap E_2 \cap E_3) \cup E_4;$$
$$E_1 \cup (E_2 \cap E_3); \qquad (E_1 \cup E_2) - \bar{E}_3; \qquad \bar{E}_1 \cup \bar{E}_2; \qquad \bar{E}_1 \cap (\bar{E}_2 \cup \bar{E}_3);$$
$$(E_1 \cup E_2 \cup E_3) - E_4; \qquad R - [(E_1 \cup E_2) \cap (E_3 \cup \bar{E}_4)].$$

1.7. The game of craps is played with a pair of ordinary 6-sided dice as follows. If the shooter rolls 7 or 11 he wins without further throwing. If he rolls 2, 3, or 12 he loses without further throwing. If he rolls 4, 5, 6, 8, 9, or 10 he must continue throwing until he gets a 7 *or* the number initially thrown. If 7 appears first he loses. If the point he initially threw appears he wins. Describe the sample space involved, and assuming true dice show that the probability the shooter wins is 244/495.

1.8 If a sample space R contains N sample points, show that the total number of events in the Boolean field generated by these N points is 2^N.

1.9 Consider the two infinite sequences of sets $E_1, E_2, \ldots$ and $F_1, F_2, \ldots$ in the xy-plane where E_n is the set of points for which $x^2 + y^2 < (1 + n)/n$ and F_n is the set of points for which $x^2 + y^2 \leqslant n/(1 + n)$. Let $\bar{E}_n$ be the complement of E_n with respect to the entire xy-plane, with a similar definition of $\bar{F}_n$. Describe, in terms of x and y, the following sets:

$$\lim_{\alpha} E_\alpha; \qquad \lim_{\alpha} \bar{E}_\alpha; \qquad \lim_{\alpha} F_\alpha; \qquad \lim_{\alpha} \bar{F}_\alpha; \qquad \lim_{\alpha} E_\alpha \cap \bar{F}_\alpha.$$

1.10 If $E_1, \ldots, E_n$ are arbitrary events in a sample space R whose complements are $\bar{E}_1, \ldots, \bar{E}_n$ show that

$$\bigcup_{\alpha=1}^{n} E_\alpha \quad \text{and} \quad \bigcap_{\alpha=1}^{n} \bar{E}_\alpha$$

are disjoint and that their union is R, and hence that

$$P\left(\bigcup_{\alpha=1}^{n} E_\alpha\right) = 1 - P\left(\bigcap_{\alpha=1}^{n} \bar{E}_\alpha\right).$$

1.11 Events $E_1, \ldots, E_n$ are such that the probability of the occurrence of any specified r of them is p_r, $r = 1, \ldots, n$. Show:

(a) That the probability of the occurrence of one or more of the events $E_1, \ldots, E_n$ is

$$\binom{n}{1}p_1 - \binom{n}{2}p_2 + \cdots + (-1)^{n-1}\binom{n}{n}p_n.$$

(b) That the probability of the occurrence of m or more of the events $E_1, \ldots, E_n$ is

$$\binom{m-1}{m-1}\binom{n}{m}p_m - \binom{m}{m-1}\binom{n}{m+1}p_{m+1} + \cdots + (-1)^{n-m}\binom{n-1}{m-1}\binom{n}{n}p_n.$$

(c) That the probability of the occurrence of exactly m of the events $E_1, \ldots, E_n$ is

$$\binom{m}{m}\binom{n}{m}p_m - \binom{m+1}{m}\binom{n}{m+1}p_{m+1} + \cdots + (-1)^{n-m}\binom{n}{m}\binom{n}{n}p_n.$$

1.12 A company manufacturing cornflakes puts a card numbered 1 or 2 or, . . . , or r at random in each package, all numbers being equally likely to be drawn. If $n(>r)$ boxes of cornflakes are purchased, show that the probability of being able to assemble at least one complete set of cards from the packages is

$$1 - \binom{r}{1}\left(1 - \frac{1}{r}\right)^n + \binom{r}{2}\left(1 - \frac{2}{r}\right)^n \cdots + (-1)^{r-1}\binom{r}{r-1}\left(1 - \frac{r-1}{r}\right)^n.$$

1.13 If an urn has N chips numbered $1, 2, \ldots, N$ and if two chips are drawn successively (without replacement) let e be any point in the sample space R generated by this operation. Let $x(e)$ be the absolute difference between the numbers on the two chips which yield e. If all points e in the sample space R are assigned equal probabilities, show that $x(e)$ is a random variable, describe its sample space and write down the formula for $P(x(e) = x')$.

1.14 In Problem 1.4 suppose all possible sequences in which the N bulbs can be tested are assigned equal probabilities. What is the probability that the sth defective bulb ($s \leqslant r$) to be found will occur with the testing of the xth bulb tested? State the range of values of x for which the required probability is positive.

1.15 Suppose a coin is thrown successively n times and that e is a sample point in the 2^n points in the sample space R generated by this operation. Let $x(e)$ be the number of heads in e. If the sample points in R are all assigned equal probabilities, show that $x(e)$ is a random variable, describe its sample space, and write down the expression for $P(x(e) = x')$.

1.16 Suppose R is a sample space whose elements e are the points inside the square with vertices (0, 0), (0, 1), (1, 0), and (1, 1) in the uv-plane. For any point e having coordinates (u, v) let $x(e) = u + v$. If E is any triangle or quadrilateral in R and if $P(E)$ = area of E, show that $x(e)$ is a random variable for which $P(x(e) \leqslant x')$ can be computed for every real x', describe its sample space, and write down the expression for $P(x(e) \leqslant x')$ as a function of x'. If F is the event in R for which $u/2 < v < u$, compute the value of $\int_F x(e)\,dP$. Compute the value of $\int_R x(e)\,dP$.

1.17 (*Continuation*) Let $y(e) = u/v$. Show that $y(e)$ is a random variable such that $P(y(e) \leqslant y')$ can be computed for every real y', and find the expression for $P(y(e) \leqslant y')$.

1.18 (*Continuation*) Show that $(x(e), y(e))$ is a two-dimensional random variable such that $P(x(e) \leqslant x', y(e) \leqslant y')$ can be computed for each real x' and y', and write down the expression for $P(x(e) \leqslant x', y(e) \leqslant y')$.

1.19 (*Continuation*) Let E_n, $n = 1, 2, \ldots$ be triangles in R for which $u > 0$, $v > 0$, $u + v \leqslant (1 + n)/2n$. What is $\lim_{n\to\infty} E_n$? Show that $\lim_{n\to\infty} P(E_n) = P\left(\lim_{n\to\infty} E_n\right) = \frac{1}{8}$. Show that the infinite sequence $E_1, E_2, \ldots$ satisfies formula (1.9.3).

1.20 Prove **1.9.1** and **1.9.2**.

1.21 Suppose $(R, \mathscr{B}, P)$ is a probability space and $x(e)$ is a non-negative random variable relative to $\mathscr{B}$. Let $I_{\delta,\alpha}$ be the interval $(\alpha\delta, (\alpha + 1)\delta)$, $\alpha = 0, 1, 2, \ldots$ where $\delta > 0$, and $I_{\delta,\alpha}^{-1}$ the set in $\mathscr{B}$ for which $x(e) \in I_{\delta,\alpha}$. For $E \in \mathscr{B}$ let

$$A(\delta) = \sum_{\alpha=1}^{\infty} \alpha\delta P(E \cap I_{\delta,\alpha}^{-1}).$$

If $A(\delta)$ is finite for some value of δ, show that $\lim_{\delta\to 0} A(\delta)$ exists and is the Lebesgue-Stieltjes integral of $x(e)$ over E.

1.22 Let $E_1, \ldots, E_n$ be arbitrary events and let $p_m(\alpha_1, \ldots, \alpha_r)$ be the probability that at least m of the events $E_{\alpha_1}, \ldots, E_{\alpha_r}$ occur. Show that

$$(k + 1 - m)\Sigma_{k+1} p_m(\alpha_1, \ldots, \alpha_{k+1}) \leqslant (n - k)\Sigma_k p_m(\alpha_1, \ldots, \alpha_k)$$

$k = 1, \ldots, n - 1$, $1 \leqslant m \leqslant k$, where Σ_i, $i = k, k + 1$, denotes summation over all $\binom{n}{i}$ possible selections of i integers from $1, \ldots, n$. (Chung (1941)).

CHAPTER 2

Distribution Functions

2.1 PRELIMINARY REMARKS

As we have seen in Section 1.8 a (one-dimensional) random variable $x(e)$ induces a probability space $(R_1, \mathscr{B}_1, P')$ associated with the real line R_1, from the basic probability space $(R, \mathscr{B}, P)$. Similarly, a k-dimensional random variable $(x_1(e), \ldots, x_k(e))$ induces a probability space $(R_k, \mathscr{B}_k, P')$ associated with Euclidean space R_k from the basic probability space $(R, \mathscr{B}, P)$.

We shall usually drop the reference to event points e in R in dealing with a random variable $x(e)$ and call it the random variable* x. If x is the (one-dimensional) random variable whose probability space is $(R_1, \mathscr{B}_1, P')$ and if E' is any set in $\mathscr{B}_1$, it will sometimes avoid ambiguity to denote the event E' by $x \in E'$ and to use the notation $P(x \in E')$ instead of $P'(E')$. In particular, if E' is the interval $(a, b]$, we shall understand that $P(x \in E')$ may be written as $P(a < x \leqslant b)$; if E' consists of a single point x', we may write $P(x = x')$, which is the probability that the value x' is *realized* by the random variable x in a given trial. If x is a k-dimensional vector random variable $(x_1, \ldots, x_k)$ and if it is desirable to indicate the dimensionality of x to avoid ambiguity we shall denote an event E' in $\mathscr{B}_k$ by $(x_1, \ldots, x_k) \in E'$, and use $P((x_1, \ldots, x_k) \in E')$ rather than $P(x \in E')$ or $P'(E')$. If E' is the k-dimensional interval $(a_1, \ldots, a_k; b_1, \ldots, b_k]$ denoted by $(a, b]_k$, it will be understood that $P((x_1, \ldots, x_k) \in E')$ can also be written as $P(a_i < x_i \leqslant b_i, i = 1, \ldots, k)$.

In discussing a k-dimensional random variable $(x_1, \ldots, x_k)$, we shall

* Ideally, it would be desirable to denote random variables by bold face letters or by some other characteristic marking. This is, however, hardly practical in a book containing many applications of random variable theory involving many different symbols (some of them classical) designating random variables. It will be made clear in the text whenever a quantity under discussion is, in fact, a random variable.

sometimes find it convenient to refer to it as "the random variables $x_1, \ldots, x_k$" and sometimes as "the k-dimensional random variable with components $x_1, \ldots, x_k$."

We shall be concerned frequently with *bounded sets* in R_k. A set E' in R_k is *bounded* if it is contained in some finite k-dimensional interval $(a, b]_k$, that is, where $a_1, \ldots, a_k$; $b_1, \ldots, b_k$ are all finite. If a random variable $(x_1, \ldots, x_k)$ has the property that $P((x_1, \ldots, x_k) \in (a, b]_k) = 1$, where $(a, b]_k$ is finite, then $(x_1, \ldots, x_k)$ is said to be *bounded with probability* 1.

2.2 DISTRIBUTION FUNCTIONS OF ONE-DIMENSIONAL RANDOM VARIABLES

Suppose x is a one-dimensional random variable whose probability space* is $(R_1, \mathscr{B}_1, P)$. We shall show that the allocation or distribution of probability over the sample space of x in R_1 can also be described by a *distribution function* $F(x)$ defined at each point in R_1 and having certain properties.

For any interval $(-\infty, x']$ on R_1 which, of course, belongs to $\mathscr{B}_1$, let $F(x')$ be defined as follows:

(2.2.1)
$$F(x') = P(-\infty < x \leqslant x').$$

$F(x)$ is clearly a single-valued, real, and non-negative function of x in R_1.

If $x'' > x'$ we have from **1.4.2**

(2.2.2)
$$\begin{aligned} F(x'') - F(x') &= P(-\infty < x \leqslant x'') - P(-\infty < x \leqslant x') \\ &= P(x' < x \leqslant x'') \geqslant 0. \end{aligned}$$

Hence $F(x)$ is a nondecreasing function of x.

If we denote the interval $(-\infty, \alpha]$ by E_α, then we have the following contracting sequence of sets

$$E_{-1} \supset E_{-2} \supset \cdots$$

for which it is clear that $\lim\limits_{\alpha \to \infty} E_{-\alpha} = \phi$. Hence we have from **1.4.5**

$$F(-\infty) = \lim_{\alpha \to \infty} F(-\alpha) = \lim_{\alpha \to \infty} P(E_{-\alpha}) = P\left(\lim_{\alpha \to \infty} E_{-\alpha}\right) = P(\phi) = 0,$$

that is,

(2.2.3)
$$F(-\infty) = 0.$$

* From now on, unless otherwise indicated, E (not E') will denote a set in $\mathscr{B}_1$, or more generally in $\mathscr{B}_k$, and we shall drop the dash on P.

Similarly, we have the expanding sequence

$$E_1 \subset E_2 \subset \cdots$$

for which $\lim_{\alpha \to \infty} E_\alpha = R_1$. Therefore from **1.4.5**, we have

$$F(+\infty) = \lim_{\alpha \to \infty} F(\alpha) = \lim_{\alpha \to \infty} P(E_\alpha) = P\left(\lim_{\alpha \to \infty} E_\alpha\right) = P(R_1) = 1,$$

that is,

$$F(+\infty) = 1. \tag{2.2.4}$$

Now consider a decreasing sequence of real numbers $x_1, x_2, \ldots$ such that $\lim_{\alpha \to \infty} x_\alpha = x'$. Then $E_{x_1}, E_{x_2}, \ldots$ is a contracting sequence of sets having $E_{x'}$ as their limit. Again by applying **1.4.5**, we have

$$\lim_{\alpha \to \infty} F(x_\alpha) = \lim_{\alpha \to \infty} P(E_{x_\alpha}) = P\left(\lim_{\alpha \to \infty} E_{x_\alpha}\right) = P(E_{x'}) = F(x')$$

that is,

$$F(x' + 0) = F(x'). \tag{2.2.5}$$

In other words, dropping dashes, the function $F(x)$ is *continuous on the right* at each value of x.

The reader should observe that (2.2.5) is purely a consequence of defining $F(x')$ as the probability contained in the half-closed interval $(-\infty, x']$. If we had chosen to define $F(x')$ as $P(-\infty < x < x')$, then we would have had $F(x' - 0) = F(x')$, that is, $F(x)$ would have been continuous on the left at each value of x. Hence the definition of $F(x')$ as the probability contained in the half-closed interval $(-\infty, x']$ rather than open interval $(-\infty, x')$ and having consequence (2.2.5) should be viewed as a convention.

Thus, if x is a random variable having probability space $(R_1, \mathscr{B}_1, P)$, there exists a function $F(x)$ defined by (2.2.1) (dropping the dashes), at every point x in R_1, and having properties (2.2.2) through (2.2.5).

Conversely, if a function $F(x)$ defined by (2.2.1) and having properties (2.2.2) through (2.2.5) is given, there exists a probability space $(R_1, \mathscr{B}_1, P)$. For we may consider as our initial class $\mathscr{F}_0$ of sets in R_1 the class of half-open intervals $(-\infty, x]$ for all real numbers x, and take the probability assigned to $(-\infty, x]$ as $F(x)$. This initial class of sets generates a Borel field $\mathscr{B}(\mathscr{F}_0)$ which is, by definition, the class $\mathscr{B}_1$. Utilizing properties (2.2.2) through (2.2.5) of $F(x)$, it can be shown that a unique probability measure can be constructed from $F(x)$ on the Borel field generated by $\mathscr{F}_0$.

Summarizing, we have the following result:

2.2.1 *The probability space* $(R_1, \mathscr{B}_1, P)$ *of a random variable* x *uniquely determines a single-valued, real, and non-negative function* $F(x)$ *defined by* (2.2.1) *for every point* x *in* R_1 *having the following properties:*

$$\begin{aligned} &(a)\ F(x'') - F(x') \geqslant 0, \qquad \text{if } x'' > x' \\ &(b)\ F(-\infty) = 0 \\ &(c)\ F(+\infty) = 1 \\ &(d)\ F(x + 0) = F(x). \end{aligned} \tag{2.2.6}$$

Conversely, a function $F(x)$ *having these properties uniquely determines a probability space* $(R_1, \mathscr{B}_1, P)$ *with* $P(E_x) = F(x)$.

$F(x)$ is called the *distribution function* (d.f.) or the *cumulative distribution function* (c.d.f.) of the (one-dimensional) random variable x. We shall ordinarily use the latter term. If one thinks of a total probability of 1 being distributed along the x-axis then $F(x)$ is simply the fraction of the probability lying on $(-\infty, x]$.

Thus, we have two alternative ways of describing the probabilities associated with a (one-dimensional) random variable x. One is by means of a probability space $(R_1, \mathscr{B}_1, P)$ and the other is by means of c.d.f. $F(x)$ defined at every point in R_1, satisfying the conditions expressed in (2.2.6). The c.d.f. description is more convenient in the analysis of random variables for most purposes, and will be used almost entirely from now on. It should be noted that if we are given a basic sample space R and a random variable $x(e)$ defined on points e in R there exists a c.d.f. $F(x)$ of this random variable defined by

$$P(x(e) \leqslant x) = F(x).$$

Conversely, suppose we are given a c.d.f. $F(x)$ with no reference to a basic sample space R. We can always define a random variable x whose c.d.f. is $F(x)$ by considering the real axis as the basic sample space R. The sample points e will then be the real numbers and for any given real number x', we assign our random variable $x(e)$ the value x', that is, $x(x') = x'$. Then we have $P(x(e) \leqslant x') = F(x')$.

The probability $P(x \in E)$ where E is any set in $\mathscr{B}_1$ exists and can be determined from $F(x)$. For instance, the probability $P(x' < x \leqslant x'')$ is determined from $F(x)$ by formula (2.2.2). Other useful probability statements concerning the random variable x which can be verified by the reader, using methods similar to those involved in obtaining (2.2.5), are the following:

$$P(x = x') = F(x') - F(x' - 0) \tag{2.2.7}$$

$$P(x' < x < x'') = F(x'' - 0) - F(x') \tag{2.2.8}$$

$$P(x' \leqslant x < x'') = F(x'' - 0) - F(x' - 0) \tag{2.2.9}$$

$$P(x' \leqslant x \leqslant x'') = F(x'') - F(x' - 0). \tag{2.2.10}$$

Note that if x is a bounded random variable, then there exist finite numbers a, b with $a < b$ where a is the largest number for which $F(a) = 0$ and b is the smallest number for which $F(b) = 1$; $b - a$ is called the *range* of x.

2.3 COMMON TYPES OF ONE-DIMENSIONAL RANDOM VARIABLES

Most one-dimensional random variables which arise in mathematical statistics belong to one of two types: the *discrete* type and the *continuous* type. The distribution function $F(x)$ and, of course, the probability measure $P(E)$, for these two types of random variables can be defined in terms of alternative, if not more primitive, functions as we shall see presently.

(a) The Discrete Type

In the *discrete* type the c.d.f. $F(x)$ is a step-function, that is, its value changes only at a finite or a countably infinite number of points $x^{(1)}$, $x^{(2)}, \ldots$ in R_1, having no finite limit point, at which *jumps* or *saltuses* of size $p(x^{(1)}), p(x^{(2)}), \ldots$ occur. The saltus $p(x^{(\alpha)})$ is given by (2.2.7) as

$$p(x^{(\alpha)}) = P(x = x^{(\alpha)}) = F(x^{(\alpha)}) - F(x^{(\alpha)} - 0). \tag{2.3.1}$$

At all other points x' in R_1, $p(x') = P(x = x') = 0$. $F(x)$ can be expressed in terms of these saltuses as follows:

$$F(x) = \sum_{x^{(\alpha)} \leqslant x} p(x^{(\alpha)}), \tag{2.3.2}$$

where the summation extends over all values of α for which $x^{(\alpha)} \leqslant x$. Letting $x \to +\infty$ in (2.3.2) it is seen that we must have

$$F(+\infty) = \sum_{\alpha} p(x^{(\alpha)}) = 1. \tag{2.3.3}$$

Since the total probability 1 is distributed among the points $x^{(1)}$, $x^{(2)}, \ldots$ it is customary to refer to these points as *probability points* or *mass points*. This set of points comprises the sample space of a random variable x having a c.d.f. given by (2.3.2).

In dealing with a discrete random variable x it will be clear from the context what the mass points of x are. Hence there will be no ambiguity if we drop α and simply write $p(x)$. The function $p(x)$ will be called the *probability function* (p.f.) of x.

We may summarize as follows:

2.3.1 *The c.d.f. $F(x)$ of a discrete random variable x is uniquely determined by the p.f. $p(x)$ and conversely.*

In dealing with problems involving a discrete random variable, it is usually more convenient to work with the p.f. $p(x)$ than with the c.d.f. $F(x)$.

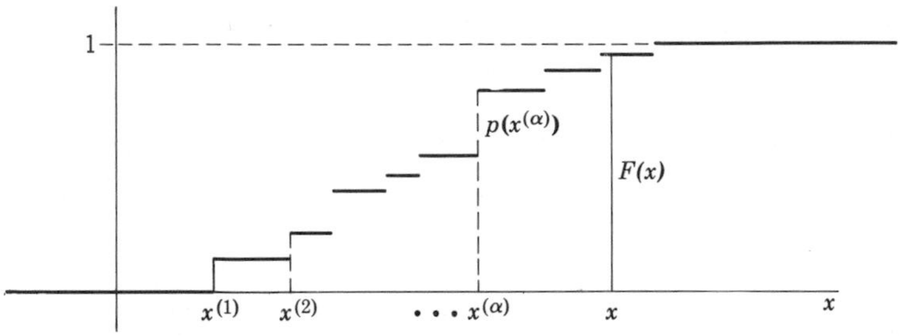

FIG. 2.1 Graph of the c.d.f. $F(x)$ of a one-dimensional discrete random variable x.

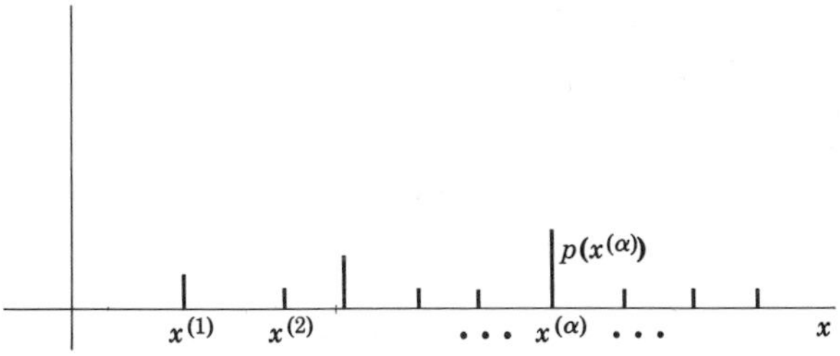

FIG. 2.2 Graph of p.f. $p(x)$ corresponding to the c.d.f. $F(x)$ graphed in Fig. 2.1.

The c.d.f. $F(x)$ of a one-dimensional discrete random variable x can be represented graphically as the graph of a step-function whose jumps of magnitude $p(x^{(1)}), p(x^{(2)}), \dots$ occur at the mass points $x^{(1)}, x^{(2)}, \dots$ respectively, as shown in Fig. 2.1.

The p.f. $p(x)$ of the random variable whose c.d.f. is represented in Fig. 2.1 can be represented graphically by vertical line segments of lengths $p(x^{(1)})$, $p(x^{(2)}), \dots$ located at the mass points $x^{(1)}$, $x^{(2)}, \dots$, respectively, and zero elsewhere, as shown in Fig. 2.2.

If $F(x)$ has only one saltus, say $p(x^{(1)})$, occurring at the mass point $x^{(1)}$, then x is called a *degenerate* (one-dimensional) random variable, and we shall denote its c.d.f. by

$$\varepsilon(x - x^{(1)}) = \begin{cases} 1 & x \geqslant x^{(1)} \\ 0 & x < x^{(1)}. \end{cases} \tag{2.3.4}$$

It should be noted that the c.d.f. $F(x)$ given by (2.3.1) can be expressed in terms of the ε-function (2.3.4) as follows:

$$F(x) = \sum_{\alpha} p(x^{(\alpha)}) \cdot \varepsilon(x - x^{(\alpha)}). \tag{2.3.5}$$

Examples. Examples of one-dimensional discrete random variables are plentiful in elementary probability theory. For instance, if x is the random variable denoting the number of dots occurring in a throw of a single "true" 6-sided die, then the mass points are $x^{(1)} = 1, \ldots, x^{(6)} = 6$ and probabilities are assigned so that the p.f. is given by $p(x) = \frac{1}{6}$, $x = 1, \ldots, 6$ and $p(x) = 0$ for all other values of x. If x is the number of aces occurring in a single hand of 13 cards dealt from a "well-shuffled" pack of 52 ordinary playing cards, then $x^{(1)} = 0$, $x^{(2)} = 1, \ldots, x^{(5)} = 4$ and probabilities are assigned so that the p.f. is given by

$$p(x) = \frac{\binom{4}{x}\binom{48}{13 - x}}{\binom{52}{13}}, \quad x = 0, 1, 2, 3, 4$$

and $p(x) = 0$ for all other values of x.

Further examples of important discrete random variables will be discussed in Chapter 6.

(b) The Continuous Type

For the *continuous* type of random variable there exists a Lebesgue-measurable function $f(x) \geqslant 0$ such that

$$F(x') = \int_{-\infty}^{x'} f(y)\, dy, \tag{2.3.6}$$

for all $x' \in R_1$. In this case dF/dx exists and

$$\frac{dF}{dx} = f(x) \tag{2.3.7}$$

except possibly for a set of values of x of probability 0. As a matter of fact a function $f(x)$ exists which satisfies (2.3.6) and (2.3.7) if and only if $F(x)$ is *absolutely continuous*, and a random variable having such a c.d.f. is sometimes called an *absolutely continuous* random variable.

Sometimes we have occasion to deal with a random variable x having merely a *continuous* c.d.f. $F(x)$, in which case $F(x)$ would not be assumed to satisfy (2.3.6) and (2.3.7). This more general type of random variable should not be confused with the case in which $F(x)$ is *absolutely continuous*. The occasions in which $F(x)$ is merely continuous rather than absolutely continuous are rare, however, and it is customary to drop the adjective "absolutely" and in such a case say that x is a *continuous* random variable. At any rate it will always be clear from the context of a situation which type of continuity is involved. It will be seen, for instance, that a good deal of the sampling theory underlying order statistics and nonparametric inference in Chapters 8, 11, and 13, holds for the case where $F(x)$ is merely continuous. Consider the expression

$$\frac{F(x'') - F(x')}{x'' - x'}. \tag{2.3.8}$$

The ratio (2.3.8) is non-negative and represents the average probability per unit length contained in $(x', x'']$. If the limit of the ratio exists as $x'' \to x'$, we obtain $f(x') \geqslant 0$ which may be thought of as a density of probability at the point $x = x'$. Accordingly, we shall call $f(x)$ the *probability density function* (p.d.f.) of the random variable x. It may be useful to summarize as follows:

2.3.2 *The c.d.f. $F(x)$ of a continuous random variable x is uniquely determined by its p.d.f. $f(x)$ in accordance with* (2.3.6). *Conversely, $f(x)$ is determined by $F(x)$ in accordance with* (2.3.7) *except possibly for a set of values of x of probability* 0.

If (2.3.7) holds, then for any (Borel) set E on R_1, we have

$$P(x \in E) = \int_E f(x)\, dx. \tag{2.3.9}$$

It follows at once from (2.3.9) that if $E = R_1$, then $\int_{R_1} f(x)\, dx = 1$.

It is sometimes convenient to use ordinary differential notation and write

$$P(x' < x < x' + dx) = f(x')\, dx \tag{2.3.10}$$

understanding, of course, the usual caution which must be exercised in dealing with differentials. The quantity $f(x)\, dx$ (dropping the dash) is called the *probability element* (p.e.) of x. If $F(x)$ is merely continuous, it is still convenient to denote $P(x' < x < x' + dx)$ by $dF(x')$ and refer to it as the p.e. of x at $x = x'$.

For any number p on the interval (0, 1), the pth *quantile* or *fractile* (100pth *percentile*) $\underline{x}_p$ of the continuous random variable x having c.d.f. $F(x)$ is defined by the smallest number $\underline{x}_p$ for which

$$F(\underline{x}_p) = p. \tag{2.3.11}$$

In particular, $\underline{x}_{0.5}$ is the *median* of x, and $\underline{x}_{0.25}$ and $\underline{x}_{0.75}$ are the *lower quartile* and *upper quartile*, respectively, of x. In the case of a discrete random variable x, if there is at least one value of x for which $F(x) = p$, the pth quantile is the smallest of such values. Thus quantiles in this case are defined only at the mass points of the random variable.

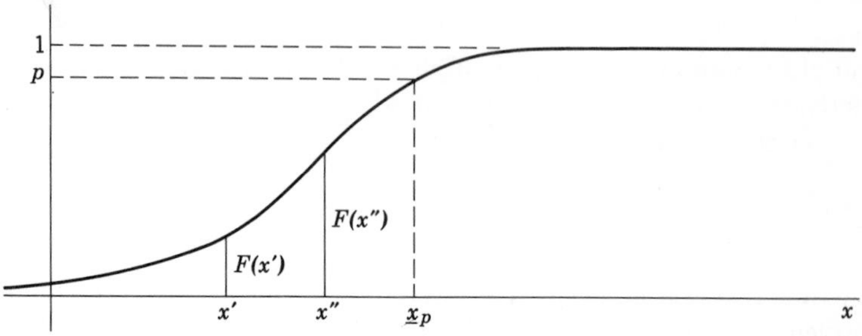

FIG. 2.3 Graph of the c.d.f. $F(x)$ of a one-dimensional continuous random variable x.

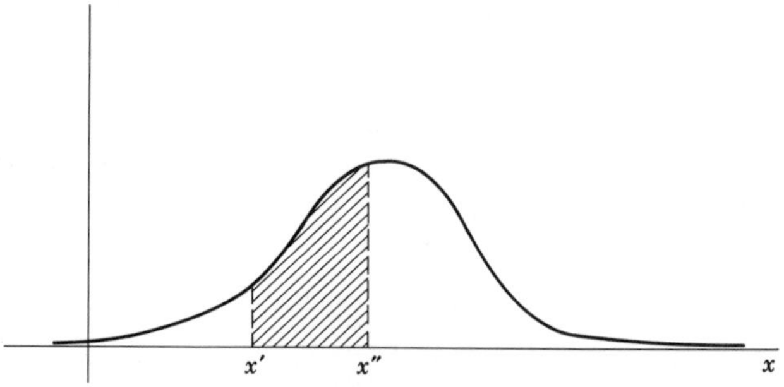

FIG. 2.4 Graph of the p.d.f. $f(x)$ of the c.d.f. $F(x)$ represented in Fig. 2.3.

In dealing with problems involving a continuous random variable x, it is usually more convenient to work with the p.d.f. $f(x)$ than with the c.d.f. $F(x)$.

The c.d.f. $F(x)$ of a one-dimensional continuous random variable and the corresponding p.d.f. $f(x)$ are represented in Figs. 2.3 and 2.4 respectively. In Fig. 2.3 the value of $P(x' < x \leqslant x'')$ is represented by the difference between the two ordinates $F(x')$ and $F(x'')$, whereas in Fig. 2.4 the same probability is represented by the shaded area.

Example. We shall deal with several important special probability distributions of the continuous type in Chapter 7. It may be useful, however, to give the following simple example. Suppose the probability that a "random" point lies inside any circle of radius δ which in turn lies within a given circle C of radius r is δ^2/r^2, and we are interested in the distance of the "random" point from the center of circle C. If we set up a random variable x to denote the distance between the "random" point and the center of the given circle and for a given x' define $F(x')$ as $P(x \leqslant x')$, then the c.d.f. of x (assuming the point falls in C) is given by

$$F(x) = \begin{cases} 1, & x > r \\ \dfrac{x^2}{r^2}, & 0 < x \leqslant r \\ 0 & x \leqslant 0 \end{cases}$$

and the p.d.f. of x is given by

$$f(x) = \begin{cases} \dfrac{2x}{r^2}, & 0 < x \leqslant r \\ 0 & x \leqslant 0, \quad x > r. \end{cases}$$

It can be shown that the most general form of $F(x)$ is a convex combination of a discrete c.d.f. $F_1(x)$ and a continuous (not necessarily absolutely continuous) c.d.f. $F_2(x)$, that is, $F(x) = aF_1(x) + bF_2(x)$, where a and b are non-negative and $a + b = 1$.

2.4 DISTRIBUTION FUNCTIONS OF TWO-DIMENSIONAL RANDOM VARIABLES

(a) General Properties

Suppose (x_1, x_2) is the two-dimensional random variable having probability space $(R_2, \mathscr{B}_2, P)$. Let $E_{(x_1,x_2)}$ be the interval $(-\infty, -\infty;\ x_1, x_2]$ in R_2. Let

(2.4.1) $$F(x_1', x_2') = P(E_{(x_1',x_2')}) = P(-\infty < x_1 \leqslant x_1', -\infty < x_2 \leqslant x_2').$$

$F(x_1, x_2)$ is clearly a single-valued, real, and non-negative function of (x_1, x_2) in R_2.

Any interval I_2 of the form $(x_1', x_2';\ x_1'', x_2'']$ belongs to $\mathscr{B}_2$ since

(2.4.2) $$I_2 = (E_{(x_1'',x_2'')} - E_{(x_1',x_2'')}) - (E_{(x_1'',x_2')} - E_{(x_1',x_2')}).$$

Furthermore, the probability that $(x_1, x_2) \in I_2$ is seen to be

(2.4.3) $$P((x_1, x_2) \in I_2) = F(x_1'', x_2'') - F(x_1', x_2'') - F(x_1'', x_2') + F(x_1', x_2').$$

It will be convenient to call the expression on the right of (2.4.3) $\Delta_{I_2}^2 F(x_1, x_2)$, the *second difference of* $F(x_1, x_2)$ *over* I_2. We then have

(2.4.4) $$P((x_1, x_2) \in I_2) = \Delta_{I_2}^2 F(x_1, x_2) \geqslant 0.$$

Figure 2.5 relates to the various quantities in $\Delta_{I_2}^2 F(x_1, x_2)$. $F(x_1', x_2'')$ and $F(x_1'', x_2')$ are the probabilities contained in the infinite regions (including their boundaries) shaded with vertical and horizontal lines, respectively. $F(x_1', x_2')$ is the probability contained in the doubly-shaded region (including upper and right boundaries) and $\Delta^2_{I_2}F(x_1, x_2)$ is the probability contained in the nonshaded rectangular region (including the upper and right boundaries).

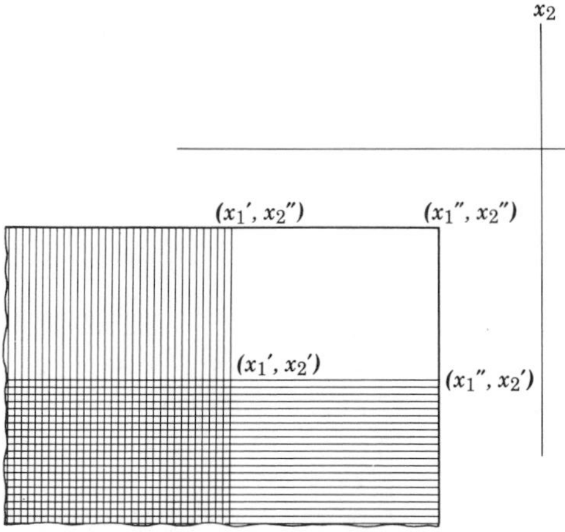

FIG. 2.5 Diagram relating to the formula for $\Delta^2_{I_2}F(x_1, x_2)$.

Now consider the sets

$$E_{(-\alpha, x_2')}, \qquad \alpha = 1, 2, \ldots$$

which is clearly a contracting sequence of sets whose limit $\lim_{\alpha\to\infty} E_{(-\alpha, x_2')} = \phi$. Therefore, it follows from **1.4.5** that

$$\lim_{\alpha\to\infty} F(-\alpha, x_2') = \lim_{\alpha\to\infty} P(E_{(-\alpha,x_2')}) = P\left(\lim_{\alpha\to\infty} E_{(-\alpha,x_2')}\right) = P(\phi) = 0.$$

Hence, dropping the dash,

$$F(-\infty, x_2) = 0. \tag{2.4.5}$$

Similarly,

$$F(x_1, -\infty) = 0. \tag{2.4.5a}$$

Now consider the sets

$$E_{(\alpha,\alpha)}, \qquad \alpha = 1, 2, \ldots.$$

This is an expanding sequence such that $\lim_{\alpha\to\infty} E_{(\alpha,\alpha)} = R_2$. Hence we have from **1.4.5**

$$\lim_{\alpha\to\infty} F(\alpha, \alpha) = \lim_{\alpha\to\infty} P(E_{(\alpha,\alpha)}) = P\left(\lim_{\alpha\to\infty} E_{(\alpha,\alpha)}\right) = P(R_2) = 1,$$

that is,

$$F(+\infty, +\infty) = 1. \tag{2.4.6}$$

By argument similar to that used in establishing (2.2.5) it can be shown that $F(x_1, x_2)$ is continuous on the right in each variable; that is, at each point (x_1, x_2) in R_2,

$$F(x_1 + 0, x_2) = F(x_1, x_2 + 0) = F(x_1, x_2). \tag{2.4.7}$$

From (2.4.7) it can be verified that $F(x_1 + 0, x_2 + 0) = F(x_1, x_2)$.

Hence, if (x_1, x_2) is a random variable having probability space $(R_2, \mathscr{B}_2, P)$, a unique function $F(x_1, x_2)$ exists, defined by (2.4.1) (dropping dashes) at every point (x_1, x_2) in R_2, and having properties (2.4.4) through (2.4.7).

Conversely, as in the one-dimensional case, a function $F(x_1, x_2)$ defined by (2.4.1) and having properties (2.4.4) through (2.4.7) uniquely determines a probability space $(R_2, \mathscr{B}_2, P)$. Summarizing, we have the two-dimensional analogue of **2.2.1**:

2.4.1 *The probability space $(R_2, \mathscr{B}_2, P)$ of a two-dimensional random variable (x_1, x_2) uniquely determines a single-valued, real, and non-negative function $F(x_1, x_2)$ defined by* (2.4.1) *at each point (x_1, x_2) in R_2 having the following properties;*

$$\begin{aligned} &(a)\ \Delta^2_{I_2} F(x_1, x_2) \geqslant 0 \\ &(b)\ F(-\infty, x_2) = F(x_1, -\infty) = 0 \\ &(c)\ F(+\infty, +\infty) = 1 \\ &(d)\ F(x_1 + 0, x_2) = F(x_1, x_2 + 0) = F(x_1, x_2). \end{aligned} \tag{2.4.8}$$

Conversely, a function $F(x_1, x_2)$ having these properties uniquely determines a probability space $(R_2, \mathscr{B}_2, P)$ such that $P(E_{(x_1,x_2)}) = F(x_1, x_2)$.

$F(x_1, x_2)$ is called the *distribution function* or *cumulative distribution function* of the two-dimensional random variable (x_1, x_2). It is sometimes convenient to say that $F(x_1, x_2)$ is the c.d.f. of the *two random variables*, x_1, and x_2; $F(x_1, x_2)$ is also referred to as a *bivariate* c.d.f.

As in the case of a one-dimensional random variable, we have two alternative schemes for describing the distribution of probabilities associated with a two-dimensional random variable: a probability space and a c.d.f. We will use the latter almost entirely.

A variety of two-dimensional analogues of formulas (2.2.7) through (2.2.10) can be set up and verified by the reader. In particular, we have

$$\begin{aligned} P(x_1 = x_1', x_2 = x_2') = {} & F(x_1', x_2') - F(x_1' - 0, x_2') \\ & - F(x_1', x_2' - 0) + F(x_1' - 0, x_2' - 0). \end{aligned} \tag{2.4.9}$$

(b) Marginal Distributions

Consider the sets

$$E_{(x_1', \alpha)}, \qquad \alpha = 1, 2, \ldots .$$

They constitute an expanding sequence such that

$$\lim_{\alpha\to\infty} E_{(x_1',\alpha)} = E_{(x_1',\,+\infty)} = E_{x_1'}$$

where $E_{x_1'}$ is the event $x_1 \leqslant x_1'$ in R_2. Hence by applying **1.4.5** we have

$$\lim_{\alpha\to\infty} F(x_1', \alpha) = \lim_{\alpha\to\infty} P(E_{(x_1',\alpha)}) = P\left(\lim_{\alpha\to\infty} E_{(x_1',\alpha)}\right) = P(E_{x_1'}).$$

That is,

$$F(x_1', +\infty) = P(E_{x_1'}) = P(x_1 \leqslant x_1'). \tag{2.4.10}$$

Dropping the dashes, let

$$F(x_1, +\infty) = F_1(x_1). \tag{2.4.11}$$

It can be verified by the reader that $F_1(x_1)$ satisfies all the conditions (2.2.6) (*a*) through (*d*) of the c.d.f. of a one-dimensional random variable. In fact, $F_1(x_1)$ is the c.d.f. of the component x_1 of the random variable (x_1, x_2) and is called the *marginal* c.d.f. of x_1, or simply the c.d.f. of x_1.

Similarly,

$$F_2(x_2) = F(+\infty, x_2) \tag{2.4.12}$$

is the marginal c.d.f. of x_2. If one thinks of a total probability of 1 being distributed in the (x_1, x_2) plane in accordance with the c.d.f. $F(x_1, x_2)$, and if this probability is orthogonally projected onto the x_1-axis $R_1^{(1)}$, then $F_1(x_1)$ is the amount of probability lying in $(-\infty, x_1]$ on the x_1-axis. A similar interpretation holds, of course, for $F_2(x_2)$.

It follows from **2.2.1** that the marginal c.d.f.'s $F_1(x_1)$ and $F_2(x_2)$ determine probability spaces $(R_1^{(1)}, \mathscr{B}_1^{(1)}, P^{(1)})$ and $(R_1^{(2)}, \mathscr{B}_1^{(2)}, P^{(2)})$ respectively.

(c) Statistically Independent Random Variables

If (x_1, x_2) is a two-dimensional random variable whose probability space is $(R_2, \mathscr{B}_2, P)$ and if the components x_1 and x_2 are one-dimensional random variables having probability spaces $(R_1^{(1)}, \mathscr{B}_1^{(1)}, P^{(1)})$ and $(R_1^{(2)}, \mathscr{B}_1^{(2)}, P^{(2)})$ respectively, then x_1 and x_2 are said to be *statistically independent* (see Section 1.6) if for every set E in $R_2 = R_1^{(1)} \times R_1^{(2)}$ of the form $E^{(1)} \times E^{(2)}$ where $E^{(1)}$ and $E^{(2)}$ are (Borel) sets in $R_1^{(1)}$ and $R_1^{(2)}$ respectively, we have $P(E) = P^{(1)}(E^{(1)}) \cdot P^{(2)}(E^{(2)})$.

Dropping the adjective "statistically" and referring merely to *independent* random variables will not cause ambiguity.

Independence of x_1 and x_2 can be more usefully expressed in terms of c.d.f.'s as follows:

2.4.2 *If (x_1, x_2) is a random variable, having c.d.f. $F(x_1, x_2)$ a necessary and sufficient condition for x_1 and x_2 to be independent is that*

$$F(x_1, x_2) = F_1(x_1) \cdot F_2(x_2), \tag{2.4.13}$$

where $F_1(x_1)$ and $F_2(x_2)$ are the marginal c.d.f.'s of x_1 and x_2.

To establish **2.4.2** we denote, as usual, the sample spaces of x_1 and x_2 by $R_1^{(1)}$ and $R_1^{(2)}$, respectively. Then

$$R_2 = R_1^{(1)} \times R_1^{(2)}. \tag{2.4.14}$$

Let $E_{x_1'}$ and $E_{x_2'}$ be the events $x_1 \leqslant x_1'$ and $x_2 \leqslant x_2'$ in $R_1^{(1)}$ and $R_1^{(2)}$, respectively. Then

$$E_{(x_1', x_2')} = E_{x_1'} \times E_{x_2'}. \tag{2.4.15}$$

If x_1 and x_2 are independent, we have

$$P(E_{(x_1', x_2')}) = P^{(1)}(E_{x_1'}) \cdot P^{(2)}(E_{x_2'}), \tag{2.4.16}$$

that is,

$$F(x_1', x_2') = F_1(x_1') \cdot F_2(x_2'). \tag{2.4.17}$$

Conversely, suppose $F(x_1, x_2)$, $F_1(x_1)$, and $F_2(x_2)$ are c.d.f.'s such that (2.4.17) holds for every point (x_1', x_2') in R_2, that is to say, (2.4.16) holds for every set $E_{(x_1', x_2')} = E_{x_1'} \times E_{x_2'}$. Then it follows from Section 1.6 that a probability space $(R_2, \mathscr{B}_2, P)$ is uniquely determined in R_2 which has the property that for any set E in $\mathscr{B}_2$ of form $E^{(1)} \times E^{(2)}$, we have $P(E) = P^{(1)}(E^{(1)}) \cdot P^{(2)}(E^{(2)})$ where $(R_1^{(1)}, \mathscr{B}_1^{(1)}, P^{(1)})$ and $(R_1^{(2)}, \mathscr{B}_1^{(2)}, P^{(2)})$ are probability spaces determined by $F_1(x_1)$ and $F_2(x_2)$ respectively, which is equivalent to the statement that x_1 and x_2 are independent.

2.5 COMMON TYPES OF TWO-DIMENSIONAL RANDOM VARIABLES

In probability and statistics, most two-dimensional random variables are of three types: *discrete*, *continuous*, and *mixed*, although the mixed type occurs much less frequently than the other two. It is worthwhile to discuss these in some detail.

(a) The Discrete Type

For a *discrete* random variable (x_1, x_2), $F(x_1, x_2)$ is a step-function such that the right-hand side of (2.4.9) is zero except at a finite or countably infinite number of *mass points* $(x_1^{(\alpha)}, x_2^{(\alpha)})$, $\alpha = 1, 2, \ldots$, having no finite limit point in R_2. At these mass points we have

$$P(x_1 = x_1^{(\alpha)}, x_2 = x_2^{(\alpha)}) = p(x_1^{(\alpha)}, x_2^{(\alpha)}) \tag{2.5.1}$$

and furthermore

$$\sum_{\alpha} p(x_1^{(\alpha)}, x_2^{(\alpha)}) = 1. \tag{2.5.2}$$

Conversely, if we are given a sequence of points $(x_1^{(\alpha)}, x_2^{(\alpha)})$ such that $p(x_1^{(\alpha)}, x_2^{(\alpha)}) > 0$, $\alpha = 1, 2, \ldots$, and $p(x_1, x_2) = 0$ for all other points in R_2, then $F(x_1, x_2)$ is the step-function defined by

$$F(x_1, x_2) = \sum p(x_1^{(\alpha)}, x_2^{(\alpha)}) \tag{2.5.3}$$

the summation extending over all values of α for which $x_i^{(\alpha)} \leqslant x_i$, $i = 1, 2$.

The random variable (x_1, x_2) is *degenerate* if there is only one mass point $(x_1^{(1)}, x_2^{(1)})$, in which case $p(x_1^{(1)}, x_2^{(1)}) = 1$. We may write

$$F(x_1, x_2) = \varepsilon(x_1 - x_1^{(1)}, x_2 - x_2^{(1)}) = \begin{cases} 1, & x_i \geqslant x_i^{(1)}, i = 1, 2 \\ 0, & \text{otherwise.} \end{cases} \tag{2.5.4}$$

Of course, it is possible for only one of the components of (x_1, x_2), say x_1, to be a degenerate random variable with a c.d.f. as defined in (2.3.4). Unless otherwise indicated we shall assume that neither component of (x_1, x_2) is degenerate.

There will be no ambiguity if we drop α and call $p(x_1, x_2)$ the *probability function* (p.f.) of (x_1, x_2).

Then we have the result that

2.5.1 *The c.d.f. $F(x_1, x_2)$ of a discrete random variable (x_1, x_2) is uniquely determined by the p.f. $p(x_1, x_2)$, and conversely.*

In dealing with a discrete random variable (x_1, x_2) it is usually more convenient to deal with $p(x_1, x_2)$ than $F(x_1, x_2)$.

In general, if E is any set in R_2, we have

$$P((x_1, x_2) \in E) = \sum_{(x_1^{(\alpha)}, x_2^{(\alpha)}) \in E} p(x_1^{(\alpha)}, x_2^{(\alpha)}).$$

In particular, the marginal c.d.f. $F_1(x_1)$ for any value of x_1, say x_1', is given by

$$F_1(x_1') = P(x_1 \leqslant x_1') = \sum_{(x_1^{(\alpha)}, x_2^{(\alpha)}) \in E_{x_1'}} p(x_1^{(\alpha)}, x_2^{(\alpha)}), \tag{2.5.5}$$

where $E_{x_1'}$ is the set in R_2 for which $x_1 \leqslant x_1'$. $F_2(x_2)$ is similarly defined.

The marginal c.d.f.'s $F_1(x_1)$ and $F_2(x_2)$ are discrete one-dimensional c.d.f.'s, which have marginal p.f.'s $p_1(x_1)$ and $p_2(x_2)$, respectively.

The reader can readily verify the following statement:

2.5.2 *If* (x_1, x_2) *is a discrete random variable with p.f.* $p(x_1, x_2)$, *a necessary and sufficient condition for* x_1 *and* x_2 *to be independent is that* $p(x_1, x_2) = p_1(x_1) \cdot p_2(x_2)$.

Example. Examples of two-dimensional discrete random variables are plentiful in elementary probability theory. For instance, if x_1 denotes the number of aces and x_2 the number of kings occurring in a hand of 13 bridge cards, the mass points (x_1^α, x_2^α), $\alpha = 1, 2, \ldots, 25$ of the random variable (x_1, x_2) are (0, 0), (0, 1), (1, 0), . . . , (4, 4). Under conditions of "perfect" shuffling; that is, assuming the $\binom{52}{13}$ possible hands are all assigned equal probabilities the p.f. of (x_1, x_2) is

$$p(x_1, x_2) = \frac{\binom{4}{x_1}\binom{4}{x_2}\binom{44}{13 - x_1 - x_2}}{\binom{52}{13}}$$

for $(x_1, x_2) = (0, 0), (0, 1), \ldots, (4, 4)$ and, of course, $p(x_1, x_2) = 0$ for all other points in R_2. The marginal p.f. of x_1 is given by

$$p_1(x_1) = \sum_{x_2=0}^{4} p(x_1, x_2) = \frac{\binom{4}{x_1}\binom{48}{13 - x_1}}{\binom{52}{13}}$$

which, of course, is the p.f. of the random variable x_1, the number of aces obtained. A similar expression holds for $p_2(x_2)$. Note that x_1 and x_2 are not statistically independent since $p(x_1, x_2) \neq p_1(x_1) \cdot p_2(x_2)$.

Further cases of two-dimensional discrete random variables will arise in Chapter 6.

(b) The Continuous Type

In the case of a *continuous* two-dimensional random variable (x_1, x_2), there exists a Lebesgue-measurable function $f(x_1, x_2) \geqslant 0$ such that

$$F(x_1, x_2) = \int_{-\infty}^{x_1} \int_{-\infty}^{x_2} f(y_1, y_2)\, dy_1\, dy_2 \tag{2.5.6}$$

for all $(x_1, x_2) \in R_2$ in which case $\dfrac{\partial^2 F(x_1, x_2)}{\partial x_1\, \partial x_2}$ exists and

$$\frac{\partial^2 F(x_1, x_2)}{\partial x_1\, \partial x_2} = f(x_1, x_2) \tag{2.5.7}$$

at all points in R_2 except possibly for a set of probability 0.

Conditions on $f(x_1, x_2)$ under which (2.5.7) is valid are two-dimensional analogues of those stated for the case of a one-dimensional random variable; the details are left to the reader.

Now let us examine the concept of probability density for the two-dimensional random variable (x_1, x_2).

If we set up the ratio

$$\frac{\Delta^2_{I_2} F(x_1, x_2)}{(x_1'' - x_1')(x_2'' - x_2')}, \qquad x_1'' > x_1', \qquad x_2'' > x_2', \tag{2.5.8}$$

where the expression in the numerator is defined in (2.4.4), it is clear that this ratio represents the average probability per unit area contained in I_2. If the limit of the ratio exists as $x_1'' \to x_1'$ and $x_2'' \to x_2'$ this limit is $f(x_1', x_2') \geq 0$, that is, a non-negative density at (x_1', x_2'). If at a given point $f(x_1, x_2)$ is given by (2.5.7) we shall say that the *probability density of* (x_1, x_2) *exists* at that point. We shall call $f(x_1, x_2)$ the *probability density function* (p.d.f.) of (x_1, x_2), and $f(x_1, x_2)\, dx_1\, dx_2$ the *probability element* (p.e.) of (x_1, x_2). Summarizing,

2.5.3 *The c.d.f. $F(x_1, x_2)$ of a continuous random variable (x_1, x_2) is uniquely determined by the p.d.f. $f(x_1, x_2)$ in accordance with* (2.5.6). *Conversely, $f(x_1, x_2)$ is determined by $F(x_1, x_2)$ in accordance with* (2.5.7) *except possibly for a set of points in R_2 having probability* 0.

If E is any (Borel) set in R_2, we have $P(E)$ given by the following Lebesgue integral:

$$P((x_1, x_2) \in E) = \iint_E f(y_1, y_2)\, dy_1\, dy_2. \tag{2.5.9}$$

In particular, we have for the marginal c.d.f. of

$$F_1(x_1') = P(x_1 \leq x_1') = \int_{-\infty}^{x_1'} \int_{-\infty}^{\infty} f(y_1, y_2)\, dy_2\, dy_1. \tag{2.5.10}$$

When we consider the integral in (2.5.10) as an iterated integral, the function

$$f_1(x_1) = \int_{-\infty}^{\infty} f(x_1, y_2)\, dy_2 \tag{2.5.11}$$

is called the *marginal* p.d.f. of x_1. The marginal p.d.f. of x_2 is similarly defined.

The following statement furnishes a useful criterion for statistical independence of two random variables x_1 and x_2 and can be verified by the reader.

2.5.4 *If (x_1, x_2) is a continuous random variable having p.d.f. $f(x_1, x_2)$ a necessary and sufficient condition for x_1 and x_2 to be independent is that*

$$f(x_1, x_2) = f_1(x_1) \cdot f_2(x_2).$$

Example. As a simple example of a two-dimensional continuous random variable suppose two numbers are independently "drawn" from the interval (0, 1), all numbers being "equally likely" to be drawn. Let (x_1, x_2) be the (two-dimensional) random variable such that x_1 and x_2 denote respectively the smaller and larger numbers drawn. We define the p.d.f. of (x_1, x_2) as follows: $f(x_1, x_2) = 2$ inside the triangle having vertices (0, 0), (1, 1), and (0, 1) in the x_1x_2 plane, and 0 at all other points in the plane. The reader will see that the marginal p.d.f. of x_1 is

$$f_1(x_1) = \int_{-\infty}^{\infty} f(x_1, y_2)\, dy_2 = \int_{x_1}^{1} 2\, dy_2 = 2(1 - x_1),$$

and similarly that $f_2(x_2) = 2x_2$.

Various special cases of more important two-dimensional continuous random variables will be discussed in Chapter 7.

(c) The Mixed Type

In a *mixed* random variable (x_1, x_2) one of the components is discrete and the other is continuous. More precisely, if $F(x_1, x_2)$ is the c.d.f. of (x_1, x_2), where x_1 is discrete and x_2 is continuous, and if $x_1^{(\alpha)}$, $\alpha = 1, 2, \ldots$ are the mass points of x_1 and if $p_1(x_1)$ is the p.f. of x_1, then there exist one-dimensional *conditional* p.d.f.'s $f(x_2 \mid x_1^{(\alpha)})$, $\alpha = 1, 2, \ldots$ given by the formula:

$$f(x_2 \mid x_1^{(\alpha)}) = \frac{1}{p_1(x_1^{(\alpha)})} \cdot \frac{d}{dx_2} [F(x_1^{(\alpha)}, x_2) - F(x_1^{(\alpha)} - 0, x_2)] \tag{2.5.12}$$

such that

$$F(x_1, x_2) = \sum_{x_1^{(\alpha)} \leqslant x_1} p_1(x_1^{(\alpha)}) \int_{-\infty}^{x_2} f(y \mid x_1^{(\alpha)})\, dy. \tag{2.5.13}$$

The marginal c.d.f.'s $F_1(x_1)$ and $F_2(x_2)$ are given by the following formulas:

$$F_1(x_1) = \sum_{x_1^{(\alpha)} \leqslant x_1} p_1(x_1^{(\alpha)}) \tag{2.5.14}$$

and

$$F_2(x_2) = \sum_{\alpha} p_1(x_1^{(\alpha)}) \int_{-\infty}^{x_2} f(y \mid x_1^{(\alpha)})\, dy. \tag{2.5.15}$$

We therefore have the following statement:

2.5.5 *The c.d.f.* $F(x_1, x_2)$ *of a mixed random variable* (x_1, x_2) *where* x_1 *is discrete and* x_2 *is continuous is uniquely determined by the pairs* $[p(x_1^{(\alpha)}), f(x_2 \mid x_1^{(\alpha)})]$, $\alpha = 1, 2, \ldots$ *and conversely.*

For purposes of interpretation, a mixed random variable (x_1, x_2), where x_1 is discrete with p.f. $p(x_1)$ and x_2 is continuous with conditional p.d.f. $f(x_2 \mid x_1)$, can be viewed in the following way. First, the probability 1 is partitioned into pieces of magnitude $p_1(x_1^{(\alpha)})$, $\alpha = 1, 2, \ldots$ and these pieces are placed at the mass points $x_1^{(\alpha)}$, $\alpha = 1, 2, \ldots$ respectively on the x_1-axis. Then these pieces of probability are continuously "smeared" along the vertical lines $x_1 = x_1^{(\alpha)}$, $\alpha = 1, 2, \ldots$ in such a way that the density at any point $(x_1^{(\alpha)}, x_2)$ on the αth line is $p_1(x_1^{(\alpha)})f(x_2 \mid x_1^{(\alpha)})$.

It will be noted that if the $f(x_2 \mid x_1^{(\alpha)})$ are identical for all α, then

$$f(x_2 \mid x_1^{(\alpha)}) = f_2(x_2), \tag{2.5.16}$$

where $f_2(x_2)$ is the p.d.f. of x_2, and $F(x_1, x_2) = F_1(x_1) \cdot F_2(x_2)$; that is x_1 and x_2 are independent. Conversely, if x_1 and x_2 are independent, then $F(x_1, x_2) = F_1(x_1) \cdot F_2(x_2)$ and (2.5.12) reduces to (2.5.16). Hence

2.5.6 *If* (x_1, x_2) *is a mixed random variable where* x_1 *is discrete and* x_2 *is continuous, a necessary and sufficient condition for* x_1 *and* x_2 *to be independent is that* $f(x_2 \mid x_1^{(\alpha)}) = f_2(x_2)$ *for all* α *where* $f_2(x_2)$ *is the marginal p.d.f. of* x_2.

Example. As stated earlier, examples of mixed random variables are more rare than those of the discrete or continuous type, but the following simple artificial example at this point might strengthen further the idea of a mixed random variable. Suppose a die is thrown, letting x_1 be the random variable denoting the number of dots occurring. Then if $x_1 = x_1^{(\alpha)}$, $(x_1^{(1)} = 1, \ldots, x_1^{(6)} = 6)$, suppose $x_1^{(\alpha)}$ numbers are drawn independently from a "uniform" distribution on the interval (0, 1), letting x_2 denote the largest of the $x_1^{(\alpha)}$ numbers obtained. Then (x_1, x_2) is a mixed random variable with x_1 discrete and having p.f. $p_1(x_1^{(\alpha)}) = \frac{1}{6}$, $x_1^{(1)} = 1, \ldots, x_1^{(6)} = 6$, and x_2 is continuous, such that for $0 < x_2 < 1$

$$f(x_2 \mid x_1^{(\alpha)}) = \frac{d}{dx_2}(x_2)^{x_1^{(\alpha)}} = x_1^{(\alpha)} x_2^{x_1^{(\alpha)}-1}$$

for $x_1^{(1)} = 1, \ldots, x_1^{(6)} = 6$ for $0 < x_2 < 1$, with $f(x_2 \mid x_1^{(\alpha)}) = 0$ for all values of x_2 outside (0, 1). One distribution of possible interest in this example is the marginal c.d.f. $F_2(x_2)$, which is given by

$$F_2(x_2) = \tfrac{1}{6}(x_2 + x_2^2 + \cdots + x_2^6).$$

From this we see that the p.d.f. of x_2 is

$$f_2(x_2) = \tfrac{1}{6}(1 + 2x_2 + \cdots + 6x_2^5).$$

Other examples of mixed random variables occur in Problems **8.34** and **8.35** at the end of Chapter 8.

2.6 DISTRIBUTION FUNCTIONS OF k-DIMENSIONAL RANDOM VARIABLES

(a) General Properties

The reader will now see how the results of Sections 2.4 and 2.5 for two-dimensional random variables can be extended to the case of a k-dimensional random variable. It is sufficient to outline the extension only briefly. For a detailed analysis of general multidimensional distributions the reader is referred to von Neumann (1950).

Suppose $(x_1, \ldots, x_k)$ is a k-dimensional random variable whose probability space is $(R_k, \mathscr{B}_k, P)$. Let $E_{(x_1', \ldots, x_k')}$ be the set $(-\infty, \ldots, -\infty;\ x_1', \ldots, x_k']$ in R_k. Then

(2.6.1)
$$F(x_1', \ldots, x_k') = P(E_{(x_1', \ldots, x_k')}) = P(-\infty < x_i \leqslant x_i',\ i = 1, \ldots, k);$$

$F(x_1, \ldots, x_k)$ is clearly a single-valued, real, and non-negative function of $(x_1, \ldots, x_k)$ in R_k.

Now any interval I_k in R_k of the form $(x_1', \ldots, x_k';\ x_1'', \ldots, x_k'']$ belongs to $\mathscr{B}_k$ since

(2.6.2) $$I_k = E_{(x_1'', \ldots, x_k'')} - [E_{(x_1', x_2'', \ldots, x_k'')} \cup \cdots \cup E_{(x_1'', x_2'', \ldots, x_{k-1}'', x_k')}].$$

Furthermore the probability that $(x_1, \ldots, x_k) \in I_k$ can be found in the terms of $F(x_1, \ldots, x_k)$ by extending (2.4.3) to the case of k variables. This gives

(2.6.3)
$$\begin{aligned} P((x_1, \ldots, x_k) \in I_k) = F(x_1'', \ldots, x_k'') & \\ - [F(x_1', x_2'', \ldots, x_k'') + \cdots & \\ + F(x_1'', \ldots, x_{k-1}'', x_k')] & \\ + [F(x_1', x_2', x_3'', \ldots, x_k'') + \cdots & \\ + F(x_1'', \ldots, x_{k-2}'', x_{k-1}', x_k')] & \\ + \cdots & \\ + (-1)^k F(x_1', \ldots, x_k'). & \end{aligned}$$

It is convenient to denote the expression on the right by $\Delta_{I_k}^k F(x_1, \ldots, x_k)$, the kth *difference* of $F(x_1, \ldots, x_k)$ over I_k. We then have

(2.6.4) $$P((x_1, \ldots, x_k) \in I_k) = \Delta_{I_k}^k F(x_1, \ldots, x_k) \geqslant 0.$$

By argument similar to that used in establishing (2.4.5), it follows that

(2.6.5) $$F(-\infty, x_2, \ldots, x_k) = \cdots = F(x_1, x_2, \ldots, x_{k-1}, -\infty) = 0.$$

Also, we can extend the argument leading to (2.4.6) and show that

(2.6.6) $$F(+\infty, \ldots, +\infty) = 1.$$

Furthermore, the k-dimensional analogue of (2.4.7) can be verified, that is,

$$(2.6.7) \quad F(x_1, \ldots, x_{i-1}, x_i + 0, x_{i+1}, \ldots, x_k) = F(x_1, \ldots, x_k), \quad i = 1, \ldots, k.$$

It can also be verified that the value of $F(x_1, \ldots, x_k)$ is unchanged if any set of x's, say $x_{i_1}, \ldots, x_{i_r}$, are replaced by $x_{i_1} + 0, \ldots, x_{i_r} + 0$ respectively.

Therefore, if $(x_1, \ldots, x_k)$ is a k-dimensional random variable having probability space $(R_k, \mathscr{B}_k, P)$, there exists a function $F(x_1, \ldots, x_k)$ defined at every point $(x_1, \ldots, x_k)$ in R_k by (2.6.1) and having properties (2.6.4) through (2.6.7).

Conversely, if $F(x_1, \ldots, x_k)$ is defined by (2.6.1) and having properties (2.6.4) through (2.6.7) we obtain a probability space $(R_k, \mathscr{B}_k, P)$. Summarizing we have the k-dimensional extension of **2.4.1** namely,

2.6.1 *The probability space* $(R_k, \mathscr{B}_k, P)$ *of a k-dimensional random variable,* $(x_1, \ldots, x_k)$, *uniquely determines a single-valued, real, and non-negative function* $F(x_1, \ldots, x_k)$ *defined by* (2.6.1) *at every point in* R_k *and having the following properties:*

(2.6.8) (*a*) $\Delta^k_{I_k} F(x_1, \ldots, x_k) \geqslant 0$

(*b*) $F(-\infty, x_2, \ldots, x_k) = \cdots = F(x_1, \ldots, x_{k-1}, -\infty) = 0$

(*c*) $F(+\infty, \ldots, +\infty) = 1$

(*d*) $F(x_1, \ldots, x_{i-1}, x_i + 0, x_{i+1}, \ldots, x_k) = F(x_1, \ldots, x_k)$, $i = 1, \ldots, k$.

Conversely, $F(x_1, \ldots, x_k)$ *defined by* $F(x_1, \ldots, x_k) = P(E_{(x_1, \ldots, x_k)})$ *and having these properties uniquely determines a probability space* $(R_k, \mathscr{B}_k, P)$.

The function $F(x_1, \ldots, x_k)$ is called the *distribution function* (d.f.) or *cumulative distribution function* (c.d.f.) of the k-dimensional random variable $(x_1, \ldots, x_k)$. $F(x_1, \ldots, x_k)$ is sometimes referred to as a *k-variate* c.d.f.

It will be noted that the probability that $x_1 = x_1', \ldots, x_k = x_k'$ is obtained by taking the limit of the right-hand side of (2.6.3) as $x_i'' \to x_i'$, $i = 1, \ldots, k$, that is,

$$(2.6.9) \quad P(x_1 = x_1', \ldots, x_k = x_k') = \lim_{x_i'' \to x_i',\ \text{all } i} \Delta^k_{I_k} F(x_1, \ldots, x_k).$$

(b) Marginal Distributions

The marginal c.d.f. of x_1, $F_1(x_1)$, is defined as

$$(2.6.10) \quad F_1(x_1) = F(x_1, +\infty, \ldots, +\infty),$$

the other marginal c.d.f.'s $F_i(x_i)$, $i = 2, \ldots, k$ being similarly defined. More generally, the marginal c.d.f. of $(x_1, \ldots, x_{k_1})$, $k_1 < k$, is defined by

$$(2.6.11) \qquad F_{1 \ldots k_1}(x_1, \ldots, x_{k_1}) = F(x_1, \ldots, x_{k_1}, +\infty, \ldots, +\infty),$$

a similar definition holding, of course, for any other subset of the components of $(x_1, \ldots, x_k)$.

(c) Independence of Two or More Vector Random Variables

If $(x_1, \ldots, x_{k_1})$ and $(x_{k_1+1}, \ldots, x_k)$, where $k = k_1 + k_2$, are vector random variables whose probability spaces are $(R_{k_1}, \mathscr{B}_{k_1}, P^{(1)})$ and $(R_{k_2}, \mathscr{B}_{k_2}, P^{(2)})$ respectively, then these two vector random variables are said to be *independent* (see Section 1.6) if for every set E_k in $R_k = R_{k_1}^{(1)} \times R_{k_2}^{(2)}$ of form $E_{k_1}^{(1)} \times E_{k_2}^{(2)}$ where $E_{k_1}^{(1)}$ and $E_{k_2}^{(2)}$ are Borel sets in $R_{k_1}^{(1)}$ and $R_{k_2}^{(2)}$ respectively, we have $P(E_k) = P^{(1)}(E_{k_1}) \cdot P^{(2)}(E_{k_2})$.

As in the two-dimensional case, independence can be more usefully expressed in terms of c.d.f.'s, in accordance with the following statement which can be verified by the reader.

2.6.2 *If $(x_1, \ldots, x_k)$ is a random variable having c.d.f. $F(x_1, \ldots, x_k)$, a necessary and sufficient condition for $(x_1, \ldots, x_{k_1})$ and $(x_{k_1+1}, \ldots, x_k)$ to be independent is that*

$$(2.6.12) \qquad F(x_1, \ldots, x_k) = F_{1 \ldots k_1}(x_1, \ldots, x_{k_1}) \cdot F_{k_1+1 \ldots k}(x_{k_1+1}, \ldots, x_k)$$

where the two functions on the right are the marginal c.d.f.'s of $(x_1, \ldots, x_{k_1})$ and $(x_{k_1+1}, \ldots, x_k)$, respectively.

The notion of independence can be extended in an obvious manner to the case of three or more mutually exclusive subsets of the components of $(x_1, \ldots, x_k)$. In particular, it should be observed that $x_1, \ldots, x_k$ are *mutually independent* if and only if

$$(2.6.13) \qquad F(x_1, \ldots, x_k) = F_1(x_1) \cdots F_k(x_k).$$

2.7 COMMON TYPES OF k-DIMENSIONAL RANDOM VARIABLES

As in the two-dimensional cases, there are three common types of k-dimensional random variables: *discrete, continuous,* and *mixed* random variables.

In the *discrete* type, $F(x_1, \ldots, x_k)$ is a step function such that the right side of (2.6.9) is zero at all points in R_k except at a finite or countably infinite number of points $(x_1^{(\alpha)}, \ldots, x_k^{(\alpha)})$, $\alpha = 1, 2, \ldots$ which have no finite limit points in R_k at which

$$(2.7.1) \qquad P(x_1 = x_1^{(\alpha)}, \ldots, x_k = x_k^{(\alpha)}) = p(x_1^{(\alpha)}, \ldots, x_k^{(\alpha)}) > 0$$

and such that

$$(2.7.2) \qquad \sum_{\alpha} p(x_1^{(\alpha)}, \ldots, x_k^{(\alpha)}) = 1.$$

The function $p(x_1, \ldots, x_k)$ which has the value 0 at all points in R_k except $(x_1^{(\alpha)}, \ldots, x_k^{(\alpha)})$, $\alpha = 1, 2, \ldots$ at which the value is given by (2.6.9) and denoted by (2.7.1); $p(x_1, \ldots, x_k)$ is the p.f. of $(x_1, \ldots, x_k)$. In case there exists only one value of α, say $\alpha = 1$, then $p(x_1^{(1)}, \ldots, x_k^{(1)}) = 1$ and $(x_1, \ldots, x_k)$ is a *degenerate* random variable. Extensions of (2.5.3) and **2.5.1** are straightforward.

The reader will readily see how the marginal p.f. $p_i(x_i)$, or, more generally $p_{1 \cdots k_1}(x_1, \ldots, x_{k_1})$ is defined, and how **2.6.2** can be restated in terms of p.f.'s. The definition of degeneracy can be extended, of course, to marginal distributions. It is important to note in particular that $x_1, \ldots, x_k$ are *mutually independent* if and only if

$$(2.7.3) \qquad p(x_1, \ldots, x_k) = p_1(x_1) \cdots p_k(x_k).$$

In the case of a *continuous* k-dimensional random variable $(x_1, \ldots, x_k)$ there exists a non-negative Lebesgue-measurable function $f(x_1, \ldots, x_k)$, such that

$$(2.7.4) \qquad F(x_1, \ldots, x_k) = \int_{-\infty}^{x_1} \cdots \int_{-\infty}^{x_k} f(y_1, \ldots, y_k)\, dy_1 \ldots dy_k$$

for all $(x_1, \ldots, x_k) \in R_k$, where

$$\frac{\partial^k F(x_1, \ldots, x_k)}{\partial x_1 \cdots \partial x_k}$$

exists in R_k, and

$$(2.7.5) \qquad \frac{\partial^k F(x_1, \ldots, x_k)}{\partial x_1 \cdots \partial x_k} = f(x_1, \ldots, x_k)$$

at all points in R_k except possibly for a set of probability 0.

Conditions under which (2.7.5) is valid are k-dimensional versions of these for the one- and two-dimensional cases and are left to the reader to formulate.

The k-dimensional extension of **2.5.3** is left to the reader.

The marginal p.d.f. $f_i(x_i)$ of any component x_i of $(x_1, \ldots, x_k)$, or the p.d.f. of any subset of the components, say $f_{1 \cdots k_1}(x_1, \ldots, x_{k_1})$ are defined from the corresponding c.d.f.'s in an obvious manner.

It will be seen that **2.6.2** can be restated in terms of p.d.f.'s in the case of continuous random variables. Furthermore, the components $x_1, \ldots, x_k$ of the random variable $(x_1, \ldots, x_k)$ are mutually independent if and only if

$$f(x_1, \ldots, x_k) = f_1(x_1) \cdots f_k(x_k). \tag{2.7.6}$$

If $(x_1, \ldots, x_{k_1})$ is a k_1-dimensional discrete random variable and $(x_{k_1+1}, \ldots, x_k)$, $k = k_1 + k_2$, is a k_2-dimensional continuous random variable, then $(x_1, \ldots, x_k)$ is a *mixed* k-dimensional random variable. Such a random variable is defined by *conditional* p.d.f.'s $f(x_{k_1+1}, \ldots, x_k \mid x_1, \ldots, x_{k_1})$ given by the formula

$$\begin{aligned} \text{(2.7.7)}\quad & f(x_{k_1+1}, \ldots, x_k \mid x_1^{(\alpha)}, \ldots, x_{k_1}^{(\alpha)}) \\ &= \frac{1}{p_{1\cdots k_1}(x_1^{(\alpha)}, \ldots, x_{k_1}^{(\alpha)})} \frac{\partial^{k_2}}{\partial x_{k_1+1} \cdots \partial x_k} [\Delta_{I_0(k_1)}^{k_1} F(x_1, \ldots, x_k)] \end{aligned}$$

where $p_{1\cdots k_1}(x_1, \ldots, x_{k_1})$ is the p.f. of $(x_1, \ldots, x_{k_1})$ and where

$$\Delta_{I_0(k_1)}^{k_1} F(x_1, \ldots, x_k) = \lim_{x_1' \to x_1, \ldots, x_{k_1}' \to x_{k_1}} [\Delta_{I_{k_1}}^{k_1} F(x_1, \ldots, x_k)]; \tag{2.7.8}$$

$\Delta_{I_{k_1}}^{k_1} F(x_1, \ldots, x_k)$ is the k_1th difference of $F(x_1, \ldots, x_k)$ with respect to $(x_1, \ldots, x_{k_1})$ over the k_1-dimensional interval $(x_1', \ldots, x_{k_1}';\ x_1, \ldots, x_{k_1}]$ where $x_{k_1+1}, \ldots, x_k$ are held fixed.

The k-dimensional analogues of **2.5.5** and **2.5.6** are straightforward extensions and are left as exercises for the reader.

2.8 FUNCTIONS OF RANDOM VARIABLES

(a) Functions of One Random Variable

It will be recalled from Section 2.2 that a (one-dimensional) random variable x has a probability space $(R_1, \mathscr{B}_1, P)$, or equivalently a c.d.f. $F(x)$. We often have to deal with the probability theory of a *measurable function** $g(x)$ of the random variable x, where $g(x)$ is real, single-valued and defined at every point x in R_1 except possibly for a set of probability 0, and where the set of points in R_1 for which $g(x) \leqslant y$, for every real number y, also belongs to the class of Borel sets $\mathscr{B}_1$ in R_1. For example, if x is a random variable such elementary functions as polynomials in x, $\sin x$, e^x, etc., are obviously measurable.

It follows from the fact that if $(R_1, \mathscr{B}_1, P)$ is the probability space of a random variable x, and from the definition of $g(x)$ that for any real number y, the set of values of x for which $g(x) \leqslant y$ is contained in $\mathscr{B}_1$ and hence the probability assigned to this set is provided by $(R_1, \mathscr{B}_1, P)$.

* Such functions are also called *Baire functions*.

Let us denote the probability of the inequality $g(x) \leqslant y$, for a fixed y, by $H(y)$. Then

$$H(y) = P(g(x) \leqslant y) = P(x \in E_y) \tag{2.8.1}$$

where E_y denotes the set of points x (in R_1) for which $g(x) \leqslant y$.

It can be readily verified that $H(y)$ has all the properties of a c.d.f. and hence we have:

2.8.1 *If x is a random variable and $g(x)$ is a random variable denoted by y, then $H(y)$ defined by* (2.8.1) *is the c.d.f. of y.*

Now suppose $g(x)$ is strictly monotonic and has a continuous nonvanishing derivative for all x in some open interval A. Let $y = g(x)$ and B be the image (interval) of A in the y space. Take $x' \in A$ and let $y' = g(x')$. Then for some $\Delta y > 0$ there is a unique solution of the equation $g(x) = y' + \Delta y$ which will be indicated by $x = g^{-1}(y' + \Delta y)$.

If the c.d.f. of x is $F(x)$, we have

$$H(y' + \Delta y) - H(y') = \pm[F(g^{-1}(y' + \Delta y)) - F(g^{-1}(y'))], \tag{2.8.2}$$

the plus sign holding if $g(x)$ is monotonically increasing and the minus sign if $g(x)$ is monotonically decreasing.

Dividing both sides of (2.8.2) by Δy we may write

$$\frac{H(y' + \Delta y) - H(y')}{\Delta y} = \left[\frac{F(g^{-1}(y' + \Delta y)) - F(g^{-1}(y'))}{g^{-1}(y' + \Delta y) - g^{-1}(y')}\right] \times \left[\frac{\pm(g^{-1}(y' + \Delta y) - g^{-1}(y'))}{\Delta y}\right]. \tag{2.8.3}$$

If $g(x)$ has a nonvanishing derivative at $x = x'$, $g^{-1}(y)$ has a nonvanishing derivative at $y = y'$. Taking the limit of (2.8.3) as $\Delta y \to 0$, assuming $F(x)$ has a derivative $f(x)$ for all $x \in A$ and dropping dashes, we obtain

$$\frac{dH(y)}{dy} = f(x) \cdot \left|\frac{dx}{dy}\right|$$

where x is replaced by $g^{-1}(y)$ on the right. We use the absolute value sign to simplify the rule about using the $+$ and $-$ signs. Furthermore,

$$\int_A f(x)\, dx = \int_B f(g^{-1}(y)) \left|\frac{dg^{-1}(y)}{dy}\right| dy. \tag{2.8.4}$$

Summarizing:

2.8.2 *Suppose x is a continuous random variable with* p.d.f. $f(x)$, *and* $y = g(x)$ *is strictly monotonic and has a continuous nonvanishing derivative in some open interval A. Let B be the image of A in the y space. If y is the random variable* $g(x)$, *then y is a continuous random variable whose p.d.f.* $h(y)$ *exists in B and is given by*

$$h(y) = f(g^{-1}(y))\left|\frac{dg^{-1}(y)}{dy}\right| \qquad (2.8.5)$$

where $g^{-1}(y)$ *is the solution of* $g(x) = y$ *for* x. *Furthermore,* (2.8.4) *holds.*

Sometimes we use the phraseology "the transformation $x = g^{-1}(y)$ carries the probability element $f(x)\,dx$ into the probability element $h(y)\,dy$," or more briefly, "$f(x)\,dx \to h(y)\,dy$" where $h(y)$ is given by (2.8.5). Formula (2.8.5) is, of course, the familiar rule for changing the variable of integration in the integrand (a probability density function here) of an ordinary integral.

Example. To illustrate the rule expressed by (2.8.5) suppose x is a continuous random variable having p.d.f. $f(x) = 1/a$ for x in the interval $(0, a)$ and $f(x) = 0$ otherwise, and that we wish to find the p.d.f. of the random variable x^n. In this case, $g(x) = x^n$ and $g^{-1}(y) = y^{1/n}$. Hence, applying (2.8.5),

$$h(y) = \frac{1}{na}\,y^{1/n-1}$$

for y in $(0, a^n)$ and 0 otherwise, that is, for the transformation $x = y^{1/n}$ we have

$$\frac{1}{a}\cdot dx \to \frac{1}{a}\cdot\left|\frac{dy^{1/n}}{dy}\right| dy = \frac{1}{na}\,y^{1/n-1}\,dy.$$

(b) Functions of Two Random Variables

Suppose we have a two-dimensional random variable (x_1, x_2). A function $g(x_1, x_2)$ of x_1 and x_2 is referred to as a random variable if $g(x_1, x_2)$ is real, single-valued, and defined at every point of R_2, the x_1x_2-plane, except possibly for a set of probability 0, and if for every real y the set of points of R_2 for which $g(x_1, x_2) \leqslant y$ belongs to the Borel class of sets $\mathscr{B}_2$ in R_2. If $g(x_1, x_2)$ is a random variable, let us denote it by y; let the set of points in the x_1x_2 plane for which $g(x_1, x_2) \leqslant y$ be denoted by E_y.

Since the random variable is characterized by a probability space $(R_2, \mathscr{B}_2, P)$, or alternatively by a c.d.f. $F(x_1, x_2)$, the probability that $(x_1, x_2) \in E_y$ is provided by $(R_2, \mathscr{B}_2, P)$ (and, of course, also by $F(x_1, x_2)$ for any real number y). Let $H(y)$ be this probability. We have

$$H(y) = P(g(x_1, x_2) \leqslant y) = P((x_1, x_2) \in E_y). \qquad (2.8.6)$$

It can be verified that $H(y)$ is a c.d.f., namely, the c.d.f. of the random variable $y = g(x_1, x_2)$.

Let us consider the vector function $(g_1(x_1, x_2), g_2(x_1, x_2))$ of the random variable (x_1, x_2), where $g_1(x_1, x_2)$ and $g_2(x_1, x_2)$ are real, single-valued, and defined at every point of R_2, except possibly for a set of probability 0, and where the set of points in the x_1x_2-plane for which $g_i(x_1, x_2) \leqslant y_i$, $i = 1, 2$ belongs to $\mathscr{B}_2$ for every pair of real numbers (y_1, y_2). For a given y_1 and y_2 let this set of points be denoted by $E_{(y_1, y_2)}$. Then the probability that $(x_1, x_2) \in E_{(y_1, y_2)}$ is provided by the probability space $(R_2, \mathscr{B}_2, P)$ of the random variable (y_1, y_2). Let this probability be $H(y_1, y_2)$. Then

(2.8.7)
$$H(y_1, y_2) = P(g_i(x_1, x_2) \leqslant y_i, i = 1, 2) = P((x_1, x_2) \in E_{(y_1, y_2)}).$$

It can be verified that $H(y_1, y_2)$ has all of the properties of a two-dimensional c.d.f. and hence:

2.8.3 *If (x_1, x_2) is a two-dimensional random variable and if (y_1, y_2) denotes the random variable $(g_1(x_1, x_2), g_2(x_1, x_2))$, then $H(y_1, y_2)$ as defined by (2.8.7) is the c.d.f. of (y_1, y_2).*

The reader should note that the c.d.f. of one of the components of (y_1, y_2), say y_1, is given by

$$H_1(y_1) = P(g_1(x_1, x_2) \leqslant y_1) = P((x_1, x_2) \in E_{y_1}) \tag{2.8.8}$$

where E_{y_1} is the set of points in the x_1x_2-plane for which $g_1(x_1, x_2) \leqslant y_1$. The important point about $H_1(y_1)$ is that although it is actually the marginal c.d.f. of y_1, defined by letting $y_2 \to +\infty$ in the c.d.f. $H(y_1, y_2)$, it can be obtained without first determining $H(y_1, y_2)$. In fact, (2.8.8) shows that the introduction of the random variable $y_2 = g_2(x_1, x_2)$ is irrelevant if one is interested in y_1 only.

The components of a two-dimensional random variable (x_1, x_2) are *linearly dependent* if there exist real constants c_1 and c_2, not both zero, such that the random variable $c_1x_1 + c_2x_2$ is a degenerate random variable. (Usually there is very little interest in the case in which one of the c's is zero, for in this situation one of the components x_1 or x_2 is itself degenerate.) If $c_1x_1 + c_2x_2$ is degenerate when neither c_1 nor c_2 is zero, we shall say that x_1 and x_2 are *properly linearly dependent*, which means, of course, that $P(c_1x_1 + c_2x_2 = c_3) = 1$, where c_3 is a constant, and the line $c_1x_1 + c_2x_2 = c_3$ is not parallel to either coordinate axis. If we rule out the case in which one or more of the components of (x_1, x_2) is degenerate, then clearly the only type of linear dependence which will arise is proper linear dependence. If x_1 and x_2 are not linearly dependent,they are said to be *linearly independent*.

In the particular case where $x_1 - x_2$ is degenerate such that

$$P(x_1 - x_2 = 0) = 1,$$

then x_1 and x_2 are called *equivalent random variables*. If $F(x_1, x_2)$ is the c.d.f. of two equivalent random variables x_1 and x_2, then $F_1(x_1)$ and $F_2(x_2)$ are identical. The converse of this statement is, of course, false.

Two vector random variables are equivalent if their corresponding components are equivalent random variables.

(c) Continuous Functions of Two Continuous Random Variables

An important special case arises when (x_1, x_2) is a continuous random variable, and when the transformation $y_i = g_i(x_1, x_2)$, $i = 1, 2$ is one-to-one between the x_1x_2- and y_1y_2-planes. In this case we can, under certain regularity conditions, give an explicit expression for the p.d.f. of (y_1, y_2) in terms of the p.d.f. of (x_1, x_2) and the first partial derivatives of $g_1(x_1, x_2)$ and $g_2(x_1, x_2)$. More precisely, we may make the following statement:

2.8.4 *Suppose (x_1, x_2) is a continuous random variable with p.d.f. $f(x_1, x_2)$. Let $g_i(x_1, x_2)$, $i = 1, 2$, be single-valued and have continuous first partial derivatives in some open region A in the x_1x_2-plane. Let $y_i = g_i(x_1, x_2)$, $i = 1, 2$, have a unique inverse $x_i = g_i^{-1}(y_1, y_2)$, $i = 1, 2$ for all points in A. Let B be the image of A in the y_1y_2-plane. Let J be the Jacobian defined by the determinant*

$$J = \left|\frac{\partial x_i}{\partial y_j}\right|, \qquad i, j = 1, 2, \tag{2.8.9}$$

having a non-zero value at all points in A. The p.d.f. $h(y_1, y_2)$ of (y_1, y_2) at any point $(y_1, y_2) \in B$ is given by

$$h(y_1, y_2) = f(x_1, x_2) \cdot |J|, \tag{2.8.10}$$

where, on the right of (2.8.10), *(x_1, x_2) are to be replaced by*

$$g_1^{-1}(y_1, y_2), g_2^{-1}(y_1, y_2),$$

respectively.

Furthermore,

$$\int_A f(x_1, x_2)\, dx_1\, dx_2 = \int_B f(x_1, x_2) \mid J \mid dy_1\, dy_2. \tag{2.8.11}$$

Theorem **2.8.4** is merely a statement, in terms of probability density functions, of the familiar theorem to be found in advanced calculus texts, on changing variables in a double integral. The reader interested in details of the proof may refer, for example, to Widder (1947).

Example. Suppose (x_1, x_2) is a random variable having p.d.f.

$$f(x_1, x_2) = \begin{cases} e^{-x_1-x_2}, & x_1 \geqslant 0, \quad x_2 \geqslant 0 \\ 0, & \text{otherwise,} \end{cases}$$

and that we are required to find the p.d.f. of the random variable $(x_1 + x_2, x_2/x_1)$. The transformation involved here is

$$y_1 = x_1 + x_2$$

$$y_2 = \frac{x_2}{x_1}$$

and its inverse is

$$x_1 = \frac{y_1}{1 + y_2}$$

$$x_2 = \frac{y_1 y_2}{1 + y_2}.$$

This transformation provides a one-to-one mapping between points in the first quadrant of the x_1x_2-plane and the first quadrant of the y_1y_2-plane. The absolute value of the Jacobian of the transformation for all points in the first quadrant is

$$\left| \frac{\partial(x_1, x_2)}{\partial(y_1, y_2)} \right| = \frac{y_1}{(1 + y_2)^2}$$

Hence we have for the p.d.f. of (y_1, y_2)

$$h(y_1, y_2) = \begin{cases} e^{-y_1} \dfrac{y_1}{(1 + y_2)^2}, & y_1 > 0, \quad y_2 > 0 \\ 0, & \text{otherwise.} \end{cases}$$

Incidentally, it should be noted here (see **2.5.4**) that y_1 and y_2 are independent random variables.

(d) Functions of k-Dimensional Random Variables

The preceding results extend in a straightforward way to several functions of several random variables. If $(x_1, \ldots, x_k)$ is a k-dimensional random variable, a function $g(x_1, \ldots, x_k)$, which we shall call y, is also a a random variable if it is real, single-valued, defined at all points in R_k except possibly for a set of probability 0, and if the set of points in R_k (the x-space) for which $g(x_1, \ldots, x_k) \leqslant y$, belongs to $\mathscr{B}_k$, the Borel sets of R_k for all real values of y. If, for a given y, E_y denotes this set of points, then the c.d.f. of y is given by

(2.8.12) $$H(y) = P((x_1, \ldots, x_k) \in E_y).$$

The definition of linear dependence extends to the case of a k-dimensional random variable $(x_1, \ldots, x_k)$. We have *linear dependence* among $x_1, \ldots,$ x_k if $c_1x_1 + \cdots + c_kx_k$ is a degenerate random variable for some constants $c_1, \ldots, c_k$ not all zero. If none of the c_i is zero, we have *proper linear dependence*. If $x_1, \ldots, x_n$ are not linearly dependent, they are said to be *linearly independent*.

A vector function $(g_1(x_1, \ldots, x_k), \ldots, g_{k_1}(x_1, \ldots, x_k))$ of a random variable $(x_1, \ldots, x_k)$, which we shall denote by $(y_1, \ldots, y_{k_1})$, is also a k_1-dimensional random variable if each of the components $g_1(x_1, \ldots, x_k), \ldots, g_{k_1}(x_1, \ldots, x_k)$ is real, single-valued, and defined at all points $(x_1, \ldots, x_k)$ in R_k except possibly for a set of probability 0, and if the set of points $E_{(y_1, \ldots, y_{k1})}$ in R_k for which $g_i(x_1, \ldots, x_k) \leqslant y_i$, $i = 1, \ldots, k_1$, for any set of real numbers $y_1, \ldots, y_{k_1}$ belongs to $\mathscr{B}_{k_1}$. Thus, the probability of the event $E_{(y_1, \ldots, y_{k1})}$ is provided by the probability space $(R_k, \mathscr{B}_k, P)$, or the c.d.f. $F(x_1, \ldots, x_k)$, of the random variable $(x_1, \ldots, x_k)$. The function $H(y_1, \ldots, y_{k_1})$ defined by

$$H(y_1, \ldots, y_{k_1}) = P((x_1, \ldots, x_k) \in E_{(y_1, \ldots, y_{k1})}) \tag{2.8.13}$$

is the c.d.f. of $(y_1, \ldots, y_{k_1})$.

Now suppose $k_1 = k$ and $(x_1, \ldots, x_k)$ is a continuous random variable with p.d.f. $f(x_1, \ldots, x_k)$. In some open region A of the space of the x's let $y_i = g_i(x_1, \ldots, x_k)$, $i = 1, \ldots, k$, have a unique inverse $x_i = g_i^{-1}(y_1, \ldots, y_k)$, $i = 1, \ldots, k$, where the $g_i(x_1, \ldots, x_k)$ possess continuous first derivatives such that the Jacobian $J = \left|\dfrac{\partial x_i}{\partial y_j}\right| \neq 0$ in A. Let the image of A in the space of $(y_1, \ldots, y_k)$ be denoted by B. Then $(y_1, \ldots, y_k)$ is a continuous random variable having p.d.f. at a point $(y_1, \ldots, y_k)$ in B given by

$$h(y_1, \ldots, y_k) = f(x_1, \ldots, x_k) \cdot |J|, \tag{2.8.14}$$

and furthermore,

$$\int_A f(x_1, \ldots, x_k)\, dx_1 \cdots dx_k = \int_B f(x_1, \ldots, x_k)\, |J|\, dy_1 \cdots dy_k. \tag{2.8.15}$$

These, of course, are the k-dimensional extensions of formulas (2.8.10) and (2.8.11). It is understood that $x_i = g_i^{-1}(y_1, \ldots, y_k)$, $i = 1, \ldots, k$, that the x's are to be expressed in terms of the y's on the right-hand side of (2.8.14) and (2.8.15).

2.9 CONDITIONAL DISTRIBUTION FUNCTIONS

(a) General Remarks

In Section 1.9 conditional probability was defined for events in the general sample space R. Since we are primarily interested in events defined by random variables it is useful to specialize the ideas of Section 1.10 for the case where events are described by random variables. Suppose x is a random variable with probability space $(R_1, \mathscr{B}_1, P)$ or equivalently, having c.d.f. $F(x)$. Let G be any event in $\mathscr{B}_1$ for which $P(G) > 0$, and $E_{x'}$ the event $x \leqslant x'$. Let

$$F(x' \mid G) = \frac{P(E_{x'} \cap G)}{P(G)}. \tag{2.9.1}$$

The reader can verify that $F(x \mid G)$ satisfies all of the conditions in (2.2.6) for a c.d.f.; it is called the *conditional* c.d.f. *of* x *given* $x \in G$. Conversely, by **2.2.1** $F(x \mid G)$ uniquely determines a *conditional* probability space $(R_1, \mathscr{B}_1, P)$.

In a similar manner suppose $(x_1, \ldots, x_k)$ is a k-dimensional random variable having c.d.f. $F(x_1, \ldots, x_k)$. We can define the conditional c.d.f. of $(x_1, \ldots, x_k)$, given that $(x_1, \ldots, x_k) \in G$, where G is a Borel set in R_k as

$$F(x_1, \ldots, x_k \mid G) = \frac{P(E_{(x_1, \ldots, x_k)} \cap G)}{P(G)} \tag{2.9.2}$$

where $P(G) > 0$, and $E_{(x_1, \ldots, x_k)}$ is the k-dimensional interval $(-\infty; x]_k$.

Actually, we shall be interested in (2.9.2) mainly for situations in which the event G consists of *sections* of the $(x_1, \ldots, x_k)$-space obtained by holding one or more of the components of $(x_1, \ldots, x_k)$ fixed. This immediately leads us into difficulties in situations where both numerator and denominator of the right-hand side of (2.9.2) are zero, as, for example, in case $(x_1, \ldots, x_k)$ is a k-dimensional continuous random variable with a p.d.f. $f(x_1, \ldots, x_k)$. Treatment of these difficulties is given in the following paragraphs.

(b) Conditional Distribution Functions of Two-Dimensional Random Variables

Let us consider the case of two random variables x_1, x_2, and let I_1, the cylinder set in R_2 for which $x_1'' < x_1 \leqslant x_1'$, be such that $P(I_1) > 0$. Then we have from (2.9.2)

$$F(x_1, x_2 \mid I_1) = \frac{P(E_{(x_1, x_2)} \cap I_1)}{P(I_1)}. \tag{2.9.3}$$

If $F(x_1, x_2)$ is the c.d.f. of (x_1, x_2) and $F_1(x_1)$ is the marginal c.d.f. of x_1 then $F(x_1', x_2 \mid I_1)$ may be written as

$$F(x_1', x_2 \mid I_1) = \frac{F(x_1', x_2) - F(x_1'', x_2)}{F_1(x_1') - F_1(x_1'')}. \tag{2.9.4}$$

Note that $F(x_1', x_2 \mid I_1)$, as a function of x_2, is a c.d.f. If the limit on the right exists as $x_1'' \to x_1'$, let it be denoted by $F(x_2 \mid x_1')$, that is, let

$$\lim_{x_1'' \to x_1'} F(x_1', x_2 \mid I_1) = F(x_2 \mid x_1'). \tag{2.9.5}$$

If $F(x_2 \mid x_1')$ exists for every x_2 it can be verified that $F(x_2 \mid x_1')$ considered as a function of x_2 is a c.d.f.; that is, it has the basic properties listed in (2.2.6). It is called the *conditional* c.d.f. *of* x_2 *given* $x_1 = x_1'$, or dropping the dash we may say, for brevity, when there is no possibility of ambiguity, that $F(x_2 \mid x_1)$ is the c.d.f. of the *conditional random variable* $x_2 \mid x_1$. $F(x_2 \mid x_1)$ is a c.d.f. in which x_1 essentially plays the role of a parameter; x_1 is sometimes called a *fixed* variable to emphasize that it is not a random variable in the definition of $F(x_2 \mid x_1)$. The quantity $x_2 \mid x_1$ may be regarded as a one-dimensional random variable whose c.d.f. $F(x_2 \mid x_1)$ for a fixed value of x_1, say x_1', is defined at all points in the x_1x_2-plane for which $x_1 = x_1'$ in such a way that, roughly speaking, $F(x_2' \mid x_1')$ is the amount of probability (or probability density) lying along the portion of the line $x_1 = x_1'$ for which $x_2 \leqslant x_2'$, expressed as a fraction of the probability (or probability density) lying along the entire line.

The preceding discussion provides a rather elementary approach to conditional random variables and their c.d.f.'s which is adequate for nearly all distributions arising in mathematical statistics. A more general approach can be made by use of the Radon-Nikodym theorem **1.10.1.**

There are two important special types of conditional random variables for the case of two dimensions which deserve special attention.

Type A. In this type the component x_1 of the random variable (x_1, x_2) is discrete. Thus, in (2.9.2) for $k = 2$, we assume x_1 is a discrete random variable with mass points $x_1^{(\alpha)}$, $\alpha = 1, 2, \ldots$. Let G be the set in R_2, the sample space of (x_1, x_2), for which $x_1 = x_1^{(\beta)}$. Then

$$P(G) = P(x = x_1^{(\beta)}) = p_1(x_1^{(\beta)}) > 0.$$

Using (2.9.2) for $k = 2$ we find at once

$$F(x_1^{(\beta)}, x_2 \mid x_1 = x_1^{(\beta)}) = \frac{P(E_{(x_1^{(\beta)}, x_2)} \cap (x_1 = x_1^{(\beta)}))}{P(x_1 = x_1^{(\beta)})}, \tag{2.9.6}$$

which we shall denote by $F(x_2 \mid x_1^{(\beta)})$. The function $F(x_2 \mid x_1^{(\beta)})$ is a c.d.f., and we shall refer to it as the c.d.f. of the *conditional random variable* $x_2 \mid x_1^{(\beta)}$. For a Type A conditional random variable, we may therefore define $F(x_2 \mid x_1^{(\beta)})$ directly from (2.9.2) for $k = 2$ and with G equal to the set of points in R_2 for which $x_1 = x_1^{(\beta)}$. Definition (2.9.5) also leads to the same result for $F(x_2 \mid x_1^{(\beta)})$ by choosing $x_1' = x_1^{(\beta)}$ in defining I_1.

Note that $x_2 \mid x_1^{(\beta)}$ is a random variable whose sample space is on the line in the x_1x_2-plane for which $x_1 = x_1^{(\beta)}$. The c.d.f. of $x_2 \mid x_1^{(\beta)}$ is $F(x_2 \mid x_1^{(\beta)})$. When there is no ambiguity we may drop the β and refer to $F(x_2 \mid x_1)$ as the c.d.f. of the conditional random variable $x_2 \mid x_1$.

There are two common cases of conditional random variables under Type A. The most common is that in which x_2 is discrete as well as x_1. In this case, it is evident that

$$F(x_2 \mid x_1^{(\beta)}) = \sum_{x_2^{(\gamma)} \leqslant x_2} \frac{p(x_1^{(\beta)}, x_2^{(\gamma)})}{p_1(x_1^{(\beta)})}, \tag{2.9.7}$$

where $p(x_1, x_2)$ is the p.f. of (x_1, x_2) and $p_1(x_1)$ is the marginal p.f. of x_1. It will be convenient to drop superscripts and let

$$p(x_2 \mid x_1) = \frac{p(x_1, x_2)}{p_1(x_1)}, \tag{2.9.8}$$

in which case $p(x_2 \mid x_1)$ is the p.f. of the conditional random variable $x_2 \mid x_1$. Note that $x_2 \mid x_1$ will be a degenerate random variable unless there are at least two distinct mass points having the given x_1 coordinate.

Formula (2.9.8) can be rewritten as

$$p(x_1, x_2) = p(x_2 \mid x_1) \cdot p_1(x_1) \tag{2.9.9}$$

which provides a method determining the p.f. of (x_1, x_2) in two steps, that is, by finding the p.f. of $x_2 \mid x_1$ and the p.f. of x_1 and multiplying the two p.f.'s. The (marginal) p.f. of x_2, which is often the objective in such a problem, can then be determined in the usual way from $p(x_1, x_2)$.

Examples. Suppose we wish to find the probability $p(x_2 \mid x_1)$ that a hand of bridge, *known* to contain x_1 aces, has x_2 kings. The probability $p(x_1, x_2)$ of getting x_1 aces and x_2 kings in a hand is given by

$$p(x_1, x_2) = \frac{\binom{4}{x_1}\binom{4}{x_2}\binom{44}{13 - x_1 - x_2}}{\binom{52}{13}}.$$

Now $p_1(x_1) = \sum_{x_2=0}^{4} p(x_1, x_2)$ is the probability that the hand contains x_1 aces and is given by

$$p_1(x_1) = \frac{\binom{4}{x_1} \cdot \binom{48}{13 - x_1}}{\binom{52}{13}}.$$

Therefore, applying (2.9.8), we have

$$p(x_2 \mid x_1) = \frac{\binom{4}{x_2} \cdot \binom{44}{13 - x_1 - x_2}}{\binom{48}{13 - x_1}}$$

Now consider an example to illustrate (2.9.9). Suppose x_1 is a random variable denoting the number of dots appearing when a "true" die is thrown. If $x_1 = x_1'$, let x_1' coins be tossed, and let x_2 be a random variable denoting the number of heads obtained. The interesting problem here is to determine $p_2(x_2)$, that is, the probability that the number of heads resulting from the entire experiment is x_2.

We have
$$p(x_2 \mid x_1) = \binom{x_1}{x_2}(\tfrac{1}{2})^{x_1}, \quad \text{and} \quad p_1(x_1) = \tfrac{1}{6}.$$

Hence
$$p(x_1, x_2) = \tfrac{1}{6}\binom{x_1}{x_2}(\tfrac{1}{2})^{x_1}.$$

To find $p_2(x_2)$, we sum $p(x_1, x_2)$ with respect to x_1 over $x_2, x_2 + 1, \ldots, 6$. This gives:
$$p_2(x_2) = \frac{63}{384}, \frac{120}{384}, \frac{99}{384}, \frac{64}{384}, \frac{29}{384}, \frac{8}{384}, \frac{1}{384},$$

for $x_2 = 0, 1, 2, 3, 4, 5, 6$, respectively.

The second, but less common, case of a Type A conditional random variable is that in which $F(x_2 \mid x_1^{(\beta)})$ is absolutely continuous in x_2, thus possessing a density function $f(x_2 \mid x_1^{(\beta)})$ such that

$$F(x_2 \mid x_1^{(\beta)}) = \frac{F(x_1^{(\beta)}, x_2) - F(x_1^{(\beta)} - 0, x_2)}{p_1(x_1^{(\beta)})} = \int_{-\infty}^{x_2} f(y \mid x_1^{(\beta)})\, dy. \tag{2.9.10}$$

This follows, of course, at once from (2.9.2) for $k = 2$ and for G taken as the set of points in R_2 for which $x_1 = x_1^{(\beta)}$. It can also be obtained from (2.9.5) by choosing $x_1' = x_1^{(\beta)}$.

Here $x_2 \mid x_1^{(\beta)}$ is a *continuous conditional random variable* with p.d.f. $f(x_2 \mid x_1^{(\beta)})$. Note that when expressed in terms of $F(x_1, x_2)$ the function $f(x_2 \mid x_1^{(\beta)})$ is given by (2.5.12) with $\alpha = \beta$.

TYPE B. In this type, x_1 is a continuous random variable with p.d.f. $f_1(x_1)$, and there are two useful cases. In the more common one (x_1, x_2) is a continuous random variable with p.d.f. $f(x_1, x_2)$, and (2.9.4) becomes

$$F(x_1', x_2 \mid I_1) = \frac{\displaystyle\int_{x_1''}^{x_1'}\int_{-\infty}^{x_2} f(x_1, x_2)\, dx_2\, dx_1}{\displaystyle\int_{x_1''}^{x_1'} f_1(x_1)\, dx_1}. \tag{2.9.11}$$

If $f_1(x_1)$ and $\int_{-\infty}^{x_2} f(x_1, x_2)\, dx_2$ are continuous functions of x_1 at x_1' with $f_1(x_1') > 0$, then taking limits as $x_1'' \to x_1'$, we obtain

$$F(x_2 \mid x_1') = \frac{\displaystyle\int_{-\infty}^{x_2} f(x_1', x_2)\, dx_2}{f_1(x_1')}. \tag{2.9.12}$$

The reader will note that this expression for $F(x_2 \mid x_1')$ for the continuous case is suggested by the analogous formula (2.9.7) for the discrete case. If we set

$$f(x_2 \mid x_1') = \frac{f(x_1', x_2)}{f_1(x_1')}, \tag{2.9.13}$$

then $f(x_2 \mid x_1')$ is a p.d.f., namely, that of the conditional random variable $x_2 \mid x_1'$. When there is no danger of confusion we may drop the dash from x_1' and rewrite (2.9.13) as

$$f(x_1, x_2) = f(x_2 \mid x_1) \cdot f_1(x_1), \tag{2.9.14}$$

which, of course, is the analogue of (2.9.9) for continuous random variables.

Example. Suppose x_1 is a random variable denoting a number picked "at random" from the interval (0, 1) (all numbers on the interval (0, 1) are to be regarded "equally likely in the elementary geometric sense"), and let x_2 be a random variable denoted by a number picked at random from the interval $(x_1', 1)$ where x_1' is the realized value of x_1. We wish to find the p.d.f. of x_2, that is, $f_2(x_2)$. We have

$$f_1(x_1) = \begin{cases} 1, & 0 < x_1 < 1 \\ 0, & \text{otherwise} \end{cases} \qquad f(x_2 \mid x_1) = \begin{cases} \dfrac{1}{1 - x_1}, & x_1 < x_2 < 1 \\ 0, & \text{otherwise.} \end{cases}$$

Hence

$$f(x_1, x_2) = \begin{cases} \dfrac{1}{1 - x_1}, & 0 < x_1 < x_2 < 1 \\ 0, & \text{otherwise in } R_2. \end{cases}$$

Therefore

$$f_2(x_2) = \int_0^{x_2} \frac{dx_1}{1 - x_1} = -\log(1 - x_2), \qquad 0 < x_2 < 1$$

and $f_2(x_2) = 0$ for x_2 outside (0, 1).

In the second, less frequent case of a Type B conditional random variable, x_1 is continuous, and x_2 is discrete, and the reader can verify that if $f_1(x_1)$ is continuous at x_1' with $f_1(x_1') > 0$, and if $F(x_1, x_2)$ and $F(x_1, x_2 - 0)$ possess derivatives with respect to x_1 at x_1', then application of (2.9.4) and (2.9.5) yields

$$F(x_2 \mid x_1') = \sum_{x_2^{(\beta)} \leqslant x_2} p^*(x_2^{(\beta)} \mid x_1') \tag{2.9.15}$$

where

$$p^*(x_2^{(\beta)} \mid x_1') = \frac{f^*(x_1', x_2^{(\beta)})}{f_1(x_1')} \tag{2.9.16}$$

and

$$f^*(x_1, x_2) = \frac{\partial}{\partial x_1}[F(x_1, x_2) - F(x_1, x_2 - 0)]. \tag{2.9.17}$$

Note that $p^*(x_2 \mid x_1)$ is similar in structure to $p(x_2 \mid x_1)$ as given by (2.9.8) except that $p^*(x_2 \mid x_1)$ is constructed from probability densities rather than probabilities (at mass points). Roughly speaking, we may think of breaking up the probability 1 into pieces $p_2(x_2^{(1)})$, $p_2(x_2^{(2)}), \ldots$ and then "smearing" these pieces of probability continuously along the lines $x_2 = x_2^{(1)}$, $x_2 = x_2^{(2)}, \ldots$ in accordance with the density functions (not p.d.f.'s) $f^*(x_1, x_2^{(1)}), f^*(x_1, x_2^{(2)}), \ldots$, respectively.

(c) Extensions to k-Dimensional Random Variables

The conditional c.d.f. defined by (2.9.5) can be extended in a straightforward manner to multidimensional random variables. Suppose $(x_1, \ldots, x_k)$ is a random variable with c.d.f. $F(x_1, \ldots, x_k)$. Let $F_{1\cdots k_1}(x_1, \ldots, x_{k_1})$ be the marginal c.d.f. of $(x_1, \ldots, x_{k_1})$, $k_1 < k = k_1 + k_2$. Let I_{k_1} be the cylinder set in R_k for which $x_i'' < x_i \leqslant x_i'$, $i = 1, \ldots, k_1$. The projection of this cylinder set onto R_{k_1}, the sample space of $(x_1, \ldots, x_{k_1})$, is the k_1-dimensional interval $(x_1'', \ldots, x_{k_1}''; x_1', \ldots, x_{k_1}']$ which we are also calling I_{k_1}. Let $\Delta_{I_{k_1}}^{k_1} F_{1\cdots k_1}(x_1, \ldots, x_{k_1})$ be the kth difference of $F_{1\cdots k_1}(x_1, \ldots, x_{k_1})$ over I_{k_1} defined in accordance with (2.6.3) and (2.6.4). This kth difference is simply $P(I_{k_1})$ which is assumed >0. Now let us put

$$\text{(2.9.18)} \qquad F(x_1, \ldots, x_k \mid I_{k_1}) = \frac{P(E_{(x_1, \ldots, x_k)} \cap I_{k_1})}{P(I_{k_1})}.$$

In particular, we have

$$\text{(2.9.19)} \quad F(x_1', \ldots, x_{k_1}', x_{k_1+1}, \ldots, x_k \mid I_{k_1}) = \frac{\Delta_{I k_1}^{k_1} F(x_1, \ldots, x_k)}{\Delta_{I_{k_1}}^{k_1} F_{1\ldots k_1}(x_1, \ldots, x_{k_1})}$$

where the numerator on the right is the k_1th difference of $F(x_1, \ldots, x_k)$ with respect to $(x_1, \ldots, x_{k_1})$ for fixed values of $(x_{k_1+1}, \ldots, x_k)$.

If the limit of the right-hand side of (2.9.19) exists as $x_1'' \to x_1', \ldots, x_{k_1}'' \to x_{k_1}'$, that is, if

$$\text{(2.9.20)} \qquad \lim_{x_i'' \to x_i', i=1, \ldots, k_1} F(x_1', \ldots, x_{k_1}', x_{k_1+1}, \ldots, x_k \mid I_{k_1}) = F(x_{k_1+1}, \ldots, x_k \mid x_1', \ldots, x_{k_1}')$$

exists, then dropping dashes, $F(x_{k_1+1}, \ldots, x_k \mid x_1, \ldots, x_{k_1})$ is called the c.d.f. of the *conditional random variable* $(x_{k_1+1}, \ldots, x_k \mid x_1, \ldots, x_{k_1})$. It can be verified that if it exists, $F(x_{k_1+1}, \ldots, x_k \mid x_1, \ldots, x_{k_1})$ has all of the properties (2.6.8) of a k_2-dimensional c.d.f. If the limit in (2.9.20) does not exist, the more general approach discussed in Section 1.10 is required.

In the case of a discrete random variable $(x_1 \ldots, x_k)$, if G in (2.9.2) is chosen as the section of R_k for which $x_1 = x_1', \ldots, x_{k_1} = x_{k_1}'$ so that $P(G) > 0$ then the conditional random variable $(x_{k_1+1}, \ldots, x_k \mid x_1', \ldots, x_{k_1}')$

exists and has a p.f. with the following formula, which is an extension of (2.9.8):

$$(2.9.21)\quad p(x_{k_1+1}, \ldots, x_k \mid x_1', \ldots, x_{k_1}') = \frac{p(x_1', \ldots, x_{k_1}', x_{k_1+1}, \ldots, x_k)}{p_{1\cdots k_1}(x_1', \ldots, x_{k_1}')}.$$

If we drop dashes, $p(x_1, \ldots, x_k)$ is the p.f. of $(x_1, \ldots, x_k)$ and $p_{1\cdots k_1}(x_1, \ldots, x_{k_1})$ is the marginal p.f. of $(x_1, \ldots, x_{k_1})$. Under the conditions we have stated it can also be shown that the c.d.f. of the conditional random variable $(x_{k_1+1}, \ldots, x_k \mid x_1', \ldots, x_{k_1}')$ [having p.d.f. (2.9.21)] is given by (2.8.20).

The reader can verify that if $(x_1, \ldots, x_k)$ is discrete we have the following multiplication formula for p.f.'s at mass points of the space of $(x_1, \ldots, x_k)$:

(2.9.22)

$$p(x_1, \ldots, x_k) = p(x_k \mid x_1, \ldots, x_{k-1}) \cdot p_{1\cdots k-1}(x_{k-1} \mid x_1, \ldots, x_{k-2}) \cdots p_1(x_1).$$

In case $(x_1, \ldots, x_k)$ is a continuous random variable, we have, corresponding to (2.9.21), the following p.d.f. of

$$(2.9.23)\quad f(x_{k_1+1}, \ldots, x_k \mid x_1, \ldots, x_{k_1}) = \frac{f(x_1, \ldots, x_k)}{f_{1\cdots k_1}(x_1, \ldots, x_{k_1})}$$

where the two functions on the right are p.d.f.'s as defined in Section 2.7. This p.d.f. may be derived from (2.9.20) by straightforward extension of the argument and assumptions used in establishing (2.9.12).

Corresponding to (2.9.22) we have, of course, for the continuous case, the following multiplication formula for p.d.f.'s:

(2.9.24)

$$f(x_1, \ldots, x_k) = f(x_k \mid x_1, \ldots, x_{k-1}) \cdot f_{1\cdots k-1}(x_{k-1} \mid x_1, \ldots, x_{k-2}) \cdots f_1(x_1)$$

assuming that the p.d.f.'s of all indicated conditional random variables exist.

(d) Conditional Distribution Functions in Case of Independence

Consider first the random variable (x_1, x_2). Let G be a Borel cylinder set in R_2 parallel to the x_2-axis such that $P(G) > 0$. There is no ambiguity if we refer to a point (x_1, x_2) of G by writing $x_1 \in G$. Now $E_{(x_1', x_2')}$ is the Cartesian product $E_{x_1'} \times E_{x_2'}$ where $E_{x_1'}$ is the set in $R_1^{(1)}$ for which $x_1 \leqslant x_1'$ and $E_{x_2'}$ is the set in $R_1^{(2)}$ for which $x_2 \leqslant x_2'$. Then

$$E_{(x_1', x_2')} \cap G = (E_{x_1'} \cap G) \times E_{x_2'}$$

and we have from (2.9.2)

$$F(x_1', x_2' \mid G) = \frac{P((E_{x_1'} \cap G) \times E_{x_2'})}{P(G)}. \tag{2.9.25}$$

If x_1 and x_2 are independent, (2.9.25) reduces to

$$F(x_1', x_2' \mid G) = \frac{P(E_{x_1'} \cap G)}{P(G)} \cdot P(E_{x_2'}) \tag{2.9.26}$$

and therefore we obtain the following result (dropping dashes):

2.9.1 *If (x_1, x_2) is a random variable with c.d.f. $F(x_1, x_2)$, such that x_1 and x_2 are independent, and if G is any (Borel) cylinder set in R_2 parallel to the x_2-axis for which $P(G) > 0$, then*

$$F(x_1, x_2 \mid G) = F_1(x_1 \mid G) \cdot F_2(x_2). \tag{2.9.27}$$

This means that the conditional probability that x_2 belongs to any specified set, given that $(x_1, x_2) \in G$, where G is a cylinder set parallel to the x_2-axis, does not depend on G. In particular, suppose G is the set for which $x_1 \in I_1$ used in (2.9.3). Then if x_1 and x_2 are independent, it follows from **2.4.2** that (2.9.4) reduces to

$$F(x_1', x_2 \mid I_1) = F_2(x_2). \tag{2.9.28}$$

In this case $\lim_{x_1'' \to x_1'} F(x_1', x_2 \mid I_1)$ exists except possibly for a set of values of x_2 with probability 0, and is equal to $F_2(x_2)$. Hence, we obtain the following corollary of **2.9.1**:

2.9.1a. *If x_1 and x_2 are independent then, except possibly for a set of probability* 0,

$$F(x_2 \mid x_1) = F_2(x_2). \tag{2.9.29}$$

In case x_1 and x_2 are independent it is clear that (2.9.8) reduces to

$$p(x_2 \mid x_1) = p_2(x_2) \tag{2.9.29a}$$

and (2.9.13) reduces to

$$f(x_2 \mid x_1) = f_2(x_2). \tag{2.9.29b}$$

More generally

2.9.2 *If $(x_1, \ldots, x_k)$ is a random variable with c.d.f. $F(x_1, \ldots, x_k)$, such that the random variables $(x_1, \ldots, x_{k_1})$ and $(x_{k_1+1}, \ldots, x_k)$ are independent, then if G is any Borel cylinder set in R_k defined by restricting values of $x_1, \ldots, x_{k_1}$ to any set such that $P(G) > 0$, we have*

(2.9.30)

$$F(x_1, \ldots, x_k \mid G) = F_{1 \ldots k_1}(x_1, \ldots, x_{k_1} \mid G) \cdot F_{k_1+1, \ldots, k}(x_{k_1+1}, \ldots, x_k).$$

If G is the interval for which $(x_1, \ldots, x_{k_1}) \in I_{k_1}$ used in (2.9.19) and furthermore if $(x_1, \ldots, x_{k_1})$ and $(x_{k_1+1}, \ldots, x_k)$ are independent, we have

$F(x_1, \ldots, x_k \mid I_{k_1}) = F_{1,\ldots,k_1}(x_1, \ldots, x_{k_1} \mid I_{k_1}) \cdot F_{k_1+1,\ldots,k}(x_{k_1+1}, \ldots, x_k)$ and

(2.9.19) reduces to

$$F(x_1', \ldots, x_{k_1}', x_{k_1+1}, \ldots, x_k \mid I_{k_1}) = F_{k_1+1,\ldots,k}(x_{k_1+1}, \ldots, x_k). \tag{2.9.31}$$

Taking the limits as $x_1'' \to x_1', \ldots, x_{k_1}'' \to x_{k_1}'$) which exists almost everywhere, we obtain as a corollary of **2.9.2**,

2.9.2a *If $(x_1, \ldots, x_k)$ is a random variable with c.d.f. $F(x_1, \ldots, x_k)$ such that $(x_1, \ldots, x_{k_1})$ and $(x_{k_1+1}, \ldots, x_k)$ are independent, except possibly for a set of probability* 0, *we have*

$$F(x_{k_1+1}, \ldots, x_k \mid x_1, \ldots, x_{k_1}) = F_{k_1+1,\ldots,k}(x_{k_1+1}, \ldots, x_k). \tag{2.9.32}$$

If $(x_1, \ldots, x_{k_1})$ and $(x_{k_1}, \ldots, x_k)$ are independent, it follows from (2.9.32) that (2.9.21) and (2.9.23) reduce to

$$p(x_{k_1+1}, \ldots, x_k \mid x_1, \ldots, x_{k_1}) = p_{k_1+1,\ldots,k}(x_{k_1+1}, \ldots, x_k) \tag{2.9.32a}$$

and

$$f(x_{k_1+1}, \ldots, x_k \mid x_1, \ldots, x_{k_1}) = f_{k_1+1,\ldots,k}(x_{k_1+1}, \ldots, x_k) \tag{2.9.32b}$$

respectively.

2.10 FINITE STOCHASTIC PROCESSES

A k-dimensional random variable $(x_1, \ldots, x_k)$ is sometimes referred to as a *finite stochastic process*, particularly in applications where the components of $(x_1, \ldots, x_k)$ correspond to measurements on the outcomes of a succession of physical operations in such a way, of course, that a k-dimensional c.d.f. is determined.

Example. For instance, if x_1 is a random variable denoting a number taken "at random" from the interval (0, 1), all numbers being assumed "equally likely," whereas x_2 is a number taken "at random" from $(x_1, 1)$ and so on for k numbers, we have a k-dimensional random variable $(x_1, \ldots, x_k)$, whose p.d.f. $f(x_1, \ldots, x_k)$ can be found by applying (2.9.24), that is,

$$f(x_1, \ldots, x_k) = \frac{1}{(1 - x_{k-1})} \cdot \frac{1}{(1 - x_{k-2})} \cdots \frac{1}{(1 - x_1)},$$

for $0 < x_1 < \cdots < x_k < 1$, and $f(x_1, \ldots, x_k) = 0$ otherwise. We may then refer to $(x_1, \ldots, x_k)$ as a (finite) stochastic process for describing the results of a succession of "cuts" of the interval (0, 1) in the manner described.

As he progresses through this book the reader will note many examples of finite stochastic processes. Infinite stochastic processes, that is, multidimensional random variables with infinitely many components, also occur at various places. They are defined and discussed in Chapter 4.

PROBLEMS

2.1 If $F(x)$ is the c.d.f. of a random variable x, establish formulas (2.2.7), (2.2.8), (2.2.9), and (2.2.10) by considering appropriate sequences of half-open intervals and **1.4.5.**

2.2 By considering a suitable sequence of two-dimensional half-open intervals establish (2.4.9).

2.3 Let $F_1(x)$ and $F_2(x)$ be c.d.f.'s of a discrete and a continuous random variable respectively. If a and b are non-negative numbers whose sum is 1, show that

$$aF_1(x) + bF_2(x)$$

has all of the properties of a c.d.f.

2.4 A mixed two-dimensional random variable (x_1, x_2) is such that x_1 is a discrete random variable with p.f. $p(x_1) = (\frac{1}{2})^{x_1}$, $x_1 = 1, 2, \ldots$ and x_2 is a continuous random variable such that the conditional random variable $x_2 \mid x_1$ has p.d.f. $x_1(1 - x_2)^{x_1-1}$ on the interval (0, 1). Show that the p.d.f. of the unconditional random variable x_2 is $2(1 + x_2)^{-2}$ on (0, 1).

2.5 A discrete random variable x_1 has p.f. qp^{x_1-1}, $x_1 = 1, 2, \ldots$ where $0 < p < 1, q = 1 - p$, whereas x_2 is a continuous random variable such that the p.d.f. of $x_2 \mid x_1$ is $x_1 x_2^{x_1-1}$ on (0, 1) and 0 otherwise. Determine the unconditional p.d.f. of x_2; the p.f. of $x_1 \mid x_2$; the c.d.f. of x_1; and the c.d.f. of x_2.

2.6 If $G(x_1, x_2) = [1 - (1 + x_1)^{-k} - (1 + x_2)^{-k} + (1 + x_1 + x_2)^{-k}]$, where $k > 0$, at any point (x_1, x_2) in the first quadrant of the x_1x_2-plane, and 0 elsewhere, show that $G(x_1, x_2)$ satisfies all conditions for the c.d.f. of a two-dimensional continuous random variable (x_1, x_2) and find its p.d.f.

2.7 If (x, y) is a pair of continuous random variables whose p.d.f. is $f(x, y)$ for $x > 0$, $y > 0$, and 0 elsewhere, show that

(a) the p.d.f. of $u = y/x$ is

$$\int_0^\infty x f(x, ux)\, dx,$$

(b) the p.d.f. of $v = x + y$ is

$$\int_0^v f(x, v - x)\, dx,$$

(c) the p.d.f. of $w = xy$ is

$$\int_0^\infty \frac{1}{x} f\left(x, \frac{w}{x}\right) dx.$$

2.8 Suppose the random variables (x_1, x_2) have p.d.f. $(2\pi)^{-1}e^{-\frac{1}{2}(x_1^2+x_2^2)}$ at every point (x_1, x_2) of the x_1x_2-plane. If y_1 and y_2 are random variables related to x_1 and x_2 as follows:

$$x_1 = y_1 \cos y_2$$
$$x_2 = y_1 \sin y_2$$

where the p.d.f. of y_1 is $y_1e^{-\frac{1}{2}y_1^2}$ for $y_1 > 0$, and 0 for $y_1 \leqslant 0$; and the p.d.f. of y_2 is $1/(2\pi)$ for $0 < y_2 < 2\pi$, and 0 otherwise, show that y_1 and y_2 are independent.

2.9 A three-dimensional random variable (x_1, x_2, x_3) has p.d.f. 6 in the tetrahedron having vertices (0, 0, 0), (0, 0, 1), (0, 1, 0), (1, 0, 0) and 0 outside. Find the c.d.f. and p.d.f. of: (x_1, x_2); x_1; $x_1 \mid x_2$; $x_1 + x_2$; and $x_1 + x_2 + x_3$.

2.10 If (x_1, x_2, x_3) are non-negative random variables having p.d.f. $e^{-(x_1+x_2+x_3)}$, find the p.d.f. of (u, x_2, x_3) where $u = x_1 + x_2 + x_3$. From the p.d.f. of (u, x_2, x_3) find the p.d.f. of

(a) u,
(b) $u \mid x_2$,
(c) $u \mid x_2, x_3$,
(d) $(x_2, x_3 \mid u)$.

2.11 If $(x_1, \ldots, x_k)$ are independent random variables having c.d.f. $F(x_1, \ldots, x_k)$, shows that in (2.6.4)

$$\Delta_{I_k}^k F(x_1, \ldots, x_k) = \Delta F_1(x_1) \cdots \Delta F_k(x_k),$$

where $$\Delta F_i(x_i) = F_i(x_i'') - F_i(x_i'),$$

and $F_i(x_i)$ is the marginal c.d.f. of x_i.

2.12 If $(x_1, \ldots, x_k)$ is a k-dimensional random variable having a p.d.f. which is symmetric in $x_1, \ldots, x_k$, show that $P(x_1 < x_2 < \cdots < x_k) = 1/k!$

2.13 A k-dimensional random variable $(x_1, \ldots, x_k)$ is known to have a p.d.f. of form

$$C(A + x_1 + \cdots + x_k)^{-k-B}$$

for $x_1 > 0, \ldots, x_k > 0$ and 0 elsewhere, where A and B are positive. Determine C. Find the p.d.f. of the marginal distribution of $(x_1, \ldots, x_r)$, $r < k$.

2.14 If $(x_1, \ldots, x_k)$ are independent random variables with identical continuous c.d.f.'s $F(x_1), \ldots, F(x_k)$, and if

$$u = \min(x_1, \ldots, x_k)$$
$$v = \max(x_1, \ldots, x_k),$$

show that the c.d.f. of (u, v) is

$$[F(v)]^k - g(u, v)[F(v) - F(u)]^k.$$

where $$g(u, v) = 1 \quad \text{for } u < v, \qquad = 0 \text{ for } u \geqslant v.$$

Find the marginal c.d.f.'s of u and of v. Also if $F(x)$ has a derivative $f(x)$ find the p.d.f.'s of (u, v), u, and v.

2.15 A jar has N chips numbered $1, 2, \ldots, N$. A chip is drawn, its number denoted by a random variable x_1, and it is replaced. A second chip is drawn, its number denoted by a random variable x_2, and it is replaced. And so on until k chips are drawn and replaced. Assigning equal probabilities to all N numbers at each step and mutual independence of $x_1, \ldots, x_k$, write down the c.d.f. of the k-dimensional random variable $(x_1, \ldots, x_k)$. If $y + 1$ is the smallest integer which exceeds every number drawn find the c.d.f. and then the p.f. of y.

2.16 (*Continuation*) Suppose the k ($<N$) chips are drawn successively *without* replacement, the probability of drawing any chip remaining in the jar at any stage being assigned so as to be equal to that of drawing any other chip remaining at that stage. How many mass points does $(x_1, \ldots, x_k)$ have? Determine the p.f. of $(x_1, \ldots, x_k)$. Show that the p.f. of the marginal distribution of any set of r ($r < k$) of the x's is the same as that of any other set of r of the x's.

2.17 If $(x_1, y_1), \ldots, (x_k, y_k)$, $k \geqslant 2$ are independent two-dimensional random variables with identical c.d.f.'s $F(x_1, y_1), \ldots, F(x_k, y_k)$ (and identical p.d.f.'s $f(x_1, y_1), \ldots, f(x_k, y_k)$), and if

$$u = \max(x_1, \ldots, x_k)$$
$$v = \max(y_1, \ldots, y_k),$$

show that the p.d.f. of (u, v) is

$$k(k-1)F^{k-2}(u, v)\frac{\partial F}{\partial u}\cdot\frac{\partial F}{\partial v} + kF^{k-1}(u, v)f(u, v).$$

2.18 (*Continuation*) If $w = \max(x_1 + y_1, \ldots, x_k + y_k)$ show that the p.d.f. of w is

$$k\left[\int_{-\infty}^{\infty}\int_{-\infty}^{w-x} f(x, y)\,dy\,dx\right]^{k-1}\int_{-\infty}^{\infty} f(x, w - x)\,dx.$$

2.19 Show that if each of the random variables $x_1, \ldots, x_k$ is independent of the remaining ones, then they are mutually independent.

2.20 If $F(x_1, \ldots, x_k)$ is a k-dimensional c.d.f. with marginal c.d.f.'s $F_1(x_1) \cdots F_k(x_k)$ show that

$$F(x_1, \ldots, x_k) \leqslant [F_1(x_1) \cdots F_k(x_k)]^{1/k}.$$

CHAPTER 3

Mean Values and Moments of Random Variables

3.1 INTRODUCTION

In Section 1.8 we have defined, for a given probability space $(R, \mathscr{B}, P)$ the Lebesgue-Stieltjes integral of a random variable $x(e)$ with respect to P over a set $E \in \mathscr{B}$.

Let E' be a Borel set in R_1, and let E be the set in R for which $x(e) \in E'$. Then if $F(x)$ is the c.d.f. of $x(e)$ we may write the Lebesgue-Stieltjes integral (1.8.5) of $x(e)$ over E in either of the two following equal forms:

(3.1.1) $$\int_E x(e)\, dP(e) = \int_{E'} x\, dF(x).$$

Similarly, let E_k' be a Borel set in R_k, and let E be the set in R for which $(x_1(e), \ldots, x_k(e)) \in E_k'$. If $F(x_1, \ldots, x_k)$ is the c.d.f. of $(x_1(e), \ldots, x_k(e))$, and if $g(x_1, \ldots, x_k)$ is measurable relative to $\mathscr{B}_k$ [thus making $g(x_1(e), \ldots, x_k(e))$ measurable relative to $\mathscr{B}$], we can then write

(3.1.2) $$\int_E g(x_1(e), \ldots, x_k(e))\, dP(e) = \int_{E_k'} g(x_1, \ldots, x_k)\, dF(x_1, \ldots, x_k).$$

Suppose E_k' in (3.1.2) is the set of points in R_k for which $g(x_1, \ldots, x_k) \in E_1'$ where E_1' is a Borel set in R_1. Then if $H(y)$ is the c.d.f. of $g(x_1, \ldots, x_k)$ we can further write

(3.1.3) $$\int_{E_k'} g(x_1, \ldots, x_k)\, dF(x_1, \ldots, x_k) = \int_{E_1'} y\, dH(y).$$

3.2 MEAN VALUE OF A RANDOM VARIABLE

If in (3.1.1) we take R_1 as the set E', then the resulting integral defines the *mean value* of the random variable x, that is, we write

(3.2.1) $$\mathscr{E}(x) = \int_{-\infty}^{\infty} x\, dF(x).$$

In case x is a discrete random variable with p.f. $p(x)$ as defined in Section 2.3(a), $\mathscr{E}(x)$ reduces to the following sum:

(3.2.2) $$\mathscr{E}(x) = \sum_{\alpha} x^{(\alpha)} p(x^{(\alpha)}),$$

where the $x^{(\alpha)}$ are the mass points of the random variable x.

If x is a continuous random variable with p.d.f. $f(x)$ we have

(3.2.3) $$\mathscr{E}(x) = \int_{-\infty}^{\infty} x\, dF(x) = \int_{-\infty}^{\infty} xf(x)\, dx.$$

More generally, if in (3.1.2) we take for E_k' the entire space R_k we obtain

(3.2.4) $$\mathscr{E}(g(x_1, \ldots, x_k)) = \int_{R_k} g(x_1, \ldots, x_k)\, dF(x_1, \ldots, x_k).$$

Furthermore, it follows from (3.1.3) that if $E_k' = R_k$, then $E_1' = R_1$ and we have

(3.2.5) $$\mathscr{E}(g(x_1, \ldots, x_k)) = \int_{-\infty}^{\infty} y\, dH(y).$$

If we denote $g(x_1, \ldots, x_k)$ by y, then the integral in (3.2.4) is simply $\mathscr{E}(y)$ and we have

(3.2.6) $$\mathscr{E}(g(x_1, \ldots, x_k)) = \mathscr{E}(y).$$

If $(x_1, \ldots, x_k)$ is a discrete random variable, then (3.2.4) reduces to a sum and if $(x_1, \ldots, x_k)$ is continuous, (3.2.4) is a k-dimensional integral over R_k.

If $g(x_1, \ldots, x_k)$ is a random variable, which for the moment we may write as y, mean values when they exist have the following useful properties:

3.2.1 *If c is a constant*

$$\mathscr{E}(cy) = c\mathscr{E}(y).$$

3.2.2 $$\mathscr{E}(ay_1 + by_2) = a\mathscr{E}(y_1) + b\mathscr{E}(y_2).$$

3.2.3 *If $m \leqslant y \leqslant M$, then $m \leqslant \mathscr{E}(y) \leqslant M$.*

3.2.4 *If $y_1 \leqslant y_2$, then $\mathscr{E}(y_1) \leqslant \mathscr{E}(y_2)$.*

3.2.5 $$|\mathscr{E}(y)| \leqslant \mathscr{E}(|y|).$$

Suppose (x_1, x_2) is a random variable with c.d.f. $F(x_1, x_2)$ such that x_1 and x_2 are independent and consider the mean value of the product $x_1 x_2$. It can be verified that

$$(3.2.7)\quad \mathscr{E}(x_1 x_2) = \int_{-\infty}^{\infty} x_1\, dF_1(x_1) \cdot \int_{-\infty}^{\infty} x_2\, dF_2(x_2) = \mathscr{E}(x_1) \cdot \mathscr{E}(x_2).$$

Thus we have the result that

3.2.6 *If (x_1, x_2) is a random variable whose components x_1 and x_2 are independent, then*

$$\mathscr{E}(x_1 x_2) = \mathscr{E}(x_1) \cdot \mathscr{E}(x_2).$$

A similar result holds, of course, for k mutually independent random variables.

3.3 MOMENTS OF ONE-DIMENSIONAL RANDOM VARIABLES

There are many problems in mathematical statistics in which it is difficult, or at least not feasible, to determine completely the c.d.f. of a random variable. In such cases it is often possible to describe the distribution of the random variable incompletely, although usefully, by moments and certain functions of moments of the random variable.

Suppose x is a random variable having c.d.f. $F(x)$. The *mean value* of x, as defined in (3.2.1), is usually denoted by $\mathscr{E}(x)$ or $\mu(x)$, that is

$$(3.3.1)\qquad \mathscr{E}(x) = \mu(x) = \int_{-\infty}^{\infty} x\, dF(x).$$

The *variance** of x is defined as the mean value of the random variable $(x - \mu(x))^2$, that is,

$$(3.3.2)\qquad \sigma^2(x) = \mathscr{E}(x - \mu(x))^2.$$

The quantity $\sigma(x)$, the positive square root of $\sigma^2(x)$, is called the *standard deviation* of x. The ratio $[\sigma(x)]/\mu(x)$ is called the *coefficient of variation* of x.

If we use concepts and language of elementary mechanics, the mean of the random variable x can be interpreted as the *center of gravity* in R_1 of the probability distribution of x. The variance of x can be interpreted as the *moment of inertia* of the same probability distribution about the center of gravity, and is an indication of the amount by which the probability mass spreads (or concentrates) about the center of gravity.

* Sometimes the notations ave (x) is used for $\mathscr{E}(x)$, and var (x) is used for $\sigma^2(x)$.

When no ambiguity arises as to what random variable is involved, we shall write μ rather than $\mu(x)$ and σ rather than $\sigma(x)$.

Making use of (3.3.1) and the properties of mean values expressed in **3.2.1** and **3.2.2**, we obtain

$$\sigma^2 = \mathscr{E}(x^2 - 2x\mu + \mu^2) = \mathscr{E}(x^2) - \mu^2. \tag{3.3.3}$$

If we take the mean value $\mathscr{E}(x - a)^2$ where a is an arbitrary constant, we have

$$\mathscr{E}(x - a)^2 = \mathscr{E}[(x - \mu) + (\mu - a)]^2 = \sigma^2 + (\mu - a)^2 \tag{3.3.4}$$

from which the following statement can be made:

3.3.1 *The value of the constant a which minimizes $\mathscr{E}(x - a)^2$ is μ and the minimum value of $\mathscr{E}(x - a)^2$ is σ^2.*

A useful inequality for the amount of probability in the "tails" of a probability distribution is provided by *Chebyshev's* (1867) *inequality* stated as follows:

3.3.2 *If x is a random variable having mean μ and variance $\sigma^2 > 0$, then*

$$P(|x - \mu| \geqslant \lambda\sigma) \leqslant \frac{1}{\lambda^2} \tag{3.3.5}$$

where λ is any positive constant.

To prove this statement we first cut the x-axis R_1 into three disjoint intervals:

$$I = (-\infty, \mu - \lambda\sigma], \quad I' = (\mu - \lambda\sigma, \mu + \lambda\sigma), \quad I'' = [\mu + \lambda\sigma, +\infty).$$

We can write the variance of x as

$$\sigma^2 = \int_I (x - \mu)^2 \, dF(x) + \int_{I'} (x - \mu)^2 \, dF(x) + \int_{I''} (x - \mu)^2 \, dF(x). \tag{3.3.6}$$

Dropping the middle term on the right-hand side of (3.3.6), we have

$$\sigma^2 \geqslant \int_I (x - \mu)^2 \, dF(x) + \int_{I''} (x - \mu)^2 \, dF(x). \tag{3.3.7}$$

By replacing x by $\mu - \lambda\sigma$ in $(x - \mu)^2$ in the first integral on the right and x by $\mu + \lambda\sigma$ in $(x - \mu)^2$ in the second, the inequality is preserved and we have

$$\sigma^2 \geqslant \sigma^2\lambda^2 P(|x - \mu| \geqslant \lambda\sigma) \tag{3.3.8}$$

which is equivalent to inequality (3.3.5).

The rth *moment* $\mu_r'(x)$, where r is a positive integer, is defined if it exists as the mean value of the random variable x^r, that is

$$\mu_r'(x) = \mathscr{E}(x^r). \tag{3.3.9}$$

For convenience we shall occasionally let $r = 0$, in which case $\mu_0'(x) = 1$. The question of whether $\mu_r'(x)$ exists and is finite is simply a matter of whether the random variable x^r is integrable over R_1. The rth *central moment* $\mu_r(x)$ is defined as the mean value of $(x - \mu)^r$, that is,

$$\mu_r(x) = \mathscr{E}[(x - \mu)^r]. \tag{3.3.10}$$

If there is no ambiguity about what random variable we are talking about, we shall denote $\mu_r'(x)$ and $\mu_r(x)$ by μ_r' and μ_r, respectively.

It is clear that the μ_r, $r = 1, 2, \ldots$ can be expressed as polynomials in the μ_r' and conversely. Noting that $\mu_1' = \mu$, we have

$$\begin{aligned} \mu_1 &= 0 \\ \mu_2 &= \mu_2' - \mu^2 \\ \mu_3 &= \mu_3' - 3\mu_2'\mu + 2\mu^3 \\ &\cdot \\ &\cdot \\ &\cdot \end{aligned} \tag{3.3.11}$$

and conversely,

$$\begin{aligned} \mu_1' &= \mu \\ \mu_2' &= \mu_2 + \mu^2 \\ \mu_3' &= \mu_3 + 3\mu_2\mu + \mu^3 \\ &\cdot \\ &\cdot \\ &\cdot \end{aligned} \tag{3.3.12}$$

Note that $\mu_2 = \sigma^2$.

The rth *absolute moment* $\nu_r'(x)$ is defined as

$$\nu_r'(x) = \mathscr{E}(|x|^r). \tag{3.3.13}$$

We shall ordinarily write $\nu_r'(x)$ as ν_r'. We define the rth *absolute central moment* as

$$\nu_r(x) = \mathscr{E}(|x - \mu|^r). \tag{3.3.13a}$$

More generally, we can similarly express the variance, and higher moments of any random variable $g(x)$, in terms of the c.d.f. of x. For instance, making use of (3.3.1) and (3.3.2) the *mean* and *variance* of $g(x)$ are defined as

$$\mu(g(x)) = \mathscr{E}[g(x)] \quad \text{and} \quad \sigma^2(g(x)) = \mathscr{E}\{(g(x))^2 - [\mu(g(x))]^2\}. \tag{3.3.14}$$

In certain kinds of problems involving discrete random variables, it is often convenient to determine the moments μ_r' by first evaluating *factorial moments*. If we let

$$x^{[r]} = x(x - 1) \cdots (x - r + 1) \tag{3.3.15}$$

the rth *factorial moment* is defined as

(3.3.16) $$\mu'_{[r]}(x) = \mathscr{E}(x^{[r]}),$$

and if no ambiguity arises we shall often write $\mu'_{[r]}(x)$ as $\mu'_{[r]}$. Expressed in terms of the ordinary moments, we have

(3.3.17)
$$\begin{aligned}
\mu'_{[1]} &= \mu'_1 \\
\mu'_{[2]} &= \mu'_2 - \mu'_1 \\
\mu'_{[3]} &= \mu'_3 - 3\mu'_2 + 2\mu'_1 \\
&\cdot \\
&\cdot \\
&\cdot
\end{aligned}$$

and conversely,

(3.3.18)
$$\begin{aligned}
\mu'_1 &= \mu'_{[1]} \\
\mu'_2 &= \mu'_{[2]} + \mu'_{[1]} \\
\mu'_3 &= \mu'_{[3]} + 3\mu'_{[2]} + \mu'_{[1]} \\
&\cdot \\
&\cdot \\
&\cdot
\end{aligned}$$

Note that we have defined the various kinds of moments as "moments of a random variable x." They are sometimes referred to as "moments of the probability distribution of x." If $F(x)$ is the c.d.f. of x, these various kinds of moments are sometimes referred to as "moments of $F(x)$."

3.4 MOMENTS OF TWO-DIMENSIONAL RANDOM VARIABLES

Suppose (x_1, x_2) is a two-dimensional random variable with c.d.f. $F(x_1, x_2)$. The *mean values* of x_1 and x_2 are

(3.4.1) $$\mu(x_1) = \mathscr{E}(x_1), \qquad \mu(x_2) = \mathscr{E}(x_2),$$

and are usually denoted by μ_1 and μ_2. The variances are

(3.4.2) $$\sigma^2(x_1) = \mathscr{E}(x_1 - \mu_1)^2, \qquad \sigma^2(x_2) = \mathscr{E}(x_2 - \mu_2)^2,$$

and are usually denoted by σ_1^2 and σ_2^2.

There is an ambiguity in the notation of (3.3.10) and (3.4.1) when μ_1 is written for $\mu(x_1)$ and μ_2 for $\mu(x_2)$. This ambiguity, however, is unimportant since $\mu_1 = 0$, and $\mu_2 = \mu'_2 - \mu^2$ in the sense of (3.3.10), and this will be clearly indicated in any discussion in which these symbols may appear.

The *covariance* of x_1 and x_2, written cov (x_1, x_2), is defined as

$$\text{cov}\,(x_1, x_2) = \mathscr{E}[(x_1 - \mu_1)(x_2 - \mu_2)], \tag{3.4.3}$$

which can be simplified to

$$\text{cov}\,(x_1, x_2) = \mathscr{E}(x_1 x_2) - \mathscr{E}(x_1) \cdot \mathscr{E}(x_2). \tag{3.4.4}$$

Note that cov $(x_1, x_2) =$ cov (x_2, x_1).

If x_1 and x_2 are independent, it follows from **3.2.6** that $\mathscr{E}(x_1 x_2) = \mathscr{E}(x_1) \cdot \mathscr{E}(x_2)$, and hence we obtain the following result:

3.4.1 *If x_1 and x_2 are independent, then*

$$\text{cov}\,(x_1, x_2) = 0. \tag{3.4.5}$$

Remark. It should be noted that the condition cov $(x_1, x_2) = 0$ does not imply that x_1 and x_2 are independent. For example, suppose (x_1, x_2) is a random variable satisfying the functional relation $x_2 = \cos x_1$ with probability 1, and the p.d.f. of x_1 is given by

$$f_1(x_1) = \begin{cases} \dfrac{1}{2\pi}, & -\pi < x_1 < +\pi \\ 0, & \text{otherwise.} \end{cases}$$

It is found that cov $(x_1, x_2) = 0$, yet x_2 can be exactly (functionally) determined from x_1 with probability 1.

The *correlation coefficient* $\rho(x_1, x_2)$ between x_1 and x_2 is defined as

$$\rho(x_1, x_2) = \frac{\text{cov}\,(x_1, x_2)}{\sigma_1 \sigma_2} \tag{3.4.6}$$

and will be written as ρ_{12} or as ρ when no ambiguity arises as to which random variables are involved. If $\rho(x_1, x_2) = 0$, x_1 and x_2 are said to be *uncorrelated*.

Suppose we take the mean value of the random variable

$$\left[t\left(\frac{x_1 - \mu_1}{\sigma_1}\right) + \left(\frac{x_2 - \mu_2}{\sigma_2}\right)\right]^2$$

where t is a real constant. We have

$$\mathscr{E}\left[t\left(\frac{x_1 - \mu_1}{\sigma_1}\right) + \left(\frac{x_2 - \mu_2}{\sigma_2}\right)\right]^2 = t^2 + 2t\rho + 1 \geqslant 0. \tag{3.4.7}$$

The condition for $t^2 + 2t\rho + 1 \geqslant 0$, for all real t, is that $\rho^2 - 1 \leqslant 0$. Therefore

3.4.2 *The correlation coefficient ρ satisfies the condition $-1 \leqslant \rho \leqslant 1$.*

If x_1 and x_2 are properly linearly dependent [see Section 2.8(*b*)] then $P(x_2 = \beta_0 + \beta_1 x_1) = 1$, where $\beta_1 \neq 0$, from which it follows that

$\sigma^2(x_2 - \beta_0 - \beta_1 x_1) = 0$ and that $\rho = +1$ or -1, depending on whether β_1 is a positive or a negative number. Conversely, if $\rho = \pm 1$, then the random variable

$$\mp\left(\frac{x_1 - \mu_1}{\sigma_1}\right) + \left(\frac{x_2 - \mu_2}{\sigma_2}\right) \tag{3.4.8}$$

is degenerate with mean value equal to 0. This means that if $\rho = \pm 1$

$$P\left(\mp\left(\frac{x_1 - \mu_1}{\sigma_1}\right) + \left(\frac{x_2 - \mu_2}{\sigma_2}\right) = 0\right) = 1. \tag{3.4.9}$$

Therefore

3.4.3 *A necessary and sufficient condition for the nondegenerate random variables x_1 and x_2, having finite variances, to be (properly) linearly dependent is that $\rho^2 = 1$.*

The *moments* $\mu'_{r_1 r_2}$ and *central moments* $\mu_{r_1 r_2}$ are defined as follows:

$$\mu'_{r_1 r_2} = \mathscr{E}(x_1^{r_1} x_2^{r_2}) \tag{3.4.10}$$

$$\mu_{r_1 r_2} = \mathscr{E}[(x_1 - \mu_1)^{r_1}(x_2 - \mu_2)^{r_2}]. \tag{3.4.11}$$

In this notation, the *means* of x_1 and x_2 are μ'_{10} and μ'_{01}, and the *variances* of x_1 and x_2 are μ_{20} and μ_{02}, respectively. The *covariance* of x_1 and x_2 is μ_{11}, and the *correlation coefficient* is $\rho_{12} = \mu_{11}/\sqrt{\mu_{20}\mu_{02}}$.

Absolute moments $\nu'_{r_1 r_2}$ and *factorial moments* $\mu'_{[r_1][r_2]}$ are defined as

$$\nu'_{r_1 r_2} = \mathscr{E}(|x_1|^{r_1} \cdot |x_2|^{r_2}), \tag{3.4.12}$$

$$\nu_{r_1 r_2} = \mathscr{E}(|x_1 - \mu_1|^{r_1} \cdot |x_2 - \mu_2|^{r_2}), \tag{3.4.13}$$

$$\mu'_{[r_1][r_2]} = \mathscr{E}(x_1^{[r_1]} x_2^{[r_2]}). \tag{3.4.14}$$

3.5 MOMENTS OF k-DIMENSIONAL RANDOM VARIABLES

The extension of the foregoing definitions to k-dimensional random variables is straightforward. Thus, if $(x_1, \ldots, x_k)$ is a k-dimensional random variable having c.d.f. $F(x_1, \ldots, x_k)$ the mean value of x_i is given by (3.2.6) with $g(x_1, \ldots, x_k) = x_i$, which reduces to (3.2.5) where $H(y)$ is the marginal c.d.f. of x_i, or stated more briefly,

$$\mu_i = \mathscr{E}(x_i) = \int_{R_k} x_i \, dF(x_1, \ldots, x_k) = \int_{-\infty}^{\infty} x_i \, dF_i(x_i). \tag{3.5.1}$$

The *variance* σ_i^2 of x_i is similarly defined from the marginal c.d.f. of x_i in accordance with definition (3.3.2) and the covariance between x_i and x_j from the marginal c.d.f. $F_{ij}(x_i, x_j)$ in accordance with definition (3.4.3).

The set of all variances and covariances among the k components of the random variable $(x_1, \ldots, x_k)$ form a $k \times k$ symmetric matrix, called the *covariance matrix*

(3.5.2) $$\|\sigma_{ij}\|, \qquad i, j = 1, \ldots, k,$$

σ_{ii} being the variance of x_i, and σ_{ij}, $i \neq j$, being the covariance between x_i and x_j. If there is any possibility of ambiguity as to what random variables are involved, we sometimes write the covariance matrix as $\|\sigma(x_i, x_j)\|$.

It is convenient to say that the random variable $(x_1, \ldots, x_k)$ has *mean* $(\mu_1, \ldots, \mu_k)$ and *covariance matrix* $\|\sigma_{ij}\|$.

We shall not only make considerable use of the covariance matrix (3.5.2) but also of the *inverse of the covariance matrix*, namely, $\|\sigma_{ij}\|^{-1}$ which will be denoted by

(3.5.3) $$\|\sigma^{ij}\|$$

where σ^{ij} can be formally expressed as σ^{ij} = (cofactor of σ_{ij} in $\|\sigma_{ij}\|)/|\sigma_{ij}|$, and $|\sigma_{ij}|$ is the determinant of the matrix $\|\sigma_{ij}\|$. Note that $\|\sigma^{ij}\|$ will exist *only* if $|\sigma_{ij}| \neq 0$. Note also that if $\|\sigma_{ij}\|$ has an inverse $\|\sigma^{ij}\|$, then $|\sigma^{ij}| = \dfrac{1}{|\sigma_{ij}|}$. Similar statements hold for the *correlation matrix* $\|\rho_{ij}\|$

It will be recalled that the necessary and sufficient condition stated in **3.4.3** for the nondegenerate random variables x_1 and x_2 to be properly linearly dependent is that $\rho^2 = 1$. An equivalent necessary and sufficient condition can be stated as

$$\begin{vmatrix} \sigma_{11} & \sigma_{12} \\ \sigma_{21} & \sigma_{22} \end{vmatrix} = 0.$$

It is useful to have a criterion for linear dependence in the case of a k-dimensional random variable. As stated in Section 2.8(d), the components of the random variable $(x_1, \ldots, x_k)$ are properly linearly dependent if there exists a set of constants $c_1, \ldots, c_k$ all different from zero, such that $c_1x_1 + \cdots + c_kx_k$ is a degenerate random variable. A criterion for proper linear dependence is provided by the following extension of Theorem **3.4.3**.

3.5.1 *A necessary and sufficient condition for the components of the random variable $(x_1, \ldots, x_k)$ to be properly linearly dependent is that the matrix $\|\sigma_{ij}\|$ be of rank $k - 1$.*

First consider the *sufficiency* of the condition.

Since $\|\sigma_{ij}\|$ is of rank $k - 1$, $|\sigma_{ij}| = 0$ and all principal minors of $|\sigma_{ij}|$ are positive, from which it follows that there exists a set of real constants c_i, $i = 1, \ldots, k$ all different from zero, such that $\sum_{i=1}^{k} \sigma_{ij} c_i = 0$ for $j = 1, \ldots, k$. Multiplying these equations by c_j and summing with respect to j, we have

$$\sum_{i,j=1}^{k} \sigma_{ij} c_i c_j = 0 \tag{3.5.4}$$

which may be written as

$$\mathscr{E}\left[\sum_{i=1}^{k} (x_i - \mu_i) c_i\right]^2 = 0. \tag{3.5.5}$$

However, this is the variance of the random variable $\sum_{i=1}^{k} c_i x_i$. Since the variance is zero, then $\sum_{i=1}^{k} c_i x_i$ is degenerate and, by definition, $x_1, \ldots, x_k$ are properly linearly dependent since none of the c_i are zero.

Now let us show that the condition is *necessary*. If $x_1, \ldots, x_k$ are properly linearly dependent, then by definition, there exists a degenerate random variable $\sum_{i=1}^{k} x_i c_i$ where the c_i are real constants all different from zero. Therefore,

$$\sigma^2\left(\sum_{i=1}^{k} c_i x_i\right) = \sum_{i,j=1}^{k} \sigma_{ij} c_i c_j = 0. \tag{3.5.6}$$

Now consider $\psi(c_1, \ldots, c_k) = \sum_{i,j=1}^{k} \sigma_{ij} c_i c_j$ as a function of the c_i. Since $\psi(c_1, \ldots, c_k)$ cannot be negative, it obviously has a minimum of zero at a point (in the space of the c's) for which no c_i is zero by hypothesis. But the values of the c_i for which $\psi(c_i, \ldots, c_k)$ has a minimum satisfy the equations

$$\frac{\partial \psi}{\partial c_i} = 0, \qquad i = 1, \ldots, k. \tag{3.5.7}$$

These equations reduce to

$$\sum_{j=1}^{k} \sigma_{ij} c_j = 0, \qquad i = 1, \ldots, k \tag{3.5.8}$$

Since the c's are not zero we must have $|\sigma_{ij}| = 0$. It is seen that the vanishing of any principal minor of $|\sigma_{ij}|$ contradicts the hypothesis that none of the c's are 0. Hence $\| \sigma_{ij} \|$ is of rank $k - 1$, which completes the proof for **3.5.1**.

If, for real $c_1, \ldots, c_k$, the quadratic form $\sum_{i,j=1}^{k} \sigma_{ij} c_i c_j$ vanishes (that is, has a minimum of zero only for $c_1 = \cdots = c_k = 0$, the quadratic form is said to be *positive definite* and its matrix $\|\sigma_{ij}\|$ is said to be a *positive definite matrix.*

A useful criterion of positive definiteness of a covariance matrix may be stated as follows:

3.5.2 *If $(x_1, \ldots, x_k)$ is a k-dimensional random variable with no degenerate components, and having covariance matrix $\|\sigma_{ij}\|$, a necessary and sufficient condition for $\|\sigma_{ij}\|$ to be positive definite is that there exist no linear dependence among the components $x_1, \ldots, x_k$.*

The proof is left to the reader.

The following property of quadratic forms will be useful in later sections.

3.5.3 *If $\sum_{i,j=1}^{k} \sigma_{ij}c_ic_j$ is positive definite, the quadratic form $\sum_{i,j=1}^{k} \sigma^{ij}c_ic_j$ is also positive definite.*

The proof of this statement is straightforward and is therefore omitted.

The various higher moments $\mu'_{r_1 \cdots r_k}$, $\mu_{r_1 \cdots r_k}$, $\nu'_{r_1 \cdots r_k}$, $\nu_{r_1 \cdots r_k}$ and $\mu'_{[r_1] \cdots [r_k]}$ pertaining to k random variables $(x_1, \ldots, x_k)$ are defined by obvious extensions of (3.4.10), (3.4.11), (3.4.12), (3.4.13), and (3.4.14).

3.6 MEANS, VARIANCES, AND COVARIANCES OF LINEAR FUNCTIONS OF RANDOM VARIABLES

The most frequently occurring type of function of several random variables is a linear function. The following facts about the mean and variance of such a function are useful:

3.6.1 *If $(x_1, \ldots, x_k)$ is a k-dimensional random variable having mean $(\mu_1, \ldots, \mu_k)$ and covariance matrix $\|\sigma_{ij}\|$ then the mean and variance of the linear function*

(3.6.1)
$$L = \sum_{i=1}^{k} c_i x_i$$

where $c_1, \ldots, c_k$ are constants, are

(3.6.2)
$$\mathscr{E}(L) = \sum_{i=1}^{k} c_i \mu_i$$

and

(3.6.3)
$$\sigma^2(L) = \sum_{i,j=1}^{k} \sigma_{ij} c_i c_j.$$

The proof is left to the reader.

In case the x_i are independent we have, by **3.4.1**, $\sigma_{ij} = 0$, $i \neq j$, and hence the following important corollary of **3.6.1**.

3.6.1*a* *If* $x_1, \ldots, x_k$ *are uncorrelated random variables with variances* $\sigma_1^2, \ldots, \sigma_k^2$ *the variance of* $L\left(= \sum_{i=1}^{k} c_i x_i\right)$ *is*

$$\sigma^2(L) = \sum_{i=1}^{k} c_i^2 \sigma_i^2. \tag{3.6.4}$$

More generally

3.6.2 *If* $L_p = \sum_{i=1}^{k} c_{ip} x_i$, $p = 1, \ldots, s$, *are linear functions of the random variables referred to in* **3.6.1**, *then the* L_p *have means*

$$\mathscr{E}(L_p) = \sum_{i=1}^{k} c_{ip}\mu_i, \qquad p = 1, \ldots, s \tag{3.6.5}$$

and covariance matrix

$$\|\sigma(L_p, L_q)\| = \left\| \sum_{i,j=1}^{k} \sigma_{ij} c_{ip} c_{jq} \right\|. \tag{3.6.6}$$

If $x_1, \ldots, x_k$ are uncorrelated random variables with variances $\sigma_1^2, \ldots, \sigma_k^2$, the covariance matrix $\| \sigma(L_p, L_q) \|$ reduces to $\left\| \sum_{i=1}^{k} \sigma_i^2 c_{ip} c_{iq} \right\|$.

3.7 MEAN VALUES OF CONDITIONAL RANDOM VARIABLES

(a) Case of Two Variables

Suppose $x_2 \mid x_1$ is a conditional random variable whose c.d.f. is $F(x_2 \mid x_1)$ as defined in Section 2.9(*b*). The mean value of $x_2 \mid x_1$, if it exists, is defined by

$$\mu(x_2 \mid x_1) = \mathscr{E}(x_2 \mid x_1) = \int_{-\infty}^{\infty} x_2 \, dF(x_2 \mid x_1). \tag{3.7.1}$$

The quantity $\mu(x_2 \mid x_1)$, considered as a function of x_1, is called the *regression function* of x_2 on x_1. Graphically, it represents the locus of the center of gravity of the conditional random variable $x_2 \mid x_1$ as a function of x_1. In particular, if

$$\mu(x_2 \mid x_1) = \beta_0 + \beta_1 x_1 \tag{3.7.2}$$

we have a *linear regression function* of x_2 on x_1, and β_0 and β_1 are called *regression coefficients*.

More generally, the mean value of the conditional random variable $g(x_2) \mid x_1$ is expressed by

$$\mathscr{E}(g(x_2) \mid x_1) = \int_{-\infty}^{\infty} g(x_2) \, dF(x_2 \mid x_1). \tag{3.7.3}$$

In particular, the variance of $x_2 \mid x_1$ is given by

$$\sigma^2(x_2 \mid x_1) = \mathscr{E}[(x_2 - \mu(x_2 \mid x_1))^2 \mid x_1]. \tag{3.7.4}$$

The quantities $\mu(x_2 \mid x_1)$ and $\sigma^2(x_2 \mid x_1)$ are sometimes referred to by the terms *conditional mean* and *conditional variance* of x_2 given x_1. $\sigma^2(x_2 \mid x_1)$ is also sometimes called the *residual variance* of x_2 on x_1.

The following statement, which is a corollary of **3.3.1**, gives some useful information about the relationship between $\sigma^2(x_2 \mid x_1)$ and $\mu(x_2 \mid x_1)$:

3.7.1 *If $x_2 \mid x_1$ is a conditional random variable as defined in Section* 2.9(*b*) *and if $u(x_1)$ is a real and single-valued function of x_1, the value of $u(x_1)$ which minimizes $\mathscr{E}[(x_2 - u(x_1))^2 \mid x_1]$ is given by $u(x_1) = \mu(x_2 \mid x_1)$. Furthermore, the minimum of $\mathscr{E}[(x_2 - u(x_1))^2 \mid x_1]$ is $\sigma^2(x_2 \mid x_1)$.*

If $\sigma^2(x_2 \mid x_1) = 0$ for all values of x_1 at which $x_2 \mid x_1$ is defined, then it is clear that we have $P[(x_2 = \mu(x_2 \mid x_1)) \mid x_1] = 1$. This means, of course, that if the value of the random variable x_1 is known (or given) for an event, then the value of x_2 for that event is $\mu(x_2 \mid x_1)$ with probability 1.

If we put $g(x_2) = x_2^r$ in (3.7.3) we obtain the rth moment of the conditional random variable $x_2 \mid x_1$. The rth central moment and factorial moment of $x_2 \mid x_1$ are similarly defined.

For the actual determination of the mean value of a function $g(x_1, x_2)$ of a random variable (x_1, x_2) in some of the simpler cases, it is sometimes convenient to proceed by iterated integration. In this case conditional random variables play an important role, which we express without proof as follows:

3.7.2 *Suppose (x_1, x_2) is a random variable with* c.d.f. $F(x_1, x_2)$. *Let $x_2 \mid x_1$ and $x_1 \mid x_2$ be conditional random variables in the sense of Section* 2.9(*b*) *with* c.d.f.'s $F(x_2 \mid x_1)$ *and* $F(x_1 \mid x_2)$, *respectively. Then, if $g(x_1, x_2)$ is a random variable,*

$$\int_{R_2} g(x_1, x_2)\, dF(x_1, x_2) = \int_{-\infty}^{\infty} \left[\int_{-\infty}^{\infty} g(x_1, x_2)\, dF(x_2 \mid x_1)\right] dF_1(x_1) \tag{3.7.5}$$

$$= \int_{-\infty}^{\infty} \left[\int_{-\infty}^{\infty} g(x_1, x_2)\, dF(x_1 \mid x_2)\right] dF_2(x_2),$$

or more briefly,

$$\mathscr{E}(g(x_1, x_2)) = \mathscr{E}_{(x_1)}[\mathscr{E}_{(x_2)}(g(x_1, x_2) \mid x_1)] = \mathscr{E}_{(x_2)}[\mathscr{E}_{(x_1)}(g(x_1, x_2) \mid x_2)]. \tag{3.7.5a}$$

The reader can readily verify **3.7.2** for the simpler, but common types of conditional random variables referred to in Section 2.9.

Extension of **3.7.1** and **3.7.2** to more general conditional random variables can be made by using the Radon-Nikodym theorem **1.10.1**.

(b) Case of Several Variables

If $x_k \mid x_1, \ldots, x_{k-1}$ is a conditional random variable with c.d.f. $F(x_k \mid x_1, \ldots, x_{k-1})$ in the sense of Section 2.9(*c*), the mean value of the conditional random variable $g(x_k) \mid x_1, \ldots, x_{k-1}$ is defined by

$$(3.7.6) \qquad \mathscr{E}(g(x_k) \mid x_1, \ldots, x_{k-1}) = \int_{-\infty}^{\infty} g(x_k)\, dF(x_k \mid x_1, \ldots, x_{k-1}).$$

In particular, the mean and variance of $x_k \mid x_1, \ldots, x_{k-1}$ are defined by

$$(3.7.7) \qquad \mu(x_k \mid x_1, \ldots, x_{k-1}) = \mathscr{E}(x_k \mid x_1, \ldots, x_{k-1})$$

and

$$(3.7.8) \qquad \sigma^2(x_k \mid x_1, \ldots, x_{k-1}) = \mathscr{E}[(x_k - \mu(x_k \mid x_1, \ldots, x_{k-1}))^2 \mid x_1, \ldots, x_{k-1}].$$

The reader will note that **3.7.1** can be extended at once to the random variable $x_k \mid x_1, \ldots, x_{k-1}$.

In general, if $x_{k_1+1}, \ldots, x_k \mid x_1, \ldots, x_{k_1}$, $k = k_1 + k_2$, is a k_2-dimensional conditional random variable with c.d.f. $F(x_{k_1+1}, \ldots, x_k \mid x_1, \ldots, x_{k_1})$, the mean value of the random variable $g(x_{k_1+1}, \ldots, x_k) \mid x_1, \ldots, x_{k_1}$ is given by

$$(3.7.9) \qquad \mathscr{E}[g(x_{k_1+1}, \ldots, x_k) \mid x_1, \ldots, x_{k_1}] = \int_{R_{k_2}} g(x_{k_1+1}, \ldots, x_k)\, dF(x_{k_1+1}, \ldots, x_k \mid x_1, \ldots, x_{k_1}).$$

It is now evident how one defines such quantities as the covariance between two conditional random variables, say $x_k \mid x_1, \ldots, x_{k-2}$ and $x_{k-1} \mid x_1, \ldots, x_{k-2}$, joint moments, etc.

The quantity $\mu(x_k \mid x_1, \ldots, x_{k-1})$, as a function of $x_1, \ldots, x_{k-1}$, is called the *regression function* of x_k on $x_1, \ldots, x_{k-1}$. In the particular case where

$$(3.7.10) \qquad \mu(x_k \mid x_1, \ldots, x_{k-1}) = \beta_0 + \beta_1 x_1 + \cdots + \beta_{k-1} x_{k-1},$$

x_k has a *linear regression function* on $x_1, \ldots, x_{k-1}$.

Finally, the reader should note that **3.7.2** can be extended so that we have as a generalization of (3.7.5),

$$(3.7.11) \qquad \int_{R_k} g(x_1, \ldots, x_k)\, dF(x_1, \ldots, x_k) = \int_{R_{k_1}} \left[\int_{R_{k_2}} g(x_1, \ldots, x_k)\, dF(x_{k_1+1}, \ldots, x_k \mid x_1, \ldots, x_{k_1}) \right] \cdot dF_{1 \cdots k_1}(x_1, \ldots, x_{k_1}),$$

or written as a completely iterated integral

$$(3.7.12)\quad \int_{R_k} g(x_1, \ldots, x_k)\, dF(x_1, \ldots, x_k)$$
$$= \int_{-\infty}^{\infty} \cdots \int_{-\infty}^{\infty} g(x_1, \ldots, x_k)\, dF(x_k \mid x_1, \ldots, x_{k-1})$$
$$\cdot\, dF_{1\cdots k-1}(x_{k-1} \mid x_1, \ldots, x_{k-2}) \cdots dF_1(x_1).$$

There are, of course, $k!$ possible orders of iteration.

We point out that formulas (3.7.6) through (3.7.12) can be given meaning under more general conditions on $F(x_1, \ldots, x_k)$ than those imposed in Section 2.9(*c*), by use of the Radon-Nikodym theorem.

(c) The Correlation Ratio

The variance $\sigma^2(x_2 \mid x_1)$ of the conditional random variable $x_2 \mid x_1$ provides some information as to how well the x_2 component of a sample point in R_2 can be determined when we are given the value of the x_1 component of the sample point. If $\sigma^2(x_2 \mid x_1) = 0$, the determination holds with probability 1. If $\sigma^2(x_2 \mid x_1) \neq 0$, some notion of how good the determination is can be obtained by taking the ratio $[\sigma^2(x_2 \mid x_1)]/\sigma^2(x_2)$ where, of course, $\sigma^2(x_2)$ is assumed $\neq 0$. However, in general, this ratio depends on x_1. A more useful criterion is sometimes obtained by taking the mean value of this ratio with respect to x_1. Denoting this mean value by $\eta^2_{2\cdot 1}$, we have

$$(3.7.13)\quad \eta^2_{2\cdot 1} = \frac{\mathscr{E}[\sigma^2(x_2 \mid x_1)]}{\sigma^2(x_2)}$$

where

$$(3.7.14)\quad \mathscr{E}[\sigma^2(x_2 \mid x_1)] = \int_{-\infty}^{\infty} \sigma^2(x_2 \mid x_1)\, dF_1(x_1).$$

The quantity $\eta^2_{2\cdot 1}$ is called the *correlation ratio* of x_2 on x_1. It is evident that $0 \leqslant \eta^2_{2\cdot 1} \leqslant 1$. We have $\eta^2_{2\cdot 1} = 0$ only when $\sigma^2(x_2 \mid x_1) = 0$, that is, only when $P[(x_2 = \mu(x_2 \mid x_1)) \mid x_1] = 1$. At the other extreme $\eta^2_{2\cdot 1} = 1$ only when x_1 and x_2 are uncorrelated.

The *multiple correlation ratio* $\eta^2_{k\cdot 12\cdots(k-1)}$ is defined as

$$(3.7.15)\quad \eta^2_{k\cdot 12\cdots(k-1)} = \frac{\mathscr{E}[\sigma^2(x_k \mid x_1, \ldots, x_{k-1})]}{\sigma^2(x_k)}$$

where

$$(3.7.16)\quad \mathscr{E}[\sigma^2(x_k \mid x_1, \ldots, x_{k-1})]$$
$$= \int_{R_{k-1}} \sigma^2(x_k \mid x_1, \ldots, x_{k-1})\, dF_{1\cdots(k-1)}(x_1, \ldots, x_{k-1}).$$

As in the case of the correlation ratio $\eta^2_{2\cdot 1}$, we have $0 \leqslant \eta^2_{k\cdot 12\cdots(k-1)} \leqslant 1$. The value 0 occurs only if $P[(x_k = \mu(x_k \mid x_1, \ldots, x_{k-1})) \mid x_1, \ldots, x_{k-1}] = 1$. The value 1 occurs only if x_k and $(x_1, \ldots, x_{k-1})$ are uncorrelated.

3.8 LEAST SQUARES LINEAR REGRESSION

(a) Case of Two Variables

Suppose we consider only linear functions of form $\beta_0 + \beta_1 x_1$ for the function $u(x_1)$ referred to in **3.7.1** and determine the value of β_0 and β_1 for which $\mathscr{E}(x_2 - \beta_0 - \beta_1 x_1)^2$ is a minimum. As usual, we assume that neither x_1 nor x_2 is degenerate, that is, that σ_1^2 and σ_2^2 are both positive.

Denoting this function by $\psi(\beta_0, \beta_1)$, we may write it as follows:

$$\psi(\beta_0, \beta_1) = \mathscr{E}[(x_2 - \mu_2) - (\beta_0 - \mu_2 + \beta_1\mu_1) - \beta_1(x_1 - \mu_1)]^2 \tag{3.8.1}$$

where $\mu_1 = \mathscr{E}(x_1)$ and $\mu_2 = \mathscr{E}(x_2)$. The values of β_0 and β_1 which minimize $\psi(\beta_0, \beta_1)$ will be given by the equations

$$\frac{\partial \psi}{\partial \beta_0} = 0, \qquad \frac{\partial \psi}{\partial \beta_1} = 0. \tag{3.8.2}$$

Using the notation of Section 3.4, and simplifying these equations, we obtain

$$\begin{aligned} \beta_0 - \mu_2 + \beta_1\mu_1 &= 0 \\ \rho\sigma_1\sigma_2 - \beta_1\sigma_1^2 &= 0. \end{aligned} \tag{3.8.3}$$

Denoting the solution by β_0^*, β_1^* we find

$$\begin{aligned} \beta_0^* &= \mu_2 - \rho\frac{\sigma_2}{\sigma_1}\mu_1 \\ \beta_1^* &= \rho\frac{\sigma_2}{\sigma_1}. \end{aligned} \tag{3.8.4}$$

Therefore:

3.8.1 *The linear function $u(x_1)$ which minimizes $\mathscr{E}(x_2 - u(x_1))^2$ is given by*

$$u(x_1) = \mu_2 + \rho\frac{\sigma_2}{\sigma_1}(x_1 - \mu_1). \tag{3.8.5}$$

The line whose equation is

$$x_2 = \mu_2 + \rho\frac{\sigma_2}{\sigma_1}(x_1 - \mu_1) \tag{3.8.6}$$

or, written more symmetrically,

$$\frac{(x_2 - \mu_2)}{\sigma_2} = \rho\frac{(x_1 - \mu_1)}{\sigma_1} \tag{3.8.7}$$

is called the *least squares regression line of x_2 on x_1*.

It can be verified that:

3.8.2 *The minimum value of* $\mathscr{E}(x_2 - \beta_0 - \beta_1 x_1)^2$, *which occurs for the values of* β_0 *and* β_1 *given by* (3.8.4), *and which is denoted by* $\sigma_{2\cdot 1}^2$ *is given by*

$$\sigma_{2\cdot 1}^2 = \sigma_2^2(1 - \rho^2). \tag{3.8.8}$$

The quantity $\sigma_{2\cdot 1}^2$ is called the *least squares residual variance of* x_2 *on* x_1. It is zero if and only if $\rho = \pm 1$, that is, if and only if x_1 and x_2 (both assumed nondegenerate) are properly linearly dependent, as pointed out in **3.4.3**.

By interchanging x_1 and x_2 one obtains the least squares regression line of x_1 on x_2.

$$\frac{(x_1 - \mu_1)}{\sigma_1} = \rho \frac{(x_2 - \mu_2)}{\sigma_2}. \tag{3.8.9}$$

Since neither x_1 nor x_2 is degenerate, it is clear that the two regression lines (3.8.7) and (3.8.9) will coincide if and only if $\rho = \pm 1$.

A question which arises here is whether there are special two-dimensional probability distributions which have the property that the *least squares* regression functions are identical with the actual regression functions. The answer is in the affirmative, and one of the most important distributions in mathematical statistics which has this property is the two-dimensional *normal* or *Gaussian* distribution which will be discussed in Section 7.3. This distribution, as we shall see, also has the property that $\sigma^2(x_2 \mid x_1)$ does not depend on x_1 and, in fact, is equal to $\sigma_{2\cdot 1}^2$.

A criterion for indicating how effectively the least squares regression line can be used for determining the value of x_2 of an event when x_1 is given is to compare the variance of x_2 about the least squares regression function of x_2 on x_1 with the variance of x_2, ignoring x_1, that is, to compare $\sigma_{2\cdot 1}^2$ with σ_2^2. The ratio of these two quantities which we write as follows:

$$\eta_{2\cdot 1(L)}^2 = \frac{\sigma_{2\cdot 1}^2}{\sigma_2^2} = 1 - \rho^2 \tag{3.8.10}$$

is called the *linear correlation ratio*. If x_1 is linearly related to x_2, that is, if x_1 and x_2 are random variables which are properly linearly dependent, then $\rho^2 = 1$ and $\eta_{2\cdot 1(L)}^2 = 0$. Conversely, if $\rho^2 = 1$ (and $\eta_{2\cdot 1(L)}^2 = 0$), x_1 and x_2 are linearly dependent. At the other extreme, if x_1 and x_2 are uncorrelated, then $\rho = 0$ and $\eta_{2\cdot 1(L)}^2 = 1$ and x_1 contains no information for determining x_2.

(b) Case of Several Variables

Now consider linear functions of the form

$$\beta_0 + \beta_1 x_1 + \cdots + \beta_{k-1} x_{k-1} \tag{3.8.11}$$

and let us determine $\beta_0, \ldots, \beta_{k-1}$ so as to minimize the quantity

$$\psi(\beta_0, \beta_1, \ldots, \beta_{k-1}) = \mathscr{E}[x_k - \beta_0 - \beta_1 x_1 - \cdots - \beta_{k-1}x_{k-1}]^2. \tag{3.8.12}$$

As in the two-dimensional case, it will be convenient to write (3.8.12) as

$$\begin{aligned}\psi = \mathscr{E}[(x_k - \mu_k) - (\beta_0 - \mu_k + \beta_1\mu_1 + \cdots + \beta_{k-1}\mu_{k-1})\\ - \beta_1(x_1 - \mu_1) - \cdots - \beta_{k-1}(x_{k-1} - \mu_{k-1})]^2.\end{aligned} \tag{3.8.13}$$

The values of the β's which minimize (3.8.13) are given by the solution of the equations

$$\frac{\partial \psi}{\partial \beta_\alpha} = 0, \qquad \alpha = 0, 1, \ldots, k - 1. \tag{3.8.14}$$

The first equation $\partial\psi/\partial\beta_0 = 0$ reduces to

$$\beta_0 - \mu_k + \beta_1\mu_1 + \cdots + \beta_{k-1}\mu_{k-1} = 0. \tag{3.8.15}$$

Making use of this result in the remaining $k - 1$ equations in (3.8.14), we find that they reduce to

$$\begin{aligned}\beta_1\sigma_{11} + \cdots + \beta_{k-1}\sigma_{1k-1} &= \sigma_{1k}\\ \cdots\cdots\cdots&\cdots\cdots\\ \beta_1\sigma_{k-1,1} + \cdots + \beta_{k-1}\sigma_{k-1,k-1} &= \sigma_{k-1,k}\end{aligned} \tag{3.8.16}$$

where $\|\sigma_{ij}\|$, $i, j = 1, \ldots, k$ is the covariance matrix of the components of the random variable $(x_1, \ldots, x_k)$ as defined in Section 3.5. Hence, it is evident that:

3.8.3 *If $|\sigma_{pq}| \neq 0$, $p, q = 1, \ldots, k - 1$, and if the values of $\beta_0, \beta_1, \ldots, \beta_{k-1}$ which minimize ψ are denoted by $\beta_0^*, \beta_1^*, \ldots, \beta_{k-1}^*$ they are given by*

$$\begin{aligned}\beta_0^* &= \mu_k - \beta_1^*\mu_1 - \cdots - \beta_{k-1}^*\mu_{k-1}\\ \beta_p^* &= \sum_{q=1}^{k-1} \sigma_{(k)}^{pq}\sigma_{qk}, \qquad p = 1, \ldots, k - 1,\end{aligned} \tag{3.8.17}$$

where $\|\sigma_{(k)}^{pq}\|$ is the inverse of the covariance matrix $\|\sigma_{pq}\|$, $p, q = 1, \ldots, k - 1$.

It will be remembered from **3.5.1** that $|\sigma_{pq}| \neq 0$ implies that $x_1, \ldots, x_{k-1}$ are not linearly dependent.

Finally, we may state that the *least squares linear regression hyperplane of x_k on $x_1, \ldots, x_{k-1}$* has as its equation

$$x_k = \mu_k + \sum_{p=1}^{k-1} \beta_p^*(x_p - \mu_p). \tag{3.8.18}$$

Equation (3.8.18) can also be written in more easily remembered determinantal form as

$$\text{(3.8.19)} \qquad \begin{vmatrix} \sigma_{11} & \cdots & \sigma_{1k} \\ \cdot & & \cdot \\ \cdot & & \cdot \\ \cdot & & \cdot \\ \sigma_{k-1,1} & \cdots & \sigma_{k-1,k} \\ (x_1 - \mu_1) & \cdots & (x_k - \mu_k) \end{vmatrix} = 0$$

as the reader will see by expanding the determinant with respect to the bottom row.

Now consider the problem of finding the value of the minimum of $\psi(\beta_0, \ldots, \beta_{k-1})$ with respect to the β's. Substituting the values of the β^*'s from (3.8.17) into (3.8.13) and denoting the minimum value of ψ by $\sigma^2_{k\cdot 12\cdots(k-1)}$ we have

(3.8.20)

$$\begin{aligned} \sigma^2_{k\cdot 12\cdots(k-1)} &= \mathscr{E}[(x_k - \mu_k) - \beta_1^*(x_1 - \mu_1) - \cdots - \beta_{k-1}^*(x_{k-1} - \mu_{k-1})]^2 \\ &= \sigma_{kk} - 2\sum_{p=1}^{k-1} \sigma_{pk}\beta_p^* + \sum_{p,q=1}^{k-1} \sigma_{pq}\beta_p^*\beta_q^*. \end{aligned}$$

Substituting the value of β_p^* from (3.8.17) into the last two terms of (3.8.20), we have

$$\text{(3.8.21)} \qquad \sum_{p=1}^{k-1} \sigma_{pk}\beta_p^* = \sum_{p,q=1}^{k-1} \sigma_{(k)}^{pq}\sigma_{pk}\sigma_{qk}$$

$$\text{(3.8.22)} \qquad \sum_{p,q=1}^{k-1} \sigma_{pq}\beta_p^*\beta_q^* = \sum_{p,q=1}^{k-1}\sum_{r,s=1}^{k-1} \sigma_{pq}\sigma_{(k)}^{ps}\sigma_{(k)}^{qr}\,\sigma_{rk}\sigma_{sk} = \sum_{r,s=1}^{k-1} \sigma_{(k)}^{rs}\sigma_{rk}\sigma_{sk},$$

since $\sum_{p=1}^{k-1} \sigma_{pq}\sigma_{(k)}^{ps} = \delta_{qs}$, where δ_{qs} is the Kronecker delta, which has the value 1 if $q = s$ and 0 if $q \neq s$. The extreme right member of (3.8.22) is, of course, the same as the right side of (3.8.21). Therefore, we have,

$$\text{(3.8.23)} \qquad \sigma^2_{k\cdot 12\cdots(k-1)} = \sigma_{kk} - \sum_{p,q=1}^{k-1} \sigma_{(k)}^{pq}\sigma_{pk}\sigma_{qk}.$$

But

$$\text{(3.8.24)} \qquad \sigma_{kk} - \sum_{p,q=1}^{k-1} \sigma_{(k)}^{pq}\sigma_{pk}\sigma_{qk} = \frac{|\sigma_{ij}|}{|\sigma_{pq}|}$$

where $i, j = 1, \ldots, k$ and $p, q = 1, \ldots, k-1$ as will be seen by performing a bordered expansion [see Bôcher (1907), for example] of the determinant $|\sigma_{ij}|$ by the kth row and kth column. Therefore, we finally obtain the following result:

3.8.4 *The minimum value of* $\psi(\beta_0, \beta_1, \ldots, \beta_{k-1})$, *denoted by* $\sigma^2_{k\cdot 12\cdots(k-1)}$, *is given by*

$$\sigma^2_{k\cdot 12\cdots(k-1)} = \frac{|\sigma_{ij}|}{|\sigma_{pq}|}. \tag{3.8.25}$$

The quantity $\sigma^2_{k\cdot 12\cdots(k-1)}$ is called the *least squares residual variance of* x_k *on* $x_1, \ldots, x_{k-1}$; it is the variance of x_k around the least squares regression plane whose equation is (3.8.18).

The correlation coefficient between the random variable x_k and the least squares regression function $\mu_k + \sum_{p=1}^{k-1} \beta_p^* (x_p - \mu_p)$ is called the *multiple correlation coefficient* between x_k and $(x_1, \ldots, x_{k-1})$. It is denoted by $\rho_{k\cdot 12\cdots(k-1)}$ and is expressed in terms of the elements of $\|\sigma_{ij}\|$ by the following formula

$$\rho^2_{k\cdot 12\cdots(k-1)} = 1 - \frac{|\sigma_{ij}|}{\sigma_{kk}|\sigma_{pq}|}. \tag{3.8.26}$$

To establish this formula we determine the variances of x_k and of $\mu_k + \sum_{p=1}^{k-1} \beta_p^* (x_p - \mu_p)$ and the covariance between them. We have

$$\sigma^2(x_k) = \sigma_{kk}$$

$$\sigma^2\left[\mu_k + \sum_{p=1}^{k-1} \beta_p^*(x_p - \mu_p)\right] = \sum_{p,q=1}^{k-1} \sigma_{pq}\beta_p^*\beta_q^*$$

$$\text{cov}\left[x_k, \mu_k + \sum_{p=1}^{k-1} \beta_p^*(x_p - \mu_p)\right] = \sum_{p=1}^{k-1} \sigma_{pk}\beta_p^*.$$

Applying the definition (3.4.6) of the correlation coefficient, and using (3.8.21) and (3.8.22) and positive square roots, we obtain

$$\rho_{k\cdot 12\cdots(k-1)} = +\sqrt{\sum_{p,q=1}^{k-1} \sigma^{pq}_{(k)}\sigma_{pk}\sigma_{qk}/\sigma_{kk}}. \tag{3.8.27}$$

Making use of (3.8.24) in (3.8.27), we obtain formula (3.8.26).

From (3.8.25) and (3.8.26) it is evident that

$$\sigma^2_{k\cdot 12\cdots(k-1)} = \sigma_{kk}(1 - \rho^2_{k\cdot 12\cdots(k-1)}). \tag{3.8.28}$$

The *linear correlation ratio* is defined by

$$\eta^2_{k\cdot 12\cdots(k-1),(L)} = \frac{\sigma^2_{k\cdot 12\cdots(k-1)}}{\sigma_{kk}} = 1 - \rho^2_{k\cdot 12\cdots(k-1)}. \tag{3.8.29}$$

As with the linear correlation ratio for two random variables, $\sigma^2_{k\cdot 12\cdots(k-1)}$

will be zero if and only if $\rho^2_{k\cdot 12\cdots(k-1)} = 1$, that is, if and only if the probability contained in the regression plane of x_k on $(x_1, \ldots, x_{k-1})$ is unity.

PROBLEMS

3.1 If x is a random variable whose first absolute central moment ν_1 exists show that the mean μ of x is finite, and that for $\lambda > 0$

$$P(|x - \mu| < \lambda\nu_1) \geqslant 1 - \frac{1}{\lambda}.$$

3.2 If $g(x)$ is a measurable function of a random variable x show that $P(|g(x)| \geqslant \lambda) \leqslant \mathscr{E}(|g(x)|)/\lambda$, for $\lambda > 0$.

3.3 Show that if the first r moments $\mu_1', \ldots, \mu_r'$ (and central moments $\mu_1, \ldots, \mu_r$) of a random variable x exist, then

$$\mu_r' = \sum_{i=0}^{r} \binom{r}{i} \mu_{r-i}\mu^i$$

and
$$\mu_r = \sum_{i=0}^{r} (-1)^i \binom{r}{i} \mu_{r-i}'\mu^i.$$

3.4 If x is any discrete random variable whose mass points are $0, 1, 2, \ldots, r$, show that all factorial moments higher than $\mu_{[r]}'$ are zero.

3.5 If the moments $\mu_{2r}, \mu_{2r+1}, \mu_{2r+2}$ of a random variable exist, show that

$$(\mu_{2r+1})^2 \leqslant \mu_{2r}\mu_{2r+2}.$$

3.6 Prove **3.5.2** and **3.5.3**.

3.7 If $(x_1, \ldots, x_k)$ is a k-dimensional random variable having means $(\mu_1, \ldots, \mu_k)$ and (positive definite) covariance matrix $\|\sigma_{ij}\|$ with inverse $\|\sigma^{ij}\|$, show that

$$P\left(\sum_{i,j=1}^{k} \sigma^{ij}(x_i - \mu_i)(x_j - \mu_j) < \lambda^2\right) \geqslant 1 - \frac{k}{\lambda^2}.$$

3.8 Prove **3.6.1**.

3.9 If $(x_1 \ldots, x_k)$ are independent random variables having zero means and unit variances show that

$$P\left(\sum_{i=1}^{k} x_i^2 \geqslant \lambda k\right) \leqslant \frac{1}{\lambda}.$$

3.10 If x is a random variable with mean μ and variance σ^2 and has c.d.f. $F(x)$ show that

$$F(x) \begin{cases} \leqslant \dfrac{1}{1 + \left(\dfrac{x - \mu}{\sigma}\right)^2}, & \text{if } x < \mu \\[2ex] \geqslant \dfrac{1}{1 + \left(\dfrac{\sigma}{x - \mu}\right)^2}, & \text{if } x > \mu, \end{cases}$$

[see Cramér (1946)].

3.11 If the $2r$th moments of two random variables x and y are finite, show that the $2r$th moment of $x + y$ is also finite.

3.12 If x is a random variable having mean μ and variance σ^2 and a p.d.f. with absolute maximum at c show that

$$P(|x - c| \geqslant \lambda\sqrt{\sigma^2 + (\mu - c)^2}) \leqslant \left(\frac{2}{3\lambda}\right)^2,$$

a result due to Gauss.

3.13. (*Continuation*) Show that

$$P(|x - \mu| \geqslant \lambda\sigma) \leqslant \frac{4(1 + \delta^2)}{9[\lambda - |\delta|]^2}$$

where $\delta = (\mu - c)/\sigma$ and $\lambda > |\delta|$. [Cramér (1946)].

3.14 If $x_1, \ldots, x_n$ are independent random variables having means all equal to μ and variances all equal to σ^2, show by Chebyshev's inequality that for any $\delta > 0$

$$\lim_{n \to \infty} P\left(\left|\frac{x_1 + \cdots + x_n}{n} - \mu\right| < \delta\right) = 1.$$

3.15 If $(x_1, \ldots, x_k)$ is a k-dimensional random variable such that the correlation coefficient between each pair of components is ρ, show that $-\dfrac{1}{k-1} \leqslant \rho \leqslant +1$.

3.16 Suppose $(x_1, \ldots, x_m, y_1, \ldots, y_n)$ is an $(m + n)$-dimensional random variable such that the variances of all components are unity, $\text{cov}(x_i, x_j) = \rho_1$, $\text{cov}(y_p, y_q) = \rho_2$, and $\text{cov}(x_i, y_q) = \rho_3$. If $u = x_1 + \cdots + x_m$, and $v = y_1 + \cdots + y_n$, show that the correlation coefficient between u and v is

$$\rho(u, v) = \frac{\sqrt{mn}\,\rho_3}{\sqrt{1 + (m-1)\rho_1} \cdot \sqrt{1 + (n-1)\rho_2}}.$$

3.17 If $x_1, \ldots, x_p, y_1, \ldots, y_q, z_1, \ldots, z_r$ are random variables having unit variances and zero covariances, show that the correlation coefficient between u and v where $u = x_1 + \cdots + x_p + y_1 + \cdots + y_q$, $v = x_1 + \cdots + x_p + z_1 + \cdots + z_r$ is given by

$$\rho(u, v) = \frac{p}{\sqrt{(p + q)(p + r)}}.$$

3.18 Suppose $(x_1, \ldots, x_n)$ are random variables such that x_1 and the conditional random variables $x_2 | x_1, \ldots, x_n | x_{n-1}$ are independent. If

$$\mathscr{E}(x_1) = \mu, \mathscr{E}(x_\xi \mid x_{\xi-1}) = x_{\xi-1}, \quad \xi = 2, \ldots, n$$

and if

$$\mathscr{E}(x_1 - \mu)^2 = \mathscr{E}[(x_\xi - x_{\xi-1})^2 \mid x_{\xi-1}] = \sigma^2, \quad \xi = 2, \ldots, n$$

show that the unconditional mean and variance of x_n are μ and $n\sigma^2$, respectively.

3.19 If x is a random variable whose first $2k$ moments $\mu_1, \ldots, \mu_{2k}$ exist, show that the matrix $\|\mu_{i+j}\|$, $i, j, = 1, \ldots, k$, is positive definite.

3.20 If x is a non-negative random variable, show that $\mathscr{E}(1/x) \geqslant 1/\mathscr{E}(x)$.

3.21 Suppose $(x_1, \ldots, x_m)$ is an m-dimensional random variable representing scores of a student (taken "at random" from grade G) on m questions of an examination in subject A and $(y_1, \ldots, y_n)$ is an n-dimensional random variable representing his scores on an examination in subject B. Let $T_1 = x_1 + \cdots + x_m$ and $T_2 = y_1 + \cdots + y_n$ be the total "scores" on the examinations in A and B, respectively. Let $\sigma_{1,m}^2$ and $\sigma_{1,m}^2 \rho_{1,m}$ be the average of the variances of the x's and the average of the covariances of all pairs of x's, respectively. Let $\sigma_{2,n}^2$, $\sigma_{2,n}^2 \rho_{2,n}$ be the corresponding quantities for the y's. Let $\sigma_{1,m}\sigma_{2,n}\rho_{3,m,n}$ be the average of the covariances of all x, y pairs. If, as $m, n \to \infty$, we have $\sigma_{1,m}^2 \to \sigma_1^2$, $\sigma_{2,n}^2 \to \sigma_2^2$, $\rho_{1,m} \to \rho_1$, $\rho_{2,m} \to \rho_2$, $\rho_{3,m,n} \to \rho_3$ where σ_1^2, σ_2^2, ρ_1, ρ_2, ρ_3 are all positive, show that

$$\lim_{m,n\to\infty} \rho(T_1, T_2) = \frac{\rho_3}{\sqrt{\rho_1\rho_2}}.$$

(Hence, if T and T^* are scores of the student on two *very long* examinations in the same subject (A or B), $\rho(T, T^*) \cong 1$).

3.22 Suppose (x_1, x_2, x_3) is a three-dimensional random variable having (finite) covariance matrix $\|\sigma_{ij}\|$, and correlation matrix $\|\rho_{ij}\|$, $i, j = 1, 2, 3$. Let $\beta_{02}^* + \beta_{12}^* x_1$ be the least squares regression line of x_2 on x_1, and $\beta_{03}^* + \beta_{13}^* x_1$ the least squares regression line of x_3 on x_1. Let $y_2 = x_2 - \beta_{02}^* - \beta_{12}^* x_1$ and $y_3 = x_3 - \beta_{03}^* - \beta_{13}^* x_1$. The correlation coefficient between y_2 and y_3 denoted by $\rho_{23\cdot 1}$ is called the *partial correlation coefficient between x_2 and x_3 with x_1 held constant.* Show that

$$\rho_{23\cdot 1} = \frac{\rho_{23} - \rho_{12}\rho_{13}}{\sqrt{(1-\rho_{13}^2)(1-\rho_{12}^2)}}.$$

3.23 Generalizing the preceding problem, let $(x_1, \ldots, x_k, x_{k+1})$ be a $(k+1)$-dimensional random variable with finite covariance matrix $\|\sigma_{ij}\|$ and correlation matrix $\|\rho_{ij}\|$, $i, j = 1, \ldots, k$.

Let $\beta_{0k}^* + \beta_{1k}^* x_1 + \cdots + \beta_{k-1,k}^* x_{k-1}$ be the least squares regression "plane" of x_k on $x_1, \ldots, x_{k-1}$ and $\beta_{0\ k+1}^* + \beta_{1\ k+1}^* x_1 + \cdots + \beta_{k-1\ k+1}^* x_{k-1}$ the least squares regression "plane" of x_{k+1} on $x_1, \ldots, x_{k-1}$.

Let $y_1 = x_k - \beta_{0k}^* - \beta_{1k}^* x_1 - \cdots - \beta_{k-1\ k}^* x_{k-1}$
and $y_2 = x_{k-1} - \beta_{0k+1}^* - \beta_{1k+1}^* x_1 - \cdots - \beta_{k-1\ k+1}^* x_{k-1}$.

The *partial correlation coefficient* $\rho_{k,k+1\cdot 12\cdots(k-1)}$ between x_k *and* x_{k+1} *with* $x_1, \ldots, x_{k-1}$ *held constant*, is the ordinary correlation coefficient between y_1 and y_2. Show that

$$\rho_{k,k+1\cdot 12\cdots(k-1)} = \frac{\Delta_{k\ k+1}}{\sqrt{\Delta_{kk}\,\Delta_{k+1\ k+1}}}$$

where Δ_{pq}, $p, q = k, k+1$ is the minor one obtains by deleting the pth row and qth column of the determinant

$$\begin{vmatrix} 1 & \rho_{12} & \cdots & & \rho_{1,k+1} \\ \rho_{21} & 1 & & & \rho_{2,k+1} \\ \cdot & \cdot & & & \cdot \\ \cdot & & \cdot & & \cdot \\ \cdot & & & \cdot & \cdot \\ & & & 1 & \rho_{k,k+1} \\ \rho_{k+1,1} & \rho_{k+1,2} & \cdots & \rho_{k+1,k} & 1 \end{vmatrix}$$

3.24 If $(x_1, \ldots, x_k)$ is a k-dimensional random variable such that the correlation coefficient between each pair of components is ρ, show that the multiple correlation coefficient between any component and the remaining components is

$$\rho \frac{\sqrt{k-1}}{\sqrt{1+(k-2)\rho}}.$$

3.25 In the preceding problem suppose the correlation coefficient between x_1 and each of the components $x_2, \ldots, x_k$ is ρ_1, whereas the correlation coefficient between each pair of the components $(x_2, \ldots, x_k)$ is ρ. Show that $\rho_1\rho \geqslant 0$ and that the multiple correlation coefficient between x_1 and $x_2, \ldots, x_k$ is

$$\frac{\sqrt{(k-1)\rho_1^2}}{\sqrt{1+(k-2)\rho}}.$$

3.26 In Problem **3.25** suppose the correlation coefficient between x_1 and x_r is ρ_r, $r = 2, \ldots, k$ while $x_2, \ldots, x_k$ are mutually independent. Show that the multiple correlation coefficient between x_1 and $x_2, \ldots, x_k$ is $\sqrt{\rho_2^2 + \cdots + \rho_k^2}$.

3.27 Reduce to simplest forms the equations of the least squares linear regression hyperplanes of x_1 on $x_2, \ldots, x_k$ in Problems **3.25** and **3.26**.

3.28 Verify that the equations given by (3.8.18) and (3.8.19) are equivalent.

CHAPTER 4

Sequences of Random Variables

4.1 DEFINITION OF A STOCHASTIC PROCESS

An important class of problems in mathematical statistics is the determination of limiting distribution functions of certain functions of n random variables as $n \to \infty$. More precisely, if $(x_1, \ldots, x_n)$ is an n-dimensional random variable and $g_n(x_1, \ldots, x_n)$ is a function of $(x_1, \ldots, x_n)$ which is itself a random variable, the problem is to determine the limiting c.d.f. of $g_n(x_1, \ldots, x_n)$ as $n \to \infty$, or at least certain properties of the c.d.f., if such a c.d.f. exists. Thus, it will be convenient to deal with random variables having infinitely many components. Such random variables are called *stochastic processes.* A stochastic process with a countably infinite number of components is often referred to as a *sequence* of random variables.

More precisely stated, a *stochastic process* is a family of random variables $\{x_\alpha; \alpha \in A\}$ where the range A may be an interval on the real axis, or a sequence of points on the real axis such as a sequence of integers, or even a more general set of points, such that every finite collection of components $(x_{\alpha_1}, \ldots, x_{\alpha_m})$ is a set of random variables with a specified c.d.f. $F_{\alpha_1, \ldots, \alpha_m}(x_{\alpha_1}, \ldots, x_{\alpha_m})$. The c.d.f.'s of the collection of all finite sets of random variables must be consistent, of course, in the sense that the c.d.f. of any finite set of random variables, say $(x_{\alpha_1}, \ldots, x_{\alpha_m})$, must be identically the same c.d.f. as the marginal c.d.f. of this set obtained from the c.d.f. of any finite set of random variables which contains $(x_{\alpha_1}, \ldots, x_{\alpha_m})$ as a subset.

4.2 PROBABILITY MEASURE FOR A STOCHASTIC PROCESS

Since we shall deal mostly with stochastic processes $(x_1, x_2, \ldots)$ with a countably infinite number of components, there will be no ambiguity if we denote by R_∞ the product space $R_1^{(1)} \times R_1^{(2)} \times \cdots$ where $R_1^{(1)}, R_1^{(2)} \cdots$ are (one-dimensional) sample spaces of $x_1, x_2, \ldots$, respectively. Then any countably infinite sequence of real numbers $(b_1, b_2, \ldots)$ is a point in R_∞ and, conversely, any point in R_∞ is a sequence of real numbers. Let b denote such a point. Thus, every realization of a stochastic process $(x_1, x_2, \ldots)$ specified by the values of a countably infinite number of components is a sample point in R_∞. We may refer to R_∞ as the sample space of such a stochastic process.

Consider any finite set of coordinates $(b_{\alpha_1}, \ldots, b_{\alpha_m})$ of b; it represents a point b' in the m-dimensional Euclidean space $R_m^{(\alpha_1, \ldots, \alpha_m)} = R_1^{(\alpha_1)} \times \cdots \times R_1^{(\alpha_m)}$. The point b' is the *projection* of b in R_∞ into $R_m^{(\alpha_1, \ldots, \alpha_m)}$ which we may write as

$$b' = h_{\alpha_1, \ldots, \alpha_m}(b). \tag{4.2.1}$$

The set of all points E in R_∞ which project into a given set E' in $R_m^{(\alpha_1, \ldots, \alpha_m)}$ is called a *cylinder set* and may be written as

$$E = h^{-1}_{\alpha_1, \ldots, \alpha_m}(E'). \tag{4.2.2}$$

If E' is a Borel set in $R_m^{(\alpha_1, \ldots, \alpha_m)}$, the cylinder set E in R_∞ corresponding to E' (the preimage of E' in R_∞) is called a *Borel cylinder set*. The reader should note that E' is simply the projection of E in R_∞ into $R_m^{(\alpha_1, \ldots, \alpha_m)}$. If we extend the definition of a Borel cylinder set to the case where $m = \infty$, it is noted that R_∞ itself is a Borel cylinder set. Furthermore if, for a finite m, E is a Borel cylinder set, so is $\bar{E}$.

Now suppose E' and E'' are Borel sets in $R_m^{(\alpha_1, \ldots, \alpha_m)}$ and $R_n^{(\beta_1, \ldots, \beta_n)}$ and let E_1 and E_2 be corresponding cylinder sets in R_∞. Consider the set $E_1 \cup E_2$ in R_∞. This is a cylinder set in R_∞ corresponding to a (presently to be defined) set of points E_* in $R_s^{(\gamma_1, \ldots, \gamma_s)}$, where $\gamma_1, \ldots, \gamma_s$ is the set of distinct integers among the integers $\alpha_1, \ldots, \alpha_m, \beta_1, \ldots, \beta_n$. $R_m^{(\alpha_1, \ldots, \alpha_m)}$ and $R_n^{(\beta_1, \ldots, \beta_n)}$ are then marginal sample spaces of $R_s^{(\gamma_1, \ldots, \gamma_s)}$, and $E_* = E_*' \cup E_*''$, where E_*' is the cylinder set in $R_s^{(\gamma_1, \ldots, \gamma_s)}$ corresponding to E' in $R_m^{(\alpha_1, \ldots, \alpha_m)}$ and E_*'' is the cylinder set in $R_s^{(\gamma_1, \ldots, \gamma_s)}$ corresponding to E'' in $R_n^{(\beta_1, \ldots, \beta_n)}$. The set $E_1 \cap E_2$ is the cylinder set in R_∞ corresponding to $E_*' \cap E_*''$ in $R_s^{(\gamma_1, \ldots, \gamma_s)}$. It should be noted that E_1 and E_2 are disjoint if and only if $E_*' \cap E_*'' = \phi$.

If E' and E'' are Borel sets in $R_m^{(\alpha_1, \ldots, \alpha_m)}$ and $R_n^{(\beta_1, \ldots, \beta_n)}$ respectively, whose preimages in $R_s^{(\gamma_1, \ldots, \gamma_s)}$ are E_*' and E_*'', then $E_*' \cup E_*''$ is also a Borel set in $R_s^{(\gamma_1, \ldots, \gamma_s)}$, and since $E_1 \cup E_2$ is the corresponding cylinder set in R_∞ this latter set is therefore a Borel cylinder set. Hence

4.2.1 *The class of Borel cylinder sets in R_∞ is a Boolean field $\mathscr{F}$ of sets.*

However, we are interested primarily in the Borel extension of this field which we shall call $\mathscr{B}_\infty$. That is, the Borel field $\mathscr{B}_\infty$ is defined as the smallest

Borel class of sets in R_∞ which contains the Boolean field $\mathscr{F}$. $\mathscr{B}_\infty$ is called *the class of Borel sets* of R_∞ and includes all Borel cylinder sets in R_∞.

Now suppose E' is a Borel set in the space $R_m^{(\alpha_1,\ldots,\alpha_m)}$, and let its preimage in R_∞ be E. To assign a probability to the Borel cylinder set E we adopt the usual rule, namely,

$$P(E) = P_{\alpha_1,\ldots,\alpha_m}(E') \tag{4.2.3}$$

where $P_{\alpha_1,\ldots,\alpha_m}(E')$ is the probability associated with E' as computed from the c.d.f. of $(x_{\alpha_1}, \ldots, x_{\alpha_m})$ where $(x_{\alpha_1}, \ldots, x_{\alpha_m})$ are the indicated components from the stochastic process $(x_1, x_2, \ldots)$. The question arises as to how probabilities are assigned to sets in $\mathscr{B}_\infty$ which are not Borel cylinder sets. This question is answered by the following extension theorem due to Kolmogorov (1933a):

4.2.2 *Let $(x_{\alpha_1}, \ldots, x_{\alpha_m})$ be any finite collection of components from a stochastic process $(x_1, x_2, \ldots)$ and let $R_m^{(\alpha_1,\ldots,\alpha_m)}$ be the sample space of $(x_{\alpha_1}, \ldots x_{\alpha_m})$. Then if E is the Borel cylinder set in R_∞ corresponding to any Borel set E' in $R_m^{(\alpha_1,\ldots,\alpha_m)}$, let the probability associated with E be assigned in accordance with* (4.2.3). *Then there exists a unique probability measure on $\mathscr{B}_\infty$, the Borel sets of R_∞, whose restriction to the cylinder sets in R_∞ is the set function defined by* (4.2.3).

The proof of this theorem is omitted. The reader who is interested in the proof is referred to Doob (1953) and Kolmogorov (1933a).

In setting up a stochastic process with a countably infinite number of components as a mathematical model in an application, there is often a physical process of some sort, just as with a finite stochastic process, which "generates" the sequence of components $x_1, x_2, \ldots$ in such a way that the conditions of **4.2.2** are assumed to be satisfied.

Examples. If a "true" die is rolled repeatedly, we can set up a very simple stochastic process $(x_1\ x_2, \ldots)$ where x_α is a random variable denoting the number of dots obtained on the αth throw, whereas $(x_{\alpha_1}, \ldots, x_{\alpha_r})$ has as its p.f. the function $p(x_{\alpha_1}, \ldots, x_{\alpha_r}) = 1/6^r$ at each of the 6^r mass points of the r-dimensional random variable $(x_{\alpha_1}, \ldots, x_{\alpha_r})$ for any choice of $\alpha_1, \ldots \alpha_r$ and any finite r.

If we let x_1 be a random variable having a rectangular distribution on (0, 1), x_2 a random variable having a rectangular distribution on $(x_1, 1)$, and in general x_α a random variable having a rectangular distribution on $(x_{\alpha-1}, 1)$, $\alpha = 1, 2, \ldots$, then $(x_1, x_2, \ldots)$ is an example of an important kind of a stochastic process called a *Markov chain of order* 1. Note that for any α the p.d.f. of the conditional random variable $x_\alpha \mid x_{\alpha-1}$ namely, $1/(1 - x_{\alpha-1})$ is exactly the same as the p.d.f. of the conditional random variable $x_\alpha \mid x_{\alpha-1}, x_{\beta_1}, \ldots, x_{\beta_r}$ where $\beta_1, \ldots, \beta_r$ is any set of $r < \alpha - 2$ of the integers $1, 2, \ldots, \alpha - 2$. A *Markov chain of order k* is a stochastic process $(x_1, x_2, \ldots)$ which has the property that for any α the conditional random variable $x_\alpha \mid x_{\alpha-1}, \ldots, x_{\alpha-k}$ has the same distribution as the conditional random variable $x_\alpha \mid x_{\alpha-1}, \ldots, x_{\alpha-k}, x_{\beta_1}, \ldots, x_{\beta_r}$ where $\beta_1, \ldots, \beta_r$ is any set of $r < \alpha - k$ of the integers $1, \ldots, \alpha - k - 1$. For discussion of the general theory of Markov chains or *Markov processes* the reader is referred to the books by Doob (1953), Feller (1957) and Loève (1955).

When we have a stochastic process $(x_1, x_2, \ldots)$ and occasion to discuss the probabilities associated with Borel sets in the (Euclidean) spaces of finite collections of x's, we can always consider the preimages of these sets in R_∞ and hence we can always talk about sets in R_∞ and their associated probabilities. For instance, instead of talking about the (Borel) set in the $x_m x_n$-plane for which $|x_m - x_n| < c$ we can talk about the (Borel cylinder) set in R_∞ for which $|x_m - x_n| < c$. The probabilities assigned to the two sets are equal by definition. In some situations it is usually a convenience to be able to refer to preimage events in R_∞ rather than to events in specific finite dimensional spaces corresponding to various finite collections of x's from $(x_1, x_2, \ldots)$. R_∞ is a fixed sample space in which events regarding $(x_1, x_2, \ldots)$ occur, whereas the sample space of $(x_1, \ldots, x_n)$ changes with n. There are, however, other reasons than convenience in considering R_∞. Indeed, as will be seen later, it is only in R_∞ that certain problems and theorems can even be formulated.

4.3 CONVERGENCE IN PROBABILITY

(a) Some Criteria for Convergence in Probability

Suppose $(x, x_1, x_2, \ldots)$ is a stochastic process such that for arbitrary $\varepsilon > 0$

$$\lim_{n\to\infty} P(|x_n - x| \geqslant \varepsilon) = 0. \tag{4.3.1}$$

Then $(x_1, x_2, \ldots)$ is said to *converge stochastically*, or to *converge in probability*, to the random variable x, and we sometimes denote this type of convergence briefly by writing

$$p \lim_{n\to\infty} x_n = x. \tag{4.3.1a}$$

If x is a degenerate random variable such that $P(x = x_0) = 1$, then the stochastic process $(x_1, x_2, \ldots)$ converges in probability to the constant x_0.

One of the simplest and most important examples of the convergence of a stochastic process to a constant is the *weak law of large numbers* stated as follows:

4.3.1 *Let $(x_1, x_2, \ldots)$ be a sequence of independent random variables having 0 means and variances $\sigma_1^2, \sigma_2^2, \ldots$. Let $c_n^2 = \sigma_1^2 + \cdots + \sigma_n^2$ and $\lim_{n\to\infty} c_n^2/n^2 = 0$. Let $\bar{x}_n = (x_1 + \cdots + x_n)/n$. Then the stochastic process $(\bar{x}_1, \bar{x}_2, \ldots)$ converges in probability to 0.*

To prove this we note that $\mathscr{E}(\bar{x}_n) = 0$ and $\mathscr{E}(\bar{x}_n^2) = c_n^2/n^2$. Hence it follows by Chebyshev's inequality (3.3.5) that for an arbitrary $\varepsilon > 0$

$$P(|\bar{x}_n| < \varepsilon) \geqslant 1 - \frac{c_n^2}{n^2\varepsilon^2}.$$

Thus, if $c_n^2/n^2 \to 0$ as $n \to \infty$, we have $\lim_{n\to\infty} P(|\bar{x}_n| < \varepsilon) = 1$, which concludes the argument.

Actually, **4.3.1** holds if the covariance between each pair of components of $(x_1, x_2, \ldots)$ is zero.

If the c.d.f.'s of the components of $(x_1, x_2, \ldots)$ are $F_1(x)$, $F_2(x), \ldots$, and if $\lim_{n\to\infty} F_n(x) = F(x)$ at every point of continuity of a c.d.f. $F(x)$, then $(x_1, x_2, \ldots)$ *converges in distribution* to $F(x)$.

Several theorems about convergence in probability and convergence in distribution follow. They are useful for later chapters.

One of the simplest criteria for convergence in probability can be stated as follows:

4.3.2 If $(x, x_1, x_2, \ldots)$ *is a stochastic process, then* $(x_1, x_2, \ldots)$ *converges in probability to the random variable* x *if*

(4.3.2) $$\lim_{n\to\infty} \mathscr{E}(x_n - x)^2 = 0.$$

It follows from Chebyshev's inequality stated in **3.3.2** that

(4.3.3) $$P(|x_n - x| \geqslant \varepsilon) \leqslant \frac{\mathscr{E}(x_n - x)^2}{\varepsilon^2},$$

and since (4.3.2) is assumed, we have $\lim_{n\to\infty} P(|x_n - x| \geqslant \varepsilon) \leqslant 0$ which implies (4.3.1). If $(x, x_1, x_2, \ldots)$ is a stochastic process which satisfies (4.3.2) with $\mathscr{E}(x_n^2) < +\infty$, $n = 1, 2, \ldots$, and $\mathscr{E}(x^2) < +\infty$, then $(x_1, x_2, \ldots)$ is said to *converge in the mean to* x. This type of convergence is sometimes denoted briefly by writing

(4.3.2*a*) $$\underset{n\to\infty}{\text{l.i.m.}}\, x_n = x.$$

4.3.3 *If* $(x, x', x_1, x_2, \ldots)$ *is a stochastic process such that* $(x_1, x_2, \ldots)$ *converges in probability to each of the random variables* x *and* x', *then* x *and* x' *are equivalent random variables.*

To prove this we first note that

(4.3.4) $$P\left(|x - x'| > \frac{1}{N}\right) = P\left(|(x_n - x') - (x_n - x)| > \frac{1}{N}\right).$$

Now let E, E', E'' be the sets of points in the sample space R_∞ of $(x, x', x_1, x_2, \ldots)$ for which $|(x_n - x') - (x_n - x)| > \frac{1}{N}$, $|x_n - x| > \frac{1}{2N}$, and $|x_n - x'| > \frac{1}{2N}$, respectively. Then it will be seen that $E \subset (E' \cup E'')$. Hence $P(E) \leqslant P(E') + P(E'')$, that is,

$$(4.3.5) \quad P\left(|x - x'| > \frac{1}{N}\right) \leqslant P\left(|x_n - x| > \frac{1}{2N}\right) + P\left(|x_n - x'| > \frac{1}{2N}\right).$$

By letting $n \to \infty$ both members on the right have limits equal to zero by hypothesis. Therefore

$$P\left(|x - x'| > \frac{1}{N}\right) = 0.$$

Denoting the set in R_∞ for which $|x - x'| > \frac{1}{N}$ by G_N, we have $G_1 \subset G_2 \subset \cdots$ and $\lim_{N\to\infty} G_N = G$, where G is the set of all points in R_∞ for which $x \neq x'$. Therefore, by **1.4.6** for the case of sets in $\mathscr{B}_\infty$, we have

$$(4.3.6) \quad P(x \neq x') = P(G_1 \cup G_2 \cup \cdots) \leqslant P(G_1) + P(G_2) + \cdots.$$

But $P(G_1) = P(G_2) = \cdots = 0$. Therefore, $P(x \neq x') = 0$, and hence x and x' are equivalent random variables [see Section 2.8(b)].

Another useful result is that convergence in probability implies convergence in distribution, or more precisely stated:

4.3.4 *Suppose* $(x, x_1, x_2, \ldots)$ *is a stochastic process with components having* c.d.f.'s $F(x), F_1(x), F_2(x), \ldots$, *respectively. If* $(x_1, x_2, \ldots)$ *converges in probability to the random variable* x, *then the sequence of* c.d.f.'s $F_1(x), F_2(x), \ldots$ *converges to* $F(x)$ *at every point of continuity of* $F(x)$.

As usual, let R_∞ be the sample space of the stochastic process $(x, x_1, x_2, \ldots)$. Suppose $F(x)$ is continuous at $x = x_0$ and let x' be a constant such that $x' < x_0$. Now consider the three events for which: $x \leqslant x'$; $x_n \leqslant x_0$; and $|x_n - x| > (x_0 - x')$ in R_∞. The first event is contained in the union of the second and third. Therefore we have

$$(4.3.7) \quad F(x') = P(x \leqslant x') \leqslant P(x_n \leqslant x_0) + P(|x_n - x| > (x_0 - x')).$$

Now take limits as $n \to \infty$. Since $(x_1, x_2, \ldots)$ converges in probability to the random variable x, we have $\lim_{n\to\infty} P(|x_n - x| > (x_0 - x')) = 0$. Therefore,

$$(4.3.8) \quad F(x') \leqslant \liminf_{n\to\infty} F_n(x_0).$$

Similarly, by taking $x'' > x_0$, we find

$$F(x'') \geqslant \lim_{n\to\infty} \sup F_n(x_0) \tag{4.3.9}$$

and hence

$$F(x') \leqslant \lim_{n\to\infty} \inf F_n(x_0) \leqslant \lim_{n\to\infty} \sup F_n(x_0) \leqslant F(x''). \tag{4.3.10}$$

But since $F(x)$ is continuous at $x = x_0$, we have $\lim_{x'\to x_0} F(x') = \lim_{x''\to x_0} F(x'') = F(x_0)$. Therefore, we have

$$\lim_{n\to\infty} F_n(x_0) = F(x_0) \tag{4.3.11}$$

which completes the argument for **4.3.4**.

(b) Convergence of Functions of Components in Stochastic Processes

If $(x, x_1, x_2, \ldots)$ is a stochastic process such that $(x_1, x_2, \ldots)$ converges in probability to the random variable x we sometimes need to know what conditions will insure the convergence in probability of $(g(x_1), g(x_2), \ldots)$ to $g(x)$. We shall consider several theorems relating to this and similar questions which will be used in later chapters.

4.3.5 *Suppose $(x, x_1, x_2, \ldots)$ is a stochastic process such that $(x_1, x_2, \ldots)$ converges in probability to the random variable x. Let $g(x)$ be a continuous function of x on R_1. Then*

$$p \lim_{n\to\infty} g(x_n) = g(x). \tag{4.3.12}$$

If x is a degenerate random variable, namely, a constant c, then (4.3.12) simply states that $(g(x_1), g(x_2), \ldots)$ converges in probability to the constant $g(c)$.

To prove **4.3.5** we note that $g(x)$ is uniformly continuous on any closed interval, say $[-M, M]$. Since x is a random variable we can, for an arbitrary $\varepsilon > 0$, choose M so that

$$P(|x| > M) < \frac{\varepsilon}{2}. \tag{4.3.13}$$

For such a choice of ε and M there exists a $\delta(\varepsilon, M)$ such that if

(i) $|x| \leqslant M$

(ii) $|x_n - x| < \delta(\varepsilon, M)$

then

(iii) $|g(x_n) - g(x)| < \varepsilon.$

If we denote the sets in R_∞ for which (i), (ii), and (iii) hold by E_1, E_2, and E_3, respectively, then it is seen that $\bar{E}_3 \subset (\bar{E}_1 \cup \bar{E}_2)$, from which we have

$P(\bar{E}_3) \leqslant P(\bar{E}_1) + P(\bar{E}_2)$, that is,

(4.3.14) $P(|g(x_n) - g(x)| \geqslant \varepsilon) \leqslant P(|x_n - x| \geqslant \delta(\varepsilon, M)) + P(|x| > M).$

But there is an $n(\varepsilon, \delta, M)$ such that for $n > n(\varepsilon, \delta, M)$ we have

(4.3.15) $$P(|x_n - x| \geqslant \delta(\varepsilon, M)) < \frac{\varepsilon}{2}.$$

Since M has been chosen so as to satisfy (4.3.13), we finally obtain

(4.3.16) $$P(|g(x_n) - g(x)| \geqslant \varepsilon) < \varepsilon$$

for $n > n(\varepsilon, \delta, M)$, which is equivalent to (4.3.12), thus establishing **4.3.5**.

It should be noted that **4.3.5** can be stated in more general form by requiring that $g(x)$ be continuous on a closed interval I where $P(x \in I) = 1$. Since this would involve only minor modifications of the argument it is left to the reader.

Sometimes we have to deal with convergence problems involving stochastic processes whose components are vectors. For instance, if $(x, y;\ x_1, y_1;\ x_2, y_2; \ldots)$ is such a stochastic process, we might be concerned with conditions under which $(g(x_1, y_1), g(x_2, y_2), \ldots)$ converges in probability to $g(x, y)$. In this case R_∞ is the sample space of $(x, y;\ x_1, y_1;\ x_2, y_2; \ldots)$.

If $(z;\ x_1, y_1;\ x_2, y_2; \ldots)$ is a stochastic process such that for an arbitrary $\varepsilon > 0$
$$\lim_{n\to\infty} P(|x_n - y_n| > \varepsilon) = 0,$$
while one of the sequences $(x_1, x_2, \ldots)$ or $(y_1, y_2, \ldots)$ converges in probability to the random variable z then it can be readily verified that the other sequence also converges in probability to the random variable z. In such a situation it is convenient to say that $(x_1, x_2, \ldots)$ and $(y_1, y_2 \ldots)$ *converge in probability together to the random variable* z. Similarly, if the above limit holds and one of the two sequences converges in distribution to $F(x)$, we shall say that both sequences *converge in distribution together to* $F(x)$.

We may state the extension of **4.3.5** to the case of several sequences of random variables as follows:

4.3.6 *Suppose* $(x^{(1)}, \ldots, x^{(k)};\ x_1^{(1)}, \ldots, x_1^{(k)};\ x_2^{(1)}, \ldots, x_2^{(k)}; \ldots)$ *is a k-dimensional vector stochastic process which converges in probability to the k-dimensional random variable* $(x^{(1)}, \ldots, x^{(k)})$. *Let* $g(x^{(1)}, \ldots, x^{(k)})$ *be a continuous function of* $(x^{(1)}, \ldots, x^{(k)})$ *in* R_k. *Then*

(4.3.17) $$p\lim_{n\to\infty} g(x_n^{(1)}, \ldots, x_n^{(k)}) = g(x^{(1)}, \ldots, x^{(k)}).$$

The proof of this theorem is similar to that of **4.3.5** and is left to the reader.

Again it is to be noted that we can relax the conditions of **4.3.6** by requiring that $g(x^{(1)}, \ldots, x^{(k)})$ be continuous in a closed k-dimensional interval I_k for which $P((x^{(1)}, \ldots, x^{(k)}) \in I_k) = 1$.

The following corollary of **4.3.6** will be particularly useful in later chapters:

4.3.6a *If* $(x, c;\ x_1, y_1;\ x_2, y_2; \ldots)$ *is a vector stochastic process such that* $(x_1, y_1;\ x_2, y_2; \ldots)$ *converges in probability to* (x, c) *where* x *has* c.d.f. $F(x)$ *and where* c *is a positive constant, then at every point of continuity of* $F(x)$,

(4.3.18) $$\lim_{n\to\infty} P(x_n + y_n \leqslant z) = F(z - c)$$

(4.3.19) $$\lim_{n\to\infty} P(x_n y_n \leqslant z) = F\left(\frac{z}{c}\right)$$

(4.3.20) $$\lim_{n\to\infty} P\left(\frac{x_n}{y_n} \leqslant z\right) = F(cz)$$

(4.3.21) $$\lim_{n\to\infty} P(ax_n + by_n \leqslant z) = F\left(\frac{z - bc}{a}\right)$$

where a *and* b *are constants and* $a > 0$.

The following theorem will be of interest later:

4.3.7 *Suppose* $(x_1^{(1)}, \ldots, x_1^{(k)}; y_1^{(1)}, \ldots, y_1^{(k)};\ x_2^{(1)}, \ldots, x_2^{(k)}; y_2^{(1)}, \ldots, y_2^{(k)}; \ldots)$ *is a vector stochastic process such that*

(4.3.22) $$|y_n^{(i)} - c^{(i)}| \leqslant |x_n^{(i)} - c^{(i)}|, \quad i = 1, \ldots, k; \quad n = 1, 2, \ldots$$

and $(x_1^{(1)}, \ldots, x_1^{(k)};\ x_2^{(1)}, \ldots, x_2^{(k)}; \ldots)$ *converges in probability to the vector constant* $(c^{(1)}, \ldots, c^{(k)})$. *Let* $g(x^{(1)}, \ldots, x^{(k)})$ *be a single-valued function defined at all points in* R_k *and continuous in some open k-dimensional rectangular interval containing* $(c^{(1)}, \ldots, c^{(k)})$. *Then*

(4.3.23) $$p \lim_{n\to\infty} g(y_n^{(1)}, \ldots, y_n^{(k)}) = g(c^{(1)}, \ldots, c^{(k)}).$$

The proof of this theorem involves no particular difficulties and is omitted.

It should be pointed out that the conditions in **4.3.7** can be relaxed by assuming $g(x^{(1)}, \ldots, x^{(k)})$ to be defined only in some closed k-dimensional interval I_k containing $(c^{(1)}, \ldots, c^{(k)})$ for which $P(x^{(1)}, \ldots, x^{(k)} \in I_k) = 1$, and such that $g(x^{(1)}, \ldots, x^{(k)})$ is continuous in I_k.

Suppose $(g(x, \theta), g(x_1, \theta), g(x_2, \theta), \ldots)$ is a stochastic process for every θ in some interval (θ', θ''). If for an arbitrary $\varepsilon > 0$ there is an n_ε such that for each θ in (θ', θ'')

(4.3.24) $$P(|g(x_n, \theta) - g(x, \theta)| > \varepsilon) < \varepsilon$$

for $n > n_\varepsilon$, then $(g(x_1, \theta), g(x_2, \theta), \ldots)$ is said to converge *in probability to* $g(x, \theta)$ *uniformly with respect to* θ *in* (θ', θ''). This notion can be extended in an obvious manner to situations where $g(x, \theta)$ is degenerate and also to those where x_n is a vector random variable.

Finally, we shall find the following theorem useful in later chapters, particularly Chapters 12 and 13:

4.3.8 *Let* $(x_1, x_2, \ldots)$ *be a stochastic process depending on a parameter* θ *such that for each value of* θ *in some interval* (θ', θ''), $f_n(x_1, \ldots, x_n, \theta)$, $n = 1, 2, \ldots,$ *and* $\theta_n^*(x_1, \ldots, x_n)$, $n = 1, 2, \ldots,$ *are sequences of random variables converging in probability respectively to* $g(\theta)$ *and* θ *uniformly with respect to* θ *in* (θ', θ''), *where* $g(\theta)$ *is continuous in* (θ', θ''). *Then* $f_n(x_1, \ldots, x_n, \theta_n^*)$, $n = 1, 2, \ldots,$ *converges in probability to* $g(\theta)$.

Consider the stochastic process $(x_1, x_2, \ldots)$ at any point θ_0 in (θ', θ''). Then the sequence $(\theta_1^*(x_1), \theta_2^*(x_1, x_2), \ldots)$ converges in probability to θ_0. Thus for an arbitrary $\varepsilon > 0$ such that $(\theta_0 - \varepsilon, \theta_0 + \varepsilon)$ is in (θ', θ'') there is an n_ε such that we have (writing θ_n^* for $\theta_n^*(x_1, \ldots, x_n)$) for $n > n_\varepsilon$.

$$P(\theta_0 - \varepsilon < \theta_n^* < \theta_0 + \varepsilon) > 1 - \varepsilon. \tag{4.3.25}$$

Now let $g_u(\theta_0)$ be the least upper bound and $g_l(\theta_0)$ the greatest lower bound of $g(\theta)$ for values of θ in $(\theta_0 - \varepsilon, \theta_0 + \varepsilon)$. Then since $(f_1(x_1, \theta), f_2(x_1, x_2, \theta), \ldots)$ converges in probability to $g(\theta)$ uniformly with respect to θ in (θ', θ''), there exists, for an arbitrary $\varepsilon' > 0$, an $n_{\varepsilon'}$ such that for $n > n_{\varepsilon'}$, and for any θ in $(\theta_0 - \varepsilon, \theta_0 + \varepsilon)$

$$P(g_l(\theta_0) - \varepsilon' < f_n(x_1, \ldots, x_n, \theta) < g_u(\theta_0) + \varepsilon') > 1 - \varepsilon'. \tag{4.3.26}$$

Now for any $n > \max(n_\varepsilon, n_{\varepsilon'})$ let E_n be the set in R_∞, the sample space of $(x_1, x_2, \ldots)$ for which both sets of inequalities indicated in (4.3.25) and (4.3.26) hold. It is evident that $P(E_n) > 1 - \varepsilon - \varepsilon'$. But any point in E_n will satisfy the following inequality:

$$g_l(\theta_0) - \varepsilon' < f_n(x_1, \ldots, x_n, \theta_n^*) < g_u(\theta_0) + \varepsilon'. \tag{4.3.27}$$

Hence for $n > \max(n_\varepsilon, n_{\varepsilon'})$ the probability that (4.3.27) is satisfied for any θ_n^* in $(\theta_0 - \varepsilon, \theta_0 + \varepsilon)$ exceeds $1 - \varepsilon - \varepsilon'$. But since ε and ε' are arbitrary and since $g(\theta)$ is continuous in (θ', θ''), the differences $g_n(\theta_0) - g(\theta_0)$ and $g(\theta_0) - g_l(\theta_0)$ can be made arbitrarily small. Hence $f_n(x_1, \ldots, x_n, \theta^*)$ converges in probability to $g(\theta_0)$. But θ_0 is any point in (θ', θ''), thus concluding the argument for **4.3.8.**

4.4 ALMOST CERTAIN CONVERGENCE

(a) Definition

Let $(x_1, x_2, \ldots)$ be a stochastic process with associated probability space $(R_\infty, \mathscr{B}_\infty, P)$, and let E be the set of all sequences $(b_1, b_2, \ldots)$ which converge. To see which points belong to E let E_{npN} be the set in R_∞ for which

$$|b_{n+j} - b_n| < \frac{1}{N}, \qquad j = 1, 2, \ldots, p \tag{4.4.1}$$

where N is a positive integer.

E_{npN}, for each n, p, and N, is a Borel cylinder set in R_∞ which, of course, belongs to $\mathscr{B}_\infty$. The set $E_{(N)} = \bigcup_{n=1}^{\infty} \bigcap_{p=1}^{\infty} E_{npN}$ also belongs to $\mathscr{B}_\infty$ and is a decreasing sequence for $N = 1, 2, \ldots$. The set of points E mentioned above is the limit of this decreasing sequence, that is,

$$E = \lim_N E_{(N)} \tag{4.4.2}$$

which also belongs to $\mathscr{B}_\infty$, and hence has a probability $P(E)$. This probability may be referred to as the *probability of convergence* of the stochastic process $(x_1, x_2, \ldots)$. E is called the *convergence set* in R_∞.

Now suppose $P(E) = 1$ and let x be a random variable which has the value $\lim_{n\to\infty} b_n$ if $(b_1, b_2, \ldots)$ belongs to E and has the value 0 if $(b_1, b_2, \ldots)$ does not belong to E. We then say that the sequence of random variables *almost certainly* converges to the random variable x.

Remark. The reader will find it helpful to interpret the preceding remarks in terms of events in the original basic probability space $(R, \mathscr{B}, P)$. Denoting, as usual, a sample point in R by e, the stochastic process (all components being measurable with respect of $\mathscr{B}$) may be written as $(x_1(e), x_2(e), \ldots)$. Thus, for each sample point e in R we have a countably infinite sequence of random variables which maps e into a point in R_∞. If there is a random variable $x(e)$ (measurable with respect to $\mathscr{B}$) such that $\lim_{n\to\infty} x_n(e) = x(e)$ for every point e in E' the preimage (in R) of the set E (in R_∞), we say that $(x_1(e), x_2(e), \ldots)$ converge to $x(e)$ with probability $P(E')$. E' is the *convergence set in R*. If $P(E') = P(E) = 1$, we say that $(x_1(e), x_2(e), \ldots)$ *almost certainly converges* or *converges with probability* 1 to $x(e)$.

(b) Relation between Almost Certain Convergence and Convergence in Probability

It will be noted that almost certain convergence is a stronger type of convergence than convergence in probability. In fact

4.4.1 *If $(x, x_1, x_2, \ldots)$ is a stochastic process such that $(x_1, x_2, \ldots)$ almost certainly converges to x, the sequence converges in probability to x.*

For if E is in the convergence set of $(x_1, x_2, \ldots)$, then for any $\varepsilon > 0$ and n, E is in the set in R_∞ defined by the limit of the following expanding sequence of sets in R_∞: $[|x_{n+j} - x| < \varepsilon, j = 0, 1, 2 \ldots]$, $n = 1, 2, \ldots$. But if in R_∞, E_n and F_n denote the events $[|x_{n+j} - x| < \varepsilon, j = 0, 1, 2, \ldots]$ and $[|x_n - x| < \varepsilon]$ respectively, then $E_n \subset F_n$. Hence

$$P(|x_n - x| < \varepsilon) \geqslant P(|x_{n+j} - x| < \varepsilon, j = 0, 1, 2, \ldots). \tag{4.4.3}$$

Taking limits as $n \to \infty$, we obtain

$$\begin{aligned}\lim_{n\to\infty} P(|x_n - x| < \varepsilon) &\geqslant \lim_{n\to\infty} P(|x_{n+j} - x| < \varepsilon, j = 0, 1, 2, \ldots) \\ &= P(\lim_{n\to\infty} E_n) = P(E) = 1.\end{aligned} \tag{4.4.4}$$

that is,

$$\lim_{n\to\infty} P(|x_n - x| < \varepsilon) = 1 \tag{4.4.5}$$

which is equivalent to (4.3.1), thus concluding the argument for **4.4.1.**

4.5 KOLMOGOROV'S INEQUALITY

The following interesting extension of notions in Chebyshev's inequality to a set of inequalities is due to Kolmogorov (1928):

4.5.1 *Suppose $x_1, \ldots, x_n$ is a set of independent random variables having 0 means and variances $\sigma_1^2, \ldots, \sigma_n^2$. Let $c_n^2 = \sigma_1^2 + \cdots + \sigma_n^2$. The probability that all of the inequalities*

$$|x_1 + \cdots + x_\alpha| < \lambda c_n, \ \alpha = 1, \ldots, n \tag{4.5.1}$$

hold is at least $1 - 1/\lambda^2$.

To prove this result let F_α be the event in the sample space R_n for which

$$|x_1 + \cdots + x_\alpha| \geqslant \lambda c_n, \quad \alpha = 1, \ldots, n.$$

Let $E_1 = F_1$, $E_2 = \bar{F}_1 \cap F_2$, $E_3 = \bar{F}_1 \cap \bar{F}_2 \cap F_3, \ldots, E_n = \bar{F}_1 \cap \bar{F}_2 \cap \cdots \cap \bar{F}_{n-1} \cap F_n$, and let G_n be the complement of $E_1 \cup E_2 \cup \cdots \cup E_n$. G_n is the set in R_n for which *all* of the inequalities in (4.5.1) hold. $E_1, \ldots, E_n, G_n$ are disjoint and their union is the entire sample space R_n. Thus,

$$P(G_n) = 1 - [P(E_1) + \cdots + P(E_n)]. \tag{4.5.2}$$

Let z_α be a random variable having the value 1 at each point in E_α and 0 at each point in $\bar{E}_\alpha$, $\alpha = 1, \ldots, n$, and z_{n+1} a random variable similarly defined for G_n. We then have

(4.5.3) $$\mathscr{E}(x_1 + \cdots + x_n)^2 = \mathscr{E}[z_1(x_1 + \cdots + x_n)^2] + \cdots \\ \cdots + \mathscr{E}[z_n(x_1 + \cdots + x_n)^2] + \mathscr{E}[z_{n+1}(x_1 + \cdots + x_n)^2].$$

Any function g_α of $(x_1, \ldots, x_\alpha)$ is independent of the remaining x's. Hence

$$\mathscr{E}(g_\alpha \cdot x_\beta) = \mathscr{E}(g_\alpha) \cdot \mathscr{E}(x_\beta) = 0,\ \beta = \alpha + 1, \ldots, n.$$

We have therefore

(4.5.4) $$\mathscr{E}[z_\alpha(x_1 + \cdots + x_n)^2] = \mathscr{E}[z_\alpha(x_1 + \cdots + x_\alpha)^2] \\ + \mathscr{E}[z_\alpha(x_{\alpha+1} + \cdots + x_n)^2] \geqslant \mathscr{E}[z_\alpha(x_1 + \cdots + x_\alpha)^2].$$

But $(x_1 + \cdots + x_\alpha)^2 \geqslant \lambda^2 c_n^2$ at each point in E_α. Therefore we have for $\alpha = 1, \ldots, n$

(4.5.5) $$\mathscr{E}[z_\alpha(x_1 + \cdots + x_n)^2] \geqslant \lambda^2 c_n^2 P(E_\alpha).$$

Using the fact that $\mathscr{E}(x_1 + \cdots + x_n)^2 = c_n^2$, applying (4.5.5) to the right-hand side of (4.5.3) for $\alpha = 1, \ldots, n$, and dropping $\mathscr{E}[z_{n+1}(x_1 + \cdots + x_n)^2]$, we obtain

(4.5.6) $$c_n^2 \geqslant \lambda^2 c_n^2 [P(E_1) + \cdots + P(E_n)].$$

It follows from (4.5.6) and (4.5.2) that

(4.5.7) $$P(G_n) \geqslant 1 - \frac{1}{\lambda^2}$$

which completes the proof of **4.5.1**.

4.6 THE STRONG LAW OF LARGE NUMBERS

In Section 4.3 we referred to the weak law of large numbers which states that under certain conditions the mean of a sequence of independent random variables with 0 means converges in probability to 0. The question arises as to whether we can make a probability statement that *all* means in the sequence of means from some point sufficiently far out in the sequence are arbitrarily close to zero. This is answered by the *strong law of large numbers*, which may be stated as follows:

4.6.1 *Let* $(x_1, x_2, \ldots)$ *be a sequence of independent random variables having* 0 *means and variances* $(\sigma_1^2, \sigma_2^2, \ldots)$ *such that* $\sum_{i=1}^{\infty} \frac{\sigma_i^2}{i^2} < +\infty$. *Let* $\bar{x}_n = (x_1 + \cdots + x_n)/n$. *Then* $(\bar{x}_1, \bar{x}_2, \ldots)$ *converges almost certainly to* 0. *That is, for arbitrary* $\delta > 0$ *and* $\varepsilon > 0$, *there is an* $N_{\delta,\varepsilon}$ *such that*

(4.6.1) $$P(|\bar{x}_n| < \varepsilon, n = N, N+1, \ldots, N+k) \geqslant 1 - \delta$$

for all $N > N_{\delta,\varepsilon}$ *and for every* k.

To prove **4.6.1** we follow a line of argument similar to that used by Feller (1957). Let $E_N, E_{N+1}, \ldots, E_{N+k}$ be the events $|\bar{x}_n| < \varepsilon$, $n = N, N+1, \ldots, N+k$, respectively. Then the probability in (4.6.1) is

$$P(E_N \cap E_{N+1} \cap \cdots \cap E_{N+k}). \tag{4.6.2}$$

But this probability has the value

$$1 - P(\bar{E}_N \cup \bar{E}_{N+1} \cup \cdots \cup \bar{E}_{N+k}). \tag{4.6.3}$$

Thus, for $N > N_{\delta,\varepsilon}$ and all k we must show that

$$P(\bar{E}_N \cup \bar{E}_{N+1} \cup \cdots \cup \bar{E}_{N+k}) \leqslant \delta. \tag{4.6.4}$$

Let us partition the positive integers into sets $I_1, I_2, \ldots$ where I_α is the set $\{2^{\alpha-1}+1, 2^{\alpha-1}+2, \ldots, 2^\alpha\}$. Let F_α be the event for which at least one of the inequalities $\{|\bar{x}_n| < \varepsilon, n \in I_\alpha\}$ fails. Then for some α and β we have

$$F_\alpha \cup F_{\alpha+1} \cup \cdots \cup F_{\alpha+\beta} \supset \bar{E}_N \cup \bar{E}_{N+1} \cup \cdots \cup \bar{E}_{N+k}. \tag{4.6.5}$$

Also, we have

$$P(F_\alpha \cup F_{\alpha+1} \cup \cdots \cup F_{\alpha+\beta}) \leqslant P(F_\alpha) + \cdots + P(F_{\alpha+\beta}). \tag{4.6.6}$$

Thus, it is enough for $\sum_{\alpha=1}^{\infty} P(F_\alpha)$ to converge. Now F_α is the event that for at least one n in I_α we have $|\bar{x}_n| \geqslant \varepsilon$, which may be written

$$|x_1 + \cdots + x_n| \geqslant n\varepsilon$$

which implies that

$$|x_1 + \cdots + x_n| > \left(\frac{\varepsilon \cdot 2^{\alpha-1}}{c_{2^\alpha}}\right) c_{2^\alpha},$$

where $c_{2^\alpha}^2$ is defined in **4.5.1** as $\sigma_1^2 + \cdots + \sigma_{2^\alpha}^2$.
But it follows from Kolmogorov's inequality that

$$P(F_\alpha) \leqslant \frac{4c_{2^\alpha}^2}{\varepsilon^2 2^{2\alpha}}.$$

Hence

$$\sum_{\alpha=1}^{\infty} P(F_\alpha) \leqslant \frac{4}{\varepsilon^2} \sum_{\alpha=1}^{\infty} 2^{-2\alpha} \sum_{i=1}^{2^\alpha} \sigma_i^2. \tag{4.6.7}$$

Reversing the order of summation with respect to i and α and observing that $\sum 2^{-2\alpha}$ over all positive integers α for which $2^\alpha \geqslant i$ does not exceed $2i^{-2}$, we obtain

$$\sum_{\alpha=1}^{\infty} P(F_\alpha) \leqslant \frac{8}{\varepsilon^2} \sum_{i=1}^{\infty} \frac{\sigma_i^2}{i^2}.$$

Hence $\sum_{\alpha=1}^{\infty} P(F_\alpha)$ converges if $\sum_{i=1}^{\infty} \sigma_i^2/i^2$ does, in which case the right-hand side of (4.6.6) can be made $<\delta$ for all values of β by choosing α sufficiently large. It then follows from (4.6.5) and (4.6.6) that for sufficiently large N (4.6.4) and (4.6.1) hold, thus completing the argument for **4.6.1**.

It should be noted that if all components in $(x_1, x_2, \ldots)$ have equal (finite) variances the condition $\sum_{i=1}^{\infty} \frac{\sigma_i^2}{i^2} < +\infty$ is automatically satisfied. As a matter of fact it will be seen that $(\bar{x}_1, \bar{x}_2, \ldots)$ converges almost certainly to 0 if $x_1, x_2, \ldots$ are independent and identically distributed with 0 means. See Khintchine (1929).

PROBLEMS

4.1 If $(x, y;\ x_1, y_1;\ x_2, y_2; \ldots)$ is a stochastic process such that $(x_1, x_2, \ldots)$ and $(y_1, y_2, \ldots)$ converge in probability to the random variables x and y, respectively, where x and y are equivalent, show that for an arbitrary $\varepsilon > 0$,

$$\lim_{n\to\infty} P(|x_n - y_n| < \varepsilon) = 1.$$

4.2 (*Continuation*) If $\lim_{n\to\infty} \mathscr{E}(x_n - y_n)^2 = 0$ and if $(x_1, x_2, \ldots)$ converges in probability to the random variable x show that $(y_1, y_2, \ldots)$ also converges in probability to x.

4.3 If $(x_1, x_2, \ldots)$ are independent random variables having identical c.d.f.'s $F(x)$ where

$$F(x) = \begin{cases} 0 & x \leqslant 0 \\ x & 0 < x \leqslant 1, \\ 1 & x > 1 \end{cases}$$

let $(u_1, u_2, \ldots)$ be the stochastic process where

$$u_n = \max(x_1, \ldots, x_n).$$

Let $v_n = n(1 - u_n)$. Show that the sequence $(v_1, v_2, \ldots)$ converges in distribution to $F(v)$, where $F(v) = 1 - e^{-v}$ for $v > 0$, and $F(v) = 0$, for $v \leqslant 0$.

4.4 (*Continuation*) If $g(x)$ is any differentiable single-valued continuous and increasing function of x over $(0, \infty)$ and if $w_n = g(n(1 - u_n))$ show that $(w_1, w_2, \ldots)$ converges in distribution to $H(w)$, where $\frac{dH(w)}{dw}$ is given by

$$e^{-g^{-1}(w)} \frac{d}{dw} g^{-1}(w)$$

for w on the interval $(g(0), g(+\infty))$ and 0 elsewhere, where $g^{-1}(w)$ is the inverse of $g(x)$.

4.5 A discrete random variable x_n which arises in the theory of extreme observations, is known to have c.d.f.

$$F_n(x) = 1 - \frac{n(n-1)\cdots(n-r+1)}{(n+nx)(n+nx-1)\cdots(n+nx-r+1)}$$

where $r \leqslant n + 1$, the mass points of x_n being $1/n$, $2/n$, ... Show that the sequence of random variables $(x_1, x_2, \ldots)$ converges in distribution to $F(x)$, where

$$F(x) = \begin{cases} 0, & x \leqslant 0 \\ 1 - \dfrac{1}{(1 + x)^r}, & x > 0. \end{cases}$$

4.6 Referring to **4.3.6a**, show that if d is any constant $\neq 0$

$$\lim_{n \to \infty} P(x_n + ny_n \leqslant d) = 0.$$

4.7 Suppose $(x_1, x_2, \ldots)$ is a stochastic process such that $\mathscr{E}(x_n) = 0$, $\sigma^2(x_n) = \sigma^2$, $n = 1, 2, \ldots$, and $\text{cov}\,(x_i, x_j) = \rho\sigma^2$, $i \neq j = 1, 2, \ldots$ where σ^2 is finite and ρ lies on $(0, 1)$. Let

$$u_k = \frac{1}{k}(x_1 + x_3 + \cdots + x_{2k-1})$$

and

$$v_k = \frac{1}{k}(x_2 + x_4 + \cdots + x_{2k})$$

$k = 1, 2, \ldots$.
Show that the vector stochastic process $(u_1, v_1; u_2, v_2; \ldots)$ converges in distribution to (u, v) where u and v are equivalent random variables having 0 means and variances $\sigma^2\rho$.

4.8 Suppose $(x_1, \ldots, x_n)$ are n independent random variables all having means 0 and variances 1. Consider the set of inequalities

$$|\bar{x}_\alpha| < \lambda \frac{1}{\sqrt{\alpha}}, \quad \alpha = 1, \ldots, n,$$

where

$$\bar{x}_\alpha = \frac{1}{\alpha}(x_1 + \cdots + x_\alpha).$$

Let G_n be the event that all inequalities are satisfied. Then $\bar{G}_n$ is the event that at least one inequality fails. If the αth inequality in the set is the first to fail given that $\bar{G}_n$ occurs, show by methods similar to those used in dealing with Kolmogorov's inequality, that for any $\lambda > 0$,

$$P(\bar{G}_n) \leqslant \frac{n}{n + (\lambda^2 - 1)\mathscr{E}(\alpha \mid \bar{G}_n)},$$

where $\mathscr{E}(\alpha \mid \bar{G}_n)$ is the conditional mean value of α given that $\bar{G}_n$ occurs, and where α is the random variable denoting the first of the above inequalities to fail.

4.9 (*Continuation*) Consider the inequalities

$$x_1^2 + \cdots + x_\alpha^2 < \alpha\lambda^2, \quad \alpha = 1, \ldots, n.$$

Show that if $\bar{G}_n$ is the event that at least one of the inequalities fails, then $P(\bar{G}_n)$ satisfies the same inequality as in the preceding problem.

4.10 If $(x_1, \ldots, x_n)$ are independent and positive random variables whose means are all equal to μ, show that for any $\lambda > 0$, the probability that all of the inequalities $(x_1 x_2 \cdots x_\alpha) < \lambda\mu^\alpha$, $\alpha = 1, \ldots, n$ are satisfied in at least $1 - 1/\lambda$.

4.11 Extend the strong law of large numbers to the case where the sequence of independent random variables $(x_1, x_2, \ldots)$ have (finite) means $\mu_1, \mu_2, \ldots$ as well as (finite) variances $\sigma_1^2, \sigma_2^2, \ldots$.

4.12 If $(x_1, \ldots, x_n)$ are n independent random variables all with zero medians, show that

$$\mathscr{E}(|x_1 + \cdots + x_n|) \geqslant \varphi(n)\bar{d}$$

where

$$\bar{d} = \frac{1}{n}[\mathscr{E}\,|x_1| + \cdots + \mathscr{E}\,|x_n|]$$

and

$$\varphi(2k+1) = \varphi(2k+2) = \frac{(2k+1)!}{2^{2k}(k!)^2}$$

a result due to Tukey (1946).

4.13 Suppose $(x_1, \ldots, x_k)$ is a k-dimensional random variable with mean $(0, \ldots, 0)$ and covariance matrix $\|\sigma_{ij}\|$ where $\sigma_{ii} = \sigma^2$ and $\sigma_{ij} = \sigma^2\rho$. Show that

$$P(\max_{\alpha} |x_\alpha| \geqslant t\sigma) \leqslant \frac{\sigma^2}{kt^2}[(k-1)\sqrt{1-\rho} + \sqrt{1+(k-1)\rho}],$$

a result due to Olkin and Pratt (1958).

4.14 In **4.3.6a** show that the four statements (4.3.18), (4.3.19), (4.3.20), and (4.3.21) are true if the two stochastic processes $(x_1, x_2, \ldots)$ and $(y_1, y_2, \ldots)$ converge respectively to the random variable x and the positive constant c.

4.15 Prove that if $(x_1, x_2, \ldots)$ is a sequence of random variables with p.d.f.'s $(f_1(x), f_2(x), \ldots)$ and if $\lim_{\alpha \to \infty} f_\alpha(x) = f(x)$ for all x in R_1 except possibly for a set of probability 0 where $f(x)$ is a p.d.f., then

$$\lim_{\alpha \to \infty} \int_E f_\alpha(x)\,dx = \int_E f(x)\,dx$$

uniformly for all sets $E \subset R_1$; [This result is due to Scheffé (1947). It should be noted that the conditions in Scheffé's theorem are stronger than those in **4.3.4**. A discussion and comparison of various conditions for convergence of distributions have been given by Robbins (1948)].

4.16 If $(x, x_1, x_2, \ldots)$ is a stochastic process such that $(x_1, x_2, \ldots)$ converges in probability to the random variable x and if $g_1(x), g_2(x), \ldots$ is a sequence of continuous functions which converge uniformly to $g(x)$ over a bounded interval show that $(g_1(x_1), g_2(x_2), \ldots)$ is a stochastic process which converges in probability to $g(x)$.

4.17 Let $x_1, \ldots, x_n$ be independent random variables and let $y_1 = x_1$, $y_2 = x_1 + x_2, \ldots, y_n = x_1 + \cdots + x_n$, if $F_\xi(y_\xi)$ denotes the c.d.f. of y_ξ, $\xi = 1, \ldots n$, and $F(y_1, \ldots, y_n)$ the c.d.f. of $(y_1, \ldots, y_n)$ show that

$$F(y_1, \ldots, y_n) \geqslant F_1(y_1) \cdots F_n(y_n),$$

a result due to Robbins (1954).

4.18 Prove **4.3.7.**

CHAPTER 5

Characteristic Functions and Generating Functions

5.1 CASE OF A ONE-DIMENSIONAL RANDOM VARIABLE

One of the most important classes of problems in mathematical statistics is the determination of distribution functions of measurable functions of random variables, that is, functions of random variables which are random variables themselves. A few methods were presented in Section 2.8 for dealing with these problems. These methods, however, are often technically difficult or tedious to carry out in specific cases. Some situations, particularly those involving linear functions of independent random variables, can often be handled in an elegant manner by making use of the characteristic function of the particular function of the random variables under consideration. The characteristic function and related devices are also useful in some cases for such tasks as generating moments and cumulants of distributions and testing independence of two or more functions of random variables. This chapter will be devoted to these methods and their application.

The *characteristic function* $\varphi(t)$ of a random variable x having c.d.f. $F(x)$ is defined as

$$\varphi(t) = \mathscr{E}(e^{\mathrm{i}tx}) = \int_{-\infty}^{\infty} e^{\mathrm{i}tx}\, dF(x) \tag{5.1.1}$$

where $\mathrm{i} = \sqrt{-1}$ and t is real.* It is sometimes convenient to say that $\varphi(t)$ is the characteristic function corresponding to $F(x)$ or more briefly, the characteristic function of x.

Since
$$e^{\mathrm{i}tx} = \cos tx + \mathrm{i} \sin tx$$

* Throughout this book we denote $\sqrt{-1}$ by i to avoid confusion in various sections with the use of italic i for indexes of summation.

$\cos tx$ and $\sin tx$ are both integrable over R_1 for any real t, then $\varphi(t)$ is a complex number whose real and imaginary parts are finite for every value of t.

The *moment-generating function* $\psi(t)$ is defined as

$$\psi(t) = \mathscr{E}(e^{tx}) = \varphi\left(\frac{t}{i}\right). \tag{5.1.2}$$

We are usually interested in $\psi(t)$ for values of t in some neighborhood of $t = 0$. But $\psi(t)$ does not exist for every c.d.f. and all values of t as does $\varphi(t)$.

The *factorial moment-generating function* $\theta(t)$, if it exists, is obtained by replacing t by $\log t$ in (5.1.2), thus yielding

$$\theta(t) = \psi(\log t) = \mathscr{E}(t^x). \tag{5.1.3}$$

If the random variable x is discrete so that the mass points of x are positive integers, then $\theta(t)$, if it exists, is also called the *probability generating function* of the random variable x. For if $\theta(t)$ is expanded into a series in t, the coefficient of t^x is $p(x)$, the p.f. of x.

If the rth moment $\mu_r'(x)$ exists, we can differentiate (5.1.1) h times, $0 < h \leqslant r$, with respect to t and obtain

$$\varphi^{(h)}(t) = i^h \int_{-\infty}^{\infty} x^h e^{itx}\, dF(x), \tag{5.1.4}$$

from which we find the hth moment $\mu_h'(x)$ to be

$$\mu_h'(x) = \frac{\varphi^{(h)}(0)}{i^h}, \qquad 0 < h \leqslant r. \tag{5.1.5}$$

For convenience we define $\varphi^{(0)}(t)$ to be $\varphi(t)$.

Similarly, if $\psi(t)$ exists for values of t in a neighborhood of zero, we have

$$\mu_h'(x) = \psi^{(h)}(0). \tag{5.1.6}$$

If all moments including the $2s$th moment of x exists, we may write

$$\varphi(t) = \int_{-\infty}^{\infty} \left[1 + \sum_{h=1}^{2s-1} \frac{(itx)^h}{h!} + \frac{(itx)^{2s}}{(2s)!}(\cos t'x + i \sin t''x)\right] dF(x) \tag{5.1.7}$$

where t' and t'' are numbers in the interval $(0, t)$. Assuming x to be non-degenerate, then $\mu_{2s}' \neq 0$ and we may write

$$\varphi(t) = 1 + \sum_{h=1}^{2s-1} \frac{(it)^h}{h!}\mu_h' + \frac{(it)^{2s}\mu_{2s}'}{(2s)!} g_1(t, s) \tag{5.1.8}$$

where $g_1(t, s)$ is a complex function whose real and imaginary components are respectively, the mean values $\mathscr{E}(x^{2s} \cos t'x/\mu_{2s}')$ and $\mathscr{E}(x^{2s} \sin t''x/\mu_{2s}')$. We note that $|g_1(t, s)| \leqslant 1$. Furthermore, since t' and t'' both lie in the interval $(0, t)$ we have $\lim_{t\to 0} g_1(t, s) = 1$.

In a similar manner, if all moments including the $(2s-1)$th moment exist we may write

$$\varphi(t) = 1 + \sum_{h=1}^{2s-2} \frac{(it)^h}{h!} \mu_h' + \frac{(it)^{2s-1}}{(2s-1)!} \nu_{2s-1}' \, g_2(t, s) \tag{5.1.9}$$

where $g_2(t, s)$ is a complex function whose real and complex components are $\mathscr{E}(x^{2s-1} \cos t_1 x/\nu_{2s-1}')$ and $\mathscr{E}(x^{2s-1} \sin t_2 x/\nu_{2s-1}')$ respectively, where t_1 and t_2 are numbers in the interval $(0, t)$, and ν_{2s-1}' is the $(2s-1)$th absolute moment as defined by (3.3.13), which is $\neq 0$, if x is nondegenerate. Again, we have $|g_2(t, s)| \leqslant 1$. Furthermore, $\lim_{t\to 0} g_2(t, s) = \dfrac{\mu_{2s-1}'}{\nu_{2s-1}'}$, which, of course, is finite.

No matter whether the highest moment is odd or even, we may summarize as follows:

5.1.1 *If the* r*th moment of a random variable x exists, then $\varphi(t)$ can be expanded in a neighborhood of $t = 0$ as follows:*

$$\varphi(t) = 1 + \sum_{h=1}^{r} \frac{(it)^h}{h!} \mu_h' + o(t^r) \tag{5.1.10}$$

where

$$\lim_{t\to 0} \frac{o(t^r)}{t^r} = 0.$$

If $\varphi(t)$ can be expanded as stated in (5.1.10), we may also expand $\log \varphi(t)$ as follows:

$$\log \varphi(t) = \sum_{h=1}^{r} \frac{\kappa_h}{h!} (it)^h + o(t^r). \tag{5.1.11}$$

The quantities κ_h are called *semi-invariants* or *cumulants* of the c.d.f. $F(x)$, originally defined and studied by Thiele (1903).

Note that any semi-invariant κ_h is a polynomial in the moments $\mu_1', \mu_2', \ldots,$ and vice versa. The first few semi-invariants are as follows:

$$\begin{aligned} \kappa_1 &= \mu_1' = \mu \\ \kappa_2 &= \mu_2' - (\mu_1')^2 = \sigma^2 \\ \kappa_3 &= \mu_3' - 3\mu_2'\mu_1' + 2(\mu_1')^3 \\ &\vdots \end{aligned} \tag{5.1.12}$$

and conversely, we have

$$\begin{aligned} \mu_1' &= \kappa_1 \\ \mu_2' &= \kappa_2 + \kappa_1^2 \\ \mu_3' &= \kappa_3 + 3\kappa_1\kappa_2 + \kappa_1^3 \\ &\vdots \end{aligned} \tag{5.1.13}$$

There are many problems in probability theory and its applications, especially to sampling theory, in which it is quite easy to find the characteristic function of a function of the components of a multidimensional random variable, particularly a linear function. The question which arises, of course, is how to find the c.d.f. of the random variable from its characteristic function. An answer to this question can be found in many cases in the following theorem due to Lévy (1925):

5.1.2 *Let x be a random variable having characteristic function $\varphi(t)$ and c.d.f. $F(x)$. Then if $F(x)$ is continuous at $x = x' \pm \delta$, $\delta > 0$, we have*

$$(5.1.14)\quad F(x' + \delta) - F(x' - \delta) = \lim_{A\to\infty} \frac{1}{\pi}\int_{-A}^{A} \frac{\sin \delta t}{t}\, e^{-itx'}\varphi(t)\, dt.$$

Furthermore, if $\int_{-\infty}^{\infty} |\varphi(t)|\, dt < +\infty$, a p.d.f. $f(x)$ exists at $x = x'$ and

$$(5.1.15)\qquad f(x') = \frac{1}{2\pi}\int_{-\infty}^{+\infty} e^{-itx'}\varphi(t)\, dt.$$

To prove this theorem we replace $\varphi(t)$ in (5.1.14) by its integral expression in (5.1.1) and write

$$(5.1.16)\qquad G(x', A, \delta) = \frac{1}{\pi}\int_{-A}^{A} \frac{\sin \delta t}{t}\, e^{-itx'}\int_{-\infty}^{\infty} e^{itx}\, dF(x)\, dt.$$

Since $\left|\frac{\sin \delta t}{t}\, e^{-it(x'-x)}\right| < \delta$ we can invert the order of integration in (5.1.16), obtaining

$$(5.1.17)\qquad G(x', A, \delta) = \int_{-\infty}^{\infty} m(x, x', A, \delta)\, dF(x)$$

where

$$\begin{aligned}(5.1.18)\quad m(x, x', A, \delta) &= \frac{1}{\pi}\int_{-A}^{A} \frac{\sin \delta t}{t}\, e^{it(x-x')}\, dt \\ &= \frac{2}{\pi}\int_{0}^{A} \frac{\sin \delta t}{t}\cos [t(x - x')]\, dt \\ &= \frac{1}{\pi}\int_{0}^{A} \frac{\sin (x - x' + \delta)t}{t}\, dt \\ &\quad - \frac{1}{\pi}\int_{0}^{A} \frac{\sin (x - x' - \delta)t}{t}\, dt.\end{aligned}$$

Taking limits, we have

$$(5.1.19)\qquad \lim_{A\to\infty} G(x', A, \delta) = \int_{-\infty}^{\infty} \lim_{A\to\infty} m(x, x', A, \delta)\, dF(x).$$

But making use of the fact that $\lim_{V\to\infty}\int_0^V \frac{\sin u}{u}\,du = \frac{\pi}{2}$, it will be seen that

$$\lim_{A\to\infty} m(x, x', A, \delta) = \begin{cases} 0, & x < x' - \delta, \quad x > x' + \delta \\ \frac{1}{2}, & x = x' - \delta, \quad x = x' + \delta \\ 1, & x' - \delta < x < x' + \delta. \end{cases} \tag{5.1.20}$$

Therefore, since $F(x)$ is continuous at $x = x' \pm \delta$, it follows, upon using these values of $\lim_{A\to\infty} m(x, x', A, \delta)$ in (5.1.19), that

$$\lim_{A\to\infty} G(x', A, \delta) = \int_{x'-\delta}^{x'+\delta} dF(x) = F(x' + \delta) - F(x' - \delta)$$

which establishes (5.1.14).

Now if we divide (5.1.14) by 2δ, we have

$$\frac{F(x' + \delta) - F(x' - \delta)}{2\delta} = \frac{1}{2\pi}\int_{-\infty}^{\infty} \frac{\sin \delta t}{\delta t} e^{-itx'} \varphi(t)\, dt. \tag{5.1.21}$$

If $F(x)$ has a derivative $f(x')$ at $x = x'$, since $\left|\frac{\sin \delta t}{\delta t} e^{-itx'} \varphi(t)\right| < |\varphi(t)|$, and if $\int_{-\infty}^{\infty} |\varphi(t)|\, dt < +\infty$, we have

$$\lim_{\delta\to 0}\left[\frac{F(x' + \delta) - F(x' - \delta)}{2\delta}\right] = \frac{1}{2\pi}\int_{-\infty}^{\infty} \lim_{\delta\to 0}\left(\frac{\sin \delta t}{\delta t}\right) e^{-itx'}\varphi(t)\, dt.$$

Since $\lim_{\delta\to 0} (\sin \delta t/\delta t) = 1$, we therefore obtain

$$f(x') = \frac{1}{2\pi}\int_{-\infty}^{\infty} e^{-itx'}\varphi(t)\, dt$$

which is formula (5.1.15).

If two random variables x_1 and x_2 have identical c.d.f.'s, then it is evident that their characteristic functions are identical. Conversely, suppose the random variables have identical characteristic functions $\varphi(t)$. Then if $F_1(x)$ and $F_2(x)$ are the c.d.f's of the two random variables it follows from (5.1.14) that if $x' \pm \delta$ is *any* interval such that $F_1(x)$ and $F_2(x)$ are continuous at the end points $x' \pm \delta$, then

$$F_1(x' + \delta) - F_1(x' - \delta) = F_2(x' + \delta) - F_2(x' - \delta)$$

which, together with the fact that $F_1(x)$ and $F_2(x)$ are c.d.f.'s, and with our convention of making all c.d.f.'s continuous on the right, implies that $F_1(x) \equiv F_2(x)$.

Thus, we have the following result:

5.1.3 *If x_1 and x_2 are random variables having* c.d.f.'s $F_1(x)$ *and* $F_2(x)$ *respectively, and characteristic functions* $\varphi_1(t)$ *and* $\varphi_2(t)$ *respectively, a necessary and sufficient condition for* $F_1(x) \equiv F_2(x)$ *is that* $\varphi_1(t) \equiv \varphi_2(t)$.

This one-to-one correspondence between c.d.f.'s and characteristic functions is highly useful in probability theory since it provides a basis by which c.d.f.'s may be identified from their corresponding characteristic functions.

Example. Suppose $x_1, \ldots, x_k$ are statistically independent random variables having p.d.f.

$$f(x_1, \ldots, x_k) = \begin{cases} e^{-\sum_1^k x_i}, & x_i \geqslant 0, \quad i = 1, \ldots, k \\ 0, & \text{otherwise,} \end{cases}$$

and let $L = \sum_1^k x_i$. Let us find the moments and p.d.f. of L.

The characteristic function of L is

$$(5.1.22) \qquad \varphi(t) = \int_0^\infty \cdots \int_0^\infty e^{-\sum_i x_i + it\sum_i x_i}\, dx_1 \cdots dx_k = (1 - it)^{-k}.$$

The rth moment $\mu_r'(L)$ is given by applying (5.1.5) for

$$\mu_r'(L) = \frac{\varphi^{(r)}(0)}{i^r} = \frac{\Gamma(k + r)}{\Gamma(k)}.$$

The p.d.f. of L is, by (5.1.15),

$$(5.1.23) \qquad f(L) = \frac{1}{2\pi} \int_{-\infty}^{\infty} e^{-itL}(1 - it)^{-k}\, dt.$$

The integral can be evaluated by contour integration in the complex plane by making the transformation

$$z = -L(1 - it)$$

which gives

$$L^{k-1}e^{-L} \cdot H$$

where H is the Hankel integral given by

$$(5.1.24) \qquad H = \frac{i}{2\pi} \int_{-L+i\infty}^{-L-i\infty} e^{-z}(-z)^{-k}\, dz$$

which has the value $1/\Gamma(k)$. [See, for example, Whittaker and Watson (1927)]. The p.d.f. of L is therefore given by

$$(5.1.25) \qquad \begin{aligned} f(L) &= \frac{1}{\Gamma(k)} L^{k-1}e^{-L}, \qquad && L \geqslant 0 \\ &= 0, && L < 0. \end{aligned}$$

5.2 CASE OF A k-DIMENSIONAL RANDOM VARIABLE

Suppose $(x_1, \ldots, x_k)$ is a k-dimensional random variable having c.d.f. $F(x_1, \ldots, x_k)$. The *characteristic function* $\varphi(t_1, \ldots, t_k)$ of $(x_1, \ldots, x_k)$ is defined as

$$(5.2.1)\quad \varphi(t_1, \ldots, t_k) = \mathscr{E}\left[\exp\left(i \sum_i^k t_i x_i\right)\right]$$

$$= \int_{R_k} \exp\left(i \sum_1^k t_i x_i\right) dF(x_1, \ldots, x_k).$$

The *moment-generating function* $\psi(t_1, \ldots, t_k)$ is defined by

$$(5.2.2)\qquad \psi(t_1, \ldots, t_k) = \varphi\left(\frac{t_1}{i}, \ldots, \frac{t_k}{i}\right).$$

Similarly, the *factorial moment-generating function* $\theta(t_1, \ldots, t_k)$ is defined by replacing t_i by $\log t_i$, $i = 1, \ldots, k$, in $\psi(t_1, \ldots, t_k)$.

If the joint moment $\mu'_{r_1 \cdots r_k}$ exists, it can be obtained by differentiating $\varphi(t_1, \ldots, t_k)$ as follows:

$$(5.2.3)\qquad \mu'_{r_1 \cdots r_k} = \frac{\varphi^{(r_1 + \cdots + r_k)}(0, \ldots, 0)}{i^{(r_1 + \cdots + r_k)}}.$$

If $\mu'_{r_1 \cdots r_k}$ exists, then all joint moments $\mu'_{h_1 \cdots h_k}$, $0 < h_i \leqslant r_i$, $i = 1, \ldots, k$ exist.

It should be observed that the characteristic function of any subset of the components of the random variable $(x_1, \ldots, x_k)$ is obtained by setting equal to zero the t's corresponding to the random variables not included in the subset. For instance, the characteristic function of the random variable $(x_1, \ldots, x_{k_1})$, $k_1 < k$, is $\varphi(t_1, \ldots, t_{k_1}, 0, \ldots, 0)$.

The extension of Lévy's theorem to the case of k random variables is straightforward and may be stated as follows:

5.2.1 *Let $(x_1, \ldots, x_k)$ be a k-dimensional random variable with characteristic function $\varphi(t_1, \ldots, t_k)$ and* c.d.f. *$F(x_1, \ldots, x_k)$. Let I_k be the interval $x_i' - \delta_i < x_i \leqslant x_i' + \delta_i$, $i = 1, \ldots, k$ in R_k, $\delta_i > 0$, and let $F(x_1, \ldots, x_k)$ be continuous on the boundary of the closed interval $x_i' - \delta_i \leqslant x_i \leqslant x_i' + \delta_i$, $i = 1, \ldots, k$. Then*

$$(5.2.4)\quad P((x_1, \ldots, x_k) \in I_k) = \lim_{A \to \infty} \frac{1}{\pi^k} \int_{-A}^{A} \cdots \int_{-A}^{A} \prod_{i=1}^{k} \left[\frac{\sin \delta_i t_i}{t_i} e^{-it_i x_i'}\right] \cdot \varphi(t_1, \ldots, t_k)\, dt_1 \cdots dt_k.$$

Furthermore, if $\int_{R_k} |\varphi(t_1, \ldots, t_k)| \, dt_1 \cdots dt_k < +\infty$, *a p.d.f.* $f(x_1, \ldots, x_k)$ *exists at* $(x_1', \ldots, x_k')$ *and*

$$(5.2.5)\quad f(x_1', \ldots, x_k') = \left(\frac{1}{2\pi}\right)^k \int_{-\infty}^{\infty} \cdots \int_{-\infty}^{\infty} \exp\left(-i \sum_1^k t_i x_i'\right) \cdot \varphi(t_1, \ldots, t_k) \, dt_1 \cdots dt_k.$$

The proof of **5.2.1** is similar to that of **5.1.2** and is omitted.

Theorem **5.1.3** on the one-to-one correspondence between c.d.f.'s and characteristic functions can be extended to the case of a k-dimensional random variable without new difficulties.

5.3 CHARACTERISTIC FUNCTIONS OF INDEPENDENT RANDOM VARIABLES

Characteristic functions are sometimes useful in determining whether two random variables are independent without having to determine first the distribution function of the two random variables. The essential result here may be stated as follows:

5.3.1 *If* (x_1, x_2) *is a two-dimensional random variable, a necessary and sufficient condition for* x_1 *and* x_2 *to be independent is that*

$$(5.3.1)\qquad \varphi(t_1, t_2) = \varphi(t_1, 0) \cdot \varphi(0, t_2)$$

where $\varphi(t_1, t_2)$ *is the characteristic function of* (x_1, x_2).

Note that $\varphi(t_1, 0)$ is the characteristic function of x_1 and $\varphi(0, t_2)$ is the characteristic function of x_2. It is convenient to denote $\varphi(t_1, 0)$ and $\varphi(0, t_2)$ by $\varphi_1(t_1)$ and $\varphi_2(t_2)$ respectively.

To see that the condition is *necessary*, we assume x_1 and x_2 to be independent, that is, $F(x_1, x_2) = F_1(x_1) F_2(x_2)$. Then we have

$$(5.3.2)\qquad \begin{aligned} \varphi(t_1, t_2) &= \int_{R_2} e^{it_1x_1 + it_2x_2} \, dF(x_1, x_2) \\ &= \int_{R_2} e^{it_1x_1 + it_2x_2} \, d[F_1(x_1)F_2(x_2)] \\ &= \int_{-\infty}^{\infty} e^{it_1x_1} \, dF_1(x_1) \int_{-\infty}^{\infty} e^{it_2x_2} \, dF_2(x_2) \end{aligned}$$

that is,

$$\varphi(t_1, t_2) = \varphi_1(t_1) \cdot \varphi_2(t_2).$$

Now consider the *sufficiency*. Assuming (5.3.1) to hold, we have, by applying formula (5.2.4) for $k = 2$,

(5.3.3)

$$P(x_i' - \delta_i < x_i \leq x_i' + \delta_i;\ i = 1, 2)$$

$$= \lim_{A\to\infty} \frac{1}{\pi^2} \int_{-A}^{A} \int_{-A}^{A} \prod_{i=1}^{2} \left[\frac{\sin \delta_i t_i}{t_i} e^{-it_ix_i'}\right] \varphi_1(t_1)\varphi_2(t_2)\, dt_1\, dt_2$$

$$= \prod_{i=1}^{2} \left[\lim_{A\to\infty} \frac{1}{\pi} \int_{-A}^{A} \frac{\sin \delta_i t_i}{t_i} e^{-it_ix_i'} \varphi_i(t_i)\, dt_i\right].$$

That is,

(5.3.4) $$P((x_1, x_2) \in I_2) = P(x_1 \in I_1^{(1)}) \cdot P(x_2 \in I_1^{(2)})$$

where $I_1^{(i)}$ is the interval $x_i' - \delta_i < x_i \leq x_i' + \delta_i$, $i = 1, 2$, and I_2 is the Cartesian product $I_1^{(1)} \times I_1^{(2)}$. But we know from **2.4.2** that (5.3.4) implies that

$$F(x_1, x_2) = F_1(x_1) \cdot F_2(x_2)$$

that is, independence of x_1 and x_2.

The extension of **5.3.1** to any (finite) number of random variables is straightforward and is left to the reader.

Another useful property of characteristic functions of independent random variables is their particular form in the case of linear functions of random variables. The essential result on this matter, and which can be readily verified by the reader may be stated as follows:

5.3.2 *Suppose L is a linear function of k independent random variables* $x_1, \ldots, x_k$ *defined as follows:*

(5.3.5) $$L = \sum_{i=1}^{k} c_i x_i.$$

Let $\varphi(t)$ *be the characteristic function of L and* $\varphi_i(t_i)$ *that of* x_i, $i = 1, \ldots, k$. *Then we have*

(5.3.6) $$\varphi(t) = \prod_{i=1}^{k} \varphi_i(c_i t).$$

Suppose x_1 and x_2 are independent random variables have c.d.f.'s $F(x_1; \theta_1)$ and $F(x_2; \theta_2)$, respectively, where θ_1 and θ_2 are values of a parameter θ. Let L denote the random variable $x_1 + x_2$. Then if the c.d.f. of L is $F(L; \theta_1 + \theta_2)$, the c.d.f. $F(x; \theta)$ is said to be *reproductive with respect to* θ. We may similarly speak of a p.f., a p.d.f. or a distribution depending on θ as being reproductive with respect to θ. Reproductivity is an important property which will be used frequently in later chapters.

The notion of reproductivity can be extended without difficulty to the case where either or both the random variable x and the parameter θ is multi-dimensional. Characteristic functions provide a useful criterion for determining whether a c.d.f. $F(x; \theta)$ is reproductive which can be stated as follows:

5.3.3 *Suppose x_1 and x_2 are independent random variables having c.d.f.'s $F(x_1; \theta_1)$ and $F(x_2; \theta_2)$ and characteristic functions $\varphi(t; \theta_1)$ and $\varphi(t; \theta_2)$, respectively, where θ_1* and *θ_2 are parameters. Then a necessary and sufficient condition for $F(x; \theta)$ to be reproductive with respect to θ is that*

$$\varphi(t; \theta_1)\varphi(t; \theta_2) = \varphi(t; \theta_1 + \theta_2). \tag{5.3.7}$$

The proof of this is straightforward and is left to the reader.

5.4 CHARACTERISTIC FUNCTIONS OF A SEQUENCE OF RANDOM VARIABLES

As we have stated in Section 4.1, one of the important problems in probability theory and its application to sampling theory and related topics is that of the convergence in probability of a sequence of random variables. Since c.d.f.'s are uniquely determined by characteristic functions, the problem of convergence in probability of a sequence of random variables can often be more easily handled by dealing with the convergence of the corresponding sequence of characteristic functions than by dealing directly with c.d.f.'s of the random variables. The fundamental principle involved here is expressed by the following theorem due to Lévy (1937) and Cramér (1937).

5.4.1 *Let $(x_1, x_2, \ldots)$ be a sequence of random variables. Let $\varphi_1(t)$, $\varphi_2(t), \ldots$ be the corresponding sequence of characteristic functions. A necessary and sufficient condition for $(x_1, x_2, \ldots)$ to converge in distribution to a* c.d.f. *$F(x)$ is that for every value of t, the sequence $\varphi_1(t), \varphi_2(t), \ldots$ converge to a limit $\varphi(t)$, which is continuous at $t = 0$. Under these conditions $\varphi(t)$ is identical with the characteristic function corresponding to $F(x)$.*

First, let us show that the condition is *necessary*. If $F_1(x), F_2(x), \ldots$ are the c.d.f.'s of $(x_1, x_2, \ldots)$ and $F(x)$ is the limit c.d.f., then we assume that the sequence $F_1(x), F_2(x), \ldots$ converges to the c.d.f. $F(x)$ for every x, and we must show that $\lim_{n\to\infty} \varphi_n(t) = \varphi(t)$ for every t. Now, we may write

$$\varphi_n(t) = \int_{-\infty}^{\infty} \cos tx \, dF_n(x) + i\int_{-\infty}^{\infty} \sin tx \, dF_n(x). \tag{5.4.1}$$

Since cos tx is bounded on $(-\infty, +\infty)$ for every t and since $F_1(x)$, $F_2(x)$, . . . converges to the c.d.f. $F(x)$ for every x, we have by the Helly-Bray theorem, [see Loève (1955), for instance],

$$\lim_{n\to\infty} \int_{-\infty}^{\infty} \cos tx \, dF_n(x) = \int_{-\infty}^{\infty} \cos tx \, dF(x), \tag{5.4.2}$$

and similarly,

$$\lim_{n\to\infty} \int_{-\infty}^{\infty} \sin tx \, dF_n(x) = \int_{-\infty}^{\infty} \sin tx \, dF(x). \tag{5.4.3}$$

But (5.4.2) and (5.4.3) taken together are equivalent to the statement

$$\lim_{n\to\infty} \int_{-\infty}^{\infty} e^{itx} \, dF_n(x) = \int_{-\infty}^{\infty} e^{itx} \, dF(x).$$

that is,

$$\lim_{n\to\infty} \varphi_n(t) = \varphi(t).$$

Now consider the *sufficiency* of the condition. We assume that $\lim_{n\to\infty} \varphi_n(t) = \varphi(t)$, for every t, and that $\varphi(t)$ is continuous at $t = 0$. Now it can be shown [see Cramér (1946) for instance] that the sequence of c.d.f.'s $F_1(x)$, $F_2(x)$, . . . contains a subsequence $F_{n_1}(x)$, $F_{n_2}(x)$, . . . which converges to a nondecreasing function $F(x)$ continuous on the right. Our problem now is to show that the function $F(x)$ to which the subsequence $F_{n_1}(x)$, $F_{n_2}(x)$, . . . converges, and which is nondecreasing and continuous on the right, satisfies the remaining conditions for a c.d.f., namely, $F(-\infty) = 0$ and $F(+\infty) = 1$. It is clear that $0 \leqslant F(x) \leqslant 1$.

Now it can be shown by argument similar to that used in establishing (5.1.14) that for $c > 0$

$$c\left[\frac{1}{c}\int_0^c F_{n_i}(y)\,dy - \frac{1}{c}\int_{-c}^0 F_{n_i}(y)\,dy\right] = \frac{1}{\pi}\int_{-\infty}^{\infty} \frac{1-\cos ct}{t^2}\,\varphi_{n_i}(t)\,dt. \tag{5.4.4}$$

In the case of the first integral on the left of (5.4.4) it will be noted that if y is regarded as a continuous random variable having p.d.f. $1/c$ in the interval $(0, c)$, then $F_{n_i}(y)$ will be a random variable which is a function of the parameter n_i. Since $F_{n_1}(x)$, $F_{n_2}(x)$, . . . converges to $F(x)$, and since all these functions lie on the interval $[0, 1]$, we have

$$\lim_{i\to\infty} \frac{1}{c}\int_0^c F_{n_i}(y)\,dy = \frac{1}{c}\int_0^c F(y)\,dy.$$

Similarly,

$$\lim_{i\to\infty} \frac{1}{c}\int_{-c}^0 F_{n_i}(y)\,dy = \frac{1}{c}\int_{-c}^0 F(y)\,dy.$$

By similar considerations it can be shown that we may take the limit as $t \to \infty$ under the integral on the right (5.4.4). We obtain

$$(5.4.5) \quad c\left[\frac{1}{c}\int_0^c F(y)\,dy - \frac{1}{c}\int_{-c}^0 F(y)\,dy\right] = \frac{1}{\pi}\int_{-\infty}^{\infty} \frac{1-\cos ct}{t^2}\,\varphi(t)\,dt.$$

Setting $t = u/c$ and dividing both sides of (5.4.5) by c, we obtain

$$(5.4.6) \quad \frac{1}{c}\int_0^c F(y)\,dy - \frac{1}{c}\int_{-c}^0 F(y)\,dy = \frac{1}{\pi}\int_{-\infty}^{\infty} \frac{1-\cos u}{u^2}\,\varphi\left(\frac{u}{c}\right)du.$$

If we let $c \to \infty$, then since $F(y)$ is nondecreasing, the limit of the left-hand side of (5.4.6) is $F(+\infty) - F(-\infty)$. Since $\varphi(t)$ is continuous at $t = 0$, we have $\lim_{c\to\infty} \varphi(u/c) = \varphi(0)$ for every u. But $\varphi_1(t), \varphi_2(t), \ldots$ converges to $\varphi(t)$ for every t. Hence $\lim_{n\to\infty} \varphi_n(0) = \varphi(0)$. But $\varphi_n(0) = 1$ for every n. Therefore $\varphi(0) = 1$, and we finally obtain from (5.4.6) by letting $c \to \infty$

$$(5.4.7) \qquad F(+\infty) - F(-\infty) = \frac{1}{\pi}\int_{-\infty}^{\infty} \frac{1-\cos u}{u^2}\,du = 1.$$

Since $F(x)$ is non-negative, nondecreasing and cannot exceed 1, we must have $F(+\infty) = 1$ and $F(-\infty) = 0$. Hence $F(x)$, the limit of $F_{n_1}(x)$, $F_{n_2}(x), \ldots$, is a c.d.f. and its characteristic function is $\varphi(t)$, the limit of $\varphi_{n_1}(t), \varphi_{n_2}(t), \ldots$. Now if there is any other subsequence of $F_1(x), F_2(x), \ldots$ which converges to a nondecreasing function, let this function be $F^*(x)$. Then it can be shown by argument similar to that used above that $F^*(x)$ is a c.d.f. whose characteristic function is identical with $\varphi(t)$. Then since $F(x)$ and $F^*(x)$ are c.d.f.'s with identical characteristic functions, it follows from **5.1.3** that $F(x) \equiv F^*(x)$. This means that every convergent subsequence of $F_1(x), F_2(x), \ldots$ converges to the c.d.f. $F(x)$, which, of course, is equivalent to the statement that $F_1(x), F_2(x), \ldots$ converges to the c.d.f. $F(x)$ which concludes the argument for **5.4.1**.

The following corollary to **5.4.1** is useful in connection with problems of establishing convergence in probability to a constant:

5.4.1a *A necessary and sufficient condition for a sequence of random variables $x_1, x_2, \ldots$ having characteristic functions $\varphi_1(t), \varphi_2(t), \ldots$ to converge in probability to the constant c, is that* $\lim_{n\to\infty} \varphi_n(t) = e^{ict}$.

The k-dimensional analogues of **5.4.1** and **5.4.1a** are straightforward and are left to the reader for formulation and proof.

5.5 DETERMINATION OF DISTRIBUTION FUNCTIONS FROM MOMENTS

(a) Determination of a c.d.f. by a Moment-Sequence

In Section 5.1 it was seen that moments of distribution functions can be found, if they exist, by differentiating characteristic functions and (5.1.10) shows the relationship between the characteristic functions and those moments which do exist. Sometimes, however, the moments of the c.d.f. of a random variable can be found more effectively by other methods than by differentiating characteristic functions. There are various situations which occur in later chapters where moment-sequences of random variables and functions of random variables are easily determined one way or another. But the basic question is this: Under what conditions does a moment-sequence $\mu_0', \mu_1', \mu_2', \ldots$, of a random variable x determine the c.d.f. of x uniquely? A useful sufficient criterion [see Cramér (1946)] may be stated as follows:

5.5.1 *Let $F(x)$ be a c.d.f. with moments μ_r', $r = 0, 1, 2, \ldots$, all of which are finite. If the series $\sum_{r=0}^{\infty} \frac{\mu_r'}{r!} c^r$ is absolutely convergent for some $c > 0$ then $F(x)$ is the only c.d.f. having these moments.*

From the definition (5.1.1) of the characteristic function belonging to $F(x)$ and following a line of argument similar to that by which (5.1.10) was established, we may write

$$(5.5.1)\quad \varphi(t + u)$$
$$= \int_{-\infty}^{\infty} \left[1 + \sum_{r=1}^{n-1} \frac{(iu)^r}{r!} x^r + \frac{(iu)^n}{n!} x^n(\cos u'x + i \sin u''x)\right] e^{itx}\, dF(x)$$

where u' and u'' are (real) numbers in the interval $(0, u)$. If the nth moment exists, we may apply (5.1.4) and obtain

$$(5.5.2)\qquad \varphi(t + u) = \sum_{r=0}^{n-1} \frac{u^r}{r!} \varphi^{(r)}(t) + \frac{\nu_n' u^n}{n!} q$$

where q is a complex function such that $|q| \leqslant 1$, and where ν_n' is the nth absolute moment of x. If n is even, then $\nu_n' = \mu_n'$ and it follows from the hypothesis of our theorem that the remainder term $(\nu_n' u^n/n!)\, q \to 0$ as $n \to \infty$ if $|u| < c$. Now consider the remainder term for n odd. We note that for any real λ

$$(5.5.3)\quad \mathscr{E}[\lambda\, |x|^{(n-1)/2} + |x|^{(n+1)/2}]^2 = \lambda^2 \nu_{n-1}' + 2\lambda \nu_n' + \nu_{n+1}' \geqslant 0$$

from which we must have

$$\nu_n'^2 \leqslant \nu_{n-1}' \cdot \nu_{n+1}'. \tag{5.5.4}$$

We may therefore write

$$\frac{\nu_n'}{n!}\, u^n \leqslant \left[\left(\frac{\nu_{n-1}'}{(n-1)!}\, u^{n-1}\right)\left(\frac{\nu_{n+1}'}{(n+1)!}\, u^{n+1}\right)\left(\frac{n+1}{n}\right)\right]^{1/2}. \tag{5.5.5}$$

Now $\nu_{n-1}' = \mu_{n-1}'$ and $\nu_{n+1}' = \mu_{n+1}'$, and hence the right-hand side of this inequality vanishes as $n \to \infty$, which implies that as $n \to \infty$ (n odd) the remainder term of (5.5.2) vanishes. Therefore we have

$$\varphi(t + u) = 1 + \sum_{r=1}^{\infty} \frac{u^r}{r!}\, \varphi^{(r)}(t) \tag{5.5.6}$$

the series being convergent at least for $|u| < c$. Putting $t = 0$ and using (5.1.5), we have

$$\varphi(u) = 1 + \sum_{r=1}^{\infty} \frac{\mu_r'}{r!}\, (iu)^r, \tag{5.5.7}$$

which means that for $|u| < c$, $\varphi(u)$ is uniquely determined by the moments μ_r'. By the process of analytic continuation it can be shown that formula (5.5.7) holds for *all* values of u. For all derivatives of $\varphi(u)$ exist, for example, at $u = \pm\frac{1}{2}c$ and can be determined from (5.5.7), and hence for $|u| < c$, $\varphi(\frac{1}{2}c + u)$ and $\varphi(-\frac{1}{2}c + u)$ can be uniquely expressed by the series (5.5.6) by replacing t by $+\frac{1}{2}c$ and $-\frac{1}{2}c$ respectively. This is equivalent to stating that (5.5.7) is valid for $|u| < \frac{3}{2}c$. Continuing this procedure step-by-step it is clear that (5.5.7) holds for *all* values of u. Thus, under the hypotheses of our theorem the characteristic function is uniquely determined by the moment-sequence and the characteristic function, in turn, uniquely determines the c.d.f. $F(x)$ as stated in **5.5.1.**

The following corollary of **5.5.1** is useful in determining distribution functions defined over finite intervals, given the moment-sequence of such distributions:

5.5.1a *If x is a bounded random variable, then its c.d.f. $F(x)$ is uniquely determined by its moments μ_r', $r = 0, 1, 2, \ldots$.*

If x is bounded, then there are finite numbers a, b, $a < b$, such that $F(a) = 0$, and $F(b) = 1$. If M denotes the largest of $|a|$ and $|b|$, then we have

$$|\mu_r'| \leqslant \int_a^b |x|^r\, dF(x) = \nu_r' \leqslant M^r \tag{5.5.8}$$

and

$$\left|\sum_{r=0}^{\infty} \frac{\mu_r' c^r}{r!}\right| \leqslant \sum_{r=0}^{\infty} \frac{|\mu_r'| c^r}{r!} \leqslant \sum_{r=0}^{\infty} \frac{\nu_r' c^r}{r!} \leqslant \sum_{r=0}^{\infty} \frac{|Mc|^r}{r!} = e^{|Mc|} \tag{5.5.9}$$

which is finite for all values of c. Therefore the sufficient condition in **5.5.1** is satisfied, thus establishing **5.5.1a.**

Extended versions of **5.5.1** and **5.5.1a** for the unique determination of a c.d.f. of a k-dimensional random variable $(x_1, \ldots, x_k)$ from the moments $\mu'_{r_1, \ldots, r_k}$ can be stated. Formulation and proof of the extensions are left to the reader.

(b) Determination of the Limit of a Sequence of c.d.f.'s by Moments

Suppose $(x_1, x_2, \ldots)$ is a sequence of random variables and the moment-sequence of each component is given. A question arising is whether there exist simple criteria based on these moments for determining whether $(x_1, x_2, \ldots)$ converges in distribution to some random variable x. A simple theorem due to Kendall and Rao (1950) concerning the convergence in distribution of subsequences from $(x_1, x_2, \ldots)$ to a c.d.f. using only the second moments may be stated as follows:

5.5.2 *Let $\mu'_2(x_n)$ be the second moment of x_n in the sequence of random variables $(x_1, x_2, \ldots)$.* If

$$\mu'_2(x_n) < \mathrm{K} < +\infty \tag{5.5.10}$$

for $n = 1, 2, \ldots$, then there is a subsequence of $(x_1, x_2, \ldots)$ which converges in distribution.

To prove this let $F_n(x)$ be the c.d.f. of x_n. We have for any $x_0 > 0$

$$\mathrm{K} > \mu'_2(x_n) = \int_{-\infty}^{\infty} x^2 \, dF_n(x) \geqslant x_0^2 \int_{-\infty}^{-x_0} dF_n(x) + x_0^2 \int_{x_0}^{\infty} dF_n(x). \tag{5.5.11}$$

Therefore, we may write

$$\frac{\mathrm{K}}{x_0^2} > F_n(-x_0) + 1 - F_n(x_0), \quad n = 1, 2, \ldots \tag{5.5.12}$$

For a given $\varepsilon > 0$, we can therefore choose $x_0 > 0$ so that $1 - [F_n(x) - F_n(-x)] < \varepsilon$ for $x > x_0$ and for all n. A subsequence of $(x_1, x_2, \ldots)$ can be found, as pointed out in the proof of **5.4.1**, whose c.d.f.'s converge to a nondecreasing function $G(x)$ at all of its points of continuity. Then clearly for $x > x_0$ we have $1 - [G(x) - G(-x)] < \varepsilon$, which implies that $G(-\infty) = 0$ and $G(+\infty) = 1$. Therefore $G(x)$ is a c.d.f. which completes the argument that a subsequence of $(x_1, x_2, \ldots)$ converges in distribution to some random variable x.

Now suppose we have a complete moment-sequence $\mu'_r(x_n) = \mu'_{r,n}$, $r = 0, 1, 2, \ldots$ for each component x_n in the sequence of random variables $(x_1, x_2, \ldots)$ such that $\mu'_{r,n}$ are all finite and $\lim_{n \to \infty} \mu'_{r,n} = \mu'_r$, $r = 0, 1, 2, \ldots$ where the limit-sequence μ'_r, $r = 0, 1, 2, \ldots$ uniquely determines a c.d.f. $F(x)$. What conditions will guarantee that the sequence of random variables $(x_1, x_2, \ldots)$ converges in distribution to a random variable x having c.d.f. identical with $F(x)$?

An answer to this question due to Kendall and Rao (1950) can be stated as follows:

5.5.3 *Let $(x_1, x_2, \ldots)$ be a sequence of random variables. Let the* rth *moment of x_n be $\mu'_{r,n}$ and finite for all n and r. Let* $\lim_{n\to\infty} \mu'_{r,n} = \mu'_r$, *where μ'_r is finite for all r. Then if $(x_1, x_2, \ldots)$ converges in distribution to $F(x)$, $\mu'_0, \mu'_1, \ldots$ is the moment-sequence of $F(x)$. Conversely, if this moment-sequence uniquely determines a c.d.f. $F(x)$, it is the limiting c.d.f. of $(x_1, x_2, \ldots)$.*

To establish **5.5.3** we first assume that $(x_1, x_2, \ldots)$ converges in distribution to a c.d.f. $F(x)$ and we must then show that $\mu'_r = \lim_{n\to\infty} \mu'_{r,n}$, $r = 1, 2, \ldots$, is the moment-sequence of $F(x)$. That is, if $F_n(x)$ is the c.d.f. of x_n, $n = 1, 2, \ldots$, we must show that

$$\lim_{n\to\infty} \left| \int_{-\infty}^{\infty} x^r \, dF_n(x) - \int_{-\infty}^{\infty} x^r \, dF(x) \right| = 0, \qquad \text{for } r = 1, 2, \ldots. \tag{5.5.13}$$

Note that for any $K > 0$

$$\left| \int_{-\infty}^{\infty} x^r \, dF_n(x) - \int_{-\infty}^{\infty} x^r \, dF(x) \right| \leqslant A_1 + A_2 + A_3,$$

where

$$\begin{aligned} A_1 &= \left| \int_{-K}^{K} x^r \, dF_n(x) - \int_{-K}^{K} x^r \, dF(x) \right| \\ A_2 &= \left| \int_{E_K} x^r \, dF_n(x) \right| \\ A_3 &= \left| \int_{E_K} x^r \, dF(x) \right|, \end{aligned} \tag{5.5.14}$$

and where E_K is the set of values of x for which $|x| > K$. It follows from Schwarz' inequality that

$$A_2^2 \leq \int_{E_K} x^{2r} \, dF_n(x) \cdot \int_{E_K} dF_n(x), \tag{5.5.15}$$

both integrals being non-negative. Since $\mu'_{2r,n} \to \mu'_{2r}$ (finite) as $n \to \infty$, there exists a constant $M_r^2 > 0$ which bounds the first integral for all n and K. Since $F_n(x) \to F(x)$ as $n \to \infty$, the second integral on the right, and hence A_2 can be made arbitrarily small for all n by choosing K sufficiently large.

Since μ'_r is finite, A_3 can be made arbitrarily small by choosing K sufficiently large.

Since $F_n(x) \to F(x)$ as $n \to \infty$ and both $\mu'_{r,n}$ and μ'_r are finite, it is evident that for any fixed K, A_1 can be made arbitrarily small by choosing n sufficiently large.

Therefore we conclude that (5.5.13) holds and hence that $\lim_{n\to\infty} \mu'_{r,n} = \mu'_r$, $r = 1, 2, \ldots$; $\mu'_1, \mu'_2, \ldots$ being the moment-sequence of $F(x)$, the limit of the sequence of c.d.f.'s of $(x_1, x_2, \ldots)$.

Now consider the converse statement in **5.5.3.** We assume that $\mu'_1, \mu'_2, \ldots$ uniquely determine a c.d.f. $F(x)$ and we must show that $F_1(x), F_2(x), \ldots$ converge to a c.d.f. which is $F(x)$. We know from **5.5.2** that every convergent subsequence of $F_1(x), F_2(x), \ldots$ converges to some c.d.f. and from the argument given in the first part of the present theorem we know that the limiting c.d.f.'s for these subsequences must all have the same moment-sequence, namely, $\mu'_1, \mu'_2, \ldots$. But this moment-sequence is assumed to determine a c.d.f. uniquely. Therefore these limiting c.d.f.'s are all identical to $F(x)$, namely, the c.d.f. having the moment-sequence $\mu'_1, \mu'_2, \ldots$.

Finally, it should be noted that the condition that $\mu'_{r,n}$ be finite for all n and r can be replaced by the condition that $\mu'_{r,n}$ be finite for all r and all n greater than some integer n^* possibly depending on r without affecting the conclusions of **5.5.3.**

PROBLEMS

5.1 If a random variable x has a p.d.f. (or p.f.) which is symmetric about the vertical axis, show that the characteristic function $\varphi(t)$ of x takes on only real values.

5.2 A random variable x has characteristic function

$$\varphi(t) = \frac{1}{1 + t^2}.$$

Show that the p.d.f. of x is $\frac{1}{2}e^{-|x|}$ for any value of x in R_1.

5.3 The characteristic function of a random variable x is

$$\frac{e^{it}(1 - e^{nit})}{n(1 - e^{it})}$$

Show that x is a discrete random variable with p.f. $p(x) = 1/n$ for $x = 1, \ldots, n$.

5.4 If $e^{-\frac{1}{2}t^2}$ is the characteristic function of a random variable x, show that the p.d.f. of x is $(2\pi)^{-\frac{1}{2}}e^{-\frac{1}{2}x^2}$.

5.5 Prove **5.2.1.**

5.6 If $x_1, \ldots, x_k$ are independent random variables whose c.d.f.'s are all $F(x)$ show that the characteristic function of $x_1 + \cdots + x_k$ is $[\varphi(t)]^k$ where $\varphi(t)$ is the characteristic function of a random variable x having c.d.f. $F(x)$.

5.7 A random variable x is said to have a *Cauchy* (1853) *distribution* if its p.d.f. is

$$f(x) = \frac{k}{\pi[k^2 + (x - \mu)^2]},$$

for $-\infty < x < +\infty$, where k and μ are real constants and $k > 0$. Show that the characteristic function of x is given by

$$\varphi(t) = e^{i\mu t - k|t|},$$

and hence that if $x_1, \ldots, x_n$ are independent random variables all having this Cauchy distribution, the random variable $(x_1 + \cdots + x_n)/n$ also has the same Cauchy distribution.

5.8 If $x_1, \ldots, x_n$ are independent random variables all having the same p.d.f. in R_1, namely, $\frac{1}{\sqrt{2\pi}} e^{-\frac{1}{2}x^2}$ show, by using characteristic functions, that the random variable $\frac{1}{\sqrt{n}}(x_1 + \cdots + x_n)$ also has the same p.d.f.

5.9 Suppose x is a random variable denoting the number of times a die must be thrown in order to obtain one ace. Determine the probability generating function $\theta(t)$ of x. Find the mean and variance of x from $\theta(t)$.

5.10 Suppose x is a random variable denoting the number of times a "true" coin must be thrown in order to get k heads, If $F(x; k)$ is the c.d.f. of x show that $F(x; k)$ is reproductive with respect to k.

5.11 If x is a random variable denoting the number of spades in a hand of 13 ordinary playing cards, show that the probability generating function $\theta(t)$ of x is the coefficient of v^{13} in $(1 + tv)^{13}(1 + v)^{39} \Big/ \binom{52}{13}$. From $\theta(t)$ find the mean and variance of x.

5.12 A random variable x has

$$\mu'_r = \frac{k}{k + r}, \qquad r = 1, 2, \ldots$$

where $k > 0$. Show that the p.d.f. of x is given by

$$\begin{aligned} f(x) &= kx^{k-1} && \text{for } 0 \leqslant x \leqslant 1 \\ &= 0 && \text{otherwise} \end{aligned}$$

and that the distribution of x is unique.

5.13 If x is a random variable having as its rth moment,

$$\mu'_r = \frac{(k + r)!}{k!}, \quad k \text{ being a positive integer,}$$

show that its p.d.f. is

$$f(x) = \begin{cases} 0 & x \leqslant 0 \\ \dfrac{x^k}{k!} e^{-x} & x > 0 \end{cases}$$

and that the distribution of x is unique.

5.14 If (x_1, x_2) is a bounded two-dimensional random variable such that for $r, s = 1, 2, \ldots, \mathscr{E}(x_1^r x_2^s)\ \mathscr{E}(x_1^r),\ \mathscr{E}(x_2^s)$ all exist, and $\mathscr{E}(x_1^r x_2^s) = \mathscr{E}(x_1^r) \cdot \mathscr{E}(x_2^s)$, show that x_1 and x_2 are independent.

5.15 Let x be a random variable denoting the number of dots which appear when n "true" dice are thrown simultaneously. Show that the moment generating function $\psi(t)$ of x is $(e^t/6)^n(1 - e^t)^{-n}(1 - e^{6t})^n$ and that the distribution of x is reproductive with respect to n.

5.16 In a sequence of random variables $(x_1, x_2, \ldots), x_n$ has p.d.f. $n(1 - x)^{n-1}$ on (0, 1) and 0 otherwise. Let $y_n = n\,x_n$. Show, by using characteristic functions, that the sequence $(y_1, y_2 \ldots)$ converges in distribution to a random variable y having p.d.f. e^{-y} on $(0, \infty)$, and 0 otherwise.

5.17 In a sequence of random variables $(x_1, x_2, \ldots)$, x_n has characteristic function

$$\varphi_n(t) = \frac{\sin nt}{nt}.$$

Show that the p.d.f. of x_n is $1/2n$ on $(-n, +n)$ and 0 elsewhere, and hence that even though the sequence of characteristic functions converges to a limit $\varphi(t)$, the sequence of c.d.f.'s does not converge to a c.d.f. What condition of **5.4.1** is violated? [Cramér (1946)].

5.18 If $(x_1, \ldots, x_k)$ are independent discrete random variables such that the p.f. of x_i is

$$p_i(x) = \frac{\mu_i^x e^{-\mu_i}}{x!}, \qquad x = 0, 1, 2, \ldots, \qquad \mu_i > 0,$$

and if z is the random variable $x_1 + \cdots + x_k$, show by using characteristic functions that the p.f. of z is given by

$$p(z) = \frac{\left(\sum_1^k \mu_i\right)^z e^{-\sum_1^k \mu_i}}{z!}, \qquad z = 0, 1, 2, \ldots$$

5.19 A sequence of non-negative random variables $(x_1, x_2, \ldots)$ is such that the rth moment of x_n is given by

$$\mu'_{r,n} = \frac{r!\, n^r(n - r - 1)!}{(n - 1)!},$$

$n > r = 1, 2, \ldots$. Show that the sequence of random variables converges in distribution to a random variable x having p.d.f.

$$f(x) = \begin{cases} 0 & x \leqslant 0 \\ e^{-x} & x > 0. \end{cases}$$

Show that this limiting distribution is uniquely determined by its moments.

5.20 Consider a *stochastic branching process*, where an object can generate other objects, each of which can generate others, and so on, such as occurs for males (or females) in successive generations, or for neutrons in a nuclear fission process.

Let x_1 be a random variable denoting the number of objects an initial (zeroth-generation) object produces. Let x_2 be a random variable denoting the number of objects each first generation object produces. In general let x_{n+1} be a random variable denoting the number of objects each nth generation object produces. Assume the p.f.'s of $x_1, x_2, \ldots$ are all identical, namely, $p(x)$, $x = 0, 1, 2, \ldots$. Let y_n be a random variable denoting the total number of objects produced in the nth generation. Let $\theta_n(t)$ be the probability generating function of y_n. [Note that $\theta_0(t) = t$ and $\theta_1(t) = \sum_{x=0}^{\infty} t^x p(x)$.] Assuming all probability generating functions converge, show that

$$\theta_{n+1}(t) = \theta_1[\theta_n(t)].$$

If we denote by μ and σ^2 the mean and variance of y_1, show that

$$\begin{aligned} \mathscr{E}(y_n) &= \mu^n \\ \sigma^2(y_n) &= n\sigma^2, \qquad \text{if } \mu = 1 \\ &= \sigma^2\mu^{n-1}(\mu^n - 1)/(\mu - 1), \qquad \text{if } \mu \neq 1, \end{aligned}$$ [Harris (1948)]

5.21 (*Continuation*). If the random variables $x_1, x_2, \ldots, x_n$ have different p.f.'s, say $p_1(x), p_2(x), \ldots, p_n(x)$ and corresponding probability generating functions $\theta_1^*(t), \theta_2^*(t), \ldots$, respectively. Show that the probability generating function $\theta_n(t)$ of y_n is given by

$$\theta_n(t) = \theta_1^*[\theta_2^*\{\cdots \theta_n^*(t) \cdots\}],$$

and hence that

$$\mathscr{E}(y_n) = \mu_1 \mu_2 \cdots \mu_n,$$

where $\mu_1, \ldots, \mu_n$ are the means of $x_1, \ldots x_n$, respectively.

5.22 Establish (5.4.4).

5.23 Suppose x is a random variable with p.f. $p(x)$, and c.d.f. $F(x)$, and whose sample space is a set of non-negative integers. Let $\theta^*(t) = \sum_{x=0}^{\infty} F(x)t^x$, the *generating function* for $F(x)$. If $\theta(t)$ is the probability generating function of x, show that for $|t| < 1$

$$\theta^*(t) = \theta(t)/(1 - t).$$

5.24 If $x_1, \ldots, x_k$ are independent random variables whose sample spaces are non-negative integers and whose probability generating functions are $\theta_1(t), \ldots, \theta_k(t)$, show that the probability generating function of $x_1 + \cdots + x_n$ is $\theta_1(t) \cdots \theta_k(t)$.

CHAPTER 6

Some Special Discrete Distributions

The purpose of this chapter is to present some of the more important discrete probability distributions of mathematical statistics, not only to provide certain basic information about the distributions themselves but to illustrate some of the concepts, principles, and methods of Chapters 1 through 5 more fully than was done by the examples in those chapters. The distributions discussed in this chapter, and some of their main properties, will also be used at various points throughout the remainder of the book. Moreover, their discussion will show that, in spite of the general principles and methods introduced in earlier chapters, the study of special distributions often requires special methods and devices.

6.1 THE HYPERGEOMETRIC DISTRIBUTION

(a) The Case of One Random Variable

The probability function $p(x)$ used at the end of Section 2.3(a) is a simple case of a *hypergeometric distribution.* More generally, suppose we have a collection Π of elements such that each element belongs either to class C or to class $\bar{C}$. Let Np be the number of elements belonging to C and Nq the number belonging to $\bar{C}$ where $p + q = 1$. Suppose a set of $n(\leqslant N)$ elements is taken from Π and we let x be a random variable denoting the number of C's in the set. We wish to find the p.f. of x. There are $\binom{N}{n}$ possible sets of n elements in Π, of which $\binom{Np}{x} \cdot \binom{Nq}{n-x}$ will contain exactly x C's.

It should be noted that for this situation the basic sample space R discussed in Chapter 1 consists of $\binom{N}{n}$ sample points, each sample point being a set of n elements from Π. The event for which $x \leqslant x'$ consists of all sample points in R for which the number of C's is $\leqslant x'$.

Assigning equal probabilities to all sample points in R, we have for the p.f. of x the following:

$$p(x) = \frac{\binom{Np}{x}\binom{Nq}{n-x}}{\binom{N}{n}}, \tag{6.1.1}$$

the mass points of x (sample space of x) in R_1 being the integers satisfying both inequalities $0 \leqslant x \leqslant Np$ and $0 \leqslant n - x \leqslant Nq$.

This is called the *hypergeometric probability function.* It will be convenient to refer to a distribution having this p.f. as the *hypergeometric distribution* $H(N, n; p)$.

To see that the sum of $p(x)$ over this range of values of x is unity consider the identity

$$(u + v)^{A+B} \equiv (u + v)^A(u + v)^B$$

where A and B are positive integers. Expanding both sides of this expression, we have

$$\sum_{r=0}^{A+B}\binom{A+B}{r}u^r v^{A+B-r} \equiv \left[\sum_{s=0}^{A}\binom{A}{s}u^s v^{A-s}\right]\left[\sum_{r=s}^{B+s}\binom{B}{r-s}u^{r-s}v^{B-r+s}\right].$$

Since this is an identity in u and v, the coefficient of $u^r v^{A+B-r}$ on the left must be identical with the coefficient of $u^r v^{A+B-r}$ on the right. Therefore

$$\binom{A+B}{r} = \sum_{s}\binom{A}{s}\binom{B}{r-s} \tag{6.1.2}$$

where $\sum_s$ denotes summation over the integers s, satisfying both inequalities $0 \leqslant s \leqslant A$ and $0 \leqslant r - s \leqslant B$.

It is now clear from (6.1.2) that

$$\sum_{x}\binom{Np}{x}\binom{Nq}{n-x} = \binom{N}{n} \tag{6.1.2a}$$

and hence the sum of $p(x)$ over the sample space of x stated in (6.1.1) is unity.

This hypergeometric distribution is an example of a distribution for which the characteristic function is virtually worthless as a device for finding moments. But we can determine the factorial moments $\mu'_{[r]}$ in a reasonably painless fashion as follows.

First, we note that

$$\sum_{x} x^{[r]}\binom{Np}{x}\binom{Nq}{n-x} = (Np)^{[r]}\sum_{x}\binom{Np-r}{x-r}\binom{Nq}{n-x}, \tag{6.1.3}$$

the range of x being identically the same as that in (6.1.2*a*).

By (6.1.2) the sum on the right has the value $\binom{N-r}{n-r}$. When we divide (6.1.3) by $\binom{N}{n}$, the expression on the left defines $\mu'_{[r]}$ and its value is therefore given by

$$\mu'_{[r]} = \frac{(Np)^{[r]}\binom{N-r}{n-r}}{\binom{N}{n}}. \tag{6.1.4}$$

In particular, we have

$$\mu'_{[1]} = np, \quad \mu'_{[2]} = \frac{np(n-1)(Np-1)}{(N-1)},$$

and the following mean and variance

$$\mu(x) = np, \qquad \sigma^2(x) = \left(\frac{N-n}{N-1}\right)npq. \tag{6.1.5}$$

The hypergeometric distribution is one of the most important distributions in elementary probability theory. It is basic in connection with the evaluation of probabilities in problems arising in most card games, in drawing samples from lots of mass-produced articles containing defectives and other finite populations, etc.

The p.f. and the c.d.f. of the hypergeometric distribution (6.1.1) have been extensively tabulated by Lieberman and Owen (1961).

(b) The k-Variate Case

Probability distribution (6.1.1) can be generalized to k random variables.

The example at the end of Section 2.5(*a*) is a special case of a two-dimensional or bivariate hypergeometric distribution. More generally, suppose Π is a collection of N elements, such that Np_i elements belong to class C_i, $i = 1, \ldots, k+1$, where $p_1 + \cdots + p_{k+1} = 1$. This means that the classes $C_1, \ldots, C_{k+1}$, are *mutually exclusive*, that is, every element of Π belongs to one and only one of these classes. Now suppose a set of n elements is taken from Π. Let $x_1, \ldots, x_{k+1}$ be random variables denoting the number of elements in the set belonging to $C_1, \ldots, C_{k+1}$, respectively. These random variables are linearly dependent since their sum is n. We may consider $x_1, \ldots, x_k$ as k linearly independent random variables, and write x_{k+1} as $n - x_1 - \cdots - x_k$.

There are $\binom{N}{n}$ possible sample points of n elements which could be formed from Π, that is, $\binom{N}{n}$ points in the basic sample space R. Of this number of sample points, exactly $\binom{Np_1}{x_1} \cdots \binom{Np_{k+1}}{x_{k+1}}$ contain

x_1 C_1's, . . . , x_{k+1} C_{k+1}'s. Assigning all of these $\binom{N}{n}$ sample points equal probabilities, we obtain the following p.f. of $(x_1, \ldots, x_k)$:

$$p(x_1, \ldots, x_k) = \frac{\binom{Np_1}{x_1} \cdots \binom{Np_{k+1}}{x_{k+1}}}{\binom{N}{n}}, \tag{6.1.6}$$

where $x_{k+1} = n - x_1 - \cdots - x_k$. This is the *k-dimensional* or *k-variate hypergeometric probability function.* The mass points (sample points) of this distribution consist of those points in R_k with coordinates which are positive integers or zero in the *simplex* $x_i \geqslant 0$, $i = 1, \ldots, k, \sum_{i=1}^{k} x_i \leqslant n$ for which all the inequalities, $0 \leqslant x_i \leqslant Np_i$, $i = 1, \ldots, k + 1$, are satisfied. It will be convenient to refer to a probability distribution having p.f. (6.1.6) as the *k-variate hypergeometric distribution* $H(N, n; p_1, \ldots, p_k)$.

The means, variances, covariances, and higher moments of (6.1.6) can be determined from factorial moments. Denoting the factorial moment $\mathscr{E}(x_1^{[r_1]} \cdots x_k^{[r_k]})$ by $\mu'_{[r_1]\cdots[r_k]}$, we find by a straightforward extension of the method used in arriving at (6.1.4)

$$\mu'_{[r_1]\cdots[r_k]} = \frac{(Np_1)^{[r_1]} \cdots (Np_k)^{[r_k]} \binom{N - r_1 - \cdots - r_k}{n - r_1 - \cdots - r_k}}{\binom{N}{n}}, \tag{6.1.7}$$

where $0 \leqslant r_i \leqslant Np_i$, $i = 1, \ldots, k$, and $r_1 + \cdots + r_k \leqslant n$. From (6.1.7) it can be verified that

$$\mu(x_i) = np_i, \qquad \sigma^2(x_i) = np_i(1 - p_i)\left(\frac{N - n}{N - 1}\right), \tag{6.1.8}$$

$$\operatorname{cov}(x_i, x_j) = -np_ip_j\left(\frac{N - n}{N - 1}\right).$$

6.2 THE BINOMIAL DISTRIBUTION

Suppose an "operation" or "trial," when performed, will have one of two possible outcomes: C or $\bar{C}$. For instance, in rolling a die we might denote by C the occurrence of an ace, and by $\bar{C}$ the occurrence of any other face. Suppose p is the probability of a C and q the probability of a $\bar{C}$. Then $p + q = 1$.

If the "trial" is performed n times, we will obtain a sequence of n letters consisting of C's and $\bar{C}$'s. There are 2^n possible sequences; these sequences form the basic sample space R, the sample points themselves being these 2^n possible sequences. Let x be a random variable whose value for any given sequence is equal to the number of C's in that sequence. If we assume that the n "trials" in the sequence are statistically independent, the probability which would be assigned to any sequence for which $x = x'$ is

$$p^{x'}q^{n-x'} \tag{6.2.1}$$

as may be seen by replacing C by p and $\bar{C}$ by q in the sequence and multiplying the p's and q's in the resulting sequence. There are $\binom{n}{x'}$ sequences for which $x = x'$, and hence we obtain

$$P(x = x') = \binom{n}{x'} p^{x'}q^{n-x'}, \qquad x' = 0, 1, \ldots, n. \tag{6.2.2}$$

Hence x is a discrete random variable whose p.f. $p(x)$ is given by the right-hand side of (6.2.2) (dropping dashes). It will be noted that $p(x)$ is the general term in the expansion of the binomial $(q + p)^n$ which means that the sum of $p(x)$ given by (6.2.2) over the sample space of the random variable x is $(q + p)^n$, which, of course, is unity since $p + q = 1$. Accordingly, a distribution having p.f. (6.2.2) is called the *binomial distribution* and will be referred to as $Bi(n, p)$. (We use $Bi(a, b)$ for binomial distribution notation and $Be(a, b)$ for beta distribution notation to be introduced in Chapter 7.) Sequences of independent trials, each resulting in one of two possible outcomes, with constant probabilities from trial to trial are sometimes called *Bernoulli trials* after J. Bernoulli (1713) who first studied them.

The characteristic function of (6.2.2) is

$$\varphi(t) = \mathscr{E}(e^{itx}) = \sum_{x=0}^{n} \binom{n}{x} e^{itx}p^x q^{n-x} = (q + pe^{it})^n, \tag{6.2.3}$$

from which the moments μ_r' can be found by applying (5.1.5). In particular, we have

$$\mu_1' = \frac{\varphi'(0)}{i} = np; \qquad \mu_2' = \frac{\varphi''(0)}{i^2} = np(q + np) \tag{6.2.4}$$

from which we find the mean and variance of (6.2.2) to be

$$\mu(x) = np, \qquad \sigma^2(x) = npq. \tag{6.2.5}$$

If we consider n as a parameter in (6.2.3) and write $\varphi(t)$ as $\varphi(t; n)$, then $\varphi(t; n)$ satisfies (5.3.7), which means that if x_1 and x_2 are two independent random variables having binomial distributions $Bi(n_1, p)$ and $Bi(n_2, p)$ respectively, then $x = x_1 + x_2$ has the binomial distribution $Bi(n_1 + n_2, p)$, an intuitively obvious result. Hence

6.2.1 *The binomial distribution $Bi(n, p)$ is reproductive with respect to n.*

It should be pointed out that the binomial distribution (6.2.2) is a limiting form of the hypergeometric distribution (6.1.1) as $N \to \infty$. For (6.1.1) may be expressed as

(6.2.6) $p(x) =$

$$\binom{n}{x} \frac{p\left(p-\frac{1}{N}\right)\cdots\left(p-\frac{x-1}{N}\right)q\left(q-\frac{1}{N}\right)\cdots\left(q-\frac{n-x-1}{N}\right)}{\left(1-\frac{1}{N}\right)\cdots\left(1-\frac{n-1}{N}\right)} \underset{N\to\infty}{\longrightarrow} \binom{n}{x} q^{n-x}p^{x}.$$

Hence, holding x and n fixed we have the result that:

6.2.2 *The limit of the p.f. of the hypergeometric distribution $H(N, n; p)$ as $N \to \infty$ is the p.f. of the binomial distribution $Bi(n, p)$.*

Since the numbers of mass points in the binomial and hypergeometric distributions are finite, **6.2.2** implies that if $P_H(x \in E)$ and $P_B(x \in E)$ are probabilities of a fixed event E computed from $H(N, n; p)$ and $Bi(n; p)$ respectively, then $\lim_{N\to\infty} P_H(x \in E) = P_B(x \in E)$. Expressed in terms of sequences of random variables: if x_N is a random variable having the hypergeometric distribution $H(N, n; p)$, then the sequence of random variables $(x_1, x_2, \ldots)$ converges in distribution to the random variable x having the binomial distribution $Bi(n, p)$.

The binomial distribution is one of the most important distributions in statistics. Both the p.f. (6.2.2) and the c.d.f. form have been widely tabulated, the most extensive tabulations being those by Army Ordnance Corps (1952), Harvard Computation Laboratory (1955), National Bureau of Standards (1949), and by Romig (1953).

6.3 THE MULTINOMIAL DISTRIBUTION

Now suppose each "operation" or "trial" will result in one and only one of mutually exclusive events $C_1, \ldots, C_{k+1}$ and let the probabilities corresponding to these events be $p_1, \ldots, p_{k+1}$, all > 0, and $\sum_i p_i = 1$. Let n independent trials be made and let $(x_1, \ldots, x_{k+1})$ be a $(k + 1)$-dimensional random variable whose components $x_1, \ldots, x_{k+1}$ denote the numbers of trials resulting in $C_1, \ldots, C_{k+1}$, respectively. The components $x_1, \ldots, x_{k+1}$ are linearly dependent since $x_1 + \cdots + x_{k+1} = n$. The basic sample space R consists of $(k + 1)^n$ sample points, each point being a sequence of n selections from $C_1, \ldots, C_{k+1}$ with repetitions allowed. The probability associated with any sample point consisting of x_1 C_1's, . . . , x_{k+1} C_{k+1}'s is

(6.3.1) $$p_1^{x_1} \cdots p_{k+1}^{x_{k+1}}.$$

The number of such sample points in R is evidently

(6.3.2) $$\frac{n!}{x_1! \cdots x_{k+1}!}.$$

Since the amount of probability at each of these sample points is given by (6.3.1), it follows that the p.f. of the random variable $(x_1, \ldots, x_k)$ is given by

(6.3.3) $$p(x_1, \ldots, x_k) = \frac{n!}{x_1! \cdots x_{k+1}!} p_1^{x_1} \cdots p_{k+1}^{x_{k+1}}$$

where it is to be understood that $x_{k+1} = n - x_1 - \cdots - x_k$ and $p_{k+1} = 1 - p_1 - \cdots - p_k$. It will be noted that the expression for $p(x_1, \ldots, x_k)$ is the general term in the expansion of the multinomial $(p_1 + \cdots + p_{k+1})^n$. Accordingly, a distribution having p.f. (6.3.3) is known as a *k-variate multinomial distribution*, which will be conveniently referred to as $M(n; p_1, \ldots, p_k)$. The mass points of this distribution, that is, the sample space of $(x_1, \ldots, x_k)$, are the lattice points in R_k contained in the simplex $x_i \geqslant 0$, $i = 1, \ldots, k$, $\sum_{i=1}^{k} x_i \leqslant n$, that is, all points in the simplex whose coordinates are positive integers or zero.

The characteristic function of the multinomial distribution is

$$\begin{aligned}\varphi(t_1, \ldots, t_k) &= \sum e^{it_1x_1 + \cdots + it_kx_k} p(x_1, \ldots, x_k) \\ &= \sum \frac{n!}{x_1! \cdots x_{k+1}!} (p_1e^{it_1})^{x_1} \cdots (p_ke^{it_k})^{x_k}(p_{k+1})^{x_{k+1}}\end{aligned}$$

where the summation is performed over all points in the sample space of $(x_1, \ldots, x_k)$. But this is merely the sum of all of the terms in the expansion of the multinomial $(p_1e^{it_1} + \cdots + p_ke^{it_k} + p_{k+1})^n$. Therefore,

(6.3.4) $$\varphi(t_1, \ldots, t_k) = (p_1e^{it_1} + \cdots + p_ke^{it_k} + p_{k+1})^n.$$

By applying (5.2.3) one can find the joint moments of $x_1, \ldots, x_k$. In particular, we find

(6.3.5) $$\begin{gathered}\mu(x_i) = np_i, \qquad \sigma^2(x_i) = np_i(1 - p_i) \\ \operatorname{cov}(x_i, x_j) = -np_ip_j.\end{gathered}$$

If n is considered as a parameter in (6.3.4), it can be seen that

6.3.1 *The multinomial distribution $M(n; p_1, \ldots, p_k)$ is reproductive with respect to n.*

6.4 THE POISSON DISTRIBUTION

In the binomial distribution (6.2.2) suppose, for a fixed x, we let $n \to \infty$ and $p \to 0$ through sequences of values such that $np = \mu$, for a fixed μ. Then the limit of (6.2.2) is the following probability function

$$p(x) = \frac{\mu^x e^{-\mu}}{x!}. \tag{6.4.1}$$

This is the p.f. of what is known as the *Poisson distribution*, named after its discoverer Poisson (1837). It will be convenient to denote the Poisson distribution having p.f. (6.4.1) by $Po(\mu)$. [We insert o in $Po(\mu)$ to avoid confusion with probability notation $P(\mu)$.] Since n is allowed to increase without limit in obtaining (6.4.1), we can fix x at any positive integer desired. Thus, the mass points of x are 0, 1, 2, ... To establish (6.4.1), note that the right-hand side of (6.2.2) can be written as . . . .

$$\frac{\frac{n}{n}\left(1 - \frac{1}{n}\right)\left(1 - \frac{2}{n}\right)\cdots\left(1 - \frac{x-1}{n}\right)}{x!}(np)^x\left(1 - \frac{np}{n}\right)^{n-x}. \tag{6.4.2}$$

Allowing $n \to \infty$ and $p \to 0$ so that $np = \mu$, we see that the limit of (6.4.2) is the distribution given in (6.4.1). It can be readily verified that the sum of the probabilities given by (6.4.1) for $x = 0, 1, 2, \ldots$ is unity.

The characteristic function of the Poisson distribution is

$$\varphi(t) = \sum_{x=0}^{\infty} \frac{e^{itx}\mu^x e^{-\mu}}{x!} = e^{-\mu(1-e^{it})} \tag{6.4.3}$$

from which one finds

$$\mu(x) = \mu, \qquad \sigma^2(x) = \mu. \tag{6.4.4}$$

Considering μ as a parameter in the characteristic function (6.4.3) it can be seen that the characteristic function satisfies (5.3.7) and hence

6.4.1 *The Poisson distribution $Po(\mu)$ is reproductive with respect to μ.*

Remark. Since the binomial distribution is an approximation to the hypergeometric distribution for large N, and the Poisson distribution is an approximation to the binomial distribution for large n and small p, one might expect the relatively simple Poisson distribution to be an approximation to the hypergeometric distribution under certain conditions. As a matter of fact, roughly speaking, this is the case if (i) n is large, (ii) p small (with $np = \mu$), and (iii) N is much larger than n. These conditions are reasonably well fulfilled in such important applications as the sampling of lots of mass-produced articles, in which case N is the size of the lot, n the size of a sample of articles drawn from the lot, and p is the fraction of defective articles in the lot.

Like the binomial distribution, the Poisson distribution is a very important one in applications of probability and statistics. There are many problems not only in the distribution of numbers of defectives in samples of industrial products but in the distribution of bacterial colonies per unit volume of a culture or liquid, number of telephone calls initiated per unit time, number of bomb fragments striking a target per unit area, etc., in which the Poisson distribution arises. In all these situations we essentially have a large number n of independent or nearly independent "trials," a small probability p of a "success" in any given trial, and we ask for the probability of getting x "successes" in n trials.

The p.f. and c.d.f. of the Poisson distribution have been extensively tabulated by Molina (1942).

6.5 DISCRETE WAITING-TIME DISTRIBUTIONS

(a) The Hypergeometric Case

Suppose Π is a set of elements consisting of Np C's and Nq $\bar{C}$'s where $p > 0$, $q > 0$, and $p + q = 1$. Let elements be drawn successively until exactly k C's are drawn. We wish to determine the probability that exactly x elements will have to be drawn to accomplish this objective. The event points of our basic sample space R will consist of all possible sequences of C's and $\bar{C}$'s which could be produced by successively drawing all N elements from Π until Π is exhausted. There are

$$\binom{N}{Np} \tag{6.5.1}$$

sample points in R.

For each sample point, the value of our random variable x is equal to the number of elements drawn in order to obtain exactly k C's. For any value x' satisfying both inequalities $0 \leqslant k - 1 \leqslant x' - 1$ and $0 \leqslant Np - k \leqslant N - x'$, the number of sample points for which $x = x'$ is seen to be

$$\binom{x'-1}{k-1}\binom{N-x'}{Np-k} \tag{6.5.2}$$

by combinatorial analysis. If all sample points in R are assigned equal probabilities, namely, $1\bigg/\binom{N}{Np}$, then we have

$$P(x = x') = \frac{\binom{x'-1}{k-1}\binom{N-x'}{Np-k}}{\binom{N}{Np}} \tag{6.5.3}$$

which (dropping dashes) is therefore the p.f. $p(x)$ of the random variable x.

The distribution having (6.5.3) as its p.f. will be called the *hypergeometric waiting-time distribution.* A specified value x' of the random variable x is essentially the number of trials we have to *wait through* in order to obtain k C's.

To verify that

$$p(x) = \binom{x-1}{k-1}\binom{N-x}{Np-k}\Big/\binom{N}{Np} \tag{6.5.4}$$

is a p.f., where the sample space of x is the set of integers $k, k+1, \ldots, Nq+k$, $1 \leqslant k \leqslant Np$, we must show that

$$\sum_{x=k}^{Nq+k}\binom{x-1}{k-1}\binom{N-x}{Np-k} = \binom{N}{Np}. \tag{6.5.5}$$

Note that for $|t| < 1$, we have

$$(1-t)^{-k} = \sum_{x=k}^{\infty}\binom{x-1}{k-1}t^{x-k}$$

$$(1-t)^{-(Np-k+1)} = \sum_{y=-\infty}^{Nq+k}\binom{N-y}{Np-k}t^{Nq-y+k} \tag{6.5.6}$$

and

$$(1-t)^{-(Np+1)} = \sum_{z=0}^{\infty}\binom{Np+z}{Np}t^{z}. \tag{6.5.7}$$

It will be seen that the coefficient of t^{Nq} in the expansion of $(1-t)^{-(Np+1)}$ given by (6.5.7) is $\binom{N}{Np}$. By multiplying the two equations in (6.5.6) we obtain

$$(1-t)^{-(Np+1)} = \sum_x\sum_y\binom{x-1}{k-1}\binom{N-y}{Np-k}t^{Nq+(x-y)}. \tag{6.5.8}$$

The coefficient of t^{Nq} in the expansion of $(1-t)^{-(Np+1)}$ given by (6.5.8) is the sum of the coefficients $\binom{x-1}{k-1}\binom{N-y}{Np-k}$ over all *possible* pairs of integers (x, y) for which $x = y$, that is, by

$$\sum_{x=k}^{Nq+k}\binom{x-1}{k-1}\binom{N-x}{Np-k}. \tag{6.5.9}$$

Since one expansion of $(1-t)^{-(Np+1)}$ gives $\binom{N}{Np}$ as the coefficient of t^{Nq}, whereas the other gives (6.5.9), we therefore obtain (6.5.5), and hence $p(x)$ as defined in (6.5.4) is a p.f.

The distribution whose p.f. is given by (6.5.4) is another example of a discrete distribution whose characteristic function is not useful for finding moments. But the moments can be found by a method similar to that used in Section 6.1. In this case it is perhaps simpler to evaluate $\mathscr{E}[(x - k)^{[r]}]$, the rth *factorial moment of x with respect to k*, and then find the ordinary moments from these factorial moments. We have

$$(6.5.10)\quad \mathscr{E}[(x-k)^{[r]}]$$
$$= \sum_{x=k}^{Nq+k} (x-k)^{[r]} \binom{x-1}{k-1}\binom{N-x}{Np-k}\bigg/\binom{N}{Np}$$
$$= (k+r-1)^{[r]}\left\{\sum_{x=k+r}^{Nq+k}\binom{x-1}{k+r-1}\binom{N-x}{Np-k}\right\}\bigg/\binom{N}{Np}.$$

If we use the fact that the sum of the numerator on the right of (6.5.4) over the sample space of x has the value $\binom{N}{Np}$ it is evident that the sum inside { } has the value $\binom{N}{Np+r}$. Therefore we obtain

$$(6.5.11)\quad \mathscr{E}[(x-k)^{[r]}] = (k+r-1)^{[r]}\binom{N}{Np+r}\bigg/\binom{N}{Np}.$$

From this expression for $r = 1, 2$ one finds the mean of x to be

$$(6.5.12)\quad \mu(x) = \frac{k(N+1)}{Np+1}$$

and the variance to be

$$(6.5.13)\quad \sigma^2(x) = \frac{Nq(Nq-1)k(k+1)}{(Np+2)(Np+1)} + \frac{k(2k+1)(N+1)}{Np+1} - \frac{k^2(N+1)^2}{(Np+1)^2} - k(k+1).$$

(b) The Binomial Case

Suppose "trials" are performed successively and the outcome of each trial is either a C or a $\bar{C}$, the trials being independent. Let $P(C) = p$ and $P(\bar{C}) = q$ where $p + q = 1$. Our basic sample space R is the set of all possible sequences (sample points) of C's and $\bar{C}$'s which could occur in an indefinitely long sequence of trials. Let x be a random variable defined for each sample point as the number of trials performed in order to obtain exactly k C's.

Let E be the event in R of getting $k - 1$ C's in the first $x - 1$ trials and F the event of getting a C on the xth trial. If G is the event of having to make exactly x trials to get exactly k C's, then $G = E \cap F$, and $P(G) = P(E \cap F) = P(E) \cdot P(F \mid E)$. But $P(F \mid E) = p$ and

$$(6.5.14)\quad P(E) = \binom{x-1}{k-1} p^{k-1} q^{x-k}.$$

Hence, if $p(x)$ is the p.f. of x, we have $p(x) = P(G)$, that is,

$$p(x) = \binom{x-1}{k-1} p^k q^{x-k}, \quad x = k, k+1, \ldots. \tag{6.5.15}$$

Note that the expression for $p(x)$ is the $(x - k + 1)$th term obtained in the expression $p^k(1 - q)^{-k}$ if $(1 - q)^{-k}$ is expanded into a series in powers of q. The sum of $p(x)$ over the values $x = k, k + 1, \ldots$ is $p^k(1 - q)^{-k} = p^k p^{-k} = 1$.

The distribution having the p.f. defined by (6.5.15) will be called the *binomial waiting-time distribution.* The p.f. $p(x)$ is simply the probability that one must wait through x (independent) trials in order to obtain k C's. The distribution is also called the *negative binomial distribution* and the *Pascal distribution.*

The characteristic function of (6.5.15) is

$$\varphi(t) = \sum_{x=k}^{\infty} \binom{x-1}{k-1} (pe^{it})^k (qe^{it})^{x-k} = (pe^{it})^k (1 - qe^{it})^{-k}$$

which reduces to

$$\varphi(t) = p^k(e^{-it} - q)^{-k}. \tag{6.5.16}$$

By the usual differentiation procedure it can be verified that

$$\mu(x) = \frac{k}{p}, \qquad \sigma^2(x) = \frac{kq}{p^2}. \tag{6.5.17}$$

Note that the mean and variance here are the limiting forms of those in (6.5.12) and (6.5.13), as $N \to \infty$. In fact, the p.f. of the binomial waiting-time distribution (6.5.15) is, as one would expect, the limiting form of the p.f. of the hypergeometric waiting-time distribution (6.5.3) as $N \to \infty$. The reader will find it instructive to verify these statements.

We have considered only the simplest kind of a discrete waiting-time distribution. Problems arise in which we wish to know not only the probability function of the number of trials required to obtain a specified number of C's but the probability function of the number of trials required to obtain specified numbers of C's and of $\bar{C}$'s. Or in the case of more than two classes of outcomes, we may wish to know the probability function of the number of trials required to obtain specified numbers in each class. Problems of this type are much more complicated than those we have discussed. Some of them have been treated by Girshick, Mosteller, and Savage (1946), Haldane (1945), Laplace (1814), and McCarthy (1947).

6.6 DISTRIBUTIONS IN THE THEORY OF RUNS

The theory of runs plays an important role in certain problems of nonparametric statistical inference as we shall see in Chapter 14. But the main distributions in the theory of runs can be conveniently introduced here. The results which will be presented below are due primarily to Mood (1940).

(a) Runs of Two Kinds of Elements

Consider a sample space R whose sample points consist of the set of all $\binom{n}{n_1}$ permutations of n_1 C's and n_2 $\bar{C}$'s, where $n_1 + n_2 = n$. Any particular sample point will be a sequence of C's and $\bar{C}$'s which will consist of alternating *runs* of C's and $\bar{C}$'s. The *length* of a run is the number of elements in it. Let r_{1j} denote the number of runs of C's of length j, and r_{2j} the number of runs of $\bar{C}$'s of length j. For instance, if the sequence is

$$CCC\bar{C}\bar{C}CC\bar{C}CC\bar{C}\bar{C}C\bar{C}$$

we have $n_1 = 8$, $n_2 = 6$, $r_{11} = 1$, $r_{12} = 2$, $r_{13} = 1$, $r_{21} = 2$, $r_{22} = 2$, all other r's being zero.

It will be seen from the definition of n_1, n_2, r_{1j} and r_{2j} that $\sum_j jr_{1j} = n_1$ and $\sum_j jr_{2j} = n_2$. Let $r_1 = \sum_j r_{1j}$ and $r_2 = \sum_j r_{2j}$ be the total numbers of runs of C's and of $\bar{C}$'s respectively. For any assignment of probabilities to the sample points in R we can set up random variables r_{1j} and r_{2j} which will have specific values as defined above for any given sample point in R. Our problem is to find the p.f. of these random variables when all sample points in R are assigned equal probabilities. For a given set of values of the r_{1j}, the number of ways of arranging the r_1 runs of C's is

$$\frac{r_1!}{r_{11}! \cdots r_{1n_1}!}, \tag{6.6.1}$$

and similarly, the number of ways of arranging the r_2 runs of $\bar{C}$'s is

$$\frac{r_2!}{r_{21}! \cdots r_{2n_2}!}. \tag{6.6.2}$$

Note that r_1 and r_2 cannot differ from each other by more than unity; for if they do, at least two runs of one kind of element would have to be adjacent, contrary to the definition of a run. If $r_1 = r_2$, a given arrangement of runs of C's can be fitted into a given arrangement of runs of $\bar{C}$'s in two ways, such that the entire sequence of C's and $\bar{C}$'s will begin with either a run of C's or a run of $\bar{C}$'s.

If we define the function $\gamma(r_1, r_2)$ to be the number of ways of arranging r_1 indistinguishable objects of one kind and r_2 indistinguishable objects of a second kind so that no two objects of the same kind appear together, then we have

$$\gamma(r_1, r_2) = \begin{cases} 0 & , \quad |r_1 - r_2| > 1 \\ 1 & , \quad |r_1 - r_2| = 1 \\ 2 & , \quad r_1 = r_2. \end{cases} \tag{6.6.3}$$

The total number of ways of getting r_{1j} runs of C's of lengths $j = 1, \ldots, n_1$ and of getting r_{2j} runs of $\bar{C}$'s of lengths $j = 1, \ldots, n_2$ is therefore

$$\frac{r_1!\, r_2!\, \gamma(r_1, r_2)}{r_{11}! \cdots r_{1n_1}!\, r_{21}! \cdots r_{2n_2}!}. \tag{6.6.4}$$

But there are $\binom{n}{n_1}$ possible arrangements of n_1 C's and n_2 $\bar{C}$'s, that is, sample points in R. If these arrangements are all assigned equal probabilities, then the $(n_1 + n_2)$-dimensional random variable $(r_{ij};\, j = 1, \ldots, n_i,\ i = 1, 2)$ will have the following p.f.:

$$p(\{r_{ij}\}) = \frac{r_1!\, r_2!}{r_{11}! \cdots r_{1n_1}!\, r_{21}! \cdots r_{2n_2}!} \cdot \frac{n_1!\, n_2!}{n!} \cdot \gamma(r_1, r_2). \tag{6.6.5}$$

If we are interested only in the p.f. of the runs of C's, that is the r_{1j}, we must take the marginal distribution of $(r_{11}, \ldots, r_{1n_1})$ in (6.6.5), that is, we must sum $p(\{r_{ij}\})$ with respect to $r_{21}, \ldots, r_{2n_2}$. This means that we must sum (6.6.2) for all $r_{21}, \ldots, r_{2n_2}$ such that $\sum_j jr_{2j} = n_2$ and $\sum_j r_{2j} = r_2$. In order to do this we make use of the following identity in s which holds for values of s near zero:

$$(s + s^2 + \cdots)^{r_2} \equiv s^{r_2}(1 - s)^{-r_2} \equiv s^{r_2} \sum_{t=0}^{\infty} \frac{(r_2 + t - 1)!}{(r_2 - 1)!\, t!} s^t. \tag{6.6.6}$$

Now the coefficient of s^{n_2} in the first expression of (6.6.6) is the sum of (6.6.2) with respect to the $r_{21}, \ldots, r_{2n_2}$ subject to the restrictions $\sum_j jr_{2j} = n_2$ and $\sum_j r_{2j} = r_2$. But the coefficient of s^{n_2} from the first expression in (6.6.6) must equal that of s^{n_2} in the last expression in (6.6.6), which is

$$\frac{(n_2 - 1)!}{(r_2 - 1)!\, (n_2 - r_2)!}. \tag{6.6.7}$$

Therefore the p.f. of $(r_{1j}, j = 1, \ldots, n_1)$ and r_2 is

$$p(\{r_{1j}\}, r_2) = \frac{r_1!}{r_{11}! \cdots r_{1n_1}!} \cdot \frac{(n_2 - 1)!}{(r_2 - 1)!\, (n_2 - r_2)!} \cdot \frac{n_1!\, n_2!}{n!} \cdot \gamma(r_1, r_2). \tag{6.6.8}$$

Finally, to obtain the p.f. of the r_{1j} we must sum (6.6.8) with respect to r_2. Making use of (6.6.3), we have, after some simplification,

(6.6.9)
$$\sum_{r_2=1}^{n_2} \frac{(n_2-1)!}{(r_2-1)!(n_2-r_2)!} \cdot \gamma(r_1, r_2) = \frac{(n_2+1)!}{r_1!\,(n_2-r_1+1)!} = \binom{n_2+1}{r_1}.$$

Therefore, we have the following result for the p.f. of $(r_{1j}, j = 1, \ldots, n_1)$

(6.6.10)
$$p(\{r_{1j}\}) = \frac{r_1!}{r_{11}! \cdots r_{1n_1}!} \cdot \binom{n_2+1}{r_1} \Big/ \binom{n}{n_1}.$$

A similar result holds, of course, for the p.f. of the r_{2j}.

Another important distribution is that of r_1 and r_2. The p.f. of this distribution is obtained by summing (6.6.8) with respect to the r_{1j} subject to the conditions $\sum_j jr_{1j} = n_1$ and $\sum_j r_{1j} = r_1$. The procedure here is similar to that used in summing (6.6.5) to obtain (6.6.8) and it yields

(6.6.11)
$$p(r_1, r_2) = \frac{\binom{n_1-1}{r_1-1}\binom{n_2-1}{r_2-1}}{\binom{n}{n_1}} \gamma(r_1, r_2).$$

The p.f. of r_1, *the total number of runs* of C's, is found (by summing (6.6.11) with respect to r_2) to be

(6.6.12)
$$p(r_1) = \frac{\binom{n_1-1}{r_1-1}\binom{n_2+1}{r_1}}{\binom{n}{n_1}}.$$

A similar expression holds for the p.f. of r_2.

To verify that the sum of $p(r_1)$ over the sample space of r_1 is unity we make use of the relation

(6.6.13)
$$\sum_{i=0}^{B} \binom{A}{K+i}\binom{B}{i} = \binom{A+B}{K+B}$$

obtained by equating the coefficients of s^K in the following identity

(6.6.14)
$$(1+s)^A\left(1+\frac{1}{s}\right)^B \equiv \frac{(1+s)^{A+B}}{s^B}.$$

It follows from (6.6.13) that

$$\sum_{r_1=1}^{n_1} \binom{n_1-1}{r_1-1}\binom{n_2+1}{r_1} = \binom{n}{n_1}$$

which is equivalent to the statement

$$\sum_{r_1=1}^{n_1} p(r_1) = 1$$

The easiest way to find the moments of r_1 is by means of factorial moments. For the gth factorial moment $\mu'_{[g]}$ we have

$$(6.6.15)\quad \mu'_{[g]} = \mathscr{E}(r_1^{[g]}) = \frac{(n_2+1)^{[g]}}{\binom{n}{n_1}} \sum_{r_1=g}^{n_1} \binom{n_1-1}{r_1-1}\binom{n_2+1-g}{r_1-g}.$$

But it follows from (6.6.13) that

$$\sum_{r_1=g}^{n_1} \binom{n_1-1}{r_1-1}\binom{n_2+1-g}{r_1-g} = \binom{n-g}{n_1-g}.$$

Therefore

$$(6.6.16)\qquad \mu'_{[g]} = (n_2+1)^{[g]} \frac{\binom{n-g}{n_1-g}}{\binom{n}{n_1}},$$

from which we find

$$(6.6.17)\qquad \mu(r_1) = \frac{n_1(n_2+1)}{n}, \qquad \sigma^2(r_1) = \frac{(n_2+1)^{[2]}(n_1)^{[2]}}{n(n)^{[2]}}.$$

Similar formulas exist for the factorial moment, the mean and variance of r_2.

By similar methods applied to (6.6.11) one can find that the general factorial moment of $(r_1 - 1)$ and $(r_2 - 1)$ is given by

(6.6.18)

$$\mathscr{E}\left[(r_1-1)^{[g_1]}(r_2-1)^{[g_2]}\right] = \frac{(n_1-1)^{[g_1]}(n_2-1)^{[g_2]}}{\binom{n}{n_1}} \binom{n-g_1-g_2}{n_1-g_2}.$$

In certain kinds of problems, the random variable $u = r_1 + r_2$, the *total number of runs*, is important.

The mean and variance of u can be obtained at once from (6.6.17) and (6.6.18) by using formulas (3.4.2) and (3.4.3). We have

$$(6.6.19)\qquad \mu(u) = \mu(r_1) + \mu(r_2) = \frac{2n_1n_2}{n} + 1$$

$$\sigma^2(u) = \sigma^2(r_1) + \sigma^2(r_2) + 2\operatorname{cov}(r_1, r_2) = \frac{2n_1n_2(2n_1n_2 - n)}{n^2(n-1)}.$$

As a matter of fact the p.f. of u can be found from (6.6.11) by summing $p(r_1, r_2)$ along the line $u = r_1 + r_2$ in the r_1r_2-plane. It will be seen that if u is even there is only *one* point at which $p(r_1, r_2)$ has a value different from zero, and if u is odd there are *two* such points. The p.f. of u can be written down at once as

(6.6.20)

$$p(u) = \begin{cases} \dfrac{2\dbinom{n_1 - 1}{\frac{1}{2}u - 1}\dbinom{n_2 - 1}{\frac{1}{2}u - 1}}{\dbinom{n}{n_1}}, & \text{if } u \text{ is even} \\[2ex] \dfrac{\dbinom{n_1 - 1}{\frac{1}{2}u - \frac{1}{2}}\dbinom{n_2 - 1}{\frac{1}{2}u - \frac{3}{2}} + \dbinom{n_1 - 1}{\frac{1}{2}u - \frac{3}{2}}\dbinom{n_2 - 1}{\frac{1}{2}u - \frac{1}{2}}}{\dbinom{n}{n_1}}, & \text{if } u \text{ is odd.} \end{cases}$$

This result was originally obtained by Stevens (1939). Various tabulations concerning $p(u)$ have been made by Swed and Eisenhart (1943).

Throughout the foregoing discussion n_1 and n_2 have been fixed. If we allow them to be random variables, then the distributions whose p.f.'s are given by (6.6.5), (6.6.10), (6.6.11), and (6.6.12) are conditional distributions for fixed values of n_1 and n_2. If the p.f. of n_1 and n_2 is $p^*(n_1, n_2)$, then the product of each of these four p.f's by $p^*(n_1, n_2)$ would give the p.f. of the r's involved in that distribution *and* n_1 and n_2. To obtain the p.f.of the r's in any case we would have to sum the product for that case over the range of values of n_1 and n_2. In particular, if we consider a given set of independent trials, each trial resulting in an C or $\bar{C}$, then n_1 and n_2 would be linearly dependent random variables satisfying the condition $n_1 + n_2 = n$ such that n_1 would have a binomial distribution $Bi(n, p)$, where p is the probability of getting a C in a single trial. In this case

$$p^*(n_1, n_2) = \binom{n}{n_1} p^{n_1} q^{n_2} \tag{6.6.21}$$

where $n_2 = n - n_1$. To find the gth factorial moment of r_1 for instance, we would multiply the expression on the right of (6.6.16) by (6.6.21) and sum with respect to n_1 from 0 to n, understanding, of course, that the terms in this sum corresponding to $n = 1, \ldots, g - 1$ would be zero.

(b) Runs of k Kinds of Elements

The preceding results can be extended in a reasonably direct manner to runs generated by several kinds of elements. Suppose there is a total of n elements consisting of n_1 C_1's, . . . , n_k C_k's where $n_1 + \cdots + n_k = n$. We let r_{ij} be a random variable denoting the number of runs of C_i of length j. Let $r_i = \sum_j r_{ij}$, the total number of runs of C_i's. Mood (1940) has shown that the p.f. of the set of r_{ij} is

$$(6.6.22) \quad p(\{r_{ij}\}) = \frac{n_1! \cdots n_k!}{n!} \left[\prod_{i=1}^{k} \frac{r_i!}{r_{i1}! \cdots r_{in_i}!} \right] \cdot \gamma(r_1, \ldots, r_k),$$

where $\gamma(r_1, \ldots, r_k)$ is the number of ways r_1 objects of one kind, r_2 objects of a second kind, and so on, can be arranged so that no two adjacent objects are of the same kind. The function $\gamma(r_1, \ldots, r_k)$ is the coefficient of $x_1^{r_1} \cdots x_k^{r_k}$ in the expansion of

$$(6.6.23) \quad (x_1 + \cdots + x_k)^k (x_2 + x_3 + \cdots + x_k)^{r_1 - 1} \cdot (x_1 + x_3 + \cdots + x_k)^{r_2 - 1} \cdots (x_1 + x_2 + \cdots + x_{k-1})^{r_k - 1}.$$

The argument for establishing (6.6.22) is a straightforward extension of that for establishing (6.6.5), and is left to the reader. Similarly, an extension of the argument leading from (6.6.5) to (6.6.11) will establish the p.f. of $(r_1, \ldots, r_k)$ as

$$(6.6.24) \quad p(\{r_i\}) = \frac{n_1! \cdots n_k!}{n!} \left[\prod_{i=1}^{k} \binom{n_i - 1}{r_i - 1} \right] \cdot \gamma(r_1, \ldots, r_k).$$

Moments can be found from (6.6.22) and (6.6.24) by procedures similar to those for the case of $k = 2$.

PROBLEMS

6.1 If x is a random variable having the hypergeometric distribution $H(N, n; p)$, show that

$$\mathscr{E}[(n - x)^{[r]}] = (Nq)^{[r]} \binom{N - r}{n - r} \Big/ \binom{N}{n}.$$

6.2 A jar of $m + n$ chips are numbered $1, 2, \ldots, m + n$. A set of n chips are drawn at random from the jar. Show that the probability is

$$\binom{m + n - x - 1}{m - 1} \Big/ \binom{m + n}{m}$$

that x of the chips drawn have numbers exceeding all numbers on the chips remaining in the jar. Also show that

$$\mathscr{E}[(m + n - x + r - 1)^{[r]}] = (m + r - 1)^{[r]} \binom{m + n + r}{m + r} \Big/ \binom{m + n}{n}.$$

From this determine the mean and variance of the random variable x.

6.3 If $(x_1, \ldots, x_k)$ is a k-dimensional random variable having the k-variate hypergeometric distribution $H(N, n; p_1, \ldots, p_k)$, show that the marginal distribution of $(x_1, \ldots, x_{k_1})$, $k_1 < k$ is the hypergeometric distribution

$$H(N, n; p_1, \ldots, p_{k_1}).$$

6.4 (*Continuation*) Show that the conditional random variable

$$x_k \mid x_1, \ldots, x_{k-1}$$

has the one-dimensional hypergeometric distribution

$$H\left(N(p_k + p_{k+1}), n - x_1 - \cdots - x_{k-1}; \frac{p_k}{p_k + p_{k+1}}\right).$$

6.5 (*Continuation*) Show that $x_1 + \cdots + x_k$ has the one-dimensional hypergeometric distribution $H(N, n; p_1 + \cdots + p_k)$.

6.6 Show by formally summing $p(x_1, \ldots, x_k)$ as defined in (6.1.6) over the entire sample space that the sum is unity.

6.7 Prove **6.2.1.**

6.8 Prove **6.3.1.**

6.9 Prove **6.4.1.**

6.10 If $n \to \infty$ and $p_1, \ldots, p_k$ each $\to 0$ so that $np_1 = \mu_1, \ldots, np_k = \mu_k$, show that the limit of the p.f. of $(x_1, \ldots, x_k)$ in (6.3.3) is the product of p.f.'s of independent Poisson distributions $Po(\mu_1), \ldots, Po(\mu_k)$.

6.11 Show that the binomial waiting-time distribution whose p.f. is given by (6.5.15) is reproductive with respect to k.

6.12 If $(x_1, \ldots, x_k)$ has the k-variate multinomial distribution

$$M(n; p_1, \ldots, p_k),$$

show by use of characteristic functions that the marginal distribution of $(x_1, \ldots, x_{k_1})$, $k_1 < k$, is the k_1-variate multinomial distribution

$$M(n; p_1, \ldots, p_{k_1}).$$

6.13 (*Continuation*) Show that the distribution of the conditional random variable $x_k \mid x_1, \ldots, x_{k-1}$ is the binomial distribution

$$Bi\left(n - x_1 - x_2 - \cdots - x_{k-1}; \frac{p_k}{p_k + p_{k+1}}\right).$$

6.14 (*Continuation*) Show that $x_1 + \cdots + x_{k_1}$, $k_1 \leqslant k$, has the binomial distribution $Bi(n; p_1 + \cdots + p_{k_1})$.

6.15 Show that the limit of the hypergeometric waiting-time distribution given by (6.5.3) as $N \to \infty$ is the binomial waiting-time distribution given by (6.5.15).

6.16 Verify formula (6.6.12).

6.17 Verify formula (6.6.18).

6.18 If x is a random variable having the Poisson distribution $Po(\mu)$ and if the conditional random variable $y \mid x$ has the binomial distribution $Bi(x, p)$, show that the unconditional distribution of y is the Poisson distribution $Po(\mu p)$.

6.19 If, in the binomial waiting-time distribution (6.5.15), y is the new random variable $x - k$, show that the limiting distribution of y as $k \to \infty$ and $q \to 0$ so that $kq \to \mu$, is the Poisson distribution $Po(\mu)$.

6.20 If x has the Poisson distribution $Po(\mu)$ show that the c.d.f. of x is given by

$$\frac{1}{x!}\int_{\mu}^{\infty} e^{-z}z^x\,dz.$$

6.21 *Neyman's* (1939) *type A contagious distribution.* Suppose x is a random variable having Poisson distribution $Po(\mu_1)$ and y is a random variable such that the conditional random variable $y \mid x$ has the Poisson distribution $Po(\mu_2 x)$. Show that the characteristic function of the unconditional random variable y is given by

$$\varphi(t) = \exp\{-\mu_1[1 - e^{-\mu_2(1-e^{it})}]\}$$

and that

$$\mathscr{E}(y) = \mu_1\mu_2$$

$$\sigma^2(y) = \mu_1\mu_2(1 + \mu_2).$$

6.22 If x is a random variable having the binomial distribution $Bi(n, p)$ show that the c.d.f. of x is given by

$$(n - x)\binom{n}{x}\int_0^q y^{n-x-1}(1 - y)^x\,dy.$$

6.23 *The bloodtesting problem.* Persons in a large population are given blood tests in groups of k persons at a time as follows: Blood specimens are taken from k persons and pooled. If a test on the pooled blood is negative, a single test is sufficient for the k persons. If the test is positive, each person's blood is tested separately. If q is the probability of a negative test on a single person picked at random, show that the value of k which minimizes the expected value of the total number of tests is the positive integer for which the interval

$$\left(\sqrt[k]{\frac{1}{k(k+1)(1-q)}},\ \sqrt[k]{\frac{q}{k(k-1)(1-q)}}\right)$$

contains q. If q is near 1, show that an approximation to k is the solution of the equation

$$k^2 g^k \log g + 1 = 0. \qquad \text{[Dorfman (1943).]}$$

6.24 Suppose a jar has n chips numbered $1, 2, \ldots, n$. A person draws a chip, returns it, draws another, returns it, and so on until he gets a chip which has been drawn before and then stops. Let x be a random variable denoting the number of drawings required to accomplish this objective. Show that the p.f. of x is given by

$$p(x) = (x - 1)!\binom{n}{x-1}\frac{(x-1)}{n^k}, \qquad x = 2, \ldots, n + 1.$$

Verify that $\sum_{x=2}^{n+1} p(x) = 1$.

Show that $$\mathscr{E}(x) = 2 + \left(1 - \frac{1}{n}\right) + \left(1 - \frac{1}{n}\right)\left(1 - \frac{2}{n}\right) + \cdots + \left(1 - \frac{1}{n}\right)\left(1 - \frac{2}{n}\right)\cdots\left(1 - \frac{n-1}{n}\right).$$

6.25 In the binomial waiting-time problem of Section 6.5, suppose x is the number of trials required in order to obtain k successive successes. Show that the m.g.f. of x is
$$\psi(t) = (pe^t)^k(1 - pe^t)(1 - e^t + p^k q e^{(k+1)t})^{-1}$$
and that
$$\mathscr{E}(x) = \frac{1 - p^k}{p^k q}.$$

6.26 *The problem of class-size distributions—the multinomial case.* In the multinomial distribution having p.f. (6.3.3) suppose
$$p_1 = \cdots = p_{k+1} = \frac{1}{k+1}.$$

Let $r_0, r_1, \ldots, r_n$ be the numbers of the components of $(x_1, \ldots, x_{k+1})$ which are $0, 1, \ldots, n$, respectively. Show that the p.f. of $(r_0, r_1, \ldots, r_n)$ is given by
$$\frac{n!\,(k+1)!\,(k+1)^{-n}}{(0!)^{r_0}(1!)^{r_1}\cdots(n!)^{r_n} r_0!\,r_1!\cdots r_n!}$$
$(r_0, r_1, \ldots, r_n)$ being subject to the conditions
$$r_0 + r_1 + \cdots + r_n = k + 1$$
$$r_1 + 2r_2 + \cdots + nr_n = n.$$
Furthermore, show that
$$\mathscr{E}[r_0^{[s_0]} r_1^{[s_1]} \cdots r_n^{[s_n]}] = \frac{n!\,(k+1)!\,(k+1)^{-n}(k+1-B)^{n-A}}{(0!)^{s_0}(1!)^{s_1}\cdots(n!)^{s_n}\,(n-A)!\,(k+1-B)!}$$
where $A = s_1 + 2s_2 + \cdots + ns_n$ and $B = s_0 + s_1 + \cdots + s_n$. From this expression find the value of $\mathscr{E}(r_t)$, $\sigma^2(r_t)$ and cov (r_t, r_u). [Tukey (1949b).]

6.27 *The problem of class-size distributions—the hypergeometric case.* Suppose a deck of $M(k + 1)$ cards has M cards of each of $k + 1$ different "suits" and after thorough shuffling suppose a "hand" of n cards is dealt. Let r_0 be the number of 0-cards (blank) suits, in the "hand," r_1 the number of 1-card suits in the "hand," r_2 the number of 2-card suits in the "hand," and so on. Show that the p.f. of the $(n + 1)$-dimensional random variable $(r_0, r_1, \ldots, r_n)$ is given by
$$\frac{(M!)^{k+1}(k+1)!\binom{M(k+1)}{n}^{-1}}{r_0!\,r_1!\cdots r_n!\,[0!\,(M-0)!]^{r_0}[1!\,(M-1)!]^{r_1}\cdots[n!\,(M-n)!]^{r_n}}.$$

Furthermore, show that

$$\mathscr{E}[r_0^{[s_0]} r_1^{[s_1]} \cdots r_n^{[s_n]}] = \frac{(M!)^B (k+1)! \binom{M(k+1-B)}{n-A}}{(k+1-B)! \binom{M(k+1)}{n} [0!\,(M-0)!]^{s_0} \cdots [n!\,(M-n)!]^{s_n}},$$

where $A = s_1 + 2s_2 + \cdots + ns_n$ and $B = s_0 + s_1 + \cdots + s_n$. From this expression find the value of $\mathscr{E}(r_t)$, $\sigma^2(r_t)$, and cov (r_t, r_u). [Tukey (1949*b*).]

6.28 *The card-matching problem for two decks of cards.* Let A be a deck of N cards, each card belonging to one and only one of the "suits" $S_1, \ldots, S_k$, the numbers of cards belonging to these suits being $m_1, \ldots, m_k$, respectively. Similarly let B be a second deck of N cards having $n_1, \ldots, n_k$ cards belonging to suits $S_1, \ldots, S_k$, respectively. Deck A is shuffled and the cards are dealt face up along a line. Similarly, deck B is shuffled and the cards are dealt face up along a line immediately below the line formed by the cards in deck A. Let x be a random variable denoting the number of pairs of cards in the two decks which match in suit. Assigning all possible permutations of cards in each suit equal probabilities, show that the p.f. of x is given by the coefficient of

$$e^{tx} a_1^{m_1} \cdots a_k^{m_k} b_1^{n_1} \cdots b_k^{n_k}$$

in the expansion of

$$\Phi = \frac{1}{M}\left(\sum_{i,j=1}^{k} a_i b_j e^{\delta_{ij} t}\right)^N$$

where

$$\delta_{ij} = 0, \quad i \neq j \quad \text{and} \quad 1, i = j$$

and

$$M = \frac{(N!)^2}{m_1! \cdots m_k!\, n_1! \cdots n_k!}.$$

Hence show that the rth moment μ'_r of x is given by the coefficient of $a_1^{m_1} \cdots a_k^{m_k} b_1^{n_1} \cdots b_k^{n_k}$ in the expansion of

$$\left[\frac{\partial^r \Phi}{\partial t^r}\right]_{t=0}.$$

In particular, show that

$$\mathscr{E}(x) = \sum_{i=1}^{k} \frac{m_i n_i}{N}$$

$$\sigma^2(x) = \frac{1}{N^2(N-1)}\left[\left(\sum_{i=1}^{k} m_i n_i\right)^2 - N \sum_{i=1}^{k} (m_i^2 n_i + m_i n_i^2) + N^2 \sum_{i=1}^{k} m_i n_i\right].$$

[Battin (1942), and Kaplansky and Riordan (1945).]

CHAPTER 7

Some Special Continuous Distributions

In this chapter we present some of the more important continuous probability distributions of mathematical statistics together with some of their properties. Many of these results will be used in subsequent chapters.

7.1 THE RECTANGULAR DISTRIBUTION

The simplest continuous distribution, and one which we shall find it useful to define here, is that which has the following p.d.f.

$$(7.1.1)\qquad f(x) = \begin{cases} \dfrac{1}{\omega}, & \mu - \dfrac{\omega}{2} \leqslant x \leqslant \mu + \dfrac{\omega}{2} \\[2ex] 0, & x < \mu - \dfrac{\omega}{2}, \qquad x > \mu + \dfrac{\omega}{2}. \end{cases}$$

It will be convenient to call the probability distribution having this p.d.f. the *rectangular distribution* $R(\mu, \omega)$. This distribution has the following mean and variance

$$(7.1.2)\qquad \mathscr{E}(x) = \mu, \qquad \sigma^2(x) = \frac{\omega^2}{12}.$$

The parameter ω is called the *range* of the distribution.

It will be noted that if x is a random variable having the rectangular distribution $R(\mu, \omega)$, then

$$y = \frac{x - \mu + \frac{1}{2}\omega}{\omega}$$

is a random variable having the rectangular distribution $R(\frac{1}{2}, 1)$, which has the nonzero part of its p.d.f. on the interval [0, 1].

An important case of a random variable having the rectangular distribution $R(\frac{1}{2}, 1)$ is embodied in the following statement:

7.1.1 *If x is a random variable having a continuous c.d.f. $F(x)$ then the random variable $y = F(x)$ has the rectangular distribution $R(\frac{1}{2}, 1)$.*

This follows at once from the fact that the c.d.f. of y is

$$H(y) = P(F(x) \leqslant y) = \begin{cases} 1, & y > 1 \\ y, & 0 < y \leqslant 1 \\ 0, & y \leqslant 0 \end{cases}$$

which is the c.d.f. of the rectangular distribution $R(\frac{1}{2}, 1)$.

7.2 THE NORMAL DISTRIBUTION

The most important distribution function of a continuous random variable is the *normal* or *Gaussian distribution*. Its p.d.f. may be written as

(7.2.1) $$f(x) = \frac{1}{\sqrt{2\pi}\,\sigma} e^{-\frac{1}{2}(x-\mu)^2/\sigma^2}$$

for $-\infty < x < +\infty$, where μ and σ^2 are *parameters*. It will be shown presently that these two parameters are actually the *mean* and *variance*,

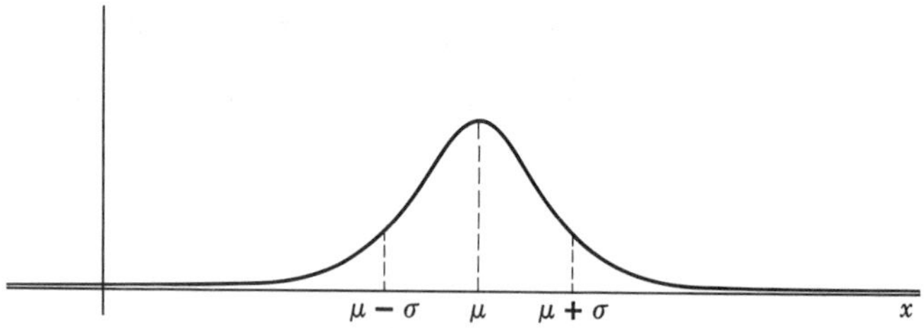

FIG. 7.1 Graph of normal p.d.f. (7.2.1)

respectively, of the normal distribution function. It will be convenient to refer to the distribution having p.d.f. (7.2.1) as the normal distribution $N(\mu, \sigma^2)$, or simply the distribution $N(\mu, \sigma^2)$ since the normal distribution occurs so frequently. The graph of (7.2.1) as shown in Fig. 7.1 is symmetrical with respect to $x = \mu$ with maximum ordinate $1/\sqrt{2\pi}\sigma$ at $x = \mu$. The graph has inflection points at $x = \mu \pm \sigma$.

The most convenient form of the normal distribution for tabulation is that corresponding to a random variable y, where $y = (x - \mu)/\sigma$. The p.d.f. of y is the *standardized form* $N(0, 1)$ of the normal distribution. Note that

$$P(x \leqslant x') = P(y \leqslant y') = \frac{1}{\sqrt{2\pi}} \int_{-\infty}^{y'} e^{-\frac{1}{2}y^2}\, dy$$

where $y' = (x' - \mu)/\sigma$.

The c.d.f. $\Phi(x)$ of the standardized form of the normal distribution defined by

(7.2.1*a*) $$\Phi(x) = \frac{1}{\sqrt{2\pi}} \int_{-\infty}^{x} e^{-\frac{1}{2}y^2}\, dy,$$

is widely tabulated, an excellent table being that prepared by the National Bureau of Standards (1942). Greenwood and Hartley (1961) give an extensive index of the various tabulations.

First we show that the integral of the normal distribution function over the entire x-axis is unity.

By putting $y = (x - \mu)/\sigma$, we can write

(7.2.2) $$\frac{1}{\sqrt{2\pi}\sigma} \int_{-\infty}^{\infty} e^{-\frac{1}{2}(x-\mu)^2/\sigma^2}\, dx = \frac{1}{\sqrt{2\pi}} \int_{-\infty}^{\infty} e^{-\frac{1}{2}y^2}\, dy.$$

Let I denote the integral on the right without the constant $1/\sqrt{2\pi}$. Our problem is to show that $I = \sqrt{2\pi}$. Now

(7.2.3) $$I^2 = \int_{-\infty}^{\infty} \int_{-\infty}^{\infty} e^{-\frac{1}{2}(y_1^2+y_2^2)}\, dy_1\, dy_2.$$

Applying the following transformation to polar coordinates

$$y_1 = r \cos \theta$$
$$y_2 = r \sin \theta$$

we have

(7.2.3*a*) $$I^2 = \int_0^{2\pi} \int_0^{\infty} re^{-\frac{1}{2}r^2}\, dr\, d\theta = 2\pi$$

and hence $I = \sqrt{2\pi}$.

Now consider the characteristic function of (7.2.1). We have

(7.2.4) $$\varphi(t) = \frac{1}{\sqrt{2\pi}\sigma} \int_{-\infty}^{\infty} e^{itx-\frac{1}{2}(x-\mu)^2/\sigma^2}\, dx$$

which can be written as

(7.2.4*a*) $$\varphi(t) = e^{it\mu-\frac{1}{2}\sigma^2t^2} \frac{1}{\sqrt{2\pi}\sigma} \int_{-\infty}^{\infty} e^{-\frac{1}{2}(x-\mu-i\sigma^2t)^2/\sigma^2}\, dx.$$

The integral in (7.2.4*a*) is the integral of the function $e^{-\frac{1}{2}z^2/\sigma^2}$, where z is a complex variable, in the complex plane along a line parallel to the real axis, namely, the line $y = -i\sigma^2t$, which can be shown to be equal to the integral of the same function along the real axis, that is, the integral $\int_{-\infty}^{\infty} e^{-\frac{1}{2}x^2/\sigma^2}\, dx$. But this integral has the value $\sigma\sqrt{2\pi}$. Therefore,

7.2.1 *The characteristic function of the distribution $N(\mu, \sigma^2)$ is*

(7.2.5) $$\varphi(t) = e^{it\mu-\frac{1}{2}\sigma^2t^2}.$$

By the usual differentiation procedure we find the mean and variance of the distribution $N(\mu, \sigma^2)$ to be

$$\mathscr{E}(x) = \mu, \qquad \sigma^2(x) = \sigma^2. \tag{7.2.6}$$

Treating (μ, σ^2) as a vector parameter in the characteristic function (7.2.5) it is evident that the characteristic function satisfies (5.3.7) and hence

7.2.2 *The distribution $N(\mu, \sigma^2)$ is reproductive with respect to (μ, σ^2).*

In fact, it has a stronger property than merely being reproductive which may be stated as follows:

7.2.3 *Suppose x_1 and x_2 are independent random variables having distributions $N(\mu_1, \sigma_1^2)$ and $N(\mu_2, \sigma_2^2)$, respectively. Let $L = c_1x_1 + c_2x_2$, where c_1 and c_2 are real constants, not both 0. Then the distribution of L is $N(c_1\mu_1 + c_2\mu_2, c_1^2\sigma_1^2 + c_2^2\sigma_2^2)$.*

A quick way to verify this statement is to set up the characteristic function of L. We have by (5.3.6)

$$\varphi(t) = \varphi_1(c_1t) \cdot \varphi_2(c_2t)$$

where $\varphi_1(t)$ and $\varphi_2(t)$ are the characteristic functions of x_1 and x_2 respectively. Thus,

$$\varphi_j(c_jt) = e^{itc_j\mu_j - \frac{1}{2}c_j^2\sigma_j^2t^2}, \qquad j = 1, 2,$$

and we have

$$\varphi(t) = \exp\left[i(c_1\mu_1 + c_2\mu_2)t - \tfrac{1}{2}(c_1^2\sigma_1^2 + c_2^2\sigma_2^2)t^2\right] \tag{7.2.7}$$

which, by (7.2.5), is seen to be the characteristic function of the normal distribution $N(c_1\mu_1 + c_2\mu_2, c_1^2\sigma_1^2 + c_2^2\sigma_2^2)$. Applying **5.1.3** completes the argument for theorem **7.2.3**. This theorem can be extended immediately to the case of k independent random variables having normal distributions. This extension is left as an exercise for the reader.

7.3 THE BIVARIATE NORMAL DISTRIBUTION

The normal distribution occupies a position of such importance in the theory of probability and statistics that it will be useful to discuss the two-dimensional case in some detail before proceeding to the k-variate case. The most convenient form of the p.d.f. of the *bivariate normal distribution* is

$$f(x_1, x_2) = \frac{\sqrt{|\sigma^{ij}|}}{2\pi} e^{-\frac{1}{2}Q(x_1, x_2)} \tag{7.3.1}$$

where (x_1, x_2) is any point in R_2, and

$$(7.3.2)\quad Q(x_1, x_2) = \sigma^{11}(x_1 - \mu_1)^2 + 2\sigma^{12}(x_1 - \mu_1)(x_2 - \mu_2) + \sigma^{22}(x_2 - \mu_2)^2.$$

We shall presently show that μ_1 and μ_2 are the *means* of x_1 and x_2 respectively, and that the matrix $\|\sigma^{ij}\|$, $i, j = 1, 2$, where $\sigma^{12} = \sigma^{21}$, is the inverse of the covariance matrix $\|\sigma_{ij}\|$ of x_1 and x_2 as defined in (3.5.3). Needless to say, we assume that $\|\sigma_{ij}\|$ is positive definite, which implies that $\|\sigma^{ij}\|$ is positive definite, that is, $Q(x_1, x_2)$ is a positive definite quadratic form.

We shall find it convenient to refer to a bivariate normal distribution having p.d.f. (7.3.1) as $N(\{\mu_i\}, \|\sigma_{ij}\|)$, $i, j = 1, 2$.

First, let us verify that

$$(7.3.3)\qquad \int_{R_2} f(x_1, x_2)\, dx_1\, dx_2 = 1$$

which is equivalent to showing that

$$(7.3.4)\qquad \int_{-\infty}^{\infty}\int_{-\infty}^{\infty} e^{-\frac{1}{2}Q(x_1,x_2)}\, dx_1\, dx_2 = \frac{2\pi}{\sqrt{|\sigma^{ij}|}}.$$

Putting $y_1 = x_1 - \mu_1$ and $y_2 = x_2 - \mu_2$ we can write the left-hand side of (7.3.4) as

(7.3.5)

$$\int_{-\infty}^{\infty}\int_{-\infty}^{\infty} \exp\{-\tfrac{1}{2}[\sqrt{\sigma^{11}}(y_1 + \sigma^{12}/\sigma^{11}\, y_2)]^2 - \tfrac{1}{2}[y_2\sqrt{|\sigma^{ij}|/\sigma^{11}}]^2\}\, dy_1\, dy_2.$$

Making the transformation

$$(7.3.6)\qquad z_1 = \sqrt{\sigma^{11}}\left(y_1 + \frac{\sigma^{12}}{\sigma^{11}}\, y_2\right)$$

$$z_2 = \sqrt{\frac{|\sigma^{ij}|}{\sigma^{11}}}\, y_2$$

which has the Jacobian

$$\left|\frac{\partial(y_1, y_2)}{\partial(z_1, z_2)}\right| = \left|\frac{\partial(z_1, z_2)}{\partial(y_1, y_2)}\right|^{-1} = \frac{1}{\sqrt{|\sigma^{ij}|}}$$

we find that the left-hand side of (7.3.4) reduces to

$$\frac{1}{\sqrt{|\sigma^{ij}|}}\int_{-\infty}^{\infty}\int_{-\infty}^{\infty} e^{-\frac{1}{2}(z_1^2+z_2^2)}\, dz_1\, dz_2.$$

But we know from (7.2.3) and (7.2.3*a*) that this double integral has value 2π. Therefore (7.3.4) is established.

To show that μ_1 and μ_2 are the means of x_1 and x_2 note that by differentiating both sides of (7.3.3) with respect to μ_1 and μ_2 we obtain

$$\mathscr{E}[\sigma^{11}(x_1 - \mu_1) + \sigma^{12}(x_2 - \mu_2)] = 0$$
$$\mathscr{E}[\sigma^{21}(x_1 - \mu_1) + \sigma^{22}(x_2 - \mu_2)] = 0$$

which can be written as

(7.3.7) $$\sigma^{11}\mathscr{E}(x_1 - \mu_1) + \sigma^{12}\mathscr{E}(x_2 - \mu_2) = 0$$
$$\sigma^{21}\mathscr{E}(x_1 - \mu_1) + \sigma^{22}\mathscr{E}(x_2 - \mu_2) = 0.$$

These are homogeneous linear equations in $\mathscr{E}(x_1 - \mu_1)$ and $\mathscr{E}(x_2 - \mu_2)$ and, since $|\sigma^{ij}| \neq 0$, will have the unique solution

(7.3.8) $$\mathscr{E}(x_1 - \mu_1) = 0, \qquad \mathscr{E}(x_2 - \mu_2) = 0.$$

But formulas (7.3.8) imply that

(7.3.9) $$\mu_1 = \mathscr{E}(x_1), \qquad \mu_2 = \mathscr{E}(x_2)$$

that is, *μ_1 and μ_2 are the means of x_1 and x_2 respectively.*

Now consider the variances and covariances. If we differentiate both sides of (7.3.4) with respect to σ^{11} and then multiply both sides by $-\sqrt{|\sigma^{ij}|}/\pi$ we obtain

$$\frac{\sqrt{|\sigma^{ij}|}}{2\pi}\int_{-\infty}^{\infty}\int_{-\infty}^{\infty}(x_1 - \mu_1)^2 e^{-\frac{1}{2}Q(x_1,x_2)}\,dx_1\,dx_2 = \frac{\sigma^{22}}{|\sigma^{ij}|}.$$

But the left-hand side is the expression for $\mathscr{E}(x_1 - \mu_1)^2$ the variance of x_1. Denoting this variance by σ_{11}, we have

$$\sigma_{11} = \frac{\sigma^{22}}{|\sigma^{ij}|}.$$

Denoting the covariance between x_1 and x_2 by σ_{12} and the variance of x_2 by σ_{22} we obtain, in a similar manner,

$$\sigma_{12} = -\frac{\sigma^{12}}{|\sigma^{ij}|}, \qquad \sigma_{22} = \frac{\sigma^{11}}{|\sigma^{ij}|}.$$

Thus, we find the *covariance matrix* $\|\sigma_{ij}\|$, expressed in terms of the parameters σ^{11}, σ^{12}, σ^{22} in (7.3.2), to be

(7.3.10) $$\|\sigma_{ij}\| = \|\sigma^{ij}\|^{-1}.$$

Hence, we have the following result:

7.3.1 *The constants μ_1 and μ_2 in (7.3.2) are the means of x_1 and x_2, while the matrix $\|\sigma^{ij}\|$ is the inverse of the covariance matrix $\|\sigma_{ij}\|$ of x_1 and x_2.*

If we write the variances of x_1 and x_2 as σ_1^2 and σ_2^2 and the covariance as $\sigma_1\sigma_2\rho$, where ρ is the correlation coefficient between x_1 and x_2, then the bivariate normal p.d.f. (7.3.1) may be written in the commonly used alternative form

(7.3.11) $$f(x_1, x_2) = \frac{1}{2\pi\sigma_1\sigma_2\sqrt{1-\rho^2}} e^{-\frac{1}{2}Q(x_1,x_2)}$$

where

(7.3.12)
$$Q(x_1, x_2) = \frac{1}{(1-\rho^2)}\left[\frac{(x_1-\mu_1)^2}{\sigma_1^2} + \frac{(x_2-\mu_2)^2}{\sigma_2^2} - 2\rho\frac{(x_1-\mu_1)(x_2-\mu_2)}{\sigma_1\sigma_2}\right].$$

It should be noted that the normal p.d.f. is constant on any ellipse of the form $Q(x_1, x_2) =$ constant, and that the greatest value of $f(x_1, x_2)$ occurs at the *center of gravity* of the distribution, namely, (μ_1, μ_2).

Now let us determine the characteristic function of (x_1, x_2), that is,

(7.3.13) $$\varphi(t_1, t_2) = \int_{-\infty}^{\infty}\int_{-\infty}^{\infty} e^{it_1x_1+it_2x_2} f(x_1, x_2)\, dx_1\, dx_2$$

Again letting $y_1 = x_1 - \mu_1$ and $y_2 = x_2 - \mu_2$ we find that

(7.3.14) $$\varphi(t_1, t_2) = \frac{e^{i\mu_1t_1+i\mu_2t_2}}{2\pi\sqrt{|\sigma_{ij}|}}\int_{-\infty}^{\infty}\int_{-\infty}^{\infty} e^{-\frac{1}{2}Q'(y_1,y_2)} dy_1\, dy_2$$

where $Q'(y_1, y_2) = \sigma^{11}y_1^2 + \sigma^{22}y_2^2 + 2\sigma^{12}y_1y_2 - 2it_1y_1 - 2it_2y_2$.

But $Q'(y_1, y_2)$ can be written as

(7.3.15) $$Q'(y_1, y_2) = z_1^2 + z_2^2 + (\sigma_{11}t_1^2 + 2\sigma_{12}t_1t_2 + \sigma_{22}t_2^2)$$

where

$$z_1 = \sqrt{\sigma^{11}}\left(y_1 + \frac{\sigma^{12}}{\sigma^{11}}y_2 - \frac{it_1}{\sigma^{11}}\right)$$

$$z_2 = \sqrt{\frac{|\sigma^{ij}|}{\sigma^{11}}}\left(y_2 + \frac{i\sigma^{12}t_1 - i\sigma^{11}t_2}{|\sigma^{ij}|}\right).$$

The Jacobian of the transformation by which the y's and z's are related is

$$\left|\frac{\partial(y_1, y_2)}{\partial(z_1, z_2)}\right| = 1/\sqrt{|\sigma^{ij}|}.$$

Utilizing (7.2.3) and (7.2.3*a*), we find from (7.3.14) that

7.3.2 *The characteristic function of the bivariate normal distribution* (7.3.1) [*or* (7.3.11)] *is*

(7.3.16) $$\varphi(t_1, t_2) = \exp\left[i(\mu_1t_1 + \mu_2t_2) - \tfrac{1}{2}(\sigma_{11}t_1^2 + \sigma_{22}t_2^2 + 2\sigma_{12}t_1t_2)\right].$$

The reader can verify that the bivariate normal distribution is reproduced with respect to the vector of means and the covariance matrix.

We should remark that z_1 and z_2 are both complex variables, and that each of the integrals $\int e^{-\frac{1}{2}z_1^2}\,dz_1$ and $\int e^{-\frac{1}{2}z_2^2}\,dz_2$, obtained when the y's in (7.3.14) are transformed to z's, is actually taken along a line in the complex plane parallel to the real axis. But each integral has the same value as if taken along the real axis.

It should be particularly noted that the matrix of the quadratic form in t_1 and t_2 in the characteristic function is the covariance matrix $\|\sigma_{ij}\|$ while the matrix of the quadratic form in x_1 and x_2 in the p.d.f. (7.3.1) is the inverse of the covariance matrix, namely $\|\sigma^{ij}\|$.

The reader will find it instructive to verify by applying the usual differentiation procedure to (7.3.16) that the means of x_1 and x_2 are μ_1 and μ_2 and the covariance matrix of x_1 and x_2 is $\|\sigma_{ij}\|$.

If we put $t_2 = 0$ in (7.3.16), we obtain the characteristic function of x_1 namely,

(7.3.17) $$\varphi(t_1, 0) = e^{i\mu_1 t_1 - \frac{1}{2}\sigma_{11}t_1^2}$$

which by **7.2.1** implies that

7.3.3 *The marginal distribution of x_1 in the distribution $N(\{\mu_i\}, \|\sigma_{ij}\|)$, $i, j = 1, 2$, is the distribution $N(\mu_1, \sigma_{11})$.*

A similar statement holds, of course, for the marginal distribution of x_2.

More generally, if $L = c_1x_1 + c_2x_2$ where c_1 and c_2 are not both zero, the characteristic function of L is given by

(7.3.18)
$$\varphi(c_1t, c_2t) = \exp\left[i(c_1\mu_1 + c_2\mu_2)t - \tfrac{1}{2}(\sigma_{11}c_1^2 + 2\sigma_{12}c_1c_2 + \sigma_{22}c_2^2)t^2\right]$$

and hence we have the following result:

7.3.4 *If x_1 and x_2 are random variables having the distribution $N(\{\mu_i\}, \|\sigma_{ij}\|)$ $i, j = 1, 2$, then $L = c_1x_1 + c_2x_2$ has the distribution*

$$N\left(\sum_{i=1}^{2} c_i\mu_i, \sum_{i,j=1}^{2} \sigma_{ij}c_ic_j\right).$$

Note that **7.2.3** is a special case of **7.3.4** obtained by putting the covariance $\sigma_{12} = 0$ and denoting σ_{11} and σ_{22} by σ_1^2 and σ_2^2 respectively.

Now consider the conditional random variable $x_2 \mid x_1$, where (x_1, x_2) has a bivariate normal distribution. The marginal p.d.f. of x_1 is

(7.3.19) $$f_1(x_1) = \frac{1}{\sqrt{2\pi}\sigma_1} e^{-\frac{1}{2}(x_1-\mu_1)^2/\sigma_1^2}$$

where $\sigma_1^2 = \sigma_{11} = \sigma^2(x_1)$.

Using form (7.3.11) of the bivariate normal p.d.f. we have from (2.9.13)

$$f(x_2 \mid x_1) = \frac{1}{\sqrt{2\pi}\sigma_2\sqrt{1-\rho^2}} e^{-\frac{1}{2}Q(x_1, x_2) + \frac{1}{2}(x_1 - \mu_1)^2/\sigma_1^2}$$

where $Q(x_1, x_2)$ is given by (7.3.12). After some algebraic simplification, we find that

(7.3.20) $f(x_2 \mid x_1)$

$$= \frac{1}{\sqrt{2\pi}\sigma_2\sqrt{1-\rho^2}} \exp\left\{-\frac{1}{2\sigma_2^2(1-\rho^2)}\left[(x_2 - \mu_2) - \rho\frac{\sigma_2}{\sigma_1}(x_1 - \mu_1)\right]^2\right\}$$

from which we can make the following statement:

7.3.5 *In the bivariate normal distribution having p.d.f.* (7.3.11) *the conditional random variable* $x_2 \mid x_1$ *has the distribution* $N\left(\mu_2 + \rho\frac{\sigma_2}{\sigma_1}(x_1 - \mu_1), \sigma_2^2(1-\rho^2)\right)$.

It can be seen from considerations of symmetry that a similar statement is true for the conditional distribution of $x_1 \mid x_2$.

It should be noted that the mean value of $x_2 \mid x_1$, that is, the regression function of x_2 on x_1 in (7.3.20), is a linear function of x_1, namely,

$$\mu(x_2 \mid x_1) = \mu_2 + \rho\frac{\sigma_2}{\sigma_1}(x_1 - \mu_1).$$

Hence the equation of the *regression line of* x_2 *on* x_1 is

$$x_2 = \mu_2 + \rho\frac{\sigma_2}{\sigma_1}(x_1 - \mu_1) \qquad (7.3.21)$$

which, it should be observed, is exactly the same as the least squares regression line of x_2 on x_1 given by (3.8.6). A similar statement holds regarding the regression function of x_1 on x_2. This substantiates a remark made in Section 3.8(*a*) which can now be stated more precisely as follows:

7.3.6 *The bivariate normal distribution has the property that its two regression functions are linear and are identically the same as the regression lines provided by the method of least squares. Furthermore, the variances of the conditional random variables* $x_2 \mid x_1$ *and* $x_1 \mid x_2$ *are identically the same as the least squares residual variances* $\sigma_{2\cdot 1}^2$ *and* $\sigma_{1\cdot 2}^2$ *respectively, defined in* (3.8.8).

7.4 THE *k*-VARIATE NORMAL DISTRIBUTION

(a) Structure of the p.d.f. of the *k*-Variate Normal Distribution

The multivariate or *k*-variate normal distribution and its properties are straightforward extensions of the case for $k = 2$ treated in Section 7.3.

The p.d.f. of the k-variate normal distribution is

$$f(x_1, \ldots, x_k) = \frac{\sqrt{|\sigma^{ij}|}}{(2\pi)^{\frac{1}{2}k}} e^{-\frac{1}{2}Q(x_1, \ldots, x_k)} \tag{7.4.1}$$

where $(x_1, \ldots, x_k)$ is any point in R_k, and

$$Q(x_1, \ldots, x_k) = \sum_{i,j=1}^{k} \sigma^{ij}(x_i - \mu_i)(x_j - \mu_j) \tag{7.4.2}$$

$\|\sigma^{ij}\|$ being the *inverse of the covariance matrix* $\|\sigma_{ij}\|$ and μ_i the *means* of the x_i.

We assume that $\|\sigma^{ij}\|$ is positive definite, which implies that $|\sigma^{ij}| \neq 0$, and hence $|\sigma_{ij}| \neq 0$. The distribution having p.d.f. (7.4.1) will be called the *k-variate normal distribution* and will be denoted by $N(\{\mu_i\}, \|\sigma_{ij}\|)$, $i, j = 1, \ldots, k$. The sample space of the random variable $(x_1, \ldots, x_k)$ is the entire k-dimensional space R_k.

If we let $y_i = x_i - \mu_i$, $i = 1, \ldots, k$ verification that the integral of $(x_1, \ldots, x_k)$ over R_k equals unity is equivalent to showing that

$$\int_{R_k} \exp\left(-\tfrac{1}{2}\sum_{i,j=1}^{k} \sigma^{ij}y_iy_j\right) dy_1 \cdots dy_k = \frac{(2\pi)^{\frac{1}{2}k}}{\sqrt{|\sigma^{ij}|}}. \tag{7.4.3}$$

We can write

$$\sum_{i,j=1}^{k} \sigma^{ij}y_iy_j = \sigma^{11}\left(y_1 + \frac{\sum_{j=2}^{k}\sigma^{1j}y_j}{\sigma^{11}}\right)^2 + \sum_{i,j=2}^{k}\left(\sigma^{ij} - \frac{\sigma^{1i}\sigma^{1j}}{\sigma^{11}}\right)y_iy_j.$$

Letting

$$z_1 = \sqrt{\sigma^{11}}\left(y_1 + \frac{1}{\sigma^{11}}\sum_{j=2}^{k}\sigma^{1j}y_j\right)$$

$$\sigma_{(1)}^{ij} = \sigma^{ij} - \frac{\sigma^{1i}\sigma^{1j}}{\sigma^{11}}, \qquad i, j = 2, \ldots, k$$

we have

$$\sum_{i,j=1}^{k} \sigma^{ij}y_iy_j = z_1^2 + \sum_{i,j=2}^{k} \sigma_{(1)}^{ij}y_iy_j.$$

Continuing this process and setting up the notation

$$\sigma_{(p)}^{ij} = \sigma_{(p-1)}^{ij} - \frac{\sigma_{(p-1)}^{pi}\sigma_{(p-1)}^{pj}}{\sigma_{(p-1)}^{pp}}, \quad i, j = p+1, \ldots, k;\ p = 1, \ldots, k-1$$

with

$$\sigma_{(0)}^{ij} = \sigma^{ij}, \qquad i, j = 1, \ldots, k,$$

we can write

$$\sum_{i,j=1}^{k} \sigma^{ij}y_iy_j = \sum_{i=1}^{k} z_i^2$$

where

(7.4.4) $$z_i = \sqrt{\sigma^{ii}_{(i-1)}}\left(y_i + \frac{1}{\sigma^{ii}_{(i-1)}} \sum_{j=i+1}^{k} \sigma^{ij}_{(i-1)} y_j\right), \qquad i = 1, \ldots, k.$$

Since $\|\sigma^{ij}\|$ is a positive definite matrix, it can be verified that the quantities $\sigma^{11}, \sigma^{22}_{(1)}, \ldots, \sigma^{kk}_{(k-1)}$ are all positive.

The procedure we have just outlined actually exhibits a linear transformation for reducing the positive definite quadratic form in the exponent of (7.4.3) to a sum of squares and is known as Lagrange's method. There is, of course, a family of linear transformations which will yield such a reduction.

The Jacobian of the transformation (7.4.4) is

$$\left|\frac{\partial(y_1, \ldots, y_k)}{\partial(z_1, \ldots, z_k)}\right| = 1 \Big/ \left|\frac{\partial(z_1, \ldots, z_k)}{\partial(y_1, \ldots, y_k)}\right| = \frac{1}{\sqrt{\sigma^{11}\sigma^{22}_{(1)} \cdots \sigma^{kk}_{(k-2)}}}$$

and hence the left side of (7.4.3) reduces to

(7.4.5) $$\frac{1}{\sqrt{\sigma^{11}\sigma^{22}_{(1)} \cdots \sigma^{kk}_{(k-1)}}} \int_{-\infty}^{\infty} \cdots \int_{-\infty}^{\infty} \exp\left(-\tfrac{1}{2}\sum_{i=1}^{k} z_i^2\right) dz_1 \cdots dz_k.$$

By making use of (7.2.3*a*) it is seen that the k-fold integral in (7.4.5) has the value $(2\pi)^{\frac{1}{2}k}$. Our problem, therefore, reduces to showing that

(7.4.6) $$\sigma^{11}\sigma^{22}_{(1)} \cdots \sigma^{kk}_{(k-1)} = |\sigma^{ij}|.$$

To establish (7.4.6), we evaluate the determinant $|\sigma^{ij}|$ as follows:

(7.4.7) $$|\sigma^{ij}| = \begin{vmatrix} \sigma^{11} & \sigma^{12} & \cdots & \sigma^{1k} \\ \sigma^{21} & \sigma^{22} & \cdots & \sigma^{2k} \\ \cdot & \cdot & \cdot & \cdot \\ \cdot & \cdot & \cdot & \cdot \\ \cdot & \cdot & \cdot & \cdot \\ \sigma^{k1} & \sigma^{k2} & \cdots & \sigma^{kk} \end{vmatrix} = \sigma^{11} \begin{vmatrix} 1 & \sigma^{12} & \cdots & \sigma^{1k} \\ \dfrac{\sigma^{21}}{\sigma^{11}} & \sigma^{22} & \cdots & \sigma^{2k} \\ \cdot & \cdot & \cdot & \cdot \\ \cdot & \cdot & \cdot & \\ \cdot & \cdot & \cdot & \cdot \\ \dfrac{\sigma^{k1}}{\sigma^{11}} & \sigma^{k2} & \cdots & \sigma^{kk} \end{vmatrix}.$$

Multiplying the first column of the determinant on the extreme right by σ^{12} and subtracting from the second column, the first column by σ^{13} and subtracting from the third, and so on, we obtain

(7.4.8) $$|\sigma^{ij}| = \sigma^{11} \begin{vmatrix} \sigma^{22}_{(1)} & \sigma^{23}_{(1)} & \cdots & \sigma^{2k}_{(1)} \\ \sigma^{32}_{(1)} & \sigma^{33}_{(1)} & \cdots & \sigma^{3k}_{(1)} \\ \cdot & \cdot & & \cdot \\ \cdot & \cdot & & \cdot \\ \cdot & \cdot & & \cdot \\ \sigma^{k2}_{(1)} & \sigma^{k3}_{(1)} & \cdots & \sigma^{kk}_{(1)} \end{vmatrix}.$$

Continuing the process described above, we find

$$|\sigma^{ij}| = \sigma^{11}\sigma^{22}_{(1)} \cdots \sigma^{kk}_{(k-1)}$$

which establishes (7.4.6).

To see that the μ_i are actually the means of the random variables having p.d.f. (7.4.1), we proceed as in the two-dimensional case, by taking the first partial derivatives of

$$\frac{\sqrt{|\sigma^{ij}|}}{(2\pi)^{\frac{1}{2}k}} \int_{R_k} e^{-\frac{1}{2}Q(x_1, \ldots, x_k)} \, dx_1 \cdots dx_k = 1$$

with respect to μ_i, $i = 1, \ldots, k$. This gives the following equations:

$$\mathscr{E}\left[\sum_{j=1}^{k} \sigma^{ij}(x_j - \mu_j)\right] = 0, \qquad i = 1, \ldots, k$$

which can be expressed as

$$\sum_{j=1}^{k} \sigma^{ij}\mathscr{E}(x_j - \mu_j) = 0. \tag{7.4.9}$$

Since we have assumed $\|\sigma_{ij}\|$ and hence $\|\sigma^{ij}\|$ to be positive definite, we have $|\sigma^{ij}| \neq 0$, and hence the only solution of (7.4.9) is

$$\mathscr{E}(x_i - \mu_i) = 0, \quad \text{or} \quad \mathscr{E}(x_i) = \mu_i, \quad i = 1, \ldots, k, \tag{7.4.10}$$

that is, $\mu_1, \ldots, \mu_k$ in (7.4.1) *are the means of* $x_1, \ldots, x_k$, *respectively*.

To verify that $\|\sigma^{ij}\|$ is the inverse of the covariance matrix, we use the relation

$$\int_{R_k} e^{-\frac{1}{2}Q(x_1, \ldots, x_k)} \, dx_1 \cdots dx_k = \frac{(2\pi)^{\frac{1}{2}k}}{\sqrt{|\sigma^{ij}|}}. \tag{7.4.11}$$

Differentiating both sides of (7.4.11) with respect to σ^{ij}, and then multiplying by $-(1 + \delta_{ij}) \dfrac{\sqrt{|1\sigma^{ij}|}}{(2\pi)^{\frac{1}{2}k}}$, where δ_{ij} is the Kronecker δ, we obtain

$$\mathscr{E}[(x_i - \mu_i)(x_j - \mu_j)] = \sigma_{ij}, \tag{7.4.12}$$

a formula which actually holds for $i = j$ as well as $i \neq j$. Therefore, we have the following result:

7.4.1 *The constants* $\mu_1, \ldots, \mu_k$ *of the quadratic form in* (7.4.1) *are the means of* $x_1, \ldots x_k$, *while the matrix* $\|\sigma^{ij}\|$ *of the quadratic form is the inverse of the covariance matrix* $\|\sigma_{ij}\|$.

If one denotes the variance of x_i by σ_i^2 and the covariance between x_i and x_j by $\sigma_i\sigma_j\rho_{ij}$, then (7.4.1) can, of course, be written in a form, although cumbersome, which is the k-variate analogue of (7.3.12). But this is left as an exercise for the reader.

(b) Characteristic Function of the k-Variate Normal Distribution

Now let us consider the characteristic function of the k-variate normal distribution. We have

$$(7.4.13)\quad \varphi(t_1, \ldots, t_k) = \exp\left\{i\sum_{i=1}^{k}\mu_i t_i\right\}\cdot\frac{\sqrt{|\sigma^{ij}|}}{(2\pi)^{\frac{1}{2}k}}$$

$$\cdot\int_{R_k}\exp\left\{-\frac{1}{2}\left[Q(x_1, \ldots, x_k) - 2i\sum_{i=1}^{k}(x_i - \mu_i)t_i\right]\right\}dx_1\cdots dx_k$$

where $Q(x_1, \ldots, x_k)$ is given by (7.4.2).

To evaluate the integral (7.4.13) we carry out the k-variate version of "completing the square" which was performed in passing from (7.2.4) to (7.2.4*a*). For this purpose, putting $y_i = x_i - \mu_i$, $i = 1, \ldots, k$, we observe that

$$(7.4.14)\quad \sum_{i,j=1}^{k}\sigma^{ij}\left[y_i - i\sum_{h=1}^{k}\sigma_{ih}t_h\right]\left[y_j - i\sum_{g=1}^{k}\sigma_{gj}t_g\right]$$

$$\equiv \sum_{i,j=1}^{k}\sigma^{ij}y_iy_j - i\sum_{i,j,h=1}^{k}\sigma^{ij}\sigma_{ih}t_hy_j - i\sum_{i,j,g=1}^{k}\sigma^{ij}\sigma_{gj}t_gy_i - \sum_{i,j,g,h=1}^{k}\sigma^{ij}\sigma_{ih}\sigma_{gj}t_gt_h.$$

Since $\sigma^{ij} = \sigma^{ji}$ and, of course, $\sigma_{ij} = \sigma_{ji}$, the quantities $\sum_{i=1}^{k}\sigma^{ij}\sigma_{ih}$ and $\sum_{j=1}^{k}\sigma^{ij}\sigma_{gj}$ are Kronecker deltas δ_{jh} and δ_{ig}, respectively, and hence the second member of (7.4.14) reduces to

$$Q(x_1, \ldots, x_k) - 2i\sum_{i=1}^{k}t_iy_i - \sum_{i,j=1}^{k}\sigma_{ij}t_it_j.$$

Therefore

$$(7.4.15)\quad Q(x_1, \ldots, x_k) - 2i\sum_{i=1}^{k}t_iy_i = Q'(x_1, \ldots, x_k) + \sum_{i,j=1}^{k}\sigma_{ij}t_it_j$$

where $Q'(x_1, \ldots, x_k)$ is the quadratic form which constitutes the first member of (7.4.14). We may therefore write (7.4.13) as

$$(7.4.16)\quad \varphi(t_1, \ldots, t_k) = \exp\left(i\sum_{i=1}^{k}\mu_it_i - \tfrac{1}{2}\sum_{i,j=1}^{k}\sigma_{ij}t_it_j\right)\cdot H$$

where

$$H = \frac{\sqrt{|\sigma^{ij}|}}{(2\pi)^{\frac{1}{2}k}}\int_{R_k}e^{-\frac{1}{2}Q'(x_1, \ldots, x_k)}\,dx_1\cdots dx_k.$$

$Q'(x_1, \ldots, x_k)$ is a quadratic form with matrix $\|\sigma^{ij}\|$ in the complex variables $(x_i - \mu_i) + iB_i$, $i = 1, \ldots, k$, where the B_i are real, whereas $Q(x_1, \ldots, x_k)$ is the quadratic form obtained by setting the $B_i = 0$ in $Q'(x_1, \ldots, x_k)$. Thus, if we denote $x_i - \mu_i + iB_i$ by y_i', it is clear that

$Q'(x_1, \ldots, x_k)$ will be reduced to a sum of squares $\sum_{i=1}^{k} z_i^2$ by the transformation (7.4.4) with y_i replaced by y_i'. The z_i in this case will, of course, be complex variables, and the integration in the case of each z_i is parallel to the real axis of that complex variable. But as we have pointed out in the cases of the one- and two-dimensional normal distributions, such integrals have the same values as the integrals of the same functions along the corresponding real axes. Hence H has the same value as that obtained by replacing $Q'(x_1, \ldots, x_k)$ by $Q(x_1, \ldots, x_k)$, namely, 1. Therefore

7.4.2 *The characteristic function of the distribution* $N\{(\mu_i\}, \|\sigma_{ij}\|)$ $i, j = 1, \ldots, k$ is

$$\varphi(t_1, \ldots, t_k) = \exp\left(i\sum_{i=1}^{k} \mu_i t_i - \tfrac{1}{2}\sum_{i,j=1}^{k} \sigma_{ij} t_i t_j\right). \tag{7.4.17}$$

If in this characteristic function we put $t_{k_1+1} = \cdots = t_k = 0$ we have

$$\varphi(t_1, \ldots, t_{k_1}, 0, \ldots, 0) = \exp\left(i\sum_{i=1}^{k_1} \mu_i t_i - \tfrac{1}{2}\sum_{i,j=1}^{k_1} \sigma_{ij} t_i t_j\right). \tag{7.4.18}$$

Hence :

7.4.3 *If* $(x_1, \ldots, x_k)$ *is a vector random variable having the k-variate normal distribution* $N(\{\mu_i\}, \|\sigma_{ij}\|)$, $i, j = 1, \ldots, k$, *the marginal distribution of* $(x_1, \ldots, x_{k_1})$, $(k_1 < k)$, *is the* k_1*-variate normal distribution* $N(\{\mu_i\}, \|\sigma_{ij}\|)$, $i, j = 1, \ldots, k_1$.

(c) Distribution of Linear Functions of Normal Variables

If in (7.4.17) we put $t_i = c_i t$, $i = 1, \ldots, k$ where the c_i are not all zero, we obtain, as stated by (5.3.6), the c.f. $\varphi(t)$ of the linear function $L = \sum_{i=1}^{k} c_i x_i$ as follows:

$$\varphi(t) = \exp\left[i\left(\sum_{i=1}^{k} c_i\mu_i\right)t - \tfrac{1}{2}\left(\sum_{i=1}^{k} \sigma_{ij} c_i c_j\right)t^2\right] \tag{7.4.19}$$

and making use of **7.2.1** we have the following result:

7.4.4 *If* $(x_1, \ldots, x_k)$ *has the k-variate distribution* $N(\{\mu_i\}, \|\sigma_{ij}\|)$, $i, j, \ldots, k$, *then* $L = c_1x_1 + \cdots + c_kx_k$ *has the distribution*

$$N\left(\sum_{i=1}^{k} c_i\mu_i, \sum_{i,j=1}^{k} \sigma_{ij} c_i c_j\right).$$

Similarly, it can be shown that

7.4.5 *More generally, if* $L_p = c_{p1}x_1 + \cdots + c_{pk}x_k$, $p = 1, \ldots, s$, $s \leqslant k$, *are linearly independent random variables*, where $(x_1, \ldots, x_k)$ *has* *the distribution* $N(\{\mu_i\}, \|\sigma_{ij}\|)$, $i, j = 1, \ldots, k$, *then* $(L_1, \ldots, L_s)$ *have the s-dimensional distribution*

$$N\left(\left\{\sum_{i=1}^{k} c_{pi}\mu_i\right\}, \left\|\sum_{i,j=1}^{k} \sigma_{ij}c_{pi}c_{qj}\right\|\right), \quad p, q = 1, \ldots, s.$$

(d) Conditional Distributions from k-Variable Normal Distribution

Finally, we consider the p.d.f. of the conditional random variable $x_k \mid x_1, \ldots, x_{k-1}$ in the k-variable normal distribution. By definition

$$f(x_k \mid x_1, \ldots, x_{k-1}) = \frac{f(x_1, \ldots, x_k)}{f_{12\cdots(k-1)}(x_1, \ldots, x_{k-1})} \tag{7.4.20}$$

where $f(x_1, \ldots, x_k)$ is the p.d.f. (7.4.1) and $f_{12\cdots(k-1)}(x_1, \ldots, x_{k-1})$ is the marginal p.d.f. of $(x_1, \ldots, x_{k-1})$, that is,

$$f_{12\cdots(k-1)}(x_1, \ldots, x_{k-1}) = \frac{\sqrt{|\sigma_{(k)}^{pq}|}}{(2\pi)^{\frac{1}{2}(k-1)}} \exp\left[-\frac{1}{2}\sum_{p,q=1}^{k-1} \sigma_{(k)}^{pq}(x_p - \mu_p)(x_q - \mu_q)\right]. \tag{7.4.21}$$

where $\|\sigma_{(k)}^{pq}\|$ is the inverse of the matrix $\|\sigma_{pq}\|$, $p, q = 1, \ldots, k-1$. Note that in the case of the matrix $\|\sigma_{ij}\|$ and its inverse $\|\sigma^{ij}\|$, we have $i, j = 1, \ldots, k$.

Now putting $y_i = x_i - \mu_i$, $i = 1, \ldots, k$, it can be verified that

$$\sum_{i,j=1}^{k} \sigma^{ij}y_iy_j \equiv \sigma^{kk}\left[y_k - \sum_{p=1}^{k-1} \beta_p^* y_p\right]^2 + \sum_{p,q=1}^{k-1} \sigma_{(k)}^{pq}y_py_q \tag{7.4.22}$$

where

$$\beta_p^* = \sum_{q=1}^{k-1} \sigma_{(k)}^{pq}\sigma_{qk}, \qquad p = 1, \ldots, k-1. \tag{7.4.23}$$

Also we have

$$\frac{|\sigma^{ij}|}{|\sigma_{(k)}^{pq}|} = \frac{|\sigma_{pq}|}{|\sigma_{ij}|} = \sigma^{kk} \tag{7.4.24}$$

where $i, j = 1, \ldots, k$ and $p, q = 1, \ldots, k-1$.

Substituting the expressions for $f(x_1, \ldots, x_k)$ and for $f_{1\cdots(k-1)}(x_1, \ldots, x_{k-1})$ into (7.4.20) and making use of (7.4.22) and (7.4.24), we obtain

$$f(x_k \mid x_1, \ldots, x_{k-1}) = \frac{\sqrt{\sigma^{kk}}}{\sqrt{2\pi}} \exp\left\{-\frac{\sigma^{kk}}{2}\left[(x_k - \mu_k) - \sum_{p=1}^{k-1} \beta_p^*(x_p - \mu_p)\right]^2\right\}. \tag{7.4.25}$$

Therefore, we have the following result:

7.4.6 *If* $(x_1, \ldots, x_k)$ *is a vector random variable having the k-variate distribution* $N(\{\mu_i\}, \|\sigma_{ij}\|)$, $i, j = 1, \ldots, k$, *then the distribution of the conditional random variable* $x_k \mid x_1, \ldots, x_{k-1}$ *is*

$$N\left(\mu_k + \sum_{p=1}^{k-1} \beta_p^*(x_p - \mu_p), \frac{1}{\sigma^{kk}}\right) \tag{7.4.26}$$

where the β_p^* *are given by* (7.4.23).

It should be noted that $\mu(x_k \mid x_1, \ldots, x_{k-1})$ from (7.4.25) is linear in $x_1, \ldots, x_{k-1}$ and is identical with the least squares regression plane of x_k on $x_1, \ldots, x_{k-1}$ given by expression (3.8.18). Furthermore, $\sigma^2(x_k \mid x_1, \ldots, x_{k-1})$ in (7.4.25) has a value, namely, $1/\sigma^{kk}$, which is identical with that of the least squares residual variance of x_k on $x_1, \ldots, x_{k-1}$ as given in (3.8.25).

Thus, we have established the important fact that

7.4.7 *In the case of a k-variate normal distribution the mean and variance of the conditional random variable* $x_k \mid x_1, \ldots, x_{k-1}$ *are identical, respectively, with the least squares regression function and residual variance of* x_k *on* $x_1, \ldots, x_{k-1}$.

7.5 THE GAMMA DISTRIBUTION

In general, the *gamma function* $\Gamma(g)$ is defined for complex numbers g, whose real part is positive, by the definite integral

$$\Gamma(g) = \int_0^\infty x^{g-1} e^{-x}\, dx. \tag{7.5.1}$$

Actually, we shall be interested only in real and positive values of g.

If we integrate by parts it follows that

$$\Gamma(g) = (g - 1)\Gamma(g - 1).$$

from which it is evident that if g is a positive integer

$$\Gamma(g) = (g - 1)!.$$

If $g > 0$, but not an integer, we have

$$\Gamma(g) = (g - 1)(g - 2) \cdots \delta\Gamma(\delta)$$

where $0 < \delta < 1$. In particular, $\Gamma(\frac{1}{2}) = \sqrt{\pi}$. This can be verified by noting that $I = \sqrt{2}\Gamma(\frac{1}{2})$, where I is defined in (7.2.3).

In problems of statistical theory, the most important values of g, as we shall see in later chapters, are (positive) multiples of $\frac{1}{2}$.

A distribution function which arises frequently in problems dealing with sums of squares and quadratic forms of random variables having normal distributions is the *gamma distribution* which for $\mu > 0$ has p.d.f.

$$f(x) = \frac{x^{\mu-1}e^{-x}}{\Gamma(\mu)} \tag{7.5.2}$$

for $x > 0$ and $f(x) = 0$ otherwise. It will be convenient to refer to a distribution having this p.d.f. as the *gamma distribution* $G(\mu)$. It is also called a *Pearson Type III distribution* (see Pearson (1906)).

The rth moment μ_r' of (7.5.2) is

$$\mu_r' = \frac{1}{\Gamma(\mu)}\int_0^\infty x^{\mu+r-1}e^{-x}\,dx = \frac{\Gamma(\mu+r)}{\Gamma(\mu)} \tag{7.5.3}$$

from which we find

$$\mathscr{E}(x) = \mu, \qquad \sigma^2(x) = \mu \tag{7.5.4}$$

that is, *the mean and variance of the gamma distribution are both equal to* μ. This, property, it will be recalled, is also true of the Poisson distribution (6.4.1), which, of course, is a discrete distribution.

By putting $\mu = p + 1$ in (7.5.2), the function

$$I(u, p) = \frac{1}{\Gamma(p+1)}\int_0^{u\sqrt{p+1}} x^p e^{-x}\,dx$$

is called the *incomplete gamma function*.

Karl Pearson (1922) has tabulated it for combinations of values of u and p, with u ranging from 0 to 12 and p ranging from 0 to 50.

Now consider the characteristic function of the gamma distribution $G(\mu)$,

$$\varphi(t) = \frac{1}{\Gamma(\mu)}\int_0^\infty x^{\mu-1}e^{-x(1-it)}\,dx. \tag{7.5.5}$$

Putting $x(1 - it) = y$, we see that

$$\varphi(t) = (1 - it)^{-\mu}. \tag{7.5.6}$$

If we regard μ as a parameter in the gamma distribution it is evident that the characteristic function satisfies (5.3.7) and hence

7.5.1 *The gamma distribution $G(\mu)$ is reproductive with respect to μ.*

As we have already stated, one of the most important applications of the gamma distribution arises in dealing with the problem of finding the distribution of certain quadratic forms of normally distributed random variables. In the general case of k normally distributed random variables having p.d.f. (7.4.1) the quadratic form in which we are interested is the one in the exponent of the p.d.f. itself, namely, $Q(x_1, \ldots, x_k)$, as defined by (7.4.2).

Consider the characteristic function of $\frac{1}{2}Q(x_1, \ldots, x_k)$. It is defined by

$$\varphi(t) = \frac{\sqrt{|\sigma^{ij}|}}{(2\pi)^{\frac{1}{2}k}}\int_{R_k} e^{-\frac{1}{2}(1-it)Q(x_1,\ldots,x_k)}\,dx_1\cdots dx_k. \tag{7.5.7}$$

If we make the transformation $y_i = \sqrt{1 - it}\,(x_i - \mu_i)$, $i = 1, \ldots, k$, (7.5.7) reduces to

$$(7.5.8)\qquad \varphi(t) = \frac{\sqrt{|\sigma^{ij}|}}{(2\pi)^{\frac{1}{2}k}}(1 - it)^{-\frac{1}{2}k}\int_{R_k} \exp\left(-\tfrac{1}{2}\sum_{i,j=1}^{k} \sigma^{ij}y_iy_j\right) dy_1 \cdots dy_k.$$

Making use of (7.4.3), we find that

$$(7.5.9)\qquad \varphi(t) = (1 - it)^{-\frac{1}{2}k}$$

which is the characteristic function of the gamma distribution $G(k/2)$. Therefore,

7.5.2 *If $(x_1, \ldots, x_k)$ is a vector random variable having the k-variate distribution $N(\{\mu_i\}, \|\sigma_{ij}\|)$, $i, j = 1, \ldots, k$, then $\frac{1}{2}Q(x_1, \ldots, x_k)$ has the gamma distribution $G(k/2)$.*

Another type of problem in which the gamma distribution arises relates to *continuous waiting-time distributions.* It will be recalled from Section 6.5 that (6.5.14) represents the probability of having to wait through x trials before obtaining k C's. Now suppose we think of C's as events spaced in time, and consider how long we must wait in order to obtain k C's. Let $(0, t)$ be the time interval during which exactly $k - 1$ C's occurred, and let $(t, t + \Delta t)$ be an increment of time during which the kth C occurs. If we cut up the interval $(0, t)$ into intervals each of length Δt, there will be $t/\Delta t$ of such intervals. Now suppose the probability of a C occurring in a specified time interval Δt is $\lambda\,\Delta t + o(\Delta t)$ where λ is a constant, namely, the *average number of C's per unit time.* Furthermore, assume that the occurrence of any number of C's in a time interval I is independent of the occurrence of any number of C's in a time interval I' which does not overlap I. We want the probability of having to wait through the time interval $(t, t + \Delta t)$ in order for the last one of k C's to occur. This probability is the product of the probability of $k - 1$ C's occurring among the $t/\Delta t$ intervals into which we have cut $(0, t)$ and the probability of the kth C occurring during $(t, t + \Delta t)$. Except for terms of order $(\Delta t)^2$ and higher, this probability is obtained from (6.5.15) by putting $x = t/\Delta t$ and $p = \lambda\,\Delta t$. This gives

$$(7.5.10)\qquad \binom{\dfrac{t}{\Delta t}}{k-1}(\lambda\,\Delta t)^k(1 - \lambda\,\Delta t)^{(t/\Delta t)-k+1}$$

$$= t\,\frac{(t - \Delta t)(t - 2\,\Delta t)\cdots(t - (k-2)\,\Delta t)}{(k-1)!}\,\lambda^{k-1}(1 - \lambda\,\Delta t)^{(t/\Delta t)-k+1}\lambda\,\Delta t.$$

Dividing by Δt and allowing $\Delta t \to 0$, we obtain the p.d.f.

$$f(t) = \frac{\lambda(\lambda t)^{k-1}e^{-\lambda t}}{(k-1)!}, \tag{7.5.11}$$

for $t > 0$ and $f(t) = 0$ otherwise.

This is the p.d.f. of the waiting time required in order to obtain the occurrence of exactly k C's, under the assumptions we have made.

It is clear from (7.5.11) that λt *is a random variable having the gamma distribution* $G(k)$.

Finally, it will be useful to state the following result, leaving verification to the reader.

7.5.3 *If x is a random variable having the rectangular distribution $R(\frac{1}{2}, 1)$, then the random variable $y = -\log x$ has the gamma distribution $G(1)$.*

7.6 THE BETA DISTRIBUTION

Another continuous distribution which occurs frequently in statistical theory is the distribution defined by the following p.d.f.

$$f(x) = \frac{\Gamma(\nu_1 + \nu_2)}{\Gamma(\nu_1)\Gamma(\nu_2)} x^{\nu_1 - 1}(1 - x)^{\nu_2 - 1} \tag{7.6.1}$$

for $0 < x < 1$, and $f(x) = 0$ elsewhere, and where ν_1 and ν_2 are positive and real. We shall refer to a distribution having p.d.f. (7.6.1) as the *beta distribution* $Be(\nu_1, \nu_2)$.

To verify that the integral of $f(x)$ over the interval (0, 1) is unity we proceed as follows. Making use of (7.5.1) we have

$$\Gamma(\nu_1)\Gamma(\nu_2) = \int_0^\infty \int_0^\infty x_1^{\nu_1 - 1} x_2^{\nu_2 - 1} e^{-x_1 - x_2}\, dx_1\, dx_2. \tag{7.6.2}$$

Applying the transformation

$$x_1 = r^2 \cos^2 \theta$$
$$x_2 = r^2 \sin^2 \theta$$

which has Jacobian

$$\left| \frac{\partial(x_1, x_2)}{\partial(r, \theta)} \right| = 4r^3 \sin\theta \cos\theta,$$

we obtain

$$\Gamma(\nu_1)\Gamma(\nu_2) = 4\int_0^{\frac{1}{2}\pi} \int_0^\infty (\cos\theta)^{2\nu_1 - 1}(\sin\theta)^{2\nu_2 - 1} r^{2\nu_1 + 2\nu_2 - 1} e^{-r^2}\, d\theta\, dr. \tag{7.6.3}$$

But applying the transformation $r = \sqrt{y}$ to

$$2\int_0^\infty r^{2\nu_1 + 2\nu_2 - 1} e^{-r^2}\, dr$$

we find that this integral reduces to the definite integral which defines $\Gamma(\nu_1 + \nu_2)$. Therefore (7.6.3) reduces to the following:

$$\text{(7.6.4)} \qquad \frac{\Gamma(\nu_1)\Gamma(\nu_2)}{\Gamma(\nu_1 + \nu_2)} = 2\int_0^{\frac{1}{2}\pi} (\cos\theta)^{2\nu_1 - 1}(\sin\theta)^{2\nu_2 - 1}\, d\theta.$$

Finally, by making the change of variable $\cos\theta = \sqrt{x}$ in (7.6.4) we obtain

$$\text{(7.6.5)} \qquad \frac{\Gamma(\nu_1)\Gamma(\nu_2)}{\Gamma(\nu_1 + \nu_2)} = \int_0^1 x^{\nu_1 - 1}(1 - x)^{\nu_2 - 1}\, dx$$

which completes verification of the fact that the integral of $f(x)$ in (7.6.1) over the interval (0, 1) is unity.

The c.d.f. $F(x)$ of the beta distribution $Be(\nu_1, \nu_2)$ designated as $I_x(\nu_1, \nu_2)$ and called the *incomplete beta function* has been tabulated under the direction of Karl Pearson (1934) in *The Tables of the Incomplete Beta Function* for $x = 0.01$ to 1.00 and for $\nu_1, \nu_2 = 0.05$ to 50.

The function of ν_1 and ν_2 defined by the definite integral on the right of (7.6.5) is called the *Beta Function* of ν_1 and ν_2 and is classically written as $B(\nu_1, \nu_2)$.

The beta distribution is another important example of a continuous distribution for which the characteristic function is awkward in determining moments. The simplest way to find the moments of (7.6.1) is by direct evaluation. Thus, if μ_r' is the rth moment of (7.6.1) we have

$$\text{(7.6.6)} \qquad \mu_r' = \frac{\Gamma(\nu_1 + \nu_2)}{\Gamma(\nu_1)\Gamma(\nu_2)} \int_0^1 x^{\nu_1 + r - 1}(1 - x)^{\nu_2 - 1}\, dx.$$

Using (7.6.5), we obtain the following expression for the rth moment

$$\text{(7.6.7)} \qquad \mu_r' = \frac{\Gamma(\nu_1 + \nu_2)\Gamma(\nu_1 + r)}{\Gamma(\nu_1 + \nu_2 + r)\Gamma(\nu_1)},$$

from which we find

$$\text{(7.6.8)} \qquad \mu(x) = \frac{\nu_1}{\nu_1 + \nu_2}, \qquad \sigma^2(x) = \frac{\nu_1\nu_2}{(\nu_1 + \nu_2)^2(\nu_1 + \nu_2 + 1)}.$$

There are various situations in which a beta distribution arises in mathematical statistics including the theory of order statistics and certain statistical tests. We shall discuss these situations in later chapters. One of the most important may be stated as follows:

7.6.1 *If x_1 and x_2 are independent random variables having gamma distributions $G(\nu_1)$ and $G(\nu_2)$ respectively, then the random variable*

$$\text{(7.6.9)} \qquad u = \frac{x_1}{x_1 + x_2}$$

has the beta distribution $Be(\nu_1, \nu_2)$.

Verification of this statement is straightforward. The p.e. of x_1 and x_2 is

$$\frac{1}{\Gamma(\nu_1)\Gamma(\nu_2)} x_1^{\nu_1-1}x_2^{\nu_2-1}e^{-x_1-x_2}\,dx_1\,dx_2. \tag{7.6.10}$$

If we apply the transformation

$$u = \frac{x_1}{x_1 + x_2} \tag{7.6.11}$$

$$v = x_2$$

to (7.6.10), we obtain as the p.e. of u and v

$$\frac{1}{\Gamma(\nu_1)\Gamma(\nu_2)} u^{\nu_1-1}v^{\nu_1+\nu_2-1}(1-u)^{-(\nu_1+1)}e^{-v/(1-u)}\,du\,dv. \tag{7.6.12}$$

The distribution of u is the marginal distribution of u in the distribution having p.e. (7.6.12). Integrating (7.6.12) with respect to v from 0 to ∞, we obtain the result that u has the beta distribution $Be(\nu_1, \nu_2)$.

Remarks. Before leaving the subject of beta distributions it will be useful to present two important formulas concerning gamma functions due to Legendre and Stirling, respectively. These formulas can be obtained by performing some elementary analysis on beta distributions.

Legendre's Duplication Formula for Gamma Functions

Suppose we put $\nu_1 = \nu_2 = \nu$ in (7.6.5). Then

$$\frac{\Gamma^2(\nu)}{\Gamma(2\nu)} = \int_0^1 x^{\nu-1}(1-x)^{\nu-1}\,dx = 2\int_0^{\frac{1}{2}} (x - x^2)^{\nu-1}\,dx. \tag{7.6.13}$$

Making the change of variable $y = 4(x - x^2)$ for $0 < x < \frac{1}{2}$, that is $x = (1 - \sqrt{1-y})/2$, we find

$$\frac{\Gamma^2(\nu)}{\Gamma(2\nu)} = 2^{1-2\nu}\int_0^1 y^{\nu-1}(1-y)^{-\frac{1}{2}}\,dy = 2^{1-2\nu}\frac{\Gamma(\nu)\Gamma(\frac{1}{2})}{\Gamma(\nu+\frac{1}{2})}. \tag{7.6.14}$$

Remembering that $\Gamma(\frac{1}{2}) = \sqrt{\pi}$, we obtain *Legendre's duplication formula for gamma functions*:

$$\Gamma(2\nu) = \frac{2^{2\nu-1}}{\sqrt{\pi}}\Gamma(\nu)\Gamma(\nu+\tfrac{1}{2}) \tag{7.6.15}$$

which we shall need in later chapters.

Stirling's Formula for Large Factorials

By using certain limit properties of a beta distribution we can establish by elementary analysis a highly useful approximation for $\Gamma(g)$ for large (real) values of g.

In (7.6.1) let $\nu_1 = g$ and $\nu_2 = n + 1$, and make the change of variable $x = y/n$. Then

$$\frac{\Gamma(g + n + 1)}{n^g\Gamma(g)\Gamma(n + 1)} \int_0^n y^{g-1}\left(1 - \frac{y}{n}\right)^n dy = 1 \tag{7.6.16}$$

for all values of n. Hence the limit of the left-hand side of (7.6.16) as $n \to \infty$ is 1. It is to be noted that the integrand in (7.6.16) converges uniformly to $y^{g-1}e^{-y}$ as $n \to \infty$ on any finite interval $(0, K)$. By taking n sufficiently large, we can make the difference

$$\int_0^{\min(n,K)} y^{g-1}\left(1 - \frac{y}{n}\right)^n dy - \int_0^K y^{g-1} e^{-y}\, dy$$

arbitrarily near zero. But K can be chosen at the outset so as to make

$$\int_K^\infty y^{g-1} e^{-y}\, dy$$

arbitrarily small. Hence

$$\lim_{n\to\infty} \int_0^n y^{g-1}\left(1 - \frac{y}{n}\right)^n dy = \int_0^\infty y^{g-1} e^{-y}\, dy = \Gamma(g). \tag{7.6.17}$$

Therefore

$$\lim_{n\to\infty} \frac{\Gamma(n + 1)n^g}{\Gamma(g + n + 1)} = 1.$$

which can be rewritten as

$$\Gamma(g) = \lim_{n\to\infty} \frac{n!\, n^g}{g(g + 1)\cdots(g + n)}. \tag{7.6.18}$$

Now let

$$S_n(g) = g \log n + \sum_{\alpha=1}^{n} \log \alpha - \sum_{\alpha=0}^{n} \log (g + \alpha). \tag{7.6.19}$$

Differentiating (7.6.19) with respect to g we find

$$S_n'(g) = \log n - \sum_{\alpha=0}^{n} \frac{1}{g + \alpha}. \tag{7.6.20}$$

Then

$$\lim_{n\to\infty} S_n(g) = \log \Gamma(g) \quad \text{and} \quad \lim_{n\to\infty} S_n'(g) = \frac{\Gamma'(g)}{\Gamma(g)}.$$

Now it will be seen from elementary calculus considerations that

$$-B_n(g) < -\sum_{\alpha=0}^{n} \frac{1}{g + \alpha} + \frac{1}{2g} + \frac{1}{2(g + n)} < -C_n(g) \tag{7.6.21}$$

where

$$B_n(g) = \tfrac{1}{2}\int_{\frac{1}{2}}^{n+\frac{1}{2}} \left(\frac{1}{g + x - 1} + \frac{1}{g + x}\right) dx, \qquad C_n(g) = \int_{\frac{1}{2}}^{n+\frac{1}{2}} \frac{dx}{g + x - \frac{1}{2}}.$$

Denoting $\log n - \frac{1}{2}(1/g + 1/(g + n))$ by $A_n(g)$, and adding $A_n(g)$ to all members of the inequality (7.6.21) we get

$$A_n(g) - B_n(g) < S_n'(g) < A_n(g) - C_n(g). \tag{7.6.22}$$

Evaluating $A_n(g)$, $B_n(g)$, and $C_n(g)$ and taking limits of (7.6.21) as $n \to \infty$, we obtain

(7.6.23) $$\log g - \frac{1}{2g} + \tfrac{1}{2}\log\left(1 - \frac{1}{4g^2}\right) < \frac{\Gamma'(g)}{\Gamma(g)} < \log - \frac{1}{2g}$$

from which it is evident that

(7.6.24) $$\frac{\Gamma'(g)}{\Gamma(g)} = \log g - \frac{1}{2g} + O\left(\frac{1}{g^2}\right).$$

Integrating with respect to g, we find

(7.6.25) $$\log \Gamma(g) = g \log g - g - \tfrac{1}{2}\log g + C + O\left(\frac{1}{g}\right)$$

where C is a constant independent of g. We can evaluate C by making use of formula (7.6.15). Replacing ν by g and taking logarithms, we have

(7.6.26)
$$\log \Gamma(2g) = (2g - 1)\log 2 + \log \Gamma(g) + \log \Gamma(g + \tfrac{1}{2}) - \tfrac{1}{2}\log \pi.$$

Replacing $\log \Gamma(2g)$, $\log \Gamma(g)$, and $\log \Gamma(g + \frac{1}{2})$ by the expressions which would be given for them by (7.6.25) and allowing $g \to \infty$, we find $C = \frac{1}{2}\log 2\pi$.
Finally, we obtain

(7.6.27) $$\Gamma(g) = \sqrt{2\pi}\, g^{g-\frac{1}{2}}\, e^{-g}\left(1 + O\left(\frac{1}{g}\right)\right).$$

Inequality (7.6.21) is not sharp enough to give us the coefficients of the various powers of $1/g$ in $O(1/g)$ in (7.6.27).

Actually $O(1/g) = 1/12g + 1/288g^2 + O(1/g^3)$, but to establish this, stronger methods of analysis are required than those used here, for example, see Whittaker and Watson (1927) or Cramér (1946). For large values of g, the asymptotic formula

(7.6.28) $$\Gamma(g) \cong \sqrt{2\pi}\, g^{g-\frac{1}{2}}\, e^{-g}$$

is sufficient for most purposes. If g is a positive integer, and since $\Gamma(g + 1) = g!$, we have *Stirling's formula for large factorials*

(7.6.29) $$g! \cong \sqrt{2\pi g}\, g^g\, e^{-g}.$$

7.7 THE DIRICHLET DISTRIBUTION

The k-variate analogue of (7.6.1) is the distribution having the p.d.f.

(7.7.1) $$f(x_1, \ldots, x_k) = \frac{\Gamma(\nu_1 + \cdots + \nu_{k+1})}{\Gamma(\nu_1)\cdots\Gamma(\nu_{k+1})} x_1^{\nu_1-1}\cdots x_k^{\nu_k-1}(1 - x_1 - \cdots - x_k)^{\nu_{k+1}-1},$$

at any point in the *simplex*: $S_k : \left\{(x_1, \ldots, x_k) : x_i \geqslant 0,\ i = 1, \ldots, k,\ \sum_{i=1}^{k} x_i \leqslant 1\right\}$, in R_k and zero outside, and where the ν_i are all real and positive. We shall refer to a distribution having p.d.f. (7.7.1) as the *k-variate Dirichlet distribution* $D(\nu_1, \ldots, \nu_k; \nu_{k+1})$. Note that if $k = 1$, $D(\nu_1; \nu_2)$ is identical with $Be(\nu_1, \nu_2)$, that is, (7.7.1) reduces to the p.d.f. of the beta distribution $Be(\nu_1, \nu_2)$. The Dirichlet distribution is basic to the probability theory of order statistics as we shall see in Section 8.7.

To verify that the integral of $f(x_1, \ldots, x_k)$ over the simplex S_k is unity we apply the transformation

$$\begin{aligned} x_1 &= \theta_1 \\ x_2 &= \theta_2(1 - x_1) \\ &\cdot \\ &\cdot \\ &\cdot \\ x_k &= \theta_k(1 - x_1 - \cdots - x_{k-1}) \end{aligned}$$

to (7.7.1), and obtain

$$(7.7.2)\quad f(x_1, \ldots, x_k)\, dx_1 \cdots dx_k$$
$$= \frac{\Gamma(\nu_1 + \cdots + \nu_{k+1})}{\Gamma(\nu_1) \cdots \Gamma(\nu_{k+1})} \theta_1^{\nu_1 - 1}(1 - \theta_1)^{\nu_2 + \cdots + \nu_{k+1} - 1}\theta_2^{\nu_2 - 1}$$
$$\cdot (1 - \theta_2)^{\nu_3 + \cdots + \nu_{k+1} - 1} \cdots$$
$$\cdot\, \theta_k^{\nu_k - 1}(1 - \theta_k)^{\nu_{k+1} - 1}\, d\theta_1 \cdots d\theta_k$$

where the range of the θ's is the *k-dimensional unit cube* $\{(\theta_1, \ldots, \theta_k)$: $0 \leqslant \theta_i \leqslant 1, i = 1, \ldots, k\}$. Making use of (7.6.5) we have

$$(7.7.3)\quad \int_{S_k} f(x_1, \ldots, x_k)\, dx_1 \cdots dx_k$$
$$= \frac{\Gamma(\nu_1 + \cdots + \nu_{k+1})}{\Gamma(\nu_1) \cdots \Gamma(\nu_{k+1})} \cdot \frac{\Gamma(\nu_1)\Gamma(\nu_2 + \cdots + \nu_{k+1})}{\Gamma(\nu_1 + \cdots + \nu_{k+1})} \cdots \frac{\Gamma(\nu_k)\Gamma(\nu_{k+1})}{\Gamma(\nu_k + \nu_{k+1})}$$

where it is to be noted that the right-hand side telescopes to unity.

The integral

$$(7.7.4)\quad \int_{S_k} x_1^{\nu_1 - 1} \cdots x_k^{\nu_k - 1}(1 - x_1 - \cdots - x_k)^{\nu_{k+1} - 1}\, dx_1 \cdots dx_k$$

which therefore has the value

$$(7.7.5)\quad \frac{\Gamma(\nu_1) \cdots \Gamma(\nu_{k+1})}{\Gamma(\nu_1 + \cdots + \nu_{k+1})}$$

was first investigated by Dirichlet (1839) and is known as the *Dirichlet integral*, and by analogy with the terminology introduced for the beta distribution (7.6.1) we may properly call the distribution having p.d.f. (7.7.1) the *Dirichlet distribution* $D(\nu_1, \nu_2, \ldots, \nu_k; \nu_{k+1})$.

It can be verified that the general moment $\mu'_{r_1 \cdots r_k}$ of the k-variate Dirichlet distribution has the following value

$$(7.7.6)\qquad \mu'_{r_1 \cdots r_k} = \frac{\Gamma(\nu_1 + r_1) \cdots \Gamma(\nu_k + r_k)\Gamma(\nu_1 + \cdots + \nu_{k+1})}{\Gamma(\nu_1 + \cdots + \nu_{k+1} + r_1 + \cdots + r_k)\Gamma(\nu_1) \cdots \Gamma(\nu_k)}$$

from which we find the means, variances, and covariances of the x's to be

$$\mu(x_i) = \frac{\nu_i}{\nu_1 + \cdots + \nu_{k+1}}, \qquad i = 1, \ldots, k$$

$$(7.7.7)\qquad \sigma^2(x_i) = \frac{\nu_i(\nu_1 + \cdots + \nu_{k+1} - \nu_i)}{(\nu_1 + \cdots + \nu_{k+1})^2(\nu_1 + \cdots + \nu_{k+1} + 1)}, \qquad i = 1, \ldots, k$$

$$\sigma(x_i, x_j) = -\frac{\nu_i \nu_j}{(\nu_1 + \cdots + \nu_{k+1})^2(\nu_1 + \cdots + \nu_{k+1} + 1)}, \qquad i \neq j = 1, \ldots, k.$$

A k-variate version of **7.6.1** can be stated as follows:

7.7.1 *Suppose $x_1, \ldots, x_{k+1}$ are independent random variables having gamma distributions $G(\nu_1), \ldots, G(\nu_{k+1})$.* Let

$$(7.7.8)\qquad y_i = \frac{x_i}{x_1 + \cdots + x_{k+1}}, \qquad i = 1, \ldots, k.$$

Then $(y_1, \ldots, y_k)$ has the k-variate Dirichlet distribution $D(\nu_1, \ldots, \nu_k; \nu_{k+1})$.

The argument for **7.7.1** is straightforward and will be left as an exercise for the reader. Suppose we put $r_{k_1+1} = \cdots = r_k = 0$ in the general moment (7.7.6), where $k_1 < k$. The resulting quantity is the general moment $\mu'_{r_1 \cdots r_{k_1}}$, of the marginal distribution of $(x_1, \ldots, x_{k_1})$ of the k-variate Dirichlet distribution $D(\nu_1, \ldots, \nu_k; \nu_{k+1})$ and it has the value

$$(7.7.9)\qquad \mu'_{r_1 \cdots r_{k_1}} = \frac{\Gamma(\nu_1 + r_1) \cdots \Gamma(\nu_{k_1} + r_{k_1})\Gamma(\nu_1 + \cdots + \nu_{k+1})}{\Gamma(\nu_1 + \cdots + \nu_{k+1} + r_1 + \cdots + r_{k_1})\Gamma(\nu_1) \cdots \Gamma(\nu_{k_1})}.$$

But this is the general moment of the k_1-variate Dirichlet distribution $D(\nu_1, \ldots, \nu_{k_1}; \nu_{k_1+1} + \cdots + \nu_{k+1})$ which by the multidimensional version of **5.5.1a** is a uniquely determined distribution. Thus

7.7.2 *If $(x_1, \ldots, x_k)$ is a vector random variable having the k-variate Dirichlet distribution $D(\nu_1, \ldots, \nu_k; \nu_{k+1})$, then the marginal distribution of $(x_1, \ldots, x_{k_1})$, $k_1 < k$, is the k_1-variate Dirichlet distribution $D(\nu_1, \ldots, \nu_{k_1}; \nu_{k_1+1} + \cdots + \nu_{k+1})$.*

We are now able to write down the p.d.f. of the conditional random variable $x_k \mid x_1, \ldots, x_{k-1}$ for a k-variate Dirichlet distribution by taking the ratio of the p.d.f. of $D(\nu_1, \ldots, \nu_k; \nu_{k+1})$ to the p.d.f. of $D(\nu_1, \ldots, \nu_{k-1}; \nu_k + \nu_{k+1})$. This gives, after some simplification, the following expression for the p.e. of $x_k \mid x_1, \ldots, x_{k-1}$:

(7.7.10) $$dF(x_k \mid x_1, \ldots, x_{k-1}) = \frac{\Gamma(\nu_k + \nu_{k+1})}{\Gamma(\nu_k)\Gamma(\nu_{k+1})}\left(\frac{x_k}{1 - x_1 - \cdots - x_{k-1}}\right)^{\nu_k - 1} \cdot \left(1 - \frac{x_k}{1 - x_1 - \cdots - x_{k-1}}\right)^{\nu_{k+1}-1} d\left(\frac{x_k}{1 - x_1 - \cdots - x_{k-1}}\right).$$

It is evident from (7.7.10) that:

7.7.3 *If $(x_1, \ldots, x_k)$ is a vector random variable having the k-variate Dirichlet distribution $D(\nu_1, \ldots, \nu_k; \nu_{k+1})$, the conditional random variable $x_k \mid x_1, \ldots, x_{k-1}$ has the property that*

$$\frac{x_k}{1 - x_1 - \cdots - x_{k-1}} \,\Bigg|\, x_1, \ldots, x_{k-1}$$

has the beta distribution $Be(\nu_k, \nu_{k+1})$.

Referring to (7.6.8) it is to be noted that the mean and variance of $x_k \mid x_1, \ldots, x_{k-1}$ are given by

$$\mu(x_k \mid x_1, \ldots, x_{k-1}) = \frac{\nu_k}{\nu_k + \nu_{k+1}}(1 - x_1 - \cdots - x_{k-1})$$

(7.7.11) $$\sigma^2(x_k \mid x_1, \ldots, x_{k-1}) = \frac{\nu_k \nu_{k+1}}{(\nu_k + \nu_{k+1})^2(\nu_k + \nu_{k+1} + 1)}(1 - x_1 - \cdots - x_{k-1})^2.$$

Another useful property of the k-variate Dirichlet distribution may be stated as follows:

7.7.4 *If $(x_1, \ldots, x_k)$ is a vector random variable having the k-variate Dirichlet distribution $D(\nu_1, \ldots, \nu_k; \nu_{k+1})$, the sum $x_1 + \cdots + x_k$ has the beta distribution $Be(\nu_1 + \cdots + \nu_k, \nu_{k+1})$.*

This follows from the fact that the rth moment of $1 - (x_1 + \cdots + x_k)$ is

$$\frac{\Gamma(\nu_{k+1} + r)\Gamma(\nu_1 + \cdots + \nu_{k+1})}{\Gamma(\nu_1 + \cdots + \nu_{k+1} + r)\Gamma(\nu_{k+1})}, \qquad r = 1, 2, \ldots.$$

which, by **5.5.1a**, implies that $1 - (x_1 + \cdots + x_k)$ has the beta distribution $Be(\nu_{k+1}, \nu_1 + \cdots + \nu_k)$. Therefore $x_1 + \cdots + x_k$ has the beta distribution $Be(\nu_1 + \cdots + \nu_k, \nu_{k+1})$.

More generally **7.7.4** can be extended as follows:

7.7.5 *If $(x_1, \ldots, x_k)$ is a vector random variable having the k-variate Dirichlet distribution $D(\nu_1, \ldots, \nu_k; \nu_{k+1})$, then the random variable $(z_1, \ldots, z_s)$ where $z_1 = x_1 + \cdots + x_{k_1}$, $z_2 = x_{k_1+1} + \cdots + x_{k_1+k_2}$, $\ldots, z_s = x_{k_1+\cdots+k_{s-1}+1} + \cdots + x_{k_1+\cdots+k_s}$ and $k_1 + \cdots + k_s \leqslant k$, has the s-variate Dirichlet distribution $D(\nu_{(1)}, \ldots, \nu_{(s)}; \nu_{(s+1)})$ where $\nu_{(1)} = \nu_1 + \cdots + \nu_{k_1}, \ldots, \nu_{(s)} = \nu_{k_1+\cdots+k_{s-1}+1} + \cdots + \nu_{k_1+\cdots+k_s}$, $\nu_{(s+1)} = \nu_{k_1+\cdots+k_s+1} + \cdots + \nu_{k+1}$.*

For we have by direct evaluation from (7.7.1)

$$(7.7.12)\quad \mathscr{E}[x_{k_1+1}^{r_{k_1+1}} \cdots x_k^{r_k}(1 - z_1 - x_{k_1+1} - \cdots - x_k)^{r_{(1)}}]$$

$$= G \cdot \frac{\Gamma(\nu_{(1)})\Gamma(\nu_{k_1+1} + r_{k_1+1}) \cdots \Gamma(\nu_k + r_k)\Gamma(\nu_{k+1} + r_{(1)})}{\Gamma(\nu_{(1)} + \nu_{k_1+1} + \cdots + \nu_k + \nu_{k+1} + r_{k_1+1} + \cdots + r_k + r_{(1)})}$$

where

$$(7.7.13)\qquad G = \frac{\Gamma(\nu_{(1)} + \nu_{k_1+1} + \cdots + \nu_{k+1})}{\Gamma(\nu_{(1)})\Gamma(\nu_{k_1+1}) \cdots \Gamma(\nu_{k+1})}.$$

But the right-hand side of (7.7.12) is the value of

$$\mathscr{E}[x_{k_1+1}^{r_{k_1+1}} \cdots x_k^{r_k}(1 - z_1 - x_{k_1+1} - \cdots - x_k)^{r_{(1)}}]$$

computed from the p.d.f.

$$(7.7.14)\quad f(z_1, x_{k_1+1}, \ldots, x_k)$$

$$= G z_1^{\nu_{(1)}-1} x_{k_1+1}^{\nu_{k_1+1}-1} \cdots x_k^{\nu_k-1}(1 - z_1 - x_{k_1+1} - \cdots - x_k)^{\nu_{k+1}-1}$$

which is uniquely determined, since $(z_1, x_{k_1+1}, \ldots, x_k)$ is a bounded random variable.

We can now repeat the process to bring in z_2 and so on, finally obtaining

$$(7.7.15)\quad f(z_1, \ldots, z_s)$$

$$= \frac{\Gamma(\nu_{(1)} + \cdots + \nu_{(s+1)})}{\Gamma(\nu_{(1)}) \cdots \Gamma(\nu_{(s+1)})} z_1^{\nu_{(1)}-1} \cdots z_s^{\nu_{(s)}-1}(1 - z_1 - \cdots - z_s)^{\nu_{(s+1)}-1}$$

which establishes **7.7.5**.

It will be useful, particularly in Section 8.7, in connection with applications to order statistics, to introduce another distribution closely related to the k-variate Dirichlet distribution. Let $(x_1, \ldots, x_k)$ be random variables having the k-variate Dirichlet distribution $D(\nu_1, \ldots, \nu_k; \nu_{k+1})$. Let

$$(7.7.16)\qquad \begin{aligned} y_1 &= x_1 \\ y_2 &= x_1 + x_2 \\ &\cdot \\ &\cdot \\ &\cdot \\ y_k &= x_1 + \cdots + x_k. \end{aligned}$$

Since the Jacobian of this transformation is unity, we have the following p.d.f. for

$$(7.7.17)\quad f(y_1, \ldots, y_k) = \frac{\Gamma(\nu_1 + \cdots + \nu_{k+1})}{\Gamma(\nu_1) \cdots \Gamma(\nu_{k+1})} y_1^{\nu_1 - 1} \cdot (y_2 - y_1)^{\nu_2 - 1} \cdots (y_k - y_{k-1})^{\nu_k - 1}(1 - y_k)^{\nu_{k+1} - 1}$$

where the range of the y's is the region $0 < y_1 < \cdots < y_k < 1$. It will be convenient to call the distribution having p.d.f. (7.7.17) the *ordered k-variate Dirichlet distribution* $D^*(\nu_1, \ldots, \nu_k; \nu_{k+1})$. Note that when $k = 1$, (7.7.17) reduces to the p.d.f. of the beta distribution $Be(\nu_1, \nu_2)$.

Sometimes we are interested in the marginal distribution of a subset of s of the y's and the following result will be useful in problems of order statistics:

7.7.6 *If $(y_1, \ldots, y_k)$ is a vector random variable having the ordered k-variate Dirichlet distribution $D^*(\nu_1, \ldots, \nu_k; \nu_{k+1})$, then the marginal distribution of $(y_{k_1}, y_{k_1+k_2}, \ldots, y_{k_1+\cdots+k_s})$ is the ordered s-variate Dirichlet distribution $D^*(\nu_{(1)}, \ldots, \nu_{(s)}; \nu_{(s+1)})$ where $\nu_{(1)}, \ldots, \nu_{(s+1)}$ are defined in* **7.7.5**.

For the random variables $y_{k_1}, \ldots, y_{k_1+\cdots k_s}$, which are defined in (7.7.16), have the same distribution as the random variables $z_1, z_1 + z_2, \ldots, z_1 + \cdots + z_s$ defined in **7.7.5**. Since $(z_1, \ldots, z_s)$ has the s-variate Dirichlet distribution $D(\nu_{(1)}, \ldots, \nu_{(s)}; \nu_{(s+1)})$, it follows by definition that $(z_1, z_1 + z_2, \ldots, z_1 + \cdots + z_s)$ has the ordered s-variate Dirichlet distribution $D^*(\nu_{(1)}, \ldots, \nu_{(s)}; \nu_{(s+1)})$.

It should be noted that **7.7.6** can also be established by the direct integration of (7.7.17) with respect to all y's except $y_{k_1}, y_{k_1+k_2}, \ldots, y_{k_1+\cdots+k_s}$. We would integrate successively with respect to $y_1, \ldots, y_{k_1-1}$ over the region $0 < y_1 < \cdots < y_{k_1}$, then successively with respect to $y_{k_1+1}, \ldots, y_{k_1+k_2-1}$ over the region $y_{k_1} < y_{k_1+1} < \cdots < y_{k_1+k_2}$, and so on.

7.8 DISTRIBUTIONS INVOLVED IN THE ANALYSIS OF VARIANCE

Three distributions closely related to gamma and beta distributions, namely, the chi-square distribution, the "Student" distribution, and the Snedecor distribution, are fundamental in the analysis of variance and in other statistical procedures based on normally distributed random variables, which are discussed in later chapters. It will be convenient to state these distributions here in their basic forms as further examples of important continuous distributions.

(a) The Chi-Square Distribution

If we make the change of variable

$$x = \chi^2/2$$

in the p.e. of the gamma distribution $G(\mu)$, we obtain

$$dF_{2\mu}(\chi^2) = \frac{\left(\frac{\chi^2}{2}\right)^{\mu-1} e^{-\frac{1}{2}\chi^2}}{2\Gamma(\mu)} \, d(\chi^2). \tag{7.8.1}$$

This is the p.e. of the *chi-square distribution with* 2μ *degrees of freedom*, and when a random variable has such a distribution we shall say it has the *chi-square distribution* $C(2\mu)$. In most statistical applications of this distribution 2μ is a positive integer. Values of χ^2_α for which $\int_{\chi^2_\alpha}^{\infty} dF_{2\mu}(\chi^2) = \alpha$, for various values of α from 0.001 to 0.99, with $2\mu = 1, 2, \ldots 30$ have been tabulated by Fisher (1925*a*). More extensive tables are given in *Biometrika Tables for Statisticians*, edited by E. S. Pearson and Hartley (1954).

It will be noted that the characteristic function of the chi-square distribution $C(2\mu)$ is

$$\varphi(t) = (1 - 2it)^{-\mu} \tag{7.8.2}$$

which is obtained by replacing t by $2t$ in (7.5.6).

It is evident from (7.8.2) that

7.8.1 *The chi-square distribution* $C(2\mu)$ *is reproductive with respect to* μ.

Since $\chi^2/2$ has the gamma distribution $G(\mu)$ it follows from (7.5.3) that the rth moment of χ^2 is

$$\mu'_r = \frac{2^r\Gamma(\mu + r)}{\Gamma(\mu)}.$$

The mean and variance of the chi-square distribution $C(2\mu)$ are therefore

$$\mathscr{E}(\chi^2) = 2\mu, \qquad \sigma^2(\chi^2) = 4\mu.$$

The chi-square notation was introduced by K. Pearson (1900), and is rather awkward, but since it is deeply embedded in statistical literature we shall use it.

Returning to the problem of the distribution of $Q(x_1, \ldots, x_k)$, given in terms of the gamma distribution by **7.5.2**, this can be expressed in chi-square notation as follows:

7.8.2 *If* $(x_1, \ldots, x_k)$ *is a vector random variable having the k-variate distribution* $N(\{\mu_i\}, \|\sigma_{ij}\|)$, $i, j = 1, \ldots, k$, *then* $\sum_{i,j=1}^{k} \sigma^{ij}(x_i - \mu_i)(x_j - \mu_j)$ *has the chi-square distribution* $C(k)$.

An important corollary of **7.8.2** is:

7.8.2a *If* x *is a random variable having the distribution* $N(0, 1)$, *then* x^2 *has the chi-square distribution* $C(1)$.

The reason for using the term *degrees of freedom* with reference to 2μ in (7.8.1) becomes apparent when we see that the number of degrees of freedom mentioned in **7.8.2** refers to the number of random variables involved in the (nondegenerate) normal distribution considered. That the choice of this term is a reasonable one will become clearer as we discuss some applications of the chi-square distribution in later chapters.

(b) The "Student" Distribution

The mathematical essentials of this distribution and how it arises from other random variables, may be stated as follows:

7.8.3 *Suppose* u *is a random variable having the distribution* $N(0, 1)$ *and* v *is a random variable having the chi-square distribution* $C(k)$. *If* u *and* v *are independent, the random variable*

$$t = \frac{u}{\sqrt{v/k}} \tag{7.8.3}$$

has the p.d.f.

$$f_k(t) = \frac{\Gamma\left(\frac{k+1}{2}\right)}{\Gamma\left(\frac{k}{2}\right)\sqrt{\pi k}}\left(1 + \frac{t^2}{k}\right)^{-\frac{1}{2}(k+1)}. \tag{7.8.4}$$

This is the p.d.f. of the "*Student*" *distribution with k degrees of freedom.* For brevity we shall call this distribution the "*Student*" *distribution* $S(k)$. Various applications of this distribution are discussed in Sections 8.4 and 10.4.

To establish the distribution (7.8.4) we apply the transformation

$$\begin{aligned} s &= v \\ t &= \frac{u}{\sqrt{v/k}} \end{aligned} \tag{7.8.5}$$

to the p.e. of u and v, namely,

$$(7.8.6)\qquad \frac{1}{2\sqrt{2\pi}\Gamma\left(\frac{k}{2}\right)}\left(\frac{v}{2}\right)^{\frac{1}{2}(k-2)} e^{-\frac{1}{2}(u^2+v)}\,du\,dv$$

and then take the marginal distribution of t.

The Jacobian of the transformation (7.8.5) is $\sqrt{s/k}$ and hence the p.e. of s and t is

$$(7.8.7)\qquad \frac{1}{2\sqrt{\pi k}\Gamma\left(\frac{k}{2}\right)}\left(\frac{s}{2}\right)^{\frac{1}{2}(k-1)} e^{-\frac{1}{2}[1+(t^2/k)]s}\,ds\,dt.$$

Taking the marginal p.e. of t (that is, integrating from 0 to ∞ with respect to s), we obtain the p.d.f. of t as that given by (7.8.4).

Values of t_α for which $P(|t| > t_\alpha) = \alpha$ have been tabulated by Fisher (1925*a*) for various values of α from 0.1 to 0.99 with $k = 1, 2, \ldots, 30$. A tabulation is also given in Pearson and Hartley's (1954) *Biometrika Tables for Statisticians.*

Note that all odd moments of the distribution having p.d.f. (7.8.4) which exist are 0. As for the even moments which exist, we have

$$\mu'_{2r} = \mu_{2r} = k^r \mathscr{E}\left(\frac{u^{2r}}{v^r}\right)$$

and since u and v are independent

$$\mathscr{E}\left(\frac{u^{2r}}{v^r}\right) = \mathscr{E}(u^{2r}) \cdot \mathscr{E}(v^{-r}).$$

But u^2 and v have independent chi-square distributions $C(1)$ and $C(k)$, respectively. Therefore,

$$(7.8.8)\qquad \mu'_{2r} = \mu_{2r} = k^r \frac{\Gamma(\frac{1}{2}+r)\Gamma\left(\frac{k}{2}-r\right)}{\Gamma(\frac{1}{2})\Gamma\left(\frac{k}{2}\right)}$$

which shows that μ'_{2r} exists if and only if $-1 < 2r < k$.

The *mean* and *variance* of the "Student" distribution $S(k)$ are defined for $t > 1$ and $t > 2$ respectively, and have values

$$\mathscr{E}(t) = 0, \qquad \sigma^2(t) = \frac{k}{k-2}\,.$$

The following statement gives the relationship between the "Student" distribution and the beta distribution:

7.8.4 *If t is a random variable having the "Student" distribution $S(k)$, then* $x = \dfrac{1}{1 + \dfrac{t^2}{k}}$ *has the beta distribution* $Be(\frac{1}{2}k, \frac{1}{2})$.

(c) The Snedecor Distribution

The mathematical essentials of this distribution and how it arises may be stated as follows:

7.8.5 *Suppose u and v are independent random variables having chi-square distributions $C(k_1)$ and $C(k_2)$, respectively. The random variable*

$$\mathscr{F} = \frac{u}{k_1}\bigg/\frac{v}{k_2}$$

has the p.e.

(7.8.9)

$$dF_{k_1,k_2}(\mathscr{F}) = \frac{\Gamma(\frac{1}{2}k_1 + \frac{1}{2}k_2)}{\Gamma(\frac{1}{2}k_1)\Gamma(\frac{1}{2}k_2)}(k_1/k_2)^{\frac{1}{2}k_1}\mathscr{F}^{\frac{1}{2}k_1-1}[1 + (k_1\mathscr{F}/k_2)]^{-\frac{1}{2}(k_1+k_2)}\,d\mathscr{F}.$$

This is the p.e. of the *Snedecor distribution with k_1, k_2 degrees of freedom*. For the sake of brevity we shall refer to this distribution as the *Snedecor distribution $S(k_1, k_2)$*. Applications of this distribution are discussed in Sections 10.4 and 10.6.

The Snedecor distribution can be established by applying the transformation

$$\mathscr{F} = \frac{u}{k_1}\bigg/\frac{v}{k_2} \qquad \mathscr{G} = v \tag{7.8.10}$$

to the p.e. of u and v

$$\frac{(\frac{1}{2}u)^{\frac{1}{2}k_1-1}(\frac{1}{2}v)^{\frac{1}{2}k_2-1}}{4\Gamma(\frac{1}{2}k_1)\Gamma(\frac{1}{2}k_2)}e^{-\frac{1}{2}(u+v)}du\,dv \tag{7.8.11}$$

and taking the marginal distribution of $\mathscr{F}$. This gives as the p.e. of $\mathscr{F}$ and $\mathscr{G}$

$$\frac{\left(\dfrac{k_1}{k_2}\right)^{\frac{1}{2}k_1}}{2\Gamma(\frac{1}{2}k_1)\Gamma(\frac{1}{2}k_2)}\mathscr{F}^{\frac{1}{2}k_1-1}\left(\frac{\mathscr{G}}{2}\right)^{\frac{1}{2}(k_1+k_2)-1}e^{-\frac{1}{2}[1+(k_1\mathscr{F}/k_2)]\mathscr{G}}d\mathscr{F}\,d\mathscr{G}. \tag{7.8.12}$$

Integrating with respect to $\mathscr{G}$ from 0 to ∞, we obtain (7.8.9) as the p.e. of $\mathscr{F}$.

The rth moment of the Snedecor distribution is

$$(7.8.13)\qquad \mu'_r = \left(\frac{k_2}{k_1}\right)^r \mathscr{E}(u^r)\mathscr{E}(v^{-r}) = \left(\frac{k_2}{k_1}\right)^r \frac{\Gamma(\frac{1}{2}k_1 + r)\Gamma(\frac{1}{2}k_2 - r)}{\Gamma(\frac{1}{2}k_1)\Gamma(\frac{1}{2}k_2)}$$

which exists only for $-k_1 < 2r < k_2$. For the mean and variance of $\mathscr{F}$ we have

$$(7.8.14)\qquad \mu(\mathscr{F}) = \frac{k_2}{k_2 - 2}, \qquad \sigma^2(\mathscr{F}) = \frac{2k_2^2(k_1 + k_2 - 2)}{k_1(k_2 - 2)^2(k_2 - 4)}.$$

There is also a connection between the Snedecor and beta distributions which may be stated as follows:

7.8.6 *If $\mathscr{F}$ is a random variable having the Snedecor distribution $S(k_1, k_2)$, then the random variable $[1 + k_1\mathscr{F}/k_2]^{-1}$ has the beta distribution $Be(\frac{1}{2}k_2, \frac{1}{2}k_1)$, and $\dfrac{k_1\mathscr{F}}{k_2 + k_1\mathscr{F}}$ has the beta distribution $Be(\frac{1}{2}k_1, \frac{1}{2}k_2)$.*

The Snedecor distribution is, from a practical point of view, a slightly more convenient form of one originally suggested by Fisher (1924) for use in the analysis of variance. The random variable z originally proposed by Fisher is related to Snedecor's $\mathscr{F}$ as follows:

$$(7.8.15)\qquad z = \tfrac{1}{2}\log_e \mathscr{F}.$$

Fisher (1925*a*) has tabulated values of z_α for which $P(z > z_\alpha) = \alpha$ for $\alpha = 0.01$, 0.05 and for $k_1 = 1, 2, 3, 4, 5, 6, 8, 12, 24, \infty$, and for $k_2 = 1, 2, \ldots, 30, 60, \infty$. Snedecor (1937) has tabulated the values of $\mathscr{F}_\alpha$ corresponding to z_α (where $z_\alpha = \frac{1}{2}\log \mathscr{F}_\alpha$) for $\alpha = 0.01$, 0.05 and for $k_1 = 1, 2, \ldots, 12, 14, 16, 20, 24, 30, 40, 50, 75, 100, 200, 500, \infty$, and $k_2 = 1, 2, \ldots, 30, 32, 34, 36, 38, 40, 42, 44, 46, 48, 50, 55, 60, 65, 70, 80, 100, 125, 150, 200, 400, 1000, \infty$. For subsequent tabulations see Greenwood and Hartley (1961).

If $\mathscr{F}$ has the Snedecor distribution $S(k_1, k_2)$, it follows from **7.8.6** that $\mathscr{F}_\alpha$ can also be determined from Karl Pearson's (1934) *Tables of the Incomplete Beta Function*, since $P(\mathscr{F} > \mathscr{F}_\alpha) = I_{x_\alpha}(\frac{1}{2}k_1, \frac{1}{2}k_2)$ and $\mathscr{F}_\alpha = \dfrac{k_2 x_\alpha}{k_1(1 - x_\alpha)}$.

PROBLEMS

7.1 Prove **7.2.2**.

7.2 If x is a random variable having the normal distribution $N(\mu, \sigma^2)$, show that

$$\mathscr{E}(|x - \mu|) = \sqrt{(2/\pi)}\sigma.$$

7.3 If x is a continuous random variable with a unique median $x_{0.5}$ show that $\mathscr{E}(|x - c|)$ is minimized for $c = x_{0.5}$.

7.4 Suppose x_1 is a random variable having the normal distribution $N(\mu, \sigma^2)$, and $x_2 \mid x_1$ is a conditional random variable having the normal distribution $N(x_1, \sigma^2)$. Show that (x_1, x_2) has the two-dimensional normal distribution $N(\mu, \mu; \|\sigma_{ij}\|)$ where

$$\|\sigma_{ij}\| = \begin{Vmatrix} \sigma^2 & \sigma^2 \\ \sigma^2 & 2\sigma^2 \end{Vmatrix}.$$

7.5 Let $f(x_1, x_2)$ be the p.d.f. of the circular normal bivariate distribution $N(\{0\}, \|\delta_{ij}\|)$, $i, j = 1, 2$. Show that the integral of $f(x_1, x_2)$ over the square $(-k, -k)$, $(-k, +k)$, $(k, -k)$, (k, k) is less than the integral of $f(x_1, x_2)$ over the circle $x_1^2 + x_2^2 < 4k^2/\pi$ and hence that

$$\frac{1}{\sqrt{2\pi}} \int_0^k e^{-\frac{1}{2}t^2}\, dt < \tfrac{1}{2}\sqrt{1 - e^{-2k^2/\pi}}.$$

a result due to Williams (1946).

7.6 Suppose $f(x_1, x_2)$ is the p.d.f. of the two-dimensional normal distribution $N(\{0\}, \|\sigma_{ij}\|)$ where

$$\|\sigma_{ij}\| = \begin{Vmatrix} 1 & \rho \\ \rho & 1 \end{Vmatrix}.$$

Let $F_1(x)$ and $F_2(x)$ be the c.d.f.'s of the marginal distributions of x_1 and x_2, respectively. Show that the correlation coefficient of the random variables $F_1(x_1)$, $F_2(x_2)$ is $(6/\pi) \sin^{-1} (\frac{1}{2}\rho)$.

7.7 If (x_1, x_2) is a two-dimensional random variable having the distribution $N(\{0\}, \|\sigma_{ij}\|)$, where

$$\|\sigma_{ij}\| = \begin{Vmatrix} 1 & \rho \\ \rho & 1 \end{Vmatrix},$$

show that the ratio

$$z = \frac{x_1}{x_2}$$

has p.d.f.

$$\frac{\sqrt{1 - \rho^2}}{\pi[1 - 2\rho z + z^2]}.$$

[Fieller (1932).]

7.8 If x_1 and x_2 are independent random variables each having distribution $R(\frac{1}{2}, 1)$, show that $\sqrt{-2 \log x_1} \cos 2\pi x_2$ and $\sqrt{-2 \log x_1} \sin 2\pi x_2$ are independent random variables each having the distribution $N(0, 1)$. [Box and Muller (1958).]

7.9 If x_1 and x_2 are independent random variables having gamma distributions $G(k_1)$ and $G(k_2)$ respectively, show that $x_1 + x_2$ and $x_1/(x_1 + x_2)$ are independent random variables (having the gamma distribution $G(k_1 + k_2)$ and beta distribution $Be(k_1, k_2)$, respectively).

7.10 If x_1 and x_2 are independent random variables with gamma distributions $G(\nu)$ and $G(\nu + \frac{1}{2})$ show that the random variable $y = 2\sqrt{x_1 x_2}$ has the gamma distribution $G(2\nu)$.

7.11 A two-dimensional random variable has the p.d.f.

$$\frac{1}{\Gamma(k_1)\Gamma(k_2)} x_1^{k_1-1}(x_2 - x_1)^{k_2-1}e^{-x_2}$$

in the x_1x_2-plane where $0 < x_1 < x_2 < \infty$, and 0 elsewhere. Show that the marginal distributions of x_1 and x_2 are gamma distributions $G(k_1)$ and $G(k_1 + k_2)$.

7.12 Suppose x is a continuous random variable such that for some positive integer k, kx has the gamma distribution $G(k)$. Suppose $y \mid x$ is a discrete conditional random variable which has the Poisson distribution $Po(x)$. Show that the unconditional distribution of y is a binomial waiting-time distribution.

7.13 If x_1 and x_2 are independent random variables having beta distributions $Be(\nu_1, \nu_2)$ and $Be(\nu_1 + \frac{1}{2}, \nu_2)$ show that $\sqrt{x_1 x_2}$ has the beta distribution $Be(2\nu_1, 2\nu_2)$.

7.14 If $x_1, \ldots, x_k$ are independent random variables each having the rectangular distribution $R(\frac{1}{2}, 1)$ show that $-\log(x_1 \cdots x_k)$ has the gamma distribution $G(k)$.

7.15 If u is a random variable having the chi-square distribution $C(k)$ show, by the use of characteristic functions, that the limiting distribution of $(u - k)/\sqrt{2k}$ as $k \to \infty$ is the distribution $N(0, 1)$.

7.16 Suppose x is a random variable with c.d.f. $F(x)$ and characteristic function $\varphi(t)$. If $[\varphi(t)]^{1/n}$ is a characteristic function of some random variable for every positive integer n then x is said to be *infinitely divisible*, that is, for each n, $F(x)$ is the distribution of a sum of n independent random variables, each having $[\varphi(t)]^{1/n}$ as its characteristic function. Show that a random variable having a Poisson, gamma, or normal distribution is infinitely divisible. [See Gnedenko and Kolmogorov (1954) for a treatment of infinitely divisible random variables.]

7.17 Show that the p.d.f. of the Student t-distribution given by (7.8.4) converges to the p.d.f. of the distribution $N(0, 1)$ for every t as $k \to \infty$.

7.18 If $x_1, \ldots, x_k$ are independent random variables all having the distribution $N(\mu, \sigma^2)$ and if $c_1, \ldots, c_k$ are (real) constants whose sum is 1 and whose sum of squares is 1, show that $c_1x_1 + \cdots + c_kx_k$ also has the distribution $N(\mu, \sigma^2)$. Show that no set of values of the c's exist, which are all positive, satisfying these conditions.

7.19 If $x_1, \ldots, x_k$ are independent random variables, each having the distribution $N(0, 1)$, and if $y_1, \ldots, y_s$, $s \leqslant k$, are new random variables defined by $y_p = c_{p1}x_1 + \cdots + c_{pk}x_k$, $p = 1, \ldots, s$ where $\sum_{i=1}^{k} c_{pi}c_{qi} = 0$ for $p \neq q$, and 1 for $p = q$, show that $y_1, \ldots, y_s$ are independent, each having the distribution $N(0, 1)$.

7.20 Show by appropriate differentiations of the characteristic function (7.4.17) of the k-dimensional random variable $(x_1, \ldots, x_k)$ having the distribution $N(\{\mu_i\}, \|\sigma_{ij}\|)$ that $\mathscr{E}(x_i) = \mu_i$ and $\mathscr{E}(x_i - \mu_i)(x_j - \mu_j) = \sigma_{ij}$.

7.21 Show (by the use of characteristic functions) that if $(x_1, \ldots, x_k)$ and $(x_1', \ldots, x_k')$ are independent k-dimensional random variables having distributions $N(\{\mu_i\}, \|\sigma_{ij}\|)$ and $N(\{\mu_i'\}, \|\sigma_{ij}'\|)$, then the k-dimensional random variable $(x_1 + x_1', \ldots, x_k + x_k')$ has the normal distribution $N(\{\mu_i + \mu_i'\}, \|\sigma_{ij} + \sigma_{ij}'\|)$. That is, the k-dimensional normal distribution is reproductive with respect to its vector of means and its covariance matrix.

7.22 Suppose μ is the "true" length of a standard bar, and let x_1 be a random variable denoting the length of a first-generation copy of the bar. Let x_2 be a random variable denoting the length of a copy of x_1, that is, a second-generation copy of the bar and so on, until x_k is a random variable denoting the length of a kth generation copy of the bar. If $(x_1 - \mu), (x_2 - x_1), \ldots, (x_k - x_{k-1})$ are independent, each having the normal distribution $N(0, 1)$, show that the distribution of $(x_1, \ldots, x_k)$ is a k-dimensional normal distribution $N(\{\mu_i\}, \|\sigma_{ij}\|)$ where $\mu_i = \mu$ and

$$\|\sigma_{ij}\| = \left\| \begin{matrix} 1 & 1 & 1 & \cdots & 1 & 1 \\ 1 & 2 & 2 & \cdots & 2 & 2 \\ 1 & 2 & 3 & \cdots & 3 & 3 \\ \cdot & \cdot & \cdot & & \cdot & \cdot \\ \cdot & \cdot & \cdot & & \cdot & \cdot \\ \cdot & \cdot & \cdot & & \cdot & \cdot \\ 1 & 2 & 3 & \cdots & k-1 & k-1 \\ 1 & 2 & 3 & \cdots & k-1 & k \end{matrix} \right\|$$

7.23 If $(x_1, \ldots, x_k)$ is a vector random variable having the k-dimensional spherical normal distribution $N(\{\mu_i\}, \|\delta_{ij}\sigma^2\|)$ where $\delta_{ij} = 1$ for $i = j$, and 0 for $i \neq j$, and if

$$r = [(x_1 - \mu_1)^2 + \cdots + (x_k - \mu_k)^2]^{1/2},$$

show that

$$\mathscr{E}(r) = \sqrt{2}\sigma \frac{\Gamma(\frac{1}{2}k + \frac{1}{2})}{\Gamma(\frac{1}{2}k)}.$$

7.24 Assume that a certain class of objects has a "mortality law" such that the probability of an object drawn at random from the class "expiring" during the time interval $(t, t + dt)$ (in suitable time units) is

$$\frac{t^{k-1}e^{-t}}{\Gamma(k)}\, dt.$$

As soon as the object "expires" it is replaced by another of the same type. As soon as the second one "expires" it is replaced by a third of the same type, and so on. Show that the probability that the rth object in such a sequence "expires" during $(t, t + dt)$ is

$$\frac{t^{rk-1}e^{-t}}{\Gamma(rk)}\, dt.$$

7.25 (*Continuation*) *Renewal process with fixed general "mortality law."* If $f(t)\, dt$ is the probability that an object in an initial (zeroth) generation has to be replaced by a successor during $(t, t + dt)$, and if $g_n(\tau)\, d\tau$ is the probability that an object in the nth generation has to be replaced by a successor during $(\tau, \tau + d\tau)$, and assuming the objects in any generation have the same "mortality law" as the initial generation, show that

$$g_{n+1}(t) = \int_0^t g_n(\tau)f(t - \tau)\, d\tau,$$

and hence that if a "steady state" replacement law $g(t)$ exists it must satisfy the integral equation $g(t) = \int_0^t g(\tau)f(t - \tau)\, d\tau$. [Lotka (1939) and Smith (1958)].

7.26 Suppose $x_1, \ldots, x_{k+1}$ are independent random variables all having the distribution $N(0, 1)$. Let

$$y_i = \frac{x_i^2}{x_1^2 + \cdots + x_{k+1}^2}, \qquad i = 1, \ldots, k.$$

Show that $(y_1, \ldots, y_k)$ has the k-dimensional Dirichlet distribution $D(\frac{1}{2}, \ldots, \frac{1}{2}; \frac{1}{2})$.

7.27 Prove **7.7.1**.

7.28 Let $y_{k_2} = k_1\tau$, where τ is a random variable having the Snedecor distribution $S(k_1, k_2)$. Show that the sequence of random variables $(y_1, y_2, \ldots)$ converges in distribution to a random variable having the chi-square distribution with k_1 degrees of freedom.

7.29 Suppose $(x_1, \ldots, x_k)$ is a k-dimensional random variable whose p.d.f. is of form $g(a_1x_1 + \cdots + a_kx_k)$ over the region where $x_1 > 0, \ldots, x_k > 0$, and 0 elsewhere, and where $a_1, \ldots, a_k$ are positive constants. Show that the p.d.f. of the random variable $y = a_1 x_1 + \cdots + a_kx_k$ is

$$\frac{1}{(a_1 \cdots a_k)\Gamma(k)} y^{k-1}g(y), \qquad y \geqslant 0, \quad \text{and 0 for } y < 0.$$

7.30 If $(x_1, \ldots, x_k)$ is a k-dimensional random variable having p.d.f. of form $g(y)$ where $y = \sum_{i,j} a_{ij}x_ix_j$, the matrix of constants, $\|a_{ij}\|$ being symmetric and positive definite, show that the random variable y has p.d.f.

$$\frac{\pi^{k/2}}{\sqrt{|a_{ij}|}\, \Gamma(k/2)} y^{(k/2)-1}g(y), \qquad y \geqslant 0, \quad \text{and 0 for } y < 0.$$

7.31 If $(x_1, \ldots, x_k)$ is a k-dimensional random variable having the distribution $N(\{\mu_i\}; \|\sigma_{ij}\|)$ show that the conditional p.d.f. $f(x_1, \ldots, x_s \mid x_{s+1}, \ldots, x_k)$ is the p.d.f. of the s-dimensional distribution $N(\mu_1^*, \ldots, \mu_s^*; \|\sigma_{pq}^*\|)$, where the value of μ_p^* is given by the equation

$$\begin{vmatrix} (\mu_p^* - \mu_p) & \sigma_{p\,s+1} & \cdots & \sigma_{pk} \\ (x_{s+1} - \mu_{s+1}) & \sigma_{s+1\,s+1} & \cdots & \sigma_{s+1k} \\ \vdots & \vdots & & \vdots \\ (x_k - \mu_k) & \sigma_{k\,s+1} & \cdots & \sigma_{kk} \end{vmatrix} = 0,$$

$p = 1, \ldots, s,$ and where

$$\sigma^*_{pq} = \frac{\begin{vmatrix} \sigma_{pq} & \sigma_{p\,s+1} & \cdots & \sigma_{pk} \\ \sigma_{s+1\,q} & \sigma_{s+1\,s+1} & \cdots & \sigma_{s+1\,k} \\ \vdots & \vdots & & \vdots \\ \sigma_{kq} & \sigma_{k\,s+1} & \cdots & \sigma_{kk} \end{vmatrix}}{\begin{vmatrix} \sigma_{s+1\,s+1} & \cdots & \sigma_{s+1\,k} \\ \vdots & & \vdots \\ \sigma_{k\,s+1} & \cdots & \sigma_{kk} \end{vmatrix}},$$

$p, q = 1, \ldots, s.$

7.32 *Poisson process.* Consider the following approach to (7.5.11). If we denote by E the event of k C's occurring during $(0, t + \Delta t)$, then E is the union of $k + 1$ mutually exclusive events $E_0, E_1, \ldots, E_k$, where E_i is the event of i C's occurring during $(t, t + \Delta t)$ and $k - i$ C's occurring during $(0, t)$. Therefore

$$P(E) = P(E_0) + P(E_1) + \cdots + P(E_k).$$

If we assume that the occurrence of any number of C's in a time interval I is independent of the occurrence of any number of C's in a time interval I' where $I \cap I' = \phi$, then the probabilities of the occurrence of $0, 1, \ldots, k$ C's during $(t, t + \Delta t)$ are

$$1 - \lambda\Delta t + o(\Delta t),\ \lambda\Delta t + o(\Delta t), \ldots, (\lambda\Delta t)^k + o((\Delta t)^k),$$

respectively. Then if we denote $P(E)$ by $f_k(t + \Delta t)$ we have

$$\begin{aligned} P(E_0) &= f_k(t)(1 - \lambda\Delta t) + o(\Delta t) \\ P(E_1) &= f_{k-1}(t)(\lambda\Delta t) + o(\Delta t), \end{aligned}$$

while $P(E_2), \ldots, P(E_k)$ are of order of magnitude $(\Delta t)^2$ or smaller. We therefore obtain

$$f_k(t + \Delta t) = f_k(t)(1 - \lambda\Delta t) + f_{k-1}(t)\lambda\Delta t + o(\Delta t).$$

From this we obtain the differential equations

$$f'_k(t) = \lambda f_{k-1}(t) - \lambda f_k(t)$$

for $k = 1, 2, \ldots$. (Note that for $k = 1, f_{-1}(t) = 0$.) Show that the solutions of this set of differential equations gives for $f_k(t)$ the expression (7.5.11). Note that $\{f_k(t): t > 0\}$ is an example of a stochastic process with a continuous parameter t. This particular stochastic process is called the *Poisson process*.

7.33 *Yule's* (1924) *birth process.* Suppose we have a population of objects which can generate (or give "birth" to) new objects, and that objects do not disappear (or "die") from the population. Let $f_k(t)$ be the probability of k objects in the population at time t and consider the probability of the event E of there still being k at time $t + \Delta t$. Then E is composed of disjoint events $E_0, E_1, \ldots, E_k$ where E_i is the event that there are $k - i$ objects in the population at time t and i were generated during $(t, t + \Delta t)$. By making assumptions of independence of occurrence of births in two nonoverlapping time intervals similar to the assumptions made regarding occurrence of C's in nonoverlapping time intervals in Problem 7.32, show that $f_k(t)$ satisfies the system of differential equations

$$f'_k(t) = (k - 1)\lambda f_{k-1}(t) - k\lambda f_k(t), k = 1,2, \ldots .$$

If $k = m$ at $t = 0$ and $f_m(0) = 1$, $f_k(0) = 0$, $k = m, m + 1, \ldots$, show that the solution of the system of differential equations is given by

$$f_k(t) = \binom{k - 1}{m - 1} e^{-m\lambda t}(1 - e^{-\lambda t})^{k-m}.$$

7.34 *Simple queueing*. Suppose the average number of arrivals of customers per unit time at a service counter is λ, whereas the average number of departures of customers (after receiving service) per unit time is μ. Let $f_k(t)$ be the probability of there being a *queue** of k customers at the counter at time t. Then $f_k(t + \Delta t)$ is the probability that there will still be a line of k customers at time $t + \Delta t$. If E is the event of there being a line of k customers at time $t + \Delta t$ and if $E_{ij(k-i+j)}$ is the event of there being $k - i + j$ customers in the line at time t and also i arrivals and j departures (after being served) during $(t, t + \Delta t)$, then the collection of events $\{E_{ij(k-i+j)}\}$ are disjoint and their union is E where (i, j) range over pairs of non-negative integers such that $k - i + j \geqslant 0$. Hence we have

$$f_k(t + \Delta t) = P(E) = \sum_{i,j} P(E_{ij(k-i+j)}).$$

Suppose we now make assumptions of independence concerning arrivals and also departures in disjoint time intervals similar to those concerning the occurrence of C's in disjoint time intervals involved in Problem 7.32. Under these assumptions the only events which have nonnegligible probabilities are

$$E_{00(k)}, E_{10(k-1)}, E_{01(k+1)}.$$

Their probabilities are: $f_k(t)[1 - (\lambda + \mu)\Delta t] + 0(\Delta t)$, $f_{k-1}(t)\lambda\Delta t + 0(\Delta t)$, and $f_{k+1}(t)\mu\Delta t + 0(\Delta t)$, the probabilities of all other events being of order $(\Delta t)^2$ or higher. Show that $f_k(t)$ satisfies the following system of differential equations:

$$f'_k(t) = [\lambda f_{k-1}(t) + \mu f_{k+1}(t) - (\lambda + \mu)f_k(t)], \quad k = 1, 2 \ldots$$

[For a detailed treatment of waiting-line (queueing) theory the reader is referred to Feller (1957), Kendall (1951, 1953) and Morse (1958). Morse's book has a substantial bibliography.]

7.35 (*Continuation*) For the steady state case in which $f'_k(t) = 0$ show that the solution of the resulting difference equation, namely,

$$\lambda f_{k-1} + \mu f_{k+1} - (\lambda + \mu)f_k = 0,$$

where $f_{-1} = 0$, is given by $f_k = \rho^k f_0$, where $\rho = \lambda/\mu$.
If the facilities will only accommodate a queue of length n, then show that

$$f_0 = \frac{(1 - \rho)}{(1 - \rho^{n+1})}$$

* By *queue* here we mean those customers who are either waiting to be served or are being served. Some authors use the term to mean only those waiting to be served.

Also show that $\mathscr{E}(k)$ the mean length of the queue (in the steady state case) is given by

$$\mathscr{E}(k) = \frac{\rho - (n+1)\rho^{n+1} + n\rho^{n+2}}{(1-\rho)(1-\rho^{n+1})}.$$

Also show that the mean length of the queue for the case $\rho = 1$ (average rates of arrivals and of departures being equal) is $\frac{1}{2}n$, and the mean length of queue for $\rho < 1$ for indefinitely large queueing accommodations ($n = \infty$) is $\rho/(1-\rho)$. Furthermore, show that the mean number $\mathscr{E}(l)$ of number of customers l *waiting to be served* is given by

$$\mathscr{E}(l) = \sum_{i=1}^{n} (i-1) f_i = \frac{\rho^2 - n\rho^{n+1} + (n-1)\rho^{n+2}}{(1-\rho)(1-\rho^n)}.$$

CHAPTER 8

Sampling Theory

8.1 DEFINITION OF A RANDOM SAMPLE

Suppose x is a one-dimensional random variable with c.d.f. $F(x)$ in R_1. *A random sample of size n from a population with c.d.f.* $F(x)$ is defined as the n-dimensional random variable $(x_1, \ldots, x_n)$ with c.d.f.

$$\prod_{\xi=1}^{n} F(x_\xi) \tag{8.1.1}$$

in the sample space $R_n = R_1^{(1)} \times \cdots \times R_1^{(n)}$, where $R_1^{(\xi)}$ is the one-dimensional sample space (the real line) of x_ξ, $\xi = 1, \ldots, n$. It should be especially noted that the elements or components $x_1, \ldots, x_n$ of the sample are mutually independent and all have the same c.d.f. In mathematical statistics the notion of a random sample as defined above was originally introduced by Fisher (1915), although he did not actually use the term "sample space" in referring to R_n. The term "sample space" was used in this sense however until the 1940's, and since then it has been used in the wider sense as stated in Chapter 1. The word "random" has been used mainly for descriptive effect, and will usually be omitted. We shall frequently abbreviate further and denote the sample $(x_1, \ldots, x_n)$ by O_n. For the sake of brevity we usually say that $(x_1, \ldots, x_n)$ (or O_n) is a sample of size n from $F(x)$. Note that a sample of size n from $F(x)$ is a simple example of a finite stochastic process, and is sometimes called *simple random sampling.*

If x is a random variable of the discrete type, with p.f. $p(x)$, then the sample has a p.f.

$$\prod_{\xi=1}^{n} p(x_\xi) \tag{8.1.2}$$

in R_n, and if x is a random variable of the continuous type the sample has a p.e.

$$f(x_1) \cdots f(x_n)\, dx_1 \cdots dx_n \tag{8.1.3}$$

in R_n. For the sake of brevity we often refer to $(x_1, \ldots, x_n)$ as a sample from $p(x)$ in the discrete case, and as a sample from $f(x)$ in the continuous case.

Remark. It is convenient to think of x_1 as the random variable denoting the value of x in the population obtained in the first "drawing," x_2 that obtained in the second "drawing," and so on. For example, if we throw a *true* die n times successively, we can regard x_ξ as a random variable denoting the number of dots appearing at the ξth throw, $\xi = 1, \ldots, n$. Our sample O_n in this case consists of the independent random variables $x_1, \ldots, x_n$ each having the p.f. defined by

$$p(x) = \tfrac{1}{6}, \qquad x = 1, \ldots, 6.$$

The p.f. of the sample O_n is

$$p(x_1) \cdots p(x_n) = (\tfrac{1}{6})^n$$

defined at each of 6^n mass points in R_n whose coordinates correspond to the 6^n *possible* sequences of faces which could be obtained in throwing a die n times.

In the theory of sampling we are usually interested in the distribution function of one or more functions of the n random variables comprising the sample; for example, the *sample sum* $z = \sum_{\xi=1}^{n} x_\xi$, the *sample mean* $\bar{x} = \frac{1}{n} \sum_{\xi=1}^{n} x_\xi$, the *sample variance* $s^2 = \frac{1}{n-1} \sum_{\xi=1}^{n} (x_\xi - \bar{x})^2$, the smallest sample element $\min(x_1, \ldots, x_n)$, the largest sample element $\max(x_1, \ldots, x_n)$, etc. In general, if $g(x_1, \ldots, x_n)$ is such a function which is itself a random variable, we are interested in determining its distribution function. Such a function $g(x_1, \ldots, x_n)$ is called a *statistic*, whose c.d.f., say $H(y)$, is a special case of (2.8.12), that is,

$$H(y) = P(g(x_1, \ldots, x_n) \leqslant y) = \int_{g^{-1}(y)} dF(x_1) \cdots dF(x_n) \tag{8.1.4}$$

where $g^{-1}(y)$ is the set in R_n for which $g(x_1, \ldots, x_n) \leqslant y$. The function $H(y)$ is the c.d.f. of $g(x_1, \ldots, x_n)$.

Similarly, if $g_i(x_1, \ldots, x_n)$, $i = 1, \ldots, s \leqslant n$ are s (functionally independent) statistics, we are interested in determining the c.d.f. of these statistics. The c.d.f. of the $g_i(x_1, \ldots, x_n)$ is defined by

$$\begin{aligned} H(y_1, \ldots, y_s) &= P(g_i(x_1, \ldots, x_n) \leqslant y_i,\ i = 1, \ldots, s) \\ &= \int_{g^{-1}(y_1, \ldots, y_n)} dF(x_1) \cdots dF(x_n), \end{aligned} \tag{8.1.5}$$

where $g^{-1}(y_1, \ldots, y_n)$ is the set in R_n for which

$$g_i(x_1, \ldots, x_n) \leqslant y_i, \qquad i = 1, \ldots, s.$$

The random variable x may be a k-dimensional random variable $(x_1, \ldots, x_k)$ with c.d.f. $F(x_1, \ldots, x_k)$ in R_k, in which case the sample O_n is a nk-dimensional random variable $(x_{1\xi}, \ldots, x_{k\xi}; \xi = 1, \ldots, n)$ with c.d.f.

$$\prod_{\xi=1}^{n} F(x_{1\xi}, \ldots, x_{k\xi}) \tag{8.1.6}$$

in $R_{nk} = R_k^{(1)} \times \cdots \times R_k^{(n)}$, where $R_k^{(\xi)}$ is the sample space of $(x_{1\xi}, \ldots, x_{k\xi})$, $\xi = 1, \ldots, n$. Again, the sampling distribution problem is to determine the distribution function of one or more functions of the nk random variables $x_{i\xi}$, $i = 1, \ldots, k$; $\xi = 1, \ldots, n$. For example, we may be interested in dealing with such statistics as *sample sums* $z_i = \sum_{\xi=1}^{n} x_{i\xi}$, with *sample means* $\bar{x}_i = \frac{1}{n}\sum_{\xi=1}^{n} x_{i\xi}$ or with the elements

$$s_{ij} = \frac{1}{n-1}\sum_{\xi=1}^{n}(x_{i\xi} - \bar{x}_i)(x_{j\xi} - \bar{x}_j)$$

of the *sample covariance matrix*, etc.

Sometimes we have to deal with sampling theory of functions of two or more random samples. For instance, suppose O_{n_1}: $(x_{11}, \ldots, x_{1n_1})$ and O_{n_2}: $(x_{21}, \ldots, x_{2n_2})$ are samples of size n_1 and n_2 respectively, from populations having c.d.f.'s $F_1(x)$ and $F_2(x)$ respectively. This means that we consider $(x_{11}, \ldots, x_{1n_1}, x_{21}, \ldots, x_{2n_2})$ as an $(n_1 + n_2)$-dimensional random variable having c.d.f.

$$\prod_{\xi_1=1}^{n_1} F_1(x_{1\xi_1}) \cdot \prod_{\xi_2=1}^{n_2} F_2(x_{2\xi_2}) \tag{8.1.7}$$

in $R_{n_1+n_2}$. The particular sampling problems with which we shall be concerned will be the determination of the distribution of one or more functions of the components of this $(n_1 + n_2)$-dimensional random variable, or at least the determination of certain properties of such a distribution. Similar remarks hold for three or more samples.

In the sampling distribution problems of mathematical statistics, we are usually interested in relatively simple statistics, such as averages, sums of squares, ratios, covariances, etc. Simple and explicit expressions exist for the p.f. or the p.d.f. of such sampling distributions only if population distributions have certain special forms, as we shall see in subsequent sections. However, one can determine means, variances, covariances, and some of the lower moments of these statistics for rather general population distributions by applying the results of Sections 3.3, 3.4, and 3.5. We shall consider some of the more important results of this type in the next section.

8.2 MEANS AND VARIANCES OF MEAN, VARIANCE, AND OTHER SYMMETRIC FUNCTIONS OF A SAMPLE

(a) Mean and Variance of Sample Mean

Suppose $(x_1, \ldots, x_n)$ is a sample from a population whose distribution has mean μ and variance σ^2. Consider the problem of determining the mean value and variance of the sample mean $\bar{x}$.

We have, by definition,

$$\bar{x} = \frac{1}{n}(x_1 + \cdots + x_n). \tag{8.2.1}$$

Taking the mean value of both sides of (8.2.1), we have

$$\mathscr{E}(\bar{x}) = \frac{1}{n}[\mathscr{E}(x_1) + \cdots + \mathscr{E}(x_n)]. \tag{8.2.2}$$

But since $x_1, \ldots, x_n$ are random variables having identical c.d.f.'s, we have,

$$\mathscr{E}(x_1) = \cdots = \mathscr{E}(x_n) = \mu.$$

Substituting these values in (8.2.2) we have

$$\mathscr{E}(\bar{x}) = \mu. \tag{8.2.3}$$

Without attempting a discussion of the basic concepts and principles of statistical estimation at this stage, we remark that (8.2.3) asserts that $\bar{x}$ is an *unbiased estimator* for μ. A treatment of estimators will be presented in Chapters 10, 11, and 12.

Now consider the variance of $\bar{x}$. We have

$$\sigma^2(\bar{x}) = \mathscr{E}[\bar{x} - \mathscr{E}(\bar{x})]^2.$$

But

$$\bar{x} - \mathscr{E}(\bar{x}) = \frac{1}{n}\sum_{\xi}(x_\xi - \mu).$$

Hence

$$\sigma^2(\bar{x}) = \mathscr{E}\left[\frac{1}{n}\sum_{\xi}(x_\xi - \mu)\right]^2$$

$$= \frac{1}{n^2}\sum_{\xi}\mathscr{E}(x_\xi - \mu)^2 + \frac{1}{n^2}\sum_{\xi \neq \eta}\mathscr{E}[(x_\xi - \mu)(x_\eta - \mu)].$$

Since $x_1, \ldots, x_n$ are independent and all have the same distribution, we have

$$\begin{aligned} &\mathscr{E}(x_\xi - \mu)^2 = \sigma^2, \qquad \xi = 1, \ldots, n \\ &\mathscr{E}[(x_\xi - \mu)(x_\eta - \mu)] = 0, \qquad \xi \neq \eta. \end{aligned} \tag{8.2.4}$$

Therefore

$$\sigma^2(\bar{x}) = \frac{\sigma^2}{n}. \tag{8.2.5}$$

(b) Mean and Variance of Sample Variance

Now consider the *sample variance* which is defined as

$$s^2 = \frac{1}{n-1}\sum_{\xi=1}^{n}(x_\xi - \bar{x})^2. \tag{8.2.6}$$

We can write (8.2.6) as

$$s^2 = \frac{1}{n-1}\sum_{\xi}\left[(x_\xi - \mu) - \frac{1}{n}\sum_{\eta}(x_\eta - \mu)\right]^2 \tag{8.2.6a}$$

$$= \frac{1}{n}\sum_{\xi}(x_\xi - \mu)^2 - \frac{1}{n(n-1)}\sum_{\xi \neq \eta}(x_\xi - \mu)(x_\eta - \mu).$$

Taking mean values and using (8.2.4), we have

$$\mathscr{E}(s^2) = \sigma^2. \tag{8.2.7}$$

We remark at this point that the reason for using $n - 1$ as the divisor in (8.2.6) rather than n is to make $\mathscr{E}(s^2)$ exactly equal to σ^2, that is, to make s^2 an *unbiased estimator* for σ^2.

Carrying out similar mean value operations we find after some reduction that

$$\mathscr{E}[(s^2)^2] = \frac{\mu_4}{n} + \frac{(n-1)^2 + 2}{n(n-1)}\sigma^4 \tag{8.2.8}$$

where μ_4 is the fourth central moment of the population distribution. Substituting from (8.2.7) and (8.2.8) in

$$\sigma^2(s^2) = \mathscr{E}[(s^2)^2] - [\mathscr{E}(s^2)]^2$$

we find for the variance of s^2

$$\sigma^2(s^2) = \frac{1}{n}\left(\mu_4 - \frac{n-3}{n-1}\sigma^4\right). \tag{8.2.9}$$

We can summarize as follows:

8.2.1 *If $(x_1, \ldots, x_n)$ is a sample from a distribution with mean μ, and variance σ^2, the sample mean $\bar{x}$ has mean and variance*

$$\mathscr{E}(\bar{x}) = \mu, \qquad \sigma^2(\bar{x}) = \frac{\sigma^2}{n}.$$

Furthermore, if the fourth moment μ_4 of the population distribution is *finite, then the sample variance s^2 has mean and variance*

$$\mathscr{E}(s^2) = \sigma^2, \qquad \sigma^2(s^2) = \frac{1}{n}\left(\mu_4 - \frac{n-3}{n-1}\sigma^4\right).$$

The means and variances of higher sample moments can be carried out in a similar manner, although the tediousness of the process increases rapidly. The reader interested in further results should consult Kendall (1943).

(c) Fisher's k-Statistics

For some purposes the semi-invariants of a distribution are more convenient to deal with than the moments of the distribution. In this section we shall consider certain unbiased estimators of the semi-invariants of a distribution from a sample from the distribution.

Fisher (1928*a*) has devised functions $k_r(x_1, \ldots, x_n)$, $r = 1, 2, \ldots$, where $k_r(x_1, \ldots, x_n)$ is the most general homogeneous polynomial of degree r in $x_1, \ldots, x_n$ subject to the conditions that (i) $k_r(x_1, \ldots, x_n)$ is symmetric in $x_1, \ldots, x_n$, (ii) $\mathscr{E}(k_r) = \kappa_r$, where κ_r is the rth semi-invariant (cumulant) of the distribution from which the sample is drawn as defined in (5.1.11). Such functions are called *k-statistics*. It is sufficient to indicate briefly how the coefficients are determined by considering the first three k-statistics. The reader interested in further details about the construction and sampling theory of k-statistics and related problems should consult Craig (1928), Fisher (1928*a*), Cornish and Fisher (1937), Dwyer (1933), and Kendall (1943).

The first three k-statistics are seen to be of the form

$$\begin{aligned} k_1 &= a_1 \sum_{\xi} x_\xi \\ k_2 &= a_{21} \sum_{\xi} x_\xi^2 + a_{22} \sum_{\xi \neq \eta} x_\xi x_\eta \qquad (8.2.10) \\ k_3 &= a_{31} \sum_{\xi} x_\xi^3 + a_{32} \sum_{\xi \neq \eta} x_\xi^2 x_\eta + a_{33} \sum_{\xi \neq \eta \neq \zeta} x_\xi x_\eta x_\zeta . \end{aligned}$$

By taking mean values of these three quantities and equating them to the first three semi-invariants of the population distribution, namely

$$\begin{aligned} \kappa_1 &= \mu_1' = \mu \\ \kappa_2 &= \mu_2' - (\mu_1')^2 = \sigma^2 \qquad (8.2.11) \\ \kappa_3 &= \mu_3' - 3\mu_1'\mu_2' + 2(\mu_1')^3 \end{aligned}$$

where μ_1', μ_2', and μ_3' are the first three moments of the population distribution, we determine the constants. Inserting the values of the constants thus found in (8.2.10) we have as the first three k-statistics

$$k_1 = \frac{1}{n} z_1 = \bar{x}$$

$$(8.2.12) \qquad k_2 = \frac{1}{n(n-1)}(nz_2 - z_1^2) = s^2$$

$$k_3 = \frac{1}{n(n-1)(n-2)}(n^2 z_3 - 3nz_1z_2 + 2z_1^3),$$

where
$$z_i = \sum_{\xi} x_{\xi}^i, \qquad i = 1, 2, 3.$$

Note that the first two k-statistics are simply $\bar{x}$ and s^2 whose mean values are μ and σ^2, these being the first two semi-invariants κ_1 and κ_2.

(d) Mean and Variance of Certain Symmetric Functions of a Sample

It will be noted that the sample mean $\bar{x}$, the sample variance s^2, the sample moments, and the sample k-statistics are all symmetric functions of the elements of a sample $(x_1, \ldots, x_n)$. Now let us consider the problem of determining the mean and variance of a more general class of symmetric functions of a sample.

Suppose $(x_1, \ldots, x_r)$ is a sample of size r from a c.d.f. $F(x)$, and let $g(x_1, \ldots, x_r)$ be a function of $(x_1, \ldots, x_r)$ such that

$$(8.2.13) \qquad \mathscr{E}(g^i(x_1, \ldots, x_r)) = \begin{cases} \theta_1 & i = 1 \\ \theta_2 & i = 2 \end{cases}$$

where θ_1 and θ_2 are both finite. For a sample $(x_1, \ldots, x_n)$ of size n, $n \geqslant r$ let

$$(8.2.14) \qquad g_0(x_{\eta_1}, \ldots, x_{\eta_r}) = \frac{1}{r!} \sum\nolimits_p g(x_{\xi_1}, \ldots, x_{\xi_r})$$

where $\eta_1, \ldots, \eta_r$ is a selection of r of the integers $1, \ldots, n$ with $\eta_1 < \cdots < \eta_r$ and where $\sum_p$ denotes summation over all $r!$ permutations $(\xi_1, \ldots, \xi_r)$ of $(\eta_1, \ldots, \eta_r)$. We note that $g_0(x_{\eta_1}, \ldots, x_{\eta_r})$ is a symmetric function of $(x_{\eta_1}, \ldots, x_{\eta_r})$. Let

$$(8.2.15) \qquad Q_{[r]}(x_1, \ldots, x_n) = \binom{n}{r}^{-1} \sum\nolimits_c g_0(x_{\eta_1}, \ldots, x_{\eta_r})$$

where $\sum_c$ denotes summation over all $\binom{n}{r}$ selections of $(\eta_1, \ldots, \eta_r)$ out of $(1, \ldots, n)$. $Q_{[r]}(x_1, \ldots, x_n)$ is a symmetric function of the sample components $(x_1, \ldots, x_n)$. Now $\mathscr{E}(g(x_{\xi_1}, \ldots, x_{\xi_r}))$, which is assumed to be finite, has the same value, namely θ_1, for all permutations $(x_{\xi_1}, \ldots, x_{\xi_r})$ of the elements of the sample $(x_1, \ldots, x_n)$. Furthermore, it is evident that

$$(8.2.16) \qquad \mathscr{E}(g_0(x_{\eta_1}, \ldots, x_{\eta_r})) = \frac{1}{r!} \sum\nolimits_p \mathscr{E}(g(x_{\xi_1}, \ldots, x_{\xi_r})) = \theta_1$$

and also

$$(8.2.17) \qquad \mathscr{E}(Q_{[r]}(x_1, \ldots, x_n)) = \binom{n}{r}^{-1} \sum\nolimits_c \mathscr{E}(g_0(x_{\eta_1}, \ldots, x_{\eta_r})) = \theta_1.$$

Now consider the variance of $Q_{[r]}(x_1, \ldots, x_n)$. We have

$$(8.2.18) \qquad \sigma^2(Q) = \mathscr{E}(Q^2_{[r]}(x_1, \ldots, x_n)) - \theta_1^2.$$

But

$$(8.2.19) \qquad \mathscr{E}(Q^2_{[r]}(x_1, \ldots, x_n)) = \binom{n}{r}^{-2} \sum\nolimits_c' \mathscr{E}(g_0(x_{\eta_1}, \ldots, x_{\eta_r})g_0(x_{\zeta_1}, \ldots, x_{\zeta_r}))$$

where $\eta_1, \ldots, \eta_r$ and $\zeta_1, \ldots, \zeta_r$ are two selections from the integers $1, \ldots, n$ with $\eta_1 < \cdots < \eta_r$ and $\zeta_1 < \cdots < \zeta_r$, and where $\sum_c'$ denotes summation over all pairs of such selections. It is evident that if $\eta_1, \ldots, \eta_r$ and $\zeta_1, \ldots, \zeta_r$ is a pair of selections with j integers in common, then $\mathscr{E}(g_0(x_{\eta_1}, \ldots, x_{\eta_r})g_0(x_{\zeta_1}, \ldots, x_{\zeta_r}))$ has the same value, say φ_j, for every such pair of selections. It follows from Schwarz' inequality that φ_j, $j = 0, 1, \ldots, r$ are all finite if θ_1 and θ_2 in (8.2.13) are finite. It is evident from combinatorial considerations that there are $\binom{n}{r}\binom{r}{j}\binom{n-r}{r-j}$ such pairs of selections, and j can take on the values $0, 1, \ldots, r$. Therefore we have

$$(8.2.20) \qquad \mathscr{E}(Q^2_{[r]}(x_1, \ldots, x_n)) = \binom{n}{r}^{-1} \sum_{j=0}^{r} \binom{r}{j}\binom{n-r}{r-j}\varphi_j$$

Thus, we obtain

$$(8.2.21) \qquad \sigma^2(Q_{[r]}) = \binom{n}{r}^{-1} \sum_{j=0}^{r} \binom{r}{j}\binom{n-r}{r-j}\varphi_j - \theta_1^2.$$

If $g(x_{\xi_1}, \ldots, x_{\xi_r})$ is chosen so that $\theta_1 = 0$, then since $\varphi_0 = \theta_1^2$, $\sigma^2(Q_{[r]})$ reduces to

$$(8.2.22) \qquad \sigma^2(Q_{[r]}) = \binom{n}{r}^{-1} \sum_{j=1}^{r} \binom{r}{j}\binom{n-r}{r-j}\varphi_j,$$

a result due to Hoeffding (1948*a*).

Summarizing we have the following result:

8.2.2 *Suppose* $(x_1, \ldots, x_n)$ *is a sample from an arbitrary distribution, and let* $g(x_1, \ldots, x_r)$ *be a function which has finite first and second moments* θ_1 *and* θ_2. *For a sample* $(x_1, \ldots, x_n)$, $n \geqslant r$, *from the same distribution, let* $Q_{[r]}(x_1, \ldots, x_n)$ *be defined according to* (8.2.15). *Then* $Q_{[r]}(x_1, \ldots, x_n)$ *is an unbiased estimator for* θ_1 *which has variance given by* (8.2.21).

Hoeffding has shown that if θ_1 and θ_2 are finite, $\theta_2 > \theta_1^2$, the limiting distribution of $\sqrt{n}(Q_{[r]} - \theta_1)/\sigma(Q_{[r]})$, as $n \to \infty$, is $N(0, 1)$. He extends this result as well as **8.2.2** to the case where $Q_{[r]}(x_1, \ldots, x_n)$ is a vector.

8.3 SAMPLING THEORY OF SAMPLE SUMS AND MEANS

(a) An Iterative Method

We shall first consider an iterative method for finding the distribution function of a sample sum or mean. It furnishes a direct approach to the problem which is fairly straightforward in some specific cases.

Consider the case of samples of size 2. If we let $G_2(z)$ be the c.d.f. of z, we have

(8.3.1)
$$G_2(z) = P(x_1 + x_2 \leqslant z) = \int_E d(F(x_1)F(x_2))$$

where E is the event in R_2 for which $x_1 + x_2 \leqslant z$. If we apply **3.7.2** for the case in which x_1 and x_2 are independent, and put $g(x_1, x_2) = 1$ for all points in E and 0 for all points in $\bar{E}$, we can write

(8.3.2)
$$\int_E d(F(x_1)F(x_2)) = \int_{-\infty}^{\infty} \left[\int_{-\infty}^{z-x_1} dF(x_2)\right] dF(x_1) = \int_{-\infty}^{\infty} F(z - x_1)\, dF(x_1).$$

Hence, for samples of size 2, we have

(8.3.3)
$$G_2(z) = \int_{-\infty}^{\infty} F(z - x_1)\, dF(x_1).$$

Extending the argument to samples of size n, and denoting the c.d.f. of z by $G_n(z)$, one finds similarly, that $G_n(z)$ can be expressed as an iterated integral as follows:

(8.3.4)
$$G_n(z) = \int_{-\infty}^{\infty} \cdots \int_{-\infty}^{z-x_1-\cdots-x_{n-3}} \int_{-\infty}^{z-x_1-\cdots-x_{n-2}} F(z - x_1 - \cdots - x_{n-1}) \cdot dF(x_{n-1})\, dF(x_{n-2}) \cdots dF(x_1).$$

The c.d.f. of the sample mean $\bar{x}$, say $H_n(\bar{x})$, is given by the relation $H_n(\bar{x}) \equiv G_n(n\bar{x})$. Therefore,

8.3.1 *If* $(x_1, \ldots, x_n)$ *is a sample of size n from the c.d.f.* $F(x)$, *then* $G_n(z)$, *the c.d.f. of the sample sum* z, *is given by* (8.3.4), *and* $H_n(\bar{x})$, *the c.d.f. of the sample mean* $\bar{x}$ *is given by* $H_n(\bar{x}) \equiv G_n(n\bar{x})$.

The c.d.f. $G_n(z)$ is sometimes referred to as the *convolution* of the c.d.f.'s $F(x_1), \ldots, F(x_n)$. The process applies in a straightforward manner so as to produce the convolution of n independent random variables with arbitrary distributions $F_1(x_1), \ldots, F_n(x_n)$. The convolution in this case, of course, is the c.d.f. of the sum of the n independent random variables.

In the case of a sample from a population having a discrete distribution with p.f. $p(x)$, it can be verified that the p.f. of z, say $p_n(z)$, satisfies the following equation:

$$p_n(z) = \sum_x p(z - x)p_{n-1}(x) \tag{8.3.4a}$$

where $p_1(x) \equiv p(x)$.

Similarly, if we have a sample from population having a continuous distribution with p.d.f. $f(x)$, the p.d.f. of z, say $f_n(z)$, is given by

$$f_n(z) = \int_{-\infty}^{\infty} f(z - x)f_{n-1}(x)\,dx \tag{8.3.4b}$$

where $f_1(x) \equiv f(x)$.

Example. An interesting special case of (8.3.4*b*) is that in which $f(x)$ is the p.d.f. of the rectangular distribution $R(\frac{1}{2}, 1)$. In this case we have

$$f_1(z) = f(z) = \begin{cases} 0 & z \leqslant 0 \\ 1 & 0 < z \leqslant 1 \\ 0 & z > 1 \end{cases}$$

$$f_2(z) = \begin{cases} 0 & z \leqslant 0 \\ z & 0 < z \leqslant 1 \\ z - 2(z-1) & 1 < z \leqslant 2 \\ 0 & z > 2 \end{cases}$$

and by mathematical induction it follows that

$$f_n(z) = \frac{1}{(n-1)!} \cdot \left[z^{n-1} - \binom{n}{1}(z-1)^{n-1} + \binom{n}{2}(z-2)^{n-1} - \cdots + (-1)^k\binom{n}{k}(z-k)^{n-1}\right]$$

for $k < z \leqslant k + 1$, $k = 0, 1, \ldots, n - 1$, $f_n(z) = 0$ for $z \leqslant 0$ and $z > n$. The p.f. of $\bar{x}$ is seen to be $nf_n(n\bar{x})$ for $k/n < \bar{x} \leqslant (k + 1)/n$, $k = 0, 1, \ldots, n - 1$. This result is due to Laplace (1814).

(b) Application of Characteristic Functions

Characteristic functions furnish a simple and powerful method of determining sampling distribution functions of sums and means in many cases.

If we refer to **5.3.2** it is evident that we have the following corollary to that theorem:

8.3.2 *If $(x_1, \ldots, x_n)$ is a sample from the c.d.f. $F(x)$ and if the characteristic function of $F(x)$ is $\varphi(t)$, then the characteristic function of the sample sum z is*

$$[\varphi(t)]^n$$

and the characteristic function of the sample mean $\bar{x}$ is

$$\left[\varphi\left(\frac{t}{n}\right)\right]^n.$$

The c.d.f. (and p.d.f. in the continuous case) of z or $\bar{x}$ can be determined, theoretically at least, from their characteristic functions by the application of **5.1.2**. In actual applications, except in certain special cases, the evaluation of the integral on the right of (5.1.14) or of (5.1.15) is complicated. Irwin (1930) has made extensive use of this technique in determining distributions of sample means from various distributions. However, in many important special cases which arise in sampling theory, we can determine the distribution functions of z and $\bar{x}$ by utilizing **5.3.3** concerning the reproductive property of c.d.f.'s. For suppose $(x_1, \ldots, x_n)$ is a sample from a c.d.f. $F(x; \theta)$ having characteristic function $\varphi(t; \theta)$. If z is the sample sum then we know from **8.3.2** that the characteristic function of z is $[\varphi(t; \theta)]^n$. But if $\varphi(t; \theta)$ satisfies the criterion of reproductivity with respect to θ as expressed by (5.3.7) then

$$[\varphi(t; \theta)]^n = \varphi(t; n\theta). \tag{8.3.5}$$

This implies that z has the c.d.f. $F(z; n\theta)$. By setting $z = n\bar{x}$ where $\bar{x}$ is the sample mean, it is seen that the c.d.f. of $\bar{x}$ is $F(n\bar{x}; n\theta)$.

We may summarize as follows:

8.3.3 *Suppose $(x_1, \ldots, x_n)$ is a sample from the c.d.f. $F(x; \theta)$, and the characteristic function of $F(x; \theta)$ is $\varphi(t; \theta)$. Then if $[\varphi(t; \theta)]^n = \varphi(t; n\theta)$ the c.d.f. of the sampling distribution of the sample sum z and sample mean $\bar{x}$ are $F(z; n\theta)$ and $F(n\bar{x}; n\theta)$ respectively.*

It should be noted that an extension of this statement holds for sampling from a population having a k-variate distribution. In this case z is a vector $(z_1, \ldots, z_k)$ where $z_i = \sum_{\xi=1}^{n} x_{i\xi}$. The parameter θ can also be a vector with several components.

The following corollaries of **8.3.3** give information about the sampling theory of sample sums for certain important special distributions, which will be useful in later chapters.

8.3.3a *If* $(x_1, \ldots, x_n)$ *is a sample from the binomial distribution* $Bi(m, p)$, *the sampling distribution of* z *is the binomial distribution* $Bi(mn, p)$.

For the characteristic function of the binomial distribution $B(m, p)$ is $(q + e^{it}p)^m$ from which it follows that the characteristic function of z is $(q + e^{it}p)^{mn}$ and hence the sampling distribution of z is $B(mn, p)$. Note that for $m = 1$, we are sampling from the *binomial population* having p.f.

$$p(x) = \begin{cases} q, & x = 0 \\ p, & x = 1, \end{cases} \tag{8.3.6}$$

which means that the binomial distribution having p.f. (6.2.2) is essentially the distribution of a sample sum $z = \sum_{\xi=1}^{n} x_\xi$, each x_ξ being 0 or 1 with probabilities q or p, respectively.

8.3.3b *If* $(x_{1\xi}, \ldots, x_{k\xi};\ \xi = 1, \ldots, n)$ *is a sample from the multinomial distribution* $M(m; p_1, \ldots, p_k)$, *the sampling distribution of the vector of sample sums* $(z_1, \ldots, z_k)$ *is the multinomial distribution* $M(mn;\ p_1, \ldots, p_k)$.

The verification of this statement is similar to that for **8.3.3a**. Note that for $m = 1$, the multinomial distribution $M(n; p_1, \ldots, p_k)$ is itself a sampling distribution of sample sums from the k-variate multinomial distribution $M(1; p_1, \ldots, p_k)$ having p.f.

$$p(x_1, \ldots, x_k) = p_1^{x_1} \cdots p_k^{x_k}(1 - p_1 - \cdots - p_k)^{1-x_1-\cdots-x_k} \tag{8.3.7}$$

where $x_1, \ldots, x_k$ are random variables such that $x_1 + \cdots + x_k \leqslant 1$, each x being 0 or 1.

8.3.3c *If* $(x_1, \ldots, x_n)$ *is a sample from the Poisson distribution* $Po(\mu)$, *the sampling distribution of* z *is the Poisson distribution* $Po(n\mu)$,

For the characteristic function of the Poisson distribution $Po(\mu)$ is $e^{-\mu(1-e^{it})}$ from which it follows that the characteristic function of z is $e^{-n\mu(1-e^{it})}$. But this is the characteristic function of the Poisson distribution $Po(n\mu)$, which implies that z has the Poisson distribution $Po\,(n\mu)$.

8.3.3d *If* $(x_1, \ldots, x_n)$ *is a sample from the normal distribution* $N(\mu, \sigma^2)$ *the sampling distribution of* z *is* $N(n\mu, n\sigma^2)$. *The sampling distribution of* $\bar{x}$ *is* $N(\mu, \sigma^2/n)$.

For we recall from Section 7.2 that the characteristic function of the normal distribution $N(\mu, \sigma^2)$ is $e^{i\mu t - \frac{1}{2}\sigma^2 t^2}$ whereas that of z is $e^{i(n\mu)t - \frac{1}{2}(n\sigma^2)t^2}$.

But the latter is the characteristic function of $N(n\mu, n\sigma^2)$, which is therefore the sampling distribution of z. The characteristic function of $\bar{x}$ is obtained by replacing t by t/n in the characteristic function of z, which gives $e^{i\mu t - \frac{1}{2}(\sigma^2/n)t^2}$. This is the characteristic function of $N(\mu, \sigma^2/n)$, which is therefore the sampling distribution of $\bar{x}$.

The corresponding statement for the k-variate normal distribution is:

8.3.3e *If* $(x_{1\xi}, \ldots, x_{k\xi};\ \xi = 1, \ldots, n)$ *is a sample from the k-variate normal distribution* $N(\{\mu_i\}, \|\sigma_{ij}\|)$, *the sampling distribution of the vector of sample sums* $(z_1, \ldots, z_k)$ *is* $N(\{n\mu_i\}, \|n\sigma_{ij}\|)$. *The sampling distribution of the vector of sample means* $(\bar{x}_1, \ldots, \bar{x}_k)$ *is* $N(\{\mu_i\}, \|\sigma_{ij}/n\|)$.

The verification of **8.3.3e** is similar to that for **8.3.3d** and is left as an exercise for the reader.

Finally, we have the following further corollary of **8.3.3** for the gamma distribution which is left to the reader for verification:

8.3.3f *If* $(x_1, \ldots, x_n)$ *is a sample from the gamma distribution* $G(\mu)$ *the sampling distribution of* z *is the gamma distribution* $G(n\mu)$.

If the distribution function of the population from which the sample $(x_1, \ldots, x_n)$ is drawn is *not* reproductive, the characteristic function is not very useful, in general, for finding the distribution function of the sample sum or mean.

Sometimes we have to deal with linear functions of means of samples from normal populations. The following statement gives a useful result concerning the distribution of such a linear function:

8.3.4 *Suppose* $(x_{11}, \ldots, x_{1n_1}), \ldots, (x_{k1}, \ldots, x_{kn_k})$ *are k (independent) samples from* $N(\mu_1, \sigma_1^2), \ldots, N(\mu_k, \sigma_k^2)$, *respectively. Let* $\bar{x}_1, \ldots, \bar{x}_k$ *be means of these samples, respectively, and let* $c_1, \ldots, c_k$ *be constants, not all zero. Then* $c_1\bar{x}_1 + \cdots + c_k\bar{x}_k$ *has as its sampling distribution* $N\left(\sum_{i=1}^{k} c_i\mu_i, \sum_{i=1}^{k} \frac{c_i^2\sigma_i^2}{n_i}\right)$.

Verification of this statement by using characteristic functions is straightforward and is left as an exercise for the reader. Note particularly that if $c_1 = 1$, $c_2 = -1$, and $c_3 = \cdots = c_k = 0$, we obtain the corollary that

8.3.4a *The sampling distribution of the difference* $\bar{x}_1 - \bar{x}_2$ *between the means of two (independent) samples* $(x_{11}, \ldots, x_{1n_1})$ *and* $(x_{21}, \ldots, x_{2n_2})$ *from* $N(\mu_1, \sigma_1^2)$ *and* $N(\mu_2, \sigma_2^2)$ *respectively, is* $N(\mu_1 - \mu_2, \sigma_1^2/n_1 + \sigma_2^2/n_2)$.

8.4 SAMPLING THEORY OF CERTAIN QUADRATIC FORMS IN SAMPLES FROM A NORMAL DISTRIBUTION

The essential facts concerning the sampling theory of sums and means of samples from normal distributions have been given in **8.3.3d**, **8.3.3e**, **8.3.4**, and **8.3.4a**. The present section will deal with the sampling theory of sums of squares, sample variances, and other quadratic forms of samples from one-dimensional normal distributions. The corresponding sampling theory for samples from k-variate normal distributions is more complicated; it belongs to *normal multivariate analysis* and will be treated in Chapter 18.

One of the basic results in the sampling theory of sums of squares of sample elements is the following:

8.4.1 *If* $(x_1, \ldots, x_n)$ *is a sample from the normal distribution* $N(\mu, \sigma^2)$ *the sampling distribution of the sum of squares* $\frac{1}{\sigma^2}\sum_{\xi=1}^{n}(x_\xi - \mu)^2$ *is the chi-square distribution* $C(n)$.

To verify this statement it is sufficient to note that if the p.d.f. of the normal distribution $N(\mu, \sigma^2)$ is inserted in (8.1.3) then $\frac{1}{\sigma^2}\sum_{\xi=1}^{n}(x_\xi - \mu)^2$ is the quadratic form which appears in the exponent of the p.d.f. of the sample. It follows from **7.8.2** that this quadratic form has the chi-square distribution $C(n)$. The result given in **8.4.1** was originally obtained by Helmert (1876*a*).

The sum of squares which appears in the definition (8.2.6) of the sample variance s^2 is $\sum_{\xi=1}^{n}(x_\xi - \bar{x})^2$, where $\bar{x}$ is the sample mean. The sampling theory of this sum of squares in samples from a normal distribution can be stated as follows:

8.4.2 *If* $(x_1, \ldots, x_n)$ *is a sample from the normal distribution* $N(\mu, \sigma^2)$, *then* $\sqrt{n}(\bar{x} - \mu)/\sigma$ and $(n - 1)s^2/\sigma^2$ *are (statistically) independent and their sampling distributions are the normal distribution* $N(0, 1)$ *and the chi-square distribution* $C(n - 1)$ *respectively.*

We have already seen from **8.3.3d** that $\bar{x}$ has the distribution $N(\mu, \sigma^2/n)$, which is equivalent to the statement that $\sqrt{n}(\bar{x} - \mu)/\sigma$ has the distribution $N(0, 1)$. To establish **8.4.2** it remains to show that $\sqrt{n}(\bar{x} - \mu)/\sigma$ and $(n - 1)s^2/\sigma^2$ are independent and that the latter has the chi-square distribution $C(n-1)$. To do this, let us set up the characteristic function of these two quantities, namely,

(8.4.1)

$$\varphi(t_1, t_2) = \mathscr{E}\left[\exp\left(it_1(\bar{x}-\mu)\frac{\sqrt{n}}{\sigma} + it_2\frac{(n-1)s^2}{\sigma^2}\right)\right]$$

$$= \left(\frac{1}{\sqrt{2\pi}\sigma}\right)^n \int_{R_n} \exp\left\{-\frac{1}{2}\left[Q_0(x_1, \ldots, x_n) - 2it_1(\bar{x}-\mu)\frac{\sqrt{n}}{\sigma}\right]\right\} \cdot dx_1 \cdots dx_n$$

where

(8.4.2) $$Q_0(x_1, \ldots, x_n) = \frac{1}{\sigma^2}\sum_{\xi=1}^{n}(x_\xi - \mu)^2 - \frac{2it_2}{\sigma^2}\sum_{\xi=1}^{n}(x_\xi - \bar{x})^2.$$

Putting

(8.4.3) $$\tau^{\xi\xi} = \frac{1}{\sigma^2}\left[1 - 2it_2\left(1 - \frac{1}{n}\right)\right], \qquad \xi = 1, \ldots, n$$

$$\tau^{\xi\eta} = \tau^{\eta\xi} = \frac{2it_2}{n\sigma^2}, \qquad \xi \neq \eta = 1, \ldots, n$$

we have

(8.4.4) $$Q_0(x_1, \ldots, x_n) = \sum_{\xi,\eta=1}^{n} \tau^{\xi\eta}(x_\xi - \mu)(x_\eta - \mu).$$

By using (7.4.15), we note that

(8.4.5)

$$Q_0(x_1, \ldots, x_n) - \frac{2it_1}{\sigma}(\bar{x}-\mu)\sqrt{n} = Q_0'(x_1, \ldots, x_n) + \left(\sum_{\xi,\eta=1}^{n}\tau_{\xi\eta}\right)\frac{t_1^2}{n\sigma^2}$$

where

(8.4.6) $$Q_0'(x_1, \ldots, x_n) = Q_0(x_1 - ig_1, \ldots, x_n - ig_n)$$

and

$$\|\tau_{\xi\eta}\| = \|\tau^{\xi\eta}\|^{-1}; \qquad g_\xi = \frac{t_1}{\sqrt{n}\sigma}\sum_{\eta=1}^{n}\tau_{\xi\eta} \qquad \xi = 1, \ldots, n.$$

It will be observed that the matrix $\|\tau^{\xi\eta}\|$ is of the form

(8.4.7) $$\begin{Vmatrix} a & b \cdots & b \\ b & a \cdots & b \\ \cdot & \cdot\ \cdot & \cdot \\ \cdot & \cdot \quad \cdot & \cdot \\ \cdot & \cdot \qquad \cdot & \cdot \\ b & b \cdots & a \end{Vmatrix}$$

which has $(a-b)^{n-1}[a+(n-1)b]$ as its determinant, and

$$(8.4.8) \qquad \begin{Vmatrix} A & B \cdots & B \\ B & A \cdots & B \\ \cdot & \cdot\ \cdot & \cdot \\ \cdot & \cdot \quad \cdot & \cdot \\ \cdot & \cdot \qquad \cdot & \cdot \\ B & B \cdots & A \end{Vmatrix}$$

as its inverse, where

$$(8.4.9) \quad A = \frac{a+(n-2)b}{(a-b)[a+(n-1)b]} \qquad B = \frac{-b}{(a-b)[a+(n-1)b]}.$$

Therefore, we have

$$(8.4.10) \qquad \tau_{\xi\xi} = \frac{\sigma^2\left(1 - 2i\,\dfrac{t_2}{n}\right)}{(1-2it_2)}, \qquad \xi = 1, \ldots, n$$

$$\tau_{\xi\eta} = \frac{-\sigma^2\left(2i\,\dfrac{t_2}{n}\right)}{(1-2it_2)}, \qquad \xi \neq \eta = 1, \ldots, n$$

and

$$(8.4.11) \qquad \sum_{\xi,\eta=1}^{n} \tau_{\xi\eta} = n\sigma^2.$$

Substituting from (8.4.11) into (8.4.5) and putting the resulting expression on the right of (8.4.5) in the integral in (8.4.1), we obtain

$$(8.4.12) \quad \varphi(t_1, t_2) = e^{-\frac{1}{2}t_1^2} \cdot \left(\frac{1}{\sqrt{2\pi}\sigma}\right)^n \int_{R_n} e^{-\frac{1}{2}Q_0'(x_1, \ldots, x_n)}\, dx_1 \cdots dx_n.$$

Following the same method by which it was shown that $H = 1$ in (7.4.16), it can be seen that the integral in (8.4.12) has the value $\sqrt{(2\pi)^n}\,|\tau^{\xi\eta}|^{-\frac{1}{2}}$. But

$$|\tau^{\xi\eta}| = \frac{1}{\sigma^{2n}}(1-2it_2)^{n-1}.$$

Therefore, we have

$$(8.4.13) \qquad \varphi(t_1, t_2) = \varphi_1(t_1) \cdot \varphi_2(t_2)$$

where

$$(8.4.14) \qquad \varphi_1(t_1) = e^{-\frac{1}{2}t_1^2}$$

and

$$(8.4.15) \qquad \varphi_2(t_2) = (1-2it_2)^{-\frac{1}{2}(n-1)}.$$

Since $\varphi(t_1, t_2)$ factors in the manner indicated in (8.4.13), it follows from **5.3.1** that $\sqrt{n}(\bar{x} - \mu)/\sigma$ and $(n - 1)s^2/\sigma^2$ are independent, their characteristic functions (8.4.14) and (8.4.15) being those of the distribution $N(0, 1)$ (see **7.2.1**) and chi-square distribution $C(n - 1)$ [see (7.8.2)] respectively. These distributions are uniquely determined by their characteristic functions, thus completing the argument for **8.4.2.**

The following important result follows from **8.4.2** and Section 7.8(*b*):

8.4.3 *If* $(x_1, \ldots, x_n)$ *is a sample from* $N(\mu, \sigma^2)$, *then*

$$t = \sqrt{n}(\bar{x} - \mu)/s \tag{8.4.16}$$

has the "Student" distribution $S(n - 1)$.

For if we let $u = \sqrt{n}(\bar{x} - \mu)/\sigma$ and $v = (n - 1)s^2/\sigma^2$ in (7.8.3) we have $t = \dfrac{u}{\sqrt{v/(n-1)}} = \sqrt{n}(\bar{x} - \mu)/s$. But if the sample comes from $N(\mu, \sigma^2)$ we know by **8.4.2** that u and v are independent and have as their distributions $N(0, 1)$ and the chi-square distribution $C(n - 1)$, respectively. Hence, applying **7.8.3** we obtain **8.4.3.**

Remark. The result stated in **8.4.3** was first surmised by W. S. Gosset (1908) writing under the name "Student". This result, as well as that stated in **8.4.2**, was later verified by Fisher (1926*a*), who attached the name "Student" to the ratio defined by (8.4.16) as well as to the more general distribution having p.d.f. (7.8.4). The fact that v has the chi-square distribution with $n - 1$ degrees of freedom was first established by Helmert (1876*b*).

It should be remarked that the converse of **8.4.2** is also true, which, stated in its essential form, is that if $\bar{x}$ and s^2 are the mean and variance of a sample of size n, from a p.d.f. $f(x)$, and if $\bar{x}$ and s^2 are independent then $f(x)$ is the p.d.f. of a normal distribution. This result was originally obtained by Geary (1936), and more recently by Kawata and Sakamoto (1949), and by Lukacs (1942). See Problem 8.33.

More basically Cramér (1936) has shown that if x_1 and x_2 are independent random variables whose sum has a normal distribution, then x_1 and x_2 each has a normal distribution. This result extends by induction to the case of several random variables.

The important feature of t as defined by (8.4.16) for statistical inference, as we shall see in later chapters, is that [assuming the sample comes from the distribution $N(\mu, \sigma^2)$] neither it nor its distribution depends on σ^2.

Again, let us recall that the p.d.f. of a sample $(x_1, \ldots, x_n)$ from $N(\mu, \sigma^2)$ has the sum of squares $Q = \dfrac{1}{\sigma^2}\sum_{\xi=1}^{n}(x_\xi - \mu)^2$ in its exponent and we know from **8.4.1** that Q has the chi-square distribution $C(n)$. Note that Q can be decomposed as follows:

(8.4.17) $$Q = Q_1 + Q_2$$

where

(8.4.18) $$Q_1 = \frac{1}{\sigma^2}\sum_{\xi=1}^{n}(x_\xi - \bar{x})^2, \qquad Q_2 = \frac{n}{\sigma^2}(\bar{x} - \mu)^2.$$

We have seen that $\sqrt{n}(\bar{x} - \mu)/\sigma$ has distribution $N(0, 1)$, and it follows from **7.8.2a** that $[\sqrt{n}(\bar{x} - \mu)/\sigma]^2$, that is, Q_2, has the chi-square distribution $C(1)$. Since Q_1 and $\sqrt{n}(\bar{x} - \mu)/\sigma$ are independent, so are Q_1 and Q_2. Thus Q, which has the chi-square distribution $C(n)$, has been decomposed into two components Q_1 and Q_2 which are independent and have chi-square distributions $C(n - 1)$ and $C(1)$, respectively. This leads to the question whether there are conditions under which the sum of squares appearing in the exponent of the sample p.d.f. can be decomposed into several independent components having chi-square distributions if the sample comes from a normal distribution. An answer is given in the following theorem due to Cochran (1934). It is sufficient to consider the case of sampling from the normal distribution $N(0, 1)$.

8.4.4 *If* $(x_1, \ldots, x_n)$ *is a sample from the distribution* $N(0, 1)$ *and if* $\sum_{\xi=1}^{n} x_\xi^2 \equiv \sum_{i=1}^{k} Q_i$, *where* Q_i *is a non-negative quadratic form in* $x_1, \ldots, x_n$ *whose matrix has rank* n_i, *a necessary and sufficient condition for the* Q_i *to be independently distributed according to chi-square distributions* $C(n_i)$, $i = 1, \ldots, k$, *is that* $\sum_{i=1}^{k} n_i = n$.

First, let us show that the condition is *necessary*. Thus we assume that the Q_i are independently distributed according to chi-square distributions $C(n_i)$, $i = 1, \ldots, k$ and we must show that this implies that $\sum_{i=1}^{k} n_i = n$. But it follows from the reproductive property of the chi-square distribution that $Q_1 + \cdots + Q_k$ has the chi-square distribution $C(n_1 + \cdots + n_k)$. But $Q_1 + \cdots + Q_k \equiv \sum_{\xi=1}^{n} x_\xi^2$ and since the sample is from $N(0, 1)$ it follows from **8.4.1** that $\sum_{i=1}^{k} x_\xi^2$ has the chi-square distribution $C(n)$, which must therefore be identical with $C(n_1 + \cdots + n_k)$. Hence $\sum_{i=1}^{k} n_i = n$.

To establish the *sufficiency* of the condition stated in **8.4.4** we assume that $n_1 + \cdots + n_k = n$, and we must show that there exists a nonsingular linear transformation

(8.4.19)
$$x_\xi = \sum_{\eta=1}^{n} b_{\xi\eta} y_\eta, \qquad \xi = 1, \ldots, n$$

which transforms $Q_1, \ldots, Q_k$ as follows:

(8.4.20)
$$\begin{aligned} Q_1 &= y_1^2 + \cdots + y_{n_1}^2 \\ Q_2 &= y_{n_1+1}^2 + \cdots + y_{n_1+n_2}^2 \\ &\cdots\cdots \\ Q_k &= y_{n_1+\cdots+n_{k-1}+1}^2 + \cdots + y_{n_1+\cdots+n_k}^2 \end{aligned}$$

where $n_1 + \cdots + n_k = n$, and where $y_1, \ldots, y_n$ are n independent random variables all having the distribution $N(0, 1)$. For if this can be done, such a transformation will transform the p.e. of $(x_1, \ldots, x_n)$ to the p.e. of a sample $(y_1, \ldots, y_n)$ from the distribution $N(0, 1)$. Thus, $Q_1, \ldots, Q_k$, being sums of squares of $n_1, \ldots, n_k$ mutually exclusive and independent random variables all having distributions $N(0, 1)$, would be independent and by **8.4.1** would have chi-square distributions $C(n_1), \ldots, C(n_k)$, respectively.

Since Q_1 is a non-negative quadratic form with matrix of rank n_1 there exist n_1 linearly independent linear combinations of $x_1, \ldots, x_n$ which we may denote by $y_1, \ldots, y_{n_1}$ (see Bôcher (1907) or Birkhoff and MacLane (1953), for example), namely

(8.4.21)
$$y_{\xi_1} = \sum_{\eta=1}^{n} b^{\xi_1\eta} x_\eta, \qquad \xi_1 = 1, \ldots, n_1$$

such that

(8.4.22)
$$Q_1 = y_1^2 + \cdots + y_{n_1}^2.$$

Similarly for $Q_2, \ldots, Q_k$ we have the following sets of $n_2, \ldots, n_k$ linearly independent linear combinations of $x_1, \ldots, x_n$ respectively:

(8.4.23)
$$\begin{aligned} y_{\xi_2} &= \sum_{\eta=1}^{n} b^{\xi_2\eta} x_\eta, \qquad \xi_2 = n_1 + 1, \ldots, n_1 + n_2 \\ &\vdots \\ y_{\xi_k} &= \sum_{\eta=1}^{n} b^{\xi_k\eta} x_\eta, \qquad \xi_k = n_1 + \cdots + n_{k-1} + 1, \ldots, n_1 + \cdots + n_{k-1} + n_k. \end{aligned}$$

We have shown that no linear dependence can exist among the y's associated with a single Q. We must next show that there can be no linear dependence among y's associated with two or more different Q's. If this were the case it would be possible to express a y associated with one Q, say Q^*, as a homogeneous linear combination of y's from one or more Q's different from Q^*. This means that $Q_1 + \cdots + Q_k$ is expressible as a non-negative quadratic form in at most $n - 1$ y's. Since each y is a homogeneous linear combination of $x_1, \ldots, x_n$, $Q_1 + \cdots + Q_k$ would then be a non-negative quadratic form in the x's whose matrix would be at most of rank $n - 1$. But we know that the matrix of $Q_1 + \cdots + Q_k$, being equal to that of $\sum_1^n x_\xi^2$, is of rank n. Hence, it is not possible to have linear dependence between the y's associated with one Q, say Q^*, and those associated with one or more Q's different from Q^*. Therefore the matrix

$$\|b^{\xi\eta}\|, \qquad \xi, n = 1, \ldots, n$$

is nonsingular. Hence the transformation

$$y_\xi = \sum_{\eta=1}^n b^{\xi\eta} x_\eta, \qquad \xi = 1, \ldots, n \tag{8.4.23}$$

is non-singular and has a unique inverse

$$x_\eta = \sum_{\xi=1}^n b_{\xi\eta} y_\xi, \qquad \eta = 1, \ldots, n, \tag{8.4.24}$$

which being a linear transformation, transforms $\sum_1^n x_\xi^2$, and hence $Q_1 + \cdots + Q_k$, into

$$(y_1^2 + \cdots + y_{n_1}^2) + (y_{n_1+1}^2 + \cdots + y_{n_1+n_2}^2) + \cdots + (y_{n_1+\cdots+n_{k-1}+1}^2 + \cdots + y_n^2).$$

The transformation is seen to be orthogonal and hence the y's are independent random variables all having the distribution $N(0, 1)$. Therefore, $Q_1, \ldots, Q_k$ have distributions which are identical with those of $(y_1^2 + \cdots + y_{n_1}^2), \ldots, (y_{n_1+\cdots+n_{k-1}+1}^2 + \cdots + y_n^2)$ respectively, which in turn are independent random variables with chi-square distributions $C(n_1), \ldots, C(n_k)$ respectively. This completes the proof of the sufficiency condition and concludes the proof of **8.4.4**.

Cochran's theorem has important applications in analysis of variance and normal regression theory as we shall see in Chapter 10.

8.5 SAMPLING FROM A FINITE POPULATION

(a) The One-Dimensional Case

In the definition of a random sample $(x_1, \ldots, x_n)$ given in Section 8.1, the elements of the sample were defined as *independent* random variables all having the same c.d.f. $F(x)$. The theory of sampling based on this definition is sometimes referred to as *simple random sampling from an infinite population* where the population has $F(x)$ as its c.d.f. This terminology is suggested by the fact that since the sample elements $x_1, \ldots, x_n$ are regarded as a sequence of independent random variables all having the same distribution as that of the population, we may think of $x_1, \ldots, x_n$ as describing the results of successive "drawings" from the population, the population distribution remaining *unchanged* throughout the "drawings." Unless otherwise stated, "a sample of size n" will always refer to sampling from an infinite population.

There is also a theory of *simple random sampling from a finite population* π_N, in which the population π_N consists of a finite number of N objects, $o_1, \ldots, o_N$. If we regard π_N as a basic sample space with $o_1, \ldots, o_N$ as sample points and if we let $x(o)$ be a random variable defined on each of these sample points, so that $x(o_t) = x_{ot}$, $t = 1, \ldots, N$, then $x(o)$ maps $o_1, \ldots, o_N$ onto points $x_{o1}, \ldots, x_{oN}$ in R_1. Note that $x_{o1}, \ldots, x_{oN}$ may be regarded as simply the x-values of the N objects in π_N. It is convenient, and there is no loss of generality, if we assume that the objects in π_N are labeled so that $x_{o1} \leqslant \cdots \leqslant x_{oN}$. If we assign equal probabilities $1/N$, to $o_1, \ldots, o_N$, and if we denote the random variable $x(o)$ by x, and its p.f. by $p(x)$, then the mass points (that is, sample space) of x are $x_{o1}, \ldots, x_{oN}$ and

$$p(x) = \frac{1}{N} \tag{8.5.1}$$

at each of these mass points. In case $x(o)$ has the same value for two or more elements from π_N, then $p(x)$ will be a multiple of $1/N$ at that value of $x(o)$. More specifically if $x(o) = x'$ for r elements from π_N, then $p(x') = r/N$. We may think of x as the random variable denoting the x-value of a single object "drawn at random" from π_N. It is convenient to call $p(x)$ thus defined the p.f. *of the finite population.*

Now suppose $(o_{\gamma_1}, \ldots, o_{\gamma_n})$ is some permutation of n of the N objects in π_N. There are $N!/(N-n)!$ such *n-permutations.* Now consider these n-permutations as the sample points in a basic sample space R. If $(o_{\gamma_1}, \ldots, o_{\gamma_n})$ is a sample point e in R, let $x_1(e), \ldots, x_n(e)$ be random variables such that $x_1(e) = x_{0\gamma_1}, \ldots, x_n(e) = x_{0\gamma_n}$. Denoting $(x_1(e), \ldots, x_n(e))$ by $(x_1, \ldots, x_n)$, we shall refer to this n-dimensional random variable as *a sample of size n from the finite population* π_N. We may think of x_1 as the random variable denoting the x-value of the first object "drawn at random" from π_N, x_2 that for the second object, ... x_n, that for the nth object—*the objects being drawn from* π_N *without replacement.*

It is evident that $(x_1, \ldots, x_n)$ is a discrete n-dimensional random variable. If we assign all sample points e in R the same probability, namely,

$$\frac{(N-n)!}{N!}, \tag{8.5.2}$$

then $p(x_1, \ldots, x_n)$, the p.f. of $(x_1, \ldots, x_n)$, is a multiple of $(N-n)!/N!$ at each mass point.

Actually, it is convenient in the discussion which follows if we consider the x-values of the N objects in π_N to be distinct and the objects labelled so that $x_{01} < \cdots < x_{0N}$. Then the mass points of the random variable x_ξ are $x_{01}, \ldots, x_{0N}$, $\xi = 1, \ldots, n$. Let $R_1^{(\xi)}$ be the real axis corresponding to random variable x_ξ, and let $E_1^{(\xi)}$ denote the set of points $x_{01}, \ldots, x_{0N}$ in $R_1^{(\xi)}$. Let E_n be the Cartesian product $E_1^{(1)} \times \cdots \times E_1^{(n)}$ and R_n the Cartesian product $R_1^{(1)} \times \cdots \times R_1^{(n)}$. Then the mass points of $(x_1, \ldots, x_n)$ form a set of points E_n' consisting of those points in E_n for which no two coordinates are equal. There are exactly $N!/(N-n)!$ points in E_n' and there is a one-to-one correspondence between the points in E_n' and the e-points in R. In other words the n-dimensional random variable $(x_1(e), \ldots, x_n(e))$ maps the e-points of R into the points E_n' in R_n. Thus, the p.f. of $(x_1, \ldots, x_n)$ is defined at the points in its sample space R_n as follows:

$$(8.5.3) \qquad p(x_1, \ldots, x_n) = \frac{(N-n)!}{N!}, \qquad x_1 \neq \cdots \neq x_n$$
$$= 0, \qquad \text{otherwise.}$$

The mass point of $(x_1, \ldots, x_n)$ corresponding to the objects $o_{\gamma_1}, \ldots, o_{\gamma_n}$ is $(x_{0\gamma_1}, \ldots, x_{0\gamma_n})$.

Note that the marginal p.f. of $(x_1, \ldots, x_{n-1})$ in (8.5.3), that is, the sum of $p(x_1, \ldots, x_n)$ with respect to x_n, gives

$$p_{1\cdots(n-1)}(x_1, \ldots, x_{n-1}) = (N-n+1)\frac{(N-n)!}{N!}$$
$$= \frac{(N-n+1)!}{N!}$$

at each point of $E_1^{(1)} \times \cdots \times E_1^{(n-1)}$ for which the $n-1$ coordinates are all different while $p_{1\cdots(n-1)}(x_1, \ldots, x_{n-1})$ is zero at all other points in $E_1^{(1)} \times \cdots \times E_1^{(n-1)}$. More generally, the marginal p.f. of $(x_{\xi_1}, \ldots, x_{\xi_r})$ is given by

$$(8.5.4) \qquad p_{\xi_1\cdots\xi_r}(x_{\xi_1}, \ldots, x_{\xi_r}) = \frac{(N-r)!}{N!}$$

at those points of $E_1^{(\xi_1)} \times \cdots \times E_1^{(\xi_r)}$ whose coordinates are all different, and zero at all other points in $E_1^{(\xi_1)} \times \cdots \times E_1^{(\xi_r)}$. The p.f. of any single component of $(x_1, \ldots, x_n)$, say x_ξ, is given by (8.5.1), that is,

$$(8.5.5) \qquad p_\xi(x_\xi) = \frac{1}{N},$$

the mass points of x_ξ being $x_{01}, \ldots, x_{0N}$. Thus, any component of $(x_1, \ldots, x_n)$ has the same distribution as the population distribution.

The *mean* μ_N and *variance* σ_N^2 of the finite population π_N will be defined as follows:

$$\text{(8.5.6)} \qquad \mu_N = \frac{1}{N}\sum_{t=1}^{N} x_{0t}, \qquad \sigma_N^2 = \frac{1}{N-1}\sum_{t=1}^{N}(x_{0t} - \mu_N)^2.$$

We shall ordinarily drop N and refer to μ_N and σ_N^2 as μ and σ^2. Furthermore, it is obvious that for $\xi = 1, \ldots, n$

$$\text{(8.5.7)} \qquad \mathscr{E}(x_\xi) = \mu, \qquad \sigma^2(x_\xi) = \frac{N-1}{N}\sigma^2.$$

It is worthwhile to examine the marginal p.f. (8.5.4) for $r = 2$, that is, the p.f. of any two components (x_ξ, x_η) of $(x_1, \ldots, x_n)$. We find

$$\text{(8.5.8)} \qquad p_{\xi\eta}(x_\xi, x_\eta) = \frac{1}{N(N-1)},$$

the mass points being $(x_{ot}, x_{ot'})$, $t \neq t' = 1, \ldots, N$. The covariance between x_ξ and x_η is given by

$$\text{(8.5.9)} \qquad \operatorname{cov}(x_\xi, x_\eta) = \sum_{x_\xi, x_\eta} (x_\xi - \mu)(x_\eta - \mu)p_{\xi\eta}(x_\xi, x_\eta)$$

$$= \sum_{t\neq t'=1}^{N} \frac{(x_{ot} - \mu)(x_{ot'} - \mu)}{N(N-1)}$$

$$= -\sum_{t=1}^{N} \frac{(x_{ot} - \mu)^2}{N(N-1)} + \frac{1}{N(N-1)}\left[\sum_{t=1}^{N}(x_{ot} - \mu)\right]^2.$$

Making use of (8.5.6) we find that

$$\text{(8.5.10)} \qquad \operatorname{cov}(x_\xi, x_\eta) = -\frac{\sigma^2}{N}.$$

It will be noted that the correlation coefficient between x_ξ and x_η has the value

$$\text{(8.5.11)} \qquad \rho(x_\xi, x_\eta) = -\frac{1}{N-1}.$$

It will be recalled from (8.2.4) that $\operatorname{cov}(x_\xi, x_\eta) = 0$ in the case of a sample from an infinite population. Summarizing:

8.5.1 *If $(x_1, \ldots, x_n)$ is a sample of size n from a finite population π_N, then*

$$\mathscr{E}(x_\xi) = \mu, \qquad \sigma^2(x_\xi) = \frac{N-1}{N}\sigma^2, \qquad \xi = 1, \ldots, n$$

$$\operatorname{cov}(x_\xi, x_\eta) = -\frac{\sigma^2}{N}, \quad \rho(x_\xi, x_\eta) = -\frac{1}{N-1}, \qquad \xi \neq \eta = 1, \ldots, n,$$

where μ and σ^2 are defined by (8.5.6).

The problem of determining the sampling theory of any function $g(x_1, \ldots, x_n)$ of a sample $(x_1, \ldots, x_n)$ from the finite population π_N simply amounts to finding the distribution function of $g(x_1, \ldots, x_n)$, where $(x_1, \ldots, x_n)$ is a random variable having p.f. (8.5.3). The reader should note that the preceding two pages of results hold with only minor changes if some of the $<$'s in $x_{01} < \cdots < x_{0N}$ are replaced by $=$'s.

(b) Mean Values of Sample Mean and Variance

For illustrative purposes we shall first consider two particular functions of $(x_1, \ldots, x_n)$, namely, $\bar{x}$ and s^2. Since $\bar{x} = \frac{1}{n}\sum_{\xi=1}^{n} x_\xi$ we have

$$\mathscr{E}(\bar{x}) = \mathscr{E}\left(\frac{1}{n}\sum_{\xi} x_\xi\right) = \frac{1}{n}\sum_{\xi}\mathscr{E}(x_\xi). \tag{8.5.12}$$

But $\mathscr{E}(x_\xi) = \mu$, $\xi = 1, \ldots, n$. Hence

$$\mathscr{E}(\bar{x}) = \mu. \tag{8.5.13}$$

Thus, $\bar{x}$ is an unbiased estimator for μ.

For $\sigma^2(\bar{x})$ we have

$$\sigma^2(\bar{x}) = \mathscr{E}(\bar{x} - \mu)^2 = \mathscr{E}\left[\frac{1}{n}\sum_{\xi}(x_\xi - \mu)\right]^2$$

$$= \frac{1}{n^2}\sum_{\xi}\mathscr{E}(x_\xi - \mu)^2 + \frac{1}{n^2}\sum_{\xi\neq\eta}\mathscr{E}(x_\xi - \mu)(x_\eta - \mu).$$

But $\mathscr{E}(x_\xi - \mu)^2 = \dfrac{N-1}{N}\sigma^2$, $\xi = 1, \ldots, n$, and for $\xi \neq \eta$,

$$\mathscr{E}(x_\xi - \mu)(x_\eta - \mu) = \operatorname{cov}(x_\xi, x_\eta) = -\frac{\sigma^2}{N}.$$

Hence

$$\sigma^2(\bar{x}) = \left(\frac{1}{n} - \frac{1}{N}\right)\sigma^2. \tag{8.5.14}$$

Now let us consider the mean value of s^2. We have

$$s^2 = \frac{1}{n-1}\sum_{\xi}(x_\xi - \bar{x})^2$$

$$= \frac{1}{n}\sum_{\xi}(x_\xi - \mu)^2 - \frac{1}{n(n-1)}\sum_{\xi\neq\eta}(x_\xi - \mu)(x_\eta - \mu).$$

Taking mean values, and using **8.5.1**, we find that

$$\mathscr{E}(s^2) = \sigma^2. \tag{8.5.15}$$

Hence, s^2 is an unbiased estimator for σ^2.

Summarizing, we have the following result:

8.5.2 *Suppose* $(x_1, \ldots, x_n)$ *is a sample from a finite population* π_N *having mean* μ *and variance* σ^2. *Then we have*

$$\mathscr{E}(\bar{x}) = \mu, \qquad \sigma^2(\bar{x}) = \left(\frac{1}{n} - \frac{1}{N}\right)\sigma^2, \qquad \mathscr{E}(s^2) = \sigma^2.$$

It should be noted that the mean and variance of the sample sum are given by

$$\mathscr{E}(z) = n\mu \qquad \sigma^2(z) = n^2\left(\frac{1}{n} - \frac{1}{N}\right)\sigma^2. \tag{8.5.16}$$

Finally, it should be observed that if μ_N and σ_N^2, as defined in (8.5.6), which are functions of $x_{01}, \ldots, x_{0N}$, are such that they have finite limits as $N \to \infty$, which can be denoted by μ and σ^2, then it will be seen that the limits of $\mathscr{E}(\bar{x})$ and $\sigma^2(\bar{x})$ in **8.5.2** as $N \to \infty$ are identical with the values of $\mathscr{E}(\bar{x})$ and $\sigma^2(\bar{x})$ for samples from infinite populations as given by **8.2.1.**

(c) Mean Values of Certain Symmetric Functions of a Sample from a Finite Population π_N

The basic ideas in Section 8.5(*b*) can be extended to obtain mean values of certain general symmetric functions of the sample. Thus, suppose $(x_1, \ldots, x_n)$ is a sample from a finite population π_N and consider a function $g(x_{\xi_1}, \ldots, x_{\xi_r})$, $\xi_1, \ldots, \xi_r$ all being different, $r < n$. Then it is evident from the form of the p.f. of $(x_{\xi_1}, \ldots, x_{\xi_r})$ given by (8.5.4) that

$$\mathscr{E}[g(x_{\xi_1}, \ldots, x_{\xi_r})] = \frac{(N-r)!}{N!} \sum\nolimits_t' g(x_{ot_1}, \ldots, x_{ot_r}), \tag{8.5.17}$$

where $\sum_t'$ denotes summation over all values of $t_1, \ldots, t_r$ which are all different and where each t ranges over the integers $1, \ldots, N$. Now let us form the following symmetric function of $(x_1, \ldots, x_n)$:

$$Q(x_1, \ldots, x_n) = \frac{(n-r)!}{n!} \sum\nolimits_\xi' g(x_{\xi_1}, \ldots, x_{\xi_r}), \tag{8.5.18}$$

where $\sum_\xi'$ has a meaning for $\xi_1, \ldots, \xi_r$ similar to that of $\sum_t'$ for $t_1, \ldots, t_r$.

If we take the mean value of $Q(x_1, \ldots, x_n)$, we have

$$\mathscr{E}(Q(x_1, \ldots, x_n)) = \frac{(n-r)!}{n!} \sum\nolimits_\xi' \mathscr{E}[g(x_{\xi_1}, \ldots, x_{\xi_r})]. \tag{8.5.19}$$

But $\mathscr{E}[g(x_{\xi_1}, \ldots, x_{\xi_r})]$ has the same value for all values of $\xi_1 \neq \xi_2 \neq \cdots \neq \xi_r$, namely, that given by the right-hand side of (8.5.17). Since there are $n!/(n-r)!$ such sets of values of $\xi_1, \ldots, \xi_r$, we obtain

(8.5.20)

$$\mathscr{E}(Q(x_1, \ldots, x_n)) = \frac{(N-r)!}{N!} \sum_t' g(x_{0t_1}, \ldots, x_{0t_r}) = Q(x_{01}, \ldots, x_{0N}),$$

that is, $Q(x_1, \ldots, x_n)$ is an unbiased estimator for $Q(x_{01}, \ldots, x_{0N})$.

Therefore, we have the following result:

8.5.3 *If $(x_1, \ldots, x_n)$ is a sample from a finite population π_N and if $Q(x_1, \ldots, x_n)$ is any symmetric function of $(x_1, \ldots, x_n)$ of form* (8.5.18), *then $Q(x_1, \ldots, x_n)$ is an unbiased estimator for $Q(x_{o1}, \ldots, x_{oN})$.*

Furthermore, it follows from **3.2.1** and **3.2.2** that

8.5.4 *If $Q_1(x_1, \ldots, x_n), \ldots, Q_s(x_1, \ldots, x_n)$ are any s symmetric functions of type* (8.5.18) *of a sample $(x_1, \ldots, x_n)$ from a finite population π_N, then, for any set of constants $d_1, \ldots, d_s$, $d_1 Q_1(x_1, \ldots, x_n) + \cdots d_s Q_s(x_1, \ldots, x_n)$ is an unbiased estimator of $d_1 Q_1(x_{o1}, \ldots, x_{oN}) + \cdots + d_s Q_s(x_{o1}, \ldots, x_{oN})$.*

If in (8.5.18) we take $g(x_{\xi_1}, \ldots, x_{\xi_r}) \equiv x_{\xi_1}^{c_1} \cdots x_{\xi_r}^{c_r}$, where $c_1, \ldots, c_r$ are positive integers (or zero), then we obtain a class of symmetric functions $\{Q(x_1, \ldots, x_n)\}$ such that the sample mean, sample variance, k-statistics, and other polynomials in sample moments can be expressed as linear functions of Q's in this class. This fact together with theorems **8.5.3** and **8.5.4** make the sampling theory of means, variances, higher moments, and especially k-statistics quite simple as Tukey (1950, 1956a) has shown.

Example. To illustrate let us determine the value of $\mathscr{E}(\bar{x}^3)$.

Now
$$\begin{aligned}\bar{x}^3 &= \frac{1}{n^3} \sum_{\xi_1, \xi_2, \xi_3 = 1}^{n} x_{\xi_1} x_{\xi_2} x_{\xi_3} \\ &= \frac{1}{n^3}\left[\sum_{\xi_1=1}^{n} x_{\xi_1}^3 + 3 \sum_{\xi_1 \neq \xi_2} x_{\xi_1}^2 x_{\xi_2} + \sum_{\xi_1 \neq \xi_2 \neq \xi_3} x_{\xi_1} x_{\xi_2} x_{\xi_3}\right] \\ &= \frac{1}{n^2} Q_1(x_1, \ldots, x_n) + \frac{3(n-1)}{n^2} Q_2(x_1, \ldots, x_n) \\ &\quad + \frac{(n-1)(n-2)}{n^2} Q_3(x_1, \ldots, x_n),\end{aligned}$$

where

$$Q_1 = \frac{1}{n} \sum_{\xi_1=1}^{n} x_{\xi_1}^3, \quad Q_2 = \frac{1}{n(n-1)} \sum_{\xi_1 \neq \xi_2} x_{\xi_1}^2 x_{\xi_2},$$
$$Q_3 = \frac{1}{n(n-1)(n-2)} \sum_{\xi_1 \neq \xi_2 \neq \xi_3} x_{\xi_1} x_{\xi_2} x_{\xi}$$

and which are functions of type (8.5.18).

Applying **8.5.4** we therefore have

$$\mathscr{E}(\bar{x}^3) = \frac{1}{n^2} Q_1(x_{o1}, \ldots, x_{oN}) + \frac{3(n-1)}{n^2} Q_2(x_{o1}, \ldots, x_{oN}) + \frac{(n-1)(n-2)}{n^2} Q_3(x_{o1}, \ldots, x_{oN}).$$

If we let

$$\mu_r' = \frac{1}{N} \sum_{t=1}^{N} x_{ot}^r,$$

we note that

$$Q_1(x_{o1}, \ldots, x_{oN}) = \mu_3'$$

$$Q_2(x_{o1}, \ldots, x_{oN}) = \frac{1}{N-1}(N\mu_2'\mu_1' - \mu_3')$$

$$Q_3(x_{o1}, \ldots, x_{oN}) = \frac{1}{(N-1)(N-2)}[N^2\mu_1'^3 + 2\mu_3' - 3N\mu_2'\mu_1'],$$

and hence $\mathscr{E}(\bar{x}^3)$ can be expressed as a polynomial in the first three population moments.

(d) The k-Dimensional Case

Now consider the situation in which $x(o)$, referred to in Section 8.5 (*a*), is a vector with k components, say $x_1(o), \ldots, x_k(o)$. Then each object in π_N is "measured" by k numbers. Thus, if $x_i(o_t) = x_{oit}$, we may refer to x_{oit} as the x_i *value* of o_t, and if the vectors for the N objects in π_N are all different, our vector random variable $(x_1(o), \ldots, x_k(o))$ maps the objects in π_N into N points in R_k, the coordinates of the point into which o_t is mapped being $(x_{o1t}, \ldots, x_{okt})$, $t = 1, \ldots, N$. If the probabilities associated with all of these N points are assigned equal values, namely, $1/N$, then we have a k-dimensional random variable $(x_1, \ldots, x_k)$ whose p.f. is

$$p(x_1, \ldots, x_k) = \frac{1}{N}, \tag{8.5.21}$$

the mass points being $(x_{o1t}, \ldots, x_{okt})$, $t = 1, \ldots, N$. We may regard $p(x_1, \ldots, x_k)$ as the p.f. of the finite population π_N.

Now consider the basic sample space R of the $N!/(N-n)!$ n-permutations referred to in Section 8.5(*a*) in which $(o_{\gamma_1}, \ldots, o_{\gamma_n})$ is a typical sample point e. Let $(x_{i1}(e), \ldots, x_{in}(e);\ i = 1, \ldots, k)$ be nk random variables so that $x_{i1}(e) = x_{oi\gamma_1}, \ldots, x_{in}(e) = x_{oi\gamma_n}$. If we denote $(x_{i1}(e), \ldots, x_{in}(e);\ i = 1, \ldots, k)$ by $(x_{i\xi},\ i = 1, \ldots, k,\ \xi = 1, \ldots, n)$, we shall refer to this nk-dimensional random variable as *a sample of size n from the k-variate finite population* π_N having p.f. (8.5.3). There will be $N!/(N-n)!$ mass points in R_{nk} of this nk-dimensional random variable, at each of which the assigned probability will be $(N-n)!/N!$. The mass point corresponding to $(o_{\gamma_1}, \ldots, o_{\gamma_n})$ is $(x_{oi\gamma_1}, \ldots, x_{oi\gamma_n},\ i = 1, \ldots, k)$.

The preceding setup is sufficient to provide the sampling theory of functions of our sample, the nk-dimensional random variable $(x_{i\xi};\ i = 1, \ldots, k;\ \xi = 1, \ldots, n)$, such as the vectors of sample *sums* and *means*

(8.5.22) $$z_i = \sum_{\xi=1}^{n} x_{i\xi}, \qquad \bar{x}_i = \frac{1}{n} \sum_{\xi=1}^{n} x_{i\xi}$$

and the *sample covariance matrix* $\|s_{ij}\|$ where

(8.5.23)
$$s_{ij} = \frac{1}{n-1} \sum_{\xi=1}^{n} (x_{i\xi} - \bar{x}_i)(x_{j\xi} - \bar{x}_j) \qquad i, j = 1, \ldots, k.$$

If we define the vector of *means* $(\mu_1, \ldots, \mu_k)$ of π_N as

(8.5.24) $$\mu_i = \frac{1}{N} \sum_{t=1}^{N} x_{oit}, \qquad i = 1, \ldots, k,$$

and the *covariance matrix* $\|\sigma_{ij}\|$ of π_N as

(8.5.25)
$$\sigma_{ij} = \frac{1}{N-1} \sum_{t=1}^{N} (x_{oit} - \mu_i)(x_{ojt} - \mu_j), \qquad i, j = 1, \ldots, k,$$

then it can be shown by mean value operations similar to those by which (8.5.12) through (8.5.15) were found that

(8.5.26) $$\mathscr{E}(\bar{x}_i) = \mu_i, \qquad \sigma^2(\bar{x}) = \left(\frac{1}{n} - \frac{1}{N}\right)\sigma_{ii}, \qquad i = 1, \ldots, k$$

$$\operatorname{cov}(\bar{x}_i, \bar{x}_j) = \left(\frac{1}{n} - \frac{1}{N}\right)\sigma_{ij}, \qquad i \neq j = 1, \ldots, k$$

$$\mathscr{E}(s_{ij}) = \sigma_{ij}, \qquad i, j = 1, \ldots, k.$$

8.6 MATRIX SAMPLING

(a) Second-Order Matrix Samples

Suppose u and v are independent random variables having c.d.f.'s $F_1(u)$ and $F_2(v)$. Let $x(u, v)$ be a random variable for which

$$\mu = \mathscr{E}(x(u, v))$$

(8.6.1) $$\mu_{u\cdot} = \int_{-\infty}^{\infty} x(u, v)\, dF_2(v)$$

$$\mu_{\cdot v} = \int_{-\infty}^{\infty} x(u, v)\, dF_1(u).$$

Furthermore, let $\varepsilon_{u\cdot}$, $\varepsilon_{\cdot v}$ and ε_{uv} be random variables defined as follows:

$$\begin{aligned} \varepsilon_{u\cdot} &= \mu_{u\cdot} - \mu \\ \varepsilon_{\cdot v} &= \mu_{\cdot v} - \mu \\ \varepsilon_{uv} &= x(u, v) - \mu_{u\cdot} - \mu_{\cdot v} + \mu. \end{aligned} \tag{8.6.2}$$

Then it is evident that the mean of each of these random variables is zero and also the covariance between each pair is zero.

Thus, we have the following decomposition theorem:

8.6.1 *If $x(u, v)$ is a random variable where u and v are independent random variables, then $x(u, v)$ can be decomposed as follows:*

$$x(u, v) = \mu + \varepsilon_{u\cdot} + \varepsilon_{\cdot v} + \varepsilon_{uv} \tag{8.6.3}$$

where $\varepsilon_{u\cdot}$, $\varepsilon_{\cdot v}$, and ε_{uv} have zero means and zero covariances and are defined by (8.6.2). *Furthermore,*

$$\sigma^2(x(u, v)) = \sigma^2(\varepsilon_{u\cdot}) + \sigma^2(\varepsilon_{\cdot v}) + \sigma^2(\varepsilon_{uv}). \tag{8.6.4}$$

Remark. It should be noted that a more general form of this theorem (and subsequent theorems of this section) holds if u and v are considered as independent sample points in arbitrary sample spaces $R^{(1)}$ and $R^{(2)}$ rather than random variables. In such spaces we would, of course, have probability measures and $x(u, v)$ would be a random variable measurable with respect to the Cartesian product of these two probability measures.

It will be convenient to abbreviate the notation for the variances as follows:

$$\begin{aligned} \sigma^2(x(u, v)) &= \sigma^2 \\ \sigma^2(\varepsilon_{u\cdot}) &= \sigma^2_{\cdot 0} \\ \sigma^2(\varepsilon_{\cdot v}) &= \sigma^2_{0\cdot} \\ \sigma^2(\varepsilon_{uv}) &= \sigma^2_{\cdot\cdot}. \end{aligned} \tag{8.6.5}$$

Then (8.6.4) becomes

$$\sigma^2 = \sigma^2_{\cdot 0} + \sigma^2_{0\cdot} + \sigma^2_{\cdot\cdot}. \tag{8.6.4a}$$

Now let $(u_1, \ldots, u_r)$ and $(v_1, \ldots, v_s)$ be independent samples from c.d.f.'s $F_1(u)$ and $F_2(v)$ respectively. The c.d.f. of these two samples is therefore

$$\prod_{\xi=1}^{r} F_1(u_\xi) \prod_{\eta=1}^{s} F_2(v_\eta). \tag{8.6.6}$$

Let $(x(u_\xi, v_\eta);\ \xi = 1, \ldots, r,\ \eta = 1, \ldots, s)$ be a new set of random variables, which may be regarded as an $r \times s$ rectangular array having r rows and s columns. It will be called a *second-order matrix sample*, but it should be noted that this sample depends on only $r + s$ random variables, namely, $u_1, \ldots, u_r$ and $v_1, \ldots, v_s$.

Remark. Matrix samples are fundamental in such fields as psychometrics and the design of experiments. In psychometrics, for instance, we may think of rows as representing a given population of examinees who take a certain test, and columns representing a given population of questions. Thus $x(u_\xi, v_\eta)$ would represent the score of the person for whom $u = u_\xi$ on the question for which $v = v_\eta$. Then for a test of s questions given to a group of r examinees, the scores of the r examinees on the s questions would constitute the matrix sample $x(u_\xi, v_\eta)$, $\xi = 1, \ldots, r$, $\eta = 1, \ldots, s$. Lord (1955) has considered the sampling theory of various linear and quadratic functions of the $x(u_\xi, v_\eta)$ which arise in psychometrics.

In a particular experimental design, rows may represent a population of operators of a certain kind of machine, and columns may represent a population of these machines. Thus, if in an experiment we pick r operators and s machines at random, we might let $x(u_\xi, v_\eta)$ be a random variable representing the output in an 8-hour day of the ξth operator on the ηth machine. Such an experiment would, of course, require s 8-hour days to perform. Applications of matrix sampling in the design of experiments will be discussed in Sections 10.6 and 10.7.

To shorten our notation, let

$$\begin{aligned} x_{\xi\eta} &= x(u_\xi, v_\eta) \\ x'_{\xi\eta} &= x_{\xi\eta} - \mu \\ \bar{x}_{\cdot\cdot} &= \frac{1}{rs}\sum_{\xi,\eta} x_{\xi\eta} \\ \bar{x}_{\xi\cdot} &= \frac{1}{s}\sum_{\eta} x_{\xi\eta} \\ \bar{x}_{\cdot\eta} &= \frac{1}{r}\sum_{\xi} x_{\xi\eta} \end{aligned} \tag{8.6.7}$$

where $\xi = 1, \ldots, r$, $\eta = 1, \ldots, s$.

It should be noted that for each ξ and η, (8.6.3) takes a form which we may write as

$$x_{\xi\eta} = \mu + \varepsilon_{\xi\cdot} + \varepsilon_{\cdot\eta} + \varepsilon_{\xi\eta}. \tag{8.6.8}$$

Now let

$$\begin{aligned} S_T &= \sum_{\xi,\eta} (x_{\xi\eta} - \bar{x}_{\cdot\cdot})^2 \\ S_{\cdot 0} &= \sum_{\xi,\eta} (\bar{x}_{\xi\cdot} - \bar{x}_{\cdot\cdot})^2 \\ S_{0\cdot} &= \sum_{\xi,\eta} (\bar{x}_{\cdot\eta} - \bar{x}_{\cdot\cdot})^2 \\ S_{\cdot\cdot} &= \sum_{\xi,\eta} (x_{\xi\eta} - \bar{x}_{\xi\cdot} - \bar{x}_{\cdot\eta} + \bar{x}_{\cdot\cdot})^2 \end{aligned} \tag{8.6.9}$$

where it will be seen that

$$S_T = S_{\cdot 0} + S_{0\cdot} + S_{\cdot\cdot}.$$

It is sometimes convenient to refer to $S_{\cdot 0}$, $S_{0\cdot}$ and $S_{\cdot\cdot}$ as the *row*, *column*, and *residual components* of the *total* sum of squares S_T. We shall consider the mean and variance of $\bar{x}_{\cdot\cdot}$, and the means of S_T, $S_{\cdot 0}$, $S_{0\cdot}$ and $S_{\cdot\cdot}$, all mean values being taken with respect to the c.d.f. in (8.6.6). Since

$$\mathscr{E}(x_{\xi\eta}) = \mu, \qquad \xi = 1, \ldots, r; \ \eta = 1, \ldots, s,$$

it is evident that

$$\mathscr{E}(\bar{x}_{\cdot\cdot}) = \mu.$$

In determining mean values of the S's in (8.6.9) we find in each case that the mean value can be expressed as a linear function of terms of the following four types:

$$\begin{aligned}
\text{(8.6.10)} \qquad \mathscr{E}(x'_{\xi\eta}x'_{\xi'\eta'}) &= \sigma^2_{\cdot\cdot} + \sigma^2_{\cdot 0} + \sigma^2_{0\cdot}, && \xi = \xi', \eta = \eta' \\
&= \sigma^2_{\cdot 0}, && \xi = \xi', \eta \neq \eta' \\
&= \sigma^2_{0\cdot}, && \xi \neq \xi', \eta = \eta' \\
&= 0, && \xi \neq \xi', \eta \neq \eta'.
\end{aligned}$$

Working out the mean values, we find

$$\begin{aligned}
&\mathscr{E}(S_{\cdot 0}) = s(r-1)\left(\sigma^2_{\cdot 0} + \frac{\sigma^2_{\cdot\cdot}}{s}\right) \\
\text{(8.6.11)} \qquad &\mathscr{E}(S_{0\cdot}) = r(s-1)\left(\sigma^2_{0\cdot} + \frac{\sigma^2_{\cdot\cdot}}{r}\right) \\
&\mathscr{E}(S_{\cdot\cdot}) = (r-1)(s-1)\sigma^2_{\cdot\cdot}.
\end{aligned}$$

while $\mathscr{E}(S_T)$ can be determined from the equation

$$\mathscr{E}(S_T) = \mathscr{E}(S_{\cdot 0}) + \mathscr{E}(S_{\cdot 0}) + \mathscr{E}(S_{\cdot\cdot}).$$

Similarly, the variance of $\bar{x}_{\cdot\cdot}$ is a linear function of the four types of mean values mentioned in (8.6.10), which turns out to be

$$\text{(8.6.12)} \qquad \sigma^2(\bar{x}_{\cdot\cdot}) = \frac{\sigma^2_{\cdot 0}}{r} + \frac{\sigma^2_{0\cdot}}{s} + \frac{\sigma^2_{\cdot\cdot}}{rs}.$$

Summarizing,

8.6.2 *Let* $(u_1, \ldots, u_r)$ *and* $(v_1, \ldots, v_s)$ *be independent samples from infinite populations. Let* $(x_{\xi\eta};\ \xi = 1, \ldots, r;\ \eta = 1, \ldots, s)$ *be a matrix sample whose mean* $\bar{x}_{\cdot\cdot}$ *is defined by* (8.6.7), *and whose row, column, and residual components* $S_{\cdot 0}$, $S_{0\cdot}$, *and* $S_{\cdot\cdot}$ *respectively, of the total sum of squares* S_T, *are defined by* (8.6.9). *Then* $\mathscr{E}(\bar{x}_{\cdot\cdot}) = \mu$ *and* $\sigma^2(\bar{x}_{\cdot\cdot})$ *is given by* (8.6.12), *whereas the mean values of* $S_{\cdot 0}$, $S_{0\cdot}$, *and* $S_{\cdot\cdot}$ *are given by* (8.6.11).

(b) Case of Finite Row and Column Populations

The preceding results can be carried out in a straightforward manner for the case where $(u_1, \ldots, u_r)$ and $(v_1, \ldots, v_s)$ are independent random samples from finite populations π_{N_1} and π_{N_2}. This case has been considered by Hooke (1956*a*) and Tukey (1950).

In the case of finite populations we may, without loss of generality, consider the elements in π_{N_1} as having distinct values of u, namely, $u_{o1} < \cdots < u_{oN_1}$. Similarly, we may take $v_{o1} < \cdots < v_{oN_2}$ as distinct values of v. It will be convenient to use the notation

(8.6.13)
$$x(u_{oi}, v_{oj}) = x_{oij}$$
$$x'_{oij} = x_{oij} - \mu_f$$

and corresponding to (8.6.1),

(8.6.14)
$$\mu_f = \frac{1}{N_1 N_2} \sum_{i,j} x_{oij}$$
$$\mu_{fi\cdot} = \frac{1}{N_2} \sum_j x_{oij}$$
$$\mu_{f\cdot j} = \frac{1}{N_1} \sum_i x_{oij}.$$

For the case of finite populations of rows and columns, we achieve some simplification of mean values of $S_{\cdot 0}$, $S_{0\cdot}$, $S_{\cdot\cdot}$ by defining σ_f^2, $\sigma_{f\cdot 0}^2$, $\sigma_{f0\cdot}^2$, $\sigma_{f\cdot\cdot}^2$ as follows:

(8.6.15)
$$\sigma_f^2 = \frac{1}{N_1 N_2 - 1} \sum_{i,j} x'^2_{oij}$$
$$\sigma_{f\cdot 0}^2 = \frac{1}{N_2(N_1 - 1)} \sum_{i,j} (\mu_{fi\cdot} - \mu_f)^2$$
$$\sigma_{f0\cdot}^2 = \frac{1}{N_1(N_2 - 1)} \sum_{i,j} (\mu_{f\cdot j} - \mu_f)^2$$
$$\sigma_{f\cdot\cdot}^2 = \frac{1}{(N_1 - 1)(N_2 - 1)} \sum_{i,j} (x_{oij} - \mu_{fi\cdot} - \mu_{f\cdot j} + \mu_f)^2.$$

Note, however, that the equation relating σ_f^2, $\sigma_{f\cdot 0}^2$, $\sigma_{f0\cdot}^2$ and $\sigma_{f\cdot\cdot}^2$ is

(8.6.16)
$$\frac{N_1 N_2 - 1}{N_1 N_2} \sigma_f^2 = \frac{(N_1 - 1)}{N_1} \sigma_{f\cdot 0}^2 + \frac{(N_2 - 1)}{N_2} \sigma_{f0\cdot}^2 + \frac{(N_1 - 1)(N_2 - 1)}{N_1 N_2} \sigma_{f\cdot\cdot}^2$$

but that (8.6.16) reduces to (8.6.4*a*) in the limit as $N_1, N_2 \to \infty$.

We shall be interested in the mean value and variance of $\bar{x}_{..}$ and the mean value of $S_{0\cdot}$, $S_{\cdot 0}$, and $S_{..}$. To determine these mean values we shall find it convenient to define the following quantities:

(8.6.17)
$$\begin{aligned} T_{(\cdot\cdot)} &= \sum_{i,j} (x'_{oij})^2 \\ T_{(\cdot 0)} &= \sum_{i,j\neq j'} x'_{oij}x'_{oij'} \\ T_{(0\cdot)} &= \sum_{i\neq i',j} x'_{oij}x'_{oi'j} \\ T_{(00)} &= \sum_{i\neq i',j\neq j'} x'_{oij}x'_{oi'j'} \end{aligned}$$

where it will be seen that

(8.6.18)
$$T_{(\cdot\cdot)} + T_{(\cdot 0)} + T_{(\cdot 0)} + T_{(00)} = \left[\sum_{i,j} x'_{oij}\right]^2 = 0.$$

Now $\sigma^2_{f\cdot 0}$, $\sigma^2_{f0\cdot}$ and $\sigma^2_{f\cdot\cdot}$ are linear functions of $T_{(\cdot\cdot)}$, $T_{(\cdot 0)}$, $T_{(0\cdot)}$ and $T_{(00)}$ such that, when solved for the T's and making use of (8.6.18), we obtain

(8.6.19)
$$\begin{aligned} T_{(\cdot\cdot)} &= (N_1 - 1)(N_2 - 1)\left[\sigma^2_{f\cdot\cdot} + \frac{N_1}{N_1 - 1}\,\sigma^2_{f0\cdot} + \frac{N_2}{N_2 - 1}\,\sigma^2_{f\cdot 0}\right] \\ T_{(\cdot 0)} &= (N_1 - 1)(N_2 - 1)\left[-\sigma^2_{f\cdot\cdot} - \frac{N_1}{N_1 - 1}\,\sigma^2_{f0\cdot} + N_2\sigma^2_{f\cdot 0}\right] \\ T_{(0\cdot)} &= (N_1 - 1)(N_2 - 1)\left[-\sigma^2_{f\cdot\cdot} + N_1\sigma^2_{f0\cdot} - \frac{N_2}{N_2 - 1}\,\sigma^2_{f\cdot 0}\right] \\ T_{(00)} &= (N_1 - 1)(N_2 - 1)[\sigma^2_{f\cdot\cdot} - N_1\sigma^2_{f0\cdot} - N_2\sigma^2_{f\cdot 0}]. \end{aligned}$$

As in the case of infinite row and column populations, let ($x_{\xi\eta}$; $\xi = 1, \ldots, r$; $\eta = 1, \ldots, s$,) be the matrix sample, and let $T_{..}$, $T_{\cdot 0}$, $T_{0\cdot}$, and T_{00} be defined from the $x'_{\xi\eta}$, as given by (8.6.7), similar to the manner in which $T_{(\cdot\cdot)}$, $T_{(\cdot 0)}$, $T_{(0\cdot)}$ and $T_{(00)}$ are defined from the x'_{oij}. Note that the sample random variables $T_{..}$, $T_{\cdot 0}$, $T_{0\cdot}$ and T_{00} do not satisfy (8.6.18).

It can be readily shown that

(8.6.20)
$$\begin{aligned} \mathscr{E}(T_{..}) &= \frac{rs}{N_1N_2}\,T_{(\cdot\cdot)} \\ \mathscr{E}(T_{\cdot 0}) &= \frac{rs(s - 1)}{N_1N_2(N_2 - 1)}\,T_{(\cdot 0)} \\ \mathscr{E}(T_{0\cdot}) &= \frac{rs(r - 1)}{N_1N_2(N_1 - 1)}\,T_{(0\cdot)} \\ \mathscr{E}(T_{00}) &= \frac{rs(r - 1)(s - 1)}{N_1N_2(N_1 - 1)(N_2 - 1)}\,T_{(00)} \end{aligned}$$

which, as we shall see presently, are the basic formulas in the problem of finding the mean values of $S_{\cdot 0}$, $S_{0\cdot}$, and $S_{\cdot\cdot}$.

Now $S_{\cdot 0}$, $S_{0\cdot}$, and $S_{\cdot\cdot}$ are the following linear functions of $T_{\cdot\cdot}$, $T_{\cdot 0}$, $T_{0\cdot}$, and T_{00},

$$S_{\cdot 0} = \frac{r-1}{rs}(T_{\cdot\cdot} + T_{\cdot 0}) - \frac{1}{rs}(T_{0\cdot} + T_{00})$$

$$\text{(8.6.21)} \quad S_{0\cdot} = \frac{s-1}{rs}(T_{\cdot\cdot} + T_{0\cdot}) - \frac{1}{rs}(T_{\cdot 0} + T_{00})$$

$$S_{\cdot\cdot} = \frac{(r-1)(s-1)}{rs}T_{\cdot\cdot} - \frac{(r-1)}{rs}T_{\cdot 0} - \frac{(s-1)}{rs}T_{0\cdot} + \frac{1}{rs}T_{00}.$$

Taking mean values, making use of (8.6.20), and substituting values of $T_{(\cdot\cdot)}$, $T_{(\cdot 0)}$, and $T_{(00)}$ from (8.6.19), we find the following mean values of the S's

$$\mathscr{E}(S_{\cdot 0}) = s(r-1)\left[\sigma^2_{f\cdot 0} + \left(\frac{1}{s} - \frac{1}{N_2}\right)\sigma^2_{f\cdot\cdot}\right]$$

$$\text{(8.6.22)} \quad \mathscr{E}(S_{0\cdot}) = r(s-1)\left[\sigma^2_{f0\cdot} + \left(\frac{1}{r} - \frac{1}{N_1}\right)\sigma^2_{f\cdot\cdot}\right]$$

$$\mathscr{E}(S_{\cdot\cdot}) = (r-1)(s-1)\sigma^2_{f\cdot\cdot}.$$

The value of $\mathscr{E}(S_T)$ can be determined from the equation

$$\mathscr{E}(S_T) = \mathscr{E}(S_{\cdot 0}) + \mathscr{E}(S_{0\cdot}) + \mathscr{E}(S_{\cdot\cdot}).$$

Note that these equations reduce to (8.6.11) as $N_1, N_2 \to \infty$, if $\sigma^2_{f\cdot\cdot}$, $\sigma^2_{f0\cdot}$, $\sigma^2_{f\cdot 0}$ have $\sigma^2_{\cdot\cdot}$, $\sigma^2_{0\cdot}$, $\sigma^2_{0\cdot}$ as their limits respectively.

To find the variance of $\bar{x}_{\cdot\cdot}$ we merely have to determine the mean value of $(\bar{x}_{\cdot\cdot} - \mu_f)^2$, which can be expressed as follows:

$$\text{(8.6.23)} \quad (\bar{x}_{\cdot\cdot} - \mu_f)^2 = \frac{1}{rs}[T_{\cdot\cdot} + T_{\cdot 0} + T_{0\cdot} + T_{00}].$$

Taking mean values of both sides of (8.6.23) and making use of (8.6.19) and (8.6.20) we find

(8.6.24)

$$\sigma^2(\bar{x}_{\cdot\cdot}) = \left(\frac{1}{r} - \frac{1}{N_1}\right)\sigma^2_{f\cdot 0} + \left(\frac{1}{s} - \frac{1}{N_2}\right)\sigma^2_{f0\cdot} + \left(\frac{1}{r} - \frac{1}{N_1}\right)\left(\frac{1}{s} - \frac{1}{N_2}\right)\sigma^2_{f\cdot\cdot}.$$

The reader will note that this formula reduces to (8.6.12) as $N_1, N_2 \to \infty$.

Formulation of a summary statement similar to **8.6.2** for matrix sampling in the case of finite populations of rows and columns is left to the reader.

(c) Third-Order Matrix Samples

The ideas in the preceding pages extend without major difficulties to the case of third-and higher order matrix samples. To indicate the varieties of variance components which arise in these higher order cases it should be sufficient perhaps to consider the third-order case briefly. Here we are concerned with a random variable $x(u, v, w)$, where u, v, w are independent random variables with c.d.f.'s $F_1(u)$, $F_2(v)$, $F_3(w)$.

For $x(u, v, w)$ we define the following means

$$\mu = \mathscr{E}(x(u, v, w))$$

(8.6.25)
$$\mu_{u\cdot\cdot} = \int_{R_2} x(u, v, w)\, dF_2(v)\, dF_3(w); \qquad \text{similarly for } \mu_{\cdot v\cdot}, \mu_{\cdot\cdot w}$$
$$\mu_{uv\cdot} = \int_{-\infty}^{\infty} x(u, v, w)\, dF_3(w); \qquad \text{similarly for } \mu_{u\cdot w}, \mu_{\cdot vw}$$

and random variables

(8.6.26)
$$\varepsilon_{u\cdot\cdot} = \mu_{u\cdot\cdot} - \mu; \qquad \text{similarly for } \varepsilon_{\cdot v\cdot}, \varepsilon_{\cdot\cdot w}$$
$$\varepsilon_{uv\cdot} = \mu_{uv\cdot} - \mu_{u\cdot\cdot} - \mu_{\cdot v\cdot} + \mu; \qquad \text{similarly for } \varepsilon_{u\cdot w}, \varepsilon_{\cdot vw}$$
$$\varepsilon_{uvw} = x(u, v, w) - \mu_{uv\cdot} - \mu_{u\cdot w} - \mu_{\cdot vw} + \mu_{u\cdot\cdot} + \mu_{\cdot v\cdot} + \mu_{\cdot\cdot w} - \mu.$$

Then we have the following extension of **8.6.1**:

8.6.3 *If u, v, w are independent random variables and if $x(u, v, w)$ is a random variable, then*

$$x(u, v, w) = \mu + \varepsilon_{u\cdot\cdot} + \varepsilon_{\cdot v\cdot} + \varepsilon_{\cdot\cdot w} + \varepsilon_{uv\cdot} + \varepsilon_{u\cdot w} + \varepsilon_{\cdot vw} + \varepsilon_{uvw}$$

where the ε's are random variables defined by (8.6.26) *which have zero means and zero covariances. Furthermore*

(8.6.27)
$$\sigma^2(x(u, v, w)) = \sigma^2(\varepsilon_{u\cdot\cdot}) + \sigma^2(\varepsilon_{\cdot v\cdot}) + \sigma^2(\varepsilon_{\cdot\cdot w}) + \sigma^2(\varepsilon_{uv\cdot}) + \sigma^2(\varepsilon_{u\cdot w}) + \sigma^2(\varepsilon_{\cdot vw}) + \sigma^2(\varepsilon_{uvw}).$$

It will be convenient to let

$$\sigma^2(x(u, v, w)) = \sigma^2$$
$$\sigma^2(\varepsilon_{u\cdot\cdot}) = \sigma^2_{\cdot 00}; \qquad \sigma^2(\varepsilon_{\cdot v\cdot}) = \sigma^2_{0\cdot 0}; \qquad \sigma^2(\varepsilon_{\cdot\cdot w}) = \sigma^2_{00\cdot};$$
$$\sigma^2(\varepsilon_{uv\cdot}) = \sigma^2_{\cdot\cdot 0}; \qquad \sigma^2(\varepsilon_{u\cdot w}) = \sigma^2_{\cdot 0\cdot}; \qquad \sigma^2(\varepsilon_{\cdot vw}) = \sigma^2_{0\cdot\cdot};$$
$$\sigma^2(\varepsilon_{uvw}) = \sigma^2_{\cdot\cdot\cdot}.$$

Then (8.6.27) may be written as

(8.6.28) $$\sigma^2 = \sigma^2_{\cdot 00} + \sigma^2_{0\cdot 0} + \sigma^2_{00\cdot} + \sigma^2_{\cdot\cdot 0} + \sigma^2_{\cdot 0\cdot} + \sigma^2_{0\cdot\cdot} + \sigma^2_{\cdot\cdot\cdot}$$

Now let $(u_1, \ldots, u_r)$, $(v_1, \ldots, v_s)$, $(w_1, \ldots, w_t)$ be independent samples from infinite populations having c.d.f.'s $F_1(u)$, $F_2(v)$, and $F_3(w)$. The c.d.f. of the sample is

(8.6.29) $$\prod_{\xi=1}^{r} F_1(u_\xi) \cdot \prod_{\eta=1}^{s} F_2(v_\eta) \cdot \prod_{\zeta=1}^{t} F_3(w_\zeta).$$

Our third-order matrix sample is the set of random variables

$$(x(u_\xi, v_\eta, w_\zeta);\ \xi = 1, \ldots, r; \quad \eta = 1, \ldots, s; \quad \zeta = 1, \ldots, t)$$

which may be regarded as a third-order matrix with r rows, s columns, and t layers.

For this sample we define $x_{\xi\eta\zeta}$, $x'_{\xi\eta\zeta}$, and the various sample means $\bar{x}_{\cdot\cdot\cdot}$, $\bar{x}_{\xi\cdot\cdot}$, $\bar{x}_{\cdot\eta\cdot}$, $\bar{x}_{\cdot\cdot\xi}$, $\bar{x}_{\xi\eta\cdot}$, $\bar{x}_{\xi\cdot\zeta}$, $\bar{x}_{\cdot\eta\zeta}$ by direct extension of the definitions (8.6.7). Each random variable $x_{\xi\eta\zeta}$ is a sum of random variables having the properties stated in **8.6.3**; we may write

(8.6.30) $$x_{\xi\eta\zeta} = \mu + \varepsilon_{\xi\cdot\cdot} + \varepsilon_{\cdot\eta\cdot} + \varepsilon_{\cdot\cdot\zeta} + \varepsilon_{\xi\eta\cdot} + \varepsilon_{\xi\cdot\zeta} + \varepsilon_{\cdot\eta\zeta} + \varepsilon_{\xi\eta\zeta}.$$

Furthermore, we define the sums of squares

(8.6.31)

$$S_T = \sum_{\xi,\eta,\zeta} (x_{\xi\eta\zeta} - \bar{x}_{\cdot\cdot\cdot})^2;$$

$$S_{\cdot 00} = \sum_{\xi,\eta,\zeta} (\bar{x}_{\xi\cdot\cdot} - \bar{x}_{\cdot\cdot\cdot})^2; \qquad \text{similarly for } S_{0\cdot 0} \text{ and } S_{00\cdot};$$

$$S_{\cdot\cdot 0} = \sum_{\xi,\eta,\zeta} (\bar{x}_{\xi\eta\cdot} - \bar{x}_{\xi\cdot\cdot} - \bar{x}_{\cdot\eta\cdot} + \bar{x}_{\cdot\cdot\cdot})^2; \qquad \text{similarly for } S_{\cdot 0\cdot} \text{ and } S_{0\cdot\cdot};$$

$$S_{\cdot\cdot\cdot} = \sum_{\xi,\eta,\xi} (x_{\xi\eta\zeta} - \bar{x}_{\xi\eta\cdot} - \bar{x}_{\xi\cdot\zeta} - \bar{x}_{\cdot\eta\zeta} + \bar{x}_{\xi\cdot\cdot} + \bar{x}_{\cdot\eta\cdot} + \bar{x}_{\cdot\cdot\zeta} - \bar{x}_{\cdot\cdot\cdot})^2$$

where it can be verified that

(8.6.32) $$S_T = S_{\cdot 00} + S_{0\cdot 0} + S_{00\cdot} + S_{\cdot\cdot 0} + S_{\cdot 0\cdot} + S_{0\cdot\cdot} + S_{\cdot\cdot\cdot}.$$

It is customary to refer to $S_{\cdot 00}$ as the *row component* of S_T with similar interpretations for $S_{0\cdot 0}$ and $S_{00\cdot}$; $S_{\cdot\cdot 0}$ as the *row-column interaction* component of S_T with similar statements for $S_{\cdot 0\cdot}$ and $S_{00\cdot}$; and $S_{\cdot\cdot\cdot}$ as the *residual* component of S_T.

Now it is evident that

$$\mathscr{E}(\bar{x}_{\cdot\cdot\cdot}) = \mu.$$

But to determine the mean values of the S's and the variance of $\bar{x}_{...}$, we find that the mean value in each case is a linear function of quantities like the following:

(8.6.33)

$$
\begin{aligned}
\mathscr{E}(x'_{\xi\eta\zeta}x'_{\xi'\eta'\zeta'}) &= \sigma^2; & \xi &= \xi', & \eta &= \eta', & \zeta &= \zeta' \\
&= \sigma^2_{\cdot\cdot 0} + \sigma^2_{\cdot 00} + \sigma^2_{0\cdot 0}; & \xi &= \xi', & \eta &= \eta', & \zeta &\neq \zeta' \\
&= \sigma^2_{\cdot 0\cdot} + \sigma^2_{\cdot 00} + \sigma^2_{00\cdot}; & \xi &= \xi', & \eta &\neq \eta', & \zeta &= \zeta' \\
&= \sigma^2_{0\cdot\cdot} + \sigma^2_{0\cdot 0} + \sigma^2_{00\cdot}; & \xi &\neq \xi', & \eta &= \eta', & \zeta &= \zeta' \\
&= \sigma^2_{\cdot 00}; & \xi &= \xi', & \eta &\neq \eta', & \zeta &\neq \zeta' \\
&= \sigma^2_{0\cdot 0}; & \xi &\neq \xi', & \eta &= \eta', & \zeta &\neq \zeta' \\
&= \sigma^2_{00\cdot}; & \xi &\neq \xi', & \eta &\neq \eta', & \zeta &= \zeta' \\
&= 0; & \xi &\neq \xi', & \eta &\neq \eta', & \zeta &\neq \zeta'
\end{aligned}
$$

where σ^2 is given by (8.6.28).

Evaluating mean values of the S's, we find

(8.6.34)

$$
\begin{aligned}
\mathscr{E}(S_{00\cdot}) &= rs(t-1)\left[\sigma^2_{00\cdot} + \frac{1}{s}\sigma^2_{0\cdot\cdot} + \frac{1}{r}\sigma^2_{\cdot 0\cdot} + \frac{1}{rs}\sigma^2_{\cdot\cdot\cdot}\right] \\
\mathscr{E}(S_{0\cdot 0}) &= rt(s-1)\left[\sigma^2_{0\cdot 0} + \frac{1}{t}\sigma^2_{0\cdot\cdot} + \frac{1}{r}\sigma^2_{\cdot\cdot 0} + \frac{1}{rt}\sigma^2_{\cdot\cdot\cdot}\right] \\
\mathscr{E}(S_{\cdot 00}) &= st(r-1)\left[\sigma^2_{\cdot 00} + \frac{1}{t}\sigma^2_{\cdot 0\cdot} + \frac{1}{s}\sigma^2_{\cdot\cdot 0} + \frac{1}{st}\sigma^2_{\cdot\cdot\cdot}\right] \\
\mathscr{E}(S_{0\cdot\cdot}) &= r(s-1)(t-1)\left[\sigma^2_{0\cdot\cdot} + \frac{1}{r}\sigma^2_{\cdot\cdot\cdot}\right] \\
\mathscr{E}(S_{\cdot 0\cdot}) &= s(r-1)(t-1)\left[\sigma^2_{\cdot 0\cdot} + \frac{1}{s}\sigma^2_{\cdot\cdot\cdot}\right] \\
\mathscr{E}(S_{\cdot\cdot 0}) &= t(r-1)(s-1)\left[\sigma^2_{\cdot\cdot 0} + \frac{1}{t}\sigma^2_{\cdot\cdot\cdot}\right] \\
\mathscr{E}(S_{\cdot\cdot\cdot}) &= (r-1)(s-1)(t-1)\sigma^2_{\cdot\cdot\cdot},
\end{aligned}
$$

and, of course, the value of $\mathscr{E}(S_T)$ is the sum of the quantities on the right of (8.6.34).

The variance of $\bar{x}_{...}$ is similarly a linear function of quantities in (8.6.33), and has the value

(8.6.35)

$$\sigma^2(\bar{x}_{...}) = \frac{1}{r}\sigma^2_{\cdot 00} + \frac{1}{s}\sigma^2_{0\cdot 0} + \frac{1}{t}\sigma^2_{00\cdot} + \frac{1}{rs}\sigma^2_{\cdot\cdot 0} + \frac{1}{rt}\sigma^2_{\cdot 0\cdot} + \frac{1}{st}\sigma^2_{0\cdot\cdot} + \frac{1}{rst}\sigma^2_{\cdot\cdot\cdot}.$$

We leave to the reader the task of formulating the extension of **8.6.2** to the case of third-order matrix samples for the case of infinite populations of rows, columns, and layers.

If the reader will compare the structure of the equations in (8.6.11) with those in (8.6.22) he should have no difficulty in surmising the form which (8.6.34) will take if, for the matrix sample $\{x(u_\xi, v_\eta, w_\zeta)\}$, $(u_1, \ldots, u_r)$, $(v_1, \ldots, v_s)$, and $(w_1, \ldots, w_t)$ are independent samples from finite row, column, and layer populations, respectively. Similarly, by comparing (8.6.12) with (8.6.24), he will infer the form which formula (8.6.35) for $\sigma^2(\bar{x}_{...})$ will take in the finite-population case. We leave it to the reader as an exercise to write down and verify these formulas.

(d) Balanced Incomplete Matrix Samples

Let us return to the second-order matrix sample $(x_{\xi\eta};\ \xi = 1, \ldots, r;\ \eta = 1, \ldots, s)$, which we denote by $\{x_{\xi\eta}\}$ and select a subsample $\{x_{\xi\eta}\}^*$ consisting of s' elements from every row and r' elements from every column of $\{x_{\xi\eta}\}$. If n is the sample size, then n, r, s, r', s' must be positive integers satisfying the conditions

$$n = rs' = r's. \tag{8.6.36}$$

Let the set of pairs of values of (ξ, η) for this selection be denoted by G^*. We shall call such a sample a *balanced incomplete matrix sample*. Examples in which such selections G^* can be made are: $n = 12, r = 3, s = 6, r' = 2, s' = 4$; $n = 16, r = 4, s = 8, r' = 2, s' = 4$; $n = 40, r = 5, s = 10, n' = 4, s' = 8$. The reader interested in the details of the construction of specific patterns of selection of such samples is referred to papers by Bose (1939), Connor (1952), Shrikhande (1952), and to books by Cochran and Cox (1957), Kempthorne (1952).

Let the mean of the incomplete sample, the mean of the ξth row and the ηth column be denoted by $\bar{x}^*_{..}$, $\bar{x}^*_{\xi.}$, $\bar{x}^*_{.\eta}$ respectively, and let

$$\begin{aligned}
S_T^* &= \sum_{\xi,\eta}{}^*(x_{\xi\eta} - \bar{x}^*_{..})^2 \\
S_{.0}^* &= \sum_{\xi,\eta}{}^*(\bar{x}^*_{\xi.} - \bar{x}^*_{..})^2 \\
S_{0.}^* &= \sum_{\xi,\eta}{}^*(x^*_{.\eta} - \bar{x}^*_{..})^2 \\
S_E^* &= S_T^* - S_{.0}^* - S_{0.}^*
\end{aligned} \tag{8.6.37}$$

where $\sum_{\xi,\eta}^*$ denotes summation for all $(\xi, \eta) \in G^*$. S_E^* is the *residual* sum of squares.

We shall consider the variance of $\bar{x}^*_{..}$ and the mean values of the S^*'s in the case of infinite populations of rows and columns. In this case, all these mean values are linear functions of quantities like those in (8.6.10). The evaluation of $\sigma^2(\bar{x}^*_{..})$, $\mathscr{E}(S^*_T)$, $\mathscr{E}(S^*_{\cdot 0})$ and $\mathscr{E}(S^*_{0\cdot})$ is straightforward whereas

(8.6.38) $$\mathscr{E}(S^*_E) = \mathscr{E}(S^*_T) - \mathscr{E}(S^*_{\cdot 0}) - \mathscr{E}(S^*_{0\cdot}).$$

Letting $r's = rs' = n$, and omitting the details, we find

(8.6.39)
$$\mathscr{E}(\bar{x}^*_{..}) = \mu$$
$$\sigma^2(\bar{x}^*_{..}) = \frac{1}{n}\sigma^2_{..} + \frac{1}{r}\sigma^2_{\cdot 0} + \frac{1}{s}\sigma^2_{0\cdot}$$
$$\mathscr{E}(S^*_{\cdot 0}) = (r-1)\sigma^2_{..} + (r-1)s'\sigma^2_{\cdot 0} + (r-r')\sigma^2_{0\cdot}$$
$$\mathscr{E}(S^*_{0\cdot}) = (s-1)\sigma^2_{..} + (s-s')\sigma^2_{\cdot 0} + (s-1)r'\sigma^2_{0\cdot}$$
$$\mathscr{E}(S^*_E) = (n-r-s+1)\sigma^2_{..} - (s-s')\sigma^2_{\cdot 0} - (r-r')\sigma^2_{0\cdot}..$$

Note that if $r' = r$, and $s' = s$, these equations reduce to (8.6.12) and (8.6.11).

Balanced incomplete matrix samples in their most general form from third- and higher order matrix samples must satisfy several varieties of conditions, and are consequently rather complicated. Here we shall only discuss a special third-order case, which, however, is basic in the theory of experimental designs.

In this case we consider a third-order matrix sample with r rows, r columns, and r layers, that is, $\{x_{\xi\eta\zeta}\}$, ξ, η, $\zeta = 1, \ldots, r$. Our balanced incomplete matrix sample $\{x_{\xi\eta\zeta}\}^*$ will be a selection from $\{x_{\xi\eta\zeta}\}$ consisting of one element from each combination of values of ξ, η, one element from each combination of values of ξ, ζ, and one element from each combination of values of η, ζ.

The number of elements in the selection $\{x_{\xi\eta\zeta}\}^*$ is r^2, thus comprising only a skeleton selection from $\{x_{\xi\eta\zeta}\}$. However, the balance incorporated into $\{x_{\xi\eta\zeta}\}^*$ is quite remarkable. In fact, if the elements in $\{x_{\xi\eta\zeta}\}^*$ are all projected onto the $\xi\eta$-"plane" (the row-column plane) it will be noted that in every row of the projected form of $\{x_{\xi\eta\zeta}\}^*$ one finds one and only one element from each of the r layers of $\{x_{\xi\eta\zeta}\}$. Similarly, in every column of the $\xi\eta$-"plane" one finds one and only one element from each of the r layers of $\{x_{\xi\eta\zeta}\}$. Furthermore, similar statements hold if $\{x_{\xi\eta\zeta}\}^*$ is projected onto the $\xi\zeta$-"plane" or the $\eta\zeta$-"plane," by suitable interchange of the words "row," "column," and "layer." Each of these projected forms of $\{x_{\xi\eta\zeta}\}^*$ may be called a *Latin-Square* selection of rows, columns, and layers.

Now let G^* be the set of values of (ξ, η, ζ) used in the particular selection of the incomplete sample. Needless to say, there are many choices of elements for G^*. The reader interested in the construction of sets G^* (the construction of Latin Squares) is referred to Bose (1938) and Mann (1943).

We shall be interested only in the sample mean $\bar{x}^*_{...}$, and the means $\bar{x}^*_{\xi..}$, $\bar{x}^*_{\cdot\eta\cdot}$, and $\bar{x}^*_{..\zeta}$ of the elements in the ξth row, ηth column, and ζth layer respectively, of the incomplete sample, and in

$$
\begin{aligned}
S_T^* &= \sum_{\xi,\eta,\zeta}{}^* (x_{\xi\eta\zeta} - \bar{x}^*_{...})^2 \\
S^*_{\cdot 00} &= \sum_{\xi,\eta,\zeta}{}^* (\bar{x}^*_{\xi..} - \bar{x}^*_{...})^2 \\
S^*_{0\cdot 0} &= \sum_{\xi,\eta,\zeta}{}^* (\bar{x}^*_{\cdot\eta\cdot} - \bar{x}^*_{...})^2 \qquad (8.6.40) \\
S^*_{00\cdot} &= \sum_{\xi,\eta,\zeta}{}^* (\bar{x}^*_{..\zeta} - \bar{x}^*_{...})^2 \\
S_E^* &= S_T^* - S^*_{\cdot 00} - S^*_{0\cdot 0} - S^*_{00\cdot}
\end{aligned}
$$

where $\sum_{\xi,\eta,\zeta}^*$ denotes summation over all $(\xi, \eta, \zeta) \in G^*$.

Now the variance of $\bar{x}^*_{...}$ and the mean values of the S^*'s in (8.6.40) are all linear functions of *only* the first four types of quantities listed in (8.6.33). Carrying out the mean value operations, omitting the details, we find:

$$
\begin{aligned}
\mathscr{E}(S^*_{\cdot 00}) &= (r-1)\sigma_E^2 + (r-1)\sigma^2_{\cdot 00} \\
\mathscr{E}(S^*_{0\cdot 0}) &= (r-1)\sigma_E^2 + (r-1)\sigma^2_{0\cdot 0} \qquad (8.6.41) \\
\mathscr{E}(S^*_{00\cdot}) &= (r-1)\sigma_E^2 + (r-1)\sigma^2_{00\cdot} \\
\mathscr{E}(S^*_E) &= (r-1)(r-2)\sigma_E^2 + (r-1)^2[\sigma^2_{\cdot 00} + \sigma^2_{0\cdot 0} + \sigma^2_{00\cdot}],
\end{aligned}
$$

where σ_E^2 satisfies

$$\sigma^2 = \sigma_E^2 + \sigma^2_{\cdot 00} + \sigma^2_{0\cdot 0} + \sigma^2_{00\cdot}. \qquad (8.6.42)$$

It should be noted from (8.6.27) and (8.6.28) that σ_E^2 is the sum of four components of σ^2, namely,

$$\sigma_E^2 = \sigma^2_{...} + \sigma^2_{..0} + \sigma^2_{\cdot 0\cdot} + \sigma^2_{0..}. \qquad (8.6.43)$$

8.7 SAMPLING THEORY OF ORDER STATISTICS

Suppose $(x_1, \ldots, x_n)$ is a sample from a population having a *continuous* c.d.f. $F(x)$. Let $x_1, \ldots, x_n$ be rearranged in order from least to greatest and let the ordered values be $x_{(1)}, \ldots, x_{(n)}$, where $x_{(1)} \leqslant \cdots \leqslant x_{(n)}$. These new random variables are called the *order statistics of the sample*; $x_{(k)}$ is called the kth *order statistic*. Note that since $F(x)$ is continuous, $P(x_{(\xi-1)} = x_{(\xi)}) = 0$, $\xi = 1, \ldots, n$. The intervals $(-\infty, x_{(1)}]$, $(x_{(1)}, x_{(2)}]$, $\ldots, (x_{(n)}, +\infty)$ are called *sample blocks* $B_1^{(1)}, \ldots, B_1^{(n+1)}$ respectively, and the functions $F(x_{(1)}), F(x_{(2)}) - F(x_{(1)}), \ldots, 1 - F(x_{(n)})$ of these blocks are called *coverages* $u_1, \ldots, u_{n+1}$ respectively. Since the sum of the u's is 1, we shall ordinarily omit u_{n+1}. The subscript on the B's denotes dimensionality. Sample blocks of two or more dimensions will be defined in Section 8.7(c). Note that a coverage for a given sample block is merely the amount of probability in the population distribution (*P-measure*) contained in that sample block, and is a random variable.

The sampling theory of order statistics and of coverages is of fundamental importance in *nonparametric statistical inference*, which is discussed in Chapters 11 and 14. In this section we shall consider some of the more basic results in the sampling theory of order statistics. The reader interested in more details in this field should consult the books by Fraser (1957), Gumbel (1958), and Kendall (1953). An extensive bibliography of publications in this field has been published by Savage (1962).

(a) Sampling Distributions of Order Statistics

As stated in **7.1.1**, if a random variable x has a continuous c.d.f. $F(x)$, the random variable y, where $y = F(x)$, has the rectangular distribution $R(\frac{1}{2}, 1)$. Now let $y_\xi = F(x_\xi)$, $\xi = 1, \ldots, n$. Then $(y_1, \ldots, y_n)$ is a sample of size n from the rectangular distribution $R(\frac{1}{2}, 1)$.

The p.e. of these y's is

(8.7.1) $$1 \cdot dy_1 \cdots dy_n$$

over the unit "cube" $\{(y_1, \ldots, y_n)\colon\ 0 \leqslant y_\xi \leqslant 1, \quad \xi = 1, \ldots, n\}$ in the sample space R_n and $0 \cdot dy_1 \cdots dy_n$ outside the cube. Let $y_{(1)} \leqslant \cdots \leqslant y_{(n)}$ be the *order statistics* of the sample $(y_1, \ldots, y_n)$ and consider the problem of determining the p.e. of these order statistics. The probability that two or more of the elements of the sample $(y_1, \ldots, y_n)$ are equal is zero. So we may consider only points in the unit cube whose coordinates are all distinct. Suppose $(y_1, \ldots, y_n)$ is such a point P. The p.e. associated with point P is $1 \cdot dy_1 \cdots dy_n$. But a total of $n!$ points in the unit cube are obtained by permuting the coordinates of P. If new random variables are formed by letting $y_{(1)}$ be the smallest coordinate of these $n!$ points, $y_{(2)}$ the next smallest, and so on, then the p.d.f. of the random variable $(y_{(1)}, \ldots, y_{(n)})$, whose components are the order statistics of the sample $(y_1, \ldots, y_n)$, will be given by adding the p.d.f.'s of these $n!$ points and using the order statistic notation for the random variables. This gives as the p.e. of $(y_{(1)}, \ldots, y_{(n)})$:

(8.7.2) $$n!\, dy_{(1)} \cdots dy_{(n)}$$

inside the region $\{(y_{(1)}, \ldots, y_{(n)}): \ 0 \leqslant y_{(1)} \leqslant \cdots \leqslant y_{(n)} \leqslant 1\}$, and $0 \cdot dy_{(1)} \cdots dy_{(n)}$ outside. This is the ordered n-variate Dirichlet distribution $D^*(1, \ldots, 1; 1)$ defined in (7.7.17). Therefore we have the following result:

8.7.1 *If $(x_{(1)}, \ldots, x_{(n)})$ are the order statistics of a sample from a continuous* c.d.f. *$F(x)$, the random variables $F(x_{(1)}), \ldots, F(x_{(n)})$ have the ordered n-variate Dirichlet distribution $D^*(1, \ldots, 1; 1)$.*

The distribution of any subset of one or more of the random variables $F(x_{(1)}), \ldots, F(x_{(n)})$ follows at once by application of **7.7.6**. Thus, for the case of one of these random variables we have

8.7.2 *The random variable $F(x_{(k)})$, $1 \leqslant k \leqslant n$, has the beta distribution $Be(k, n - k + 1)$.*

For the case of s of these random variables we can state the following result:

8.7.3 *The random variables $F(x_{(k_1)}), F(x_{(k_1+k_2)}), \ldots, F(x_{(k_1+\cdots+k_s)})$ have the ordered s-variate Dirichlet distribution $D^*(k_1, \ldots, k_s; n - k_1 - \cdots - k_s + 1)$.*

It should be noted that in **8.7.1**, **8.7.2**, and **8.7.3** we assumed only that $F(x)$ is continuous. Now if we make the somewhat stronger assumption that a p.d.f. $f(x)$ exists, we can determine p.e.'s of the order statistics $x_{(1)}, \ldots, x_{(n)}$ and subsets of them. Thus, corresponding to **8.7.1**, **8.7.2**, and **8.7.3**, we have the following:

8.7.1a *If $(x_{(1)}, \ldots, x_{(n)})$ are the order statistics of a sample of size n from a population having* c.d.f. *$F(x)$ and* p.d.f. *$f(x)$, then the* p.e. *of the order statistics is*

$$n!f(x_{(1)}) \cdots f(x_{(n)})\, dx_{(1)} \cdots dx_{(n)}, \tag{8.7.3}$$

the sample space of $(x_{(1)}, \ldots, x_{(n)})$ being the region in R_n for which $-\infty \leqslant x_{(1)} \leqslant \cdots \leqslant x_{(n)} \leqslant +\infty$.

8.7.2a *The* p.e. *of $x_{(k)}$, $1 \leqslant k \leqslant n$, under the conditions stated in* **8.7.1a** *is*

$$\frac{\Gamma(n+1)}{\Gamma(k)\Gamma(n-k+1)} [F(x_{(k)})]^{k-1}[1 - F(x_{(k)})]^{n-k} f(x_{(k)})\, dx_{(k)}, \tag{8.7.4}$$

the sample space of $x_{(k)}$ being R_1.

8.7.3a *The* p.e. *of* $(x_{(k_1)}, x_{(k_1+k_2)}, \ldots, x_{(k_1+\cdots+k_s)})$ *under the conditions stated in* **8.7.1a** *is*

$$\frac{\Gamma(n+1)}{\Gamma(k_1)\cdots\Gamma(k_s)\Gamma(n-k_1-\cdots-k_s+1)} \tag{8.7.5}$$
$$\cdot [F(x_{(k_1)})]^{k_1-1}[F(x_{(k_1+k_2)}) - F(x_{(k_1)})]^{k_2-1}\cdots$$
$$\cdot [1 - F(x_{(k_1+\cdots+k_s)})]^{n-k_1-\cdots-k_s} f(x_{(k_1)})\cdots$$
$$\cdot f(x_{(k_1+\cdots+k_s)})\, dx_{(k_1)}\cdots dx_{(k_1+\cdots+k_s)},$$

the sample space of $(x_{(k_1)}, x_{(k_1+k_2)}, \ldots, x_{(k_1+\cdots+k_s)})$ *being the region in* R_s *for which* $-\infty \leqslant x_{(k_1)} \leqslant \cdots \leqslant x_{(k_1+\cdots+k_s)} \leqslant +\infty$.

Formulas (8.7.3), (8.7.4), and (8.7.5) were originally obtained by Craig (1932), and they have many interesting special cases. For instance, if in (8.7.4) we put $k = 1$ we obtain the p.e. of the *smallest element in the sample*, and for $k = n$ we find the p.e. of the *largest element in the sample*. If n is odd and $k = (n+1)/2$, we obtain the p.e. of the *sample median*. If, in (8.7.5) $s = 2$, $k_1 = 1$, $k_2 = n - 1$ we obtain the p.e. of the *largest* and *smallest* elements in the sample:

$$n(n-1)[F(x_{(n)}) - F(x_{(1)})]^{n-2} f(x_{(1)}) f(x_{(n)})\, dx_{(1)}\, dx_{(n)}. \tag{8.7.6}$$

The distribution of the *sample range* w can be obtained by applying the transformation

$$x_{(1)} = v$$
$$x_{(n)} = v + w$$

to (8.7.6) and taking the marginal distribution of w, that is, integrating out the variable v.

(b) Sampling Distributions of One-Dimensional Coverages

The distributions occurring in **8.7.1**, **8.7.2**, and **8.7.3** have a wider interpretation than may be immediately apparent from the preceding statements. Suppose we make the transformation

$$\begin{aligned} y_{(1)} &= u_1 \\ y_{(2)} &= u_1 + u_2 \\ &\cdots \\ y_{(n)} &= u_1 + \cdots + u_n \end{aligned} \tag{8.7.7}$$

The p.e. of the random variables $u_1, \ldots, u_n$ is

$$n!\, du_1 \cdots du_n \tag{8.7.8}$$

in the simplex S_n: $\left\{(u_1, \ldots, u_n): u_\xi \geqslant 0,\ \xi = 1, \ldots, n, \sum_{\xi=1}^{n} u_\xi \leqslant 1\right\}$ and $0 \cdot du_1 \cdots du_n$ outside S_n. This is the p.e. of the n-variate Dirichlet distribution $D(1, \ldots, 1; 1)$ defined in (7.7.1). But note that the random variables $u_1, \ldots, u_n$ are the coverages defined in the first paragraph of Section 8.7. Furthermore, the distribution of the coverages does not depend on the population c.d.f. $F(x)$, which is assumed to be continuous. For this reason, the coverages behave in a *distribution-free* manner.

We may summarize as follows:

8.7.4 *Let* $(x_{(1)}, \ldots, x_{(n)})$ *be the order statistics of a sample from a continuous* c.d.f. $F(x)$. *Then the coverages* $u_1 = F(x_{(1)})$, $u_2 = F(x_{(2)}) - F(x_{(1)}), \ldots, u_n = F(x_{(n)}) - F(x_{(n-1)})$, *are random variables having the n-variate Dirichlet distribution* $D(1, \ldots, 1; 1)$.

Note that the distribution of the coverages given in **8.7.4** is completely symmetrical in the variables. This fact leads us to refer sometimes to the sample blocks $B_1^{(1)}, \ldots, B_1^{(n+1)}$ corresponding to the coverages $u_1, \ldots, u_{n+1}$, where $u_{n+1} = 1 - u_1 - \cdots - u_n$ respectively, as *statistically equivalent blocks*. It follows from this fact and **7.7.2** that

8.7.5 *Any* k $(k \leqslant n)$ *of the coverages listed in* **8.7.4** *have the k-variate Dirichlet distribution* $D(1, \ldots, 1; n - k + 1)$.

Applying **7.7.4** we also have

8.7.6 *The sum of any k of the coverages listed in* **8.7.4** *has the beta distribution* $Be(k, n - k + 1)$.

It will be observed that **8.7.2** is a special case of **8.7.6** since $F(x_{(n)})$ is the sum of the *first* k coverages $u_1, \ldots, u_k$.

Finally, we point out that application of **7.7.5** leads to the following result:

8.7.7 *If* $v_1, \ldots, v_s$ *are sums of* $k_1, \ldots, k_s$ *respectively, of the coverages listed in* **8.7.4**, *where no coverage belongs to more that one v, then the distribution of* $(v_1, \ldots, v_s)$ *is the s-variate Dirichlet distribution* $D(k_1, \ldots, k_s; n - k_1 - \cdots - k_s + 1)$.

(c) Sampling Distributions of Multidimensional Coverages

Thus far we have only considered order statistics in samples from a one-dimensional distribution. The results which have been stated in **8.7.4**, **8.7.5**, **8.7.6**, and **8.7.7** hold for an extension of the definition of sample blocks and coverages to two or more dimensions. Let us consider the case of two dimensions. We assume that $(x_{1\xi}, x_{2\xi}; \xi = 1, \ldots, n)$ is a sample from a continuous two-dimensional c.d.f. $F(x_1, x_2)$. We introduce an *ordering function* $h(x_1, x_2)$ such that $w = h(x_1, x_2)$ is a random variable which has a continuous c.d.f. $H(w)$. Then the random variables

$w_\xi = h(x_{1\xi}, x_{2\xi})$, $\xi = 1, \ldots, n$ constitute a sample from a population whose distribution is that of $h(x_1, x_2)$ and this sample can be ordered. Let the order statistics for the sample $(w_1, \ldots, w_n)$ be $(w_{(1)}, \ldots, w_{(n)})$. The coverages $u_1', u_2', \ldots, u_{n+1}'$, where $u_{n+1}' = 1 - u_1' - \cdots - u_n'$ and

$$u_1' = H(w_{(1)}),\ u_2' = H(w_{(2)}) - H(w_{(1)}), \ldots, u_n' = H(w_{(n)}) - H(w_{(n-1)}),$$

are random variables associated, respectively, with the two-dimensional sample blocks $B_2^{(1)}, \ldots, B_2^{(n+1)}$ into which the x_1x_2-plane is decomposed

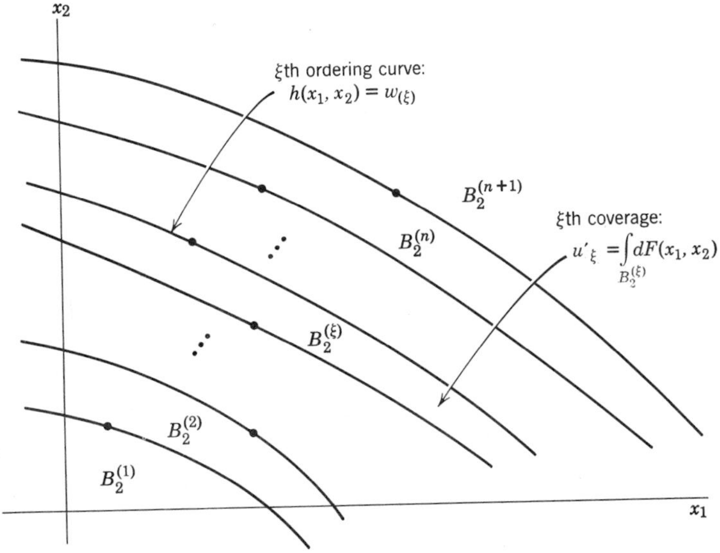

FIG. 8.1 Sample blocks and coverages for two dimensions.

by the ordering "curves" $w_{(\xi)} = h(x_1, x_2)$, $\xi = 1, \ldots, n$ as illustrated in Fig. 8.1.

It is evident from the foregoing discussion how one would define coverages $u_1', \ldots, u_{n+1}'$ for k-dimensional sample blocks $B_k^{(1)}, \ldots, B_k^{(n+1)}$ produced by a k-variate ordering function $h(x_1, \ldots, x_k)$ having a continuous c.d.f. $H(w)$. If $w_\xi = h(x_{1\xi}, \ldots, x_{k\xi})$, $\xi = 1, \ldots, n$ and if $(w_{(1)}, \ldots, w_{(n)})$ are the order statistics corresponding to $(w_1, \ldots, w_n)$ then since $(w_{(1)}, \ldots, w_{(n)})$ are order statistics from a population whose distribution is that of $h(x_1, \ldots, x_k)$, we have the following result:

8.7.8 *The coverages $u_1', \ldots, u_n'$ for the k-dimensional blocks defined by the ordering function $h(x_1, \ldots, x_n)$, have the same distribution properties as those possessed by the coverages $u_1, \ldots, u_n$ and stated in* **8.7.4, 8.7.5, 8.7.6,** *and* **8.7.7.**

The notion of coverages for two or more dimensions is more general than it may appear at first. Wald (1943) gave a method of constructing coverages based on rectangular sample blocks, which was subsequently generalized by Tukey (1947).

We shall consider Tukey's method for the case of two dimensions. Instead of using only one ordering function $h(x_1, x_2)$ for the n sample points $(x_{1\xi}, x_{2\xi})$, $\xi = 1, \ldots, n$ we can use as many as n ordering functions, $h_\eta(x_1, x_2)$, $\eta = 1, \ldots, n$. It is assumed that each of the functions $h_\eta(x_1, x_2)$

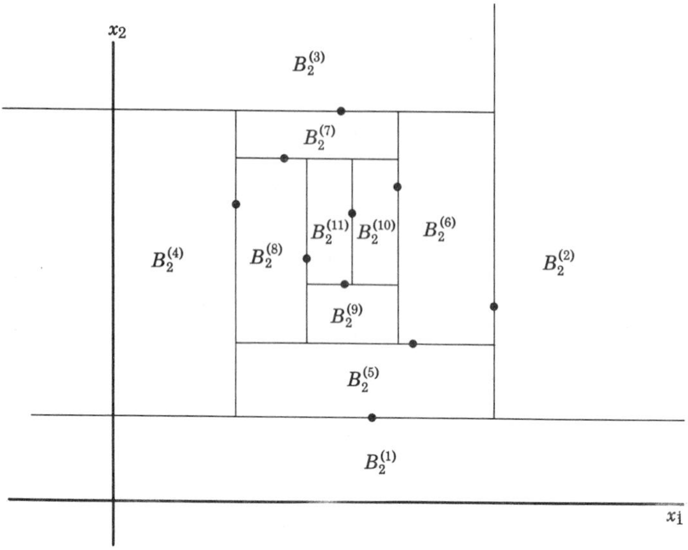

FIG. 8.2 Example of two-dimensional sample blocks generated by horizontal and vertical lines, taken alternately, as graphs of ordering functions.

has a continuous c.d.f., the c.d.f. of the population distribution being a continuous c.d.f. $F(x_1, x_2)$. Some or all of the functions can be identical, of course, if we let $w_\xi^\eta = h_\eta(x_{1\xi}, x_{2\xi})$, then for each value of η we would have a set of order statistics $w_{(\xi)}^\eta$, $\xi = 1, \ldots, n$. We now cut up the x_1x_2-plane as follows: The curve $w_{(1)}^1 = h_1(x_1, x_2)$ cuts the x_1x_2-plane into two point sets $B_2^{(1)}$ and $\bar{B}_2^{(1)}$ where $B_2^{(1)}$ is the set for which $h_1(x_1, x_2) < w_{(1)}^1$. We define our first coverage, say u_1^*, as the P-measure (the probability determined from $F(x_1, x_2)$) of the sample block $B_2^{(1)}$. The curve $w_{(2)}^2 = h_2(x_1, x_2)$ cuts the set $\bar{B}_2^{(1)}$ into two sets, $B_2^{(2)}$ and $\bar{B}_2^{(2)}$, such that on $B_2^{(2)}$, $h_2(x_1, x_2) < w_{(2)}^2$. Let coverage u_2^* be the P-measure of $B_2^{(2)}$. Continuing this process, we successively define sets $B_2^{(1)}, B_2^{(2)}, \ldots, B_2^{(n)}$ such that $B_2^{(2)} \subset \bar{B}_2^{(1)}, \ldots, B_2^{(n)} \subset \bar{B}_2^{(n-1)}$. There will, of course, be a residual set $B_2^{(n+1)}$ such that $B_2^{(1)} \cup \cdots \cup B_2^{(n)} \cup B_2^{(n+1)} = R_2$. The P-measures of

$B_2^{(1)}, \ldots, B_2^{(n)}$ are our general coverages $u_1^*, \ldots, u_n^*$. Figure 8.2 shows an example for $n = 10$ where the ordering functions $h_\eta(x_1, x_2)$, $\eta = 1, \ldots, 10$ are straight lines making angles of $(\eta - 1)90°$ with the x_1-axis.

The definition of k-variate analogues of the coverages $u_1^*, \ldots, u_n^*$ is straightforward and is left to the reader. The following statement concerning these coverages holds:

8.7.9 *The coverages $u_1^*, \ldots, u_n^*$ for the k-dimensional blocks (as defined above) have the same distribution properties as those possessed by the coverages $u_1, \ldots, u_n$ in* **8.7.4**, **8.7.5**, **8.7.6**, and **8.7.7.**

To establish **8.7.9** it will be sufficient to show that $u_1^*, \ldots, u_n^*$ have the n-variate Dirichlet distribution $D(1, \ldots, 1; 1)$ for $k = 2$. If $F(x_1, x_2)$ is the c.d.f. of the population from which the sample $(x_{1\xi}, x_{2\xi}; \xi = 1, \ldots, n)$ is drawn, the p.e. of this sample, which we denote by O_n, is

$$\prod_{\xi=1}^{n} dF(x_{1\xi}, x_{2\xi}). \tag{8.7.9}$$

Let $(x_{1*}^{(1)}, x_{2*}^{(1)})$ be the sample element among the n sample elements which yields the smallest value of $h_1(x_1, x_2)$ and let $(x_{1\xi_1}^{(1)}, x_{2\xi_1}^{(1)}; \xi_1 = 1, \ldots, n-1)$, to be denoted by $O_{n-1}^{(1)}$, be the $n - 1$ sample elements obtained by deleting $(x_{1*}^{(1)}, x_{2*}^{(1)})$ from O_n. The p.e. of $(x_{1*}^{(1)}, x_{2*}^{(1)})$ and $(x_{1\xi_1}^{(1)}, x_{2\xi_1}^{(1)}; \xi_1 = 1, \ldots, n-1)$ is

$$n \cdot dF(x_{1*}^{(1)}, x_{1*}^{(2)}) \prod_{\xi_1=1}^{n-1} dF(x_{1\xi_1}^{(1)}, x_{2\xi_1}^{(1)}). \tag{8.7.10}$$

Now let u_1'' be the coverage associated with the set $B_2^{(1)}$ in the x_1x_2-plane for which $h_1(x_1, x_2) < h_1(x_{1*}^{(1)}, x_{2*}^{(1)})$. It follows from **8.7.2** that the distribution of u_1'' is given by the beta distribution $Be(1, n)$, that is, the p.e. of u_1'' is

$$n(1 - u_1'')^{n-1}\, du_1'' \tag{8.7.11}$$

where

$$u_1'' = \int_{B_2^{(1)}} dF(x_1, x_2), \qquad du_1'' = dF(x_{1*}^{(1)}, x_{2*}^{(1)}). \tag{8.7.12}$$

Now consider the $(2n - 2)$-dimensional conditional random variable $(x_{1\xi_1}^{(1)}, x_{2\xi_1}^{(2)} \mid x_{1*}^{(1)}, x_{2*}^{(2)}; \xi_1 = 1, \ldots, n)$. The p.e. of this distribution is given by the ratio of (8.7.10) to (8.7.11), which reduces to

$$\prod_{\xi_1=1}^{n-1} dF^{(1)}(x_{1\xi_1}^{(1)}, x_{2\xi_1}^{(1)}) \tag{8.7.13}$$

where

$$F^{(1)}(x_{1\xi_1}^{(1)}, x_{2\xi_1}^{(1)}) = \frac{F(x_{1\xi_1}^{(1)}, x_{2\xi_1}^{(1)})}{1 - u_1''}. \tag{8.7.14}$$

This means that for given values of $(x_{1*}^{(1)}, x_{2*}^{(1)})$, and hence for a given value of u_1'', the remaining $n - 1$ elements of the original sample behave exactly like a sample $O_{n-1}^{(1)}$ of $n - 1$ elements from a population having the c.d.f.

$$F^{(1)}(x_1, x_2) = \frac{F(x_1, x_2)}{1 - u_1''} \tag{8.7.15}$$

which is obtained by assigning zero probability to $B_2^{(1)}$ and normalizing back to unity the portion of the original population distribution contained in $\bar{B}_2^{(1)}$.

Now we repeat the process and let $(x_{1*}^{(2)}, x_{2*}^{(2)})$ be the sample element among the $n - 1$ elements of $O_{n-1}^{(1)}$ which yields the smallest value of $h_2(x_1, x_2)$ and let $(x_{1\xi_2}^{(2)}, x_{2\xi_2}^{(2)}, \xi_2 = 1, \ldots, n - 2)$, to be denoted by $O_{n-2}^{(2)}$, be the $n - 2$ elements of $O_{n-1}^{(1)}$ obtained by deleting $(x_{1*}^{(2)}, x_{2*}^{(2)})$. The p.e. of $(x_{1*}^{(2)}, x_{2*}^{(2)})$ and $(x_{1\xi_2}^{(2)}, x_{2\xi_2}^{(2)}; \xi_2 = 1, \ldots, n - 2)$ is

$$(n - 1)\, dF^{(1)}(x_{1*}^{(2)}, x_{2*}^{(2)}) \prod_{\xi_2=1}^{n-2} dF^{(1)}(x_{1\xi_2}^{(2)}, x_{2\xi_2}^{(2)}). \tag{8.7.16}$$

Let $B_2^{(2)}$ be the subset of $\bar{B}_2^{(1)}$ for which $h_2(x_1, x_2) < h_2(x_{1*}^{(2)}, x_{2*}^{(2)})$ and let $u_2'' = \int_{B_2^{(2)}} dF^{(1)}(x_1, x_2)$ be the *conditional coverage* associated with $B_2^{(2)}$ as determined from $F^{(1)}(x_1, x_2)$. It follows as a special case of **8.7.2** that the distribution of u_2'' has p.e.

$$(n - 1)(1 - u_2'')^{n-2}\, du_2'' \tag{8.7.17}$$

and the p.e. of $(x_{1\xi_2}^{(2)}, x_{2\xi_2}^{(2)} \mid x_{1*}^{(2)}, x_{2*}^{(2)}; \xi_2 = 1, \ldots, n - 2)$ is

$$\prod_{\xi_2=1}^{n-2} dF^{(2)}(x_{1\xi_2}^{(2)}, x_{2\xi_2}^{(2)}) \tag{8.7.18}$$

where

$$F^{(2)}(x_{1\xi_2}^{(2)}, x_{2\xi_2}^{(2)}) = \frac{F^{(1)}(x_{1\xi_2}^{(2)}, x_{2\xi_2}^{(2)})}{1 - u_2''}. \tag{8.7.18a}$$

Thus, for given values of $(x_{1*}^{(1)}, x_{2*}^{(1)})$ and $(x_{1*}^{(2)}, x_{2*}^{(2)})$ the elements of $O_{n-2}^{(2)}$ behave like a sample from a population having c.d.f.

$$F^{(2)}(x_1, x_2) = \frac{F^{(1)}(x_1, x_2)}{1 - u_2''}. \tag{8.7.19}$$

The distribution of u_1'', u_2'' and the elements of $O_{n-2}^{(2)}$ is the product of (8.7.11), (8.7.17), and (8.7.18).

Continuing this process we obtain a sequence of conditional coverages $u_1'', \ldots, u_n''$ having a distribution with p.e.

$$n!\,(1 - u_1'')^{n-1} \cdots (1 - u_n'')^{1-1}\, du_1'' \cdots du_n''. \tag{8.7.20}$$

But expressed in terms of the coverages u_ξ^*, $\xi = 1, \ldots, n$ referred to in **8.7.9** we have

$$
\begin{aligned}
u_1'' &= u_1^* \\
u_2'' &= \frac{u_2^*}{1 - u_1^*} \\
&\;\;\vdots \\
u_n'' &= \frac{u_n^*}{1 - u_1^* - \cdots - u_{n-1}^*}.
\end{aligned} \tag{8.7.21}
$$

Applying this transformation to (8.7.20) we obtain

$$n!\, du_1^* \cdots du_n^*. \tag{8.7.22}$$

But this is the p.e. of the n-variate Dirichlet distribution $D(1, \ldots, 1; 1)$, which concludes the argument for **8.7.9.**

Further results in the sampling theory of coverages have been obtained by Fraser (1951, 1953), by Fraser and Guttman (1956), and by Kemperman (1956). In particular, they have shown how to obtain statistically equivalent blocks by using ordering functions which can depend on the results obtained by ordering functions used earlier in the process of determining sample blocks. Extensions of the notion of sample blocks and coverages to the case of discontinuous c.d.f.'s have been considered by Tukey (1948).

8.8 ORDER STATISTICS IN SAMPLES FROM FINITE POPULATIONS

Suppose π_N is a finite population whose elements have distinct x-values, say, $x_{o1} < \cdots < x_{oN}$. Let $(x_{(1)}, \ldots, x_{(n)})$ be the order statistics of a sample of size n from π_N. We shall consider the sampling theory of the kth order statistic $x_{(k)}$. It is evident by combinatorial analysis that

$$p(x_{(k)} = x_{ot}) = \frac{\binom{t-1}{k-1}\binom{N-t}{n-k}}{\binom{N}{n}} = p_{N,n,k}(t), \qquad \text{say,} \tag{8.8.1}$$

where $t = k, k+1, \ldots, N-n+k$. Thus, $p_{N,n,k}(t)$ can be viewed either as: (i) the p.f. of the random variable $x_{(k)}$, its mass points being x_{ot}, $t = k, k+1, \ldots, N-n+k$, or (ii) the p.f. of the random variable t, that is, the *rank* of the x-value in π_N to which the kth order statistic in the sample is equal. The random variable t is simpler to handle than the random variable $x_{(k)}$ since the mass points of t are located at the successive integers $k, k+1, \ldots, N-n+k$, whereas those of $x_{(k)}$ are located at $x_{o,k}, x_{o,k+1}, \ldots, x_{o,N-n+k}$.

Moments of t of the factorial variety are quite readily found. In fact, using the notation of (3.3.15) we have

$$(8.8.2)\quad \mathscr{E}((t+r-1)^{[r]})$$

$$= \sum_{t=k}^{N-n+k} (t+r-1)^{[r]} p_{N,n,k}(t)$$

$$= \frac{(k+r-1)^{[r]}\binom{N+r}{n+r}}{\binom{N}{n}} \cdot \sum_{t=k}^{N-n+k} \frac{\binom{t+r-1}{k+r-1}\binom{N-t}{n-k}}{\binom{N+r}{n+r}}.$$

But the sum on the extreme right has the value 1. Therefore,

$$(8.8.3)\qquad \mathscr{E}((t+r-1)^{[r]}) = \frac{(k+r-1)^{[r]}\binom{N+r}{n+r}}{\binom{N}{n}}.$$

Putting $r = 1$ and 2, we find the mean and variance of t to be

$$\mathscr{E}(t) = \frac{k(N+1)}{(n+1)}$$

$$(8.8.4)\qquad \sigma^2(t) = \frac{k(N+1)(N-n)(n-k+1)}{(n+1)^2(n+2)}.$$

Summarizing,

8.8.1 *If π_N is a finite population whose elements have distinct x-values $x_{o1} < \cdots < x_{oN}$ and if $(x_{(1)}, \ldots, x_{(n)})$ are the order statistics of a sample from π_N, the p.f. of $x_{(k)}$ is given by* (8.8.1), *the mass points being x_{ot}, $t = k, k+1, \ldots, N-n+k$. The value of $\mathscr{E}((t+r-1)^{[r]})$ is given by* (8.8.3).

If we consider the infinite sequence of random variables

$$(8.8.5)\qquad u_{k,N} = \frac{t}{N}, \qquad N = k, k+1, \ldots$$

then $\lim_{N\to\infty} u_{k,N}$ is the proportion of the x-values of an infinite population which $x_{(k)}$ exceeds. It is intuitively evident that $\lim_{N\to\infty} u_{k,N}$ corresponds to $F(x_{(k)})$ for the case of sampling from an infinite population having c.d.f.

$F(x)$ and should, in accordance with **8.7.2**, have the beta distribution $Be(k, n - k + 1)$.

To see that this is actually true we proceed as follows:

$$\text{(8.8.6)} \quad \lim_{N\to\infty} \mathscr{E}\left(\frac{t}{N}\right)^r = \lim_{N\to\infty} \mathscr{E}\left(\frac{(t + r - 1)^{[r]}}{N^r}\right) = \frac{\Gamma(n + 1)\Gamma(k + r)}{\Gamma(k)\Gamma(n + r + 1)}$$

which is the rth moment of a random variable having the beta distribution $Be(k, n - k + 1)$. It now follows from **5.5.3** that the sequence of random variables (8.8.5) converges in distribution to this beta distribution. Therefore

8.8.2 *Let π_N, $N = n, n + 1, \ldots$ be a sequence of finite populations such that the elements in π_N have distinct x-values $x_{o1} < \cdots < x_{oN}$. Let $(x_{(1)}, \ldots, x_{(n)})$ be the order statistics of a sample of size n from π_N, and let $u_{k,N}$ denote the proportion of elements in π_N having x-values exceeded by $x_{(k)}$. The sequence of random variables $u_{k,N}$, $N = n, n + 1, \ldots$ converges in distribution to a random variable having the beta distribution $Be(k, n - k + 1)$.*

PROBLEMS

8.1 If $(x_1, \ldots, x_n)$ is a random sample from a population whose distribution has finite moments $\mu_2, \ldots, \mu_{2r}$ about the distribution mean, and if

$$m_r = \frac{1}{n}\sum_{\xi=1}^{n}(x_\xi - \bar{x})^r,$$

show that

$$\sigma^2(m_r) = \frac{1}{n}\left[\mu_{2r} - \mu_r^2 - r(r - 1)\mu_2\mu_{r-2}\mu_r + r\mu_{r-1}(r\mu_2\mu_{r-1} - 2\mu_{r+1})\right] + 0\left(\frac{1}{n^2}\right).$$

8.2 If $\bar{x}_1$, $\bar{x}_2$ are the sample means and s_1^2, s_2^2 are the sample variances of two independent random samples of sizes n_1 and n_2 from $N(\mu_1, \sigma^2)$ and $N(\mu_2, \sigma^2)$, respectively, show that

$$\frac{(\bar{x}_1 - \bar{x}_2) - (\mu_1 - \mu_2)}{\sqrt{\dfrac{1}{n_1} + \dfrac{1}{n_2}}\left[\dfrac{(n_1 - 1)s_1^2 + (n_2 - 1)s_2^2}{n_1 + n_2 - 2}\right]^{\frac{1}{2}}}$$

has the "Student" t distribution $S(n_1 + n_2 - 2)$.

8.3 If $(x_{11}, \ldots, x_{1n})$ and $(x_{21}, \ldots, x_{2n})$ are independent samples from $N(\mu_1, \sigma_1^2)$ and $N(\mu_2, \sigma_2^2)$, respectively, and if $\bar{x}_1$, $\bar{x}_2$ are the sample means, s_1^2, s_2^2 are the sample variances, and $r = \dfrac{1}{(n - 1)s_1s_2}\sum_{\xi=1}^{n}(x_{1\xi} - \bar{x}_1)(x_{2\xi} - \bar{x}_2)$ the correlation coefficient between the two samples, show that

$$\frac{[(\bar{x}_1 - \bar{x}_2) - (\mu_1 - \mu_2)]\sqrt{n}}{[s_1^2 + s_2^2 - 2rs_1s_2]^{\frac{1}{2}}}$$

has the "Student" distribution $S(n - 1)$.

8.4 Suppose $\bar{x}_1$ and $\bar{x}_2$ are means of independent samples of sizes n_1 and n_2 from distributions whose variances are both equal to σ^2. Let $\bar{x}$ be the mean of the two samples pooled together as a sample of size $n_1 + n_2$. Show that the variance of $\bar{x} - \bar{x}_1$ is $\sigma^2 n_2/[n_1(n_1 + n_2)]$.

8.5 If x is a random variable having the distribution $N(\mu, \sigma^2)$ and if $F(x)$ is the c.d.f. of this normal distribution show that the correlation coefficient between the random variables x and $F(x)$ is $\sqrt{3/\pi}$.

8.6 If (x, y) is a pair of random variables having an arbitrary two-dimensional normal distribution with correlation coefficient ρ and if $F(y)$ is the c.d.f. of y, show that the correlation coefficient between the random variables x and $F(y)$ is $\rho\sqrt{3/\pi}$.

8.7 Making use of **8.2.2** for $r = 2, g(x_1, x_2) \equiv x_1 x_2$, construct $Q_{[2]}(x_1, \ldots, x_n)$ for a sample of size n from a distribution having finite first and second moments μ_1' and μ_2', thus obtaining an unbiased estimator for $(\mu_1')^2$. Find the variance of this estimator.

8.8 If $(x_1, \ldots, x_n)$ is a simple random sample from a finite population π_N of size N having variance σ^2, show for any (real) constants $c_1, \ldots, c_n$ that the variance of $c_1 x_1 + \cdots + c_n x_n$ is

$$\left[(c_1^2 + \cdots + c_n^2) - \frac{1}{N}(c_1 + \cdots + c_n)^2\right]\sigma^2.$$

Hence, show that the variance of the difference between the means of two random samples of sizes n_1 and n_2, $n_1 + n_2 < N$ drawn without replacement from π_N is $\sigma^2(1/n_1 + 1/n_2)$.

8.9 Suppose samples of sizes $n_1, \ldots, n_k$ are independent samples from identical normal distributions $N(\mu, \sigma^2)$. Let $\bar{x}_1, \ldots, \bar{x}_k$ be the means and $s_1^2, \ldots, s_k^2$ the variances of these samples. Let

$$u = \frac{1}{\sigma^2}[(n_1 - 1)s_1^2 + \cdots + (n_k - 1)s_k^2]$$

$$v = \frac{1}{\sigma^2}[n_1(\bar{x}_1 - \bar{x})^2 + \cdots + n_k(\bar{x}_k - \bar{x})^2]$$

$$w = \frac{1}{\sigma^2}[n(\bar{x} - \mu)^2]$$

where $\bar{x}$ is the mean of all samples pooled together as a single sample, and $n = n_1 + \cdots + n_k$. Show by Cochran's theorem or by characteristic functions that u, v, and w are independent random variables having chi-square distributions $C(n_1 + \cdots + n_k - k)$, $C(k - 1)$, and $C(1)$, respectively.

8.10 Suppose $(y_1, \ldots, y_n)$ are independent random variables from

$$N(\mu + \beta x_1, \sigma^2), \ldots, N(\mu + \beta x_n, \sigma^2) \text{ respectively,}$$

where $\mu, \beta, x_1, \ldots, x_n$ are constants. Let $\hat{\mu}$ and $\hat{\beta}$ be the values of μ and β which minimize the sum of squares

$$\sum_{\xi=1}^{n} (y_\xi - \mu - \beta x_\xi)^2,$$

and let

$$v = \frac{1}{\sigma^2} \sum_{\xi=1}^{n} (y_\xi - \hat{\mu} - \hat{\beta} x_\xi)^2.$$

Show that $(\hat{\mu}, \hat{\beta})$ has the two-dimensional normal distribution $N(\{\mu, \beta\}; \|\sigma_{ij}\|)$, where

$$\|\sigma_{ij}\| = \left\| \begin{matrix} \dfrac{(\Sigma\, x_\xi^2)\sigma^2}{n a_n} & -\dfrac{\bar{x}\sigma^2}{a_n} \\ -\dfrac{\bar{x}\sigma^2}{a_n} & \dfrac{\sigma^2}{a_n} \end{matrix} \right\|$$

and $a_n = \sum_\xi (x_\xi - \bar{x})^2$,

and that v has the chi-square distribution $C(n - 2)$. Furthermore, show that $(\hat{\mu}, \hat{\beta})$ and v are independent.

8.11 If $(x_1, \ldots, x_n)$ is a sample from $N(\mu, \sigma^2)$ show that the characteristic function of

$$u = \frac{1}{\sigma^2} \sum_{\xi=1}^{n} (x_\xi - \mu + \delta)^2$$

is

$$e^{-\frac{1}{2}\beta^2} \sum_{r=0}^{\infty} \frac{1}{r!} \left(\frac{\beta^2}{2}\right)^r (1 - 2it)^{-\frac{1}{2}n} \exp\left(\frac{\beta^2 it}{1 - 2it}\right),$$

where $\beta^2 = n\delta^2/\sigma^2$, and hence that the p.d.f. of u is

$$f(u) = e^{-\frac{1}{2}\beta^2} \sum_{r=0}^{\infty} \frac{1}{r!} \left(\frac{\beta^2}{2}\right)^r f_{n+2r}(u),$$

where $f_{n+2r}(u)$ is the p.d.f. of the chi-square distribution $C(n + 2r)$. (The distribution of u, originally obtained by Fisher (1928*b*), is known as the *noncentral chi-square distribution* with n degrees of freedom and parameter β^2.)

8.12 If n, $\bar{x}$, s^2 are the size, mean, and variance of a sample from $N(\mu, \sigma^2)$, show that the distribution of

$$t = \frac{(\bar{x} - \mu + \delta)\sqrt{n}}{s}$$

has p.d.f.

$$f(t) = \frac{e^{-\frac{1}{2}\gamma^2}\Gamma(\frac{1}{2}n)}{\sqrt{\pi(n-1)}\,\Gamma(\frac{1}{2}n - \frac{1}{2})} \left(1 + \frac{t^2}{n-1}\right)^{-\frac{1}{2}n} g(t),$$

where

$$g(t) = \sum_{r=0}^{\infty} \left(\frac{\sqrt{2}\gamma t}{\sqrt{n-1}}\right)^r \left(1 + \frac{t^2}{n-1}\right)^{-\frac{1}{2}r} \frac{\Gamma[\frac{1}{2}(n+r)]}{r!\,\Gamma(\frac{1}{2}n)}$$

and $\gamma = \sqrt{n}\delta/\sigma$. (This distribution obtained by Tang (1938), is known as the *noncentral Student distribution* with $n - 1$ degrees of freedom and parameter γ.)

8.13 Find the finite-population forms of formulas (8.6.34) and (8.6.35) if the numbers of elements in the populations of rows, columns, and layers are N_1, N_2, and N_3, respectively.

8.14 If $(x_1, \ldots, x_n)$ is a sample from the rectangular distribution $R(\frac{1}{2}\theta, \theta)$ show that the sampling distribution of $u = \max\,(x_1, \ldots, x_n)$ has p.d.f. nu^{n-1}/θ^n on $(0, \theta)$ and 0 outside.

8.15 If a sample of size $(2n + 1)$ is taken from the rectangular distribution $R(\frac{1}{2}, 1)$ show that the median of this sample has the beta distribution

$$Be(n + 1, n + 1).$$

8.16 If $(x_1, \ldots, x_n)$ is a sample from the rectangular distribution $R(\mu, \omega)$ show that the p.d.f. of the sample range r is

$$\frac{n(n-1)}{\omega}\left(\frac{r}{\omega}\right)^{n-2}\left(1 - \frac{r}{\omega}\right),$$

for $r \in (0, \omega)$ and 0 for $r \notin (0, \omega)$.

8.17 Suppose $(x_{(1)}, \ldots, x_{(n)})$ are the order statistics of a sample of size n from a population having a continuous c.d.f. $F(x)$. Show that the covariance matrix of $(F(x_{(k_1)}), F(x_{(k_2)}))$ where $1 \leqslant k_1 < k_2 \leqslant n$ is

$$\left\|\begin{matrix} \dfrac{p_1(1-p_1)}{(n+2)} & \dfrac{p_1(1-p_2)}{(n+2)} \\[2ex] \dfrac{p_1(1-p_2)}{(n+2)} & \dfrac{p_2(1-p_2)}{(n+2)} \end{matrix}\right\|$$

where

$$p_1 = \frac{k_1}{n+1}, \qquad p_2 = \frac{k_2}{n+1}.$$

8.18 If $(x_{(1)}, \ldots, x_{(n)})$ are the order statistics of a sample from a population having a continuous c.d.f. $F(x)$, show that the mean value of the range $R = x_{(n)} - x_{(1)}$ is given by

$$\mathscr{E}(R) = \int_{-\infty}^{\infty} \{1 - F^n(x) - [1 - F(x)]^n\}\, dx$$

a result due to Tippett (1925).

8.19 (*Continuation*) Show that the c.d.f. of R is given by

$$n\int_{-\infty}^{\infty} \{F(x + R) - F(x)\}^{n-1}\, dF(x).$$

8.20 In a sample of size n from the rectangular distribution $R(\mu, \omega)$, let $x_{(1)}$ and $x_{(n)}$ be the smallest and largest order statistics of the sample. Let $r = [\frac{1}{2}(x_{(n)} + x_{(1)}) - \mu]/(x_{(n)} - x_{(1)})$ show that the p.e. of r is given by

$$g(r)\, dr = (n - 1)(1 + 2|r|)^{-n}\, dr$$

on the interval $(-\infty, +\infty)$. [Carlton (1946)].

8.21 If $x_{(1)}$, $x_{(n)}$ are the smallest and largest order statistics of a sample of size n from a continuous c.d.f. $F(x)$, show that the random variable $2n\sqrt{F(x_{(1)})(1 - F(x_{(n)}))}$ has a limit distribution as $n \to \infty$, with mean $\frac{1}{2}\pi$ and variance $4 - \pi^2/4$. [Elfving (1947).]

8.22 Let $(x_{(1)}, \ldots, x_{(n)})$ be the order statistics of a sample from the p.d.f. $f(x)$ where

$$f(x) = \begin{cases} \dfrac{1}{\sigma} e^{-(x-\mu)/\sigma}, & x > \mu \\ 0, & x \leqslant \mu \end{cases}$$

and let

$$w = \frac{1}{n-1} \sum_{\xi=2}^{n} (x_{(\xi)} - x_{(1)}).$$

Show that the p.d.f. $g(t)$ of the random variable t defined by

$$t = (x_{(1)} - \mu)/w$$

is given by

$$g(t) = \begin{cases} n[1 + nt/(n-1)]^{-n}, & t > 0 \\ 0, & t \leqslant 0, \end{cases}$$ [Guttman (1960).]

8.23 If $(x_1, \ldots, x_n)$ is an n-dimensional random variable having a p.d.f. $f(x_1, \ldots, x_n)$ which is symmetric in $x_1, \ldots, x_n$, show that the order statistics $(x_{(1)}, \ldots, x_{(n)})$ of $(x_1, \ldots, x_n)$ have p.d.f. $n!f(x_{(1)}, \ldots, x_{(n)})$, in the region for which $-\infty < x_{(1)} < \cdots < x_{(n)} < +\infty$, and 0 otherwise.

8.24 If $(x_1, \ldots, x_n)$ is a sample from the rectangular distribution $R(\frac{1}{2}, 1)$ let

$$y = (x_1 \cdots x_n)^{1/n}.$$

Show that the p.d.f. of y is

$$\frac{n^n y^{n-1}}{(n-1)!} (-\log y)^{n-1} \text{ on } (0, 1) \text{ and 0 otherwise.}$$

8.25 If m independent samples of size n are drawn from the rectangular distribution $R(\frac{1}{2}, 1)$ and if $u = \prod_{i=1}^{m} z_i$, where $z_1, \ldots, z_m$ are the largest order statistics in the m samples respectively, show that the p.d.f. of u is

$$\frac{n^m}{(m-1)!} u^{n-1} (-\log u)^{m-1}$$

on (0, 1) and 0 otherwise, a result due to Rider (1955).

8.26 If $(x_1, \ldots, x_n)$ is a sample from a gamma distribution $G(\mu)$, and if $h(x_1, \ldots, x_n)$ is a random variable such that

$$h(x_1, \ldots, x_n) = h(cx_1, \ldots, cx_n)$$

for each $c > 0$, show that $(x_1 + \cdots + x_n)$ and $h(x_1, \ldots, x_n)$ are independent. [Pitman (1937*a*)].

8.27 If $(x_1, \ldots, x_n)$ is a sample from a p.d.f. $f(x)$ and if, for any set of constants $c_1, \ldots, c_n$, not all zero, the ratio $(c_1x_1 + \cdots + c_nx_n)/(x_1 + \cdots + x_n)$ and the sum $(x_1 + \cdots + x_n)$ are independent, show that $f(x)$ is the p.d.f. of a gamma distribution. [Laha (1954)].

8.28 If $(x_{(1)}, \ldots, x_{(n)})$ are the order statistics of a sample of size n from a population having a continuous c.d.f. with finite mean μ and variance σ^2, show by use of Schwarz' inequality, that

$$\mathscr{E}(x_{(n)}) \leqslant \mu + \frac{(n-1)\sigma}{\sqrt{2n-1}},$$

a slightly different form of a result due to Hartley and David (1954).

8.29 If $(x_{(1)}, \ldots, x_{(n)})$ are order statistics in a sample of size n from a continuous c.d.f. $F(x)$, show that the random variables

$$y_i = \left[\frac{F(x_{(i)})}{F(x_{(i+1)})}\right]^i, \qquad i = 1, \ldots, n-1$$

are independently distributed according to the rectangular distribution $R(\frac{1}{2}, 1)$ [Malmquist (1950) and Renyi (1953)].

8.30 Suppose $(x_1, \ldots, x_n)$ is a sample from the rectangular distribution $R(\frac{1}{2}\omega, \omega + \delta)$ where $\omega > 0$ and $0 < \frac{1}{2}\delta < \omega$. Let I_ξ be the random interval $(x_\xi - \frac{1}{2}\delta, x_\xi + \frac{1}{2}\delta)$. Let E be the event $\bigcup_{\xi=1}^{n} I_\xi$, and I the interval $(0, \omega)$. Show that

$$\mathscr{E}(\mathrm{E} \cap I) = \omega\left[1 - \left(1 - \frac{1}{\omega + \delta}\right)^n\right]. \qquad \text{[Robbins (1944}a\text{)].}$$

8.31 If n points are taken "at random" on a line of length L, show that if $0 < d < L/(n-1)$, the probability is $(L - (n-1)d)^n/L^n$ that no two points will be closer together than the distance d. [Parzen (1960)].

8.32 If $(x_1, \ldots, x_n)$ is a sample from a normal population having mean μ and variance σ^2 and if d is the sample mean deviation about μ defined by

$$d = \frac{1}{n}\sum_{\xi=1}^{n} |x_\xi - \mu|$$

show that

$$\mathscr{E}(d) = \sqrt{2/\pi}\,\sigma$$

$$\sigma^2(d) = \frac{\sigma^2}{n}\left(1 - \frac{2}{\pi}\right).$$

8.33 Let x be a random variable with p.d.f. $f(x)$ having mean μ and variance σ^2. Let $\varphi(t)$ be the characteristic function of x. Let $\bar{x}$ and s^2 be the sample mean and variance of a sample of size n from $f(x)$. Show that if $\bar{x}$ and s^2 are independent for any n, then $\varphi(t)$ satisfies the differential equation

$$\varphi\frac{d^2\varphi}{dt^2} - \left(\frac{d\varphi}{dt}\right)^2 + \sigma^2\varphi^2 = 0,$$

that its solution, subject to the boundary conditions $\varphi(0) = 1$, and $\varphi'(0) = i\mu$ is

$$\varphi(t) = e^{i\mu t - \frac{1}{2}\sigma^2 t^2}$$

and hence that $f(x)$ is the p.d.f. of the normal distribution $N(\mu, \sigma^2)$. [Lukacs (1942)].

8.34 Suppose $x_{(k)}$ is the kth order statistic in a sample of size m from a continuous c.d.f. $F(x)$. In a second independent sample of size n from the same c.d.f. let y be the random variable denoting the number of x's in the second sample which do not exceed $x_{(k)}$. Show that the p.f. of y is

$$\frac{\binom{m}{k}\binom{n}{y}}{\binom{m+n}{k+y}} \cdot \frac{k}{(k+y)}, \qquad y = 0, 1, \ldots, n$$

and show that the rth factorial moment of y is given by

$$\mathscr{E}(y^{[r]}) = \frac{m!\, n!\, (k+r-1)!}{(k-1)!\,(m+r)!\,(n-r)!} \qquad r \leqslant n, \qquad \text{[Epstein (1954)].}$$

8.35 (*Continuation*) Suppose the second sample is drawn one element at a time until y elements are less than $x_{(k)}$. Show that the size n of the second sample is a random variable having the following p.f.

$$\frac{\binom{n-1}{y-1}\binom{m-1}{k-1}}{\binom{m+n-1}{y+k-1}} \cdot \frac{m}{(m+n)}$$

$n = y, y+1, \ldots$
and show that

$$\mathscr{E}[(n-y)^{[r]}] = \frac{(y+r-1)!\,(k-r-1)!\,(m-k+r)!}{(y-1)!\,(k-1)!\,(m-k)!},$$

$r \leqslant k-1$. [Wilks (1959*b*)].

8.36 Suppose a finite population π_N consists of N chips marked $1, 2, \ldots, N$ respectively. In a sample of size n drawn from π_N without replacement, let x be the largest number drawn in the sample. Show that the p.f. of x is

$$\binom{x-1}{n-1}\Big/\binom{N}{n}, \qquad x = n, \ldots, N,$$

and that

$$\mathscr{E}(x) = \frac{n}{n+1}(N+1), \qquad \sigma^2(x) = \frac{n(N-n)(N+1)}{(n+1)^2(n+2)}.$$

8.37 (*Continuation*) Suppose a sample of size $2n+1$ is drawn from population π_N without replacement. Let y be the median of the sample. Show that the p.f. of y is

$$\binom{y-1}{n}\binom{N-y}{n}\Big/\binom{N}{2n+1},$$

$y = n+1, \ldots, N-n$.
Show that

$$\mathscr{E}(y) = \frac{N+1}{2} \qquad \sigma^2(y) = \frac{(N-2n-1)(N+1)}{8n+12}.$$

8.38 (*Continuation*) If s and t are the k_1th and k_2th order statistics ($k_1 < k_2$) in a sample of size n drawn from π_N, show that the p.f. of (s, t) is

$$\binom{s-1}{k_1-1}\binom{t-s-1}{k_2-k_1-1}\binom{N-t}{n-k_2}\Big/\binom{N}{n}.$$

Show that

$$\mathscr{E}(s) = k_1\left(\frac{N+1}{n+1}\right)$$

$$\sigma^2(s) = \frac{k_1(N-n)(N+1)(n-k_1+1)}{(n+1)^2(n+2)}$$

$$\text{Cov}\,(s, t) = \frac{k_1(n-k_2+1)(N-n)(n+1)}{(n+1)^2(u+2)}.$$

8.39 Suppose $p(x)$ and $q(x)$ are two discrete distributions having the same mass points $x_1, \ldots, x_k$. Let O_m and O_n be independent samples of size m and n drawn from $p(x)$ and $q(x)$ respectively, and let $(m_1, \ldots, m_k)$, where $m_1 + \cdots + m_k = m$, be the numbers of components of O_m having values at $x_1, \ldots, x_k$ respectively, with a similar definition of $(n_1, \ldots, n_k)$. If $r_{m,n}$ is the correlation coefficient between $(\sqrt{m_1/m}, \ldots, \sqrt{m_k/m})$ and $(\sqrt{n_1/n}, \ldots, \sqrt{n_k/n})$, show that if $p(x) \equiv q(x)$, and for an arbitrary $\lambda > 0$,

$$P\left(r_{m,n} > 1 - \frac{\lambda}{2}\right) \geqslant 1 - \frac{k-1}{\lambda}\left(\frac{1}{\sqrt{m}} + \frac{1}{\sqrt{n}}\right)^2$$

a result due to Matusita (1957).

8.40 (*Continuation*) If $\sum_{i=1}^{k}(\sqrt{p(x_i)} - \sqrt{q(x_i)})^2 \geqslant \delta^2$, where $\delta > \lambda > 0$, show that

$$P\left(r_{m,n} < 1 - \frac{\lambda}{2}\right) \geqslant 1 - \frac{k-1}{(\delta-\lambda)^2}\left(\frac{1}{\sqrt{m}} + \frac{1}{\sqrt{n}}\right)^2$$

a result also due to Matusita (1957).

8.41 Show that the c.d.f. of z for the distribution described in Section 8.3(*a*)

$$F_n(z) = \frac{1}{n!}\left[z^n - \binom{n}{1}(z-1)^n + \binom{n}{2}(z-2)^n - \cdots + (-1)^k\binom{n}{k}(z-k)^n\right]$$

$$k < z \leqslant k+1, \qquad k = 0, 1, \ldots, n-1$$

and

$$F_n(z) = 0 \quad \text{for} \quad z \leqslant 0,$$
$$= 1 \quad \text{for} \quad z > n.$$

8.42 Suppose $(x_{(1)}, \ldots, x_{(n)})$ are the order statistics of a sample of size n from the rectangular distribution $R(\frac{1}{2}, 1)$. Let $(u_1, \ldots, u_n)$ be an n-dimensional random variable denoting the segments $(x_{(1)}, x_{(2)} - x_{(1)}, \ldots, x_{(n)} - x_{(n-1)})$ and let $v = \max(u_1, \ldots, u_n)$. Show that the c.d.f. of v is given by

$$F_n(v) = \left[1 - \binom{n}{1}(1-v)^n + \binom{n}{2}(1-2v)^n - \cdots + (-1)^k\binom{n}{k}(1-kv)^n\right]$$

$$\frac{1}{k+1} < u \leqslant \frac{1}{k}, \qquad k = n-1, n-2, \ldots, 1,$$

and

$$F_n(v) = 0 \qquad \text{for} \qquad u \leqslant 0$$
$$= 1 \qquad \text{for} \qquad u > 1.$$

(The result in the preceding problem is useful in solving this one.)

8.43 (*Continuation*). *Distribution of largest segment produced by n random points on the interval* (0, 1). The order statistics $(x_{(1)}, \ldots, x_{(n)})$ cut the interval (0, 1) into $n + 1$ disjoint segments $u_1, \ldots, u_{n+1}$ where, of course, $u_1 + \cdots + u_{n+1} = 1$. Let $w = \max(u_1, \ldots, u_{n+1})$. Show that the c.d.f. of w is given by

$$F_n(w) = \left[1 - \binom{n+1}{1}(1-w)^n + \binom{n+1}{2}(1-2w)^n - \cdots + (-1)^k\binom{n+1}{k}(1-kw)^n\right]$$

$$\frac{1}{k+1} < w \leqslant \frac{1}{k}, \qquad k = n, n-1, \ldots, 1$$

and

$$F_n(w) = 0 \qquad \text{for} \qquad w \leqslant \frac{1}{n+1}$$
$$= 1 \qquad \text{for} \qquad w > 1.$$

8.44 If $(x_{(1)}, x_{(2)}, x_{(3)}\, x_{(4)})$, are the order statistics of a sample of size four from $N(0, 1)$ and if $y = \frac{1}{2}(x_{(3)} + x_{(4)}) - \frac{1}{2}(x_{(1)} + x_{(2)})$, show that the c.d.f. $G(y)$ of y is given by

$$G(y) = 8\left[\frac{1}{\sqrt{2\pi}}\int_0^y e^{-\frac{1}{2}x^2}\,dx\right]^3 \qquad y \geqslant 0 \qquad \text{[Walsh (1946)]}$$

8.45 If $(x_{(1)}, x_{(2)})$ are the order statistics of a sample of size two from $N(\mu, \sigma^2)$ show that their mean values are $(\mu - \sigma/\sqrt{\pi}, \mu + \sigma/\sqrt{\pi})$.

CHAPTER 9

Asymptotic Sampling Theory for Large Samples

The sampling theory presented in Chapter 8 has been based on finite samples of size n. We now consider what kinds of results can be obtained about sampling distributions if we allow $n \to \infty$. In this case we have a population with c.d.f. $F(x)$ and we consider a stochastic process $(x_1, x_2, \ldots)$ such that the elements (components) are mutually independent and any set of n of them is a sample of size n from the population having c.d.f. $F(x)$. This stochastic process is sometimes called *simple random sampling from an infinite population* or *random sampling from a probability distribution.* In this chapter we shall consider some limit theorems and results which are useful in approximating distributions of various functions of samples in the case of large samples.

9.1 CONVERGENCE OF SAMPLE MEAN IN PROBABILITY

One of the simplest results concerning a sample mean for large samples is given by the following theorem due to Khintchine (1929):

9.1.1 *If* $(x_1, \ldots, x_n)$ *is a sample from a* c.d.f. $F(x)$ *for which* $\mathscr{E}(x)$ *has a finite value* μ, *then the sample mean* $\bar{x}$ *converges in probability to* μ *as* $n \to \infty$.

To prove this, we note that if the mean μ exists, then it follows from (5.1.10) that the characteristic function $\varphi(t)$ corresponding to $F(x)$ may be written as

$$\varphi(t) = 1 + i\mu t + o(t). \tag{9.1.1}$$

Making use of the fact from **8.3.2** that the characteristic function of $\bar{x}$ is $[\varphi(t/n)]^n$, we have

$$\left[\varphi\left(\frac{t}{n}\right)\right]^n = \left[1 + \frac{\mu i t}{n} + o\left(\frac{t}{n}\right)\right]^n \tag{9.1.2}$$

where $n \cdot o(t/n)$ tends to 0 as $n \to \infty$ for any t. Allowing $n \to \infty$, we have

$$\lim_{n\to\infty}\left[\varphi\left(\frac{t}{n}\right)\right]^n = e^{\mu i t} \tag{9.1.3}$$

which is the characteristic function of the degenerate c.d.f. $\varepsilon(x - \mu)$ defined by (2.3.4). Applying **5.4.1** it follows that as $n \to \infty$ the distribution of $\bar{x}$ converges to $\varepsilon(x - \mu)$, which is equivalent to stating that $\bar{x}$ converges in probability to μ as $n \to \infty$, thus completing the argument for **9.1.1.**

If it is assumed that *both* the variance σ^2 and mean μ of x exist and are finite for the population distribution in **9.1.1**, then it can be proved quite simply by Chebyshev's inequality (3.3.5) that $\bar{x}$ converges in probability to μ. For we know from **8.2.1** that $\mathscr{E}(\bar{x}) = \mu$ and $\sigma^2(\bar{x}) = \sigma^2/n$. Applying (3.3.5) we may say, for any $\varepsilon > 0$, that

$$P(|\bar{x} - \mu| \geqslant \varepsilon) = P\left(|\bar{x} - \mu| \geqslant \left(\frac{\varepsilon}{\sigma_{\bar{x}}}\right)\cdot \sigma_{\bar{x}}\right) \leqslant \frac{\sigma_{\bar{x}}^2}{\varepsilon^2} = \frac{\sigma^2}{n\varepsilon^2}. \tag{9.1.4}$$

From this it is clear that

$$\lim_{n\to\infty} P(|\bar{x} - \mu| \geqslant \varepsilon) = 0; \tag{9.1.5}$$

that is, $\bar{x}$ converges in probability to μ.

The k-dimensional analogue of **9.1.1** is straightforward. Its formulation and proof are left to the reader.

Theorem **9.1.1** is sometimes called the *weak law of large numbers* and it essentially states that 100% of the probability in the distribution of $\bar{x}$ ultimately accumulates in *any* neighborhood containing the value $\bar{x} = \mu$ as $n \to \infty$. The condition which guarantees this accumulation is the existence (finiteness) of the mean μ.

The Cauchy Distribution. An example of a distribution which, at first glance, looks fairly well behaved but which does not produce sample means with this property of convergence in probability is the *Cauchy* (1853) *distribution*, having p.d.f.

$$f(x) = \frac{\theta_2}{\pi[\theta_2^2 + (x - \theta_1)^2]} \tag{9.1.6}$$

the range of x being $(-\infty, +\infty)$, and θ_1, θ_2 being real numbers with $\theta_2 > 0$.

It can be verified that the characteristic function $\varphi(t)$ corresponding to (9.1.6) is

$$\varphi(t) = e^{\theta_1 it - \theta_2|t|}. \tag{9.1.7}$$

Hence the characteristic function of $\bar{x}$ for a sample of size n is

$$\left[\varphi\left(\frac{t}{n}\right)\right]^n = [e^{\theta_1 it/n - \theta_2|t/n|}]^n = \varphi(t). \tag{9.1.8}$$

In other words, $\bar{x}$ has exactly the same characteristic function as x, which means that the sampling distribution of $\bar{x}$ for any n is exactly the same as that of x in the population. The reader should note that $f(x)$ is symmetrical with respect to $x = \theta_1$ but that θ_1 is *not* the mean of the distribution; the mean does not exist. Nor does any higher moment. The value θ_1 is, however, both the median and the center of symmetry of the distribution, whereas θ_2 is the interquartile range, that is, the distance between the quartiles $\underline{x}_{0.25}$ and $\underline{x}_{0.75}$.

9.2 LIMITING DISTRIBUTION OF SAMPLE SUMS AND MEANS

(a) The One-Dimensional Case

If the variance of the population distribution exists as well as the mean, then we can say more about the manner in which the distribution of $\bar{x}$ behaves as $n \to \infty$. The fundamental theorem for this situation which is due to Lindeberg (1922), can be stated as follows:

9.2.1 *If z and $\bar{x}$ are the sum and mean, respectively, of a sample of size n from a distribution having finite variance σ^2 and mean μ, then*

(9.2.1)
$$\lim_{n\to\infty} P\left[\frac{z - n\mu}{\sqrt{n}\sigma} \leqslant y\right] = \lim_{n\to\infty} P\left[\frac{\sqrt{n}(\bar{x} - \mu)}{\sigma} \leqslant y\right] = \frac{1}{\sqrt{2\pi}}\int_{-\infty}^{y} e^{-\frac{1}{2}u^2}\,du.$$

Whenever (9.2.1) holds, it is convenient to say that z is *asymptotically normally distributed* according to $N(n\mu, n\sigma^2)$ for large n (or $\bar{x}$ is asymptotically normally distributed according to $N(\mu, \sigma^2/n)$ for large n).

Since $(z - n\mu)/\sqrt{n}\sigma$ is essentially an alternative notation for the random variable $\sqrt{n}(\bar{x} - \mu)/\sigma$, it will be sufficient to consider the former. Let $\varphi_n(t)$ be the characteristic function of $(z - n\mu)/\sqrt{n}\sigma$. We have

(9.2.2)
$$\varphi_n(t) = \mathscr{E}[e^{it(z-n\mu)/\sqrt{n}\sigma}] = \mathscr{E}\left[\exp\left(it\sum_{\xi=1}^{n}(x_\xi - \mu)/\sqrt{n}\sigma\right)\right] = [\varphi(t)]^n$$

where $\varphi(t)$ is the characteristic function of $(x - \mu)/\sqrt{n}\sigma$. From (5.1.10) we have

$$\varphi(t) = 1 + i\mathscr{E}\left(\frac{x - \mu}{\sqrt{n}\sigma}\right)t - \tfrac{1}{2}\mathscr{E}\left(\frac{x - \mu}{\sqrt{n}\sigma}\right)^2 t^2 + o\left(\frac{t^2}{n}\right) \tag{9.2.3}$$

where $n \cdot o(t^2/n) \to 0$ as $n \to \infty$ for any given value of $t \neq 0$; $o(t^2/n) = 0$ if $t = 0$.

Hence, for any given t, we have

$$\lim_{n\to\infty} \varphi_n(t) = \lim_{n\to\infty} \left[1 - \frac{t^2}{2n} + o\left(\frac{t^2}{n}\right)\right]^n = e^{-t^2/2} \tag{9.2.4}$$

But, we know from **7.2.1** that $e^{-t^2/2}$ is the characteristic function associated with the normal distribution $N(0, 1)$. Therefore, it follows from **5.4.1** that the c.d.f. of $(z - n\mu)/\sqrt{n}\sigma$ as $n \to \infty$ converges to the c.d.f. of the distribution $N(0, 1)$, and this is equivalent to the statement contained in (9.2.1) which completes the proof of **9.2.1**.

The following is an important corollary of **9.2.1**, known as the *De Moivre-Laplace theorem*:

9.2.1a *If z is a random variable having the binomial distribution $Bi(n, p)$, then z is asymptotically distributed according to $N(np, npq)$.*

It follows from the special case of **8.3.3a** for $m = 1$ that if a sample of size n is drawn from the binomial distribution $Bi(1, p)$, then the sample sum z has the binomial distribution $Bi(n, p)$. Then applying **9.2.1** we obtain **9.2.1a**. Result **9.2.1a** was surmised by De Moivre (1718), but it was not firmly established until a century later by Laplace (1812). Gauss (1809*b*) discovered the approximation **9.2.1** by a rather heuristic argument in connection with the theory of errors.

(b) The Central Limit Theorem

Theorem **9.2.1** is a special case of a more general result known as the *central limit theorem*, which states that under certain conditions

$$\lim_{n\to\infty} P\left[\frac{\sum_{\xi=1}^{n} (x_\xi - \mu_\xi)}{\sqrt{\sum_{\xi=1}^{n} \sigma_\xi^2}} < y\right] = \frac{1}{\sqrt{2\pi}} \int_{-\infty}^{y} e^{-\frac{1}{2}u^2}\, du \tag{9.2.5}$$

where $(x_1, x_2, \ldots)$ is a sequence of independent random variables with means $(\mu_1, \mu_2, \ldots)$ and variances $(\sigma_1^2, \sigma_2^2, \ldots)$ respectively. Various studies of conditions under which this statement holds have been made by Chebyshev (1890), Feller (1935), Lévy (1935), Lindeberg (1922), Lyapunov (1900, 1901), Markov (1900), and others. A comprehensive account of the central limit theorem and related problems has been given by Gnedenko and Kolmogorov (1954). A modern version of the central limit theorem in general form can be stated as follows:

9.2.2 *Let $(x_1, x_2, \ldots)$ be a sequence of independent random variables with c.d.f.'s $(F_1(x), F_2(x), \ldots)$, means $(\mu_1, \mu_2, \ldots)$, and variances*

$(\sigma_1^2, \sigma_2^2, \ldots)$. *Let* $z_n = x_1 + \cdots + x_n$, $\zeta_n = \mu_1 + \cdots + \mu_n$, $\tau_n^2 = \sigma_1^2 + \cdots + \sigma_n^2$, *and* $\tau_n^{*2} = \sigma_1^{*2} + \cdots + \sigma_n^{*2}$, *where, for arbitrary* $\varepsilon > 0$

(9.2.6) $$\sigma_\xi^{*2} = \int_{\mu_\xi - \varepsilon\tau_n}^{\mu_\xi + \varepsilon\tau_n} (x - \mu_\xi)^2 \, dF_\xi(x), \qquad \xi = 1, \ldots, n.$$

A necessary and sufficient condition for

(9.2.7) $$\lim_{n\to\infty} P\left(\frac{z_n - \zeta_n}{\tau_n} < y\right) = \frac{1}{\sqrt{2\pi}} \int_{-\infty}^{y} e^{-\frac{1}{2}u^2} \, du$$

is that

(9.2.8) $$\lim_{n\to\infty} \frac{\tau_n^{*2}}{\tau_n^2} = 1$$

and $\lim_{n\to\infty} \tau_n^2 = \infty$, *for every* $\varepsilon > 0$.

The proof of **9.2.2** is rather long and is omitted. The sufficiency of condition (9.2.8) was established by Lindeberg (1922), and the necessity by Feller (1935).

The reader can verify with very little difficulty that if $(x_1, x_2, \ldots)$ is a sequence of independent random variables all having identical distributions with variance σ^2, then (9.2.8) is satisfied.

(c) The k-Dimensional Case

The k-dimensional analogue of **9.2.1** can be stated as follows:

9.2.3 *Suppose* $(x_{1\xi}, \ldots, x_{k\xi}; \xi = 1, \ldots, n)$ *is a sample of size n from a k-variate distribution having finite means* μ_i, $i = 1, \ldots, k$, *and (positive definite) covariance matrix* $\|\sigma_{ij}\|$, $i, j = 1, \ldots, k$. *Let* $(z_1, \ldots, z_k)$ *be the sample sums and* $(\bar{x}_1, \ldots, \bar{x}_k)$ *the sample means, as defined in Section* 8.1. *Then* $(z_1, \ldots, z_k)$ *and* $(\bar{x}_1, \ldots, \bar{x}_k)$ *have as their asymptotic distributions the k-variate distributions* $N(\{n\mu_i\}, \|n\sigma_{ij}\|)$ *and* $N\left(\{\mu_i\}, \left\|\frac{\sigma_{ij}}{n}\right\|\right)$ *respectively. That is to say,*

(9.2.9) $$\lim_{n\to\infty} P\left(\frac{(z_i - n\mu_i)}{\sqrt{n}} < y_i,\ i = 1, \ldots, k\right)$$
$$= \lim_{n\to\infty} P((\bar{x}_i - \mu_i)\sqrt{n} < y_i,\ i = 1, \ldots, k)$$
$$= \frac{\sqrt{|\sigma^{ij}|}}{(2\pi)^{\frac{1}{2}k}} \int_{-\infty}^{y_k} \cdots \int_{-\infty}^{y_1} \exp\left(-\frac{1}{2}\sum_{i,j=1}^{k} \sigma^{ij} u_i u_j\right) du_1 \cdots du_k.$$

The argument for **9.2.3** is a direct extension of that for **9.2.1**. The main thing to do here is to set up the characteristic function $\varphi_n(t_1, \ldots, t_k)$ of the k-dimensional random variable $[(z_1 - n\mu_1)/\sqrt{n}, \ldots, (z_k - n\mu_k)/\sqrt{n}]$ and to show that

$$\lim_{n\to\infty} \varphi_n(t_1, \ldots, t_k) = \exp\left(-\frac{1}{2}\sum_{i,j=1}^{k} \sigma_{ij}t_it_j\right) \tag{9.2.10}$$

which, by the k-dimensional analogue of **5.4.1** implies (9.2.9). The details of establishing (9.2.10) are straightforward and are left to the reader.

One of the most important applications of **9.2.3** is the k-dimensional extension of De Moivre's Theorem **9.2.1a**, which can be stated as follows:

9.2.3a *If $(x_{1\xi}, \ldots, x_{k\xi}, \xi = 1, \ldots, n)$ is a sample of size n from the multinomial distribution $M(1; p_1, \ldots, p_k)$, then the sample sums $(z_1, \ldots, z_k)$ have, as their asymptotic distribution for large n, the distribution $N(\{np_i\}, \|n(p_i\delta_{ij} - p_ip_j)\|$* where *$\delta_{ij}$ is the Kronecker delta.*

The proof of this statement as a corollary of **9.2.3** amounts to verifying from the p.f. of the multinomial distribution $M(1; p_1, \ldots, p_k)$ given by (6.3.3) that

$$\mathscr{E}(x_i) = p_i, \qquad \sigma^2(x_i, x_j) = \sigma_{ij} = p_i\delta_{ij} - p_ip_j \tag{9.2.11}$$

and is straightforward.

9.3 ASYMPTOTIC DISTRIBUTION OF FUNCTIONS OF SAMPLE MEANS

(a) General Case

It is sometimes important to know something about the asymptotic distribution of some function of a sample mean $\bar{x}$, say $g(\bar{x})$, for large n. A useful result on this problem can be stated as follows:

9.3.1 *Suppose $(x_1, \ldots, x_n)$ is a sample from a distribution having mean μ and variance σ^2, both finite. Let $g(x)$ be a function which has a first derivative $g'(x)$ in some neighborhood of the point $x = \mu$ such that $g'(\mu) \neq 0$. Then $g(\bar{x})$ has $N(g(\mu), [\sigma g'(\mu)]^2/n)$ as its asymptotic distribution for large n.*

To prove this statement, let $V(\mu)$ be a neighborhood of $x = \mu$ such that $g'(x)$ exists for all values of x in $V(\mu)$. Since the c.d.f. of the population from which the sample is drawn has finite mean μ and variance σ^2, it follows from **4.6.1** that for arbitrary $\varepsilon > 0$, there exists an n_ε such that for $n > n_\varepsilon$

$$P(\bar{x} \in V(\mu), \quad \text{for all } n > n_\varepsilon) > 1 - \varepsilon. \tag{9.3.1}$$

For any point $x = \bar{x}$ in $V(\mu)$, we may write

$$g(\bar{x}) = g(\mu) + g'(x^*) \cdot (\bar{x} - \mu) \tag{9.3.2}$$

where x^* is a random variable such that $|x^* - \mu| \leqslant |\bar{x} - \mu|$. But (9.3.2) can be written as

$$(g(\bar{x}) - g(\mu))\sqrt{n} = g'(x^*)(\bar{x} - \mu)\sqrt{n}. \tag{9.3.3}$$

Let the random variable on the left be denoted by u_n and that on the right by v_n. The probability that (9.3.3) holds exceeds $1 - \varepsilon$ for all $n > n_\varepsilon$. Since ε is arbitrary, the sequence of random variables $(u_1, u_2, \ldots)$ and the sequence $(v_1, v_2, \ldots)$ thus form a stochastic process such that $(u_1, u_2, \ldots)$ and $(v_1, v_2, \ldots)$ converge in distribution together if one of them converges in distribution. Since σ^2 is finite, we know from **9.2.1** that the sequence of random variables $\sqrt{n}(\bar{x} - \mu)$, $n = 1, 2, \ldots$ converges in distribution to the normal distribution $N(0, \sigma^2)$. Since $g'(x)$ exists at all points of $V(\mu)$, it is continuous in $V(\mu)$ and hence at $x = \mu$. It follows from **4.3.7** that $g'(x^*)$ converges in probability to $g'(\mu)$.

Therefore

$$\lim_{n\to\infty} P(v_n \leqslant w) = P(g'(\mu)s \leqslant w) \tag{9.3.4}$$

where s has the distribution $N(0, \sigma^2)$, which implies that

$$\lim_{n\to\infty} P(u_n \leqslant w) = P(t \leqslant w) \tag{9.3.5}$$

where t has the distribution $N(0, (\sigma g'(\mu))^2)$. But (9.3.5) is equivalent to the statement that the asymptotic distribution of $g(\bar{x})$ for large n is $N(g(\mu), [\sigma g'(\mu)]^2/n)$, which concludes the proof of **9.3.1**.

The extension of **9.3.1** to the k-dimensional case can be stated as follows:

9.3.1a *Suppose* $(x_{1\xi}, \ldots, x_{k\xi}, \ \xi = 1, \ldots, n)$ *is a sample from a k-dimensional distribution with finite means* $\{\mu_i\}$ *and positive definite covariance matrix* $\|\sigma_{ij}\|$, $i, j = 1, \ldots, k$. *Let* $g(x_1, \ldots, x_k)$ *be a function which possesses first derivatives* $\partial g/\partial x_i = g_i$, say, $i = 1, \ldots, k$ *at all points in some neighborhood of* $(\mu_1, \ldots, \mu_k)$, *and let* $g_i^0 = g_i(\mu_1, \ldots, \mu_k)$. *Then if at least one of the* g_i^0 is $\neq 0$, $g(\bar{x}_1, \ldots, \bar{x}_k)$ *has the asymptotic distribution* $N\left(g(\mu_1, \ldots, \mu_k), \frac{1}{n}\sum_{i,j=1}^{k} \sigma_{ij} g_i^0 g_j^0\right)$ *for large n.*

The proof is straightforward and is left as an exercise for the reader.

(b) A Quadratic Function of Sample Means and Sums

It should be noted that (9.2.9) essentially states that we have a sequence of independent k-dimensional random variables

$$\left(\frac{z_1 - n\mu_1}{\sqrt{n}}, \ldots, \frac{z_k - n\mu_k}{\sqrt{n}}\right), \qquad n = 1, 2, \cdots \tag{9.3.6}$$

or equivalently a sequence of independent k-dimensional random variables

$$((\bar{x}_1 - \mu_1)\sqrt{n}, \ldots, (\bar{x}_k - \mu_k)\sqrt{n}), \qquad n = 1, 2, \cdots \tag{9.3.7}$$

which converges in distribution to the k-dimensional normal distribution $N(\{0\}, \|\sigma_{ij}\|)$. For convenience, denote the c.d.f. of the random variable (9.3.6) [or (9.3.7)] for a given n by $F_n(y_1, \ldots, y_k)$ and the c.d.f. of the distribution $N(\{0\}, \|\sigma_{ij}\|)$ by $\Phi(y_1, \ldots, y_k)$, the form of which is given in (9.2.9).

In **7.8.2** it was shown that if $(x_1, \ldots, x_k)$ has the distribution $N(\{\mu_i\}, \|\sigma_{ij}\|)$, then $\sum_{i,j=1}^{k} \sigma^{ij}(x_i - \mu_i)(x_j - \mu_j)$ has the chi-square distribution $C(k)$. If we set up the quadratic form

(9.3.8)
$$Q_n = \sum_{i,j=1}^{k} \sigma^{ij}\left(\frac{z_i - n\mu_i}{\sqrt{n}}\right)\left(\frac{z_j - n\mu_j}{\sqrt{n}}\right) = n \sum_{i,j=1}^{k} \sigma^{ij}(\bar{x}_i - \mu_i)(\bar{x}_j - \mu_j),$$

the sequence of random variables Q_n, $n = 1, 2, \ldots$ converges in distribution to the chi-square distribution $C(k)$ as $n \to \infty$. For it follows from **4.3.4** and **4.3.6** that since the stochastic process (9.3.6) converges in distribution to the k-dimensional normal distribution $N(\{0\}, \|\sigma_{ij}\|)$, then $Q_1, Q_2, \ldots$ converges in distribution to the chi-square distribution $C(k)$. We can summarize as follows:

9.3.2 *If $(x_{1\xi}, \ldots, x_{k\xi};\ \xi = 1, \ldots, n)$ is a sample from a k-dimensional distribution, with finite means μ_i, $i = 1, \ldots, k$, and finite, positive definite, covariance matrix $\|\sigma_{ij}\|$, then Q_n as given by* (9.3.8) *has the chi-square distribution $C(k)$ as its asymptotic distribution for large n, that is,*

$$\lim_{n\to\infty} P(Q_n < y) = \frac{1}{2\Gamma(\frac{1}{2}k)} \int_0^y \left(\frac{u}{2}\right)^{\frac{1}{2}k-1} e^{-\frac{1}{2}u}\, du. \tag{9.3.9}$$

It should be noted that Q_n is a function of $\bar{x}_1, \ldots, \bar{x}_k$, say $Q_n(\bar{x}_1, \ldots, \bar{x}_k)$, which violates the condition in **9.3.1a** that at least one of its first derivatives evaluated at $(\mu_1, \ldots, \mu_k)$ is $\neq 0$. Yet Q_n has a well-behaved limiting distribution as $n \to \infty$, namely, $C(k)$.

(c) Pearson's Chi-Square Goodness-of-Fit Criterion

An important special case of **9.3.2** is that in which the k-dimensional population being sampled has the multinomial distribution $M(1; p_1, \ldots, p_k)$ for which the means and covariance matrix are given by (9.2.11). In this case it can be verified that

$$\|\sigma^{ij}\| = \|p_i\delta_{ij} - p_ip_j\|^{-1} = \left\|\frac{\delta_{ij}}{p_i} + \frac{1}{p_{k+1}}\right\|. \tag{9.3.10}$$

Substituting $\|\delta_{ij}/p_i + 1/p_{k+1}\|$ for σ^{ij} and p_i for μ_i in (9.3.8), we find that Q_n can be reduced as follows:

$$Q_n = \sum_{i=1}^{k+1} \frac{(z_i - np_i)^2}{np_i}, \quad \text{where } z_{k+1} = \sum_{\xi=1}^{n} x_{k+1\xi} = n - \sum_{i=1}^{k} z_i. \tag{9.3.11}$$

Therefore we have the following corollary of **9.3.2**, recalling from **8.3.3b** that if a sample of size n is drawn from the multinomial distribution $M(1; p_1, \ldots, p_k)$, the sample sums $(z_1, \ldots, z_k)$ have the distribution $M(n; p_1, \ldots, p_k)$:

9.3.2a *If $(z_1, \ldots, z_k)$ is a k-dimensional random variable having the multinomial distribution $M(n; p_1, \ldots, p_k)$, then*

$$Q_n = \sum_{i=1}^{k+1} \frac{(z_i - np_i)^2}{np_i} \tag{9.3.12}$$

is asymptotically distributed according to the chi-square distribution $C(k)$ for large n.

Note that (9.3.12) is the sum of squares of discrepancies between the sample "frequencies" z_i and their mean values, weighted inversely by the mean values, and is the well-known *chi-square goodness of fit criterion* originally introduced by K. Pearson (1900). It may be regarded as an index of the extent to which the z_i depart collectively from their respective mean values, and the significance of the magnitude of this index must be established for a particular (large) sample in terms of probability computed from its asymptotic distribution, namely, the chi-square distribution $C(k)$. Further consideration of this problem leads us into the theory of testing statistical hypotheses, which is treated in Chapter 13.

9.4 ASYMPTOTIC EXPANSION OF DISTRIBUTION OF SAMPLE SUM

Theorem **9.2.1** contains a statement of the limiting form of the distribution of $(z - n\mu)/\sqrt{n}\sigma$ as $n \to \infty$. One problem which arises here is the determination, for large values of n, of a higher degree of approximation to the distribution of $(z - n\mu)/\sqrt{n}\sigma$ than that provided by the distribution $N(0, 1)$. We shall examine this problem for populations having p.d.f.'s.

Suppose the central moments $\mu_1, \mu_2, \mu_3, \ldots$ of a continuous c.d.f. $F(x)$ exist and are finite. Then if $\varphi(t)$ is the characteristic function of $(x - \mu)/\sqrt{n}\sigma$, we have

$$\varphi(t) = 1 - \frac{t^2}{2n} + \sum_{j=3}^{\infty} \frac{(it)^j \alpha_j}{j!\,(\sqrt{n})^j} \tag{9.4.1}$$

where $\alpha_j = \mu_j/\sigma^j$. But the characteristic function of $(z - n\mu)/\sqrt{n}\sigma$, namely, $\varphi_n(t)$, is given by

$$\varphi_n(t) = [\varphi(t)]^n. \tag{9.4.2}$$

Taking logarithms, we find

$$\log \varphi_n(t) = -\frac{t^2}{2} + n \sum_{j=3}^{\infty} \frac{(it)^j (\kappa_j^*)}{j!\,(\sqrt{n})^j} \tag{9.4.3}$$

where the κ_j^* are semi-invariants of the distribution of $(x - \mu)/\sigma$ in the population. Therefore, we have

$$\varphi_n(t) = e^{-\frac{1}{2}t^2} \exp\left[n \sum_{j=3}^{\infty} \frac{(it)^j (\kappa_j^*)}{j!\,(\sqrt{n})^j}\right], \tag{9.4.4}$$

which can be written as

$$\varphi_n(t) = e^{-\frac{1}{2}t^2}\left[1 + \sum_{j=1}^{\infty} \frac{u_j(it)}{(\sqrt{n})^j}\right] \tag{9.4.5}$$

where $u_j(it)$ is a polynomial of degree $3j$ in (it) whose coefficients are functions of the κ_j^*'s but do not depend on n. The lowest power of (it) in $u_j(it)$ is $j + 2$.

If we let

$$F_n(x) = P\left(\frac{(z - n\mu)}{\sqrt{n}\sigma} \leqslant x\right) \tag{9.4.6}$$

and put $x' = (x + y)/2$, $\delta = (x - y)/2$ in (5.1.14), then for any x and y where $x > y$, we have

$$F_n(x) - F_n(y) = \frac{1}{\pi} \int_{-\infty}^{\infty} \frac{\sin\left(\frac{x - y}{2}\right)t}{t}\, e^{-it[\frac{1}{2}(x+y)]} \varphi_n(t)\, dt. \tag{9.4.7}$$

Substituting the expression for $\varphi_n(t)$ from (9.4.5) and simplifying, we obtain

$$F_n(x) - F_n(y) = \frac{1}{\pi} \int_{-\infty}^{\infty} \frac{(e^{-itx} - e^{-ity})}{2(-it)} \cdot e^{-\frac{1}{2}t^2}\left[1 + \sum_{j=1}^{\infty} \frac{u_j(it)}{(\sqrt{n})^j}\right] dt. \tag{9.4.8}$$

But it can be verified without particular difficulty that

$$\frac{1}{\pi}\int_{-\infty}^{\infty}\frac{(e^{-itx}-e^{-ity})}{2(-it)}e^{-\frac{1}{2}t^2}\,dt=\Phi(x)-\Phi(y) \tag{9.4.9}$$

where $\Phi(x)$ is the c.d.f. of the distribution $N(0, 1)$. All other terms in (9.4.7) are of the following form, except for a constant multiplier,

$$\frac{1}{2\pi}\int_{-\infty}^{\infty}(-it)^j(e^{-itx}-e^{-ity})e^{-\frac{1}{2}t^2}\,dt$$

which has the value

(9.4.10)

$$\frac{1}{2\pi}\frac{d^j}{dx^j}\int_{-\infty}^{\infty}e^{-itx-\frac{1}{2}t^2}\,dt-\frac{1}{2\pi}\frac{d^j}{dy^j}\int_{-\infty}^{\infty}e^{-ity-\frac{1}{2}t^2}\,dt=\Phi^{(j+1)}(x)-\Phi^{(j+1)}(y)$$

where

$$\Phi^{(j+1)}(x)=\frac{d^{j+1}}{dx^{j+1}}\Phi(x)=\frac{d^j}{dx^j}\left(\frac{1}{\sqrt{2\pi}}e^{-\frac{1}{2}x^2}\right). \tag{9.4.11}$$

Hence, we obtain for (9.4.7)

$$F_n(x)-F_n(y)=\Phi(x)-\Phi(y)+\sum_{j=1}^{\infty}\frac{[u_j^*(x)-u_j^*(y)]}{\sqrt{n^j}} \tag{9.4.12}$$

where $u_j^*(x)$ and $u_j^*(y)$ are the functions one obtains by replacing $(it)^p$ in $u_j(it)$ by $\Phi^{(p+1)}(x)$ and $\Phi^{(p+1)}(y)$ respectively.

If we let $y \to -\infty$ in (9.4.12), we obtain as the asymptotic expansion of $F_n(x)$,

$$F_n(x)=\Phi(x)+\sum_{j=1}^{\infty}\frac{u_j^*(x)}{\sqrt{n^j}}. \tag{9.4.13}$$

If $F_n(x)$ has p.d.f. $f_n(x)$, we find, by taking the derivative of (9.4.13) with respect to x that

$$f_n(x)=\Phi^{(1)}(x)+\sum_{j=1}^{\infty}\frac{u_j^{*\prime}(x)}{\sqrt{n^j}}, \tag{9.4.14}$$

where $u_j^{*\prime}(x)$ is the first derivative of $u_j^*(x)$.

As a matter of fact, even if $F_n(x)$ has no p.d.f. (that is, if $F_n(x)$ were a discrete c.d.f.), (9.4.13) can be formally established for points of continuity of $F_n(x)$. However, in this case the expression given by (9.4.14) would be meaningless.

We shall not write out the general expression for the function $u_j^*(x)$.

But it is of some interest to write out the expressions on the right-hand sides of (9.4.13) and (9.4.14) to terms of order $n^{-\frac{3}{2}}$. These are

$$(9.4.15)\quad F_n(x) = \Phi(x) - \frac{1}{\sqrt{n}}\left(\frac{\alpha_3}{3!}\right)\Phi^{(3)}(x) + \frac{1}{n}\left[\frac{1}{4!}(\alpha_4 - 3)\Phi^{(4)}(x) + \frac{10}{6!}\alpha_3^2\Phi^{(6)}(x)\right] - \frac{1}{n^{\frac{3}{2}}}\left[\frac{1}{5!}(\alpha_5 - 10\alpha_3)\Phi^{(5)}(x) + \frac{35}{7!}\alpha_3(\alpha_4 - 3)\Phi^{(7)}(x) + \frac{280}{9!}\alpha_3^3\Phi^{(9)}(x)\right] + o\left(\frac{1}{n^{\frac{3}{2}}}\right)\cdots$$

and

$$(9.4.16)\quad f_n(x) = \Phi^{(1)}(x) - \frac{1}{\sqrt{n}}\left(\frac{\alpha_3}{3!}\right)\Phi^{(4)}(x) + \frac{1}{n}\left[\frac{1}{4!}(\alpha_4 - 3)\Phi^{(5)}(x) + \frac{10}{6!}\alpha_3^2\Phi^{(7)}(x)\right] - \frac{1}{n^{\frac{3}{2}}}\left[\frac{1}{5!}(\alpha_5 - 10\alpha_3)\Phi^{(6)}(x) + \frac{35}{7!}\alpha_3(\alpha_4 - 3)\Phi^{(8)}(x) + \frac{280}{9!}\alpha_3^3\Phi^{(10)}(x)\right] + o\left(\frac{1}{n^{\frac{3}{2}}}\right)\cdots.$$

We may summarize these results, which were originally obtained by Edgeworth (1905), as follows:

9.4.1 *If* $(x_1, \ldots, x_n)$ *is a sample from a continuous* p.d.f. *with finite moments* $\mu_1, \mu_2, \mu_3, \ldots$, *then the* c.d.f. $F_n(x)$ *of* $(z - n\mu)/\sqrt{n}\sigma$ *can be expanded in the form* (9.4.13), *the explicit expansion to terms of order* $n^{-\frac{3}{2}}$ *being given by* (9.4.15).

The quantities α_3 ($= \mu_3/\sigma^3$) and $\alpha_4 - 3$ ($= \mu_4/\sigma^4 - 3$), usually denoted by γ_1 and γ_2 respectively, are regarded as indices of *skewness* and *kurtosis* respectively, of a distribution function having mean μ, variance σ^2, and third and fourth central moments μ_3 and μ_4. These two constants play an important role in the degree to which the c.d.f. $F_n(x)$ can be approximated by the c.d.f. $\Phi(x)$ of the distribution $N(0, 1)$. It will be noted from an inspection of (9.4.15) that, in general, $F_n(x)$ is approximated by $\Phi(x)$ except for terms of order $1/\sqrt{n}$. But the following corollary of **9.4.1** gives conditions under which higher orders of approximation hold:

9.4.1a *If the skewness of the distribution from which* $(x_1, \ldots, x_n)$ *is drawn is zero,* $F_n(x)$ *is approximated by* $\Phi(x)$ *except for terms of order*

$1/n$; if both the skewness and kurtosis are zero, $F_n(x)$ is approximated by $\Phi(x)$ except for terms of order $1/\sqrt{n^3}$.

It should be pointed out that Lyapunov (1901) was the pioneer on the problem of determining higher degrees of approximation to the distribution of $\bar{x}$ in large samples than that provided by the normal distribution. Cramér (1937) has shown that the remainder term in (9.4.13) and (9.4.14) is of the same order as the first term neglected. Esseen (1944) has made more recent investigations of the accuracy of such asymptotic expansions. Asymptotic expansions in powers of $1/\sqrt{n}$ have been established for other statistics than sample means by Cramér (1937), Hsu (1945*a*, 1945*b*), Chung (1946), and others. An expository article on asymptotic approximations to distributions with an extensive bibliography has been published by Wallace (1958).

9.5 LIMITING DISTRIBUTIONS OF LINEAR FUNCTIONS IN LARGE SAMPLES FROM LARGE FINITE POPULATIONS

In the preceding section we have considered the limiting distributions of sums and means of samples from infinite populations. In this section we shall consider limiting distributions of linear functions of elements of samples from finite populations as sample size and population size increase indefinitely. A general theorem of basic importance in dealing with this problem due to Wald and Wolfowitz (1944) can be stated as follows:

9.5.1 *Let $(a_{N1}, \ldots, a_{NN})$ and $(x_{N1}, \ldots, x_{NN})$, $N = 1, 2, \ldots$ be two sets of sequences of real numbers such that for $r = 3, 4, \ldots$ and large N*

(9.5.1)
$$\frac{m_{r,N}(a)}{[m_{2,N}(a)]^{r/2}} = O(1)$$

and

(9.5.2)
$$\frac{m_{r,N}(x)}{[m_{2,N}(x)]^{r/2}} = O(1)$$

where

(9.5.3)
$$m_{r,N}(a) = \frac{1}{N}\sum_{i=1}^{N}(a_{Ni} - \bar{a}_N)^r, \qquad \bar{a}_N = \frac{1}{N}\sum_{i=1}^{N} a_{Ni}$$

with similar definitions of $m_{r,N}(x)$ and $\bar{x}_N$. For each value of N let $(x_1, \ldots, x_N)$ be a vector random variable whose sample space is the set of all $N!$ permutations of $(x_{N1}, \ldots, x_{NN})$. Let

(9.5.4)
$$L_N = \sum_{i=1}^{N} a_{Ni}x_i.$$

Then

$$\mathscr{E}(L_N) = N\bar{a}_N\bar{x}_N \tag{9.5.5}$$

and

$$\sigma^2(L_N) = \frac{N^2}{N-1} m_{2,N}(a) \cdot m_{2,N}(x). \tag{9.5.6}$$

Furthermore;

$$\lim_{N\to\infty} P\left(\frac{L_N - \mathscr{E}(L_N)}{\sigma(L_N)} \leqslant z\right) = \frac{1}{\sqrt{2\pi}} \int_{-\infty}^{z} e^{-\frac{1}{2}u^2}\, du. \tag{9.5.7}$$

The proof of **9.5.1** is straightforward but tedious. The procedure is to first set

$$a'_{Ni} = \frac{a_{Ni} - \bar{a}_N}{\sqrt{m_{2,N}(a)}}, \qquad x'_{Ni} = \frac{x_{Ni} - \bar{x}_N}{\sqrt{m_{2,N}(x)}}.$$

Then the sets of sequences $(a'_{N1}, \ldots, a'_{NN})$, $(x'_{N1}, \ldots, x'_{NN})$, $N = 1, 2, \ldots$ satisfy conditions similar to (9.5.1) and (9.5.2). If $(x'_1, \ldots, x'_N)$ is the same permutation of $(x'_{N1}, \ldots, x'_{NN})$ as $(x_1, \ldots, x_N)$ is of $(x_{N1}, \ldots, x_{NN})$ and if we let $L'_N = \sum_{i=1}^{N} a'_{Ni}x'_i$, then we have $\mathscr{E}(L'_N) = 0$ and

$$\frac{L'_N}{\sigma(L'_N)} = \frac{L_N - \mathscr{E}(L_N)}{\sigma(L_N)}. \tag{9.5.8}$$

Omitting detailed moment computations, which are given by Wald and Wolfowitz (1944), it is found that for any positive integer k

$$\begin{aligned} \mathscr{E}(L'^s_N) &= \frac{(2k)!}{2^k k!} N^k + o(N^k), \qquad s = 2k. \\ &= o(N^k), \qquad s = 2k+1, \end{aligned} \tag{9.5.9}$$

and hence

$$\begin{aligned} \lim_{N\to\infty} \mathscr{E}\left(\frac{L'_N}{\sigma(L'_N)}\right)^s = \lim_{N\to\infty} \mathscr{E}\left(\frac{L_N - \mathscr{E}(L_N)}{\sigma(L_N)}\right)^s &= \frac{(2k)!}{2^k k!}, \qquad s = 2k \\ &= 0, \qquad s = 2k+1. \end{aligned} \tag{9.5.10}$$

Thus as $N \to \infty$ the sth moment of $[L_N - \mathscr{E}(L_N)]/\sigma(L_N)$ converges to the sth moment of a random variable having the distribution $N(0, 1)$. Therefore it follows from **5.5.3** that the limiting distribution of $[L_N - \mathscr{E}(L_N)]/\sigma(L_N)$ as $N \to \infty$ is $N(0, 1)$.

If, in the definition of L_N as given in (9.5.4), we choose $a_{Ni} = 1/n$, $i = 1, \ldots, n$ and $a_{Ni} = 0$, $i = n+1, \ldots, N$ then L_N is the mean of a sample of size n from a finite population π_N whose N elements have $x_{N1}, \ldots, x_{NN}$ as their x-values. Furthermore, $\mathscr{E}(L_N) = \mu_N$ where $\mu_N = \frac{1}{N}\sum_{i=1}^{N} x_{Ni}$ and $\sigma^2(L_N) = (1/n - 1/N)\sigma_N^2$, where μ_N and σ_N^2 are the mean and variance of π_N as defined in Section 8.5. In this case, (9.5.1) reduces to the condition that the limit of

$$\frac{\left(\frac{N}{n} - 1\right)^{r-1} - (-1)^{r-1}}{\left(\frac{N}{n}\right)\left(\frac{N}{n} - 1\right)^{\frac{1}{2}r-1}} \tag{9.5.11}$$

be finite for $r \geqslant 3$ as $N, n \to \infty$. It is seen that this occurs if $\lim\limits_{N, n \to \infty} \frac{N}{n} = c$ where $+1 < c < +\infty$. Therefore we have the following corollary of **9.5.1**:

9.5.1a *Suppose $\{\pi_N, N = 1, 2, \ldots\}$ is a sequence of finite populations such that π_N has mean μ_N and variance σ_N^2, and let $\bar{x}_n$ be the mean of a random sample of size n from π_N. Then if $(x_{N1}, \ldots, x_{NN}, N = 1, 2, \ldots)$ satisfies (9.5.2) for large N and if $N, n \to \infty$ in such a way that* $\lim N/n = c$ *where* $+1 < c < +\infty$ *then*

$$\lim P\left(\frac{(\bar{x}_n - \mu_N)}{\left[\left(\frac{1}{n} - \frac{1}{N}\right)\sigma_N^2\right]^{\frac{1}{2}}} < z\right) = \frac{1}{\sqrt{2\pi}} \int_{-\infty}^{z} e^{-\frac{1}{2}u^2}\, du\,. \tag{9.5.12}$$

9.6 ASYMPTOTIC DISTRIBUTIONS CONCERNING ORDER STATISTICS

(a) Limiting Distributions of Sums of Coverages

In **8.7.2** it was shown that if $(x_{(1)}, \ldots, x_{(n)})$ are the order statistics of a sample from continuous c.d.f. $F(x)$, the random variable $y_{(k)} = F(x_{(k)})$ has the c.d.f.

$$D_n(y_{(k)}) = \frac{\Gamma(n+1)}{\Gamma(k)\Gamma(n-k+1)} \int_0^{y_{(k)}} x^{k-1}(1-x)^{n-k}\, dx. \tag{9.6.1}$$

In fact, it should be borne in mind from **8.7.6** that (9.6.1) is the c.d.f. of the sum of *any* k of the $n + 1$ coverages $F(x_{(1)})$, $F(x_{(2)}) - F(x_{(1)}), \ldots,$ $1 - F(x_{(n)})$. $F(x_{(s+k)}) - F(x_{(s)})$ is such a sum where $0 < s < n - k$. Now consider the random variable $w_k = ny_{(k)}$. If we denote by $H_n(w_k)$ the c.d.f. of w_k, we have

$$H_n(w_k) = D_n\left(\frac{w_k}{n}\right) = \int_0^{w_k} h_n(y)\, dy \tag{9.6.2}$$

where

$$h_n(y) = \frac{\Gamma(n+1)}{n^k\Gamma(k)\Gamma(n-k+1)}\, y^{k-1}\left(1 - \frac{y}{n}\right)^{n-k}.$$

It is evident that as $n \to \infty$ the sequence $h_n(y)$, $n = k, k + 1, \ldots$ converges to the function $\frac{1}{\Gamma(k)} y^{k-1}e^{-y}$ uniformly over the interval $(0, w_k)$, and hence

$$\lim_{n\to\infty} H_n(w_k) = \frac{1}{\Gamma(k)} \int_0^{w_k} y^{k-1}e^{-y}\, dy. \tag{9.6.3}$$

Therefore we have the result that

9.6.1 *If $(x_{(1)}, \ldots, x_{(n)})$ are the order statistics of a sample from a population having a continuous c.d.f. $F(x)$, then for a fixed k, $nF(x_{(k)})$, $n = k, k + 1, \ldots$, is a sequence of random variables converging in distribution to the gamma distribution $G(k)$.*

It should be noted that **9.6.1** holds when $F(x_{(k)})$ is replaced by the sum of any k of the coverages $F(x_{(1)})$, $F(x_{(2)}) - F(x_{(1)})$, $\ldots$, $1 - F(x_{(n)})$, for example $1 - F(x_{(n-k+1)})$ or $F(x_{(s+k)}) - F(x_{(s)})$. Also, the statement is true if $F(x_{(k)})$ is replaced by any k coverages determined by a sample of size n from a population having a continuous multidimensional c.d.f. [see Section 8.7(c)].

Theorem **9.6.1** can be extended to the case of two or more sums of fixed numbers of coverages. In the case of two such sums the analogous result can be stated as follows:

9.6.2 *Suppose $(x_{(1)}, \ldots, x_{(n)})$ are the order statistics of a sample from a continuous* c.d.f. *$F(x)$. Then if k_1 and k_2 are fixed integers, the sequence of pairs of random variables $(nF(x_{(k_1)}), nF(x_{(k_1+k_2)}))$, $n = m, m + 1, \ldots$ where $m \geqslant k_1 + k_2$ converges in distribution to a random variable having* p.d.f.

$$f(w_1, w_2) = \frac{1}{\Gamma(k_1)\Gamma(k_2)} w_1^{k_1-1}(w_2 - w_1)^{k_2-1}e^{-w_2}$$

in the region $0 < w_1 < w_2 < \infty$, and $f(w_1, w_2) = 0$ elsewhere in the w_1w_2-plane.

The proof of this statement is similar to the argument used in establishing **9.6.1** and is left as an exercise for the reader. It should be particularly noted that **9.6.2** holds if $F(x_{(k_1)})$ is replaced by the sum of any k_1 coverages and $F(x_{(k_1+k_2)})$ by the sum of any $k_1 + k_2$ coverages which includes the first k_1 coverages.

Now let us consider the limiting distribution of $y_{(k)}$ as $n \to \infty$, not for a fixed k, but for an increasing sequence of values of k and n such that $k/n = p_n = p + O(1/n)$. Consider the sequence of random variables

$v_n = (y_{(np_n)} - p)\sqrt{n}$, $n = m, \; m+1, \ldots; m \geqslant np_n$, where $y_{(np_n)} = F(x_{(np_n)})$. Denoting the c.d.f. of v_n by $H_n^*(v)$, we have

$$H_n^*(v) = D_n\left(p + \frac{v}{\sqrt{n}}\right) \tag{9.6.4}$$

where $D_n(y_{(np_n)})$ is defined in (9.6.1). Now $D_n(p + v/\sqrt{n})$ can be expressed as

$$\int_{-p\sqrt{n}}^{v} h_n^*(z)\, dz \tag{9.6.5}$$

where

$$\begin{aligned} h_n^*(z) &= \frac{\Gamma(n+1)}{\sqrt{n}\Gamma(k)\Gamma(n-k+1)}\left(p + \frac{z}{\sqrt{n}}\right)^{k-1}\left(1 - p - \frac{z}{\sqrt{n}}\right)^{n-k} \\ &= \frac{\Gamma(n+1)(p + z/\sqrt{n})^{-1}}{\sqrt{n}\Gamma(p_n n)\Gamma((1-p_n)n+1)}[p^{p_n}(1-p)^{1-p_n}]^n \\ &\quad \cdot \left[\left(1 + \frac{z}{p\sqrt{n}}\right)^{p_n}\left(1 - \frac{z}{(1-p)\sqrt{n}}\right)^{1-p_n}\right]^n. \end{aligned} \tag{9.6.6}$$

But since $p_n = p + O(1/n)$, we have

(9.6.7)
$$\left(1 + \frac{z}{p\sqrt{n}}\right)^{p_n}\left(1 - \frac{z}{(1-p)\sqrt{n}}\right)^{1-p_n} = \left(1 - \frac{z^2}{2np(1-p)} + \varphi(z, n)\right)$$

where $\varphi(z, n)$ is such that $\lim_{n\to\infty} n \cdot \varphi(z, n) = 0$ for any value of z. By making use of Stirling's approximation (7.6.27) for $\Gamma(g)$ for large values of g we find that

(9.6.8)
$$\frac{\Gamma(n+1)}{\sqrt{n}\Gamma(p_n n)\Gamma((1-p_n)n+1)}[p^{p_n}(1-p)^{1-p_n}]^n = \sqrt{\frac{p}{2\pi(1-p)}} + O\left(\frac{1}{n}\right).$$

Finally we obtain

(9.6.9)
$$h_n^*(z) = \left(p + \frac{z}{\sqrt{n}}\right)^{-1}\left(\sqrt{\frac{p}{2\pi(1-p)}} + O\left(\frac{1}{n}\right)\right)\left(1 - \frac{z^2}{2np(1-p)} + \varphi(z, n)\right)^n$$

which converges uniformly to the function

$$h^*(z) = \frac{1}{\sqrt{2\pi p(1-p)}} e^{-z^2/2p(1-p)} \tag{9.6.10}$$

in any interval $(-K, v)$ as $n \to \infty$. For any $\varepsilon > 0$ we can choose K and n_1 so that for $n > n_1$

(9.6.11) $\displaystyle\int_{-\infty}^{-K} h^*(z)\, dz < \frac{\varepsilon}{3}$ and $\left|\displaystyle\int_{-K}^{v} h_n^*(z)\, dz - \int_{-p\sqrt{n}}^{v} h_n^*(z) \ldots\right|$

Furthermore, there exists an $n_2 > K^2/p^2$ such that for any v and $n > n_2$,

$$\left| \int_{-K}^{v} h_n^*(z)\, dz - \int_{-K}^{v} h^*(z)\, dz \right| < \frac{\varepsilon}{3}. \tag{9.6.12}$$

Therefore, for $n > \max(n_1, n_2)$,

$$\left| \int_{-\infty}^{v} h^*(z)\, dz - \int_{-p\sqrt{n}}^{v} h_n^*(z)\, dz \right| < \varepsilon \tag{9.6.13}$$

and hence

$$\lim_{n\to\infty} P\left(\frac{[F(x_{(np_n)}) - p]\sqrt{n}}{\sqrt{p(1-p)}} < w\right) = \frac{1}{\sqrt{2\pi}} \int_{-\infty}^{w} e^{-\frac{1}{2}u^2}\, du. \tag{9.6.14}$$

Summarizing, we have the following result:

9.6.3 *If $(x_{(1)}, \ldots, x_{(n)})$ are the order statistics of a sample from a continuous c.d.f. $F(x)$, and np_n is an integer such that $p_n = p + O(1/n)$, where $0 < p < 1$, then for large n, $F(x_{(np_n)})$ is asymptotically distributed according to $N\left(p, \dfrac{p(1-p)}{n}\right)$.*

It is to be noted that **9.6.3** holds if $F(x_{(np_n)})$ is replaced by the sum of any np_n coverages in one or more dimensions.

The statement can be extended without significant difficulties to the case of several sums of coverages. It will be sufficient to state the result for two sums, leaving the proof to the reader.

9.6.4 *If $(x_{(1)}, \ldots, x_{(n)})$ are the order statistics of a sample from a population having a continuous c.d.f. $F(x)$ and if np_{1n} and np_{2n} are integers such that $p_{1n} = p_1 + O(1/n)$, $p_{2n} = p_2 + O(1/n)$ where $0 < p_1 < p_2 < 1$, then for large n, the two-dimensional random variable $(F(x_{(np_{1n})}), F(x_{(np_{2n})}))$ has $N\left(\{p_i\}; \left\|\dfrac{\sigma_{ij}}{n}\right\|\right)$ as its asymptotic distribution for large n, where $\sigma_{11} = p_1(1 - p_1)$, $\sigma_{12} = \sigma_{21} = p_1(1 - p_2)$, and $\sigma_{22} = p_2(1 - p_2)$.*

(b) Limiting Distributions of Order Statistics

The limiting distributions obtained in **9.6.1**, **9.6.2**, **9.6.3**, and **9.6.4** referred directly to distribution properties of sums of coverages and only implicitly to large-sample distribution properties of order statistics themselves. Since the population distribution $F(x)$ was assumed to be continuous, it is possible to make some statements about the limiting distribution of the order statistics themselves. For instance, in **9.6.1**, if $F^{-1}(y)$ is the inverse of $F(x)$, suppose we consider a sequence of values of n for which $F^{-1}(k/n)$ has a unique inverse. Since $F(x)$ is continuous and defined for all real values of x, it is evident that there exists an infinite sequence of such values of n, say $n_1, n_2, \ldots$. Then it follows from **9.6.1** that for fixed k,

$$\text{(9.6.15)} \qquad \lim_{i\to\infty} P\left(x_{(k)} \leqslant F^{-1}\left(\frac{u}{n_i}\right)\right) = \frac{1}{\Gamma(k)} \int_0^u y^{k-1} e^{-y}\, dy$$

and hence for large n,

$$\text{(9.6.16)} \qquad P(x_{(k)} \leqslant v) \cong \frac{1}{\Gamma(k)} \int_0^{nF(v)} y^{k-1} e^{-y}\, dy.$$

Formulas analogous to (9.6.15) and (9.6.16) for $x_{(n-k'+1)}$ for fixed k' can be similarly established. Asymptotic results of this type, particularly for $k = 1$ (the smallest order statistic) and $k' = 1$ (the largest order statistic) have been considered in detail by Dodd (1923), Fisher and Tippett (1928), Fréchet (1927), Gumbel (1935, 1958), and Smirnov (1935).

Similarly, from **9.6.3**, we may write

$$\text{(9.6.17)} \quad \lim_{i\to\infty} P\left(x_{(n_i p_{n_i})} \leqslant F^{-1}\left(p + \frac{w\sqrt{p(1-p)}}{\sqrt{n_i}}\right)\right) = \frac{1}{\sqrt{2\pi}} \int_{-\infty}^{w} e^{-\frac{1}{2}y^2}\, dy$$

and for large n,

$$\text{(9.6.18)} \qquad P(x_{(np_n)} \leqslant v) \cong \frac{1}{\sqrt{2\pi}} \int_{-\infty}^{T} e^{-\frac{1}{2}y^2}\, dy$$

where

$$T = \frac{(F(v) - p)\sqrt{n}}{\sqrt{p(1-p)}}.$$

Formulas (9.6.16) and (9.6.18) can be extended without further important difficulties to two or more order statistics. In the two-dimensional case the results follow directly from **9.6.2** and **9.6.4**.

It is evident from (9.6.14) that as $n \to \infty$, $F(x_{(np_n)})$ converges in probability to the constant p. Now suppose $F(x) = p$ has a unique solution, that is, that a unique pth quantile $\underline{x}_p$ exists. Then since $F(x)$ is continuous, $x_{(np_n)}$ will converge in probability to $\underline{x}_p$. Therefore we have the following result:

9.6.5 *If in addition to the assumptions of* **9.6.3** *we add the condition that a unique* pth *quantile* $\underline{x}_p$ *exists, then as* $n \to \infty$, $x_{(np_n)}$ *converges in probability to* $\underline{x}_p$.

A similar statement can be made for the case of the two order statistics $x_{(np_{1n})}$ and $x_{(np_{2n})}$ or any larger specified number of similar order statistics.

Now suppose $F(x)$ has a derivative $f(x)$ in some neighborhood $V(\underline{x}_p)$ of the point $x = \underline{x}_p$ such that $f(\underline{x}_p) > 0$. Then a unique pth quantile $\underline{x}_p$ exists and hence, by **9.6.5** $x_{(np_n)}$ converges in probability to $\underline{x}_p$. Now if $x_{(np_n)}$ is any point in $V(\underline{x}_p)$ we may write

$$\text{(9.6.19)} \qquad F(x_{(np_n)}) = p + f(x^*)(x_{(np_n)} - \underline{x}_p)$$

where x^* is a random variable such that $|x^* - \underline{x}_p| < |x_{(np_n)} - \underline{x}_p|$. But since $x_{(np_n)}$ converges in probability to the quantile $\underline{x}_p$, as $n \to \infty$, then for an arbitrary $\varepsilon > 0$, we can choose an n_ε such that

$$(9.6.20) \qquad P(F(x_{(np_n)}) = p + f(x^*)(x_{(np_n)} - \underline{x}_p) \text{ for all } n > n_\varepsilon) > 1 - \varepsilon,$$

which amounts to stating that

$$(9.6.21) \qquad \frac{(F(x_{(np_n)}) - p)\sqrt{n}}{\sqrt{p(1-p)}}, \qquad n = m, m+1, \ldots, \qquad m \geqslant np_n$$

and

$$(9.6.22) \qquad \frac{f(x^*)(x_{(np_n)} - \underline{x}_p)\sqrt{n}}{\sqrt{p(1-p)}}, \qquad n = m, m+1, \ldots, \qquad m \geqslant np_n$$

form a stochastic process such that the two sequences converge in distribution together to the distribution $N(0, 1)$.

The fact that $F(x)$ has a derivative $f(x)$ in $V(\underline{x}_p)$ implies that $f(x)$ is continuous in $V(\underline{x}_p)$ and hence at $x = \underline{x}_p$. Therefore, by **4.3.5**, $f(x_{(np_n)})$ converges in probability to the constant $f(\underline{x}_p)$ and by **4.3.7** $f(x^*)$ also converges in probability to $f(\underline{x}_p)$ as $n \to \infty$. Finally, by applying **4.3.3** we see that the sequence of random variables (9.6.22) and the sequence

$$(9.6.22a) \qquad \frac{f(\underline{x}_p)(x_{(np_n)} - \underline{x}_p)\sqrt{n}}{\sqrt{p(1-p)}}, \qquad n = m, m+1, \ldots, \qquad m \geqslant np_n$$

constitute a stochastic process such that the two sequences converge together in distribution to the distribution $N(0, 1)$.

Summarizing, we have the following result:

9.6.6 *If in addition to the assumptions of* **9.6.3** *we add the condition that $F(x)$ has a derivative $f(x)$ in some neighborhood $V(\underline{x}_p)$ of $x = \underline{x}_p$ such that $f(\underline{x}_p) > 0$, then, for large n, $x_{(np_n)}$ is asymptotically distributed according to* $N\left(\underline{x}_p, \dfrac{p(1-p)}{nf^2(\underline{x}_p)}\right)$.

Example. An interesting special case of **9.6.6** arises for $p = \frac{1}{2}$. In this case, $x_{(np_n)}$ is the sample median and we see that, for large n, its asymptotic distribution is $N\left(\underline{x}_{0.5}, \dfrac{1}{4nf^2(\underline{x}_{0.5})}\right)$. If the population distribution is $N(\mu, \sigma^2)$, then the sample median has $N(\mu, \pi\sigma^2/2n)$ as its asymptotic distribution for large n.

A statement similar to **9.6.6** holds, of course, for the joint distribution of two or more order statistics. In the case of the two order statistics $x_{(np_{1n})}$ and $x_{(np_{2n})}$, we can say that

9.6.7 *If, in addition to the assumptions of* **9.6.4**, *we add the condition that $F(x)$ has a derivative $f(x)$ in neighborhoods of each of the points*

$x = \underline{x}_{p_1}$ and $\underline{x}_{p_2}$, $0 < p_1 < p_2 < 1$, such that $f(\underline{x}_{p_1}) > 0$ and $f(\underline{x}_{p_2}) > 0$, then, for large n, the random variable $(x_{(np_{1n})}, x_{(np_{2n})})$ is asymptotically distributed according to $N\left(\{\underline{x}_{p_i}\}, \left\|\frac{\sigma_{ij}^*}{n}\right\|\right)$, $i, j = 1, 2$, where

$$\sigma_{11}^* = \frac{p_1(1-p_1)}{f^2(\underline{x}_{p_1})}, \qquad \sigma_{12}^* = \frac{p_1(1-p_2)}{f(\underline{x}_{p_1})f(\underline{x}_{p_2})}, \qquad \sigma_{22}^* = \frac{p_2(1-p_2)}{f^2(\underline{x}_{p_2})}.$$

Theorems **9.6.6** and **9.6.7** were established by Smirnov (1935), although the formulas for the variances and covariance of $x_{(np_{1n})}$ and $x_{(np_{2n})}$ were originally established by K. Pearson (1920). Mosteller (1946) has extended **9.6.7** to the case of several order statistics.

PROBLEMS

9.1 State and prove the k-dimensional version of **9.1.1**.

9.2 If $(x_{1\xi}, \ldots, x_{k\xi}; \xi = 1, \ldots, n)$ is a sample from a k-dimensional distribution having (finite) means $(\mu_1, \ldots, \mu_k)$ and (finite) covariance matrix $\|\sigma_{ij}\|$ and if $(\bar{x}_1, \ldots, \bar{x}_k)$ is the vector of means of the sample, show that for arbitrary $\delta_i > 0$, $i = 1, \ldots, k$,

$$P(|\bar{x}_i - \mu_i| < \delta_i, i = 1, \ldots, k) \geqslant 1 - \frac{1}{n}\left(\frac{\sigma_{11}}{\delta_1^2} + \cdots + \frac{\sigma_{kk}}{\delta_k^2}\right)$$

and hence that $(\bar{x}_1, \ldots, \bar{x}_k)$ converges in probability to $(\mu_1, \ldots, \mu_k)$.

9.3 (*Continuation*) Show that for arbitrary $\delta^2 > 0$,

$$P\left(\sum_{i,j=1}^{k} \sigma^{ij}(\bar{x}_i - \mu_i)(\bar{x}_j - \mu_j) < \delta^2\right) \geqslant 1 - \frac{k}{n\delta^2}$$

where $\|\sigma^{ij}\| = \|\sigma_{ij}\|^{-1}$, and hence that $(\bar{x}_1, \ldots, \bar{x}_k)$ converges in probability to $(\mu_1, \ldots, \mu_k)$.

9.4 Prove **9.2.3**.

9.5 If $(x_1, \ldots, x_n)$ is a sample from a Poisson distribution $Po(\mu)$, show that for large n, $2\sqrt{\bar{x}}$ has as its asymptotic distribution $N(2\sqrt{\mu}, 1/n)$. Show that the same result holds if $(x_1, \ldots, x_n)$ is a sample from the gamma distribution $G(\mu)$.

9.6 If $(x_1, \ldots, x_n)$ is a sample from the binomial distribution $Bi(1, p)$ show that the asymptotic distribution of $\sin^{-1}(2\bar{x} - 1)$, for large n, is

$$N(\sin^{-1}(2p - 1), 1/n).$$

9.7 If $(x_1, \ldots, x_n)$ is a sample from the waiting-time distribution having p.f.

$$qp^{x-1}, \qquad x = 1, 2, \ldots$$

where $0 < p < 1$, $p + q = 1$, show that the asymptotic distribution of $\log[\bar{x}(1 + \sqrt{1 - 1/\bar{x}}) - \frac{1}{2}]$ for large n, is

$$N\left(\log\left[\frac{1+\sqrt{p}}{q} - \frac{1}{2}\right], \frac{1}{n}\right).$$

9.8 If $(x_1, \ldots, x_n)$ is a sample from the rectangular distribution $R(\frac{1}{2}\theta, \theta)$, show that the asymptotic distribution of $\sqrt{12}\log(2\bar{x})$ for large n, is

$$N(\sqrt{12}\log\theta, 4/n).$$

9.9 If $\bar{x}$ is the mean of a sample of size n from the rectangular distribution $R(\frac{1}{2}, 1)$, determine the expansions (9.4.15) and (9.4.16) for the c.d.f. and p.d.f. of the random variable $(\bar{x} - \frac{1}{2})\sqrt{12n}$ up to terms of order $n^{-\frac{3}{2}}$.

9.10 If $(x_1, \ldots, x_n)$ is a sample from a distribution having finite mean μ and variance σ^2, show that the sample variance s^2 converges in probability to σ^2, and also that the asymptotic distribution of $\sqrt{n}(\bar{x} - \mu)/s$, for large n, is $N(0, 1)$.

9.11 If n_1, $\bar{x}_1$, s_1^2 are the size, mean, and variance of a sample from a distribution having mean μ_1, and variance σ_1^2, whereas n_2, $\bar{x}_2$, s_2^2 are the size, mean, and variance of an independent sample from a distribution having mean μ_2 and variance σ_2^2, show that the limiting distribution of

$$\frac{(\bar{x}_1 - \bar{x}_2) - (\mu_1 - \mu_2)}{\sqrt{\dfrac{s_1^2}{n_1} + \dfrac{s_2^2}{n_2}}}$$

as $n_1, n_2 \to \infty$, is $N(0, 1)$.

9.12 If $\tilde{x}$ is the median of a sample of size n from a continuous c.d.f. $F(x)$, show that the asymptotic distribution of $F(\tilde{x})$, for large n, is $N(\frac{1}{2}, 1/(4n))$.

9.13 If $(x_{(1)}, \ldots, x_{(n)})$ are the order statistics of a sample from a continuous c.d.f. $F(x)$, show that the limiting distribution of $n\left(1 - \int_{x_{(1)}}^{x_{(n)}} dF(x)\right)$ as $n \to \infty$ is the gamma distribution $G(2)$.

9.14 If $(x_1, \ldots, x_{2n+1})$ is a sample from a distribution having p.d.f. $\lambda e^{-\lambda x}$, $x > 0$, $\lambda > 0$, show that, for large n, the median $\tilde{x}$ of the sample has

$$N(\log 2/\lambda, 1/(2\lambda^2 n))$$

as its asymptotic distribution.

9.15 If $\bar{x}$ and $\tilde{x}$ are the mean and median of a sample of size n from a population having the distribution $N(\mu, \sigma^2)$, show that the asymptotic distribution of $(\bar{x}, \tilde{x})$ for large n is

$$N\left(\mu, \mu; \left\|\frac{\sigma_{ij}}{n}\right\|\right)$$

where

$$\left\|\frac{\sigma_{ij}}{n}\right\| = \left\|\begin{matrix} \dfrac{\sigma^2}{n} & \dfrac{\sigma^2}{n} \\[2ex] \dfrac{\sigma^2}{n} & \dfrac{\pi\sigma^2}{2n} \end{matrix}\right\|.$$

9.16 (*Fisher's* (1925*a*) *transformation of the correlation coefficient*). It is known that the distribution of the sample correlation coefficient r in samples from a two-dimensional normal distribution having correlation coefficient ρ has, as its asymptotic distribution, for large n, the distribution $N(\rho, (1-\rho^2)^2/n)$. Show that the asymptotic distribution of $\frac{1}{2}\log\left(\frac{1+r}{1-r}\right)$ for large n is

$$N\left(\frac{1}{2}\log\left(\frac{1+\rho}{1-\rho}\right), \frac{1}{n}\right).$$

9.17 Prove **9.6.2**.

9.18 Prove **9.6.4**.

CHAPTER 10

Linear Statistical Estimation

10.1 INTRODUCTORY COMMENTS

In Chapters 8 and 9 we have presented some results on the theory of sampling, that is, probability theory of certain functions of the elements or components of a sample from a given c.d.f. $F(x)$. In problems of applied statistics, $F(x)$ is usually unknown and the main purpose of sampling is to acquire information on the basis of which statements or inferences can be made about $F(x)$, or some of its properties. These statements are made in terms of functions of the elements (components) of a sample and expressed as probability statements. Assumptions which can be made about $F(x)$ in advance of any sampling can range all the way from those stating that $F(x)$ satisfies only the basic properties of a c.d.f. given in **2.2.1** to a complete specification of $F(x)$ for all values of x in R_1. In general, assumptions which can be made about $F(x)$ in any given situation lie between these extremes. For instance, it might be assumed that $F(x)$ has a finite, but unknown, mean μ and variance σ^2, the problem being to devise *estimators* for μ and σ^2, as functions of the elements of a random sample from $F(x)$. Thus, in Section 8.2 it was shown that the sample mean $\bar{x}$ and the sample variance s^2 in samples from infinite populations are *unbiased estimators* for μ and σ^2, that is, $\mathscr{E}(\bar{x}) = \mu$ and $\mathscr{E}(s^2) = \sigma^2$. More generally, the problem is to make inferences about a population distribution function beyond the assumptions made concerning the distribution function by utilizing information in a sample from the distribution.

It is sufficient for many problems in applied statistics to devise from samples unbiased estimators for *parameters* of population distributions such as means, variances, covariances, and regression coefficients involved in the distributions. A class of relatively simple estimators of such parameters, known as *linear estimators*, can be devised as linear forms of the sample elements and involves no stronger assumptions about the population c.d.f. than finiteness of first- and second-order moments of the components of the sample. There are, of course, other classes of estimators, but these are considered in Chapters 11 and 12. In general, an *estimator* for a *population parameter* θ is an *observable* random variable determined from a sample, that is, one which is a *known* function of sample elements which is used in place of the *unknown true value* of the parameter which it estimates. In devising an estimator for a parameter θ, it is important to construct it from the sample so that its distribution concentrates as much as possible in some sense around the *true value* θ_0 when calculated from samples from a population in which θ actually has the value θ_0. As we shall see a fairly natural criterion for measuring this concentration in the case of linear estimators is the variance of the estimator. In this chapter we shall confine our attention to the theory of linear estimators, to criteria for constructing them, and to their application to some of the more important statistical problems. Similar consideration is given to other classes of estimators in Chapters 11 and 12.

As has been pointed out in Sections 8.2 and 8.5, the sample mean, which, of course, is a linear function of the sample elements, is an unbiased estimator for the population mean in both infinite and finite populations. But the sample mean is only one of many possible unbiased linear estimators of the population mean that we could devise. For example, if $(x_1, \ldots, x_n)$ is a sample from a population (finite or infinite) with mean μ, then it is evident that $a_1x_1 + \cdots + a_nx_n$ is an unbiased estimator for μ if $a_1, \ldots, a_n$ are known constants such that $a_1 + \cdots + a_n = 1$. Estimators of this type are called *unbiased linear estimators.* If one is to select an unbiased linear estimator for μ, one must consider what criteria to use in making the selection. One widely accepted criterion is to choose that linear estimator from all unbiased linear estimators having the smallest possible variance. Such an estimator will be called a *minimum variance linear estimator.*

Similarly, if $(x_1, \ldots, x_n)$ is a sample from a finite or infinite population having mean μ and variance σ^2, any quadratic form $\sum_{\xi,\eta} a_{\xi\eta}(x_\xi - \bar{x})(x_\eta - \bar{x})$ whose mean value is σ^2 is an *unbiased quadratic estimator* for σ^2. If a unique unbiased quadratic estimator for σ^2 exists having smallest possible variance, it will be called the *minimum variance quadratic estimator* for σ^2. However, it should be noted that any quadratic estimator is a *linear function* of quadratic terms, and much of the theory of linear estimation to be developed in this chapter applies to quadratic estimators.

Minimum variance linear estimators themselves usually have variances which can be estimated without bias by quadratic estimators.

In this chapter we shall deal with the theory of linear estimation and its application to the estimation of means and variances of population distributions. In these problems no assumptions are required about the distribution of the random variables used in the linear estimation process except finiteness of means and of the elements of the covariance matrix of these random variables. The underlying random variables for quadratic estimators are quadratic forms in the sample components.

It will be sometimes convenient to use the notation $\mathscr{E}^{-1}(\theta)$ to denote an unbiased estimator for θ. Thus, if T is an observable random variable which is an unbiased estimator for θ, that is, if

$$\mathscr{E}(T) = \theta$$

we may write

$$\mathscr{E}^{-1}(\theta) = T.$$

A simple but basic theorem in linear estimation theory which will be used repeatedly in dealing with linear and quadratic estimators is the following:

10.1.1 *Suppose $(x_1, \ldots, x_k)$ are random variables whose mean values are*

$$\mathscr{E}(x_i) = \sum_j a_{ij}\theta_j, \qquad i = 1, \ldots, k$$

where $\theta_1, \ldots, \theta_k$ are unknown parameters and where $\|a_{ij}\|$ is a nonsingular matrix whose elements are known (that is, do not depend on the parameters $\theta_1, \ldots, \theta_k$). Then

$$\mathscr{E}^{-1}(\theta_i) = \sum_j a^{ij}x_j, \qquad i = 1, \ldots, k$$

are unbiased linear estimators for $\theta_1, \ldots, \theta_k$ respectively, where $\|a^{ij}\| = \|a_{ij}\|^{-1}$.

The proof is left to the reader.

10.2 MINIMUM VARIANCE ESTIMATORS FOR THE MEAN AND VARIANCE OF A POPULATION FROM RANDOM SAMPLES

(a) Minimum Variance Linear Estimator for the Population Mean

Suppose $(x_1, \ldots, x_n)$ is a sample from a distribution having mean μ and variance σ^2. We have seen in Section 8.2 that the sample mean is an unbiased estimator for μ. We now show that $\bar{x}$ is the minimum variance linear estimator for μ. Let $\mathscr{E}^{-1}(\mu)$ be any unbiased linear estimator for μ, that is, let

(10.2.1) $$\mathscr{E}^{-1}(\mu) = a_1x_1 + \cdots + a_nx_n$$

where

(10.2.2) $$a_1 + \cdots + a_n = 1.$$

For the variance of $\mathscr{E}^{-1}(\mu)$ we have from **3.6.1a**

$$\sigma^2(\mathscr{E}^{-1}(\mu)) = (a_1^2 + \cdots + a_n^2)\sigma^2. \tag{10.2.3}$$

It is seen that $\sigma^2(\mathscr{E}^{-1}(\mu))$ has a unique minimum which occurs for

$$a_1 = \cdots = a_n = \frac{1}{n}. \tag{10.2.4}$$

But for this choice of values of the a's, $\mathscr{E}^{-1}(\mu)$ is identical with $\bar{x}$. Therefore

10.2.1 *If $(x_1, \ldots, x_n)$ is a sample of size n from a distribution having mean μ and variance σ^2, then $\bar{x}$ is the minimum variance linear estimator for μ.*

It can also be verified that the same conclusion holds if $(x_1, \ldots, x_n)$ is a sample from a finite population. The proof is left to the reader.

(b) Minimum Variance Quadratic Estimator for the Population Variance

To deal with this problem of quadratic estimators it will be convenient to consider first the following theorem from some general results on unbiased estimation theory by Halmos (1946):

10.2.2 *Let $(x_1, \ldots, x_n)$ be a sample from a* c.d.f. *$F(x)$ and let $g(x_1, \ldots, x_n)$ be any statistic having mean θ and variance $\sigma^2 < +\infty$. Let $(i_1, \ldots, i_n)$ be the ith in the set (suitably indexed) of all $n!$ permutations of the integers $(1, \ldots, n)$ and let $g_i(x_1, \ldots, x_n) = g(x_{i_1}, \ldots, x_{i_n})$. If $\bar{g} = \frac{1}{n!}\sum_{i=1}^{n!} g_i(x_1, \ldots, x_n)$ then $\mathscr{E}(\bar{g}) = \theta$ and the variance of $\bar{g}$ is smaller than that of $g(x_1, \ldots, x_n)$ unless $g(x_1, \ldots, x_n)$ is symmetric in $x_1, \ldots, x_n$ with probability* 1, *in which case $\bar{g}$ is identical with $g(x_1, \ldots, x_n)$.*

To establish **10.2.2**, we first note that since $(x_1, \ldots, x_n)$ is a random sample from $F(x)$ we have $\mathscr{E}(g_i) = \theta$, $i = 1, \ldots, n!$ Therefore $\mathscr{E}(\bar{g}) = \theta$. Furthermore, $\sigma^2(g_i) = \sigma^2$, $i = 1, \ldots, n!$, and

$$\sigma^2(\bar{g}) = \frac{1}{n!}\sigma^2 + \left(\frac{1}{n!}\right)^2 \sum_{i \neq j} \operatorname{cov}(g_i, g_j). \tag{10.2.5}$$

But $\operatorname{cov}(g_i, g_j) \leqslant \sigma^2$. Hence

$$\sigma^2(\bar{g}) \leqslant \sigma^2. \tag{10.2.6}$$

But equality in (10.2.6) holds if and only if $g_i \equiv a + bg_j$ with probability 1 in the sample space R_n, for all $i \neq j$, where a and b are constants which must have the values 0 and 1 respectively. Hence

$$g_i(x_1, \ldots, x_n) \equiv g_j(x_1, \ldots, x_n) \tag{10.2.7}$$

for all points $(x_1, \ldots, x_n)$ in the sample space R_n (except possibly for a set of zero probability) and for $i \neq j = 1, \ldots, n!$. This condition implies that $g(x_1, \ldots, x_n)$ is symmetric in $(x_1, \ldots, x_n)$. The conclusion of **10.2.2** follows at once.

The following corollary of **10.2.2**, states that under certain mild conditions the sample variance s^2 is the minimum variance quadratic estimator for the variance σ^2 of the population distribution. The proof is a straightforward application of a slightly extended version of **10.2.2** described below and is left as an exercise for the reader.

10.2.2a *If $(x_1, \ldots, x_n)$ is a sample from a* c.d.f. $F(x)$ *having mean μ and variance σ^2 and finite third and fourth moments, if $\{a_{\xi\eta}\}$ is any set of constants for which*

$$Q = \sum_{\xi,\eta=1}^{n} a_{\xi\eta}(x_\xi - \bar{x})(x_\eta - \bar{x})$$

has mean σ^2 the values of the $\{a_{\xi\eta}\}$ for which Q has minimum variance are $a_{\xi\xi} = 1/(n-1)$, $\xi = 1, \ldots, n$; $a_{\xi\eta} = 0$, $\xi \neq \eta$, in which case Q reduces to the sample variance s^2.

The reader should note that **10.2.2** holds with only minor changes in the argument if the assumption that $(x_1, \ldots, x_n)$ is a random sample from $F(x)$ is replaced by the assumption that $(x_1, \ldots, x_n)$ is a vector random variable whose c.d.f. $F(x_1, \ldots, x_n)$ is symmetric in $x_1, \ldots, x_n$. With this extended version of **10.2.2**, it is seen that the following version of **10.2.2a** states that in sampling from a finite population the sample variance s^2 is the minimum variance unbiased estimator for the population variance σ^2.

10.2.2b *If $(x_1, \ldots, x_n)$ is a sample from a finite population having variance σ^2 and if $\{a_{\xi\eta}\}$ is any set of constants for which*

$$Q = \sum_{\xi,\eta=1}^{n} a_{\xi\eta}(x_\xi - \bar{x})(x_\eta - \bar{x})$$

has mean σ^2, the values of $\{a_{\xi\eta}\}$ for which Q has minimum variance are those given in **10.2.2a**, *in which case Q reduces to the sample variance s^2.*

(c) Interval Estimators for μ and σ^2 in a Normal Distribution

If we make the further assumption that $(x_1, \ldots, x_n)$ is a sample from the distribution $N(\mu, \sigma^2)$, then we have a special case where we can obtain what are known as *interval estimators* for μ and σ^2. This case deserves special attention. We know from **8.4.3** that $\sqrt{n}(\bar{x} - \mu)/s$ has the "Student" distribution $S(n - 1)$. This fact enables us to make the following statement:

$$P(t_1 < \sqrt{n}(\bar{x} - \mu)/s < t_2) = \int_{t_1}^{t_2} f_{n-1}(t)\, dt = \gamma, \tag{10.2.8}$$

where $f_{n-1}(t)$ is given by (7.8.4), and t_1 and t_2 are chosen so that the integral in (10.2.8) has the value γ. But the statement

$$P\left(t_1 < \frac{(\bar{x} - \mu)\sqrt{n}}{s} < t_2\right) = \gamma$$

is equivalent to the statement

$$P\left(\bar{x} - t_2 \frac{s}{\sqrt{n}} < \mu < \bar{x} - t_1 \frac{s}{\sqrt{n}}\right) = \gamma. \tag{10.2.9}$$

Thus, $(\bar{x} - t_2(s/\sqrt{n}),\ \bar{x} - t_1(s/\sqrt{n}))$ is an observable random interval such that the probability is γ that it includes the point μ. The interval is called a $100\gamma\%$ *confidence interval* for μ, and γ is called the *confidence coefficient*. It is an example of a method of setting up an estimator for a parameter by using a (realizable or observable) random interval with a specified probability of including the "true" value of the parameter. Such intervals are sometimes called *interval estimators.* We shall defer a discussion of interval estimation under more general conditions until Chapters 11 and 12.

In the particular example above the length of the interval is $(t_2 - t_1)s/\sqrt{n}$ which is shortest for a given s if, for the given γ, t_1 and t_2 are chosen so that $t_2 = -t_1 = t_{n-1,\gamma}$ where $t_{n-1,\gamma}$ satisfies

$$\int_{-t_{n-1,\gamma}}^{+t_{n-1,\gamma}} f_{n-1}(t)\, dt = \gamma \tag{10.2.10}$$

in which case the $100\gamma\%$ confidence interval for μ is the following interval centered at $\bar{x}$:

$$\bar{x} \pm t_{n-1,\gamma} \frac{s}{\sqrt{n}}. \tag{10.2.11}$$

Similarly, if $(x_1, \ldots, x_n)$ is a sample from a population having the distribution $N(\mu, \sigma^2)$ we can set up an interval estimator for σ^2 from the fact that $(n - 1)s^2/\sigma^2$ has the chi-square distribution $C(n - 1)$ (see **8.4.2**). For we have

$$P\left(\chi_1^2 < \frac{(n - 1)s^2}{\sigma^2} < \chi_2^2\right) = \int_{\chi_1^2}^{\chi_2^2} dF_{n-1}(\chi^2) = \gamma \tag{10.2.12}$$

where $dF_{n-1}(\chi^2)$ is given by (7.8.1), so that the integral in (10.2.12) has the value γ. But (10.2.12) is equivalent to

$$P\left(\frac{(n-1)s^2}{\chi_2^2} < \sigma^2 < \frac{(n-1)s^2}{\chi_1^2}\right) = \gamma$$

from which it is evident that $((n-1)s^2/\chi_2^2, (n-1)s^2/\chi_1^2)$ is a $100\gamma\%$ confidence interval for σ^2. There are many ways, of course, of choosing χ_1^2, and χ_2^2 to satisfy (10.2.12). In practice, they are usually chosen so that

$$\int_0^{\chi_1^2} dF_{n-1}(\chi^2) = \int_{\chi_2^2}^{\infty} dF_{n-1}(\chi^2) = \frac{1-\gamma}{2} \tag{10.2.13}$$

although the confidence interval with shortest mean length is obtained if $(1/\chi_1^2 - 1/\chi_2^2)$ is minimized subject to the condition that (10.2.12) be satisfied.

10.3 ESTIMATORS FOR PARAMETERS IN LINEAR REGRESSION ANALYSIS

(a) Estimators for Regression Coefficients

We shall now consider a generalization of **10.2.1** which arises in estimation problems of regression analysis, experimental designs, and related problems. Suppose $y_1, \ldots, y_n$, $n > k$, are n independent random variables having variances all equal to σ^2 but with means given by the *regression function*

$$\mathscr{E}(y_\xi) = \beta_1 x_{1\xi} + \cdots + \beta_k x_{k\xi}, \tag{10.3.1}$$

$\xi = 1, \ldots, n$ where the $(x_{1\xi}, \ldots, x_{k\xi})$, $\xi = 1, \ldots, n$ are known (real) vectors but $\beta_1, \ldots, \beta_k$ are unknown (real) parameters, called *regression coefficients*, to be estimated. The parameter σ^2, usually unknown, is called the *residual variance*. It is convenient to introduce $x_1, \ldots, x_k$ and refer to them as *fixed variables* as contrasted with random variables, in which case $(x_{1\xi}, \ldots, x_{k\xi})$, $\xi = 1, \ldots, n$, is a set of n specified values of these fixed variables. It is customary to take $x_{1\xi} = 1$, $\xi = 1, \ldots, n$, but it will be convenient not to make this assignment at present. We will show that under mild conditions minimum variance estimators exist for $\beta_1, \ldots, \beta_k$, and a rather simple unbiased estimator exists for σ^2.

First, consider the estimation of $\beta_1, \ldots, \beta_k$. Let $\mathscr{E}^{-1}(\beta_i)$ be an arbitrary unbiased linear estimator for β_i, that is,

$$\mathscr{E}^{-1}(\beta_i) = \sum_\xi c_{i\xi} y_\xi, \qquad i = 1, \ldots, k. \tag{10.3.2}$$

Then

$$\mathscr{E}(\mathscr{E}^{-1}(\beta_i)) = \sum_\xi \sum_j \beta_j c_{i\xi} x_{j\xi} \equiv \beta_i, \qquad i = 1, \ldots, k, \tag{10.3.3}$$

from which it is evident that the $c_{i\xi}$ must satisfy

$$\sum_{\xi} c_{i\xi} x_{j\xi} = \delta_{ij} \tag{10.3.4}$$

where δ_{ij} is the Kronecker δ. The variance of $\mathscr{E}^{-1}(\beta_i)$ is given by

$$\sigma^2(\mathscr{E}^{-1}(\beta_i)) = \sum_{\xi} c_{i\xi}^2 \sigma^2. \tag{10.3.5}$$

Minimizing $\sigma^2(\mathscr{E}^{-1}(\beta_i)$ with respect to the $c_{i\xi}$, subject to condition (10.3.4), one finds that $c_{i\xi}$ must be of the following form

$$c_{i\xi} = \sum_j \lambda_{ij} x_{j\xi}, \qquad i = 1, \ldots, k, \qquad \xi = 1, \ldots, n, \tag{10.3.6}$$

where it will be seen that the λ_{ij} must satisfy the equations

$$\sum_{j'} \lambda_{ij'} a_{j'j} = \delta_{ij}, \tag{10.3.7}$$

where

$$a_{j'j} = \sum_{\xi} x_{j'\xi} x_{j\xi}. \tag{10.3.8}$$

If the matrix $\|a_{j'j}\|$ is nonsingular, which will be true if and only if $(x_{i1}, \ldots, x_{in})$, $i = 1, \ldots, k$ are linearly independent vectors, then it is evident that the solution of (10.3.7) is

$$\|\lambda_{ij}\| = \|a_{ij}\|^{-1} = \|a^{ij}\|. \tag{10.3.9}$$

Therefore, the minimum variance linear estimator for β_i, which we denote by b_i, is

$$b_i = \sum_j a^{ij} a_{j0} \tag{10.3.10}$$

where $a_{j0}, j = 1, \ldots, k$ are random variables defined by

$$a_{j0} = \sum_{\xi} x_{j\xi} y_{\xi} = a_{0j}. \tag{10.3.11}$$

To find the variance of b_i, we substitute λ_{ij} from (10.3.9) into (10.3.6) and, in turn, substitute $c_{i\xi}$ into (10.3.5). This gives

$$\sigma^2(b_i) = \sum_{j,j'=1}^{k} a^{ij} a^{ij'} a_{j'j} \sigma^2 = \sum_j a^{ij} \delta_{ij} \sigma^2 = a^{ii} \sigma^2. \tag{10.3.12}$$

One similarly finds the covariance matrix of the b_i, $i = 1, \ldots, k$, to be

$$\|a^{ij}\sigma^2\|. \tag{10.3.13}$$

Summarizing, we have the Markov (1900) theorem:

10.3.1 *Suppose y_{ξ}, $\xi = 1, \ldots, n$ are independent random variables with means $\sum_{i=1}^{k} \beta_i x_{i\xi}$, $\xi = 1, \ldots, n \geqslant k$ and with variances all equal to σ^2, where $(x_{i1}, \ldots, x_{in})$, $i = 1, \ldots, k$, are known and are linearly* *independent vectors. The minimum variance linear estimators of the regression coefficients β_i are b_i, $i = 1, \ldots, k$, where the b_i are defined by (10.3.10). The covariance matrix of the b_i is $\|a^{ij}\sigma^2\|$, where $a_{ij} = \sum_\xi x_{i\xi}x_{j\xi}$, $i, j = 1, \ldots, k$, and $\|a^{ij}\| = \|a_{ij}\|^{-1}$.*

If we minimize the sum of squares

$$Q = \sum_\xi (y_\xi - \beta_1 x_{1\xi} - \cdots - \beta_k x_{k\xi})^2 \tag{10.3.14}$$

with respect to $\beta_1, \ldots, \beta_k$, we note that if $\|a_{ij}\|$ is nonsingular, the *least squares estimators* for the β_i that is, the values of the β_i which minimize (10.3.14) are the b_i. The details are straightforward and are left to the reader. Hence we have the following theorem on linear estimators for regression coefficients:

10.3.2 *Under the conditions of* **10.3.1**, *the minimum variance linear estimators of the regression coefficients β_i are identically the same as the least squares estimators of the β_i.*

The minimum variance approach to the estimation of the β_i is due to Markov (1900), whereas the much earlier least squares method is due to Gauss (1809*a*). The combination of **10.3.1** and **10.3.2** is commonly called the *Gauss-Markov theorem.*

In many regression problems $x_{1\xi} = 1$, $\xi = 1, \ldots, n$, that is the mean of y_ξ is assumed to be $\beta_1 + \beta_2 x_{2\xi} + \cdots + \beta_k x_{k\xi}$. In this case, by applying **10.3.2**, we see that the minimum variance linear estimators b_i for the β_i can be expressed as follows:

$$b_1 = \bar{y} - b_2\bar{x}_2 - \cdots - b_k\bar{x}_k$$

$$b_{i'} = \sum_{j'=2}^{k} A^{i'j'}A_{j'0}, \qquad i' = 2, \ldots, k, \tag{10.3.15}$$

where

$$\bar{y} = \frac{1}{n}\sum_\xi y_\xi, \qquad \bar{x}_{i'} = \frac{1}{n}\sum_\xi x_{i'\xi}, \qquad i' = 2, \ldots, k,$$

and

$$A_{i'j'} = \sum_\xi (x_{i'\xi} - \bar{x}_{i'})(x_{j'\xi} - \bar{x}_{j'}). \qquad i', j' = 2, \ldots, k \tag{10.3.16}$$

$$A_{j'0} = \sum_\xi (x_{j'\xi} - \bar{x}_{j'})(y_\xi - \bar{y})$$

and

$$\|A^{i'j'}\| = \|A_{i'j'}\|^{-1}. \tag{10.3.17}$$

Furthermore, by putting $x_{1\xi} = 1$, $\xi = 1, \ldots, n$, in a_{1i} one finds from (10.3.13) that the covariance matrix of the $b_{i'}$, $i' = 2, \ldots, k$ is

$$\|A^{i'j'}\sigma^2\|, \qquad i', j' = 2, \ldots, k \tag{10.3.18}$$

whereas

$$\begin{aligned} \sigma^2(b_1) &= \left[\frac{1}{n} + \sum_{i',j'=2}^{k} A^{i'j'}\bar{x}_{i'}\,\bar{x}_{j'}\right]\sigma^2 \\ \sigma(b_1, b_{i'}) &= \left[\sum_{j'=2}^{k} A^{i'j'}\bar{x}_{j'}\right]\sigma^2. \end{aligned} \tag{10.3.19}$$

Remarks on Gauss-Markov Theorem and Weighing Problems. Suppose in (10.3.1) that $\beta_1, \ldots, \beta_k$ represent the "true" weights of k objects, say, $o_1, \ldots, o_k$ respectively. Consider weighing various combinations of these objects on a chemical balance (scales with right and left weighing pans). Let $x_{i\xi} = +1$ if o_i is placed on left pan, $x_{i\xi} = -1$ if o_i is placed on right pan, and $x_{i\xi} = 0$ if o_i is not weighed, $i = 1, \ldots, k$. Then $\beta_1 x_{1\xi} + \cdots + \beta_k x_{k\xi}$ is the "true" reading on the ξth weighing of the set of objects in accordance with the configuration determined by $x_{1\xi}, \ldots, x_{k\xi}$. For n different weighings the *weighing design matrix* $\|x_{i\xi}\|$, $i = 1, \ldots, k$, $\xi = 1, \ldots, n$, consists only of -1's, 0's, and $+1$'s.

If the scales are bias-free the "actual" reading y_ξ on the ξth weighing may be considered as a random variable whose mean value is $\beta_1 x_{1\xi} + \cdots + \beta_k x_{k\xi}$. If we assume that $y_1, \ldots, y_n$ are the n random variables one obtains in n weighings, and if we assume these random variables to be independent with equal variances, namely, σ^2, then the minimum variance estimators $b_1, \ldots, b_k$ of the "true" weights $\beta_1, \ldots, \beta_k$ of the objects $o_1, \ldots, o_k$ are given by (10.3.10) and the covariance matrix of these estimators is given by (10.3.13). Note that k is the smallest value of n for which it is always possible to find a weighing design matrix which will yield estimators for all β's. The problem of constructing weighing design matrices so as to provide vector estimators $(b_1, \ldots, b_k)$ for $(\beta_1, \ldots, \beta_k)$ in various "best" senses has been extensively investigated by Hotelling (1944), Kishen (1945), Mood (1946), and others.

(b) Estimator for the Residual Variance σ^2

Now we consider the problem of constructing an unbiased quadratic estimator for the residual variance σ^2. Let

$$z_\xi = y_\xi - \sum_{i=1}^{k} \beta_i x_{i\xi}, \tag{10.3.20}$$

$$\tilde{y}_\xi = \sum_{i=1}^{k} b_i x_{i\xi}. \tag{10.3.21}$$

$$\begin{aligned} S &= \sum_{\xi=1}^{n} z_\xi^2, \qquad S_1 = \sum_{\xi=1}^{n} (y_\xi - \tilde{y}_\xi)^2, \\ S_2 &= \sum_{i,j=1}^{k} a_{ij}(b_i - \beta_i)(b_j - \beta_j). \end{aligned} \tag{10.3.22}$$

It is seen that

(10.3.23) $$S = S_1 + S_2.$$

But

(10.3.24) $$\mathscr{E}(S) = n\sigma^2$$

and making use of the covariance matrix (10.3.13) we have

(10.3.25) $$\mathscr{E}(S_2) = \sum_{i,j=1}^{k} a^{ij}a_{ij}\sigma^2 = k\sigma^2.$$

Since $\mathscr{E}(S) = \mathscr{E}(S_1) + \mathscr{E}(S_2)$ we therefore obtain

(10.3.26) $$\mathscr{E}(S_1) = (n - k)\sigma^2.$$

Since S_1 is free of the parameters $\beta_1, \ldots, \beta_k$ and σ^2, and hence observable, we therefore have the following result:

10.3.3 *Under the conditions of* **10.3.1**, $S_1/(n - k)$ *is an unbiased estimator for* σ^2.

Note that the mean values of S, S_1, and S_2 are respectively, $n\sigma^2$, $(n - k)\sigma^2$, $k\sigma^2$. The numbers n, $n - k$, k are referred to as *degrees of freedom* of S, S_1, and S_2.

If S_1 in (10.3.22) is squared and summed over ξ, we find

(10.3.27) $$S_1 = a_{00} - \sum_{i,j} a_{ij}b_ib_j = a_{00} - \sum_{i,j} a^{ij}a_{i0}a_{j0},$$

that is, S_1 can be written in the relatively simple form

(10.3.27*a*) $$S_1 = \frac{|a_{i_0j_0}|}{|a_{ij}|}$$

where $a_{00} = \sum_{\xi} y_\xi^2$ and

(10.3.28) $$|a_{i_0j_0}| = \begin{vmatrix} a_{00} & a_{01} & \cdots & a_{0k} \\ a_{10} & a_{11} & \cdots & a_{1k} \\ \cdot & \cdot & \cdot & \cdot \\ \cdot & \cdot & \cdot & \cdot \\ \cdot & \cdot & \cdot & \cdot \\ a_{k0} & a_{k1} & \cdots & a_{kk} \end{vmatrix}.$$

The equivalence of (10.3.27) and (10.3.27a) is evident if one performs a bordered expansion of the determinant given in (10.3.28) by the first row and first column [for example, see Bôcher (1907)].

If some of the regression coefficients, say $\beta_{k_1+1}, \beta_{k_1+2}, \ldots, \beta_k$, have known values, we can replace y_ξ by y'_ξ in the preceding paragraphs, where $y'_\xi = (y_\xi - \beta_{k_1+1}x_{k_1+1\xi} - \cdots - \beta_k x_{k\xi})$, and carry all of the analysis through with k replaced by k_1, thus obtaining trivially modified forms of **10.3.1**, **10.3.2**, and **10.3.3**.

(c) Distributions of Regression Estimators in Normal Regression Theory

If we make the further assumption that the random variables y_ξ, $\xi = 1, \ldots, n$, are independent with distributions $N(\beta_1 x_{1\xi} + \cdots + \beta_k x_{k\xi}, \sigma^2)$, we obtain some results of considerable importance in applied statistics. Let z_ξ be defined by (10.3.20). Then the p.e. of the random variables z_ξ, $\xi = 1, \ldots, n$, is

(10.3.29)

$$dF(z_1, \ldots, z_n) = (1/\sqrt{2\pi}\sigma)^n \exp\left(-\frac{1}{2\sigma^2}\sum_\xi z_\xi^2\right) dz_1 \cdots dz_n.$$

Referring to (10.3.22) we find that S_1 and S_2 can be written as

(10.3.30)
$$S_1 = \sum_\xi \left[z_\xi - \sum_{i,j=1}^{k} a^{ij}\left(\sum_{\eta=1}^{n} x_{j\eta} z_\eta\right) x_{i\xi}\right]^2$$
$$S_2 = \sum_{\xi,\eta}\sum_{i,j} a^{ij} x_{i\xi} x_{j\eta} z_\xi z_\eta$$

which are quadratic forms in the z_ξ, $\xi = 1, \ldots, n$ having matrices of ranks $n - k$ and k respectively. Since S is a quadratic form having matrix of rank n in the z_ξ, $\xi = 1, \ldots, n$, it follows by Cochran's theorem **8.4.4** that S_1/σ^2 and S_2/σ^2 are independently distributed according to chi-square distributions $C(n - k)$ and $C(k)$ respectively. Furthermore, we know from (10.3.10) and (10.3.11) that the b_i are linear functions of the y_ξ, namely,

(10.3.31)
$$b_i = \sum_\xi \sum_j a^{ij} x_{j\xi} y_\xi, \qquad i = 1, \ldots, k$$

having means β_i, and covariance matrix given by (10.3.13). Hence, by **7.4.4**, if $\|a_{ij}\|$ is nonsingular, the b_i have the k-dimensional normal distribution $N(\{\beta_i\}, \|a^{ij}\sigma^2\|)$. As a matter of fact, S_1 and $(b_1, \ldots, b_k)$ are two independent sets of random variables under our assumption of normality. This can be established by evaluating the characteristc function $\mathscr{E}(e^{itS_1 + i[t_1 b_1 + \cdots + t_k b_k]})$, of $(S_1, b_1, \ldots, b_k)$ which turns out to be

(10.3.32)
$$(1 - 2\sigma^2 it)^{-\frac{1}{2}(n-k)} \cdot \exp\left\{-\frac{\sigma^2}{2}\sum_{i,j} a^{ij}\, t_i t_j + i \sum_i \beta_i t_i\right\}.$$

Therefore, we have the following important theorem in normal regression theory:

10.3.4 *If the random variables y_ξ, $\xi = 1, \ldots, n$, are independent with distributions $N(\beta_1 x_{1\xi} + \cdots + \beta_k x_{k\xi}, \sigma^2)$, $\xi = 1, \ldots, n$, where the*

matrix $\|a_{ij}\|$, $i, j = 1, \ldots, k$, *defined by* (10.3.8) *is nonsingular, then:*

(i) *The* b_i *defined by* (10.3.10) *are unbiased linear estimators for the regression coefficients* β_i, $i = 1, \ldots, k$, *and have the k-dimensional distribution* $N(\{\beta_i\}, \|a^{ij}\sigma^2\|)$;

(ii) S_1 *and* $(b_1, \ldots, b_k)$ *are independent sets of random variables;*

(iii) S_1/σ^2 *and* S_2/σ^2 *are independent and have chi-square distributions* $C(n - k)$ *and* $C(k)$, *respectively.*

10.4 INTERVAL AND ELLIPSOIDAL ESTIMATORS FOR THE PARAMETERS IN NORMAL REGRESSION THEORY

Since S_1 and $(b_1, \ldots, b_k)$ are two independently distributed sets of random variables under the conditions of **10.3.4**, it is evident that S_1 and any given b_i are independent. But b_i has the distribution $N(\beta_i, a^{ii}\sigma^2)$ and S_1/σ^2 has the chi-square distribution $C(n - k)$. Therefore, from **7.8.3** it follows that for any b_i,

(10.4.1)
$$\frac{(b_i - \beta_i)\sqrt{n - k}}{\sqrt{a^{ii}S_1}}$$

has the "Student" distribution $S(n - k)$, and it follows from an argument similar to that by which the confidence interval (10.2.11) was established that

(10.4.2)
$$b_i \pm (t_{n-k,\gamma})\sqrt{\frac{a^{ii}S_1}{(n - k)}}$$

is a $100\gamma\%$ confidence interval for β_i, where $t_{n-k,\gamma}$ satisfies (10.2.10) with $n - 1$ replaced by $n - k$.

If we set up the confidence intervals (10.4.2) for $i = 1, \ldots, k$ it is evident that the number of confidence intervals covering their respective β's has mean value γk. But this says little about the probability of the confidence intervals *simultaneously* covering their respective β's (unless the b's are independent, which will occur only if $\|a^{ij}\|$ is a diagonal matrix). The question then arises whether one can establish some kind of a simple random region R_γ in the k-dimensional β-space such that the probability is γ that R_γ covers the parameter point $(\beta_1, \ldots, \beta_k)$. In this instance a region can be readily found from the fact that S_1/σ^2 and S_2/σ^2 are independently distributed according to chi-square distributions $C(n - k)$ and $C(k)$, respectively. For it follows from **7.8.5** that

(10.4.3)
$$\frac{(n - k)S_2}{kS_1}$$

has the Snedecor distribution $S(k, n-k)$. Substituting for S_2 from (10.3.22) we have

$$(10.4.4)\quad P\left(\left(\frac{n-k}{kS_1}\right)\sum_{i,j=1}^{k} a_{ij}(b_i-\beta_i)(b_j-\beta_j) < \mathscr{F}_{k,n-k,\gamma}\right) = \int_0^{\mathscr{F}_{k,n-k,\gamma}} dF_{k,n-k}(\mathscr{F}) = \gamma$$

where $dF_{k,n-k}(\mathscr{F})$ is the p.e. of the Snedecor distribution $S(k, n-k)$, whose general form was given by (7.8.9). But (10.4.4) can be stated as follows:

$$(10.4.5)\quad P((\beta_1, \ldots, \beta_k) \in R_\gamma) = \gamma$$

where R_γ is the interior of a random ellipsoid in the β-space centered at $(b_1, \ldots, b_k)$ and having equation

$$(10.4.6)\quad \sum_{i,j=1}^{k} a_{ij}(\beta_i - b_i)(\beta_j - b_j) = \frac{kS_1\mathscr{F}_{k,n-k,\gamma}}{(n-k)}.$$

This ellipsoid is an example of a $100\gamma\%$ *confidence region* for the parameter point $(\beta_1, \ldots, \beta_k)$. We may refer to the ellipsoid (10.4.6) as a *region estimator* for $(\beta_1, \ldots, \beta_k)$. In a similar manner we can set up a region estimator for any subset of the β's. This is left as an exercise for the reader. For a single β, say β_i, the confidence interval estimator (10.4.2) is, of course, an interval (one-dimensional region) estimator.

The notion of a confidence ellipsoid was introduced by Hotelling (1931) in connection with his generalized "Student" distribution which is discussed in Chapter 18.

10.5 SIMULTANEOUS CONFIDENCE INTERVALS: MULTIPLE COMPARISONS

In the preceding discussion we have seen how the vector of sample regression coefficients $(b_1, \ldots, b_k)$ can be used for constructing a confidence ellipsoid for the vector of population regression coefficients $(\beta_1, \ldots, \beta_k)$. In some problems, particularly in analysis of variance problems, such as those to be discussed in subsequent sections it is desirable to establish confidence intervals which hold simultaneously for a large number of linear combinations (with known coefficients) of the components of k-dimensional normal random variables having a known or observable covariance matrix. This problem has been considered by Duncan (1952), Dwass (1959), Roy and Bose (1953), Roy (1954), Scheffé (1953), Tukey (1953), and others.

(a) A Probability Inequality for Simultaneous Confidence Intervals

First we shall consider an inequality for the probability that parameters are simultaneously contained in h respective confidence intervals. Suppose

$\mu_1, \ldots, \mu_h$ are unknown parameters and let $(\underline{\mu}_1, \bar{\mu}_1), \ldots, (\underline{\mu}_h, \bar{\mu}_h)$ be random intervals, each having confidence coefficient $1 - (1 - \gamma)/h$. Let E_i be the event that $(\underline{\mu}_i, \bar{\mu}_i)$ contains μ_i and $\bar{E}_i$ its complement, $i = 1, \ldots, h$. Then we have

$$P(\bar{E}_i) = \frac{1}{h}(1 - \gamma) \qquad i = 1, \ldots, h.$$

The probability that all events $E_1, \ldots, E_h$ occur simultaneously is $P(E_1 \cap \cdots \cap E_h)$. But

$$P(E_1 \cap \cdots \cap E_h) = 1 - P(\bar{E}_1 \cup \cdots \cup \bar{E}_h) \tag{10.5.1}$$

and

$$P(\bar{E}_1 \cup \cdots \cup \bar{E}_h) \leqslant P(\bar{E}_1) + \cdots + P(\bar{E}_h). \tag{10.5.2}$$

Hence

$$P(E_1 \cap \cdots \cap E_h) \geqslant 1 - [P(\bar{E}_1) + \cdots + P(\bar{E}_h)] \tag{10.5.3}$$

that is,

$$P(E_1 \cap \cdots \cap E_h) \geqslant \gamma. \tag{10.5.4}$$

We may summarize in the following result due to Tukey (1953):

10.5.1 *Suppose $\mu_1, \ldots, \mu_h$ are unknown parameters and $(\underline{\mu}_1, \bar{\mu}_1), \ldots, (\underline{\mu}_h, \bar{\mu}_h)$ are $100[1 - \frac{1}{h}(1 - \gamma)]\%$ confidence intervals for $\mu_1, \ldots, \mu_h$ respectively. Then the probability is at least γ that these confidence intervals simultaneously contain $\mu_1, \ldots, \mu_h$ respectively.*

We now consider the problem of the simultaneous fulfillment of large numbers of confidence intervals under some special conditions particularly applicable to the Model I analysis of variance problems to be discussed in subsequent sections.

(b) Scheffé's Method

A basic result due to Scheffé (1953) can be stated in its essentials as follows:

10.5.2 *Suppose $(u_1, \ldots, u_k)$ is a k-dimensional random variable having distribution*

$$N(\{\mu_i\}, \|\sigma^2 a_{ij}\|), \; i, j = 1, \ldots, k,$$

where $\|a_{ij}\|$ is nonsingular, symmetric and known, and σ^2 is unknown. Let v/σ^2 be a random variable independent of $(u_1, \ldots, u_k)$ and *having the chi-square distribution with m degrees of freedom. Let* $\mathcal{F}_{k,m,\gamma}$ *be the* $100\gamma\%$ *point of the Snedecor distribution* $S(k, m)$ *and* $\delta = \sqrt{kv\mathcal{F}_{k,m,\gamma}/m}$. *If* $\mathscr{C}$ *is the set of all real vectors* $(c_1, \ldots, c_k)$, *where* $c_1, \ldots, c_k$ *are not all zero, the probability is* γ *that the inequalities*

$$\sum_i c_i u_i - \delta\sqrt{\sum_{i,j} a_{ij}c_ic_j} \leq \sum_i c_i\mu_i \leq \sum_i c_iu_i + \delta\sqrt{\sum_{i,j} a_{ij}c_ic_j} \tag{10.5.5}$$

hold simultaneously for all $(c_1, \ldots, c_k)$ *in* $\mathscr{C}$.

Note that in the trivial case where $c_1, \ldots, c_k$ are all zero (which is not included in $\mathscr{C}$) (10.5.5) holds with probability 1.

To prove **10.5.2** we note that $\dfrac{1}{\sigma^2}\sum_{i,j} a^{ij}(u_i - \mu_i)(u_j - \mu_j)$ and v/σ^2 are independent random variables having chi-square distributions with k and m degrees of freedom respectively. Hence

$$\frac{m}{kv}\sum_{i,j} a^{ij}(u_i - \mu_i)(u_j - \mu_j)$$

has the Snedecor distribution $S(k, m)$. Therefore if $\mathcal{F}_{k,m,\gamma}$ is the $100\gamma\%$ point of this distribution we have

$$P\left(\sum_{i,j} a^{ij}(u_i - \mu_i)(u_j - \mu_j) < \delta^2\right) = \gamma \tag{10.5.6}$$

where

$$\delta^2 = \frac{kv}{m}\mathcal{F}_{k,m,\gamma}.$$

It will be useful from now on to make use of k-dimensional geometric concepts and terminology.

The set of points in the space of $(\mu_1, \ldots, \mu_k)$ for which

$$\sum_{i,j} a^{ij}(\mu_i - u_i)(\mu_j - u_j) < \delta^2 \tag{10.5.7}$$

is the interior of a $100\gamma\%$ confidence ellipsoid for the true parameter point $(\mu_1, \ldots, \mu_k)$ which is centered at $(u_1, \ldots, u_k)$. If we consider the set of points in the space of $(\mu_1, \ldots, \mu_k)$ contained between all possible pairs of parallel $(k - 1)$-dimensional hyperplanes tangent to this ellipsoid, this set of points constitutes the interior of the ellipsoid (10.5.6) and the probability associated with this set is, of course, γ. It now remains to show that for any particular choice of $(c_1, \ldots, c_k)$ in $\mathscr{C}$ the two parallel $(k - 1)$-dimensional hyperplanes in the space of $(\mu_1, \ldots, \mu_k)$ having equations

$$\sum_i c_i\mu_i = \sum_i c_iu_i \pm \delta\sqrt{\sum_{i,j} a_{ij}c_ic_j} \tag{10.5.8}$$

are tangent to the ellipsoid

$$\sum_{i,j} a^{ij}(\mu_i - u_i)(\mu_j - u_j) = \delta^2. \tag{10.5.9}$$

It is evident that any point $(\mu_1, \ldots, \mu_k)$ between the two hyperplanes (10.5.8) satisfies (10.5.5).

For the moment let $\mu_i - u_i = y_i$. Then (10.5.9) can be written as

$$\sum_{i,j} a^{ij} y_i y_j = \delta^2 \tag{10.5.10}$$

and the equation of an arbitrary hyperplane in the space of $(y_1, \ldots, y_k)$ as

$$\sum_i c_i y_i = d. \tag{10.5.11}$$

We must find the two values of d for which the hyperplane (10.5.11) is tangent to the ellipsoid (10.5.10). Using a Lagrange multiplier λ, we must find the stationary points in the $(y_1, \ldots, y_k)$-space of

$$\Phi = \tfrac{1}{2}\lambda\left(\delta^2 - \sum_{i,j} a^{ij} y_i y_j\right) + \sum_i c_i y_i. \tag{10.5.12}$$

Differentiating with respect to y_j we find

$$-\lambda \sum_i a^{ij} y_i + c_j = 0$$

or

$$y_i = \frac{1}{\lambda} \sum_j a_{ij} c_j. \tag{10.5.13}$$

Substituting in (10.5.10) we find

$$\lambda = \pm \frac{1}{\delta} \sqrt{\sum_{i,j} a_{ij} c_i c_j}. \tag{10.5.14}$$

From (10.5.14), (10.5.13), and (10.5.11), we find

$$d = \pm \delta \sqrt{\sum_{i,j} a_{ij} c_i c_j}.$$

Putting this value of d in (10.5.11) and using the fact that $y_i = \mu_i - u_i$ we obtain (10.5.8) as the equations of the two tangent hyperplanes for specified $(c_1, \ldots, c_k)$.

Finally, note that if we take only a finite number N of the vectors in $\mathscr{C}$, the set of points in the $(\mu_1, \ldots, \mu_k)$-space which lie between *all* N pairs of hyperplanes corresponding to these N vectors is a random k-dimensional polyhedron G_k which circumscribes the ellipsoid (10.5.9). Since G_k contains the ellipsoid, the probability contained in G_k exceeds that in the ellipsoid, namely, γ. Hence, the probability exceeds γ that any finite number N of the inequalities corresponding to N vectors in $\mathscr{C}$ are simultaneously fulfilled. For example, if we take $c_i = 1$, $c_j = -1$ and all other c's equal to 0, $i > j = 1, \ldots, k$, we obtain a subset of $N = k(k-1)/2$ vectors in $\mathscr{C}$. Then the probability exceeds γ that all of the following $k(k-1)/2$ inequalities hold simultaneously

$$(10.5.5a)\quad (u_i - u_j) - \delta\sqrt{a_{ii} + a_{jj} - 2a_{ij}} \leqslant (\mu_i - \mu_j) \leqslant (u_i - u_j) + \delta\sqrt{a_{ii} + a_{jj} - 2a_{ij}}$$
$$i > j = 1, \ldots, k.$$

(c) Tukey's Method

Results similar to those in (10.5.5a) for the case of all possible differences $\mu_i - \mu_j$, $i > j = 1, \ldots, k$, but using a "Studentized range" estimator for σ, have been obtained by Tukey (1953). Suppose $(z_1, \ldots, z_k)$ is a sample of size k from $N(\mu, d^2\sigma^2)$, where d^2 is a known positive constant and let v/σ^2 be a random variable independent of $(z_1, \ldots, z_k)$ and having the chi-square distribution with m degrees of freedom. Let R be the range of $(z_1, \ldots, z_k)$, that is, $R = \max(z_1, \ldots, z_k) - \min(z_1, \ldots, z_k)$. The random variable $\sqrt{m}R/\sqrt{v}\,d$ is called the *Studentized range* $R_{k,m}$. For a given confidence coefficient γ let $R_{k,m,\gamma}$ be the upper $100\gamma\%$ point of the distribution of $R_{k,m}$ defined as follows:

$$(10.5.15)\qquad P(R_{k,m} \leqslant R_{k,m,\gamma}) = \gamma.$$

Tukey's (1953) result can be stated as follows:

10.5.3 *Suppose $(u_1, \ldots, u_k)$ is a k-dimensional random variable having distribution $N(\{\mu_i\}, \|\sigma^2 a_{ij}\|)$ where $a_{ii} = a^2$, $a_{ij} = a_{ji} = \rho a^2$, $i \neq j = 1, \ldots, k$, and ρ being known.*

Let v/σ^2 be a random variable independent of $(u_1, \ldots, u_k)$ and having the chi-square distribution with m degrees of freedom. If $\mathscr{C}_0$ is the set of real vectors $(c_1, \ldots, c_k)$ for which $\sum_i c_i = 0$, and $c_1, \ldots, c_k$ are not all zero, then the probability is γ that the inequalities

$$(10.5.16)\qquad \textstyle\sum_i c_i u_i - D \leqslant \sum_i c_i \mu_i \leqslant \sum_i c_i u_i + D$$

hold simultaneously for all vectors in $\mathscr{C}_0$, where

$$(10.5.17)\qquad D = R_{k,m,\gamma}\sqrt{a^2 v(1-\rho)/m}\,(\tfrac{1}{2}\textstyle\sum_i |c_i|).$$

Note that for the trivial case in which $c_1, \ldots, c_k$ are all zero (which is not included in $\mathscr{C}_0$), (10.5.16) holds with probability 1.

To prove **10.5.3** let $u_i - \mu_i = y_i$, $i = 1, \ldots, k$. Then $(y_1, \ldots, y_k)$ has the distribution $N(\{0\}, \|\sigma^2 a_{ij}\|)$ where the a_{ij} are as defined in **10.5.3**. Let

$w = y_1 + \cdots + y_k$ and $z_i = y_i + hw$, where h is a constant satisfying the quadratic equation (where $-1/(k-1) < \rho < 1$ since $\|a_{ij}\|$ is positive definite)

(10.5.18) $$h^2k[1 + (k-1)\rho] + 2h[1 + (k-1)\rho] + \rho = 0.$$

Then $z_1, \ldots, z_k$ are independent, each having the distribution $N(0, \sigma^2a^2(1-\rho))$. Now consider the inequalities

(10.5.19) $$|z_i - z_j| \leqslant H, \qquad i, j = 1, \ldots, k.$$

These inequalities are all satisfied if and only if the range R of $(z_1, \ldots, z_k)$ satisfies

(10.5.20) $$R \leqslant H.$$

Recalling that v/σ^2 has a chi-square distribution with m degrees of freedom and is independent of $(z_1, \ldots, z_k)$ and using the fact that $(z_1, \ldots, z_k)$ is a sample of size k from $N(0, \sigma^2a^2(1-\rho))$ it is seen that the random variable $R\sqrt{m/[a^2v(1-\rho)]}$ is the Studentized range $R_{k,m}$. Now suppose $(c_1, \ldots, c_k)$ is a vector such that $c_1, \ldots, c_k$ are not all zero, but $\sum_i c_i = 0$. If we let $\sum_i'$ denote summation over all values of i for which $c_i > 0$, and similarly let $\sum_j''$ denote summation over all j for which $(-c_j) > 0$, then we may write

(10.5.21) $$\textstyle\sum_i' c_i = \sum_j'' (-c_j) = \frac{1}{2}\sum_i |c_i| = K.$$

Furthermore

$$\begin{aligned}\textstyle\sum_i c_iz_i &= \textstyle\sum_i' c_iz_i - \sum_j''(-c_j)z_j \\ &= \frac{1}{K}\left[\textstyle\sum_i'\sum_j'' c_i(-c_j)z_i - \sum_i'\sum_j'' c_i(-c_j)z_j\right] \\ &= \frac{1}{K}\textstyle\sum_i'\sum_j'' c_i(-c_j)(z_i - z_j).\end{aligned}$$

If $|z_i - z_j| \leqslant H$, $i, j = 1, \ldots, k$, we have

$$\begin{aligned}\left|\textstyle\sum_i c_iz_i\right| &= \frac{1}{K}\left|\textstyle\sum_i'\sum_j'' c_i(-c_j)(z_i - z_j)\right| \\ &\leqslant \frac{1}{K}\textstyle\sum_i'\sum_j'' c_i(-c_j)\,|z_i - z_j| \leqslant H\cdot K.\end{aligned}$$

That is,

(10.5.22) $$\left|\textstyle\sum_i c_iz_i\right| \leqslant H\cdot(\tfrac{1}{2}\textstyle\sum_i |c_i|)$$

which holds for all real vectors $(c_1, \ldots, c_k)$ in $\mathscr{C}_0$. But (10.5.22) holds for all vectors $(c_1, \ldots, c_k)$ in $\mathscr{C}_0$ if and only if (10.5.19) holds, which holds if and only if (10.5.20) holds. Hence

(10.5.23) $$P\left(\left|\textstyle\sum_i c_iz_i\right| \leqslant H(\tfrac{1}{2}\textstyle\sum_i|c_i|) \text{ for all } (c_1, \ldots, c_k) \text{ in } \mathscr{C}_0\right) = P(R \leqslant H).$$

If $R_{k,m}$ denotes the Studentized range defined in (10.5.15), we have

$$R_{k,m} = \sqrt{\frac{R^2 m}{a^2 v(1-\rho)}}. \tag{10.5.24}$$

Thus if we choose

$$H = R_{k,m,\gamma}\sqrt{\frac{a^2 v(1-\rho)}{m}} \tag{10.5.25}$$

where $R_{k,m,\gamma}$ is defined in (10.5.15), we have

$$P\left(R \leqslant R_{k,m,\gamma}\sqrt{\frac{a^2 v(1-\rho)}{m}}\right) = \gamma \tag{10.5.26}$$

and therefore

$$P\left(\left|\textstyle\sum_i c_i z_i\right| \leqslant D \quad \text{for all } (c_1, \ldots, c_k) \text{ in } \mathscr{C}_0\right) = \gamma \tag{10.5.27}$$

where D is given by (10.5.17).

But

$$\textstyle\sum_i c_i z_i = \sum_i c_i(y_i + hw) = \sum_i c_i(u_i - \mu_i)$$

and hence

$$P\left(\left|\textstyle\sum_i c_i u_i - \sum_i c_i \mu_i\right| \leqslant D \qquad \text{for all } (c_1, \ldots, c_k) \text{ in } \mathscr{C}_0\right) = \gamma. \tag{10.5.28}$$

This is equivalent to stating that the probability is γ that the inequalities (10.5.16) hold simultaneously for all $(c_1, \ldots, c_k)$ in $\mathscr{C}_0$, thus concluding the argument for **10.5.3**.

It should be noted that the probability exceeds γ that for any finite set of vectors in $\mathscr{C}_0$ the corresponding inequalities (10.5.16) hold simultaneously. For instance, the probability exceeds γ that the inequalities

$$(u_i - u_j) - R_{k,m,\gamma}\sqrt{\frac{a^2 v(1-\rho)}{m}} \leqslant (\mu_i - \mu_j) \leqslant (u_i - u_j) + R_{k,m,\gamma}\sqrt{\frac{a^2 v(1-\rho)}{m}},$$

for all choices of i, j, for $i > j = 1, \ldots, k$, hold simultaneously. This, of course, is equivalent to stating that

$$\left((u_i - u_j) \pm R_{k,m,\gamma}\sqrt{\frac{a^2 v(1-\rho)}{m}}\right)$$

are confidence intervals for $(\mu_i - \mu_j)$ respectively, for all $i > j = 1, \ldots, k$, such that the probability exceeds γ that all differences $(\mu_i - \mu_j)$, $i > j = 1, \ldots, k$, are simultaneously contained in their respective confidence intervals.

Dwass (1959) has formulated a generalization of the problem of simultaneous intervals which includes the results of Scheffé and Tukey as special cases. It should be pointed out that the basic idea of simultaneous confidence intervals in the special case of a confidence region for a regression line is due to Working and Hotelling (1929).

10.6 NORMAL LINEAR REGRESSION ANALYSIS IN EXPERIMENTAL DESIGNS

In this section we apply the regression theory of Section 10.3 to a fairly simple stochastic description of experimental designs, sometimes known as the *Model I* description. We shall only consider three of the simpler designs: the *two-factor*, the *three-factor*, and the *Latin Square* designs. The reader interested in fuller treatments of these and other designs and their associated statistical analyses should consult books by Fisher (1935*a*), Cochran and Cox (1957), Graybill (1961), Kempthorne (1946), Mann (1949), and Scheffé (1959). Tables of experimental designs have been prepared by Fisher and Yates (1938) and Kitagawa and Mitome (1953). A guide to the literature of experimental designs has been incorporated in a book by Greenwood and Hartley (1961).

The theory of linear regression analysis has also been applied to what is now called *response surface analysis.* The principal contributors to this type of analysis are Box (1952, 1954), Box and Hunter (1957), and Box and Wilson (1951).

(a) The Complete Two-Factor Experimental Design

In this type of experiment it is assumed that we have a set of rs independent random variables $\{x_{\xi\eta};\ \xi = 1, \ldots, r,\ \eta = 1, \ldots, s\}$, all having equal variances, say σ^2, and having mean values

$$\mathscr{E}(x_{\xi\eta}) = \mu + \mu_{\xi\cdot} + \mu_{\cdot\eta} \tag{10.6.1}$$

where

$$\sum_{\xi=1}^{r} \mu_{\xi\cdot} = \sum_{\eta=1}^{s} \mu_{\cdot\eta} = 0. \tag{10.6.2}$$

Note from (10.6.1) that for every ξ and η, $\xi(x_{\xi\eta})$ under conditions (10.6.2) is a linear function of $r + s - 1$ regression coefficients μ, $\mu_{\xi\cdot}$, $\mu_{\cdot\eta}$ of form (10.3.1) where the fixed variables all have the value 0, 1 or -1.

In this setup we may think of a rectangular array consisting of r *rows* and s *columns*, the r rows being associated with r specified *levels* or *categories* $R_1, \ldots, R_r$ of *factor* R, and the s columns being associated with s given *levels* or *categories* $C_1, \ldots, C_s$ of *factor* C. Then $x_{\xi\eta}$ is a random variable which describes the *response* or *yield* associated with the combination (R_ξ, C_η) of the R and C factors; the set $\{x_{\xi\eta}\}$ will be called *response* random variables.

Remark. To illustrate these ideas more concretely, suppose we have r operators $R_1, \ldots, R_r$ and s machines $C_1, \ldots, C_s$, $s \geqslant r$, producing piece parts of a certain kind. If we allow each operator to operate each machine for an eight-hour day and let $x_{\xi\eta}$ be the output of operator R_ξ from machine C_η, we have a two-factor experiment which could be run in s eight-hour days.

The regression coefficient μ in (10.6.1) is called the *over-all average response level*, $\mu_{\xi\cdot}$ the *differential effect* due to R_ξ and $\mu_{\cdot\eta}$ the *differential effect* due to C_η. We wish to determine minimum variance linear estimators for μ, $\mu_{\xi\cdot}$ and $\mu_{\cdot\eta}$, variances of these estimators, and also an estimator for σ^2.

From the set of random variables $\{x_{\xi\eta}\}$ we define the means $\bar{x}_{\cdot\cdot}$, $\bar{x}_{\xi\cdot}$ and $\bar{x}_{\cdot\eta}$ exactly as in (8.6.7). Furthermore, let

$$\begin{aligned} m &= \bar{x}_{\cdot\cdot} \\ m_{\xi\cdot} &= \bar{x}_{\xi\cdot} - \bar{x}_{\cdot\cdot}, \qquad \xi = 1, \ldots, r \\ m_{\cdot\eta} &= \bar{x}_{\cdot\eta} - \bar{x}_{\cdot\cdot}, \qquad \eta = 1, \ldots, s \end{aligned} \tag{10.6.3}$$

and

$$\begin{aligned} S &= \sum_{\xi,\eta} (x_{\xi\eta} - \mu - \mu_{\xi\cdot} - \mu_{\cdot\eta})^2 \\ S_{\cdot\cdot} &= \sum_{\xi,\eta} (x_{\xi\eta} - m - m_{\xi\cdot} - m_{\cdot\eta})^2 \\ S_{\cdot 0}(\mu_{\xi\cdot}) &= \sum_{\xi,\eta} (m_{\xi\cdot} - \mu_{\xi\cdot})^2 \\ S_{0\cdot}(\mu_{\cdot\eta}) &= \sum_{\xi,\eta} (m_{\cdot\eta} - \mu_{\cdot\eta})^2 \\ S_{00}(\mu) &= \sum_{\xi,\eta} (m - \mu)^2. \end{aligned} \tag{10.6.4}$$

It can be verified by elementary algebra that

$$S = S_{\cdot\cdot} + S_{\cdot 0}(\mu_{\xi\cdot}) + S_{0\cdot}(\mu_{\cdot\eta}) + S_{00}(\mu). \tag{10.6.5}$$

It should be observed that $S_{\cdot\cdot}$ is identically the same as $S_{\cdot\cdot}$ in (8.6.9), whereas $S_{\cdot 0}(0) = S_{\cdot 0}$ and $S_{0\cdot}(0) = S_{0\cdot}$ in (8.6.9). Also note that $S_{\cdot\cdot} + S_{\cdot 0}(0) + S_{0\cdot}(0) = S_T$ where S_T is given in (8.6.9).

We know from **10.3.1** that the minimum variance linear estimators for μ, $\mu_{\xi\cdot}$ and $\mu_{\cdot\eta}$ are those values which minimize S in (10.6.4). This minimization process as will be seen by examining (10.6.4) and applying **10.3.1** and **10.3.2** produces m, $m_{\xi\cdot}$ and $m_{\cdot\eta}$ as the minimum variance linear estimators for μ, $\mu_{\xi\cdot}$, and $\mu_{\cdot\eta}$ respectively. Furthermore, it follows from **10.3.3** that $S_{\cdot\cdot}/(r - 1)(s - 1)$ is an unbiased estimator for σ^2.

We leave it to the reader to verify that the covariance matrix of the estimators m, $m_{\xi\cdot}$ and $m_{\cdot\eta}$ has the following elements:

$$\sigma^2(m) = \frac{\sigma^2}{rs}, \qquad \sigma(m, m_{\xi\cdot}) = \sigma(m, m_{\cdot\eta}) = 0$$

$$\sigma^2(m_{\xi\cdot}) = \frac{r-1}{rs}\sigma^2, \qquad \sigma^2(m_{\cdot\eta}) = \frac{s-1}{rs}\sigma^2 \tag{10.6.6}$$

$$\sigma(m_{\xi\cdot}, m_{\xi'\cdot}) = \sigma(m_{\cdot\eta}, m_{\cdot\eta'}) = -\frac{\sigma^2}{rs}, \qquad \xi \neq \xi', \eta \neq \eta'$$

$$\sigma(m_{\xi\cdot}, m_{\cdot\eta}) = 0.$$

Unbiased estimators of these matrix elements, in turn, are obtained by replacing σ^2 in each instance by $S_{..}/(r-1)(s-1)$. It should be particularly noted that the covariance between m's from any two of the three sets m, $\{m_{\alpha\cdot}\}$, $\{m_{\cdot\beta}\}$ is zero.

In conclusion we have the following result:

10.6.1 *Let $\{x_{\xi\eta};\ \xi = 1, \ldots, r,\ \eta = 1, \ldots, s\}$ be a set of independent random variables all having the same variance σ^2 and means $\mathscr{E}(x_{\xi\eta}) = \mu + \mu_{\xi\cdot} + \mu_{\cdot\eta}$ where $\sum_\xi \mu_{\xi\cdot} = \sum_\eta \mu_{\cdot\eta} = 0$. Minimum variance linear estimators for μ, $\mu_{\xi\cdot}$, $\mu_{\cdot\eta}$ are m, $m_{\xi\cdot}$, $m_{\cdot\eta}$ respectively, whereas $S_{..}/(r-1)(s-1)$ is an unbiased estimator for σ^2, where $S_{..}$ is defined in* (10.6.4). *The covariance between m's from any two of the three sets of random variables m, $\{m_{\xi\cdot}\}$, $\{m_{\cdot\eta}\}$ is zero. The covariance matrix among all m's has elements given by* (10.6.6).

If the additional assumption is made that the set of random variables $\{x_{\xi\eta};\ \xi = 1, \ldots, r,\ \eta = 1, \ldots, s\}$ are independently distributed according to the normal distributions $N(\mu + \mu_{\xi\cdot} + \mu_{\cdot\eta}, \sigma^2)$, the following result can be established by argument similar to that used in arriving at **10.3.4**:

10.6.2 *Let $\{x_{\xi\eta};\ \xi = 1, \ldots, r,\ \eta = 1, \ldots, s\}$ be a set of independent random variables having distributions $N(\mu + \mu_{\xi\cdot} + \mu_{\cdot\eta}, \sigma^2)$ where $\sum_\xi \mu_{\xi\cdot} = \sum_\eta \mu_{\cdot\eta} = 0$. Then m, $\{m_{\xi\cdot}\}$ $\{m_{\cdot\eta}\}$ are three independent normally distributed sets of random variables with means μ, $\{\mu_{\xi\cdot}\}$, $\{\mu_{\cdot\eta}\}$ respectively, and with covariance matrices given by* (10.6.6), *the two latter distributions being degenerate and subject to the restrictions $\sum_\xi m_{\xi\cdot} = 0$, $\sum_\eta m_{\cdot\eta} = 0$ respectively. Furthermore, $S_{..}/\sigma^2$, $S_{\cdot 0}(\mu_{\xi\cdot})/\sigma^2$, $S_{0\cdot}(\mu_{\cdot\eta})/\sigma^2$, $S_{00}(\mu)/\sigma^2$ are independent random variables having chi-square distributions $C((r-1)(s-1))$, $C(r-1)$, $C(s-1)$, $C(1)$ respectively.*

From this result, and by using the Student distribution, we can write down confidence intervals for any of the constants μ; $\mu_{1\cdot}, \ldots, \mu_{r\cdot}$; $\mu_{\cdot 1}, \ldots, \mu_{\cdot s}$. For instance, $100\gamma\%$ confidence intervals for μ, $\mu_{\xi\cdot}$ and $\mu_{\cdot\eta}$ are

$$m \pm t_{(r-1)(s-1),\gamma} \cdot \sqrt{\frac{S_{\cdot\cdot}}{rs(r-1)(s-1)}}$$

(10.6.7) $$m_{\xi\cdot} \pm t_{(r-1)(s-1),\gamma} \cdot \sqrt{\frac{S_{\cdot\cdot}}{rs(s-1)}}$$

$$m_{\cdot\eta} \pm t_{(r-1)(s-1),\gamma} \cdot \sqrt{\frac{S_{\cdot\cdot}}{rs(r-1)}}$$

respectively. We can similarly set up confidence intervals for the differences such as $\mu_{\xi\cdot} - \mu_{\xi'\cdot}$, $\xi \neq \xi'$ or other linear functions of the μ's. Also by using **10.5.2** or **10.5.3** we can set up sets of confidence intervals for $\mu_{\xi\cdot} - \mu_{\xi'\cdot}$, $\xi > \xi' = 1, \ldots, r$, (or for $\mu_{\cdot\eta} - \mu_{\cdot\eta'}$, $\eta > \eta' = 1, \ldots, s$) which hold simultaneously with probability at least γ.

Also, confidence regions similar to those defined by (10.4.6) can be set up for simultaneous estimation of two or more of the μ's. A particularly important case arises when it is desired to estimate the $\mu_{\xi\cdot}$ (or the $\mu_{\cdot\eta}$), simultaneously. Suppose we wish to estimate the $\mu_{\xi\cdot}$ simultaneously. Since, as stated in **10.6.2**, $S_{\cdot 0}(\mu_{\xi\cdot})/\sigma^2$ and $S_{\cdot\cdot}/\sigma^2$ are independently distributed according to chi-square distributions $C(r-1)$ and $C((r-1)(s-1))$, it follows from **7.8.5** that $(s-1)S_{\cdot 0}(\mu_{\xi\cdot})/S_{\cdot\cdot}$ has the Snedecor distribution $S((r-1), (r-1)(s-1))$. Therefore, we have, using the notation of (10.4.6),

(10.6.8) $$P\left(\frac{(s-1)S_{\cdot 0}(\mu_{\xi\cdot})}{S_{\cdot\cdot}} < \mathscr{F}_{r-1,(r-1)(s-1),\gamma}\right) = \gamma.$$

But this can be stated as follows:

(10.6.9) $$P((\mu_{1\cdot}, \ldots, \mu_{r\cdot}) \in R_\gamma) = \gamma$$

where R_γ is the $(r-1)$-dimensional $100\gamma\%$ *confidence sphere* for the point $(\mu_{1\cdot}, \ldots, \mu_{r\cdot})$, having equation

(10.6.10) $$\sum_{\xi,\eta} (\mu_{\xi\cdot} - m_{\xi\cdot})^2 = \frac{S_{\cdot\cdot}}{s-1} \mathscr{F}_{r-1,(r-1)(s-1),\gamma}$$

subject to $\sum_{\xi,\eta} (\mu_{\xi\cdot} - m_{\xi\cdot}) = 0$. If we are particularly interested in the possibility that the $\mu_{\xi\cdot}$ are all zero, we see whether the sphere having equation (10.6.10) includes the point $(0, \ldots, 0)$. This reduces merely to checking whether the inequality

(10.6.11) $$\frac{(s-1)S_{\cdot 0}(0)}{S_{\cdot\cdot}} < \mathscr{F}_{r-1,(r-1)(s-1),\gamma}$$

holds. If it does hold, we say that the set of random variables $\{x_{\xi\eta}\}$ *support the statistical hypothesis* that the $\mu_{\xi\cdot}$ are all zero at the $100\gamma\%$ *confidence level*. If (10.6.11) holds, an alternative statement is to say that the $m_{\xi\cdot}$ (the estimators for the $\mu_{\xi\cdot}$) are *not significantly different from zero at the* $100(1 - \gamma)\%$ *level of significance*. The important point is that the random variable used for making the test, namely $(s - 1)S_{\cdot 0}(0)/S_{\cdot\cdot}$ is observable.

In a similar manner we test whether the set of random variables $\{x_{\xi\eta}\}$ support the statistical hypothesis that the $\mu_{\cdot\eta}$ are all zero.

It is customary to set up the constituents of the Model I description of the complete two-factor design under the assumption of normality into an *analysis of variance table* as shown in Table 10.1, remembering that $S_{\cdot 0}(0) = S_{\cdot 0}$ and $S_{0\cdot}(0) = S_{0\cdot}$.

TABLE 10.1 MODEL I ANALYSIS OF VARIANCE TABLE FOR COMPLETE TWO-FACTOR EXPERIMENTAL DESIGN

Source of Variation	Degrees of Freedom (D.F.)	Sum of Squares (S.S.)	Mean Sum of Squares (M.S.S.)	Snedecor $\mathscr{F}$-Ratio
Rows	$r - 1$	$S_{\cdot 0}$	$\dfrac{S_{\cdot 0}}{(r-1)}$	$(s-1)S_{\cdot 0}/S_{\cdot\cdot} = \mathscr{F}_{\cdot 0}$
Columns	$s - 1$	$S_{0\cdot}$	$\dfrac{S_{0\cdot}}{(s-1)}$	$(r-1)S_{0\cdot}/S_{\cdot\cdot} = \mathscr{F}_{0\cdot}$
Residuals (error)	$(r-1)(s-1)$	$S_{\cdot\cdot}$	$\dfrac{S_{\cdot\cdot}}{(r-1)(s-1)}$	
Total	$rs - 1$	S_T		

The first Snedecor ratio $\mathscr{F}_{\cdot 0}$ is to test the hypothesis that the $\mu_{\xi\cdot}$ are all zero, and the second is to test the hypothesis that the $\mu_{\cdot\eta}$ are all zero.

The arrangement of analysis of variance constituents into table form, such as Table 10.1, is due to Fisher (1925*a*), who also developed the theory and application of Model I analysis of variance procedures.

(b) The Complete Three-Factor Experimental Design

The ideas of Section 10.6(*a*) can be extended in a straightforward fashion to higher order designs, that is, experimental layouts involving three or more factors. It is perhaps worthwhile to show how the extension goes for three factors. Here we have a set of *rst* independent random variables

$\{x_{\xi\eta\zeta};\ \xi = 1, \ldots, r;\ \eta = 1, \ldots, s;\ \zeta = 1, \ldots, t\}$ whose variances are all equal to σ^2 and whose mean values are given by

$$\text{(10.6.12)} \quad \mathscr{E}(x_{\xi\eta\zeta}) = \mu + \mu_{\xi\cdot\cdot} + \mu_{\cdot\eta\cdot} + \mu_{\cdot\cdot\zeta} + \mu_{\xi\eta\cdot} + \mu_{\xi\cdot\zeta} + \mu_{\cdot\eta\zeta}$$

where

$$\sum_{\zeta} \mu_{\xi\cdot\cdot} = \sum_{\eta} \mu_{\cdot\eta\cdot} = \sum_{\zeta} \mu_{\cdot\cdot\zeta} = 0,$$
$$\sum_{\xi} \mu_{\xi\eta\cdot} = \sum_{\eta} \mu_{\xi\eta\cdot} = 0, \qquad \sum_{\xi} \mu_{\xi\cdot\zeta} = \sum_{\zeta} \mu_{\xi\cdot\zeta} = 0$$
$$\sum_{\eta} \mu_{\cdot\eta\zeta} = \sum_{\zeta} \mu_{\cdot\eta\zeta} = 0.$$

In this type of experimental layout we may think of r *rows*, s *columns*, and t *layers*, the rows being associated with r given levels $R_1, \ldots, R_r$ of a factor R, columns being associated with s given levels $C_1, \ldots, C_s$ of factor C, and layers associated with t given levels $L_1, \ldots, L_t$ of factor L. The constant $\mu_{\xi\cdot\cdot}$ is the *differential effect* due to R_ξ with similar interpretations for $\mu_{\cdot\eta\cdot}$ and $\mu_{\cdot\cdot\zeta}$, whereas $\mu_{\xi\eta\cdot}$ is the differential effect associated with (R_ξ, C_η) or the *interaction* between R_ξ and C_η, with similar interpretations for $\mu_{\xi\cdot\zeta}$ and $\mu_{\cdot\eta\zeta}$.

To estimate the various μ's we define $\bar{x}_{\cdots}$, $\bar{x}_{\xi\cdot\cdot}$, $\bar{x}_{\cdot\eta\cdot}$, $\bar{x}_{\cdot\cdot\zeta}$, $\bar{x}_{\xi\eta\cdot}$, $\bar{x}_{\xi\cdot\zeta}$, and $\bar{x}_{\cdot\eta\zeta}$ by obvious extension of (8.6.7). Also, we define

$$m = \bar{x}_{\cdots}$$
$$\text{(10.6.13)} \quad m_{\xi\cdot\cdot} = \bar{x}_{\xi\cdot\cdot} - \bar{x}_{\cdots}, \qquad \text{similarly, for } m_{\cdot\eta\cdot} \text{ and } m_{\cdot\cdot\zeta},$$
$$m_{\xi\eta\cdot} = \bar{x}_{\xi\eta\cdot} - \bar{x}_{\xi\cdot\cdot} - \bar{x}_{\cdot\eta\cdot} + \bar{x}_{\cdots}, \qquad \text{similarly for } m_{\xi\cdot\zeta} \text{ and } m_{\cdot\eta\zeta}.$$

Furthermore, let

$$S = \sum_{\xi,\eta,\zeta} (x_{\xi\eta\zeta} - \mu - \mu_{\xi\cdot\cdot} - \mu_{\cdot\eta\cdot} - \mu_{\cdot\cdot\zeta} - \mu_{\xi\eta\cdot} - \mu_{\xi\cdot\zeta} - \mu_{\cdot\eta\zeta})^2,$$
$$S_{\cdots} = \sum_{\xi,\eta,\zeta} (x_{\xi\eta\zeta} - m - m_{\xi\cdot\cdot} - m_{\cdot\eta\cdot} - m_{\cdot\cdot\zeta} - m_{\xi\eta\cdot} - m_{\xi\cdot\zeta} - m_{\cdot\eta\zeta})^2,$$

(10.6.14)

$$S_{\cdot\cdot 0}(\mu_{\xi\eta\cdot}) = \sum_{\xi,\eta,\zeta} (m_{\xi\eta\cdot} - \mu_{\xi\eta\cdot})^2, \qquad \text{similarly for } S_{\cdot 0\cdot}(\mu_{\xi\cdot\zeta}) \text{ and } S_{0\cdot\cdot}(\mu_{\cdot\eta\zeta}),$$
$$S_{\cdot 00}(\mu_{\xi\cdot\cdot}) = \sum_{\xi,\eta,\zeta} (m_{\xi\cdot\cdot} - \mu_{\xi\cdot\cdot})^2, \qquad \text{similarly for } S_{0\cdot 0}(\mu_{\cdot\eta\cdot}) \text{ and } S_{00\cdot}(\mu_{\cdot\cdot\zeta}),$$
$$S_{000}(\mu) = \sum_{\xi,\eta,\zeta} (m - \mu)^2.$$

Note that $S_{\cdots}$ is exactly as defined in (8.6.31), whereas $S_{\cdot\cdot 0}(0) = S_{\cdot\cdot 0}$, $S_{\cdot 00}(0) = S_{\cdot 00}$, etc., where $S_{\cdot\cdot 0}$, $S_{\cdot 00}$, etc., are defined in (8.6.31). Then we have the following decomposition for S:

$$\text{(10.6.15)} \quad S = S_{\cdots} + S_{\cdot\cdot 0}(\mu_{\xi\eta\cdot}) + S_{\cdot 0\cdot}(\mu_{\xi\cdot\eta}) + S_{0\cdot\cdot}(\mu_{\cdot\eta\zeta})$$
$$+ S_{\cdot 00}(\mu_{\xi\cdot\cdot}) + S_{0\cdot 0}(\mu_{\cdot\eta\cdot}) + S_{00\cdot}(\mu_{\cdot\cdot\zeta}) + S_{000}(\mu).$$

If we make use of **10.3.1** it is evident that the minimum variance linear estimators for μ, $\mu_{\xi\cdot\cdot}$, $\mu_{\cdot\eta\cdot}$, $\mu_{\cdot\cdot\zeta}$, $\mu_{\xi\eta\cdot}$, $\mu_{\xi\cdot\zeta}$, $\mu_{\cdot\eta\zeta}$ are m, $m_{\xi\cdot\cdot}$, $m_{\cdot\eta\cdot}$, $m_{\cdot\cdot\zeta}$, $m_{\xi\eta\cdot}$, $m_{\xi\cdot\zeta}$, $m_{\cdot\eta\zeta}$ respectively. Furthermore, if we use **10.3.3** we find that $S_{\cdots}/[(r-1)(s-1)(t-1)]$ is an unbiased estimator for σ^2

It is sufficient to list only the following nonzero elements in the covariance matrix of the m's since the remaining nonzero elements can be written down by considerations of symmetry:

$$\sigma^2(m) = \frac{\sigma^2}{rst}, \qquad \sigma^2(m_{\xi\cdot\cdot}) = \frac{(r-1)\sigma^2}{rst}$$

$$\sigma(m_{\xi\cdot\cdot}, m_{\xi'\cdot\cdot}) = -\frac{\sigma^2}{rst}, \qquad \xi \neq \xi'$$

$$\sigma^2(m_{\xi\eta\cdot}) = \frac{(r-1)(s-1)\sigma^2}{rst}, \tag{10.6.16}$$

$$\sigma(m_{\xi\eta\cdot}, m_{\xi\eta'\cdot}) = \frac{-(r-1)\sigma^2}{rst}, \qquad \eta \neq \eta'$$

$$\sigma(m_{\xi\eta\cdot}, m_{\xi'\eta'\cdot}) = -\frac{\sigma^2}{rst}, \qquad \xi \neq \xi', \eta \neq \eta'.$$

The covariance between m's from any two of the sets m, $\{m_{\xi\cdot\cdot}\}$, $\{m_{\cdot\eta\cdot}\}$, $\{m_{\cdot\cdot\zeta}\}$, $\{m_{\xi\eta\cdot}\}$, $\{m_{\xi\cdot\zeta}\}$, $\{m_{\cdot\eta\zeta}\}$ is zero.

Substituting $S_{\cdots}/[(r-1)(s-1)(t-1)]$ for σ^2 in (10.6.16)] provides unbiased estimators for the variances and covariances defined in (10.6.16).

Formulation of statements similar to **10.6.1** and **10.6.2** for the three-factor experiment are left to the reader.

In extending **10.6.2** to the three-factor experimental design we point out that if the random variables in the set $\{x_{\xi\eta\zeta}\}$ are independent and have normal distributions

$$N(\mu + \mu_{\xi\cdot\cdot} + \mu_{\cdot\eta\cdot} + \mu_{\cdot\cdot\zeta} + \mu_{\xi\eta\cdot} + \mu_{\xi\cdot\zeta} + \mu_{\cdot\eta\zeta}, \sigma^2) \tag{10.6.17}$$

then

$$\frac{S_{\cdots}}{\sigma^2}, \quad \frac{S_{\cdot\cdot 0}(\mu_{\xi\eta\cdot})}{\sigma^2}, \quad \frac{S_{\cdot 0\cdot}(\mu_{\xi\cdot\zeta})}{\sigma^2}, \quad \frac{S_{0\cdot\cdot}(\mu_{\cdot\eta\zeta})}{\sigma^2}, \tag{10.6.18}$$

$$\frac{S_{\cdot 00}(\mu_{\xi\cdot\cdot})}{\sigma^2}, \quad \frac{S_{0\cdot 0}(\mu_{\cdot\eta\cdot})}{\sigma^2}, \quad \frac{S_{00\cdot}(\mu_{\cdot\cdot\zeta})}{\sigma^2}, \quad \frac{S_{000}(\mu)}{\sigma^2}$$

are independently distributed according to chi-square distributions with

$$(r-1)(s-1)(t-1), \quad (r-1)(s-1), \quad (r-1)(t-1), \tag{10.6.19}$$
$$(s-1)(t-1), \quad (r-1), \quad (s-1), \quad (t-1), \quad 1$$

degrees of freedom respectively. Here we can set up confidence intervals for individual μ's similar to those given in (10.6.7) for two-factor experimental designs, and confidence regions similar to those in (10.6.10) for the simultaneous estimation of several μ's. By using **10.5.2** or **10.5.3**, we can also set up simultaneous confidence intervals for sets of differences such as $\{\mu_{\xi\cdot\cdot} - \mu_{\xi'\cdot\cdot};\ \xi > \xi' = 1, \ldots, r\}$, $\{\mu_{\xi\eta\cdot} - \mu_{\xi'\eta};\ \xi > \xi' = 1, \ldots, r$ and η fixed$\}$, and so on.

We leave it to the reader to set up a Model I analysis of variance table similar to Table 10.1 for the complete three-factor design.

(c) The Latin Square Experimental Design

The complete three-factor experimental design described above involves random variables, which, in some applications may constitute a prohibitive amount of experimentation. Reduction of the amount of experimentation can be achieved by balanced incomplete three-factor designs. The simplest of these designs and the only one we shall consider is the Latin Square design, which selects a balanced incomplete matrix sample $\{x_{\xi\eta\zeta}\}^*$ of response random variables from a three-way sample matrix $\{x_{\xi\eta\zeta}\}$ of r rows, r columns, and r layers.

In the Latin Square design we consider three factors R, C, and L and r levels of each of the three factors, $R_1, \ldots, R_r;\ C_1, \ldots, C_r;\ L_1, \ldots, L_r$. We may associate R_ξ with the ξth *row*, C_η with the ηth *column* and L_ζ with the ζth *layer* of a three-dimensional rectangular array of cells. The sample of random variables $\{x_{\xi\eta\zeta}\}^*$ for our Latin Square design is a balanced incomplete matrix sample having r^2 components from the matrix sample $\{x_{\xi\eta\zeta}\}$ associated with this $r \times r \times r$ array of cells; $\{x_{\xi\eta\zeta}\}^*$ is selected so that for each combination of values of (ξ, η) there is exactly one random variable, for each combination of values of (ξ, ζ) exactly one random variable, and for each combination of values of (η, ξ) exactly one random variable (see Section 8.6*d*). Let the set of r^2 values of (ξ, η, ξ) thus selected be G^*.

Now it is assumed that the r^2 elements in the sample $\{x_{\xi\eta\zeta}\}^*$, are independent with normal distributions $N(\mathscr{E}(x_{\xi\eta\zeta}), \sigma^2)$ where for any $(\xi, \eta, \zeta) \in G^*$, we have

(10.6.20) $$\mathscr{E}(x_{\xi\eta\zeta}) = \mu + \mu_{\xi\cdot\cdot} + \mu_{\cdot\eta\cdot} + \mu_{\cdot\cdot\zeta}$$

where

(10.6.21) $$\sum_\xi \mu_{\xi\cdot\cdot} = \sum_\eta \mu_{\cdot\eta\cdot} = \sum_\zeta \mu_{\cdot\cdot\zeta} = 0.$$

Comparing this with (10.6.12) it is to be particularly noted that we assume all second-order interactions $\mu_{\xi\eta\cdot}$, $\mu_{\xi\cdot\zeta}$, $\mu_{\cdot\eta\zeta}$ to be zero.

The Latin Square design provides relatively simple minimum variance linear estimators for the μ's in (10.6.20). To see this let $\bar{x}^*_{\cdot\cdot\cdot}$, $\bar{x}^*_{\xi\cdot\cdot}$, $\bar{x}^*_{\cdot\eta\cdot}$, $\bar{x}^*_{\cdot\cdot\zeta}$ be defined as in (8.6.40). Also, let m^*, $m^*_{\xi\cdot\cdot}$, $m^*_{\cdot\eta\cdot}$, $m^*_{\cdot\cdot\zeta}$ be defined in terms of the x's in $\{x_{\zeta\xi\eta}\}^*$ in exactly the same way that m, $m_{\xi\cdot\cdot}$, $m_{\cdot\eta\cdot}$, $m_{\cdot\cdot\zeta}$

are defined in terms of the x's of $\{x_{\xi\eta\zeta}\}$ in (10.6.13). Furthermore, let $S^*_{\cdot 00}(\mu_{\xi\cdot\cdot})$, $S^*_{0\cdot 0}(\mu_{\cdot\eta\cdot})$, $S^*_{00\cdot}(\mu_{\cdot\cdot\zeta})$, $S^*_{000}(\mu)$ be defined as in (10.6.14) with m's replaced by their corresponding m^*'s, the summation being performed for all $(\xi, \eta, \zeta) \in G^*$, of course. Also, let

$$S^* = \sum_{\xi,\eta,\zeta}{}^{*} (x_{\xi\eta\zeta} - \mu - \mu_{\xi\cdot\cdot} - \mu_{\cdot\eta\cdot} - \mu_{\cdot\cdot\zeta})^2 \tag{10.6.22}$$
$$S^*_{\cdots} = \sum_{\xi,\eta,\zeta}{}^{*} (x_{\xi\eta\zeta} - m^* - m^*_{\xi\cdot\cdot} - m^*_{\cdot\eta\cdot} - m^*_{\cdot\cdot\zeta})^2$$

where, as in (8.6.40), $\sum\limits_{\xi,\eta,\zeta}^{*}$ denotes summation over all $(\xi, \eta, \zeta) \in G^*$.

Now we have the following decomposition of S^*,

$$S^* = S^*_{\cdots} + S^*_{\cdot 00}(\mu_{\xi\cdot\cdot}) + S^*_{0\cdot 0}(\mu_{\cdot\eta\cdot}) + S^*_{00\cdot}(\mu_{\cdot\cdot\zeta}) + S^*_{000}(\mu), \tag{10.6.23}$$

where the components on the right have $(r - 1)(r - 2)$, $(r - 1)$, $(r - 1)$, $(r - 1)$, 1 degrees of freedom respectively. Applying **10.3.2**, it is evident that m^*, $m^*_{\xi\cdot\cdot}$, $m^*_{\cdot\eta\cdot}$, $m^*_{\cdot\cdot\zeta}$ are minimum variance linear estimators for μ, $\mu_{\xi\cdot\cdot}$, $\mu_{\cdot\eta\cdot}$, $\mu_{\cdot\cdot\zeta}$ respectively. Furthermore, $S^*_{\cdots}/(r - 1)(r - 2)$ is an unbiased estimator for σ^2.

The covariance matrix for the estimators m^*, $m^*_{\xi\cdot\cdot}$, $m^*_{\cdot\eta\cdot}$, $m^*_{\cdot\cdot\zeta}$ has the following nonzero elements:

$$\sigma^2(m^*) = \frac{\sigma^2}{r^2},$$
$$\sigma^2(m^*_{\xi\cdot\cdot}) = \sigma^2(m^*_{\cdot\eta\cdot}) = \sigma^2(m^*_{\cdot\cdot\zeta}) = \frac{r-1}{r^2}\sigma^2 \tag{10.6.24}$$
$$\sigma(m^*_{\xi\cdot\cdot}, m^*_{\xi'\cdot\cdot}) = \sigma(m^*_{\cdot\eta\cdot}, m^*_{\cdot\eta'\cdot}) = \sigma(m^*_{\cdot\cdot\zeta}, m^*_{\cdot\cdot\zeta'}) = -\frac{\sigma^2}{r^2}$$
$$\text{for } \xi \neq \xi', \quad \eta \neq \eta', \quad \zeta \neq \zeta'.$$

Unbiased estimators for these variances and covariances are obtained by replacing σ^2 by $S^*_{\cdots}/(r - 1)(r - 2)$.

Formulation of summary statements similar to **10.6.1** and **10.6.2** for the Latin Square design is left to the reader. We also leave it to the reader to set up a Model I analysis of variance table similar to Table 10.1 for the Latin Square.

10.7 ESTIMATION OF VARIANCE COMPONENTS FROM LINEAR COMBINATIONS OF RANDOM VARIABLES

In many statistical problems, particularly in such fields as the design of experiments, error theory, and psychological factor analysis, there are situations where only linear combinations of certain random variables having zero covariances are observable, whereas the random variables themselves are not observable. An example of such a linear combination of random variables is that given in (8.6.3), where $x(u, v)$ is an observable random variable which is expressed as a linear function of μ and the unobservable random variables $\varepsilon_{u\cdot}$, $\varepsilon_{\cdot v}$ and ε_{uv}.

In such situations the basic problem is to devise estimators for the variances of the component random variables from the observable linear combinations of these unobservable random variables.

More precisely, let x_ξ, $\xi = 1, \ldots, n$ be observable random variables such that x_ξ has the following form:

$$x_\xi = \mu + \sum_{g=1}^{m} a_{g\xi}\varepsilon_{g\xi}, \qquad \xi = 1, \ldots, n \geqslant m \tag{10.7.1}$$

where

$$\begin{aligned} &\mathscr{E}(\varepsilon_{g\xi}) = 0; \qquad \mathscr{E}(\varepsilon_{g\xi}^2) = \sigma_g^2; \qquad g = 1, \ldots, m; \ \xi = 1, \ldots, n \\ &\mathscr{E}(\varepsilon_{g\xi}\varepsilon_{g\xi'}) = \mathscr{E}(\varepsilon_{g\xi}\varepsilon_{g'\xi}) = 0, \qquad g \neq g', \qquad \xi \neq \xi', \end{aligned} \tag{10.7.2}$$

and where the $a_{g\xi}$ are known constants and $\|a_{g\xi}\|$ is of rank m. It is not assumed that the $\varepsilon_{g\xi}$ are observable, that is, they may be functions of unknown parameters.

Now it is evident that the mean of the sample, $\bar{x}$, is an unbiased estimator for μ. The variance of this estimator is

$$\sigma^2(\bar{x}) = \frac{1}{n^2}\sum_g \sum_\xi a_{g\xi}^2\sigma_g^2. \tag{10.7.3}$$

Also, if s^2 is the sample variance, we have

$$\mathscr{E}(s^2) = \frac{1}{n}\sum_g \sum_\xi a_{g\xi}^2\sigma_g^2. \tag{10.7.4}$$

Note that s^2 is an unbiased estimator for the quantity $(1/n) \sum_g\sum_\xi a_{g\xi}^2\sigma_g^2$; and hence that s^2/n is an unbiased estimator for $\sigma^2(\bar{x})$. But we wish to estimate the individual variance components $\sigma_1^2, \ldots, \sigma_m^2$ from the sample $(x_1, \ldots, x_m)$. To do this consider the total sum of squares

$$S_T = \sum_\xi (x_\xi - \bar{x})^2. \tag{10.7.5}$$

On the right we have a quadratic form in the random variables $(x_1 - \bar{x}), \ldots,$ $(x_n - \bar{x})$ which can be expressed as one in $(x_1 - \mu), \ldots, (x_n - \mu)$, having a matrix of rank $n - 1$. If $n - 1 \geqslant m$, it can be shown by the elementary theory of positive semidefinite quadratic forms that S_T can be decomposed (in many ways) as follows:

$$S_T = S_1 + \cdots + S_m \tag{10.7.6}$$

where $S_1, \ldots, S_m$ are positive semidefinite quadratic forms in $(x_1 - \bar{x})$, $\ldots, (x_n - \bar{x})$ which can be expressed as quadratic forms in $(x_1 - \mu), \ldots,$ $(x_n - \mu)$ having matrices of ranks $n_1, \ldots, n_m$ (all $\geqslant 1$) respectively, where $n_1 + \cdots + n_m = n - 1$. Then we may write for $h = 1, \ldots, m$

$$S_h = \sum_{\xi,\eta} A_{\xi\eta}^{(h)}(x_\xi - \mu)(x_\eta - \mu) \tag{10.7.7}$$

where $\|A_{\xi\eta}^{(h)}\|$ is a symmetric matrix of rank n_h whose elements are known numbers. Substituting $(x_\xi - \mu)$ and $(x_\eta - \mu)$ from (10.7.1) into (10.7.7) and taking mean values we find

$$\mathscr{E}(S_h) = \sum_g B_{hg}\sigma_g^2 \tag{10.7.8}$$

where

$$B_{hg} = \sum_\xi A_{\xi\xi}^{(h)} a_{g\xi}^2. \tag{10.7.9}$$

Now it follows from the properties of $\|A_{\xi\eta}^{(h)}\|$ together with our assumptions about $\|a_{g\xi}\|$ that $\|B_{hg}\|$ is nonsingular. Therefore, applying **10.1.1**, we have from (10.7.8) the following unbiased estimators for $\sigma_1^2, \ldots, \sigma_m^2$:

$$\mathscr{E}^{-1}(\sigma_g^2) = \sum_h B^{gh} S_h. \tag{10.7.10}$$

An unbiased estimator for $\sigma^2(\bar{x})$ is obtained by replacing σ_g^2 in (10.7.3) by $\mathscr{E}^{-1}(\sigma_g^2)$, $g = 1, \ldots, m$.

To summarize:

10.7.1 *Suppose x_ξ, $\xi = 1, \ldots, n \geqslant m$ are observable random variables such that*

$$x_\xi = \mu + \sum_{g=1}^m a_{g\xi}\varepsilon_{g\xi},$$

the $\varepsilon_{g\xi}$ being random variables whose means and covariances are all zero and such that $\mathscr{E}(\varepsilon_{g\xi}^2) = \sigma_g^2$, where μ and the σ_g^2 are unknown parameters and the $a_{g\xi}$ are known numbers such that $\|a_{g\xi}\|$ is of rank m. Let $\bar{x}$ be the mean of $(x_1, \ldots, x_n)$ and let $S_T, S_1, \ldots, S_m$ be quadratic forms in $(x_1 - \bar{x}), \ldots, (x_n - \bar{x})$ as defined in (10.7.5) *and* (10.7.6). *Then $\bar{x}$ is an unbiased estimator for μ whereas*

$$\mathscr{E}^{-1}(\sigma_g^2) = \sum_h B^{gh} S_h$$

are unbiased estimators for σ_g^2, $g = 1, \ldots, m$, where

$$\|B^{gh}\| = \|\sum_\xi A_{\xi\xi}^{(h)} a_{g\xi}^2\|^{-1}$$

$\|A_{\xi\eta}^{(h)}\|$ being the matrix of the quadratic form S_h.

In many special problems in which variance components are to be estimated, the matrix of known numbers $\|a_{g\xi}\|$ takes a special form which immediately suggests the decomposition (10.7.6) for S_T. In the next section we shall consider several applications of the preceding principles to experimental designs.

10.8 ESTIMATORS FOR VARIANCE COMPONENTS IN EXPERIMENTAL DESIGNS

In this section we shall apply the ideas of estimation of variance components to some of the simpler experimental designs. Actually, we shall consider only the complete two-factor, the balanced incomplete two-factor, the complete three-factor, and the Latin Square designs from the point of view of variance component analysis. The reader interested in this subject in greater detail than given here is referred to books by Cochran and Cox (1957), Graybill (1961), Kempthorne (1952), and Scheffé (1959). We shall consider only the problem of determining unbiased estimators of the variance components involved in these experimental designs and not the variances of these estimators. The reader interested in this latter problem is referred to papers by Hooke (1956*b*), Scheffé (1959), and Tukey (1956*b*, 1957*a*, 1957*b*). Results on confidence intervals of these variance component estimators have been obtained by Bulmer (1957), Moriguti (1954), Scheffé (1959), and others.

The stochastic description of experimental designs by using observable linear functions of random variables is sometimes called the *Model II* description as contrasted with the *Model I* description based on normal regression analysis and discussed in Section 10.6. The Model I and Model II designations for the two analyses of variance schemes were introduced by Eisenhart (1947).

(a) The Complete Two-Factor Experimental Design

The basic sampling theory required for the Model II stochastic description of a complete two-factor experimental design is provided by that of second-order matrix samples as presented in Section 8.6(*a*). We shall therefore adopt the notation of that section.

First, however, some comments are in order on the basic difference between the Model I and the Model II approaches to the complete two-factor experimental design. In the Model I description of the complete two-factor experimental design discussed in Section 10.4(*a*) it was assumed that R-factor levels $R_1, \ldots, R_r$ and C-factor levels $C_1, \ldots, C_s$ were *fixed* for the totality of all experiments described by the response random variables $x_{\xi\eta}$. In some situations, however, the r R-factor levels or the s C-factor levels or both are samples from larger populations of levels which may be finite or infinite. In this instance the regression analysis (Model I) approach is not appropriate. We shall consider the case where the populations of levels are infinite. More precisely, in the Model II approach we assume that for any R-factor level u and any C-factor level v, the response random variable $x(u, v)$ is of the form given by (8.6.3). Its variance σ^2 will be given by (8.6.4) or, in shorter notation, by (8.6.4a), that is, by

$$\sigma^2 = \sigma_{\cdot 0}^2 + \sigma_{0\cdot}^2 + \sigma_{\cdot\cdot}^2 \tag{10.8.1}$$

where $\sigma_{\cdot 0}^2$ is the *row* (R-factor) *component*, $\sigma_{0\cdot}^2$ the *column* (C-factor) *component*, and $\sigma_{\cdot\cdot}^2$ is the *residual component*.

If r R-factor levels are drawn at random, and s C-factor levels are drawn at random—both from infinite populations—our response random variables will be the matrix sample $\{x_{\xi\eta};\ \xi = 1, \ldots, r;\ \eta = 1, \ldots, s\}$ as defined in Section 8.6(a), the $x_{\xi\eta}$ being independent and expressible by (8.6.8), that is,

$$x_{\xi\eta} = \mu + \varepsilon_{\xi\cdot} + \varepsilon_{\cdot\eta} + \varepsilon_{\xi\eta} \tag{10.8.2}$$

where μ is the population mean, and where the $\varepsilon_{\xi\cdot}$, $\varepsilon_{\cdot\eta}$, and $\varepsilon_{\xi\eta}$ are random variables having zero means and zero covariances.

The variance σ^2 of $x_{\xi\eta}$ satisfies (10.8.1). The main problem of the Model II analysis is to estimate $\sigma_{\cdot 0}^2$, $\sigma_{0\cdot}^2$, and $\sigma_{\cdot\cdot}^2$ from the sample $\{x_{\xi\eta}\}$.

For this purpose, we adopt the definitions of $\bar{x}_{\cdot\cdot}$, $\bar{x}_{\xi\cdot}$, $\bar{x}_{\cdot\eta}$ as given in (8.6.7), and of $S_{\cdot 0}$, $S_{0\cdot}$, $S_{\cdot\cdot}$ given in (8.6.9). Under the assumptions of **8.6.2** we have, as given in (8.6.11),

$$\begin{aligned} \mathscr{E}(S_{\cdot 0}) &= (r - 1)(s\sigma_{\cdot 0}^2 + \sigma_{\cdot\cdot}^2) \\ \mathscr{E}(S_{0\cdot}) &= (s - 1)(r\sigma_{0\cdot}^2 + \sigma_{\cdot\cdot}^2) \\ \mathscr{E}(S_{\cdot\cdot}) &= (r - 1)(s - 1)\sigma_{\cdot\cdot}^2 \end{aligned} \tag{10.8.3}$$

Applying **10.1.1** (or **10.7.1**) we therefore obtain the following unbiased estimators for $\sigma_{\cdot 0}^2$, $\sigma_{0\cdot}^2$, and $\sigma_{\cdot\cdot}^2$:

$$\begin{aligned} \mathscr{E}^{-1}(\sigma_{\cdot 0}^2) &= \frac{1}{s(r - 1)}\left[S_{\cdot 0} - \frac{S_{\cdot\cdot}}{s - 1}\right] \\ \mathscr{E}^{-1}(\sigma_{0\cdot}^2) &= \frac{1}{r(s - 1)}\left[S_{0\cdot} - \frac{S_{\cdot\cdot}}{r - 1}\right] \\ \mathscr{E}^{-1}(\sigma_{\cdot\cdot}^2) &= \frac{S_{\cdot\cdot}}{(r - 1)(s - 1)}\,. \end{aligned} \tag{10.8.4}$$

The estimator for μ, the population mean, is $\bar{x}_{\cdot\cdot}$, whose variance is given by $\sigma_{\cdot 0}^2/r + \sigma_{0\cdot}^2/s + \sigma_{\cdot\cdot}^2/rs$ as stated in (8.6.12). An unbiased estimator for this variance is therefore provided by replacing $\sigma_{\cdot 0}^2$, $\sigma_{0\cdot}^2$, and $\sigma_{\cdot\cdot}^2$ with their respective estimators given in (10.8.4).

It is customary also to display the constituents in the Model II description in an analysis of variance table as shown in Table 10.2.

TABLE 10.2 MODEL II ANALYSIS OF VARIANCE TABLE FOR COMPLETE TWO-FACTOR EXPERIMENTAL DESIGN

Source of Variation	Degrees of Freedom (D.F.)	Sum of Squares (S.S.)	Mean sum of Squares (M.S.S.)	Mean value of M.S.S.
Rows	$r - 1$	$S_{\cdot 0}$	$\frac{S_{\cdot 0}}{(r-1)}$	$s\sigma^2_{\cdot 0} + \sigma^2_{\cdot\cdot}$
Columns	$s - 1$	$S_{0\cdot}$	$\frac{S_{0\cdot}}{(s-1)}$	$r\sigma^2_{0\cdot} + \sigma^2_{\cdot\cdot}$
Residual (error)	$(r-1)(s-1)$	$S_{\cdot\cdot}$	$\frac{S_{\cdot\cdot}}{(r-1)(s-1)}$	$\sigma^2_{\cdot\cdot}$
Total	$rs - 1$	S_T		

Finally, the reader should note that for the case of finite populations of R-factor levels and of C-factor levels, we find estimators for the finite population version of the variance components, that is, $\sigma^2_{f\cdot 0}$, $\sigma^2_{f0\cdot}$, and $\sigma^2_{f\cdot\cdot}$, by replacing equations (10.8.3) with (8.6.22) and applying **10.7.1.** A table similar to Table 10.2 can, of course, also be drawn up for this case.

(b) The Balanced Incomplete Two-Factor Experimental Design

The sampling theory involved in the Model II treatment of the balanced incomplete two-factor design is given in Section 8.6(*d*). This design selects the subsample $\{x_{\xi\eta}\}^*$ of size $n = rs' = r's$ of response random variables from the matrix sample $\{x_{\xi\eta}\}$ as defined in Section 8.6(*d*). Each element in $\{x_{\xi\eta}\}^*$ is a random variable which is itself a linear combination of random variables as stated by (10.8.2). Furthermore, the variance of each element $x_{\xi\eta}$ in $\{x_{\xi\eta}\}^*$ has the value σ^2 which, in turn, is the sum of three components as stated in (10.8.1).

The main problem of a Model II analysis of a balanced incomplete two-factor design is to devise unbiased estimators for μ, and the variance components $\sigma^2_{\cdot 0}$, $\sigma^2_{0\cdot}$, and $\sigma^2_{\cdot\cdot}$ by using the elements of the subsample $\{x_{\xi\eta}\}^*$.

The mean $\bar{x}^*_{\cdot\cdot}$ of the sample $\{x_{\xi\eta}\}^*$ is an unbiased estimator for μ which has variance

(10.8.5) $$\sigma^2(\bar{x}^*_{\cdot\cdot}) = \frac{\sigma^2_{\cdot\cdot}}{n} + \frac{\sigma^2_{\cdot 0}}{r} + \frac{\sigma^2_{0\cdot}}{s}$$

as we have seen in (8.6.39).

It will be observed from the last three equations in (8.6.39) that those equations have a unique solution for $\sigma^2_{\cdot 0}$, $\sigma^2_{0\cdot}$, and $\sigma^2_{\cdot\cdot}$ provided $r' > 1$ and $s' > 1$. Under these conditions we apply **10.1.1** (or **10.7.1**), obtaining the following estimators for $\sigma^2_{\cdot 0}$, $\sigma^2_{0\cdot}$, and $\sigma^2_{\cdot\cdot}$:

$$\mathscr{E}^{-1}(\sigma^2_{\cdot 0}) = \frac{S^*_{\cdot 0} + S^*_E}{n - s} - \mathscr{E}^{-1}(\sigma^2_{\cdot\cdot})$$

(10.8.6)
$$\mathscr{E}^{-1}(\sigma^2_{0\cdot}) = \frac{S^*_{0\cdot} + S^*_E}{n - r} - \mathscr{E}^{-1}(\sigma^2_{\cdot\cdot})$$

$$\mathscr{E}^{-1}(\sigma^2_{\cdot\cdot}) = \frac{1}{K}[AS^*_{0\cdot} + BS^*_{\cdot 0} + CS^*_E]$$

where

$$A = \frac{(r - r')}{(n - r)}, \qquad B = \frac{(s - s')}{(n - s)}$$

$$C = A + B - 1$$

$$K = n - r - s + 1.$$

An unbiased estimator for $\sigma^2(\bar{x}^*_{\cdot\cdot})$ is obtained, by substituting the estimators $\mathscr{E}^{-1}(\sigma^2_{\cdot 0})$, $\mathscr{E}^{-1}(\sigma^2_{0\cdot})$, and $\mathscr{E}^{-1}(\sigma^2_{\cdot\cdot})$ for $\sigma^2_{\cdot 0}$, $\sigma^2_{0\cdot}$, and $\sigma^2_{\cdot\cdot}$ respectively, in (10.8.5).

We leave it to the reader as an exercise to set up a Model II analysis of variance table similar to Table 10.2 for the incomplete two-factor design.

(c) The Complete Three-Factor Experimental Design

The estimation of variance components in the complete three-factor design follows from the underlying sampling theory of third-order matrix samples given in Section 8.6(*c*). We shall use the notation of that section. The sample of response random variables associated with the complete three-factor design is the matrix sample $\{x_{\xi\eta\zeta}\}$ referred to in Section 8.6(*c*). We are assuming, of course, that the r R-factor levels, the s C-factor levels, and the t L-factor levels are samples from infinite populations. Thus, each element $x_{\xi\eta\zeta}$ of the sample $\{x_{\xi\eta\zeta}\}$ is the sum of random variables having zero means and zero covariances as stated in (8.6.30), that is,

(10.8.7)
$$x_{\xi\eta\zeta} = \mu + \varepsilon_{\xi\cdot\cdot} + \varepsilon_{\cdot\eta\cdot} + \varepsilon_{\cdot\cdot\zeta} + \varepsilon_{\xi\eta\cdot} + \varepsilon_{\xi\cdot\zeta} + \varepsilon_{\cdot\eta\zeta} + \varepsilon_{\xi\eta\zeta}$$

the variance σ^2 of $x_{\xi\eta\zeta}$ being the sum of seven components as stated by (8.6.27) or (8.6.28), that is,

(10.8.8)
$$\sigma^2 = \sigma^2_{\cdot 00} + \sigma^2_{0\cdot 0} + \sigma^2_{00\cdot} + \sigma^2_{\cdot\cdot 0} + \sigma^2_{\cdot 0\cdot} + \sigma^2_{0\cdot\cdot} + \sigma^2_{\cdot\cdot\cdot}.$$

Now, applying **10.1.1**, (or **10.7.1**) we obtain estimators for these components from $S_{\cdot 00}$, $S_{0\cdot 0}$, $S_{00\cdot}$, $S_{\cdot\cdot 0}$, $S_{\cdot 0\cdot}$, $S_{0\cdot\cdot}$, and $S_{\cdots}$ as defined in (8.6.31), by solving equations (8.6.34) for the variance components. This gives

$$
\begin{aligned}
\mathscr{E}^{-1}(\sigma^2_{\cdots}) &= \frac{S_{\cdots}}{(r-1)(s-1)(t-1)} \\
\mathscr{E}^{-1}(\sigma^2_{0\cdot\cdot}) &= \frac{S_{0\cdot\cdot}}{r(s-1)(t-1)} - \frac{1}{r}\mathscr{E}^{-1}(\sigma^2_{\cdots}) \\
\mathscr{E}^{-1}(\sigma^2_{\cdot 0\cdot}) &= \frac{S_{\cdot 0\cdot}}{s(r-1)(t-1)} - \frac{1}{s}\mathscr{E}^{-1}(\sigma^2_{\cdots}) \\
\mathscr{E}^{-1}(\sigma^2_{\cdot\cdot 0}) &= \frac{S_{\cdot\cdot 0}}{t(r-1)(s-1)} - \frac{1}{t}\mathscr{E}^{-1}(\sigma^2_{\cdots}) \qquad (10.8.9)\\
\mathscr{E}^{-1}(\sigma^2_{00\cdot}) &= \frac{S_{00\cdot}}{rs(t-1)} - \frac{1}{s}\mathscr{E}^{-1}(\sigma^2_{0\cdot\cdot}) - \frac{1}{r}\mathscr{E}^{-1}(\sigma^2_{\cdot 0\cdot}) - \frac{1}{rs}\mathscr{E}^{-1}(\sigma^2_{\cdots}) \\
\mathscr{E}^{-1}(\sigma^2_{0\cdot 0}) &= \frac{S_{0\cdot 0}}{rt(s-1)} - \frac{1}{t}\mathscr{E}^{-1}(\sigma^2_{0\cdot\cdot}) - \frac{1}{r}\mathscr{E}^{-1}(\sigma^2_{\cdot\cdot 0}) - \frac{1}{rt}\mathscr{E}^{-1}(\sigma^2_{\cdots}) \\
\mathscr{E}^{-1}(\sigma^2_{\cdot 00}) &= \frac{S_{\cdot 00}}{st(r-1)} - \frac{1}{t}\mathscr{E}^{-1}(\sigma^2_{\cdot 0\cdot}) - \frac{1}{s}\mathscr{E}^{-1}(\sigma^2_{\cdot\cdot 0}) - \frac{1}{st}\mathscr{E}^{-1}(\sigma^2_{\cdots}).
\end{aligned}
$$

An unbiased estimator for μ is $\bar{x}_{\cdots}$, whose variance is given by formula (8.6.35). An unbiased estimator for $\sigma^2(\bar{x}_{\cdots})$ is, of course, obtained by replacing the variance components in (8.6.35) by their estimators in (10.8.9).

The reader will find it instructive to set up a Model II analysis of variance table similar to Table 10.2 for the complete three-factor design.

We also remark that estimation of variance components in a complete three-factor design for the case of finite populations of R-factor, C-factor, and L-factor levels is straightforward and will be left to the reader.

(d) The Latin Square Experimental Design

As we have pointed out in Section 8.6(d), and Section 10.6(c), a Latin Square design is a special case of a balanced incomplete third-order matrix sample, namely, that for $s = t = r$ and $s' = t' = r' = 1$. The sample of response random variables $\{x_{\xi\eta\zeta}\}^*$ for a Latin Square design, where $\{x_{\xi\eta\zeta}\}^*$ is defined in Section 8.6(d) is such that each $x_{\xi\eta\zeta}$ in this sample is of form (10.8.7) and its variance σ^2 consists of the seven components given in (10.8.8). However, if the total sum of squares S_T^* for the sample is decomposed into the sums of squares $S^*_{\cdot 00}$, $S^*_{0\cdot 0}$, $S^*_{00\cdot}$ and S_E^* as defined in

(8.6.40), it is only possible, as may be seen in (8.6.41), to estimate the following components in (10.8.8):

(10.8.10) $$\sigma_E^2,\ \sigma_{\cdot 00}^2,\ \sigma_{0\cdot 0}^2,\ \sigma_{00\cdot}^2$$

where $$\sigma_E^2 = \sigma_{\cdots}^2 + \sigma_{\cdot\cdot 0}^2 + \sigma_{\cdot 0\cdot}^2 + \sigma_{0\cdot\cdot}^2.$$

In other words, we can estimate only the row, column, and layer components individually, and the sum of the remaining four components. These estimates are obtained, of course, by applying **10.7.1** to (8.6.41), which gives:

(10.8.11)
$$\mathscr{E}^{-1}(\sigma_{\cdot 00}^2) = \frac{S_{\cdot 00}^*}{r-1} - \mathscr{E}^{-1}(\sigma_E^2)$$
$$\mathscr{E}^{-1}(\sigma_{0\cdot 0}^2) = \frac{S_{0\cdot 0}^*}{r-1} - \mathscr{E}^{-1}(\sigma_E^2)$$
$$\mathscr{E}^{-1}(\sigma_{00\cdot}^2) = \frac{S_{00\cdot}^*}{r-1} - \mathscr{E}^{-1}(\sigma_E^2)$$
$$\mathscr{E}^{-1}(\sigma_E^2) = \frac{S_{\cdot 00}^* + S_{0\cdot 0}^* + S_{00\cdot}^*}{(2r-1)} - \frac{S_E^*}{(2r-1)(r-1)}.$$

A Model II analysis of variance table similar to Table 10.2 for the Latin Square design is straightforward and we leave it to be set up as an exercise.

10.9 LINEAR ESTIMATORS FOR MEANS OF STRATIFIED POPULATIONS

(a) Definitions and Notation

In many practical problems of sampling from a finite population π_N, the population π_N is decomposed into m disjoint *strata* $\pi_{N_1}, \ldots, \pi_{N_m}$, where $N_1 + \cdots + N_m = N$ and where the mean and variance of π_{N_g} are μ_g and σ_g^2, $g = 1, \ldots, m$. If the mean and variance of π_N are μ and σ^2, then it can be verified that μ and σ^2 can be expressed in terms of N_g, μ_g, σ_g^2, $g = 1, \ldots, m$, as follows:

(10.9.1) $$\mu = \sum_g p_g \mu_g$$

and

(10.9.2) $$\sigma^2 = \sigma_W^2 + \sigma_B^2$$

where

(10.9.3) $$\sigma_W^2 = \sum_g \left(\frac{N_g - 1}{N-1}\right)\sigma_g^2, \qquad \sigma_B^2 = \frac{N}{N-1}\sum_g p_g(\mu_g - \mu)^2$$

and

$$p_g = \frac{N_g}{N}, \qquad g = 1, \ldots, m. \tag{10.9.4}$$

Note that σ^2 is the sum of two components, namely, σ_W^2, the *within-strata* component, and σ_B^2, the *between-strata* component.

If $O_{n_1}, \ldots, O_{n_m}$ are independent samples of sizes $n_1, \ldots, n_m$, from the strata $\pi_{N_1}, \ldots, \pi_{N_m}$ respectively, let $\bar{x}_1, \ldots, \bar{x}_m$ be the sample means, and $s_1^2, \ldots, s_m^2$ the sample variances of $O_{n_1}, \ldots, O_{n_m}$ respectively. The *strata samples* $O_{n_1}, \ldots, O_{n_m}$ taken collectively, where $n_1 + \cdots + n_m = n$, are called a *stratified sample* of size n from π_N.

The main problem here is to find optimum linear estimators for μ based on means of samples from the strata for various amounts of information given concerning N_g, μ_g, and σ_g^2, $g = 1, \ldots, m$. Actually, we shall consider the problem under two sets of conditions, namely:

(i) where the strata sizes $N_1, \ldots, N_m$ are known;
(ii) where the strata variances $\sigma_1^2, \ldots, \sigma_m^2$ are known in addition to the strata sizes.

If the population π_N is unstratified, we have already seen in Section 10.2 that the minimum variance linear estimator of μ from a simple random sample O_n from π_N is $\bar{x}$ and the variance of $\bar{x}$ is $(1/n - 1/N)\sigma^2$. This result will provide a "standard" against which the variances of estimators devised for conditions (i) and (ii) can be compared.

Stratified sampling is widely used in sample surveys in government, business, and industry and there is a great deal of literature on the subject. Here we shall only give an introduction to the mathematics of this kind of sampling. The reader interested in additional reading is referred to books by Cochran (1953), Hansen, Hurwitz, and Madow (1953), Stephan and McCarthy (1958), Sukhatme (1954), and Yates (1949).

(b) Linear Estimator for the Population Mean from a General Stratified Sample

In this case it is assumed that the strata sizes $N_1, \ldots, N_m$ are known, and the strata sample sizes $n_1, \ldots, n_m$, each $\geqslant 2$, are specified. Let $(x_{g1}, \ldots, x_{gn_g})$ be the elements in sample O_{n_g}, $g = 1, \ldots, m$. This set of strata samples is called a *general stratified sample* from π_N. Let L be a linear function of all sample elements, that is,

$$L = \sum_{g=1}^{m} \sum_{\xi=1}^{n_g} c_{g\xi} x_{g\xi}. \tag{10.9.5}$$

We wish to determine the $c_{g\xi}$ so that L is an unbiased estimator for μ having minimum variance. For L to be unbiased we require that the $c_{g\xi}$ be chosen so as to satisfy

$$\mathscr{E}(L) = \sum_{g=1}^{m} \sum_{\xi=1}^{n_g} c_{g\xi}\mu_g \equiv \sum_g p_g\mu_g$$

that is,

$$\sum_{\xi=1}^{n_g} c_{g\xi} = p_g, \qquad g = 1, \ldots, m \tag{10.9.6}$$

and also so as to minimize $\sigma^2(L)$. Since the strata samples are independent we have

$$\sigma^2(L) = \sum_{g=1}^{m} \left[\sigma_g^2\left(\sum_{\xi=1}^{n_g} c_{g\xi}^2\right) - \frac{\sigma_g^2}{N_g}\left(\sum_{\xi=1}^{n_g} c_{g\xi}\right)^2\right]. \tag{10.9.7}$$

It can be verified that the values of the $c_{g\xi}$ satisfying (10.9.6) and minimizing (10.9.7) are

$$c_{g\xi} = \frac{p_g}{n_g}, \qquad \xi = 1, \ldots, n_g, \; g = 1, \ldots, m. \tag{10.9.8}$$

When these values are substituted in (10.9.5), let the resulting *general stratified sample* estimator for μ be denoted by $\mathscr{E}_s^{-1}(\mu)$. Then

$$\mathscr{E}_s^{-1}(\mu) = p_1\bar{x}_1 + \cdots + p_m\bar{x}_m. \tag{10.9.9}$$

The variance of $\mathscr{E}_s^{-1}(\mu)$ is given by

$$\sigma^2(\mathscr{E}_s^{-1}(\mu)) = \sum_g p_g^2\left(\frac{1}{n_g} - \frac{1}{N_g}\right)\sigma_g^2, \tag{10.9.10}$$

from which it is seen that

$$\sum_g p_g^2\left(\frac{1}{n_g} - \frac{1}{N_g}\right)s_g^2 \tag{10.9.11}$$

is an unbiased estimator for $\sigma^2(\mathscr{E}_s^{-1}(\mu))$.

Summarizing,

10.9.1 *Let π_N be a finite population with disjoint strata $\pi_{N_1}, \ldots, \pi_{N_m}$ of known sizes. Let the mean and variance of π_{N_g} be μ_g and σ_g^2, $g = 1, \ldots, m$. Let $O_{n_1}, \ldots, O_{n_m}$ be a general stratified sample from π_N. Let the sample mean and variance of O_{n_g} be $\bar{x}_g$ and s_g^2, $g = 1, \ldots, m$, respectively. Then $\mathscr{E}_s^{-1}(\mu)$ given in* (10.9.9) *for the mean of π_N is a minimum-variance linear unbiased estimator for μ. The variance of $\mathscr{E}_s^{-1}(\mu)$ is given by* (10.9.10), *and an unbiased estimator for this variance is given by* (10.9.11).

In the preceding discussion it should be noted that the strata sample sizes $n_1, \ldots, n_m$, each $\geqslant 2$, are fixed. It will be further noted that $\mathscr{E}_s^{-1}(\mu)$

is an unbiased estimator for the mean of π_N for any choice of the strata sample sizes.

In the particular case where the strata sample sizes $n_1, \ldots, n_m$ are chosen to be proportional to the strata sizes, that is, if $n_g = p_g n$, $g = 1, \ldots, m$ we have a *proportional stratified sample* of size n from π_N. (In a practical situation, of course, $p_g n$ would be rounded to the nearest integer.) For this choice of n_g, $g = 1, \ldots, m$, we shall denote $\mathscr{E}_s^{-1}(\mu)$ by $\mathscr{E}_{sp}^{-1}(\mu)$. The variance of $\mathscr{E}_{sp}^{-1}(\mu)$ is

$$\text{(10.9.10}a\text{)} \qquad \sigma^2(\mathscr{E}_{sp}^{-1}(\mu)) = \left(\frac{1}{n} - \frac{1}{N}\right) \sum_g p_g \sigma_g^2,$$

whereas

$$\text{(10.9.11}a\text{)} \qquad \left(\frac{1}{n} - \frac{1}{N}\right) \sum_g p_g s_g^2$$

is an unbiased estimator for $\sigma^2(\mathscr{E}_{sp}^{-1}(\mu))$.

The *efficiency of* $\mathscr{E}_{sp}^{-1}(\mu)$ *relative to* $\bar{x}$, the mean of a simple random sample of size n from π_N, in estimating the mean μ of π_N is defined by the ratio

$$\text{(10.9.12)} \qquad \text{eff}(\mathscr{E}_{sp}^{-1}(\mu)) = \frac{\sigma^2(\bar{x})}{\sigma^2(\mathscr{E}_{sp}^{-1}(\mu))} = 1 + K$$

where

$$\text{(10.9.13)} \qquad K = \frac{\sigma_B^2}{\sigma_W^2} + o\left(\frac{1}{N}\right).$$

It is evident from (10.9.12) that the advantage of proportional stratified sampling over random sampling for linearly estimating the mean of π_N depends on the ratio σ_B^2/σ_W^2. If $\sigma_B^2 = 0$, that is, if all strata have equal means, then proportional stratified sampling offers no advantage over simple random sampling.

Hence

10.9.2 *If we choose a proportional stratified sample with* $n_g = p_g n$, $g = 1, \ldots, m$, *the resulting linear estimator* $\mathscr{E}_{sp}^{-1}(\mu)$ *has variance* (10.9.10*a*), *of which* (10.9.11*a*) *is an unbiased estimator. Furthermore, the efficiency of* $\mathscr{E}_{sp}^{-1}(\mu)$ *relative to* $\bar{x}$, *the mean of a random sample of size* n *from* π_N, *as an estimator for the mean of* p_N, *is given by* (10.9.12).

(c) Minimum Variance Linear Estimator for the Population Mean if Sizes and Variances of Population Strata are Known

If the values of $\sigma_1^2, \ldots, \sigma_m^2$ as well as $N_1, \ldots, N_m$ are known, we can make a "better" choice of strata sample sizes than those provided by proportional stratified sampling. The problem here amounts to minimizing $\sigma^2(\mathscr{E}_s^{-1}(\mu))$ as given by (10.9.10), subject to the condition that $n_1 + \cdots + n_m = n$. Allowing $n_1, \ldots, n_m$ for the moment to vary continuously, the reader can verify that the values of $n_1, \ldots, n_m$ satisfying the required conditions are as follows:

$$n_g = b_g n, \qquad g = 1, \ldots, m \tag{10.9.14}$$

where

$$b_g = p_g\sigma_g / \sum_g p_g\sigma_g . \tag{10.9.15}$$

For this optimum choice of $n_1, \ldots, n_m$ let $\mathscr{E}_s^{-1}(\mu)$ be denoted by $\mathscr{E}_{s0}^{-1}(\mu)$. Then we have, after some simplification of (10.9.10),

$$\sigma^2(\mathscr{E}_{s0}^{-1}(\mu)) = \frac{1}{n}\left(\sum_g p_g\sigma_g\right)^2 - \frac{1}{N}\sum_g p_g\sigma_g^2. \tag{10.9.10b}$$

(In a practical situation, one would, of course, choose for each n_g the positive integer nearest nb_g.)

It can be verified that

$$\sigma^2(\mathscr{E}_{s0}^{-1}(\mu)) \leqslant \sigma^2(\mathscr{E}_{sp}^{-1}(\mu)) \tag{10.9.16}$$

the equality holding if and only if $\sigma_1 = \cdots = \sigma_m$. For by comparing (10.9.10) and (10.9.10a), (10.9.16) holds if and only if

$$\left(\sum_g p_g\sigma_g\right)^2 \leqslant \sum_g p_g\sigma_g^2,$$

that is, if and only if

$$\left(\sum_g p_g\sigma_g\right)^2 \leqslant \left(\sum_g p_g\right)\left(\sum_g p_g\sigma_g^2\right) \tag{10.9.17}$$

since $\sum_g p_g = 1$, the p_g all being positive.

But (10.9.17) is a special case of the well-known Schwarz inequality, and equality holds if and only if $\sigma_1 = \cdots = \sigma_m$.

A stratified sample from π_N of size n for which n_g satisfies (10.9.14) is sometimes called an *optimum stratified sample* of size n from π_N.

Summarizing, we have the following result due to Neyman (1934):

10.9.3 *Let π_N be a finite population with disjoint strata $\pi_{N_1}, \ldots, \pi_{N_m}$ whose sizes $N_1, \ldots, N_m$ and variances $\sigma_1^2, \ldots, \sigma_m^2$ are known. Then the stratified sample of size n from π_N which provides the minimum variance linear estimator $\mathscr{E}_{s0}^{-1}(\mu)$ for the mean of π_N is that in which the strata and sample sizes are n_g, as given by* (10.9.14). *The variance of $\mathscr{E}_{s0}^{-1}(\mu)$ is given by* (10.9.10b). *$\mathscr{E}_{s0}^{-1}(\mu)$ and $\mathscr{E}_{sp}^{-1}(\mu)$ are equally efficient for estimating the mean of π_N if and only if the strata variances are all equal.*

(d) Minimum Variance Linear Estimator for the Population Mean for Fixed Total Cost

In practical situations there are often considerable variations in the cost of sampling from the various strata. Under such conditions it is sometimes desirable to choose sample sizes from the several strata so as to minimize $\sigma^2(\mathscr{E}_s^{-1}(\mu))$ in (10.9.10) subject to a fixed value of over-all cost. More precisely, suppose c_g is the cost of determining the value of each element in the sample of size n_g from π_{N_g}. If C is the total cost, then

(10.9.18) $$C = c_1 n_1 + \cdots + c_m n_m.$$

It is readily found that the minimum of $\sigma^2(\mathscr{E}_s^{-1}(\mu))$ subject to (10.9.18) occurs for

(10.9.19) $$n_g = CB_g, \qquad g = 1, \ldots, m.$$

where

(10.9.20) $$B_g = \frac{\sqrt{c_g}\, p_g \sigma_g}{c_g \cdot \sum_g (\sqrt{c_g}\, p_g \sigma_g)}.$$

It is to be emphasized that this procedure assumes strata sizes $N_1, \ldots, N_m$ and strata variances, $\sigma_1^2, \ldots, \sigma_m^2$ as well as the costs $c_1, \ldots, c_m$ are available (known).

For the choice of the n_g indicated in (10.9.19) let $\mathscr{E}_s^{-1}(\mu)$ be denoted by $\mathscr{E}_{sc}^{-1}(\mu)$. Then we have

(10.9.21) $$\sigma^2(\mathscr{E}_{sc}^{-1}(\mu)) = \frac{1}{C}\Big(\sum_g \sqrt{c_g}\, p_g \sigma_g\Big)^2 - \frac{1}{N}\sum_g p_g \sigma_g^2.$$

(e) Extension of Results to Stratified Sampling from Infinite Populations

The results summarized in **10.9.1**, **10.9.2**, and **10.9.3** can be extended in a straightforward manner to the case of stratified sampling from an infinite population. Extensions to this case are left as exercises for the reader.

10.10 LINEAR ESTIMATOR FOR MEAN OF STRATIFIED POPULATIONS IN TWO-STAGE SAMPLING

(a) Two-Stage Sampling

In situations where π_N has a large number of strata, it may not be economically feasible to draw a sample from each stratum. In this case, a *two-stage* sampling procedure may be used, in which, first, a number of strata are designated by a random process with specified probabilities, and then, elements are drawn by simple random sampling from each of these strata. The object of the sampling is to obtain a linear estimator for the mean of μ, the variance of this estimator, and an estimator for this variance.

For simplicity we shall consider the case where the size of each stratum is a multiple of an integer v, that is,

$$N_g = M_g v, \qquad g = 1, \ldots, m. \tag{10.10.1}$$

We shall refer to v as the *sampling unit*. Thus, the gth stratum may be regarded as consisting of M_g sampling units. The total number of sampling units in π_N is therefore $M_1 + \cdots + M_m = M$, say. Therefore

$$N = Mv. \tag{10.10.2}$$

We now consider a random process proposed by Wilks (1960b) for determining the number of sampling units to be drawn without replacement from each stratum of the population, subject to the condition that the total number of sampling units to be drawn is u. We assume that the process is such that each of the M sampling units may be regarded as having the same probability of being drawn. (This can be realized in practice by the use of random numbers.) Let δ_g be a random variable denoting the number of sampling units to be drawn from π_{Ng}, $g = 1, \ldots, m$. $(\delta_1, \ldots, \delta_m)$ is therefore an $(m - 1)$-dimensional random variable having the $(m - 1)$-dimensional hypergeometric distribution with p.f.

$$p(\delta_1, \ldots, \delta_m) = \frac{\binom{M_1}{\delta_1} \cdots \binom{M_m}{\delta_m}}{\binom{M}{u}} \tag{10.10.3}$$

where $\delta_1 + \cdots + \delta_m = u$.

The result of the first stage of sampling, therefore, essentially tells us that a sample of size $\delta_g v$ is to be drawn from π_{Ng}, $g = 1, \ldots, m$.

At the second stage of sampling, we draw samples of sizes $\delta_1 v, \ldots, \delta_m v$ from $\pi_{N_1}, \ldots, \pi_{N_m}$, respectively. This process provides a *two-stage sample of size uv* from π_N, which can be regarded as a stratified sample where strata to be sampled have been chosen by a random procedure. In practice where the number of strata is large many of the δ's would be 0, that is, many strata would not be sampled at all. The total number of strata sampled would be the number of nonzero δ's.

(b) Linear Estimator for Population Mean

Let $\mathscr{E}_2^{-1}(\mu)$ be the following linear estimator for μ

$$\mathscr{E}_2^{-1}(\mu) = \frac{(\delta_1 v)\bar{x}_1 + \cdots + (\delta_m v)\bar{x}_m}{uv}$$

that is,

$$\mathscr{E}_2^{-1}(\mu) = \frac{\delta_1\bar{x}_1 + \cdots + \delta_m\bar{x}_m}{u}. \tag{10.10.4}$$

We have

$$\mathscr{E}(\mathscr{E}_2^{-1}(\mu)) = \sum_g \mathscr{E}\left(\frac{\delta_g\bar{x}_g}{u}\right). \tag{10.10.5}$$

Making use of iterated mean values as stated by **3.7.2**, we have for $g = 1, \ldots, m$

$$\mathscr{E}\left(\frac{\delta_g\bar{x}_g}{u}\right) = \frac{1}{u}\mathscr{E}_{(\delta_g)}[\delta_g \cdot \mathscr{E}_{(\bar{x}_g)}(\bar{x}_g \mid \delta_g)], \tag{10.10.6}$$

where $\mathscr{E}_{(x_g)}(\bar{x}_g \mid \delta_g)$ is the mean value of the conditional random variable $\bar{x}_g \mid \delta_g$, while $\mathscr{E}_{(\delta_g)}[\quad]$ denotes unconditional mean value with respect to the random variable δ_g. But

$$\mathscr{E}_{(\bar{x}_g)}(\bar{x}_g \mid \delta_g) = \mu_g$$

and it follows from (6.1.8) that

$$\mathscr{E}_{(\delta_g)}(\delta_g) = p_g u.$$

Therefore,

$$\mathscr{E}(\mathscr{E}_2^{-1}(\mu)) = \sum_g p_g\mu_g = \mu \tag{10.10.7}$$

which shows that $\mathscr{E}_2^{-1}(\mu)$ is an unbiased estimator for μ, the mean of π_N.

The variance of $\mathscr{E}_2^{-1}(\mu)$ is defined by

$$\begin{aligned} \sigma^2(\mathscr{E}_2^{-1}(\mu)) &= \mathscr{E}\left[\sum_g \left(\frac{\delta_g\bar{x}_g}{u} - p_g\mu_g\right)\right]^2 \\ &= \mathscr{E}\left[\sum_g \frac{\delta_g}{u}(\bar{x}_g - \mu_g) + \sum_g \left(\frac{\delta_g}{u} - p_g\right)\mu_g\right]^2. \end{aligned} \tag{10.10.8}$$

Squaring this expression, and taking iterated mean values, we find after some simplification that

$$\sigma^2(\mathscr{E}_2^{-1}(\mu)) = \frac{N - uv}{u(N - v)}\left[\frac{N-1}{N}\left(\frac{\sigma_W^2}{v} + \sigma_B^2\right) - \frac{v-1}{Nv}\sum_g \sigma_g^2\right]. \tag{10.10.9}$$

By referring to (10.9.3) it will be seen that $\sigma^2(\mathscr{E}_2^{-1}(\mu))$ is a linear function of the three quantities

$$\sum_g \sigma_g^2, \qquad \sum_g p_g\sigma_g^2, \qquad \sum_g p_g(\mu_g - \mu)^2. \tag{10.10.10}$$

If we let G_1, G_2, and G_3 be functions of the two-stage sample elements defined as follows, $v \geqslant 2$:

(10.10.11)

$$G_1 = \sum_g \delta_g s_g^2; \qquad G_2 = \sum_g \frac{\delta_g}{N_g} s_g^2; \qquad G_3 = \sum_g \delta_g(\bar{x}_g - \mathscr{E}_2^{-1}(\mu))^2$$

it can be verified in a straightforward manner that $\mathscr{E}(G_1)$, $\mathscr{E}(G_2)$, and $\mathscr{E}(G_3)$ are linear functions of the three quantities in (10.10.10), such that there exists a linear function of G_1, G_2, G_3 which is an unbiased estimator for $\sigma^2(\mathscr{E}_2^{-1}(\mu))$. Actually, this unbiased estimator is obtained by replacing σ_W^2, σ_B^2 and $\sum_g \sigma_g^2$ in (10.10.9) by the following unbiased estimators respectively:

(10.10.12)

$$\mathscr{E}^{-1}(\sigma_W^2) = \frac{N}{u(N-1)}(G_1 - G_2)$$

$$\mathscr{E}^{-1}(\sigma_B^2) = \frac{Nv}{uv(N-1) + N}\left[\frac{N-2}{uv(N-1)}(G_1 - G_2) - \frac{Nu + 2v}{u^2 v} G_2 + G_3\right]$$

$$\mathscr{E}^{-1}\left(\sum_g \sigma_g^2\right) = \frac{N}{u} G_2.$$

We may summarize the preceding results as follows:

10.10.1 *Let π_N be a finite population consisting of the disjoint strata $\pi_{N_1}, \ldots, \pi_{N_m}$ where $N_g = M_g v$, $g = 1, \ldots, m$ and v is a given integer (sampling unit). Let a two-stage sample of u sampling units be drawn from π_N. Then $\mathscr{E}_2^{-1}(\mu)$ is an unbiased linear estimator for the mean of π_N. The variance of $\mathscr{E}_2^{-1}(\mu)$ is given by* (10.10.9), *whereas the right-hand side of* (10.10.9) *with σ_W^2, σ_B^2, $\sum_g \sigma_g^2$ replaced by their estimators given in* (10.10.12) *is an unbiased estimator for $\sigma^2(\mathscr{E}_2^{-1}(\mu))$, for $v \geqslant 2$.*

(c) Case of Infinite Population

For the case of sampling from an infinite stratified population having probabilities $p_1, \ldots, p_m$, means $\mu_1, \ldots, \mu_m$, and variances $\sigma_1^2, \ldots, \sigma_m^2$ of the strata $\pi_1, \ldots, \pi_m$ respectively, the situation is as follows.

The random variable $(\delta_1, \ldots, \delta_m)$ has the multinomial p.f.

$$p(\delta_1, \ldots, \delta_m) = \frac{u!}{\delta_1! \cdots \delta_m!} p_1^{\delta_1} \cdots p_m^{\delta_m} \tag{10.10.13}$$

where $\delta_1 + \cdots + \delta_m = u$.

The variance of $\mathscr{E}_2^{-1}(\mu)$ in this case has the following form:

$$\sigma^2(\mathscr{E}^{-1}(\mu)) = \frac{\sigma^2_{W'}}{uv} + \frac{\sigma^2_{B'}}{u} \tag{10.10.14}$$

where

$$\sigma^2_{W'} = \sum_g p_g \sigma^2_g, \qquad \sigma^2_{B'} = \sum_g p_g(\mu_g - \mu)^2.$$

while

$$\frac{n+v+1}{un(n+1)} G_1 + \frac{v^2}{n(n+1)} G_3 - \frac{v}{un(n+1)} \sum_g \frac{\delta_g}{p_g} s_g^2 \tag{10.10.15}$$

is an unbiased estimator for $\sigma^2(\mathscr{E}_2^{-1}(\mu))$ where $n = uv$. Note that $\sigma^2_{B'} = \sigma^2_B + O(1/N)$ and $\sigma^2_{W'} = \sigma^2_W + O(1/N)$.

(d) Minimum Variance Linear Estimator for the Population Mean for Fixed Total Cost

Suppose we hold the sample size uv fixed, say let $uv = n$, and write $\sigma^2(\mathscr{E}_2^{-1}(\mu))$ as follows:

$$\sigma^2(\mathscr{E}_2^{-1}(\mu)) = \frac{N-n}{n}\left(\frac{A+NB}{N-v} - B\right) \tag{10.10.16}$$

where

$$A = \sigma^2_{W'}, \qquad B = \sigma^2_{B'} - \frac{1}{N}\sum_g \sigma^2_g. \tag{10.10.17}$$

It is evident from (10.10.16) that since v must be a positive integer, the value of v which minimizes $\sigma^2(\mathscr{E}_2^{-1}(\mu))$ is $v = 1$, provided, of course, that $A + NB \geqslant 0$ and there are n strata. Here our two-stage sample from π_N reduces to a simple random sample of size n from π_N. In particular, it should be noted that in this instance the variance of $\mathscr{E}_2^{-1}(\mu)$ given by (10.10.9) reduces to $(1/n - 1/N)\sigma^2$, where σ^2 is given by (10.9.2).

In view of this situation it is clear that the choice of values of u and v in a given situation cannot be based on the criterion of minimizing $\sigma^2(\mathscr{E}_2^{-1}(\mu))$, since this criterion would require that $v = 1$. In a practical situation the criterion for choice of u and v is to select a pair of values of u and v so as to minimize $\sigma^2(\mathscr{E}_2^{-1}(\mu))$ for a fixed total cost. In this case the simplest assumption is that the cost per sampling unit is C_1, whereas the cost per sample element is C_2. The total cost C, therefore, will be

$$C = C_1 u + C_2 uv. \tag{10.10.18}$$

If $\sigma^2(\mathscr{E}_2^{-1}(\mu))$, as given by (10.10.9), is minimized subject to the restriction

(10.10.18), it is found by allowing u and v to vary continuously that the minimizing value of v, say $\tilde{v}$, is given by

$$\tilde{v} = \frac{N}{1 + \sqrt{\left(1 + \frac{H}{G}N\right)\left(1 + \frac{B}{A}N\right)}} \tag{10.10.19}$$

where

$$G = N\frac{C_1}{C}, \qquad H = N\frac{C_2}{C} - 1 \tag{10.10.20}$$

and where A and B are given in (10.10.17). The corresponding value of u, say $\tilde{u}$, would be found, of course, by substituting $\tilde{v}$ for v in (10.10.18) and solving for u. If we let $\mathscr{E}_{2c}^{-1}(\mu)$ denote the estimator $\mathscr{E}_2^{-1}(\mu)$ under these conditions then, of course, the value of $\sigma^2(\mathscr{E}_{2c}^{-1}(\mu))$ is given by (10.10.9) with u and v replaced by $\tilde{u}$ and $\tilde{v}$. Similarly, if these values of $\tilde{u}$ and $\tilde{v}$ are inserted in (10.10.12) for u and v, we obtain an unbiased estimator for $\sigma^2(\mathscr{E}_{2c}^{-1}(\mu))$. In the case of large N, we have

$$\begin{aligned} \tilde{v} &= \frac{\sigma_{W'}}{\sigma_{B'}}\sqrt{\frac{C_1}{C_2}} + o\left(\frac{1}{N}\right) \\ \tilde{u} &= \frac{C}{C_1 + \frac{\sigma_{W'}}{\sigma_{B'}}\sqrt{C_1C_2}} + o\left(\frac{1}{N}\right). \end{aligned} \tag{10.10.21}$$

PROBLEMS

10.1 Prove **10.2.1** for the case where $\bar{x}$ is the mean of a sample of size n from a finite population of size N.

10.2 Suppose $(x_1, \ldots, x_n)$ is an n-dimensional random variable having a distribution such that the mean of each x is μ, the variance of each x is σ^2, and the covariance of each pair of x's is $\rho\sigma^2$. Show that the minimum variance linear estimator for μ is the mean $\bar{x}$ of the n random variables $x_1, \ldots, x_n$.

10.3 If $\bar{x}$ and $\bar{x}'$ are means of independent samples of sizes n and n' from distributions having means μ and μ', and variances σ^2 and σ'^2 respectively, show that $\bar{x} - \bar{x}'$ is the minimum variance linear estimator for $\mu - \mu'$.

10.4 If $T_1, \ldots, T_k$ are unbiased estimators of a parameter θ having (non-singular) covariance matrix $\|\sigma_{ij}\|$ with inverse $\|\sigma^{ij}\|$, and if $l_1, \ldots, l_k$ are constants whose sum is 1, show that the minimum variance estimator of θ of form $\sum_{i=1}^{k} l_iT_i$, occurs for $l_i = \sum_j \sigma^{ij} \Big/ \sum_{i,j} \sigma^{ij}$, $i = 1, \ldots, k$, and that the variance of this estimator is $1 \Big/ \sum_{i,j} \sigma^{ij}$.

10.5 Show that the conclusions of **10.2.2** are still true if it is assumed that $(x_1, \ldots, x_n)$ is an n-dimensional random variable whose c.d.f. $F(x_1, \ldots, x_n)$ is a symmetric function of $x_1, \ldots, x_n$.

10.6 If $n_1, \bar{x}_1, s_1^2$ are the size, mean, and variance of a sample from the distribution $N(\mu_1, \sigma^2)$ and $n_2, \bar{x}_2, s_2^2$ are similar quantities for an independent sample from $N(\mu_2, \sigma^2)$, show that

$$(\bar{x}_1 - \bar{x}_2) \pm t_{n_1+n_2-2,\gamma}\, s\sqrt{1/n_1 + 1/n_2}$$

are $100\gamma\%$ confidence limits for $\mu_1 - \mu_2$, where

$$s^2 = \frac{(n_1 - 1)s_1^2 + (n_2 - 1)s_2^2}{n_1 + n_2 - 2} \quad \text{and} \quad t_{n_1+n_2-2,\gamma}$$

is defined by (10.2.10).

10.7 Prove **10.3.2.**

10.8 If in **10.3.1** $(x_{11}, \ldots, x_{1n}), \ldots, (x_{k1}, \ldots, x_{kn})$ are chosen as unit vectors (constants) mutually orthogonal to each other, show that the minimum variance linear estimators $\sum_\xi x_{i\xi}y_\xi$, $i = 1, \ldots, k$, for $\beta_1, \ldots, \beta_k$, respectively, all have variances equal to σ^2, and covariances equal to zero.

10.9 Referring to **10.3.4**, show that the interior of the ellipsoid

$$\sum_{i,j=1}^{k_1} a_{ij}^*(\beta_i - b_i)(\beta_j - b_j) = \frac{k_1 S_1 \mathscr{F}_{k_1,n-k,\gamma}}{(n - k)}$$

is a $100\gamma\%$ confidence region for $(\beta_1, \ldots, \beta_{k_1})$, $k_1 \leqslant k$, where

$$\|a_{ij}^*\| = \begin{Vmatrix} a^{11} & \cdots & a^{1k_1} \\ \cdot & & \cdot \\ \cdot & & \cdot \\ \cdot & & \cdot \\ a^{k_1 1} & \cdots & a^{k_1 k_1} \end{Vmatrix}^{-1}$$

and where

$$\mathscr{F}_{k_1,n-k,\gamma}$$

satisfies

$$\int_0^{\mathscr{F}_{k_1,n-k,\gamma}} dF_{k_1,\,n-k}(\mathscr{F}) = \gamma$$

$dF_{k_1,n-k}(\mathscr{F})$ being the p.e. of the Snedecor distribution $S(k_1, n - k)$.

10.10 If a sphere in R_n has center $(a_1, \ldots, a_n)$ and radius r show that the equations of the two hyperplanes (R_{n-1} spaces) parallel to the hyperplane having equation

$$\sum_{\xi=1}^{n} b_\xi x_\xi = k$$

and tangent to the sphere are

$$\sum_{\xi=1}^{n} b_\xi(x_\xi - a_\xi) = \pm r\sqrt{\sum_{\xi=1}^{n} b_\xi^2}.$$

10.11 Referring to Section 10.5, show that the probability is γ that the inequalities

$$\sum_i c_i b_i - \delta\sqrt{\sum_{i,j} a_{ij}c_ic_j} < \sum_i c_i\beta_i < \sum_i c_i b_i + \delta\sqrt{\sum_{i,j} a_{ij}c_ic_j}$$

for all possible real vectors $(c_1, \ldots, c_k)$, where $c_1, \ldots, c_k$ are not all zero, are simultaneously satisfied where

$$\delta^2 = \frac{kS_1}{(n-k)} \cdot \mathscr{F}_{k,n-k,\gamma}$$

b_i, β_i, S_1 and $\mathscr{F}_{k,n-k,\gamma}$ being defined in **10.3.4** and (10.4.4).

10.12 Suppose $(u_1, \ldots, u_k)$ has the distribution $N(\{\mu_i\}; \|\sigma_{ij}\|)$, where $(\mu_1, \ldots, \mu_k)$ are unknown and $\|\sigma_{ij}\|$ are known. Let $\chi^2_{k,\gamma}$ be the $100\gamma\%$ point of the chi-square distribution with k-degrees of freedom. Show that the probability is γ that the inequalities

$$\sum_i c_iu_i - \chi_{k,\gamma}\sqrt{\sum_{i,j}\sigma_{ij}c_ic_j} < \sum_i c_i\mu_i < \sum_i c_iu_i + \chi_{k,\gamma}\sqrt{\sum_{i,j}\sigma_{ij}c_ic_j}$$

for all possible real vectors $(c_1, \ldots, c_k)$, where $c_1, \ldots, c_k$ are not all zero, are simultaneously satisfied.

10.13 Verify formulas (10.6.5) and (10.6.6).

10.14 In Problem 8.20 show how to obtain a $100\gamma\%$ confidence interval for μ.

10.15 Verify (10.6.15) and (10.6.16).

10.16 Formulate statements corresponding to **10.6.1**, **10.6.2**, and set up a Model I analysis of variance table for the complete three-way experimental design.

10.17 Formulate statements corresponding to **10.6.1** and **10.6.2**, and set up a Model I analysis of variance table for the Latin Square experimental design.

10.18 Set up a Model II analysis of variance table for the incomplete balanced two-factor experimental design described in Section 10.8(*b*).

10.19 Verify that the variance component estimators (10.8.6) for the incomplete balanced two-factor experimental design reduce to those in (10.8.4) for the complete two-factor design if $s' = s$ and $r' = r$.

10.20 Write down the estimators corresponding to (10.8.4) and the table corresponding to Table 10.2 for the case of finite populations of N_1 R-factor levels and of N_2 C-factor levels.

10.21 Set up the Model II analysis of variance table for the complete three-factor experimental design.

10.22 Set up the Model II analysis of variance table for the Latin-Square experimental design.

10.23 In the complete two-factor Model I experimental design if $\mu_{\cdot\eta} = 0$, $\eta = 1, \ldots, s$, the experimental design may be described as s-replicates of a complete one-factor design. In this case show that $S^*_{::}/[r(s-1)]$ is an unbiased estimator for σ^2, where $S^*_{..} = S_{..} + S_{0.}(0)$. Assuming normality of distribution as in **10.6.2**, show that $100\gamma\%$ confidence intervals for μ and $\mu_{\xi.}$ in this case are

$$m \pm t_{r(s-1),\gamma}\sqrt{S^*_{..}/[r^2s(s-1)]}$$

$$m_{\xi\cdot} \pm t_{r(s-1),\gamma}\sqrt{S^*_{..}(r-1)/[r^2s(s-1)]}$$

and that

$$\sum_{\xi=1}^{r}(\mu_{\xi\cdot} - m_{\xi\cdot})^2 = \frac{s(r-1)}{r(s-1)}S^*_{..}\cdot\mathscr{F}_{(r-1),r(s-1),\gamma},$$

subject to $\sum_{\xi=1}^{r}(\mu_{\xi\cdot} - m_{\xi\cdot}) = 0$, is an $(r-1)$-dimensional spherical region which gives a $100\gamma\%$ confidence region for $(\mu_{1.}, \ldots, \mu_{r.})$.

10.24 In the complete three-factor Model I experimental design, if $\mu_{..\zeta}$, $\mu_{.\eta\zeta}$, $\mu_{\xi.\zeta}$ are all zero, we have t replicates of a complete two-factor experiment of r rows and s columns. In this case show that $S^*_{...}/[rs(t-1)]$ is an unbiased estimator for σ^2, where $S^*_{...} = S_{...} + S_{0..}(0) + S_{.0.}(0) + S_{00.}(0)$. Also show that

$$\frac{1}{\sigma^2}S^*_{...},\qquad \frac{1}{\sigma^2}S_{..0}(\mu_{\xi\eta\cdot}),\qquad \frac{1}{\sigma^2}S_{.00}(\mu_{\xi..}),\qquad \frac{1}{\sigma^2}S_{0\cdot0}(\mu_{.\eta\cdot})$$

are independently distributed according to chi-square distributions with $rs(t-1)$, $(r-1)(s-1)$, $(r-1)$, $(s-1)$ degrees of freedom, respectively. Set up a Model I analysis of variance table to test the hypotheses: (i) that all $\mu_{\xi\eta\cdot}$ are zero; (ii) that all $\mu_{\xi..}$ are zero; (iii) that all $\mu_{.\eta\cdot}$ are zero.

10.25 Referring to Section 10.6(*b*), show that the probability exceeds γ that the confidence intervals

$$(m_{\xi.} - m_{\xi'.}) \pm \delta\sqrt{2(r-2)/rs}$$

all contain $(\mu_{\xi.} - \mu_{\xi'.})$, $\xi \neq \xi' = 1, \ldots, r$ respectively, where

$$\delta^2 = \frac{S_{..}}{(s-1)}\mathscr{F}_{(r-1),(r-1)(s-1),\gamma}$$

and $\mathscr{F}_{(r-1),(r-1)(s-1),\gamma}$ is the $100\gamma\%$ point on the Snedecor distribution $S((r-1), (r-1)(s-1))$.

10.26 Suppose $\bar{x}_1, \ldots, \bar{x}_k$ are the means and $s_1^2, \ldots, s_k^2$ the sample variances of independent samples of sizes $n_1, \ldots, n_k$ from $N(\mu_1, \sigma^2), \ldots, N(\mu_k, \sigma^2)$ respectively. Show that the probability exceeds γ that all of the $\frac{1}{2}k(k-1)$ confidence intervals $(\bar{x}_i - \bar{x}_j) \pm \delta\sqrt{1/n_i + 1/n_j}$ contain $(\mu_i - \mu_j)$, $i > j = 1, \ldots, k$ respectively, where

$$\delta^2 = \frac{k}{n-k}[(n_1 - 1)s_1^2 + \cdots + (n_k - 1)s_k^2]\,\mathscr{F}_{k,n-k,\gamma}$$

and $\mathscr{F}_{k,n-k,\gamma}$ is the $100\gamma\%$ point of the Snedecor distribution $S(k, n-k)$.

10.27 Referring to the notation and definitions of Sections 10.5(*c*) and 10.6(*a*) show that the probability exceeds γ that the inequalities

$$(\bar{x}_{\xi.} - \bar{x}_{\xi'.}) - D < (\mu_{\xi.} - \mu_{\xi'.}) < (\bar{x}_{\xi.} - \bar{x}_{\xi'.}) + D$$

hold simultaneously for all $\xi \neq \xi' = 1, \ldots, r - 1$ where

$$D = R_{r-2,(r-1)(s-1),\gamma}\sqrt{\frac{(r-2)S_{..}}{rs(r-1)(s-1)}}.$$

10.28 Referring to the notation of Sections 10.5(*c*) and 14.6(*b*) show that for a fixed η the probability exceeds γ that the inequalities

$$(m_{\xi\eta\cdot} - m_{\xi'\eta\cdot}) - D < (\mu_{\xi\eta\cdot} - \mu_{\xi'\eta\cdot}) < (m_{\xi\eta\cdot} - m_{\xi'\eta\cdot}) + D$$

hold simultaneously for all choices of $\xi \neq \xi' = 1, \ldots, r - 1$ where

$$D = R_{r-2,(r-1)(s-1)(t-1),\gamma}\sqrt{\frac{(r-2)S_{...}}{rst(r-1)(t-1)}}.$$

10.29 Referring to Section 10.9(*a*) suppose neither the strata sizes nor strata means nor strata variances are known and that the following double-sampling scheme is used: The first sample is a simple random sample of size r from π_N, which yields $r_1, \ldots, r_m$ elements from $\pi_{N_1}, \ldots, \pi_{N_m}$, respectively, where $r_1 + \cdots + r_m = r$. Then a stratified sample of size n is drawn from π_N so that $n_1 = (r_1/r)n, \ldots, n_m = (r_m/r)n$ elements are drawn at random, respectively, from $\pi_{N_1}, \ldots, \pi_{N_m}$. Let

$$\bar{x}' = \sum_g \frac{r_g}{r}\bar{x}_g$$

where $\bar{x}_g$ is the mean of the sample of size n_g from π_{N_g}. Show that $\bar{x}'$ is an unbiased estimator for the mean μ of π_N. Also show that the variance of $\bar{x}'$ is given by

$$\sigma^2(\bar{x}') = \left[\frac{r(N-n-1)+n}{rnN}\right]\sigma_W^2 + \frac{N-r}{rN}\sigma_B^2 - \frac{n-r}{rnN(N-1)}\sum_{g=1}^{m}\sigma_g^2.$$

10.30 (*Continuation*) Suppose N is large enough to neglect terms of order $1/N$, then

$$\sigma^2(\bar{x}') \cong \frac{\sigma_{W'}^2}{n} + \frac{\sigma_{B'}^2}{r},$$

where $\sigma_{W'}^2$ and $\sigma_{B'}^2$ are defined in (10.10.14).
Suppose the cost of drawing and observing an element in the first sample is c_1 and that of drawing and observing an element in the second sample is c_2. For a fixed total cost $c = c_1 r + c_2 n$ determine the choices of r and n which minimize $\sigma^2(\bar{x}')$.

10.31 If $(x_1, \ldots, x_n)$ is a sample from a distribution having mean μ and variance σ^2, and if u is any unbiased linear estimator for μ show that the correlation coefficient between $\bar{x}$ and u is $\sigma(\bar{x})/\sigma(u)$.

10.32 Suppose L_1 and L_2 are two unbiased linear estimators of a population mean constructed from a sample. Let the variances of L_1 and L_2 be σ_1^2 and σ_2^2 and let the correlation coefficient between L_1 and L_2 be ρ. Determine constants $c_1 > 0$, $c_2 > 0$, and $c_1 + c_2 = 1$, so that $c_1L_1 + c_2L_2$ has minimum variance.

10.33 If n, $\bar{x}$, s^2 are the size, mean, and variance of a sample from a normal distribution show that the probability is γ that the next independent drawing from the same normal distribution gives an x in the interval $\bar{x} \pm t_{n-1,\gamma}\, s\sqrt{(n+1)/n}$ where $t_{n-1,\gamma}$ is defined by (10.2.10).

10.34 *Sheppard's corrections for moments.* Suppose $(x_1, \ldots, x_n)$ is a sample from a p.d.f. $f(x)$. Let h be a random variable independent of $(x_1, \ldots, x_n)$ and having the rectangular distribution $R(0, \delta)$. Let I_α be the interval $(h + \delta\alpha - \frac{1}{2}\delta,\ h + \delta\alpha + \frac{1}{2}\delta]$, $\alpha = \ldots, -2, -1, 0, +1, +2, \ldots$ and let n_α be the number of sample elements which fall in I_α. The quantity

$$M'_{r,\delta} = \frac{1}{n}\sum_\alpha n_\alpha(h + \alpha\delta)^r$$

is seen to be the rth moment of the sample if the sample elements are rounded to the nearest unit of size δ using an arbitrary (and "randomly placed") origin. Show that the characteristic function $\varphi(t)$ of $M'_{r,\delta}$ is given by

$$\varphi(t) = \frac{1}{\delta}\int_{-\frac{1}{2}\delta}^{+\frac{1}{2}\delta}\left[\sum_\alpha p_\alpha e^{it(h+\alpha\delta)^r/n}\right]^n dh$$

where $p_\alpha = \int_{I_\alpha} f(x)\,dx$. Using this characteristic function show that

$$\mathscr{E}(M'_{1,\delta}) = \mu'_1;\ \mathscr{E}(M'_{2,\delta}) = \mu'_2 + \delta^2/12;\ \mathscr{E}(M'_{3,\delta}) = \mu'_3 + \delta^2\mu'_1/4$$

and in general

$$\mathscr{E}(M'_{r,\delta}) = \frac{1}{\delta(r+1)}[\mathscr{E}(x + \tfrac{1}{2}\delta)^{r+1} - \mathscr{E}(x - \tfrac{1}{2}\delta)^{r+1}]$$

where $\mu'_1, \mu'_2, \ldots$ are moments of x and

$$\mathscr{E}(x \pm \tfrac{1}{2}\delta)^{r+1} = \int_{-\infty}^{+\infty}(x \pm \tfrac{1}{2}\delta)^{r+1} f(x)\,dx.$$

Thus, $M'_{1,\delta}$, $M'_{2,\delta} - \delta^2/12$, $M'_{3,\delta} - M'_{1,\delta}\,\delta^2/4, \cdots$ are unbiased estimators for $\mu'_1, \mu'_2, \mu'_3, \ldots$. [Sheppard (1898)].

10.35 Suppose $f(x_1, \ldots, x_k)$ is a function such that $0 < f(x_1, \ldots, x_k) < a_{k+1}$ for every point $(x_1, \ldots, x_k)$ in the k-dimensional interval $I_k = \{(x_1, \ldots, x_k)\colon 0 \leqslant x_i \leqslant a_i,\ i = 1, \ldots, k\}$ and such that the following integral exists

$$J = \int_0^{a_1}\cdots\int_0^{a_k} f(x_1, \ldots, x_k)\,dx_1 \cdots dx_k.$$

Let $x_{1\xi}, \ldots, x_{k+1\xi}$ be independent random variables from the rectangular distributions $R(\frac{1}{2}a_1, a_1), \ldots, R(\frac{1}{2}a_{k+1}, a_{k+1})$, $\xi = 1, \ldots, n$. Let z_ξ be a random variable with value 1 if $x_{k+1\xi} \leqslant f(x_{1\xi}, \ldots, x_{k\xi})$ and 0 if $x_{k+1\xi} > f(x_{1\xi}, \ldots, x_{k\xi})$, $\xi = 1, \ldots, n$, and let

$$\bar{z} = \frac{1}{n}\sum_{\xi=1}^n z_\xi.$$

Show that $(a_1 \cdots a_{k+1})\bar{z}$ is an unbiased estimator for J having variance not exceeding

$$\frac{(a_1 \cdots a_{k+1})^2}{4n}.$$

CHAPTER 11

Nonparametric Statistical Estimation

11.1 INTRODUCTORY REMARKS

In Chapter 10 we considered problems of devising linear and quadratic estimators for such population parameters as means, variances, covariances, and regression coefficients of finite and infinite populations. The process of linear estimation on which such estimators are based is, as we have seen, a relatively simple procedure which requires only mild assumptions about the random variables that enter into the linear estimators; namely, finiteness of means and of covariance matrices of these random variables.

In this chapter we shall consider another class of rather simple estimators for quantiles and functions of quantiles of the c.d.f. of the population under consideration. These estimators, as we shall see, do not depend on the functional form of the c.d.f. of the population from which the sample is drawn. Estimators of this type are sometimes called *nonparametric estimators* to contrast them with *parametric estimators* discussed in Chapter 12. Parametric estimators, as we shall see in Chapter 12, are encountered in dealing with the problem of estimating values of unknown parameters in a c.d.f. having a specified functional form and depending on these parameters.

The basic random variables involved in nonparametric estimation are the order statistics determined by a sample, and the coverages determined by these order statistics. The sampling theory of order statistics and coverages has been discussed in Section 8.7, and some of these results are used in the succeeding sections.

11.2 CONFIDENCE INTERVALS FOR QUANTILES

(a) Case of Small Samples

Suppose we have an infinite population with a continuous c.d.f. $F(x)$. We shall consider first the problem of determining confidence intervals for a given quantile $\underline{x}_p$ from the order statistics of a sample from $F(x)$.

More specifically, let $(x_1, \ldots, x_n)$ be a sample from $F(x)$ and let $(x_{(1)}, \ldots, x_{(n)})$ be the order statistics of the sample from $F(x)$ as defined in Section 8.7. Let $x_{(k_1)}$ and $x_{(k_1+k_2)}$ be any two order statistics of the sample. We know from **8.7.3** that the random variables $u = F(x_{(k_1)})$, $v = F(x_{(k_1+k_2)})$, which are sums of coverages, have the ordered bivariate Dirichlet distribution $D^*(k_1, k_2; n - k_1 - k_2 + 1)$ and hence the p.e. of (u, v) is

$$(11.2.1)\quad f(u, v)\,du\,dv = \frac{\Gamma(n+1)}{\Gamma(k_1)\Gamma(k_2)\Gamma(n - k_1 - k_2 + 1)} \cdot u^{k_1-1}(v-u)^{k_2-1}(1-v)^{n-k_1-k_2}\,du\,dv,$$

for $0 < u < v < 1$, and 0 elsewhere. Now consider the inequality

$$(11.2.2)\qquad F(x_{(k_1)}) < p < F(x_{(k_1+k_2)}).$$

Since $F(x)$ is continuous, (11.2.2) is satisfied if and only if

$$(11.2.3)\qquad x_{(k_1)} < \underline{x}_p < x_{(k_1+k_2)}.$$

Hence the probability that (11.2.3) holds is equal to the probability that (11.2.2) holds, and therefore we can write

$$(11.2.4)\qquad P(x_{(k_1)} < \underline{x}_p < x_{(k_1+k_2)}) = \int_p^1\int_0^p f(u, v)\,du\,dv.$$

But

$$(11.2.5)\quad \int_p^1\int_0^p f(u, v)\,du\,dv = \int_0^p\int_u^1 f(u, v)\,dv\,du - \int_0^p\int_0^v f(u, v)\,du\,dv.$$

By applying the transformations

$$\begin{aligned} u &= r \\ v &= 1 - s(1 - r) \end{aligned} \qquad \text{and} \qquad \begin{aligned} u &= rs \\ v &= s, \end{aligned}$$

respectively, to the first and second integrals on the right of (11.2.5), we find

$$(11.2.6)\quad P(x_{(k_1)} < \underline{x}_p < x_{(k_1+k_2)}) = I_p(k_1, n - k_1 + 1) - I_p(k_1 + k_2, n - k_1 - k_2 + 1)$$

where $I_p(\nu_1, \nu_2)$ is the incomplete beta function

$$(11.2.7)\qquad I_p(\nu_1, \nu_2) = \frac{\Gamma(\nu_1 + \nu_2)}{\Gamma(\nu_1)\Gamma(\nu_2)}\int_0^p x^{\nu_1-1}(1-x)^{\nu_2-1}\,dx.$$

It should be particularly observed that the probability that the interval $(x_{(k_1)}, x_{(k_1+k_2)})$ contains the quantile $\underline{x}_p$ does not depend on $F(x)$, and is therefore a confidence interval for $\underline{x}_p$ having confidence coefficient given by the right-hand side of (11.2.6).

We may summarize as follows:

11.2.1 *If* $(x_1, \ldots, x_n)$ *is a sample from a continuous c.d.f.* $F(x)$ *and if* $x_{(k_1)}$ *and* $x_{(k_1+k_2)}$ *are the* k_1th *and* $(k_1 + k_2)$th *order statistics of the sample, then* $(x_{(k_1)}, x_{(k_1+k_2)})$ *is a confidence interval for the quantile* $\underline{x}_p$ *having confidence coefficient* $I_p(k_1, n - k_1 + 1) - I_p(k_1 + k_2, n - k_1 - k_2 + 1)$.

It should be noted that since k_1 and k_2 are (positive) integers it is possible to find confidence intervals for $\underline{x}_p$ by using two order statistics $x_{(k_1)}$ and $x_{(k_1+k_2)}$ for only those values of the confidence coefficient γ which can be taken on by the right-hand side of (11.2.6). One can, of course, set up confidence intervals whose confidence coefficients are at least γ for certain ranges of values of γ. Furthermore, the two order statistics we have discussed are arbitrary; (11.2.6) holds for *any* two order statistics $x_{(k_1)}$ and $x_{(k_1+k_2)}$. Ordinarily, for a given γ, we would choose order statistics whose ranks are as close together as possible, that is, choose k_1 and k_2 so that k_2 is as small as possible. For instance, in setting up confidence intervals for the median $\underline{x}_{0.5}$, we would select the largest value of k for which

$$P(x_{(k)} < \underline{x}_{0.5} < x_{(n-k+1)}) \geqslant \gamma. \tag{11.2.8}$$

As a matter of fact Nair (1940) has tabulated the values of k for which (11.2.8) holds for $n = 6, 7, \ldots, 81$ and for $\gamma = 0.95$ and 0.99. The exact value of the probability $P(x_{(k)} < \underline{x}_{0.5} < x_{(n-k+1)})$ is $1 - 2I_{0.5}(n - k + 1, k)$, a result found independently by Thompson (1936) and Savur (1937).

(b) Case of Large Samples

Suppose $(x_1, \ldots, x_n)$ is a sample from a continuous c.d.f. $F(x)$, and let $\underline{x}_p$ be the pth quantile. If n_1 is the number of components of the sample which are less than $\underline{x}_p$, then n_1 is a random variable having the binomial distribution $Bi(n, p)$. We know from **9.2.1a** that for large n, n_1 is asymptotically distributed according to $N(np, np(1 - p))$. Hence for a given confidence coefficient γ, we have

$$\lim_{n\to\infty} P\left(-y_\gamma < \frac{n_1 - np}{\sqrt{np(1 - p)}} < +y_\gamma\right) = \gamma \tag{11.2.9}$$

where

$$\frac{1}{\sqrt{2\pi}} \int_{-y_\gamma}^{y_\gamma} e^{-\frac{1}{2}y^2}\, dy = \gamma. \tag{11.2.10}$$

Thus, for large n, an approximate $100\gamma\%$ confidence interval $(\underline{p}_\gamma, \bar{p}_\gamma)$ for p

is given by the set of all values of p satisfying the inequality in (11.2.9) for fixed n_1, n, and y_γ, that is, $\underline{p}_\gamma$ and $\bar{p}_\gamma$ are the two solutions of

$$\frac{(n_1 - np)^2}{np(1-p)} = y_\gamma^2 \tag{11.2.11}$$

for p.

Hence, we have for large n,

$$P(\underline{p}_\gamma < p < \bar{p}_\gamma) \cong \gamma \tag{11.2.12}$$

which is equivalent to

$$P(x_{([n\underline{p}_\gamma])} < \underline{x}_p < x_{([n\bar{p}_\gamma])}) \cong \gamma. \tag{11.2.13}$$

Therefore, the two order statistics $(x_{([n\underline{p}_\gamma])}, x_{([n\bar{p}_\gamma])})$ constitute an approximate $100\gamma\%$ confidence interval for the quantile $\underline{x}_p$, where $[n\underline{p}_\gamma]$ and $[n\bar{p}_\gamma]$ are the largest integers in $n\underline{p}_\gamma$ and $n\bar{p}_\gamma$ respectively.

11.3 CONFIDENCE INTERVALS FOR QUANTILE INTERVALS

Suppose $p_1 < p_2$ are any two quantiles, and $x_{(k_1)}$ and $x_{(k_1+k_2)}$ are order statistics in a sample of size n. Then by argument similar to that by which we obtained (11.2.4) we find

$$P(x_{(k_1)} < \underline{x}_{p_1} < \underline{x}_{p_2} < x_{(k_1+k_2)}) = \int_0^{p_1}\int_{p_2}^{1} f(u, v)\, du\, dv, \tag{11.3.1}$$

the left side of which may be rewritten as

$$\begin{aligned} &P((\underline{x}_{p_1}, \underline{x}_{p_2}) \subset (x_{(k_1)}, x_{(k_1+k_2)})) \\ &\quad = \frac{n!}{k_2!}\sum_{i=0}^{k_1-1}\frac{(-1)^i p_1^{i+k_1}}{i!\,(n-k_2-i)!} I_{p_1}(n - k_1 - k_2 + 1; k_1 - i). \end{aligned} \tag{11.3.2}$$

If for a given γ there exist values of k_1, k_2, and n so that the probability expressed by (11.3.2) has the value γ, then we would say that $(x_{(k_1)}, x_{(k_1+k_2)})$ is a $100\gamma\%$ *outer confidence interval* for the quantile interval $(\underline{x}_{p_1}, \underline{x}_{p_2})$.

In a similar manner a $100\gamma\%$ *inner confidence interval* $(x_{(k_1)}, x_{(k_1+k_2)})$ for $(\underline{x}_{p_1}, \underline{x}_{p_2})$ would be obtained by considering the relation

$$P(\underline{x}_{p_1} < x_{(k_1)} < x_{(k_1+k_2)} < \underline{x}_{p_2}) = \int_{p_1}^{p_2}\int_{p_1}^{v} f(u, v)\, du\, dv \tag{11.3.3}$$

where k_1, k_2, and n have values for which the probability expressed by (11.3.3) is equal to γ, and $f(u, v)$ is given by (11.2.1).

In the case of outer (inner) confidence intervals the "best" choice of k_1 and k_2 would be regarded as that for which k_2 is as small (large) as possible. For instance, if $p_1 = p$ and $p_2 = 1 - p$, $2p < 1$, it can be shown that the "best" choice of k_1 and k_2 in the sense just mentioned for both inner and outer confidence intervals would be that they be of the form $x_{(c)}$ and $x_{(n-c+1)}$ respectively.

11.4 CONFIDENCE INTERVALS FOR QUANTILES IN FINITE POPULATIONS

Suppose π_N is a finite population whose elements have distinct x-values $x_{o1} < \cdots < x_{oN}$. It was shown in Section 8.8 that the p.f. of the kth order statistic $x_{(k)}$ in a sample of size n from π_N is

$$p_{N,n,k}(t) = \frac{\binom{t-1}{k-1}\binom{N-t}{n-k}}{\binom{N}{n}}, \tag{11.4.1}$$

the mass points of $x_{(k)}$ being x_{ot}, $t = k, k+1, \ldots, N-n+k$. Now for any fixed value of t, say t'

$$P(x_{(k)} \leqslant x_{ot'}) = \sum_{t=k}^{t'} p_{N,n,k}(t). \tag{11.4.2}$$

If, for fixed N, n, t' and $\gamma > 0$, there is a largest k, say k' such that

$$\sum_{t=k'}^{t'} p_{N,n,k'}(t) \geqslant \gamma, \tag{11.4.3}$$

we would regard $x_{(k')}$ as the "best" lower $100\gamma\%$ confidence limit for $x_{ot'}$. We may regard $x_{ot'}$ as the (t'/N)th quantile of the population π_N. Except for values of N, n, t' and $1-\gamma$ which are uninterestingly small such lower confidence limits can be shown to exist.

In a similar manner the "best" upper $100\gamma\%$ confidence limit for $x_{ot'}$ is obtained by choosing the smallest k, say k'', for which

$$\sum_{t=t'}^{N-n+k''} p_{N,n,k''}(t) \geqslant \gamma. \tag{11.4.4}$$

One can also write down a "best" $100\gamma\%$ confidence interval for $x_{ot'}$, that is, simultaneous upper and lower confidence limits, but the p.f. involved here is more cumbersome than the p.f. $p_{N,n,k}(t)$, and we shall not write it out. It can be shown, however, that if in a population π_N having distinct x-values $x_{o1} < \cdots < x_{oN}$, and if $x_{(k_1)}, x_{(k_1+k_2)}$ are the indicated order statistics of a sample of size n from π_N

$$\lim_{N\to\infty} P(x_{(k_1)} < x_{o(Np)} < x_{(k_1+k_2)}) = I_p(k_1, n-k_1+1) - I_p(k_1+k_2, n-k_1-k_2+1). \tag{11.4.5}$$

Hence, for large N, $(x_{(k_1)}, x_{(k_1+k_2)})$ is a confidence interval for $x_{o(Np)}$ with coefficient given approximately by the right-hand side of (11.4.5).

11.5 TOLERANCE LIMITS

(a) Case of Small Samples

Suppose $(x_1, \ldots, x_n)$ is a sample from a continuous c.d.f. $F(x)$. Let $L_1(x_1, \ldots, x_n) < L_2(x_1, \ldots, x_n)$ be any two observable symmetric functions of $(x_1, \ldots, x_n)$ such that the distribution of $F(L_2) - F(L_1)$ does not depend on $F(x)$, and such that for $0 < \beta < 1$,

$$P[(F(L_2) - F(L_1)) \geqslant \beta] = \gamma. \tag{11.5.1}$$

Then L_1 and L_2 will be called $100\beta\%$ *distribution-free tolerance limits at probability level* γ. This concept is due to Shewhart (1931).

As pointed out by Wilks (1941, 1942), order statistics can be used for distribution-free tolerance limits whenever $F(x)$ is continuous. To see this, suppose $(x_{(1)}, \ldots, x_{(n)})$ are the order statistics of a sample from an unknown continuous c.d.f. $F(x)$. For any two order statistics $x_{(k_1)}$, $x_{(k_1+k_2)}$ the amount of probability in the population distribution contained in the interval $(x_{(k_1)}, x_{(k_1+k_2)})$, which we shall call U_{k_2}, is a random variable. It will be seen from Section 8.7(*b*) that U_{k_2} is the sum of k_2 coverages, and that

$$U_{k_2} = F(x_{(k_1+k_2)}) - F(x_{(k_1)}). \tag{11.5.2}$$

It follows from **8.7.6** that U_{k_2} has the beta distribution $Be(k_2, n - k_2 + 1)$, which, of course, does not depend on the population c.d.f. $F(x)$. Now for given values of $\beta > 0$ and $\gamma > 0$, suppose n, k_1, and k_2 exist so that

$$P(U_{k_2} \geqslant \beta) = \gamma. \tag{11.5.3}$$

Then $x_{(k_1)}$ and $x_{(k_1+k_2)}$ are $100\beta\%$ distribution-free tolerance limits at probability level γ. Note that U_{k_2} is a special form of $F(L_2) - F(L_1)$ in (11.5.1). Robbins (1944*b*) has shown that if $P(F(L_2) - F(L_1) \geqslant \beta) = \gamma$ for all absolutely continuous $F(x)$, then L_1 and L_2 are necessarily order statistics.

Suppose $k_1 = c$ and $k_2 = n - 2c + 1$, thus making the interval $(x_{(c)}, x_{(n-c+1)})$ symmetrically chosen from the order statistics. If we use incomplete beta function notation, (11.5.3) reduces to

$$1 - I_\beta(n - 2c + 1, 2c) = \gamma. \tag{11.5.4}$$

For a fixed c there may exist no sample size n for which this equation holds exactly. However, since the probability computed from the beta distribution $Be(n - 2c + 1, 2c)$ on the interval $(1 - \varepsilon, 1)$, for any $\varepsilon > 0$, can be made arbitrarily near 1 by taking n sufficiently large, there exists a *smallest* n for which

$$1 - I_\beta(n - 2c + 1, 2c) \geqslant \gamma. \tag{11.5.5}$$

For instance, if we choose $\beta = 0.99$ and $\gamma = 0.95$ and $c = 1$ (that is, $k_1 = 1, k_2 = n - 1$) we find $n = 473$. Murphy (1948) has tabulated, and has presented in graphical form, values of β for which (11.5.3) holds for $n = 1(1)10(10)100(100)500$, $\gamma = 0.90, 0.95, 0.99$, and for $n - k_2 + 1 = m = 1(1)6(2)10(5)30(10)60(20)100$. Somerville (1958) has extended Murphy's results in tabular form.

The notion of distribution-free tolerance limits for one-dimensional distributions having continuous c.d.f.'s extends to the case of multi-dimensional distributions having continuous c.d.f.'s. Thus, suppose we have a sample of size n from an infinite population having a continuous k-dimensional c.d.f. $F(x_1, \ldots, x_k)$. Suppose the k-dimensional sample space of $(x_1, \ldots, x_k)$ is cut up into $n + 1$ mutually exclusive and exhaustive sample blocks $B_k^{(1)}, \ldots, B_k^{(n+1)}$ by ordering functions as described in Section 8.7(*c*). Consider any rule for choosing some r of these sample blocks and let the union of the blocks be T_r. Let the sum of the coverages for the selected blocks be U_r, that is,

$$U_r = \int_{T_r} dF(x_1, \ldots, x_k). \tag{11.5.6}$$

Then U_r is a random variable, and it follows from **8.7.9** that U_r has the beta distribution $Be(r, n - r + 1)$. Thus, if for given $\beta > 0$ and $\gamma > 0$,

$$P(U_r \geqslant \beta) = \gamma. \tag{11.5.7}$$

U_r would be a $100\beta\%$ distribution-free *tolerance region* at probability level γ. If, for any fixed positive integer c, we choose $r = n - 2c + 1$, then (11.5.7) reduces to (11.5.4), the same equation we had for one-dimensional tolerance limits.

(b) Case of Finite Populations

If t' is chosen so that $t'/N = (1 - \beta)$ then (11.4.2) can be written as

$$P(x_{(k')} \leqslant x_{o(N(1-\beta))}) \geqslant \gamma. \tag{11.5.8}$$

in which case we may regard $x_{(k')}$ as a $100\beta\%$ *lower tolerance limit* at probability level γ, that is, the probability is at least γ that the fraction of elements in π_N having x-values exceeding $x_{(k')}$ is β.

Similarly, if $t' = \beta N$, we can rewrite (11.4.4) as

$$P(x_{(k'')} \geqslant x_{o(N\beta)}) \geqslant \gamma, \tag{11.5.9}$$

thus making $x_{(k'')}$ a $100\beta\%$ *upper tolerance limit* at probability level γ.

It is evident how a $100\beta\%$ *tolerance interval* at probability level γ would be defined.

If, for a large population π_N having distinct x-values, we let $U_{k_2,N}$ be the fraction of x-values in π_N lying in the interval $(x_{(k_1)}, x_{(k_1+k_2)})$ it can be shown that the limiting distribution of $U_{k_2,N}$, as $N \to \infty$, is $Be(k_2, n - k_2 + 1)$. Hence, for large N, the interval is an approximate $100\beta\%$ tolerance interval at probability level γ.

11.6 ONE-SIDED CONFIDENCE CONTOURS FOR A CONTINUOUS DISTRIBUTION FUNCTION

One of the fundamental problems in applying the sampling theory of order statistics to statistical inference is to construct functions $F_n^+(x)$ and $F_n^-(x)$ from the order statistics of a sample from a population having a continuous c.d.f. $F(x)$ so that for a given γ

$$\begin{aligned} &P(F_n^+(x) \geqslant F(x);\ \text{for all } x) = \gamma \\ &P(F_n^-(x) \leqslant F(x);\ \text{for all } x) = \gamma. \end{aligned} \tag{11.6.1}$$

Such functions $F_n^+(x)$ and $F_n^-(x)$ will be called, respectively, *upper* and *lower* $100\gamma\%$ *confidence contours* for $F(x)$.

An asymptotic solution for this problem for large n was originally obtained by Smirnov (1939*a*). More recently, Smirnov (1944), and also Birnbaum and Tingy (1951) have obtained a simple expression for the probabilities (11.6.1) that we shall present here.

Let $(x_{(1)}, \ldots, x_{(n)})$ be the order statistics of a sample from an infinite population having c.d.f. $F(x)$. The *empirical* c.d.f. $F_n(x)$ is constructed from the order statistics as follows:

$$F_n(x) = \begin{cases} 0 & x < x_{(1)} \\ \dfrac{\xi - 1}{n} & x_{(\xi-1)} \leqslant x < x_{(\xi)}, \qquad \xi = 2, \ldots, n \\ 1 & x \geqslant x_{(n)}. \end{cases} \tag{11.6.2}$$

For $0 < d \leqslant 1$ and for any value of x let

$$\begin{aligned} F_n^+(x, d) &= \min\ [F_n(x) + d; 1] \\ F_n^-(x, d) &= \max\ [F_n(x) - d; 0] \end{aligned} \tag{11.6.3}$$

and let

$$\begin{aligned} D_n^+(d) &= \inf_x\ (F_n^+(x, d) - F(x)) \\ D_n^-(d) &= \sup_x\ (F_n^-(x, d) - F(x)). \end{aligned} \tag{11.6.4}$$

Note that $P(D_n^+(d) > 0)$ is the probability that the graph of $F_n(x) + d$ never meets the graph of $F(x)$; and similarly $P(D_n^-(d) < 0)$ is the probability that the graph of $F_n(x) - d$ never meets that of $F(x)$.

The main result can be stated as follows:

11.6.1 *Let $(x_{(1)}, \ldots, x_{(n)})$ be the order statistics of a sample from a continuous c.d.f. $F(x)$, and let $F_n^+(x, d)$ and $F_n^-(x, d)$ be constructed from the empirical c.d.f. $F_n(x)$ as in (11.6.3). Then $P(D_n^+(d) > 0) = P(D_n^-(d) < 0) = P_n(d)$ where*

$$P_n(d) = 1 - d \sum_{i=1}^{[n(1-d)]} \binom{n}{i} \left(1 - d - \frac{i}{n}\right)^{n-i} \left(d + \frac{i}{n}\right)^{i-1} \tag{11.6.5}$$

and $[n(1 - d)]$ is the largest integer contained in $n(1 - d)$.

To establish **11.6.1** we first transform to new random variables $y_{(1)}, \ldots, y_{(n)}$ defined by

$$y_{(\xi)} = F(x_{(\xi)}), \qquad \xi = 1, \ldots, n. \tag{11.6.6}$$

Then we know from (8.7.2) that $y_{(1)}, \ldots, y_{(n)}$ have p.e.

$$n!\, dy_{(1)} \cdots dy_{(n)} \tag{11.6.7}$$

inside the region $0 < y_{(1)} < \cdots < y_{(n)} < 1$ and 0 elsewhere.

First consider $P(D_n^+(d) > 0)$. It is evident that the order statistics $x_{(1)}, \ldots, x_{(n)}$ will satisfy the inequality $D_n^+(d) > 0$ if and only if the order statistics $y_{(1)}, \ldots, y_{(n)}$ satisfy the following set of inequalities:

$$\begin{aligned} &y_{(\xi-1)} < y_{(\xi)} < \frac{\xi - 1}{n} + d, &&\xi = 1, \ldots, k + 1 \\ &y_{(\xi-1)} < y_{(\xi)} < 1, &&\xi = k + 2, \ldots, n \end{aligned} \tag{11.6.8}$$

where k is the largest integer in $n(1 - d)$ and $y_{(0)} = 0$. Similarly, $D_n^-(d) < 0$ will be satisfied if and only if

$$\begin{aligned} &\frac{\xi}{n} - d < y_{(\xi)} < y_{(\xi+1)}, &&\xi = n - k, \ldots, n \\ &0 < y_{(\xi)} < y_{(\xi+1)}, &&\xi = 1, \ldots, n - k - 1 \end{aligned} \tag{11.6.9}$$

where $y_{(n+1)} = 1$. But if we make the change of variables $y_{(\xi)} = 1 - y'_{(n-\xi+1)}$, $\xi = 1, \ldots, n$, then the order statistics $y'_{(1)}, \ldots, y'_{(n)}$ have p.e. (11.6.7) and the inequalities (11.6.9) reduce to (11.6.8). Therefore, $P(D_n^+(d) > 0) = P(D_n^-(d) < 0)$. Therefore, the common value of these probabilities, say $P_n(d)$, is obtained by integrating the p.e. (11.6.7) over the region within $0 < y_{(1)} < \cdots < y_{(n)} < 1$ determined by the inequalities (11.6.8). That is,

$$P_n(d) = n!\, G_n(k, d) \tag{11.6.10}$$

where

$$(11.6.11)\quad G_n(k, d) = \int_0^d \int_{y_{(1)}}^{1/n+d} \cdots \int_{y_{(k)}}^{k/n+d} \int_{y_{(k+1)}}^{1} \cdots \int_{y_{(n-1)}}^{1} dy_{(n)} \cdots dy_{(1)}.$$

Integrating with respect to $y_{(n)}, \ldots, y_{(k+2)}$, we find

(11.6.12)

$$G_n(k, d) = \int_0^d \int_{y_{(1)}}^{1/n+d} \cdots \int_{y_{(k)}}^{k/n+d} \frac{(1 - y_{(k+1)})^{n-k-1}}{(n-k-1)!} dy_{(k+1)} \cdots dy_{(1)}.$$

From here on the evaluation of $G_n(k, d)$ proceeds by induction. Thus, by integrating with respect to $y_{(k+2)}$ in $G_n(k + 1, d)$, one finds

$$(11.6.13)\quad G_n(k+1, d) = G_n(k, d) - \frac{\left(1 - d - \dfrac{k+1}{n}\right)^{n-k-1}}{(n-k-1)!} H_n(k, d)$$

where

$$(11.6.14)\quad H_n(k, d) = \int_0^d \int_{y_{(1)}}^{1/n+d} \cdots \int_{y_{(k)}}^{k/n+d} dy_{(k+1)} \cdots dy_{(1)}.$$

Omitting the details of a mathematical induction, we find

$$(11.6.15)\quad H_n(k, d) = -\frac{d}{(k+1)!} \sum_{i=1}^{k+1} (-1)^i \binom{k+1}{i} \left(d + \frac{k+1-i}{n}\right)^k,$$

which can be written as

$$(11.6.16)\quad H_n(k, d) = -\frac{d}{(k+1)!} \frac{\partial^k}{\partial t^k} \{e^{(d+(k+1)/n)t}[(1 - e^{-t/n})^{k+1} - 1]\}_{t=0}.$$

But we note that

$$(11.6.17)\quad \left\{\frac{\partial^k}{\partial t^k} [e^{(d/(k+1)+1/n)t} - e^{(d/(k+1))t}]^{k+1}\right\}_{t=0} = 0.$$

Therefore,

(11.6.18)

$$H_n(k, d) = \frac{d}{(k+1)!} \left\{\frac{\partial^k}{\partial t^k} e^{(d+(k+1)/n)t}\right\}_{t=0} = \frac{d}{(k+1)!} \left(d + \frac{k+1}{n}\right)^k.$$

Substituting this value for $H_n(k, d)$ in (11.6.13) and noting that

$$(11.6.19)\quad G_n(0, d) = \frac{1 - (1 - d)^n}{n!}$$

we find by induction the value of $G_n(k, d)$, which, when inserted in (11.6.10), and noting that k is the largest integer in $n(1 - d)$, yields the expression in (11.6.5) for $P_n(d)$, thus completing the argument for **11.6.1**.

Birnbaum and Tingey have tabulated values of d for which $P_n(d) = \gamma$ for $\gamma = 0.90, 0.95, 0.99, 0.999$, and $n = 5, 8, 10, 20, 40, 50$. For values of n greater than 50 the Smirnov asymptotic approximation (11.6.22) provides a good approximation for values of d.

Now let us consider the limit of $P_n(d)$ as $n \to \infty$. Putting $d = \lambda/\sqrt{n}$, we may write (11.6.5) as follows:

(11.6.20)

$$P_n\left(\frac{\lambda}{\sqrt{n}}\right) = 1 - \lambda\left\{\sum_{i=1}^{[n-\sqrt{n}\lambda]} \sqrt{n}\binom{n}{i}\left(1 - \frac{\lambda}{\sqrt{n}} - \frac{i}{n}\right)^{n-i}\left(\frac{\lambda}{\sqrt{n}} + \frac{i}{n}\right)^{i-1} \Delta y\right\}$$

where $\Delta y = 1/n$. Making use of Stirling's approximation for large factorials (7.6.29), it can be verified that the sum in { } converges to the integral

$$\frac{1}{\sqrt{2\pi}} \int_0^1 \frac{1}{\sqrt{y(1-y)}} e^{-\lambda^2/2y(1-y)}\, dy \tag{11.6.21}$$

as $n \to \infty$, which, when integrated, yields

$$\frac{1}{\lambda} e^{-2\lambda^2}.$$

Hence, we have the following result:

11.6.2 *Under the conditions of* **11.6.1**,

$$\lim_{n\to\infty} P_n\left(\frac{\lambda}{\sqrt{n}}\right) = 1 - e^{-2\lambda^2}. \tag{11.6.22}$$

This result was originally established by Smirnov (1939*a*), but the more direct derivation from (11.6.5) is due to Dempster (1955).

11.7 CONFIDENCE BANDS FOR A CONTINUOUS DISTRIBUTION FUNCTION

The problem of establishing two-sided contours, or a *confidence band*, for $F(x)$, that is, of determining the value of the probability

$$P(D_n < d) \tag{11.7.1}$$

where

$$D_n = \sup_x |F_n(x) - F(x)|$$

for arbitrary d, is more difficult than the problem of the one-sided contour discussed in the preceding section. Kolmogorov (1933*b*) gave an asymptotic solution to this problem for large n and recurrence formulas from which values of (11.7.1) can be computed for finite n. Wald and Wolfowitz (1939) have given a solution for finite n in determinant form. More recently, Massey (1950) has given a fairly simple solution to the problem if n is finite and if d is a multiple of $1/n$. His solution is in the form of recurrence formulas which we shall consider here. Massey's result can be stated as follows:

11.7.1 *Let $(x_{(1)}, \ldots, x_{(n)})$ be the order statistics of a sample of size n from a continuous c.d.f. $F(x)$. Let*

$$D_n = \sup_x |F_n(x) - F(x)| \tag{11.7.2}$$

where $F_n(x)$ is defined in (11.6.2). *Then*

$$P\left(D_n < \frac{k}{n}\right) = \frac{n!}{n^n} U(k, n), \qquad k = 1, \ldots, n-1 \tag{11.7.3}$$

where $U(j, m+1)$, $j = 1, \ldots, 2k-1$; $m = 0, 1, \ldots, n-1$, satisfy the system of equations

$$U(j, m+1) = \sum_{i=1}^{j+1} \frac{U(i, m)}{(j+1-i)!} \tag{11.7.4}$$

subject to the boundary conditions

$$\begin{aligned} U(i, m) &= 0, \qquad i \geqslant m + k \\ U(i, 0) &= 0, \qquad i = 1, \ldots, k-1 \\ U(k, 0) &= 1. \end{aligned} \tag{11.7.5}$$

To establish this result suppose $(x_1, \ldots, x_n)$ is a sample from the continuous c.d.f. $F(x)$, and for convenience let us denote the random variable $F(x_\xi)$ by y_ξ, $\xi = 1, \ldots, n$. Then y_ξ has the rectangular distribution on the interval (0, 1), that is, $R(\frac{1}{2}, 1)$. Let $G(y)$ be the c.d.f. of y and $G_n(y)$ be the empirical c.d.f. constructed from the order statistics $(y_{(1)}, \ldots, y_{(n)})$. Then, of course,

$$P\left(D_n < \frac{k}{n}\right) = P\left(\sup_y |G_n(y) - G(y)| < \frac{k}{n}\right). \tag{11.7.6}$$

We now cut up the interval (0, 1] into n equal intervals $I_\xi = ((\xi - 1)/n, \xi/n)$, $\xi = 1, \ldots, n$. Let $(r_1, \ldots, r_n)$ be a random variable (degenerate, with $r_1 + \cdots + r_n = n$) denoting the numbers of elements in the sample $(y_1, \ldots, y_n)$ falling into $I_1, \ldots, I_n$, respectively. The r's, of course, have the $(n-1)$-dimensional multinomial distribution $M(n; 1/n, \ldots, 1/n)$ whose p.f. is

$$p(r_1, \ldots, r_n) = \frac{n!}{r_1! \cdots r_n! \, n^n}. \tag{11.7.7}$$

Now $(r_1, \ldots, r_n)$ uniquely determines $G_n(y)$, and hence the value of $P(D_n < k/n)$ is determined by summing $p(r_1, \ldots, r_n)$ over all points in the sample space of $(r_1, \ldots, r_n)$ for which $|G_n(y) - G(y)| < k/n$ for all y. As we follow the graph of $G_n(y)$ from left to right, which lies completely within the band E for which $|G_n(y) - G(y)| < k/n$ for $y = 1/n, 2/n, \ldots, (m+1)/n$, the graph must pass through one of the points $(m/n, (m-k+i)/n)$, $i = 1, \ldots, 2k-1$. Let us call these points $A(i, m)$, $i = 1, \ldots, 2k-1$ respectively. Now let

$$U(i, m) = \textstyle\sum_{(i)} \frac{1}{r_1! \cdots r_m!} \tag{11.7.8}$$

where $\sum_{(i)}$ denotes summation over all sets of values of $(r_1, \ldots, r_m)$ for which the graph of $G_n(y)$ arrives at $A(i, m)$ while remaining inside the band E. Then, since $G_n(y)$ is nondecreasing, its path can reach $A(j, m+1)$ only by having passed through one of the points $A(1, m), A(2, m), \ldots, A(j+1, m)$ and having r_{m+1} take on the values $j, j-1, \ldots, 1, 0$, respectively. Therefore, we must have

$$U(j, m+1) = \frac{U(1, m)}{j!} + \frac{U(2, m)}{(j-1)!} + \cdots + \frac{U(j+1, m)}{0!} \tag{11.7.9}$$

for $j = 1, 2, \ldots, 2k-1$ and $m = 0, 1, \ldots, n-1$, where, of course, $U(i, m) = 0$ if $i \geqslant m + k$. Furthermore, it is evident that $U(i, m)$ satisfies the following boundary conditions:

$$\begin{aligned} U(i, 0) &= 0, \qquad i = 1, \ldots, k-1 \\ U(k, 0) &= 1. \end{aligned} \tag{11.7.10}$$

If the system of difference equations (11.7.9) is solved for $U(k, n)$ subject to conditions (11.7.8), we obtain $P(D_n < k/n)$ by formula (11.7.3), thus completing the argument for **11.7.1**.

Various values of $P(D_n < k/n)$ for $n = 5(5)80$ and $k = 1(1)9$ have been tabulated by Massey, and from these values he has computed by interpolation values of $P(D_n < \lambda/\sqrt{n})$ for $n = 10(10)80$ and $\lambda = 0.9(0.1)1.40$.

Kolmogorov's (1933*b*) asymptotic result referred to earlier is that under the conditions of **11.7.1**

$$\lim_{n \to \infty} P\left(D_n < \frac{\lambda}{\sqrt{n}}\right) = \sum_{i=-\infty}^{+\infty} (-1)^i e^{-2i^2\lambda^2}. \tag{11.7.11}$$

Kolmogorov's proof of this result is complicated. A simpler proof has been given by Feller (1948). Doob (1949) and Donsker (1952) have provided a proof by the use of Gaussian stochastic process theory. Darling (1957) has published an expository article covering this and related problems, which has an extensive list of references.

It should be noted that (11.7.11) implies that for arbitrary $\varepsilon > 0$,

$$\lim_{n \to \infty} P(D_n < \varepsilon) = 1, \tag{11.7.12}$$

that is, $F_n(x)$ converges in probability to $F(x)$ uniformly in x as $n \to \infty$.

PROBLEMS

11.1 If in the sample space of (u, v), whose p.e. is (11.2.1), E is the set of points for which $u < p$ and F is the set for which $v > p$, derive (11.2.6) from the basic probability law $P(E \cup F) = P(E) + P(F) - P(E \cap F)$.

11.2 In (11.2.8) show that

$$P(x_{(k)} < \underline{x}_{0.5} < x_{(n-k+1)}) = 1 - 2I_{\frac{1}{2}}(n - k + 1, k).$$

11.3 If $(x_{(1)}, \ldots, x_{(n)})$ are the order statistics of a sample from the continuous c.d.f. $F(x)$, show that the probability is $1 - n\beta^{n-1} + (n - 1)\beta^n$ that

$$F(x_{(n)}) - F(x_{(1)}) > \beta.$$

(That is, the fraction of the population contained in the sample range exceeds β with probability $1 - n\beta^{n-1} + (n - 1)\beta^n$.)

11.4 (*Continuation*) Show that for any positive values of δ_1 and δ_2 for which $\delta_1 + \delta_2 < 1$, the probability is $1 - (1 - \delta_1)^n - (1 - \delta_2)^n + (1 - \delta_1 - \delta_2)^n$ that both of the following inequalities hold

$$F(x_{(1)}) < \delta_1$$
$$1 - F(x_{(n)}) < \delta_2.$$

11.5 (*Continuation*) Show that for any fixed integer $k < n/2$ the maximum value of $P(x_{(k+r)} < \underline{x}_{0.5} < x_{(n-k+r+1)})$ over all possible values of r (positive or negative integers or zero) occurs for $r = 0$, where $F(\underline{x}_{0.5}) = 0.5$.

11.6 (*Continuation*) Show that

$$P(x_{(k)} < \underline{x}_p) = I_p(k, n - k + 1)$$

where $F(\underline{x}_p) = p$.

11.7 (*Continuation*) Show that $(x_{(k)}, x_{(n-k+1)})$ may be regarded as a confidence interval having confidence coefficient

$$\gamma_m = \sum_{t=k}^{n-k} p_m(t)$$

for the median of a further independent sample of size $2m + 1$ from the same distribution where

$$p_m(t) = \frac{(m + 1)}{(m + t + 1)} \cdot \frac{\binom{2m + 1}{m}\binom{n}{t}}{\binom{2m + n + 1}{m + t + 1}}.$$

Show that the limit of γ_m as $m \to \infty$ is $1 - 2I_{0.5}(n - k + 1, k)$.

11.8 In (11.2.6) show that if $p = 0.5$,

$$P(x_{(k_1)} < \underline{x}_{0\cdot 5} < x_{(k_1+k_2)}) = \left(\frac{1}{2}\right)^n \sum_{\xi=k_1}^{k_1+k_2-1} \binom{n}{\xi}.$$

11.9 If $(x_{(1)}, \ldots, x_{(n)})$ are the order statistics of a sample of size n from a finite population of size $2m + 1$ whose elements have different values of x, show that $(x_{(k)}, x_{(n-k+1)})$ is a confidence interval for the population median having confidence coefficient γ_m identical with that given in the preceding problem.

11.10 If a sample of size n is drawn from a continuous c.d.f. show that the probability is

$$\frac{n(n-1)}{(m+n)}\binom{m}{t}\bigg/\binom{m+n-1}{m-t+1},$$

that t x's in a further independent sample of size m will fall within the range of the first sample.

11.11 (*Continuation*) Let $U_{1,n,m}$ be the fraction of the x's in the second sample which lie within the range of the first. Show that the limiting distribution of $U_{1,n,m}$ as $m \to \infty$ is the beta distribution $Be(n-1,2)$. More generally if $x_{(k_1)}$ and $x_{(k_1+k_2)}$ are the indicated order statistics in the first sample, and if $U_{k_1,k_1+k_2,m}$ is the fraction of x's in the second sample lying in the interval $(x_{(k_1)}, x_{(k_1+k_2)})$, show that the limiting distribution of $U_{k_1,k_1+k_2,m}$ is the beta distribution $Be(k_2, n-k_2+1)$.

11.12 Suppose a sample of size n from a continuous two-dimensional c.d.f. is represented by n points in the xy-plane. Vertical lines are drawn, respectively, through the two points having the smallest and the largest x-coordinates. Horizontal lines are drawn through the two points of the remaining $n-2$ points which have the smallest and largest y-coordinates. Consider the fraction U of the population contained in the rectangle bounded by these four lines. (Note that U is the coverage of the rectangle.) Show that the probability is

$$1 - 4\binom{n}{4}\beta^{n-3}\left[\frac{1}{n-3} - 3\frac{\beta}{n-2} + 3\frac{\beta^2}{n-1} - \frac{\beta^3}{n}\right]$$

that the coverage U exceeds β.

11.13 Referring to Section 11.5(*a*) for the definition of U_{n+1-k} show that for fixed k and $n \to \infty$, $\beta \to 1$, so that $n(1-\beta) \to \lambda$,

$$\lim P(U_{n+1-k} > \beta) = 1 - e^{-\lambda}\left[1 + \lambda + \frac{\lambda^2}{2!} + \cdots + \frac{\lambda^k}{k!}\right].$$

11.14 Verify (11.4.5).

11.15 Verify that $H_n(k, d)$ is given by (11.6.18).

11.16 Show that the quantity in { } in (11.6.20) converges to the integral (11.6.21) as $n \to \infty$.

11.17 Let $\bar{x}$ and s^2 be the mean and variance of a sample of size n from $N(\mu, \sigma^2)$ and let $G(x)$ be the c.d.f. of $N(\mu, \sigma^2)$. Let (L_1, L_2) be the interval $\bar{x} \pm \lambda s$, $\lambda > 0$. Show that $G(L_2) - G(L_1)$, as a coverage, has a distribution depending only on λ and that for $\lambda = t_{n-1,\gamma}\sqrt{(n+1)/n}$, where $t_{n-1,\gamma}$ is defined by (10.2.10), $\mathscr{E}(G(L_2) - G(L_1)) = \gamma$. [See Wilks (1941) and Wald and Wolfowitz (1946).]

CHAPTER 12

Parametric Statistical Estimation

Many problems of statistical estimation deal with the problem of sampling from a c.d.f. of specified functional form $F(x; \theta)$, where θ is an unknown (real) parameter whose true value θ_0 is to be estimated from the elements of a sample $(x_1, \ldots, x_n)$ assumed to have been drawn from $F(x; \theta)$. The true value θ_0 is a point in a *parameter space* Ω of values of θ, where Ω is an open interval or region (or all) of a Euclidean space. The parameter space Ω is sometimes called the set of *admissible values* of θ. Needless to say, both x and θ can be multidimensional.

There are two important types of parametric estimators, namely, *point estimators* and *interval estimators*. In point estimation, the estimator for the true value of θ_0 of a parameter θ in $F(x; \theta)$ is an observable random variable, say $\tilde{\theta}(x_1, \ldots, x_n)$, which is a function of the sample elements $(x_1, \ldots, x_n)$, and whose distribution is, in some sense, concentrated around the true value θ_0 of θ. As in linear estimation, it will be found that the variance of the point estimator is often a reasonable criterion for measuring the concentration.

In interval estimation we devise two observable random variables $\underline{\theta}(x_1, \ldots, x_n)$, $\bar{\theta}(x_1, \ldots, x_n)$, usually abbreviated $(\underline{\theta}, \bar{\theta})$ where $\underline{\theta} < \bar{\theta}$, such that there is a specified probability that the random interval $(\underline{\theta}, \bar{\theta})$ contains θ_0 and is, in some sense, as short as possible. These ideas can be extended, of course, to the case where x or θ (or both) in $F(x; \theta)$ are multidimensional. Instead of talking about an "estimator $\tilde{\theta}$ for the true value θ_0 of a parameter θ", we will usually say "estimator $\tilde{\theta}$ for θ."

The linear estimators which have been considered in Chapter 10 are examples of point estimators. We have given specific examples of interval estimators in Sections 10.2(*c*), 10.4, 10.5, 10.6, and 11.2(*a*) which arise in sampling from normal distributions.

In this chapter we shall discuss the ideas of point and interval estimation in the more general setting of parametric estimation for samples from infinite populations and then present some of the basic results for finite samples as well as some asymptotic results for large samples.

12.1 DIFFERENTIATION OF PARAMETRIC DISTRIBUTION FUNCTIONS

(a) Case of a One-dimensional Parameter

In dealing with parametric statistical estimation, the testing of parametric statistical hypotheses, and related problems, we shall need certain properties of derivatives of distribution functions which depend on a parameter θ. It will be useful to discuss these questions briefly here.

Suppose x is a random variable which has a c.d.f. $F(x; \theta)$ where θ is a (real) one-dimensional parameter having values in a *parameter space* Ω. We shall consider x as a one-dimensional random variable for convenience, although it will be seen that all results obtained will hold, with minor changes in notation, if x is k-dimensional.

Thus, for various points $\theta_1, \theta_2, \ldots$ in Ω we have a corresponding set of c.d.f.'s $F(x; \theta_1)$, $F(x; \theta_2), \ldots$. We shall be especially interested in some open interval Ω_0 of Ω containing a particular point θ_0, the *true* value of θ. Now suppose we differentiate both sides of

(12.1.1)
$$\int_{-\infty}^{\infty} dF(x; \theta) = 1,$$

one or more times with respect to θ. We shall consider what happens with one and two differentiations. If the two differentiations are formally performed, we obtain

(12.1.2)
$$\frac{d}{d\theta}\int_{-\infty}^{\infty} dF(x; \theta) = \int_{-\infty}^{\infty}\left[\frac{\partial}{\partial\theta}\log dF(x; \theta)\right] dF(x; \theta) = 0$$

and

(12.1.3)
$$\frac{d^2}{d\theta^2}\int_{-\infty}^{\infty} dF(x; \theta) = \int_{-\infty}^{\infty}\left[\frac{\partial^2}{\partial\theta^2}\log dF(x; \theta)\right] dF(x; \theta)$$
$$+\int_{-\infty}^{\infty}\left[\frac{\partial}{\partial\theta}\log dF(x; \theta)\right]^2 dF(x; \theta) = 0.$$

To examine (12.1.2) and (12.1.3) a little more closely and for later reference, let us write, for the moment,

$$S(x; \theta) = \frac{\partial}{\partial\theta} \log dF(x; \theta) \tag{12.1.4}$$

$$S'(x; \theta) = \frac{\partial}{\partial\theta} S(x; \theta) \tag{12.1.5}$$

$$H(\theta, \theta') = \int_{-\infty}^{\infty} \log dF(x; \theta')\, dF(x; \theta) \tag{12.1.6}$$

$$A(\theta, \theta') = \int_{-\infty}^{\infty} S(x; \theta')\, dF(x; \theta) \tag{12.1.7}$$

$$B^2(\theta, \theta') = \int_{-\infty}^{\infty} [S(x; \theta')]^2\, dF(x; \theta) \tag{12.1.8}$$

$$D(\theta, \theta') = \int_{-\infty}^{\infty} S'(x; \theta')\, dF(x; \theta), \tag{12.1.9}$$

where θ is any point in Ω_0, and (θ, θ') any point in the Cartesian product set $\Omega_0 \times \Omega_0$.

First, we consider $S(x; \theta)$ and $S'(x; \theta)$. Assuming that $F(x; \theta)$ has a first derivative with respect to θ for any point θ in Ω_0 and for all x, except possibly for a set of probability zero, it is evident that $S(x; \theta)$ is defined as

$$S(x'; \theta) = \lim_{x \to x'} \frac{\dfrac{\partial}{\partial\theta}\Big[F(x'; \theta) - F(x; \theta)\Big]}{[F(x'; \theta) - F(x; \theta)]} \tag{12.1.10}$$

where $x < x'$, provided the indicated limit exists. We shall assume the limit exists for all θ in Ω_0 and all x in R_1 and that it is nonzero on a set of values of x of positive probability. $S'(x; \theta)$ is similarly defined. If for $\theta = \theta'$ x is a random variable which has c.d.f. $F(x; \theta')$, then $S(x;\theta)$ and $S'(x; \theta)$ are random variables for $\theta \in \Omega_0$.

Note that if x is a discrete random variable having p.f. $p(x; \theta)$, then

$$S(x; \theta) = \frac{\partial}{\partial\theta} \log p(x; \theta), \tag{12.1.10a}$$

and if x is a continuous random variable having p.d.f. $f(x; \theta)$

$$S(x; \theta) = \frac{\partial}{\partial\theta} \log f(x; \theta). \tag{12.1.10b}$$

Similar statements hold for $S'(x; \theta)$ in the cases of discrete and continuous random variables.

Next, let us consider $H(\theta, \theta')$. Let the x-axis be divided into disjoint intervals $(x_\alpha, x_{\alpha+1}]$, $\alpha = \cdots -1, 0, +1, \ldots$ and let I_α denote the interval

$(x_\alpha, x_{\alpha+1}]$. Then

$$P(x \in I_\alpha \mid \theta) = F(x_{\alpha+1}; \theta) - F(x_\alpha; \theta)$$

with a similar meaning for $P(x \in I_\alpha \mid \theta')$. Let $\Delta = \max_\alpha \{\text{length } I_\alpha\}$ and put

$$H_\Delta(\theta, \theta') = \sum_{\alpha=-\infty}^{\infty} \log P(x \in I_\alpha \mid \theta') \cdot P(x \in I_\alpha \mid \theta) \tag{12.1.11}$$

$$= \log \left\{ \prod_{\alpha=-\infty}^{\infty} [P(x \in I_\alpha \mid \theta')]^{P(x \in I_\alpha \mid \theta)} \right\}.$$

It is seen that every term in the upper line of (12.1.11) is negative, and hence $H_\Delta(\theta, \theta')$ is negative. In order to avoid having terms equal to $-\infty$ it is sufficient to assume that there exists no set E in the x-space for which $P(x \in E \mid \theta)$ and $P(x \in E \mid \theta')$ are not either both zero or both positive. Two distributions having c.d.f.'s $F(x; \theta)$ and $F(x; \theta')$ satisfying this condition are said to be *absolutely continuous with respect to each other.* This assumption is a little stronger than is required here since it would be sufficient for $F(x; \theta')$ to be absolutely continuous with respect to $F(x; \theta)$, that is, if no set E exists for which $P(x \in E \mid \theta') = 0$ and $P(x \in E \mid \theta) > 0$, but this generality is offset by the symmetry of $F(x; \theta)$ and $F(x; \theta')$ in this respect.

Making use of the fact that if $p_1, \ldots, p_r$ and $q_1, \ldots, q_r$ are any two sets of positive numbers

$$p_1^{q_1} \cdots p_r^{q_r} \leqslant (p_1 + \cdots + p_r)^{q_1 + \cdots + q_r}, \tag{12.1.12}$$

it will be seen that the positive quantity in { } in (12.1.11) cannot increase with successive subdivisions of the intervals in $\{I_\alpha\}$. Hence if there is a set $\{I_\alpha\}$ such that $H_\Delta(\theta, \theta')$ is finite, then the finite negative quantity $H_\Delta(\theta, \theta')$ cannot decrease as $\Delta \to 0$, and hence $\lim_{\Delta \to 0} H_\Delta(\theta, \theta')$ exists at all points (θ, θ') in $\Omega_0 \times \Omega_0$ and is nonpositive. The limit is denoted by the integral in (12.1.6). $H(\theta, \theta')$ is $\Sigma \log p(x; \theta')p(x; \theta)$ in the case of a discrete random variable with p.f. $p(x; \theta)$, and $\int_{-\infty}^{\infty} \log f(x; \theta')f(x; \theta')\, dx$ in the case of a continuous random variable with p.d.f. $f(x; \theta)$.

The function $H(\theta, \theta')$ is of basic importance in connection with information theory, entropy in statistical mechanics, and optimum estimation of parameters, and optimum statistical tests in statistical inference.

If, for a given function $g(x; \theta')$, measurable with respect to $F(x; \theta)$, there exists a non-negative function $h(x)$, measurable and having finite mean value with respect to $F(x; \theta)$ and such that

$$|g(x; \theta')| < h(x)$$

for $(\theta, \theta') \in \Omega_0 \times \Omega_0$, we shall say that $g(x, \theta')$ is *dominated by the integrable function* $h(x)$.

In order for (12.1.2) to be valid, it is sufficient for $\partial/\partial\theta' \log dF(x; \theta')$ to be dominated by some integrable function $h_1(x)$. Similarly, (12.1.3) is valid if $\partial^2/\partial\theta'^2 \log dF(x; \theta')$ and $[\partial/\partial\theta' \log dF(x; \theta')]^2$ are dominated by integrable functions $h_2(x)$ and $h_3(x)$. For general theorems on the sufficiency of conditions such as these the reader is referred to books on integration and real variable analysis such as those by McShane (1944), McShane and Botts (1959) and Saks (1937).

By expressing (12.1.2) and (12.1.3) in terms of mean values, we shall say that $F(x; \theta)$ is *regular with respect to its first θ-derivative in Ω_0 if*

$$\mathscr{E}(S(x;\theta)) = \frac{d}{d\theta}\int_{-\infty}^{\infty} dF(x;\theta) = 0, \tag{12.1.2a}$$

and *regular with respect to its second θ-derivative in* Ω_0 if $B^2(\theta, \theta) < +\infty$ and

$$\mathscr{E}(S'(x;\theta)) + \mathscr{E}(S(x;\theta))^2 = \frac{d^2}{d\theta^2}\int_{-\infty}^{\infty} dF(x;\theta) = 0. \tag{12.1.3a}$$

We return for a moment to $H(\theta, \theta')$ as defined in (12.1.6). If $F(x; \theta)$ is regular with respect to its first two θ-derivatives, it can be shown that $H(\theta, \theta')$ has first partial derivatives with respect to θ and θ' over $\Omega_0 \times \Omega_0$. Furthermore, $H(\theta, \theta')$, as a function of θ' for fixed θ, has a maximum for $\theta' = \theta$.

(b) Case of Vector Parameters

Now suppose $F(x; \theta)$ is a c.d.f. where θ is an r-dimensional parameter with (functionally independent) components $(\theta_1, \ldots, \theta_r)$, the parameter space being Ω_r. For convenience, we leave x one-dimensional; trivial modifications of notation show that results hold if x is k-dimensional. Sufficient conditions under which we can differentiate

$$\int_{-\infty}^{\infty} dF(x;\theta)$$

under the integral sign one or more times with respect to one or more of the components of θ can be developed without additional difficulties. Considering only two typical differentiations, we obtain corresponding to (12.1.2) and (12.1.3), the following:

$$\frac{\partial}{\partial\theta_p}\int_{-\infty}^{\infty} dF(x;\theta) = \int_{-\infty}^{\infty}\left[\frac{\partial}{\partial\theta_p}\log dF(x;\theta)\right] dF(x;\theta) = 0, \tag{12.1.13}$$

$$\begin{aligned}\frac{\partial^2}{\partial\theta_p\,\partial\theta_q}&\int_{-\infty}^{\infty} dF(x;\theta)\\ &= \int_{-\infty}^{\infty}\left[\frac{\partial^2}{\partial\theta_p\,\partial\theta_q}\log dF(x;\theta)\right] dF(x;\theta)\\ &\quad + \int_{-\infty}^{\infty}\left[\frac{\partial}{\partial\theta_p}\log dF(x;\theta)\right]\left[\frac{\partial}{\partial\theta_q}\log dF(x;\theta)\right] dF(x;\theta) = 0.\end{aligned} \tag{12.1.14}$$

We are interested in the validity of (12.1.13) and (12.1.14) for p, $q = 1, \ldots, r$ and for all points in an r-dimensional (Euclidean) open interval Ω_{r0}.

For discussion of validity of (12.1.13) and (12.1.14), and for later reference, we write down expressions corresponding to (12.1.4) through (12.1.9) as follows:

$$S_p(x; \theta) = \frac{\partial}{\partial \theta_p} \log dF(x; \theta) \tag{12.1.15}$$

$$S_{pq}(x; \theta) = \frac{\partial^2}{\partial \theta_p \, \partial \theta_q} \log dF(x; \theta) \tag{12.1.16}$$

$$H(\theta, \theta') = \int_{-\infty}^{\infty} \log dF(x; \theta') \, dF(x; \theta) \tag{12.1.17}$$

$$A_p(\theta, \theta') = \int_{-\infty}^{\infty} S_p(x; \theta') \, dF(x; \theta) \tag{12.1.18}$$

$$B_{pq}(\theta, \theta') = \int_{-\infty}^{\infty} S_p(x; \theta') S_q(x; \theta') \, dF(x; \theta) \tag{12.1.19}$$

$$D_{pq}(\theta, \theta') = \int_{-\infty}^{\infty} S_{pq}(x; \theta') \, dF(x; \theta), \tag{12.1.20}$$

where $p, q = 1, \ldots, r$, θ is a point in Ω_{r0} and (θ, θ') a point in $\Omega_{r0} \times \Omega_{r0}$. Since the components of θ are assumed to be functionally independent, the components of the random variable $(S_p(x; \theta), p = 1, \ldots, r)$ are linearly independent and hence the matrix $\|B_{pq}(\theta, \theta')\|$ is positive definite for (θ, θ') in $\Omega_{r0} \times \Omega_{r0}$. In the light of our discussion for the case of a one-dimensional parameter, these functions require no additional comment.

It is evident from the case of a one-dimensional parameter that for (12.1.13) to hold, it is sufficient for $(\partial/\partial\theta'_p) \log dF(x; \theta')$ to be dominated by integrable functions $h_{1p}(x)$, $p = 1, \ldots, r$. Similarly, for (12.1.14) to hold it is sufficient for

$$(\partial^2/\partial\theta'_p \partial\theta'_q) \log dF(x; \theta')$$

and

$$[(\partial/\partial\theta'_p) \log dF(x; \theta')][(\partial/\partial\theta'_q) \log dF(x; \theta')]$$

to be dominated by integrable functions $h_{2pq}(x)$ and $h_{3pq}(x)$, $p, q = 1, \ldots, r$ respectively.

Expressing (12.1.13) and (12.1.14) in terms of mean values, we shall say that $F(x; \theta)$ is *regular with respect to its first partial θ-derivatives in* Ω_{r0} if

$$\mathscr{E}(S_p(x; \theta)) = \frac{\partial}{\partial \theta_p} \int_{-\infty}^{\infty} dF(x; \theta) = 0, \tag{12.1.13a}$$

$p = 1, \ldots, r$ and *regular with respect to its second partial θ-derivatives in* Ω_{r0} if $\|B_{pq}(\theta, \theta)\|$ is finite and if

$$(12.1.14a) \quad \mathscr{E}(S_p(x;\theta)S_q(x;\theta)) + \mathscr{E}(S_{pq}(x;\theta)) = \frac{\partial^2}{\partial\theta_p\,\partial\theta_q}\int_{-\infty}^{\infty} dF(x;\theta) = 0.$$

(c) Remarks Concerning Extension to Vector Random Variables

The preceding discussion relates to the case of a one-dimensional random variable. The discussion extends to the case of a k-dimensional random variable with minor changes. The main changes lie in the definitions of $S(x;\theta)$ and $H(\theta, \theta')$ if x is a vector $(x_1, \ldots, x_k)$. In this case the simple difference $[F(x';\theta) - F(x;\theta)]$ of $F(x;\theta)$ over the interval $(x, x']$ which occurs in (12.1.10) is replaced by the difference of $F(x_1, \ldots, x_k;\theta)$ over the k-dimensional interval $(x_1, \ldots, x_k; x_1', \ldots, x_k']$. The resulting limit, which is assumed to exist, defines $(\partial/\partial\theta)\log dF(x_1, \ldots, x_k;\theta)$ and is denoted by $S(x_1, \ldots, x_k;\theta)$.

Similarly, the difference $F(x_{\alpha+1};\theta) - F(x_\alpha;\theta)$ of $F(x;\theta)$ over $(x_\alpha, x_{\alpha+1}]$ which appears in (12.1.11) is replaced by the kth difference of $F(x_1, \ldots, x_k;\theta)$ over the k-dimensional interval $(x_{1,\alpha}, \ldots, x_{k,\alpha}; x_{1,\alpha+1}, \ldots, x_{k,\alpha+1}]$ whereas $\Delta = \max_\alpha \{\Delta_\alpha\}$, where Δ_α is the largest dimension of this k-dimensional interval.

The remaining changes in notation are straightforward.

12.2 POINT ESTIMATION

(a) Definitions

Suppose $(x_1, \ldots, x_n)$ is an n-dimensional random variable from a c.d.f. $F_n(x_1, \ldots, x_n;\theta)$, where θ is a one-dimensional real parameter with parameter space Ω.

Let $\tilde{\theta}(x_1, \ldots, x_n)$, or more briefly $\tilde{\theta}$, be a function of $(x_1, \ldots, x_n)$ where $\tilde{\theta}$ itself is a random variable. If the realized (observed) value of $\tilde{\theta}$ corresponding to a realized (observed) value of $(x_1, \ldots, x_n)$ is used for θ_0, the *true* value of θ, then the random variable $\tilde{\theta}$ is called a *point estimate* or *estimator* for θ_0. This use of $\tilde{\theta}$ normally would be made, of course, only when the value of θ_0 is unknown. If, when $\theta = \theta_0$, $\mathscr{E}(\tilde{\theta}) = \theta_0$, which we may write more briefly as $\mathscr{E}(\tilde{\theta} \mid \theta_0) = \theta_0$, then $\tilde{\theta}$ is called an *unbiased estimator* for θ_0. Actually, it would be more accurate to say that $\tilde{\theta}$ is an *estimator for* θ_0 *unbiased in the mean.* If $\tilde{\theta}$ were a statistic having a c.d.f. $W(\tilde{\theta};\theta_0)$ continuous at $\tilde{\theta} = \theta_0$, such that $W(\theta_0;\theta_0) = \frac{1}{2}$, we would say that $\tilde{\theta}$ is an *estimator for* θ_0 *unbiased in the median.* Unless otherwise indicated, however, an unbiased estimator will be understood as being unbiased in the mean.

If an estimator $\tilde{\theta}$ converges in probability to θ_0 as $n \to \infty$ it is called a *consistent estimator* for θ_0.

If $\tilde{\theta}$ is an unbiased estimator for θ_0 having finite variance, and has the further property that no other unbiased estimator has a smaller variance, then $\tilde{\theta}$ is called an *efficient estimator* for θ_0.

If $\tilde{\theta}$ is a statistic such that for any other statistic $\breve{\theta}$ the distribution of the conditional random variable $\breve{\theta} \mid \tilde{\theta}$ does not depend on θ_0, then $\tilde{\theta}$ is called a *sufficient statistic* for θ_0. If also $\mathscr{E}(\tilde{\theta} \mid \theta_0) = \theta_0$ we shall say that $\tilde{\theta}$ is a *sufficient estimator* for θ_0.

The concepts of consistency, efficiency, and sufficiency are due to Fisher (1922, 1925b).

For simple random sampling, that is, where $(x_1, \ldots, x_n)$ is a random sample of size n from a c.d.f. $F(x; \theta)$, these notions of unbiasedness, consistency, sufficiency, and efficiency are of special importance. In this case

$$F_n(x_1, \ldots, x_n; \theta) = \prod_{\xi=1}^{n} F(x_\xi; \theta). \tag{12.2.1}$$

For a given sample $(x_1, \ldots, x_n)$, the quantity $dF_n = \prod_{\xi=1}^{n} dF(x_\xi; \theta)$ is called the *likelihood element* of θ for $(x_1, \ldots, x_n)$.

Most of the material in this chapter will be concerned with simple random sampling.

(b) Lower Bound of Variance of an Estimator

Let $(x_1, \ldots, x_n)$ be an n-dimensional random variable having c.d.f. $F_n(x_1, \ldots, x_n; \theta)$. If $\tilde{\theta}(x_1, \ldots, x_n)$ is an unbiased estimator for θ, then we have

$$\int_{R_n} (\tilde{\theta} - \theta)\, dF_n = 0. \tag{12.2.2}$$

If $F_n(x_1, \ldots, x_n; \theta)$ is regular with respect to its first θ-derivative in some (open) interval Ω_0 containing θ_0, the *true* value of θ, we may differentiate (12.2.2) under the integral sign and obtain

$$\int_{R_n} (\tilde{\theta} - \theta) S_n(x_1, \ldots, x_n; \theta)\, dF_n = 1 \tag{12.2.3}$$

where

$$S_n(x_1, \ldots, x_n; \theta) = \frac{\partial \log dF_n}{\partial \theta}. \tag{12.2.4}$$

Applying the Schwarz inequality to (12.2.3), we obtain

$$1 = \left[\int_{R_n} (\tilde{\theta} - \theta) S_n\, dF_n\right]^2 \leqslant \int_{R_n} (\tilde{\theta} - \theta)^2\, dF_n \cdot \int_{R_n} S_n^2\, dF_n$$

Since

$$\sigma^2(\tilde{\theta}) = \int_{R_n} (\tilde{\theta} - \theta)^2 \, dF_n \tag{12.2.5}$$

we therefore obtain

$$\sigma^2(\tilde{\theta}) \geqslant \frac{1}{\mathscr{E}(S_n^2)}, \quad \text{for } \theta_0 \in \Omega_0, \tag{12.2.6}$$

the equality holding if and only if

$$K[\tilde{\theta}(x_1, \ldots, x_n) - \theta] \equiv S_n(x_1, \ldots, x_n; \theta) \tag{12.2.7}$$

in R_n with probability 1, where K depends possibly on θ but not on $(x_1, \ldots, x_n)$.

If equality holds in (12.2.6), we say that $\tilde{\theta}$ is an *efficient estimator* for θ, in which case (12.2.7) gives the form which $\tilde{\theta}$ takes.

If $\sigma^2(\tilde{\theta})$ is denoted by $\sigma^2(\tilde{\theta} \mid \theta_0)$ when evaluated for $\theta = \theta_0$, that is, by putting $\theta = \theta_0$ in (12.2.5), with a similar meaning for $\mathscr{E}(S_n^2 \mid \theta_0)$, then

$$\sigma^2(\tilde{\theta} \mid \theta_0) \geqslant 1/[\mathscr{E}(S_n^2 \mid \theta_0)].$$

An efficient estimator of θ is usually denoted by $\hat{\theta}$.

Summarizing, we have the following result:

12.2.1 *Suppose* $(x_1, \ldots, x_n)$ *is a random variable having c.d.f.* $F_n(x_1, \ldots, x_n; \theta)$ *where* θ *is one-dimensional, and where* $F_n(x_1, \ldots, x_n; \theta)$ *is regular in its first* θ*-derivative in* Ω_0. *If* $\tilde{\theta}(x_1, \ldots, x_n)$ *is an unbiased estimator for* θ_0, *then* $\sigma^2(\tilde{\theta} \mid \theta_0) \geqslant 1/[\mathscr{E}(S_n^2 \mid \theta_0)]$ *where* S_n *is given by* (12.2.4), *the equality holding if and only if* (12.2.7) *holds with probability* 1 *at* $\theta = \theta_0$. *If an efficient estimator* $\hat{\theta}$ *exists for* θ_0 *its variance is* $1/\mathscr{E}(S_n^2 \mid \theta_0)$.

Lower bounds for $\sigma^2(\tilde{\theta} \mid \theta_0)$ without the regularity assumption have been given by Chapman and Robbins (1951) and by Kiefer (1952).

If an efficient estimator $\hat{\theta}$ exists for θ_0, and if $\tilde{\theta}$ is any other unbiased estimator for θ_0, then the efficiency of $\tilde{\theta}$ for estimating θ_0, is defined by ratio

$$\text{eff}\,(\tilde{\theta} \mid \theta_0) = \frac{\sigma^2(\hat{\theta} \mid \theta_0)}{\sigma^2(\tilde{\theta} \mid \theta_0)}. \tag{12.2.8}$$

(c) Case of Random Sampling

In this important case (12.2.1) holds and

$$S_n(x_1, \ldots, x_n; \theta) = \sum_{\xi=1}^{n} S(x_\xi; \theta) \tag{12.2.9}$$

where

$$S(x; \theta) = \frac{\partial}{\partial \theta} \log dF(x; \theta). \tag{12.2.10}$$

$\sum_{\xi=1}^{n} S(x_\xi; \theta)$ is called the *score* for θ based on $(x_1, \ldots, x_n)$.

Furthermore

$$\mathscr{E}(S_n^2) = n\mathscr{E}(S^2) = nB^2(\theta, \theta) \tag{12.2.11}$$

where $B^2(\theta, \theta)$ is given in (12.1.8). Therefore (12.2.6) specializes to

$$\sigma^2(\tilde{\theta}) \geqslant \frac{1}{nB^2(\theta, \theta)}. \tag{12.2.12}$$

This result was originally stated by Fisher (1922). It was later established by Cramér (1946), Dugué (1937), and Rao (1945).

Equality in (12.2.12) holds if and only if

$$K(\tilde{\theta} - \theta) \equiv \sum_{\xi=1}^{n} S(x_\xi; \theta) \tag{12.2.13}$$

over R_n with probability 1 where K does not depend on $(x_1, \ldots, x_n)$. Thus, if an efficient estimator $\hat{\theta}$ exists, it is the statistic $\tilde{\theta}$ which satisfies (12.2.13) and its variance is given by

$$\sigma^2(\hat{\theta}) = \frac{1}{nB^2(\theta, \theta)}. \tag{12.2.14}$$

Thus, we have the following important corollary of **12.2.1**:

12.2.1a *If* $(x_1, \ldots, x_n)$ *is a sample from the c.d.f.* $F(x; \theta)$ *which is regular with respect to its first* θ*-derivative in* Ω_0*, and if* $\tilde{\theta}$ *is an unbiased estimator for* θ_0*, then* $\sigma^2(\tilde{\theta} \mid \theta_0) \geqslant 1/[nB^2(\theta_0, \theta_0)]$ *where* $B^2(\theta, \theta)$ *is given by* (12.1.8). *Furthermore, equality holds if and only if* $\tilde{\theta}$ *satisfies* (12.2.13), *in which case the solution for* $\tilde{\theta}$*, denoted by* $\hat{\theta}$*, is an estimator for* θ_0 *with variance* $1/[nB^2(\theta_0, \theta_0)]$.

If $\tilde{\theta}$ is an arbitrary unbiased estimator for θ, and if an efficient estimator $\hat{\theta}$ exists for θ, the efficiency of $\tilde{\theta}$ in estimating θ_0 is defined by

$$\text{eff}(\tilde{\theta} \mid \theta_0) = \frac{\sigma^2(\hat{\theta} \mid \theta_0)}{\sigma^2(\tilde{\theta} \mid \theta_0)} = \frac{1}{\sigma^2(\sqrt{n}\tilde{\theta} \mid \theta_0)B^2(\theta_0, \theta_0)}. \tag{12.2.15}$$

Fisher (1922) has called $nB^2(\theta_0, \theta_0)$, the reciprocal of the variance of an efficient estimator $\hat{\theta}$, the *amount of information* contained in the sample regarding θ_0, or $B^2(\theta_0, \theta_0)$ the amount of information about θ_0 per observation from $F(x; \theta_0)$.

From this point of view the efficiency of an unbiased estimator $\tilde{\theta}$ for θ may be regarded as the fraction of information contained in $\tilde{\theta}$ for estimating θ_0 relative to that contained in an efficient estimator for estimating θ_0.

Example. Suppose $(x_1, \ldots, x_n)$ is a sample from the Poisson distribution $Po(\mu_0)$ referred to in **8.3.3c.** We have

$$S_n(x_1, \ldots, x_n; \mu) = \frac{\partial}{\partial \mu} \log \prod_{\xi=1}^{n} dF(x_\xi; \mu) = \frac{n}{\mu}\left[\frac{1}{n} \sum_{\xi=1}^{n} x_\xi - \mu\right]$$

which, according to (12.2.13), shows that $(1/n) \sum_{\xi=1}^{n} x_\xi$, that is, the sample mean, is an efficient estimator for μ_0, the true value of μ, for any given sample size. Denoting $\frac{1}{n} \sum_{\xi=1}^{n} x_\xi$ by $\hat{\mu}(x_1, \ldots, x_n)$, or $\hat{\mu}$, we find the variance of $\hat{\mu}$ by applying (12.2.14), that is,

$$\sigma^2(\hat{\mu} \mid \mu_0) = \frac{1}{n \sum_{x=0}^{\infty} \left(\frac{x}{\mu_0} - 1\right)^2 \frac{\mu_0^x e^{-\mu_0}}{x!}} = \frac{\mu_0}{n}.$$

Also, we have

$$\sigma^2(\sqrt{n}\hat{\mu} \mid \mu_0) = \mu_0,$$

and hence it follows from **12.2.1a** that it is impossible to find an unbiased estimator $\tilde{\mu}(x_1, \ldots, x_n)$ for μ_0 for which $\sigma^2(\sqrt{n}\tilde{\mu} \mid \mu_0) < \mu_0$.

(d) Lower Bound of Variance of a Biased Estimator

A generalization of **12.2.1** (see Cramér (1946), Dugué (1937) and Rao (1945)) can be obtained by considering a *biased* estimator $\tilde{\theta}$ for θ as follows. We replace $(\tilde{\theta} - \theta)$ in (12.2.2) by

$$[\tilde{\theta} - \theta - b_n(\theta)] \tag{12.2.16}$$

where $b_n(\theta)$ is the *bias* of the estimator $\tilde{\theta}$ and repeat essentially the same argument as that involved in **12.2.1a.** Then in (12.2.3) we would replace $(\tilde{\theta} - \theta)$ by $(\tilde{\theta} - \theta - b_n(\theta))$ and 1 on the right by $1 + b_n'(\theta)$, where $b_n'(\theta) \equiv d/d\theta\, b_n(\theta)$. In place of (12.2.12) we obtain

$$\sigma^2(\tilde{\theta}) \geqslant \frac{[1 + b_n'(\theta)]^2}{nB^2(\theta, \theta)} \tag{12.2.17}$$

with the equality holding if and only if

$$K[\tilde{\theta} - \theta - b_n(\theta)] \equiv \sum_{\xi=1}^{n} S(x_\xi; \theta) \tag{12.2.18}$$

over R_n with probability 1, and where K does not depend on $(x_1, \ldots, x_n)$.

(e) Properties of Sufficient Estimators

Suppose $(x_1, \ldots, x_n)$ is a random variable with p.d.f. $f_n(x_1, \ldots, x_n; \theta)$ which factors as follows:

$$f_n(x_1, \ldots, x_n; \theta) = v(\tilde{\theta}; \theta)u(x_1, \ldots, x_n \mid \tilde{\theta}) \tag{12.2.19}$$

where $v(\tilde{\theta}; \theta)$ is the p.d.f. of $\tilde{\theta}$, and $u(x_1, \ldots, x_n \mid \tilde{\theta})$ is the p.d.f. of the conditional random variable $(x_1, \ldots, x_n \mid \tilde{\theta})$ which does not depend on θ. Then $\tilde{\theta}$ is a sufficient statistic for θ. For if θ is any other statistic which does not depend on $\tilde{\theta}$, the distribution of the conditional random variable $\breve{\theta} \mid \tilde{\theta}$ is completely determined from $u(x_1, \ldots, x_n \mid \tilde{\theta})$.

Conversely, suppose $\tilde{\theta}$ is a sufficient statistic with p.d.f. $v(\tilde{\theta}; \theta)$. Let $y_2, \ldots, y_n$ be any $n - 1$ further statistics such that $\tilde{\theta}, y_2, \ldots, y_n$ have a p.d.f. as follows:

$$g(\tilde{\theta}, y_2, \ldots, y_n; \theta) = f_n(x_1, \ldots, x_n; \theta) \frac{1}{|J|}, \tag{12.2.20}$$

where J is the Jacobian of $(\tilde{\theta}, y_2, \ldots, y_n)$ with respect to $(x_1, \ldots, x_n)$. [For conditions under which (12.2.20) hold, see Section 2.8(*d*).]

Let $h(y_2, \ldots, y_n \mid \tilde{\theta}; \theta)$ be the p.d.f. of the conditional random variable $y_2, \ldots, y_n \mid \tilde{\theta}$. Then

$$h(y_2, \ldots, y_n \mid \tilde{\theta}; \theta) = \frac{g(\tilde{\theta}, y_2, \ldots, y_n; \theta)}{v(\tilde{\theta}; \theta)}, \tag{12.2.21}$$

where $v(\tilde{\theta}; \theta) \neq 0$ and is given by

$$v(\tilde{\theta}; \theta) = \int_{R_{n-1}} g(\tilde{\theta}, y_2, \ldots, y_n; \theta)\, dy_2 \cdots dy_n. \tag{12.2.22}$$

If $h(y_2, \ldots, y_n \mid \tilde{\theta}; \theta)$ does not depend on θ, in which case we will denote it by $h^*(y_2, \ldots, y_n \mid \tilde{\theta})$, then $\tilde{\theta}$ is sufficient for θ, and we have

$$g(\tilde{\theta}, y_2, \ldots, y_n; \theta) = v(\tilde{\theta}; \theta) h^*(y_2, \ldots, y_n \mid \tilde{\theta}). \tag{12.2.23}$$

Using (12.2.20), we therefore have

$$f_n(x_1, \ldots, x_n; \theta) = v(\tilde{\theta}; \theta) h^*(y_2, \ldots, y_n \mid \tilde{\theta}) \cdot |J|. \tag{12.2.24}$$

Note that $h^*(y_2, \ldots, y_n \mid \tilde{\theta}) \cdot |J|$ depends on $(\tilde{\theta}, y_2, \ldots, y_n)$, and hence on $(x_1, \ldots, x_n)$, but not on θ. Thus any statistic depending on $(x_1, \ldots, x_n)$ through $(y_2, \ldots, y_n)$ but not on θ would have a distribution which would not depend on θ.

We may therefore summarize as follows:

12.2.2 *Let $(x_1, \ldots, x_n)$ be a random variable having p.d.f. $f_n(x_1, \ldots, x_n; \theta)$. A necessary and sufficient condition for a statistic $\tilde{\theta}(x_1, \ldots, x_n)$ to be sufficient for θ is that*

$$f_n(x_1, \ldots, x_n; \theta) = v(\tilde{\theta}; \theta) w(x_1, \ldots, x_n) \tag{12.2.25}$$

where $v(\tilde{\theta}; \theta)$ is the p.d.f. *of $\tilde{\theta}$ and $w(x_1, \ldots, x_n)$ is a function of $(x_1, \ldots, x_n)$ which does not depend on θ.*

Fisher (1922) first pointed out the sufficiency of the *factorability criterion* (12.2.25). Neyman (1935) showed it is also a necessary condition.

In the special but important case of simple random sampling where $(x_1, \ldots, x_n)$ is a sample from a p.d.f. $f(x; \theta)$, we would have a corollary of **12.2.2** in which $f_n(x_1, \ldots, x_n; \theta)$ takes the special form $\prod_{\xi=1}^{n} f(x_\xi; \theta)$.

In a similar manner, one can show that if $(x_1, \ldots, x_n)$ is an n-dimensional discrete random variable with p.d.f. $p_n(x_1, \ldots, x_n; \theta)$, a necessary and sufficient condition for a statistic $\tilde{\theta}$ to be sufficient for θ is that $p_n(x_1, \ldots, x_n; \theta)$ factor as follows:

$$p_n(x_1, \ldots, x_n; \theta) = v(\tilde{\theta}; \theta)\, w(x_1, \ldots, x_n) \tag{12.2.26}$$

where $v(\tilde{\theta}; \theta)$ is the p.f. of $\tilde{\theta}$ and where $w(x_1, \ldots, x_n)$ depends on $(x_1, \ldots, x_n)$ but not on θ.

For simple random sampling from a p.f. $p(x; \theta)$, $p_n(x_1, \ldots, x_n; \theta)$ would take the form $\prod_{\xi=1}^{n} p(x_\xi; \theta)$.

Finally, we remark that **12.2.2** can be extended to the case where θ is a vector $(\theta_1, \ldots, \theta_r)$, $r \leqslant n$. A necessary and sufficient condition for $(\tilde{\theta}_1, \ldots, \tilde{\theta}_s)$ to be a *set of sufficient statistics* for $(\theta_1, \ldots, \theta_r)$, $s \geqslant r$, is that (12.2.27)

$$f_n(x_1, \ldots, x_n; \theta_1, \ldots, \theta_r) = v(\tilde{\theta}_1, \ldots, \tilde{\theta}_s; \theta_1, \ldots, \theta_r) w(x_1, \ldots, x_n)$$

where $w(x_1, \ldots, x_n)$ depends on $(x_1, \ldots, x_n)$ but not on θ.

A similar factorability criterion holds, of course, for the case in which θ is a vector $(\theta_1, \ldots, \theta_r)$ and $(x_1, \ldots, x_n)$ is a discrete random variable having p.f. $p_n(x_1, \ldots, x_n; \theta_1, \ldots, \theta_r)$.

For a random variable $(x_1, \ldots, x_n)$ having a general distribution, a generalization of the factorability criterion stated in **12.2.2** together with its extension to the case of a vector parameter has been given by Halmos and Savage (1949) by making use of the Radon-Nikodym theorem. An even more abstract treatment of necessary and sufficient conditions for the characterization of sufficient statistics has been given by Bahadur (1954).

Example. As an example of a sufficient estimator suppose $(x_1, \ldots, x_n)$ is a sample from the Poisson distribution $Po(\mu_0)$. We have for arbitrary μ on $(0, \infty)$,

$$dF(x; \mu) = \frac{\mu^x e^{-\mu}}{x!}$$

and

$$\prod_{\xi=1}^{n} dF(x_\xi; \mu) = \frac{\mu^{\sum_\xi x_\xi} e^{-n\mu}}{\prod_\xi x_\xi!}$$

which can be written as

$$\left[\frac{\mu^{\sum_\xi x_\xi} e^{-n\mu}}{\left(\sum_\xi x_\xi\right)!}\right] \cdot \left\{\frac{\left(\sum_\xi x_\xi\right)!}{\prod_\xi x_\xi!}\right\}.$$

Denoting $\frac{1}{n}\sum_\xi x_\xi$ by $\tilde{\mu}(x_1, \ldots, x_n)$, and referring to **8.3.3c**, the expression in [ ] is the p.f. of $\tilde{\mu}$, that is,

$$dV(\tilde{\mu}, \mu) = \frac{\mu^{n\tilde{\mu}} e^{-n\mu}}{(n\tilde{\mu})!}$$

whereas the expression in { } is actually the p.f. of the conditional random variable $(x_1, \ldots, x_n \mid \tilde{\mu})$, that is,

$$dU(x_1, \ldots, x_n \mid \tilde{\mu}) = \frac{(n\tilde{\mu})!}{\prod_\xi x_\xi!}.$$

It is evident that if $\tilde{\mu}^*$ is any other estimator, then the p.f. $p(\tilde{\mu}^* \mid \tilde{\mu})$ of the conditional random variable $\tilde{\mu}^* \mid \tilde{\mu}$ is obtained by summing $(n\tilde{\mu})!/\prod_\xi x_\xi!$ over those positive integral (or zero) values of the x_ξ subject to the two conditions $\tilde{\mu}^*(x_1, \ldots, x_n) = \tilde{\mu}^*$ and $\tilde{\mu}(x_1, \ldots, x_n) = \tilde{\mu}$. It is seen that $p(\tilde{\mu}^* \mid \tilde{\mu})$ does not depend on μ. Hence, $(1/n)\sum_\xi x_\xi$, is sufficient for estimating μ_0.

An important property of a sufficient estimator is that if one starts with any initial unbiased estimator of the parameter that is not a function of the sufficient estimator one can find an unbiased estimator depending on the sufficient estimator which has a smaller variance than that of the initial estimator. More precisely the situation is stated in the following theorem due to Blackwell (1947) and Rao (1945):

12.2.3 *Suppose $\tilde{\theta}$ is a sufficient statistic for θ_0 and $\breve{\theta}$ is any unbiased estimator for θ_0. Let $\mathscr{E}(\breve{\theta} \mid \tilde{\theta}) = h(\tilde{\theta})$. Then $h(\tilde{\theta})$ is an unbiased estimator for θ_0 whose variance cannot exceed that of $\breve{\theta}$.*

To prove **12.2.3** let $G(\tilde{\theta}, \breve{\theta}; \theta_0)$ be the c.d.f. of $(\tilde{\theta}, \breve{\theta})$, $V(\tilde{\theta}; \theta_0)$ the c.d.f. of $\tilde{\theta}$, and $U(\breve{\theta} \mid \tilde{\theta})$ the c.d.f. of the conditional random variable $\breve{\theta} \mid \tilde{\theta}$. Then

$$h(\tilde{\theta}) = \int_{-\infty}^{\infty} \breve{\theta}\, dU(\breve{\theta} \mid \tilde{\theta}), \tag{12.2.28}$$

thus we see that $h(\tilde{\theta})$ is an unbiased estimator for θ_0.

For the variance of $\breve{\theta}$ we have

$$\begin{aligned}\sigma^2(\breve{\theta}) &= \mathscr{E}(\breve{\theta} - \theta_0)^2 = \mathscr{E}[(h(\tilde{\theta}) - \theta_0) + (\breve{\theta} - h(\tilde{\theta}))]^2 \\ &= \sigma^2(h(\tilde{\theta})) + \mathscr{E}(\breve{\theta} - h(\tilde{\theta}))^2 + 2\mathscr{E}[(h(\tilde{\theta}) - \theta_0)(\breve{\theta} - h(\tilde{\theta}))].\end{aligned}$$

But

$$\mathscr{E}[(h(\tilde{\theta}) - \theta_0)(\breve{\theta} - h(\tilde{\theta}))]$$
$$= \int_{-\infty}^{\infty} [h(\tilde{\theta}) - \theta_0]\left\{\int_{-\infty}^{\infty} (\breve{\theta} - h(\tilde{\theta}))\, dU(\breve{\theta} \mid \tilde{\theta})\right\} dV(\tilde{\theta}; \theta_0) = 0,$$

since the quantity in { } vanishes. Therefore, since $\mathscr{E}(\breve{\theta} - h(\tilde{\theta}))^2 \geqslant 0$, we have

(12.2.29) $$\sigma^2(\breve{\theta}) \geqslant \sigma^2(h(\tilde{\theta}))$$

which completes the argument for **12.2.3**.

12.3 POINT ESTIMATION FROM LARGE SAMPLES

(a) Asymptotic Distribution of the Score

An efficient estimator of a parameter exists for samples of size n, only for certain special cases of c.d.f.'s $F(x; \theta)$. In such cases, $S_n(x_1, \ldots, x_n; \theta)$, which is given by (12.2.9), takes on the special form given in (12.2.13), and the efficient estimator $\hat{\theta}_n(x_1, \ldots, x_n)$ is therefore essentially given by solving the equation

(12.3.1) $$S_n(x_1, \ldots, x_n; \theta) = 0$$

for θ.

But suppose $S_n(x_1, \ldots, x_n; \theta)$ does not have the special form indicated by (12.2.13) but that (12.3.1) does have a solution for θ, which we can, without ambiguity, continue to call $\hat{\theta}_n(x_1, \ldots, x_n)$. What properties does this solution have as an estimator for θ_0? The answer is that under certain conditions to be developed presently, $\hat{\theta}_n(x_1, \ldots, x_n)$ is an efficient estimator for θ_0 in an asymptotic sense for large samples.

To deal with this question we first establish the following result:

12.3.1 *Suppose $(x_1, \ldots, x_n)$ is a sample from a c.d.f. $F(x; \theta_0)$, where $F(x; \theta)$ is regular with respect to its first θ-derivative in Ω_0. Then if $B^2(\theta; \theta)$, as defined in* (12.1.8), *exists and is finite, $S_n(x_1, \ldots, x_n; \theta_0)$ is asymptotically distributed according to $N(0, nB^2(\theta_0, \theta_0))$.*

This is essentially a corollary of **9.2.1**. For if we denote the random variable $S(x; \theta)$ by y, then y has a c.d.f. $G(y)$ defined by

$$G(y) = \int_{E_y} dF(x; \theta)$$

where E_y is the set of points on the x-axis for which $S(x; \theta) \leqslant y$. Since $F(x; \theta)$ is regular with respect to its first θ-derivative, it is clear that $\mathscr{E}(y) = 0$. Thus $S_n(x_1, \ldots, x_n; \theta)$ is the sum of a sample of size n from a population having zero mean and finite variance $B^2(\theta, \theta)$. Applying **9.2.1** we obtain the conclusion of **12.3.1**.

(b) Convergence of Maximum Likelihood Estimators

It is clear that under the conditions stated in **12.3.1** the random variable $(1/n)S_n(x_1, \ldots, x_n; \theta)$ converges in probability to 0. This suggests that if we set

$$S_n(x_1, \ldots, x_n; \theta) = 0 \tag{12.3.2}$$

we should obtain a sequence of sets of roots $\{\hat{\theta}(x_1, \ldots, x_n)\}$, $n = 1, 2, \ldots$, each set containing at least one root, from which a sequence can be found which converges to the true value of θ with probability 1. We shall show this is true under certain conditions. Let us assume that $S(x; \theta)$ is a continuous function of θ in Ω_0 for all values of x in R_1 except possibly for a set of probability zero. Furthermore, we shall assume that $F(x; \theta)$ is regular with respect to its first θ-derivative in Ω_0, from which it follows that $A(\theta_0, \theta)$ as defined in (12.1.7) is continuous and strictly decreasing in θ over some subinterval Ω_0' of Ω_0 which contains θ_0.

Referring to (12.2.9) and the proof of **12.3.1**, it is evident that $(1/n)S_n(x_1, \ldots, x_n; \theta)$ is the mean of a sample of size n from a population having mean $A(\theta_0, \theta)$ if θ_0 is the true value of θ. Therefore by **4.6.1**, $(1/n)S_n(x_1, \ldots, x_n; \theta)$ converges almost certainly to $A(\theta_0, \theta)$. Without loss of generality, we may take Ω_0' to be $(\theta_0 - \delta, \theta_0 + \delta)$ where $\delta > 0$. Thus $A(\theta_0, \theta)$ is monotonically decreasing over this interval and since $A(\theta_0, \theta_0) = 0$ we have $A(\theta_0, \theta_0 - \delta) > 0$ and $A(\theta_0, \theta_0 + \delta) < 0$. Therefore there exists an $n(\delta, \varepsilon)$ so that the probability exceeds $1 - \varepsilon$ that both of the following inequalities hold for all $n > n(\delta, \varepsilon)$ if θ_0 is the true value of θ:

$$\begin{aligned} &\frac{1}{n} S_n(x_1, \ldots, x_n; \theta) > 0 \qquad \text{if} \qquad \theta = \theta_0 - \delta \\ &\frac{1}{n} S_n(x_1, \ldots, x_n; \theta) < 0 \qquad \text{if} \qquad \theta = \theta_0 + \delta. \end{aligned} \tag{12.3.3}$$

Since $S(x, \theta)$ is continuous in θ over $(\theta_0 - \delta, \theta_0 + \delta)$ for all x in R_1 except for a set of probability 0, a similar statement holds for

$$\frac{1}{n}\sum_{\xi=1}^{n} S(x_\xi; \theta), \quad \text{that is,} \quad \frac{1}{n} S_n(x_1, \ldots, x_n; \theta).$$

Therefore, if θ_0 is the true value of θ, we have

(12.3.4)

$$P\left(\frac{1}{n} S_n(x_1, \ldots, x_n; \theta) = 0, \right.$$
$$\left. \text{for some } \theta \text{ in } (\theta_0 \pm \delta) \text{ for all } n > n(\delta, \varepsilon) \,\middle|\, \theta_0\right) > 1 - \varepsilon$$

which is equivalent to the statement that a sequence of roots of (12.3.2) exists which converge almost certainly to θ_0. In particular, if (12.3.2) has a unique solution $\hat{\theta}(x_1, \ldots, x_n)$ for $n = n_0, n_0 + 1, \ldots$, for some integer n_0, then the sequence $\hat{\theta}(x_1, \ldots, x_n)$, $n = n_0, n_0 + 1, \ldots$ converges almost certainly to θ_0.

Summarizing:

12.3.2 *Suppose $(x_1, \ldots, x_n)$ is a sample from the c.d.f. $F(x; \theta_0)$, where $F(x; \theta)$ is regular with respect to its first θ-derivative in Ω_0. Let $S(x; \theta)$ be continuous in θ for all values of x in R_1, except possibly for a set of probability zero. Then there exists a sequence of solutions of* (12.3.2) *which converge almost certainly to θ_0. In particular, if* (12.3.2) *has a unique solution $\hat{\theta}(x_1, \ldots, x_n)$ for $n \geqslant$ some n_0, then the sequence $\hat{\theta}(x_1, \ldots, x_n)$, $n = n_0, n_0 + 1, \ldots$ converges almost certainly to θ_0.*

In view of the fact that $S_n(x_1, \ldots, x_n; \theta)$ is the first θ-derivative of the logarithm of the likelihood element, namely, $\sum_{\xi=1}^{n} \log dF(x_\xi; \theta)$, and that $\hat{\theta}(x_1, \ldots, x_n)$ is a value of θ for which $S_n(x_1, \ldots, x_n; \theta)$ vanishes, $\hat{\theta}$, if it is unique and maximizes the likelihood element, is called the *maximum likelihood estimator* for θ_0, a term introduced by Fisher (1922). Wald (1949*b*) has shown that the solution of (12.3.2) which maximizes the likelihood is, under certain conditions, a consistent estimator for θ_0. Other detailed analyses of the consistency of maximum likelihood estimators and related problems have been made by Barankin and Gurland (1951), Huzurbazar (1948), LeCam (1956), Wald (1948, 1949*b*), and others.

(c) Asymptotic Distribution of Maximum Likelihood Estimators

The assumption of regularity of $F(x; \theta)$ with respect to its second θ-derivative is strong enough to enable us to make the following statement about the asymptotic distribution of $\hat{\theta}$ for large n:

12.3.3 *If $(x_1, \ldots, x_n)$ is a sample from the* c.d.f. *$F(x; \theta_0)$, where $F(x; \theta)$ is regular with respect to its second θ-derivative in Ω_0 and if the maximum likelihood estimator $\hat{\theta}$ for θ_0 is unique for $n \geqslant$ some n_0, and a random variable (measurable) with respect to $\prod_{\xi=1}^{n} F(x_\xi; \theta)$, as defined in* (12.2.1), *its distribution is asymptotically normal $N(\theta_0; 1/[nB^2(\theta_0, \theta_0)])$ for large n.*

Since $S(x; \theta)$ has a θ-derivative everywhere in Ω_0 and for all points x in R_1 except possibly for a set of zero probability, a similar θ-derivative statement holds for $S_n(x_1, \ldots, x_n; \theta)$ for all points $(x_1, \ldots, x_n)$ in R_n, except possibly for a set of zero probability. If $\hat{\theta}$ is unique for $n \geqslant$ some n_0, we have seen from **12.3.2** that $\hat{\theta}$ converges almost certainly to θ_0 as $n \to \infty$. Furthermore, if $\hat{\theta}(x_1, \ldots, x_n)$ is a random variable, then for arbitrary $\delta > 0$ and $\varepsilon > 0$ there is an $n(\delta, \varepsilon, n_0)$ and a set E_n in R_n defined by $|\theta_0 - \hat{\theta}(x_1, \ldots, x_n)| < \delta$ such that

(12.3.5) $P((x_1, \ldots, x_n) \in E_n \text{ for all } n > n(\delta, \varepsilon, n_0) \mid \theta_0) > 1 - \varepsilon.$

In E_n

(12.3.6)
$$\frac{1}{\sqrt{n}} S_n(x_1, \ldots, x_n; \theta_0) = \frac{1}{\sqrt{n}} S_n(x_1, \ldots, x_n; \hat{\theta}) + \frac{1}{\sqrt{n}} S_n'(x_1, \ldots, x_n; \theta^*)(\theta_0 - \hat{\theta})$$

where $\theta^*(x_1, \ldots, x_n)$ is a random variable satisfying

(12.3.7)
$$|\theta_0 - \theta^*| \leqslant |\theta_0 - \hat{\theta}|$$

and where

(12.3.8)
$$S_n'(x_1, \ldots, x_n; \theta^*) = \sum_{\xi=1}^{n} S'(x_\xi; \theta^*)$$

and

(12.3.9)
$$S'(x; \theta) = \frac{\partial}{\partial \theta} S(x; \theta).$$

But (12.3.5) and (12.3.6) together are equivalent to the statement

(12.3.10)
$$P\left(\frac{1}{\sqrt{n}} S_n(x_1, \ldots, x_n; \theta_0) = \frac{1}{\sqrt{n}} S_n(x_1, \ldots, x_n; \hat{\theta}) + \frac{1}{n} S_n'(x_1, \ldots, x_n; \theta^*) \cdot [\sqrt{n}(\theta_0 - \hat{\theta})] \text{ for all } n > n(\delta, \varepsilon, n_0) \,|\theta_0\right) > 1 - \varepsilon.$$

Since $\hat{\theta}$, by definition, satisfies (12.3.2), we see that (12.3.10) reduces to

(12.3.11)
$$P\left(\frac{1}{\sqrt{n}} S_n(x_1, \ldots, x_n; \theta_0) = \frac{1}{n} S_n'(x_1, \ldots, x_n; \theta^*)[\sqrt{n}(\theta_0 - \hat{\theta})] \text{ for all } n > n(\delta, \varepsilon, n_0) \mid \theta_0\right) > 1 - \varepsilon,$$

which implies that

(12.3.12)
$$\frac{1}{\sqrt{n}} S_n(x_1, \ldots, x_n; \theta_0), \qquad n = 1, 2, \cdots$$

and

(12.3.13)
$$\frac{1}{n} S_n'(x_1, \ldots, x_n; \theta^*)[\sqrt{n}(\theta_0 - \hat{\theta})], \qquad n = 1, 2, \cdots$$

are sequences of random variables, which converge together in distribution if either converges in distribution. Now

$$\frac{1}{n} S_n'(x_1, \ldots, x_n; \theta_0) = \frac{1}{n} \sum_{\xi=1}^{n} S'(x_\xi; \theta_0) \tag{12.3.14}$$

which is the mean of a sample of size n from a population having mean

$$\int_{-\infty}^{\infty} S'(x; \theta_0)\, dF(x; \theta_0) = -B^2(\theta_0; \theta_0) \tag{12.3.15}$$

and hence by **9.1.1** the expression on the left of (12.3.14) converges in probability to the expression given in (12.3.15). Now it follows from **4.3.7** that since $\hat{\theta}$ converges in probability to θ_0, so does θ^*. Therefore, by applying **4.3.8**, we conclude that $(1/n)\, S_n'(x_1, \ldots, x_n; \theta^*)$ converges in probability to the expression on the right of (12.3.15).

Finally, by applying **4.3.6** we conclude that the sequences of random variables given in (12.3.12) and (12.3.13) and the sequence

$$\sqrt{n}(\hat{\theta} - \theta_0)B^2(\theta_0, \theta_0), \qquad n = 1, 2, \ldots \tag{12.3.16}$$

converge together in distribution if any one of the three does. But we know from **12.3.1** that the sequence (12.3.12) converges in distribution to

$$N(0, B^2(\theta_0, \theta_0)). \tag{12.3.17}$$

Therefore, for large n, the asymptotic distribution of $\hat{\theta}$ is

$$N\left(\theta_0, \frac{1}{nB^2(\theta_0, \theta_0)}\right), \tag{12.3.18}$$

thus establishing **12.3.3**. The asymptotic distribution of $\hat{\theta}$ given by (12.3.18) was originally stated by Fisher (1922).

(d) Asymptotic Efficiency of Maximum Likelihood Estimators

Suppose $\tilde{\theta}(x_1, \ldots, x_n)$ is a biased estimator for θ_0 with bias $b_n(\theta_0)$ as defined in (12.2.16). If $\tilde{\theta}$ is a consistent estimator for θ_0 its bias $b_n(\theta_0)$ converges to zero as $n \to \infty$.

For any given n, we have seen that the lower bound of $\sigma^2(\tilde{\theta}_n \mid \theta_0)$ is given by (12.2.17). From this, it follows that

(12.3.19)

$$\sigma^2(\sqrt{n}\tilde{\theta} \mid \theta_0) = \mathscr{E}([\sqrt{n}(\tilde{\theta} - \theta_0 - b_n(\theta_0)]^2 \mid \theta_0) \geqslant \frac{[1 + b_n'(\theta_0)]^2}{B^2(\theta_0, \theta_0)}.$$

Since the inequality holds for every n, we have, assuming $b_n'(\theta_0) \to 0$ as $n \to \infty$,

(12.3.20)
$$\liminf_{n\to\infty} \sigma^2(\sqrt{n}\tilde{\theta} \mid \theta_0) \geqslant \lim_{n\to\infty}\left(\frac{[1 + b_n'(\theta_0)]^2}{B^2(\theta_0, \theta_0)}\right) = \frac{1}{B^2(\theta_0, \theta_0)}$$

that is,

(12.3.21)
$$\liminf_{n\to\infty} \sigma^2(\sqrt{n}\tilde{\theta} \mid \theta_0) \geqslant \frac{1}{B^2(\theta_0, \theta_0)}.$$

Therefore we have the following result:

12.3.5 *Let $(x_1, \ldots, x_n)$ be a sample from the* c.d.f. $F(x; \theta_0)$, *where $F(x; \theta)$ is regular with respect to its first θ-derivative and $B^2(\theta, \theta)$ exists for θ in Ω_0. If $\tilde{\theta}$ is a consistent estimator for θ_0, and if the θ-derivative of the bias of $\tilde{\theta}$ has zero as its limit as $n \to \infty$, the least upper bound of $\sigma^2(\sqrt{n}\tilde{\theta})$ as $n \to \infty$ cannot be less than $1/B^2(\theta_0, \theta_0)$.*

An important case of a consistent estimator for θ is the maximum likelihood estimator $\hat{\theta}$ under the conditions of **12.3.3**. Also we have from **12.3.3** that the variance of the limiting distribution of $\sqrt{n}(\hat{\theta} - \theta_0)$ is

(12.3.22)
$$\frac{1}{B^2(\theta_0, \theta_0)}.$$

If the variance of the limiting distribution of $\sqrt{n}(\tilde{\theta} - \theta_0)$ is $1/B^{*2}(\theta_0, \theta_0)$, we shall say that the limiting or *asymptotic efficiency* of $\tilde{\theta}$ as $n \to \infty$ is defined as

(12.3.23)
$$\text{leff}\,(\tilde{\theta} \mid \theta_0) = \frac{B^{*2}(\theta_0, \theta_0)}{B^2(\theta_0, \theta_0)}.$$

Thus

(12.3.24)
$$\text{leff}\,(\hat{\theta} \mid \theta_0) = 1.$$

Any estimator $\tilde{\theta}$ for θ_0 for which leff $(\tilde{\theta} \mid \theta_0) = 1$ is said to be an asymptotically efficient estimator for θ_0, and this is true of $\hat{\theta}$. Hence,

12.3.6 *Under the assumptions of* **12.3.3** *the maximum likelihood estimator $\hat{\theta}$ for θ_0 is consistent and asymptotically efficient.*

The reader should note the distinction between efficiency as defined by (12.2.15) and *asymptotic efficiency* which is defined by (12.3.23). The former definition holds for any n, but only when an efficient estimator exists. The latter is a property which is defined as a limit for $n \to \infty$ and holds when the asymptotically efficient estimator $\hat{\theta}$ exists, that is, under the assumptions of **12.3.3**. The asymptotic efficiency of a consistent estimator $\tilde{\theta}$ may also be viewed as the large-sample extension of Fisher's notion of amount of information in $\tilde{\theta}$ concerning θ_0.

Further detailed studies of asymptotic properties of maximum likelihood estimators have been carried out by Bahadur (1958), Barankin and Gurland (1951), Kraft and LeCam (1956), Rao (1949), Wald (1948), and others.

Example. To illustrate the asymptotic efficiency of an estimator, suppose $(x_1, \ldots, x_n)$ is a sample from the normal distribution $N(\mu, \sigma^2)$. Let $\tilde{\mu}$ be the median and $\hat{\mu}$ the mean of the sample. It follows from **9.6.6** for $p = \frac{1}{2}$ that the asymptotic distribution of $\tilde{\mu}$ for large n is $N(\mu, \pi\sigma^2/2n)$. It can be verified by the reader that an efficient estimator $\hat{\mu}$ for μ exists for any value of n and that $\hat{\mu}$ is the sample mean. Furthermore, we know from **8.3.3*d*** that the distribution of the sample mean is the normal distribution $N(\mu, \sigma^2/n)$. Therefore applying (12.3.23) the asymptotic efficiency of the sample median $\tilde{\mu}$ in estimating the true value μ_0 of μ is given by

$$\text{leff}\,(\tilde{\mu} \mid \mu_0) = \frac{\sigma^2}{(\pi/2)\sigma^2} = 0.637.$$

This means that for large n the mean of a sample of size $(2/\pi)n$ will estimate the true value of the mean μ of a normal distribution $N(\mu, \sigma^2)$ with the same degree of precision approximately as the median of a sample of size n, no matter what the true values of μ and σ^2 may be.

(e) Function of a Maximum Likelihood Estimator Having Asymptotic Variance $1/n$

Referring to **12.3.3** it will be noted that under the regularity conditions placed on $F(x; \theta)$ the maximum likelihood estimator $\hat{\theta}$ has as its asymptotic variance for large n, the quantity $1/[nB^2(\theta_0, \theta_0)]$ where θ_0 is the true value of θ. A question of considerable theoretical and practical interest is whether one can use as a new parameter ζ some function of θ, say $\zeta(\theta)$, such that the asymptotic variance of $\zeta(\hat{\theta})$ is simply $1/n$. A function of this kind can be found under certain conditions. To determine $\zeta(\theta)$ we proceed as follows. Let $\zeta(\theta)$ be a function of θ having a derivative $\zeta'(\theta)$ and unique inverse $\theta(\zeta)$ in Ω_0 and let $\theta(\zeta)$ have a derivative $\theta'(\zeta)$ with respect to ζ. Let ζ_0 be the true value of ζ, that is, $\theta(\zeta_0) = \theta_0$. Referring to (12.1.8) we may write

(12.3.25) $$B^2(\theta, \theta) = \int_{-\infty}^{\infty} (S(x; \theta))^2 \, dF(x; \theta).$$

Since

(12.3.26) $$S(x; \theta) = \frac{d}{d\theta} \log dF(x; \theta)$$

we may write

(12.3.27) $$S(x; \theta(\zeta)) = \frac{d}{d\zeta} \log dF(x; \theta(\zeta)) \cdot \zeta'(\theta).$$

Substituting in (12.3.25), we have

(12.3.28)
$$B^2(\theta(\zeta), \theta(\zeta)) = [\zeta'(\theta)]^2 \int_{-\infty}^{\infty} \left[\frac{\partial}{\partial\zeta} \log dF(x; \theta(\zeta))\right]^2 dF(x; \theta(\zeta)).$$

Requiring the maximum likelihood estimator $\hat{\zeta}$ (for ζ_0) to have an asymptotic variance of $1/n$ for large n, is equivalent to requiring the integral on the right of (12.3.28) to have the value 1. Therefore, the function $\zeta(\theta)$ must satisfy the differential equation

$$\left(\frac{d\theta}{d\zeta}\right)^2 = \frac{1}{B^2(\theta(\zeta), \theta(\zeta))}. \tag{12.3.29}$$

or

$$\zeta(\theta) = \int_{\theta_0}^{\theta} B(\theta, \theta)\, d\theta + \zeta_0, \qquad B(\theta,\theta) = +\sqrt{B^2(\theta,\theta)}. \tag{12.3.30}$$

Summarizing, we obtain the following result:

12.3.7 *If $(x_1, \ldots, x_n)$ is a sample from the* c.d.f. $F(x; \theta_0)$ *where* $F(x; \theta)$ *is regular with respect to its second θ-derivative, and if $\hat{\theta}$ is the maximum likelihood estimator for θ_0, then $\zeta(\hat{\theta})$, where $\zeta(\theta)$ is given by* (12.3.30) *and has a derivative $\zeta'(\theta)$ and a unique inverse $\theta(\zeta)$ in Ω_0, is asymptotically distributed according to the normal distribution $N(\zeta_0, 1/n)$, for large n.*

Example. Suppose $(x_1, \ldots, x_n)$ is a sample from the Poisson distribution $Po(\mu_0)$ whose p.f. is

$$p(x; \mu_0) = \frac{\mu_0^x e^{-\mu_0}}{x!}, \qquad x = 0, 1, \ldots$$

We wish to determine what function of μ, say $\zeta(\mu)$, is such that $\zeta(\hat{\mu})$ has as its asymptotic distribution the normal distribution $N(\zeta(\mu_0), 1/n)$ as $n \to \infty$. We know from the example in Section 12.2(c) that the maximum likelihood estimator $\hat{\mu}$ for μ_0 is the sample mean $\bar{x}$. It can be verified by applying formula (12.3.25) to the Poisson distribution (that is; where $dF(x; \mu) = p(x; \mu)$) that

$$B^2(\mu, \mu) = \sum_{x=0}^{\infty} \left(\frac{x}{\mu} - 1\right)^2 p(x, \mu) = \frac{1}{\mu}.$$

Applying (12.3.30) and choosing $\zeta_0 = 2\sqrt{\mu_0}$, we find

$$\zeta(\mu) = 2\sqrt{\mu}.$$

Therefore, we conclude from **12.3.7** that if $(x_1, \ldots, x_n)$ is a sample from a population having the Poisson distribution $Po(\mu_0)$, then $2\sqrt{\bar{x}}$ has, as its asymptotic distribution for large n, the normal distribution $N(2\sqrt{\mu_0}, 1/n)$.

12.4 INTERVAL ESTIMATION

(a) Definition of Confidence Interval

Suppose $(x_1, \ldots, x_n)$ is a random variable having c.d.f. $F_n(x_1, \ldots, x_n; \theta)$. Let $\underline{\theta}(x_1, \ldots, x_n)$, $\bar{\theta}(x_1, \ldots, x_n)$ be two functions (random variables) of the sample elements such that $\underline{\theta} < \bar{\theta}$. If $\underline{\theta}$ and $\bar{\theta}$ can be chosen so that, for a given γ

$$P(\underline{\theta} < \theta < \bar{\theta} \mid \theta) = \gamma \tag{12.4.1}$$

where $P(\underline{\theta} < \theta < \bar{\theta} \mid \theta)$ denotes the indicated probability as determined from $F_n(x_1, \ldots, x_n; \theta)$, then $(\underline{\theta}, \bar{\theta})$ is called a $100\gamma\,\%$ *confidence interval* for θ, whereas $\underline{\theta}$ and $\bar{\theta}$ are called *lower* and *upper confidence limits* for θ, and γ is called the *confidence coefficient*. Note that $(\underline{\theta}, \bar{\theta})$ is a two-dimensional random variable such that the probability is γ that the interval $(\underline{\theta}, \bar{\theta})$ contains the true value of θ in $F_n(x_1, \ldots, x_n; \theta)$. Examples of confidence intervals have already been given in Sections 10.2(*c*), 10.4, 10.5, and 10.6, in the case of sampling from normal distributions. The mathematical formality of a confidence interval was first introduced by Laplace (1814) in dealing with the problem of inferring the value of p in the binomial distribution (6.2.2) from an observed value of the random variable x of the distribution. Laplace regarded the confidence interval as fixed and p as a random variable. Laplace's procedure was rediscovered by Wilson (1927) who gave the correct interpretation of the interval as a random interval. The development of the modern theory and terminology of confidence intervals is due to Neyman (1937).

(b) Procedure for Constructing Confidence Intervals from Samples from Continuous c.d.f.'s

A procedure by which confidence intervals can be constructed in certain cases from a sample from a continuous c.d.f. $F(x; \theta)$ can be stated as follows:

12.4.1 *Suppose* $(x_1, \ldots, x_n)$ *is a sample from a continuous c.d.f.* $F(x; \theta)$. *Suppose* $g(x_1, \ldots, x_n; \theta)$: (i) *is defined at every point* θ *in an interval* (θ_1, θ_2) *containing* θ_0, *and at every point in the sample space* R_n, *except possibly for a set of probability zero;* (ii) *is continuous and monotonically increasing or decreasing in* θ*; and* (iii) *has a c.d.f. that does not depend on* θ. *Let* (g_1, g_2) *be an interval for which* $P(g_1 < g < g_2 \mid \theta) = \gamma$. *Then if* θ_0 *is the true value of* θ, *the solutions* $\underline{\theta}, \bar{\theta}$, *(where* $\underline{\theta} < \bar{\theta}$*), of the equations* $g(x_1, \ldots, x_n; \theta) = g_1, g_2$ *exist and* $(\underline{\theta}, \bar{\theta})$ *is a* $100\gamma\,\%$ *confidence interval for* θ_0.

To establish **12.4.1** note that since the c.d.f. of $g(x_1, \ldots, x_n; \theta)$ does not depend on θ, we can, for a given $0 < \gamma < 1$ and a given true value θ_0, choose g_1 and g_2 (in many ways, of course) so that if θ_0 is the true value of θ

$$P(g_1 < g(x_1, \ldots, x_n; \theta_0) < g_2 \mid \theta_0) = \gamma. \tag{12.4.2}$$

Since $g(x_1, \ldots, x_n; \theta)$ is continuous and monotonically increasing (or decreasing) in θ over (θ_1, θ_2), it is evident that (12.4.2) is equivalent to the statement

$$P(\underline{\theta} < \theta_0 < \bar{\theta} \mid \theta_0) = \gamma \tag{12.4.3}$$

where $\underline{\theta} < \bar{\theta}$ are the solutions of the equations $g(x_1, \ldots, x_n; \theta) = g_1, g_2$ and, of course, are random variables.

One may ask whether it is always possible to find a function $g(x_1, \ldots, x_n; \theta)$ whose c.d.f. is independent of θ, assuming that $F(x; \theta)$ is continuous in x. Such a function can always be found. In fact, the form taken by $F_n(x_1, \ldots, x_n; \theta)$ for simple random sampling, namely $\prod_{\xi=1}^{n} F(x_\xi; \theta)$, is such a function if $F(x; \theta)$ is a continuous monotonically increasing or decreasing function in θ for all points in the space of x, except possibly for a set of probability zero. For the random variables $F(x_\xi; \theta)$, $\xi = 1, 2, \ldots, n$ are independent and, as pointed out in Section 8.7*a*, each is distributed according to the rectangular distribution $R(\frac{1}{2}, 1)$. Furthermore, $-\log F(x_\xi; \theta)$ has the gamma distribution $G(1)$, and it follows from the reproductive property of the gamma distribution that $-\sum_{\xi=1}^{n} \log F(x_\xi; \theta)$ has the gamma distribution $G(n)$. Therefore

(12.4.4)

$$P\left(-\log b_2 < -\sum_{\xi=1}^{n} \log F(x_\xi; \theta) < -\log b_1 \mid \theta\right) = \frac{1}{\Gamma(n)} \int_{-\log b_2}^{-\log b_1} y^{n-1} e^{-y}\, dy.$$

If b_1 and b_2 are chosen so that the integral on the right is γ, then we have

$$P\left(b_1 < \prod_{\xi=1}^{n} F(x_\xi; \theta) < b_2 \mid \theta\right) = \gamma. \tag{12.4.5}$$

Now if $F(x; \theta)$ is continuous and monotonically increasing (or decreasing) in θ for all points x in R_1 (except possibly for a set of zero probability), then the statement holds for $\prod_{\xi=1}^{n} F(x_\xi; \theta)$. Therefore the inequalities in the probability statement (12.4.5) can be inverted and written in the form

$$P(\underline{\theta} < \theta_0 < \bar{\theta} \mid \theta_0) = \gamma \tag{12.4.6}$$

thus providing a $100\gamma\%$ confidence interval for θ_0.

The reader should observe that if the assumption of monotonicity of $g(x_1, \ldots, x_n; \theta)$ in **12.4.1** is removed, (12.4.2) is still valid, but the inversion of the inequalities in (12.4.2) yields a *random set* E in (θ_1, θ_2) where E

depends on $(x_1, \ldots, x_n)$, instead of the confidence interval $(\underline{\theta}, \bar{\theta})$. Then (12.4.3) would be replaced by $P(\theta \in E \mid \theta) = \gamma$. It should be carefully noted that what is random is E and not θ. From the viewpoint of the estimation of θ_0, a random interval $(\underline{\theta}, \bar{\theta})$ is more appealing and more useful than a random set E. The extension of the idea of interval estimation of θ_0 to set estimation of θ_0 will be useful in Section 12.7 when we consider the problem of extending interval estimation to a multidimensional parameter θ.

(c) Confidence Intervals from Samples from Discrete Distributions

In case $F(x; \theta)$ is the c.d.f. of a discrete random variable, then, of course, **12.4.1** is not applicable; that is, one cannot find confidence intervals having confidence coefficients exactly equal to γ, where γ is arbitrary. However, under certain conditions one can find confidence intervals having confidence coefficients not less than γ, although the situation is less elegant than that for the case of a continuous random variable. More specifically,

12.4.2 *Suppose $(x_1, \ldots, x_n)$ is a sample from the c.d.f. $F(x; \theta)$, where x is a discrete random variable and θ is a parameter whose space is an interval (θ_1, θ_2). Let $\tilde{\theta}$ be an estimator for θ defined at every mass point in the sample space R_n and lying in (θ_1, θ_2). Let $V(\tilde{\theta}; \theta)$ be the c.d.f. of $\tilde{\theta}$ and $V^*(\tilde{\theta}; \theta) = 1 - V(\tilde{\theta}; \theta)$. Furthermore let $V(\tilde{\theta}; \theta)$ be continuous and decreasing in θ at each mass point of $\tilde{\theta}$ so that $\lim_{\theta \to \theta_1} V(\tilde{\theta}; \theta) = 1$, $\lim_{\theta \to \theta_2} V(\tilde{\theta}; \theta) = 0$ for all $\tilde{\theta} \in (\theta_1, \theta_2)$. Let $\underline{\theta}$ and $\bar{\theta}$ be the values of θ for which $V(\tilde{\theta}; \theta) = \gamma_1$ and $V^*(\tilde{\theta}; \theta) = \gamma_1^*$, respectively where γ_1 and γ_1^* are non-negative and $0 < \gamma = 1 - \gamma_1 - \gamma_1^* < 1$. Then $(\underline{\theta}, \bar{\theta})$ is a confidence interval for θ_0 with confidence coefficient $\geqslant \gamma$.*

To establish **12.4.2**, let θ_1 be the largest value of $\tilde{\theta}$ for which $V(\tilde{\theta}; \theta_0) \leqslant \gamma_1$ and θ_2 the smallest value of $\tilde{\theta}$ for which $V^*(\tilde{\theta}; \theta_0) \leqslant \gamma_1^*$. Then

$$P(\theta_1 < \tilde{\theta} < \theta_2) \geqslant \gamma \tag{12.4.7}$$

where $\gamma = 1 - \gamma_1 - \gamma_1^*$. But since $V(\tilde{\theta}; \theta)$ is monotonically decreasing in θ and nondecreasing in $\tilde{\theta}$ and $V^*(\tilde{\theta}; \theta)$ is monotonically increasing in θ and nonincreasing in $\tilde{\theta}$, it is evident that $\theta_1 < \tilde{\theta} < \theta_2$ if and only if $V(\tilde{\theta}; \theta_0) > \gamma_1$ and $V^*(\tilde{\theta}; \theta_0) > \gamma_1^*$, that is, if and only if $\underline{\theta} \leqslant \theta_0 \leqslant \bar{\theta}$. Therefore

$$P(\underline{\theta} \leqslant \theta_0 \leqslant \bar{\theta} \mid \theta_0) \geqslant \gamma \tag{12.4.8}$$

and hence $(\underline{\theta}, \bar{\theta})$ is a confidence interval for θ_0 with confidence coefficient $\geqslant \gamma$.

Example. As an illustrative example, consider a sample $(x_1, \ldots, x_n)$ from a population having a binomial distribution $Bi(1, p_0)$. The p.f. of this distribution is

$$f(x; p_0) = p_0^x(1 - p_0)^{1-x}$$

the mass points being $x = 0, 1$, and p_0 is a number on $(0, 1)$. The sample mean $\bar{x}$ has the binomial distribution $Bi(n, p_0)$ over its sample space $0, 1/n, \ldots, n/n$, and is a consistent estimator for p_0. Furthermore, when the population being sampled has the binomial distribution $Bi(1, p)$, the c.d.f. of $\bar{x}$ is defined by

$$V(\bar{x}\,; p) = \sum_{i=0}^{n\bar{x}} \binom{n}{i} p^i(1 - p)^{n-i}$$

and $V^*(\bar{x}; p)$ is defined by

$$V^*(\bar{x}; p) = 1 - V(\bar{x}; p).$$

Now $V(\bar{x}; p)$ is monotonically decreasing in p and $V^*(\bar{x}; p)$ is monotonically increasing in p for $0 < n\bar{x} < p$ since $(d/dp)V(\bar{x}; p) < 0$ and $(d/dp)V^*(\bar{x}; p) > 0$. Furthermore $V(\bar{x}; 0) = 1$ and $V(\bar{x}; 1) = 0$ for all $\bar{x} \in (0, 1)$. Therefore, if we apply **12.4.2** it is evident that if $\underline{p}$ and $\bar{p}$ are the solutions of $V(\bar{x}; p_0) = \gamma_1$ and $V^*(\bar{x}; p_0) = \gamma_1^*$ respectively, then if p_0 is the true value of p in the population, and if $\gamma = 1 - \gamma_1 - \gamma_1^*$, we have

$$P(\underline{p} \leqslant p_0 \leqslant \bar{p} \mid p_0) \geqslant \gamma,$$

that is $(\underline{p}, \bar{p})$ is a confidence interval for p_0 with confidence coefficient $\geqslant \gamma$.

Confidence intervals of this type for p have been constructed, presented graphically and published by Clopper and Pearson (1934) for the case when $\gamma_1 = \gamma_1^* = 0.05$, 0.025, and for various values of n ranging from 10 to 1000.

A procedure similar to that used in the preceding Example has been applied by Garwood (1936) and Ricker (1937), to the problem of determining confidence intervals for the parameter μ in a Poisson distribution.

The procedure stated in **12.4.2** can be extended to the case in which the parameter space consists of a discrete set of points or some other subset of the points in the interval (θ_1, θ_2). This situation arises, for example, in setting up confidence intervals for p in the case of a random variable x having the hypergeometric distribution $H(N, n; p)$ as defined in Section 6.1(a), where N and n are known. In this case the parameter space Ω for p consists of discrete points $0, 1/N, 2/N, \ldots, 1$. Confidence intervals for p in this case have been computed by Chung and DeLury (1950) for $N = 500$, 2500, 10,000, $n/N = 0.05$, 0.10 (0.10) 0.90, and for $\gamma = 0.90$, 0.95, 0.99.

(d) Remarks Concerning Fiducial Probability and Fiducial Intervals

A notion introduced by Fisher (1935b) and related to that of confidence intervals is the concept of a fiducial interval. Suppose the parameter space of θ is the interval (θ_1, θ_2) and let s be a statistic whose sample space is (a, b). Let $v(s; \theta)$ be the p.d.f. and $V(s; \theta)$ the c.d.f. of s. If $V^*(\theta; s) = 1 - V(s; \theta)$ is continuous and strictly monotonically increasing in θ for each s in (a, b) so that $\lim_{\theta\to\theta_1} V^*(\theta; s) = 0$, and $\lim_{\theta\to\theta_2} V^*(\theta; s) = 1$, for each s in (a, b), then $G(\theta; s)$ as a function of θ on (θ_1, θ_2) has all the formal properties of a c.d.f. for each s in (a, b) and is called the *fiducial* c.d.f. of θ, based on the statistic s. Thus, if θ' and θ'', $\theta' < \theta''$, are chosen so that $V^*(\theta'; s) = \gamma_1$ and $V^*(\theta''; s) = \gamma_2$, where $\gamma_2 - \gamma_1 = \gamma$, where $0 < \gamma < 1$, we may write the following *fiducial probability statement*:

$$\text{fid } P(\theta' < \theta < \theta'' \mid s) = \gamma \tag{12.4.9}$$

so that (θ', θ'') would be called a $100\gamma\%$ *fiducial interval* for θ. If θ_0 is the true value of θ, and if we let s' and s'' be the values of s satisfying the equations $V(s; \theta_0) = 1 - \gamma_2$ and $V(s; \theta_0) = 1 - \gamma_1$, respectively, where $\gamma_2 - \gamma_1 = \gamma$, it will be seen that $P(s' < s < s'' \mid \theta_0) = \gamma$. But θ_0 lies in (θ', θ'') if and only if s lies in (s', s''). Thus, (θ', θ'') is also a $100\gamma\%$ confidence interval for θ_0.

Another procedure for generating a fiducial distribution and fiducial intervals is by means of *pivotal functions*. Thus, for the statistic s and parameter referred to above, suppose $g(s; \theta)$ is a strictly monotonically increasing function of θ for each s and having a similar property as a function of s for each θ, such that $g(s; \theta)$ has a probability element $h(g)\,dg$ which does not depend on θ except through g. If $g(s; \theta)$ has first partial derivatives with respect to s and with respect to θ, then $h(g)(\partial g/\partial s)\,ds$ is the probability element of the random variable s, and $h(g)(\partial g/\partial\theta)\,d\theta$ is the fiducial probability element of the parameter θ. Thus, let g_1 and g_2, $(g_1 < g_2)$, be numbers such that

$$P(g_1 < g(s; \theta) < g_2 \mid \theta) = \gamma. \tag{12.4.10}$$

Let s_1 and s_2 $(s_1 < s_2)$, be values of s for which $g(s; \theta) = g_1, g_2$ respectively, whereas θ_1 and θ_2, $\theta_1 < \theta_2$, are values of θ for which $g(s; \theta) = g_1, g_2$ respectively. Then $\theta \in (\theta_1, \theta_2)$ if and only if $s \in (s_1, s_2)$. Therefore

$$\int_{\theta_1}^{\theta_2} h(g)\frac{\partial g}{\partial\theta}\,d\theta = \int_{s_1}^{s_2} h(g)\frac{\partial g}{\partial s}\cdot ds$$

where the left side is the fiducial probability fid $P(\theta_1 < \theta < \theta_2)$ whereas the right side is the probability $P(s_1 < s < s_2)$. A similar treatment can be given if $g(s; \theta)$ is strictly monotonically increasing in s but decreasing in θ and vice versa, and when $g(s; \theta)$ is decreasing in both variables.

Remark. In the two procedures mentioned above for generating fiducial distributions, fiducial intervals and confidence intervals are equivalent.

There are some situations, however, when fiducial intervals are not equivalent to confidence intervals, and this has given rise to controversy as to what interpretation should be placed on a fiducial interval. This controversy centers on the meaning of the formal fiducial distribution of the difference $\mu_1 - \mu_2$ of means of two normal distributing $N(\mu_1, \sigma_1^2)$ and $N(\mu_2, \sigma_2^2)$, the statistics involved being the sample mean and sample variance of a sample from each distribution. More precisely, suppose n_1, $\bar{x}_1$, s_1^2 and n_2, $\bar{x}_2$, s_2^2 are the size, mean, and variance of independent samples from $N(\mu_1, \sigma_1^2)$ and $N(\mu_2, \sigma_2^2)$ respectively. Then

$$t_1 = \sqrt{n_1}(\bar{x}_1 - \mu_1)/s_1, \qquad t_2 = \sqrt{n_2}(\bar{x}_2 - \mu_2)/s_2 \tag{12.4.11}$$

are independent Student ratios having $n_1 - 1$ and $n_2 - 1$ degrees of freedom respectively. If $f_m(t)\,dt$ is the probability element of a Student ratio t having m degrees of freedom, the probability element of t_1 and t_2 is $f_{n_1-1}(t_1)f_{n_2-1}(t_2)\,dt_1\,dt_2$. Using the quantities denoted by t_1 and t_2 in (12.4.11) as pivotal quantities, one obtains the fiducial probability element of (μ_1, μ_2) for given n_1, $\bar{x}_1$, s_1^2, n_2, $\bar{x}_2$, s_2^2 the following

$$f_{n_1-1}(\sqrt{n_1}(\bar{x}_1 - \mu_1)/s_1)f_{n_2-1}(\sqrt{n_2}(\bar{x}_2 - \mu_2)/s_2)[\sqrt{n_1 n_2}/s_1 s_2]\,d\mu_1\,d\mu_2. \tag{12.4.12}$$

From (12.4.12) one can formally find the *fiducial probability distribution* of $(\mu_1 - \mu_2)$ and from this distribution fiducial intervals for $(\mu_1 - \mu_2)$. But it can be shown that there exists no function $g(\bar{x}_1, \bar{x}_2, s_1^2, s_2^2, \mu_1 - \mu_2)$, satisfying certain regularity conditions, whose distribution depends only on n_1, n_2, and g from which one can obtain confidence intervals for $\mu_1 - \mu_2$. Thus in this example fiducial intervals appear to be different from confidence intervals, and the question is how does one interpret fiducial intervals in this case?

This is known as the *Behrens-Fisher problem* [See Fisher (1935*b*)]. The reader interested in further details concerning this problem should consult Tukey (1957*c*) and the references at the end of his paper.

12.5 INTERVAL ESTIMATION FROM LARGE SAMPLES

(a) Interval Estimates from Likelihood Estimating Function

The asymptotic theory of interval estimation of population parameters from large samples follows fairly directly from the theory of point estimation from large samples as presented in Section 12.3. Asymptotically efficient point estimation implies, as we shall see, asymptotically shortest interval estimation in the case of large samples.

Suppose $(x_1, \ldots, x_n)$ is a sample from a distribution having c.d.f. $F(x; \theta_0)$. It will be convenient to define the *likelihood estimating function* $h_n(x_1, \ldots, x_n; \theta)$ as

$$h_n(x_1, \ldots, x_n; \theta) = \frac{S_n(x_1, \ldots, x_n; \theta)}{nB(\theta_0, \theta)}, \tag{12.5.1}$$

where $B(\theta_0, \theta) = +\sqrt{B^2(\theta_0, \theta)}$. Under the assumptions of **12.3.1**, we know $\sqrt{n}h_n(x_1, \ldots, x_n; \theta_0)$ has as its limiting distribution as $n \to \infty$, the distribution $N(0, 1)$ if the true value of θ is θ_0. If we adopt the stronger conditions of **12.3.3**, then since the probability exceeds $1 - \varepsilon$ that for all $n > n_\varepsilon$

$$\sqrt{n}h_n(x_1, \ldots, x_n; \theta_0) = h_n'(x_1, \ldots, x_n; \theta^*)[\sqrt{n}(\theta_0 - \hat{\theta})]$$

for points $(x_1, \ldots, x_n)$ in E_n as defined in (12.3.5), where $h_n'(x_1, \ldots, x_n; \theta) = (\partial/\partial\theta)\, h_n(x_1, \ldots, x_n; \theta)$ and $\hat{\theta}$ and θ^* are as defined in (12.3.7), the two sequences of random variables

$$\sqrt{n}h_n(x_1, \ldots, x_n; \theta_0) \qquad n = 1, 2, \ldots \tag{12.5.2}$$

and

$$h_n'(x_1, \ldots, x_n; \theta^*)[\sqrt{n}(\theta_0 - \hat{\theta})] \qquad n = 1, 2, \ldots \tag{12.5.3}$$

converge together in distribution to the normal distribution $N(0, 1)$. Furthermore, if $(\partial/\partial\theta)B^2(\theta_0, \theta)$ exists and is bounded over Ω_0, and if $\theta = \theta_0$, then

$$h_n'(x_1, \ldots, x_n; \theta^*) \qquad \text{and} \qquad h_n'(x_1, \ldots, x_n; \hat{\theta}) \tag{12.5.4}$$

converge in probability to $B(\theta_0, \theta_0)$ as $n \to \infty$. Therefore, we may write

$$\lim_{n \to \infty} P[-\lambda_\gamma < h_n'(x_1, \ldots, x_n; \hat{\theta})[\sqrt{n}(\theta_0 - \hat{\theta})] < +\lambda_\gamma \mid \theta_0] = \gamma$$

where $\Phi(\lambda_\gamma) - \Phi(-\lambda_\gamma) = \gamma$, from which it is seen that

(12.5.5)

$$\lim_{n \to \infty} P\left[\hat{\theta} - \frac{\lambda_\gamma}{\sqrt{n}h_n'(x_1, \ldots, x_n; \hat{\theta})} < \theta_0 < \hat{\theta} + \frac{\lambda_\gamma}{\sqrt{n}h_n'(x_1, \ldots, x_n; \hat{\theta})} \middle| \theta_0\right] = \gamma.$$

We shall say that the interval having endpoints $\hat{\theta} \pm \dfrac{\lambda_\gamma}{\sqrt{n}h_n'(x_1, \ldots, x_n; \hat{\theta})}$ is an *asymptotic* $100\gamma\%$ *confidence interval* for θ_0 for large n; its length is $\dfrac{2\lambda_\gamma}{\sqrt{n}h_n'(x_1, \ldots, x_n; \hat{\theta})}$.

(b) Interval Estimates from General Estimating Functions

We shall show that the ratio of the squared length of this interval to that of a similar interval obtained by using an arbitrary estimating function $g_n(x_1, \ldots, x_n; \theta)$, satisfying certain conditions, converges in probability to a number which cannot exceed 1.

Let $(x_1, \ldots, x_n)$ be a sample from $F(x; \theta)$ which is regular in its first two θ-derivatives, and let F_n denote $\prod_{\xi=1}^{n} F(x_\xi; \theta)$, and F_{n0} the value of F_n at $\theta = \theta_0$.

Let

(12.5.6) $$g_n(x_1, \ldots, x_n; \theta)$$

be a random variable such that:

(12.5.7) (i) $$\int_{R_n} [\sqrt{n} g_n(x_1, \cdots, x_n; \theta)]^j \, dF_n = 0, 1; \qquad j = 1, 2$$

for θ in Ω_0;

(ii) $\sqrt{n} g_n(x_1, \ldots, x_n; \theta_0)$ has $N(0, 1)$ as its limiting distribution as $n \to \infty$ if θ_0 is the true value of θ;

(iii) $g_n(x_1, \ldots, x_n; \theta)$ has a continuous θ-derivative $g_n'(x_1, \ldots, x_n; \theta)$ for θ in Ω_0 and for all points in the sample space R_n except possibly for a set of zero probability;

(iv) if θ_0 is the true value of θ, $g_n'(x_1, \ldots, x_n; \theta)$ converges in probability to $B^*(\theta_0, \theta)$ uniformly with respect to θ in Ω_0, where $B^*(\theta_0, \theta)$ is bounded on Ω_0 and $B^*(\theta_0, \theta_0) \neq 0$;

(v) if

$$A_n^*(\theta, \theta') = \int_{R_n} g_n(x_1, \ldots, x_n; \theta') \, dF_n$$

and

$$B_n^*(\theta, \theta') = \int_{R_n} g_n'(x_1, \ldots, x_n; \theta') \, dF_n,$$

then

(12.5.8) $$\begin{cases} \dfrac{\partial}{\partial \theta'} A_n^*(\theta, \theta') = B_n^*(\theta, \theta'), & \text{for } (\theta, \theta') \text{ in } \Omega_0 \times \Omega_0 \\ A_n^*(\theta, \theta) = 0, & \text{for } \theta \text{ in } \Omega_0 \\ \lim_{n \to \infty} B_n^*(\theta, \theta) = B^*(\theta, \theta), & \text{for } \theta \text{ in } \Omega_0. \end{cases}$$

Any function $g_n(x_1, \ldots, x_n; \theta)$ having the properties listed above will be called a *regular estimating function* for θ_0. It is to be noted that the function $h_n(x_1, \ldots, x_n; \theta)$ defined by (12.5.1) is a regular estimating function for θ_0.

It can be verified, by following a line of argument similar to that by which **12.3.2** was established, that if, for $n \geqslant$ some n_0, the equation $g_n(x_1, \ldots, x_n; \theta) = 0$ has a unique root, then the equations

(12.5.9) $$g_n(x_1, \ldots, x_n; \theta) = 0, \qquad n = n_0, n_0 + 1, \ldots$$

have a sequence of roots

(12.5.10) $$\tilde{\theta}(x_1, \ldots, x_n), \qquad n = n_0, n_0 + 1, \ldots$$

which converges in probability to θ_0. Furthermore the two sequences of random variables

(12.5.11) $$\sqrt{n}g_n(x_1, \ldots, x_n; \theta_0), \qquad n = n_0, n_0 + 1, \ldots$$

(12.5.12) $$g_n'(x_1, \ldots, x_n; \tilde{\theta}^*)[\sqrt{n}(\theta_0 - \tilde{\theta})], \qquad n = n_0, n_{0+1}, \ldots$$

where $\tilde{\theta}^*$ satisfies $|\theta_0 - \tilde{\theta}^*| \leqslant |\theta_0 - \tilde{\theta}|$, converge together in distribution to the normal distribution $N(0, 1)$. But it follows from **4.3.8** that

(12.5.13) $$g_n'(x_1, \ldots, x_n; \tilde{\theta}^*) \quad \text{and} \quad g_n'(x_1, \ldots, x_n; \tilde{\theta})$$

converge in probability to $B^*(\theta_0, \theta_0)$ as $n \to \infty$. Therefore, as in (12.5.5), we have the following statement:

(12.5.14)
$$\lim_{n\to\infty} P\left[\tilde{\theta} - \frac{\lambda_\gamma}{\sqrt{n}g_n'(x_1, \ldots, x_n; \tilde{\theta})} < \theta_0 < \tilde{\theta} + \frac{\lambda_\gamma}{\sqrt{n}g_n'(x_1, \ldots, x_n; \tilde{\theta})}\right] = \gamma.$$

Thus, $\tilde{\theta} \pm \dfrac{\lambda_\gamma}{\sqrt{n}g_n'(x_1, \ldots, x_n; \tilde{\theta})}$ are end points of an asymptotic $100\gamma\%$ confidence interval for θ_0 for large n, whose length is $\dfrac{2\lambda_\gamma}{\sqrt{n}g_n'(x_1, \ldots, x_n; \tilde{\theta})}$.

(c) Asymptotically Shortest Confidence Intervals

The ratio r_n of the squared length of the confidence interval in (12.5.5) to that in (12.5.14) is

(12.5.15) $$r_n = c_n^2 \cdot d_n^2$$

where

(12.5.16) $$c_n = \frac{g_n'(x_1, \ldots, x_n; \theta_0)}{h_n'(x_1, \ldots, x_n; \hat{\theta})}$$

and

(12.5.17) $$d_n = \frac{g_n'(x_1, \ldots, x_n; \tilde{\theta})}{g_n'(x_1, \ldots, x_n; \theta_0)}.$$

Now c_n and d_n are random variables. But it follows from **4.3.7** that the numerator and denominator of d_n^2 each converge in probability to $B^{*2}(\theta_0, \theta_0)$ and hence d_n^2 converges in probability to 1. Similarly, c_n^2, and hence r_n, converges in probability to

(12.5.18) $$\frac{B^{*2}(\theta_0, \theta_0)}{B^2(\theta_0, \theta_0)}$$

which we wish to show cannot exceed 1.

Since (12.5.7) holds for all n, and (12.5.8) can be used, we can differentiate (12.5.7) for $j = 1$ with respect to θ under the integral sign, and obtain at $\theta = \theta_0$

$$(12.5.19)\qquad \int_{R_n} g_n'(x_1, \ldots, x_n; \theta_0)\, dF_{n0} + \int_{R_n} \sqrt{n} g_n(x_1, \ldots, x_n; \theta_0)\left[\frac{1}{\sqrt{n}} S_n(x_1, \ldots, x_n; \theta_0)\right] dF_{n0} = 0.$$

Now the squares of these integrals are equal. But the square of the first integral is $B_n^{*2}(\theta_0, \theta_0)$. Applying Schwarz' inequality to the second integral, and using the fact that the integral in (12.5.7) for $j = 2$ is unity, we obtain

$$(12.5.20)\qquad B_n^{*2}(\theta_0, \theta_0) \leqslant \int_{R_n} \left[\frac{1}{\sqrt{n}} S_n(x_1, \ldots, x_n; \theta_0)\right]^2 dF_{n0}.$$

But the right-hand side reduces to $B^2(\theta_0, \theta_0)$ which does not depend on n. Hence

$$B_n^{*2}(\theta_0, \theta_0) \leqslant B^2(\theta_0, \theta_0)$$

and taking limits as $n \to \infty$, we have

$$(12.5.21)\qquad \frac{B^{*2}(\theta_0, \theta_0)}{B^2(\theta_0, \theta_0)} \leqslant 1.$$

Thus, the ratio of squared lengths of confidence intervals converges in probability to a number $\leqslant 1$, and we shall say that the confidence intervals defined in (12.5.5) are *asymptotically shortest* $100\gamma\%$ *confidence intervals* for θ_0 for large n.

We may summarize as follows:

12.5.1 *Suppose* $(x_1, \ldots, x_n)$ *is a random sample from the* c.d.f. $F(x; \theta_0)$, *where* $F(x; \theta)$ *is regular with respect to its second* θ*-derivative for* θ *in* Ω_0. *Then if* $g_n(x_1, \ldots, x_n; \theta)$ *is any regular estimating function for* θ_0, *asymptotic* $100\gamma\%$ *confidence limits for* θ_0 *for large* n *are provided by* (12.5.14). *Furthermore, there exists no regular estimating function for* θ_0 *which provides an asymptotically shorter* $100\gamma\%$ *confidence interval for* θ_0 *than that produced by the likelihood estimating function* $h_n(x_1, \ldots, x_n; \theta)$ *defined by* (12.5.1). *Thus the asymptotically shortest confidence interval is given by* (12.5.5).

The equivalence of the problem of asymptotically shortest confidence intervals and that of asymptotic efficiency of estimation can be seen by noting that the asymptotic efficiency of the estimator $\tilde{\theta}$ in (12.5.10) as computed by formula (12.3.23) is merely the ratio on the left of (12.5.21). That is, the square root of the asymptotic efficiency of the estimator $\tilde{\theta}$ for θ_0 as defined in (12.5.10) is equal to the limit (in probability) of the ratio of length of the asymptotic $100\gamma\,\%$ confidence interval provided by the likelihood estimating function $h_n(x_1, \ldots, x_n; \theta)$ to length of the corresponding confidence interval yielded by $g_n(x_1, \ldots, x_n; \theta)$.

The problem of asymptotically shortest confidence intervals for the case where $g_n(x_1, \ldots, x_n; \theta)$ is of the form $\sum_{\xi=1}^{n} g(x_\xi; \theta)$ was considered by Wilks (1938*b*), and for the more general case by Wald (1942).

12.6 MULTIDIMENSIONAL POINT ESTIMATION

(a) Introductory Remarks

The results obtained in Sections 12.2 through 12.5 pertain to a sample from a c.d.f. $F(x; \theta)$, in which θ is one-dimensional. As pointed out at the beginning of Section 12.1, the results of those sections remain valid with minor changes in notation if the sample is from a k-dimensional c.d.f. $F(x_1, \ldots, x_k; \theta)$ in which θ is one-dimensional.

Now, if θ is r-dimensional with components $\theta_1, \ldots, \theta_r$, the basic results we have obtained have r-dimensional versions which may not be immediately evident to the reader. We shall state some r-dimensional results without giving details of proof in all cases. The reader who is sufficiently familiar with the results already given for the case of a one-dimensional parameter should have no particular difficulty in furnishing these details.

Throughout this section, we shall consider simple random sampling from a c.d.f. $F(x; \theta)$ where θ is r-dimensional and for convenience x is taken to be one-dimensional. The results can be extended with only minor modifications of notation to the case where x is k-dimensional. The components of θ will be $(\theta_1, \ldots, \theta_r)$. The true value $(\theta_{10}, \ldots, \theta_{r0})$ of θ will be denoted by θ_0. Corresponding to the one-dimensional interval Ω_0 we will have an r-dimensional (open) rectangle Ω_{r0} containing θ_0.

An estimator $(\tilde{\theta}_1, \ldots, \tilde{\theta}_r)$ for θ_0 will be denoted by θ. An *unbiased estimator* $\tilde{\theta}$ for θ_0 is one for which $\tilde{\theta}_p$ is an unbiased estimator for θ_{p0}, $p = 1, \ldots, r$. A set of sufficient statistics for estimating θ has been defined in Section 12.2(*e*). A *consistent* estimator $\tilde{\theta}$ for θ_0 is one for which $\tilde{\theta}_p$ converges in probability to θ_{p0} as $n \to \infty$.

(b) Efficiency of a Multidimensional Estimator

We shall now show how to extend the notion of efficiency to the case of multidimensional estimators. In this extension, we shall adopt the notation of Section 12.1*b*. Furthermore, let

(12.6.1)

$$S_{pn}(x_1, \ldots, x_n; \theta) = \frac{\partial}{\partial\theta_p} \log dF_n(x_1, \ldots, x_n; \theta) = \sum_{\xi=1}^{n} S_p(x_\xi; \theta)$$

$$S_{pqn}(x_1, \ldots, x_n; \theta) = \frac{\partial}{\partial\theta_q} S_{pn}(x_1, \ldots, x_n; \theta).$$

If $\tilde{\theta}$ is an unbiased estimator of some θ in Ω_{r0}, and if c'_p, $p = 1, \ldots, r$, are arbitrary constants we may write

$$\int_{R_n} \sum_{p=1}^{r} (\tilde{\theta}_p - \theta_p)c'_p \, dF_n = 0 \tag{12.6.2}$$

which, of course, also states that $\sum_{p=1}^{r} c'_p\tilde{\theta}_p$ is an unbiased estimator for $\sum_{p=1}^{r} c'_p\theta_p$. We shall assume that the components $\tilde{\theta}_p$ are linearly independent.

If $F(x; \theta)$ is regular with respect to all of its first θ-derivatives, we can differentiate (12.6.2) with respect to θ_q, obtaining

$$\int_{R_n} \sum_{p=1}^{r} (\tilde{\theta}_p - \theta_p)c'_p \cdot S_{qn}(x_1, \ldots, x_n; \theta) \, dF_n = c'_q \tag{12.6.3}$$

for $q = 1, \ldots, r$. Multiplying both sides of (12.6.3) by c_q, summing both sides with respect to q, squaring both sides of the resulting equation, and then applying Schwarz' inequality to the squared integral on the left, we obtain for θ in Ω_{r0},

$$\left(\sum_1^r c_q c'_q\right)^2 \leqslant \left\{\int_{R_n}\left[\sum_1^r (\tilde{\theta}_p - \theta_p)c'_p\right]^2 dF_n\right\} \cdot \left\{\int_{R_n}\left[\sum_1^r S_{qn}(x_1, \ldots, x_n; \theta)c_q\right]^2 dF_n\right\}. \tag{12.6.4}$$

But the quantity in the first { } on the right is $\sigma^2\left(\sum_1^r c'_p\tilde{\theta}_p \mid \theta\right)$ and that in the second { } is found to be $n\sigma^2\left[\sum_1^r S_p(x, \theta)c_p \mid \theta\right]$. Therefore, at $\theta = \theta_0$, (12.6.4) reduces to

$$\sigma^2\left(\sum_1^r c'_p\tilde{\theta}_p \mid \theta_0\right) \geqslant \underset{(c_1, \ldots, c_r)}{\text{l.u.b.}} \frac{\left(\sum_1^r c_p c'_p\right)^2}{n\sigma^2\left[\sum_{p=1}^{r} S_p(x; \theta)c_p \mid \theta_0\right]} \tag{12.6.5}$$

which, incidentally, furnishes a lower bound for the variance of the unbiased estimator $\sum_1^r c'_p\tilde{\theta}_p$ for $\sum_1^r c'_p\theta_{p0}$.

As will be seen at the place in (12.6.4) where the Schwarz inequality was applied, the equality sign holds if and only if a linear dependence of form

$$(12.6.6) \qquad K(\theta) \sum_p (\tilde{\theta}_p - \theta_p)c'_p \equiv \sum_p S_{pn}(x_1, \ldots, x_n; \theta)c_p$$

exists for all points in the x-space R_n, except possibly for a set of probability zero, where $K(\theta)$ does not depend on $(x_1, \ldots, x_n)$ and where neither the set $c_1, \ldots, c_r$ nor the set $c'_1, \ldots, c'_r$ vanishes identically. If an unbiased estimator for θ_0 exists so that (12.6.6) is satisfied for θ in Ω_{r0}, we shall call it an *efficient estimator* and denote it by $\hat{\theta}$.

Let the covariance matrix of $\sqrt{n}\tilde{\theta}_p, p = 1, \ldots, r$, at $\theta = \theta_0$ be denoted by $\|B'^{(n)}_{pq}\|$; the covariance matrix of $S_q/\sqrt{n}$ at $\theta = \theta_0$ by $\|B_{pq}\|$ as defined in Section 12.1(*b*). The covariance of $\sqrt{n}\tilde{\theta}_p$ and $S_q/\sqrt{n}$ is δ_{pq}, where $\delta_{pq} = 1$ if $p = q$ and 0 if $p \neq q$, as will be seen from (12.6.3). We may then write (12.6.4) as

$$(12.6.7) \qquad \left(\sum_{p,q} B'^{(n)}_{pq}\, c'_p c'_q\right)\left(\sum_{p,q} B_{pq} c_p c_q\right) \geqslant \left(\sum_{p,q} \delta_{pq} c_p c'_q\right)^2.$$

with the equality holding only when (12.6.6) holds; assuming, of course, that neither the set $c_1, \ldots, c_r$ nor the set $c'_1, \ldots, c'_r$ vanishes identically.

Now the inequality in (12.6.7) holds only when the covariance matrix of the random variables $\sqrt{n}\tilde{\theta}_1, \ldots, \sqrt{n}\tilde{\theta}_r, S_1/\sqrt{n}, \ldots, S_r/\sqrt{n}$ is positive definite, which implies that

$$(12.6.8) \qquad |B'^{(n)}_{pq}| \cdot |B_{pq}| > |\delta_{pq}| = 1$$

and the equality in (12.6.7) holds only when the covariance matrix of the same random variables is made positive semidefinite by the linear dependence expressed by (12.6.6), which implies that

$$(12.6.9) \qquad |B'^{(n)}_{pq}| \cdot |B_{pq}| = 1.$$

In other words, the inequality (equality) of (12.6.7) holds only when the inequality (equality) of

$$(12.6.10) \qquad |B'^{(n)}_{pq}| \cdot |B_{pq}| \geqslant 1$$

holds. Since the components $\tilde{\theta}_1, \ldots, \tilde{\theta}_p$ are linearly independent, we know by **3.5.2** that $\|B'^{(n)}_{pq}\|$ is positive definite and hence $|B'^{(n)}_{pq}| > 0$. For similar reasons $|B_{pq}| > 0$.

These facts suggest that, as a generalization of (12.2.15), we may define the following ratio as the *efficiency* of the r-dimensional unbiased estimator $\tilde{\theta}$ of θ_0:

$$(12.6.11) \qquad \text{eff}\,(\tilde{\theta} \mid \theta_0) = \frac{1}{|B'^{(n)}_{pq}| \cdot |B_{pq}|}.$$

Thus $\text{eff}\,(\tilde{\theta} \mid \theta_0) = 1$ for an efficient estimator $\hat{\theta}$.

We may summarize the preceding results as follows:

12.6.1 *Let* $(x_1, \ldots, x_n)$ *be a sample from the* c.d.f. $F(x; \theta)$, *where* θ *is r-dimensional, and let* $F(x; \theta)$ *be regular with respect to its second* θ*-derivatives. If* $\tilde{\theta}$ *is an unbiased estimator for* θ_0 *with components not linearly dependent whose covariance matrix* $\|B'^{(n)}_{pq}\|$ *exists and is positive definite, then, for any two sets of constants* $c_1, \ldots, c_r$ *and* $c'_1, \ldots, c'_r$, *not all constants in either set being zero, inequalities* (12.6.5) *and* (12.6.10) *hold. Furthermore, in both cases equality holds if and only if* $\tilde{\theta}$ *is an efficient estimator, that is, if and only if* (12.6.6) *holds.*

12.7 MULTIDIMENSIONAL POINT ESTIMATION FROM LARGE SAMPLES

(a) Asymptotic Distribution of the Score

The score for the r-dimensional vector θ is an r-dimensional vector whose components are $S_{pn}(x_1, \ldots, x_n; \theta)$, $p = 1, \ldots, r$, as defined by (12.6.1). These components have an asymptotic r-dimensional normal distribution for large n under conditions analogous to those stated in **12.3.1.** More precisely,

12.7.1 *Suppose* $(x_1, \ldots, x_n)$ *is a sample from the c.d.f.* $F(x; \theta_0)$, *where* θ_0 *is r-dimensional. Let* $F(x; \theta)$ *be regular with respect to its first* θ*-derivatives in* Ω_{r0}. *Then if* $\|B_{pq}(\theta, \theta)\|$, $p, q = 1, \ldots, r$, *is positive definite for* θ *in* Ω_{r0}, $(S_{pn}(x_1, \ldots, x_n; \theta_0), p = 1, \ldots, r)$ *is asymptotically distributed for large n according to the r-dimensional distribution* $N(\{0\}, \|nB_{pq}\|)$ *where* $B_{pq} = B_{pq}(\theta_0, \theta_0)$.

The proof of this statement is similar to that of the one-dimensional case given by **12.3.1** and is left as an exercise for the reader.

(b) Convergence of the Maximum Likelihood Estimator

As in the one-dimensional case, if there exists no unbiased estimator $\tilde{\theta}$ for θ_0 whose components satisfy (12.6.6) for each n, it can be shown under certain conditions that there is a sequence of solutions of the equations

$$S_p(x_1, \ldots, x_n; \theta) = 0, \qquad p = 1, \ldots, r \tag{12.7.1}$$

which converges to θ_0 with probability 1.

More precisely, we have the following r-dimensional version of **12.3.2**:

12.7.2 *Suppose* $(x_1, \ldots, x_n)$ *is a sample from the c.d.f.* $F(x; \theta_0)$, *where* $\theta_0 = (\theta_{10}, \ldots, \theta_{r0})$ *is r-dimensional, and where* $F(x; \theta)$ *is regular* *with respect to its first θ-derivatives in Ω_{r0}. Let $S_p(x; \theta)$, $p = 1, \ldots, r$, be a continuous function of θ in Ω_{r0} for all values of x in R_1, except possibly for a set of zero probability. Then, there exists a sequence of solutions of* (12.7.1) *which converges almost certainly to $(\theta_{10}, \ldots, \theta_{r0})$. If the solution is a unique vector $(\hat{\theta}_1, \ldots, \hat{\theta}_r)$ for $n \geqslant$ some n_0, the sequence of vectors converges almost certainly to $(\theta_{10}, \ldots, \theta_{r0})$ as $n \to \infty$.*

The proof of **12.7.2** is a fairly straightforward extension of that given for **12.3.2** and is omitted.

(c) Asymptotic Distribution of the Maximum Likelihood Estimator

The r-dimensional version of **12.3.3** can be stated as follows:

12.7.3 *If $(x_1, \ldots, x_n)$ is a sample from the c.d.f. $F(x; \theta_0)$, where θ_0 is r-dimensional and $F(x; \theta)$ is regular with respect to its first and second θ-derivatives for θ in Ω_{r0}, and if the maximum likelihood estimator $(\hat{\theta}_1, \ldots, \hat{\theta}_r)$ satisfying* (12.7.1) *is unique for $n \geqslant$ some n_0, and measurable with respect to $\prod_{\xi=1}^{n} F(x_\xi; \theta)$, then it is asymptotically distributed for large n, according to the r-dimensional normal distribution $N(\{\theta_{p0}\}, \|nB_{pq}\|^{-1})$ where $\|B_{pq}\| = \|B_{pq}(\theta_0, \theta_0)\|$.*

The line of argument for **12.7.3** follows closely that of **12.3.3** and is omitted.

(d) Asymptotic Efficiency of the Maximum Likelihood Estimator

In r-dimensional estimation, $\tilde{\theta}$ is a *consistent estimator* for θ_0 if each component of $\tilde{\theta}$ is a consistent estimator of the corresponding component of θ_0.

If $\tilde{\theta}$ is a consistent estimator for θ_0 whose components $(\tilde{\theta}_1, \ldots, \tilde{\theta}_r)$ are such that

$$(\sqrt{n}(\tilde{\theta}_1 - \theta_0), \ldots, \sqrt{n}(\tilde{\theta}_r - \theta_0))$$

has a limiting normal r-dimensional distribution $N(\{0\}; \|B^*_{pq}\|^{-1})$, as $n \to \infty$, we define the *asymptotic efficiency* of $\tilde{\theta}$ as

$$\text{leff}\,(\tilde{\theta} \mid \theta_0) = \frac{|B^*_{pq}|}{|B_{pq}|}. \tag{12.7.2}$$

$$\lim_{n\to\infty} \sigma^2\Big(\sqrt{n} \sum_p c'_p \hat{\theta}_p\Big) = \sum_{p,q} B^{pq} c'_p c'_q \tag{12.7.3}$$

It will be seen from **12.7.3** that

$$(\sqrt{n}(\hat{\theta}_1 - \theta_{10}), \ldots, \sqrt{n}(\hat{\theta}_r - \theta_{r0}))$$

has $N(\{0\}; \|B_{pq}\|^{-1})$ as its limiting distribution as $n \to \infty$. Hence

$$\text{leff}\,(\hat{\theta} \mid \theta_0) = \frac{|B_{pq}|}{|B_{pq}|} = 1. \tag{12.7.4}$$

Therefore

12.7.4 *Under the assumptions of* **12.7.3**, *the maximum likelihood estimator* $\hat{\theta}$ *has asymptotic efficiency* 1 *for estimating* θ_0.

As in the case of efficient estimation of one-dimensional parameters, the earliest studies of efficient estimation of multidimensional parameters were made by Fisher (1922). A more recent study of the subject based on modern mathematical methods has been made by Barankin and Gurland (1951).

12.8 MULTIDIMENSIONAL CONFIDENCE REGIONS

In interval estimation of a one-dimensional parameter θ_0, as discussed in Section 12.4, the essential idea is that for a fixed confidence coefficient γ and for a sample $(x_1, \ldots, x_n)$ from a c.d.f. $F(x; \theta)$ there exist random variables $\underline{\theta} < \bar{\theta}$ such that

$$P(\underline{\theta} < \theta < \bar{\theta} \mid \theta) = \gamma.$$

A method by which $\underline{\theta}$, $\bar{\theta}$ can be constructed under certain conditions is provided by **12.4.1**. As briefly pointed out in Section 12.4(*b*), if the requirement of monotonicity of $g_n(x_1, \ldots, x_n; \theta)$ in θ over Ω is removed in **12.4.1** then there exists a random set $E_1(x_1, \ldots, x_n)$ in R_1, which may be written briefly as E_1, consisting of all real numbers y in Ω for which

$$g_1 < g_n(x_1, \ldots, x_n; y) < g_2 \tag{12.8.1}$$

and not depending on θ_0 such that

$$P(g_1 < g_n(x_1, \ldots, x_n; \theta_0) < g_2 \mid \theta_0) = P(\theta_0 \in E_1 \mid \theta_0) = \gamma. \tag{12.8.2}$$

In other words, the probability is γ that E_1 contains θ_0. That E_1 is a Borel set follows from the fact that $g(x_1, \ldots, x_n; \theta_0)$ is a random variable for all values of θ in Ω. That the probability of E_1 containing θ does not depend on θ follows from the fact that the c.d.f. of $g_n(x_1, \ldots, x_n; \theta)$ does not depend on θ. It should be noted particularly that E_1 is a random set in Ω having probability γ of containing θ_0, assuming, of course, that the sample $(x_1, \ldots, x_n)$ has been drawn from $F(x; \theta_0)$.

Therefore (12.8.2) provides an extension of the idea of interval estimation to *set estimation.* This extension indicates that we can generalize confidence interval estimation to the case of an r-dimensional parameter as follows:

12.8.1 *Suppose* $(x_1, \ldots, x_n)$ *is a sample from the c.d.f.* $F(x; \theta_0)$ *where* θ_0 *is r-dimensional. Let* $g_n(x_1, \ldots, x_n; \theta)$ *be a random variable so defined at every point* θ *in the parameter space* Ω_r *and at every point in sample space* R_n, *except possibly for a set of probability zero, that when the c.d.f. is* $F(x; \theta)$, *the c.d.f. of g does not depend on* θ. *Let* $E_r(x_1, \ldots, x_n) = E_r$, *say, be the set of points* $y = (y_1, \ldots, y_r)$ *in* Ω_r *for which* $g_1 < g_n(x_1, \ldots, x_n; y) < g_2$ *where* g_1 *and* g_2 *are chosen so that*

$$P(g_1 < g_n(x_1, \ldots, x_n; \theta) < g_2 \mid \theta) = \gamma \tag{12.8.3}$$

where γ *does not depend on* θ. *Then* E_r *is an r-dimensional random set such that for the true parameter point* θ_0, *we have*

$$P(\theta_0 \in E_r \mid \theta_0) = \gamma. \tag{12.8.4}$$

The proof of this statement is straightforward and is omitted. We shall call the set E_r a $100\gamma\,\%$ *confidence region* for θ_0. An example of a confidence region for estimating a k-dimensional parameter was given in Section 10.4.

More generally, if $g_n(x_1, \ldots, x_n; \theta)$ is a vector function with components $g_{p'n}(x_1, \ldots, x_n; \theta)$, $p' = 1, \ldots, r'$, whose c.d.f. does not depend on θ when the population c.d.f. is $F(x, \theta)$, then for any set $E^*_{r'}$ in $R_{r'}$ for which

$$P((g_{1n}, \ldots, g_{r'n}) \in E^*_{r'}) = \gamma, \tag{12.8.5}$$

the set E_r consisting of all points $y = (y_1, \ldots, y_r)$ in Ω_r for which

$$(g_{1n}(x_1, \ldots, x_n; y), \ldots, g_{r'n}(x_1, \ldots, x_n; y)) \in E^*_{r'} \tag{12.8.6}$$

is a $100\gamma\,\%$ confidence region for θ_0, that is,

$$P(\theta_0 \in E_r) = \gamma. \tag{12.8.7}$$

Ordinarily, the most appealing, interesting and useful functions $g_n(x_1, \ldots, x_n; \theta)$ are those which provide convex, or at least simply connected, confidence regions (random sets) E_r, since such sets are more likely to be easily described and used. For $r = r' = 1$, E_1 will have such properties if $g_n(x_1, \ldots, x_n; \theta)$ is continuous and monotonic in θ as stated in **12.4.1**. For more general values of r and r' the situation is more complicated.

But, in addition to the property of convexity or of being simply connected our intuition requires E_r to be "smallest" in some sense. This, however, is a complicated problem for small n. For large n, however, there is a satisfactory asymptotic solution under certain conditions which we shall consider presently.

Example. To illustrate the preceding ideas, suppose $(x_1, \ldots, x_n)$ is a sample from $N(\mu_0, \sigma_0^2)$. Then we know from **8.4.1** that the function

$$g(x_1, \ldots, x_n; \mu_0, \sigma_0) = \frac{(n-1)s^2 + n(\bar{x} - \mu_0)^2}{\sigma_0^2}$$

has the chi-square distribution $C(n)$. Therefore for a given γ, we can find χ_γ^2, so that

$$P(g < \chi_\gamma^2) = \gamma.$$

Denote by E_2 the region in the y_1y_2-half-plane with $y_2 > 0$ which lies between the two branches of the hyperbola, having the equation

$$\frac{y_2^2}{n} - \frac{(y_1 - \bar{x})^2}{\chi_\gamma^2} = \frac{(n-1)s^2}{n\chi_\gamma^2}.$$

We see that if (μ_0, σ_0) is the true parameter point, then

$$P((\mu_0, \sigma_0) \in E_2) = \gamma.$$

Now let us consider the two-dimensional vector function

$$g_{p'}(x_1, \ldots, x_n; \mu_0, \sigma_0), \qquad p' = 1, 2$$

where

$$g_1 = \frac{n(\bar{x} - \mu_0)^2}{\sigma_0^2}, \qquad g_2 = \frac{n-1}{\sigma_0^2}\, s^2.$$

It follows from **8.4.2** that g_1 and g_2 are independently distributed according to chi-square distributions $C(1)$ and $C(n-1)$ respectively. If we choose χ_1^2 and χ_2^2 so that

$$P(g_1 < \chi_1^2, g_2 > \chi_2^2) = \gamma$$

and let E_2^* be the region in the y_1y_2-plane for which

$$\frac{n(y_1 - \bar{x})^2}{y_2} < \chi_1^2, \qquad \frac{\sqrt{(n-1)}s}{y_2} > \chi_2^2$$

where $y_2 > 0$, we have

$$P((\mu_0, \sigma_0) \in E_2^*) = \gamma.$$

Thus, E_2 and E_2^* are examples of confidence regions determined by vector estimating functions having one and two components respectively. From an intuitive point of view, E_2^* is more satisfactory than E_2 since E_2^* is bounded and convex, whereas E_2 has neither of these properties. If one chooses as $g(x_1, \ldots, x_n; \mu_0, \sigma_0)$ the p.d.f. of $(\bar{x}, s^2)$, one can obtain a random region E_2^{**} which is more satisfactory than E_2^* on the criterion of being "smaller," but E_2^{**} is more complicated than E_2^* and will not be described in detail. In case of large n, there is an optimum method of determining an asymptotically smallest region for estimating (μ_0, σ_0). The general problem of asymptotically smallest confidence regions is considered in the next section.

12.9 ASYMPTOTICALLY SMALLEST CONFIDENCE REGIONS FROM LARGE SAMPLES

We shall now consider the r-dimensional extension of **12.5.1**.

It follows from **12.7.1** and **9.3.2** that if the true value of θ is θ_0, the sequence of random variables

$$(12.9.1)\quad U_n(x_1, \ldots, x_n; \theta_0) = n \sum_{p,q=1}^{r} B^{pq} h_{pn}(\theta_0) h_{qn}(\theta_0), \qquad n = 1, 2, \ldots$$

where $h_{pn}(\theta) = (1/n)S_{pn}(x_1, \ldots, x_n; \theta)$ as defined in (12.6.1) converges in distribution to the chi-square distribution $C(r)$. Under the stronger conditions of **12.7.3** we can apply the mean value theorem of differential calculus to U_n and write

$$U_n^* \equiv U_n \qquad \text{if the value of } \hat{\theta} \text{ belongs to } \Omega_{r0}$$

and

$$U_n^* \equiv 0 \qquad \text{if the value of } \hat{\theta} \text{ does not belong to } \Omega_{r0}$$

where

$$(12.9.2)\quad U_n^*(x_1, \ldots, x_n; \theta_0) = \sum_{p,q} B_{pq}^{(n)}[\sqrt{n}(\theta_{p0} - \hat{\theta}_p)][\sqrt{n}(\theta_{q0} - \hat{\theta}_q)]$$

for all points in the sample space R_n except possibly for a set of probability zero, where

$$(12.9.3)\qquad B_{pq}^{(n)} = \left[\sum_{p',q'=1}^{r} B^{p'q'}\left(\frac{\partial h_{p'n}}{\partial \theta_p}\right)\left(\frac{\partial h_{q'n}}{\partial \theta_q}\right)\right]_{\theta=\hat{\theta}^*}$$

and

$$|\theta_{p0} - \hat{\theta}_p^*| < |\theta_{p0} - \hat{\theta}_p|, \qquad p = 1, \ldots, r.$$

It can be verified from the conditions of **12.7.3** and from **4.3.7** and **4.3.8** that $B_{pq}^{(n)}$ converges in probability to $\sum_{p',q'=1}^{r} B^{p'q'} B_{p'p} B_{q'q}\,(= B_{pq})$ as $n \to \infty$ and hence that U_n and U_n^*, $n = 1, 2, \ldots$, are sequences of random variables converging together in distribution to the chi-square distribution $C(r)$. Now let us choose χ_γ^2 such that

$$(12.9.4)\qquad \lim_{n\to\infty} P(U_n < \chi_\gamma^2) = \lim_{n\to\infty} P(U_n^* < \chi_\gamma^2) = \gamma$$

and let $E_r(x_1, \ldots, x_n)$ and $E_r^*(x_1, \ldots, x_n)$ be the sets of points $y = (y_1, \ldots, y_r)$ in Ω_{r_0} such that

$$(12.9.5)\qquad U_n(x_1, \ldots, x_n; y) < \chi_\gamma^2$$

and

$$(12.9.6)\qquad U_n^*(x_1, \ldots, x_n; y) < \chi_\gamma^2$$

respectively. Therefore, if θ_0 is the true value of θ, we have

$$(12.9.7) \qquad \lim_{n\to\infty} P(\theta_0 \in E_r) = \lim_{n\to\infty} P(\theta_0 \in E_r^*) = \gamma,$$

that is, E_r and E_r^* are *asymptotically equivalent* confidence regions for estimating θ_0. But E_r^* is an ellipsoid in Ω_{r0} centered at θ_0 whose volume is

$$(12.9.8) \qquad \frac{K(r)\chi_\gamma}{\sqrt{n\,|B_{pq}^{(n)}|}}$$

approximately, for large n, where $K(r)$ is a constant depending only on r.

Now let $g_n(x_1, \ldots, x_n; \theta)$ be a vector random variable with linearly independent components $g_{pn}(x_1, \ldots, x_n; \theta)$, $p = 1, \ldots, r$, whose first θ-derivatives are

$$(12.9.9) \quad g_{pqn}(x_1, \ldots, x_n; \theta) = (\partial/\partial\theta_q)g_{pn}(x_1, \ldots, x_n; \theta), \; p, q = 1, \ldots, r.$$

Denote $g_{pn}(x_1, \ldots, x_n; \theta)$ and $g_{pqn}(x_1, \ldots, x_n; \theta)$ by $g_p(\theta)$ and $g_{pq}(\theta)$. We assume that $g_p(\theta), p = 1, \ldots, r$, and $g_{pq}(\theta)$, $p, q = 1, \ldots, r$, satisfy conditions (iii) through (v) listed under (12.5.7) for $g_n(x_1, \ldots, x_n; \theta)$ and $g_n'(x_1, \ldots, x_n; \theta)$ respectively. Let the corresponding A_n^*, B_n^* functions be denoted by $A_{pn}^*(\theta_0, \theta)$, $p = 1, \ldots, r$, and $B_{pqn}^*(\theta_0, \theta)$, $p, q = 1, \ldots, r$, where (12.5.8) holds with the obvious replacements. Let $B_{pq}^*(\theta, \theta) = \lim_{n\to\infty} B_{pqn}^*(\theta, \theta)$, and $B_{pq}^* = B_{pq}^*(\theta_0, \theta_0)$. For (12.5.7) (i) we would have for $j = 1$

$$(12.9.10) \qquad \int_{R_n} \sqrt{n}g_p(\theta)\, dF_n = 0$$

and for $j = 2$, we would have

$$(12.9.11) \qquad \int_{R_n} [\sqrt{n}g_p(\theta)][\sqrt{n}g_q(\theta)]\, dF_n = C_{pqn}(\theta).$$

Corresponding to (12.5.7) (ii), we assume that if $\theta = \theta_0$ in the population distribution, then $(\sqrt{n}g_p(\theta_0), p = 1, \ldots, r)$ is asymptotically distributed according to the distribution $N(\{0\}, \|C_{pq}\|)$, for large n, where

$$(12.9.12) \qquad C_{pq} = \lim_{n\to\infty} C_{pqn}(\theta_0),$$

and where $\|C_{pq}\|$ is positive definite.

A vector function $g(x_1, \ldots, x_n; \theta)$ satisfying the conditions in the preceding paragraph will be called a *regular estimating vector function* for θ_0. The reader will note that the likelihood estimating vector function $h_p(\theta) = (1/n)S_{pn}(x_1, \ldots, x_n; \theta)$ is a regular estimating vector function.

For such an estimating function it can be shown that if there exists a sequence of unique vector solutions $(\tilde{\theta}_1, \ldots, \tilde{\theta}_r)$, for $n \geqslant$ some n_0, of the equations

(12.9.13)

$$g_{pn}(x_1, \ldots, x_n; \theta) = 0, \qquad p = 1, \ldots, r, \qquad n = n_0, n_0 + 1, \ldots$$

and if θ_0 is the true value of θ, then the sequence of random variables $[\sqrt{n}(\tilde{\theta}_1 - \theta_{10}), \ldots, \sqrt{n}(\tilde{\theta}_r - \theta_{r0})]$, $n = n_0, n_0 + 1, \ldots$ has the r-dimensional distribution $N(\{0\}, \|D^{pq}\|)$ in the limit as $n \to \infty$ where

(12.9.14)
$$D_{pq} = \sum_{p',q'=1}^{r} C^{p'q'} B^*_{p'p} B^*_{q'q}.$$

It follows from **9.3.2** that if θ_0 is the true value of θ, then $V_n(x_1, \ldots, x_n; \theta_0)$ defined by

(12.9.15)
$$V_n = n \sum_{p,q=1}^{r} C^{pq} g_p(\theta_0) g_q(\theta_0)$$

has as its limiting distribution as $n \to \infty$ the chi-square distribution $C(r)$.

Furthermore, by an argument similar to that by which it was established that if θ_0 is the true value of θ, the limiting distribution of each of the random variables U_n and U_n^* as $n \to \infty$ is a chi-square distribution $C(r)$, it can be shown that V_n and the associated random variable $V_n^*(x_1, \ldots, x_n; \theta_0)$ defined by

(12.9.16)
$$V_n^* = \sum_{p,q=1}^{r} D_{pq}^{(n)} [\sqrt{n}(\tilde{\theta}_p - \theta_{p0})][\sqrt{n}(\tilde{\theta}_q - \theta_{q0})]$$

where

(12.9.17)
$$D_{pq}^{(n)} = \sum_{p',q'=1}^{r} C^{p'q'} g_{p'p}(\tilde{\theta}^*) g_{q'q}(\tilde{\theta}^*)$$

and where $\quad |\theta_{p0} - \tilde{\theta}_p^*| < |\theta_{p0} - \tilde{\theta}_p|, \qquad p = 1, \ldots, r$

converge together in distribution to the chi-square distribution $C(r)$ as $n \to \infty$.

The sets $E_r'(x_1, \ldots, x_n)$ and $E_r^{*\prime}(x_1, \ldots, x_n)$ defined as the sets of points $y = (y_1, \ldots, y_r)$ in Ω_{r0} for which

(12.9.18)
$$V_n(x_1, \ldots, x_n; y) < \chi_\gamma^2$$

and

(12.9.19)
$$V_n^*(x_1, \ldots, x_n; y) < \chi_\gamma^2$$

respectively, where χ_γ^2 is chosen as in (12.9.4), are asymptotically equivalent sets in Ω_{r0}, that is, the limits of $P(E_r')$ and $P(E_r^{*\prime})$ as $n \to \infty$ are equal.

But $E_r^{*\prime}$ is an ellipsoid, centered at θ_0 and having as its volume

$$\frac{K(r)\chi_\gamma}{\sqrt{n\,|D_{pq}^{(n)}|}}\,. \tag{12.9.20}$$

Now the ratio r_n of the volume of E_r^* given by (12.9.8) to that of $E_r^{*\prime}$ given by (12.9.20) converges (in probability) as $n \to \infty$ to the number r given by

$$r = \frac{\sqrt{|D_{pq}|}}{\sqrt{|B_{pq}|}}\,. \tag{12.9.21}$$

But substituting the expression for D_{pq} given by (12.9.14), we find

$$r = \frac{|B_{pq}^*|}{\sqrt{|C_{pq}|\cdot|B_{pq}|}}\,. \tag{12.9.22}$$

By differentiating (12.9.10) with respect to θ_q, we obtain the following result

$$\int_{R_n} g_{pq}(\theta)\,dF_n + \int_{R_n} g_p(\theta)S_q(\theta)\,dF_n = 0 \tag{12.9.23}$$

where $S_q(\theta)$ stands for $S_{qn}(x_1, \ldots, x_n; \theta)$.

From (12.9.23) it is clear that at $\theta = \theta_0$,

$$\int_{R_n} g_{pq}(\theta_0)\,dF_n = -\text{cov}\left(\sqrt{n}g_p(\theta_0), \frac{S_q(\theta_0)}{\sqrt{n}}\right). \tag{12.9.24}$$

But the left-hand side of (12.9.24) is $B_{pqn}^*(\theta_0, \theta_0)$. Therefore, the right-hand side is also $B_{pqn}^*(\theta_0, \theta_0)$, which we shall abbreviate to B_{pqn}^*. Therefore, the sequence of $2r$-dimensional random variables

$$\left(\sqrt{n}g_1(\theta_0), \ldots, \sqrt{n}g_r(\theta_0), \frac{1}{\sqrt{n}}S_1(\theta_0), \ldots, \frac{1}{\sqrt{n}}S_r(\theta_0)\right), \quad n = 1, 2, \ldots$$

converges in probability to a $2r$-dimensional random variable having the covariance matrix

$$\left\|\begin{array}{c|c} C_{pq} & B_{pq}^* \\ \hline B_{pq}^* & B_{pq} \end{array}\right\| \tag{12.9.25}$$

and since (12.9.25) is a covariance matrix, we have

$$|C_{pq}|\cdot|B_{pq}| \geqslant |B_{pq}^*|^2. \tag{12.9.26}$$

Therefore the limit (in probability) of the ratio of volume of confidence regions given by (12.9.22) cannot exceed 1.

Since E_r and E_r' are asymptotically equivalent to E_r^* and $E_r^{*\prime}$ respectively, this means that there exists no regular vector estimating function $g(x_1, \ldots, x_n; \theta)$ for which the $100\gamma\,\%$ confidence region for θ_0 defined by

$E_r'(x_1, \ldots, x_n)$ in (12.9.18) is asymptotically smaller than the $100\gamma\,\%$ confidence region for θ_0 similarly obtained by replacing $g_p(\theta)$ by $h_p(\theta)$.

We may summarize the preceding discussion and conclusions in the following r-dimensional extension of **12.5.1**.

12.9.1 *Suppose $(x_1, \ldots, x_n)$ is a sample from the c.d.f. $F(x; \theta_0)$, where θ_0 is r-dimensional, and $F(x; \theta)$ is regular with respect to all of its second θ-derivatives for θ in Ω_{r0}. Then if $g_{pn}(x_1, \ldots, x_n; \theta), p = 1, \ldots, r$, is any regular estimating vector function for θ_0, an asymptotic $100\gamma\,\%$ confidence region for θ_0 is provided by the set E_r' defined by the points $y = (y_1, \ldots, y_r)$ for which* (12.9.18) *holds. Furthermore, there exists no regular estimating vector function which provides an asymptotically smaller $100\gamma\,\%$ confidence region for θ_0 than that obtained by using the likelihood estimating vector function $h_{pn}(x_1, \ldots, x_n; \theta), p = 1, \ldots, r$, defined in* (12.9.1). *Thus the asymptotically smallest $100\gamma\,\%$ confidence region E_r consists of the points $y = (y_1, \ldots, y_r)$ for which* (12.9.5) *is satisfied.*

The problem of asymptotically smallest confidence regions for the case of multidimensional parameters has been discussed by Bartlett (1953), Beale (1960), and by Wilks and Daly (1939). The reader interested in further details should consult these papers.

Example. As an example of the problem of obtaining the asymptotically smallest $100\gamma\,\%$ confidence region consider a sample from a *multinomial population*, having the k-parameter p.f.

$$dF(s; \theta) = \theta_1^{s_1} \cdots \theta_{k+1}^{s_{k+1}}$$

where the random variable $(s_1, \ldots, s_{k+1})$ can take on one and only one of the $k + 1$ values $(1, 0, \ldots, 0), \ldots, (0, \ldots, 0, 1)$ and where $\theta_1 > 0, \ldots, \theta_{k+1} > 0$ with $\theta_1 + \cdots \theta_{k+1} = 1$. We may, for convenience, take $\theta_1, \ldots, \theta_k$ as the k parameters to be estimated by a confidence region. Suppose $(s_{1\xi}, \ldots, s_{k+1\xi}, \xi = 1, \ldots, n)$ is a sample from this distribution. For $h_p(\theta)$ we have

$$h_p(\theta) = \frac{1}{n}\sum_{\xi=1}^{n} \frac{\partial \log dF(s_\xi; \theta)}{\partial \theta_p} = \frac{1}{n}\sum_{\xi=1}^{n}\left(\frac{s_{p\xi}}{\theta_p} - \frac{s_{k+1\xi}}{\theta_{k+1}}\right)$$

where $\theta_{k+1} = 1 - \theta_1 - \cdots - \theta_k$. Setting $\sum_{\xi=1}^{n} s_{p\xi} = x_p$, we may write

$$h_p(\theta) = \frac{x_p}{n\theta_p} - \frac{x_{k+1}}{n\theta_{k+1}}, \qquad p = 1, \ldots, k$$

where $x_{k+1} = n - x_1 - \cdots - x_k$. Furthermore

$$B_{pq} = \mathscr{E}\left[\left(\frac{\partial \log dF(s; \theta)}{\partial \theta_p}\right)\left(\frac{\partial \log dF(s; \theta)}{\partial \theta_q}\right)\right] = \mathscr{E}\left[\left(\frac{s_p}{\theta_p} - \frac{s_{k+1}}{\theta_{k+1}}\right)\left(\frac{s_q}{\theta_q} - \frac{s_{k+1}}{\theta_{k+1}}\right)\right]$$

$$= \frac{\delta_{pq}}{\theta_p} + \frac{1}{\theta_{k+1}} \qquad p, q = 1, \ldots, k$$

where δ_{pq} is the Kronecker δ which has the value 1 if $p = q$, and 0 if $p \neq q$. The element B^{pq} in the inverse matrix is readily seen to be $B^{pq} = \delta_{pq}\theta_p - \theta_p\theta_q$. Therefore, for the quadratic form U_n in (12.9.1) we find, after some algebraic reduction,

$$U_n = n \sum_{p,q=1}^{k} B^{pq}h_p(\theta)h_q(\theta) = \sum_{p=1}^{k+1} \frac{(x_p - n\theta_p)^2}{n\theta_p}$$

which is the classical χ^2 function originally proposed by K. Pearson (1900). Thus, for large n, the 100γ % asymptotically smallest confidence region for estimating the k-dimensional parameter $\theta = (\theta_1, \ldots, \theta_k)$ is provided by the set E_k of real and positive vectors $x = (y_1, \ldots, y_k)$ for which

$$\sum_{p=1}^{k+1} \frac{(x_p - ny_p)^2}{ny_p} < \chi_\gamma^2$$

where $y_{k+1} = 1 - y_1 - \cdots y_k$ and χ_γ^2 is the 100γ % point of the chi-square distribution $C(k)$.

PROBLEMS

12.1 Show that the two distributions having p.d.f.'s

$$\begin{aligned} f_1(x; \theta) &= \frac{1}{\theta} & 0 < x \leqslant \theta \\ &= 0, & \text{otherwise} \\ f_2(x; \theta') &= \frac{1}{\theta'} & 0 < x \leqslant \theta' \\ &= 0 & \text{otherwise} \end{aligned}$$

are not absolutely continuous with respect to each other unless $\theta' = \theta$.

12.2 Suppose x is a random variable having a distribution with c.d.f.

$$F(x; \theta) = \begin{cases} 0 & x \leqslant 0 \\ 1 - e^{-\theta x} & x > 0 \end{cases}$$

where θ is any number in $(0, +\infty)$. Show that $\mathscr{E}(S(x; \theta)) = 0$ and find $\sigma^2(S(x; \theta))$. Determine $H(\theta, \theta')$ and show that for a fixed θ it has a maximum with respect to θ' at $\theta' = \theta$.

12.3 A discrete random variable x has the p.f.

$$p(x; \theta) = (1 - \theta)\theta^{x-1} \qquad x = 1, 2, \ldots$$

where θ is any number on (0, 1). Show that $\mathscr{E}(S(x; \theta)) = 0$ and find $\sigma^2(S(x; \theta))$.

12.4 If $(x_1, \ldots, x_n)$ is a sample from any distribution having finite kth moment $\mathscr{E}(x^k)$, show that $\frac{1}{n}\sum_{\xi=1}^{n} x_\xi^k$ is a consistent estimator for $\mathscr{E}(x^k)$.

12.5 If $\bar{x}$ is the mean of a sample from the binomial distribution $Bi(1, p)$ show that $\bar{x}$ is sufficient for estimating p. Also show that $\bar{x}$ is an efficient estimator for p.

12.6 Suppose $(x_1, \ldots, x_n)$ is a sample from the p.d.f.

$$\theta e^{-\theta x}, \qquad x > 0$$

where θ is a positive parameter. Show that the sample mean $\bar{x}$ is sufficient for estimating θ and that $(n - 1)/n\bar{x}$ is the only unbiased estimator for θ depending on $\bar{x}$.

12.7 (*Continuation*) Show that $\bar{x}$ is the maximum likelihood estimator for $1/\theta$ and that it is efficient.

12.8 If $\bar{x}$ and s^2 are the mean and variance of a sample of size $n > 1$ from a normal distribution $N(\mu, \sigma^2)$, show that the two-dimensional random variable $(\bar{x}, s^2)$ is sufficient for estimating the vector of parameters (μ, σ^2).

12.9 If $\hat{\theta}_1$ and $\hat{\theta}_2$ are two efficient estimators for a parameter θ show that the correlation coefficient between $\hat{\theta}_1$ and $\hat{\theta}_2$ is unity.

12.10 If $\hat{\theta}$ is an efficient estimator for θ, and $\tilde{\theta}$ is any other unbiased estimator for θ, and if σ^2 and $k\sigma^2$, $k > 1$, are the variances of $\hat{\theta}$ and $\tilde{\theta}$, show that the correlation coefficient between $\hat{\theta}$ and $\tilde{\theta}$ equals $1/\sqrt{k}$.

12.11 (*Continuation*) If $\hat{\theta}$ is an efficient estimator for θ, whereas $\tilde{\theta}_1$ and $\tilde{\theta}_2$ are inefficient but unbiased estimators with variances both equal to $k\sigma^2$, where $k > 1$, show that the correlation coefficient between $\tilde{\theta}_1$ and $\tilde{\theta}_2$ is at least $(2 - k)/k$.

12.12 Suppose $(x_1, \ldots, x_k)$ is a k-dimensional random variable having p.d.f. $f(x_1, \ldots, x_k; \theta)$, for $\theta \in \Omega_0$. Let $g(x_1, \ldots, x_n)$ be an unbiased estimator for $\psi(\theta)$. If f is regular in its first θ-derivative in Ω_0 and if $\psi(\theta)$ has a derivative with respect to θ in Ω_0, show that

$$\sigma^2(g) \geqslant \left(\frac{\partial \psi}{\partial \theta}\right)^2 \Big/ \mathscr{E}\left(\frac{\partial \log f}{\partial \theta}\right)^2$$

and that the equality holds if and only if $\partial \log f/\partial \theta \equiv C(g - \psi(\theta))$, with probability 1, where C does not depend on $(x_1, \ldots, x_k)$.

12.13 If $(x_{(1)}, \ldots, x_{(n)})$ are the order statistics of a sample from the rectangular distribution $R(\theta/2, \theta)$, show that $x_{(n)}$ is sufficient for estimating θ and that $(n - 1)x_{(n)}/n$ is an efficient estimator for θ. Show that $(x_{(n)}, x_{(n)}/\sqrt[n]{1 - \gamma})$ is a 100γ % confidence interval for θ.

12.14 If $(x_{(1)}, \ldots, x_{(n)})$ are the order statistics of a sample from the rectangular distribution $R((\theta_1 + \theta_2)/2, \theta_2 - \theta_1)$, $\theta_2 > \theta_1 > 0$, show that the two-dimensional random variable $(x_{(1)}, x_{(n)})$ provides sufficient estimators for (θ_1, θ_2). Show that

$$\frac{nx_{(1)}}{n - 1} - \frac{x_{(n)}}{n - 1}$$

and

$$\frac{nx_{(n)}}{n - 1} - \frac{x_{(1)}}{n - 1}$$

are minimum variance estimators of θ_1 and θ_2, respectively. Also show that

$$\frac{x_{(1)} + x_{(n)}}{2} \quad \text{and} \quad \frac{n + 1}{n - 1}(x_{(n)} - x_{(1)})$$

are minimum variance estimators of the midpoint $(\theta_1 + \theta_2)/2$ and range $\theta_2 - \theta_1$.

12.15 Show that the variance of every unbiased estimator for σ^2 in samples of size n from $N(\mu, \sigma^2)$ is at least $2\sigma^4/n$.

12.16 If $(x_1, \ldots, x_n)$ is a sample from the normal distribution $N(\mu, \sigma^2)$ where μ is known, show that

$$\frac{1}{n}\sqrt{\pi/2}\sum_{\xi=1}^{n}|x_\xi - \mu|$$

is an unbiased estimator for σ with asymptotic efficiency $1/(\pi - 2)$ for large n.

12.17 If $(x_1, \ldots, x_n)$ is a sample from the Cauchy distribution having p.d.f.

$$\frac{1}{\pi[1 + (x - \mu)^2]}$$

show that the asymptotic efficiency of the sample median for estimating μ in large samples is $8/\pi^2$.

12.18 If $(x_1, \ldots, x_n)$ is a sample from the distribution having p.d.f.

$$\frac{\lambda^{k+1}x^k e^{-\lambda x}}{\Gamma(k + 1)}, \qquad x > 0$$

where $\lambda > 0$ and k is a known constant, show that the maximum likelihood estimator $\hat{\lambda}$ for λ is $(k + 1)/\bar{x}$. Show that this estimator is biased but consistent and that its asymptotic distribution for large n is $N(\lambda, \lambda^2/[n(k + 1)])$.

12.19 (*Continuation*) Show that $\sqrt{(k + 1)}\log\hat{\lambda}$ is a function of $\hat{\lambda}$ whose asymptotic distribution for large n is normal with variance $1/n$.

12.20 If $(x_1, \ldots, x_n)$ is a sample from the binomial distribution $Bi(1, p)$, show that the maximum likelihood estimator $\hat{p}$ for p is $\bar{x}$ and that its asymptotic distribution for large samples is $N(p, [p(1 - p)]/n)$. Show that $\sin^{-1}(2\hat{p} - 1)$, has an asymptotic normal distribution with variance $1/n$ in large samples.

12.21 If $\bar{x}$ is the mean of a sample of size n from the binomial distribution $Bi(1, p)$, show that the asymptotically shortest $100\gamma\,\%$ confidence interval for large n is given by the two values of p which satisfy

$$\frac{(\bar{x} - p)\sqrt{n}}{\sqrt{p(1 - p)}} = \pm y_\gamma$$

where $P(-y_\gamma < u < +y_\gamma) = \gamma$, u being a random variable having the distribution $N(0, 1)$.

12.22 If $\bar{x}$ is the mean of a sample of size n from the Poisson distribution $Po(\mu)$ show that asymptotically shortest $100\gamma\%$ confidence interval for μ for large n is given by the two values of μ which satisfy

$$\frac{(\bar{x} - \mu)\sqrt{n}}{\sqrt{\mu}} = \pm y_\gamma$$

where y_γ is defined in the preceding problem.

12.23 If $(x_{1\xi}, \ldots, x_{k\xi}, \xi = 1, \ldots, n)$ is a sample of size n from the multinomial distribution $M(1; p_1, \ldots, p_k)$, show that the maximum likelihood estimators for $(p_1, \ldots, p_k)$ are $(\bar{x}_1, \ldots, \bar{x}_k)$ where $\bar{x}_i = \frac{1}{n} \sum_{\xi=1}^{n} x_{i\xi}$, and that the covariance matrix of these estimators is

$$\left\| \frac{1}{n} (\delta_{ij} p_i - p_i p_j) \right\|$$

where $\delta_{ij} = 1$, $i = j$, and 0, $i \neq j$.

12.24 If $(x_{1\xi}, \ldots, x_{k\xi}, \xi = 1, \ldots, n)$ is a sample of size n from the k-dimensional normal distribution $N(\{\mu_i\}, \|\sigma_{ij}\|)$, show that the maximum likelihood estimator for the vector $(\mu_1, \ldots, \mu_k)$ is the vector of sample component means $(\bar{x}_1, \ldots, \bar{x}_k)$ and the maximum likelihood estimators of the matrix $\|\sigma_{ij}\|$ is

$$\left\| \frac{n-1}{n} s_{ij} \right\|$$

where

$$s_{ij} = \frac{1}{n-1} \sum_{\xi=1}^{n} (x_{i\xi} - \bar{x}_i)(\bar{x}_{j\xi} - \bar{x}_j).$$

12.25 Suppose $(y_1, \ldots, y_n)$ are independent random variables having the normal distributions

$$N(\beta_1 x_{11} + \cdots + \beta_k x_{k1}, \sigma^2), \ldots, N(\beta_1 x_{1n} + \cdots + \beta_k x_{kn}, \sigma^2)$$

respectively, where $x_{11}, \ldots, x_{k1}, \ldots, x_{1n}, \ldots, x_{kn}$ are constants. Let

$$a_{pq} = \sum_{\xi=1}^{n} x_{p\xi} x_{q\xi}, \qquad p, q = 1, \ldots, k$$

$$a_{p0} = \sum_{\xi=1}^{n} x_{p\xi} y_{\xi}, \qquad p = 1, \ldots, k.$$

Show that the maximum likelihood estimators for $(\beta_1, \ldots, \beta_k)$ are

$$\left(\sum_{p=1}^{k} a_{p0} a^{p1}, \ldots, \sum_{p=1}^{k} a_{p0} a^{pk} \right),$$

where $\|a^{pq}\| = \|a_{pq}\|^{-1}$, $p, q = 1, \ldots, k$, and $\|a_{pq}\|$ is assumed to be nonsingular. Also show that the maximum likelihood estimator for σ^2 is

$$(1/n)\, |a_{p_0 q_0}| / |a_{pq}|, \qquad p_0, q_0 = 0, 1, \ldots, k; \qquad p, q = 1, \ldots, k,$$

where $a_{00} = \sum_{\xi=1}^{n} y_\xi^2$.

12.26 Suppose $(x_1, \ldots, x_n)$ has p.d.f. $f_n(x_1, \ldots, x_n; \theta_0)$ and $\tilde{\theta}$ is an unbiased estimator for θ_0. Show that for any $\delta > 0$

$$\sigma^2(\tilde{\theta} \mid \theta_0) \geqslant 1/[\mathscr{E}(g^2 \mid \theta_0)],$$

where

$$g = \frac{f_n(x_1, \ldots, x_n; \theta_0 + \delta) - f_n(x_1, \ldots, x_n; \theta_0)}{\delta f_n(x_1, \ldots, x_n; \theta_0)}$$

and hence that

$$\sigma^2(\tilde{\theta} \mid \theta_0) \geqslant 1 \Big/ \left[\inf_{\delta} \mathscr{E}(g^2 \mid \theta_0) \right],$$

a result due to Chapman and Robbins (1951).

12.27 Suppose $(x_1, \ldots, x_n)$ is a random variable having p.d.f. $f_n(x_1, \ldots, x_n; \theta)$ such that (12.2.25) holds, that is, such that a sufficient statistic $\tilde{\theta}$ exists for θ, and suppose $f_n(x_1, \ldots, x_n; \theta)$ and $v(\tilde{\theta}; \theta)$ have second partial derivatives $\partial^2 f_n/(\partial x_\xi\, \partial\theta)$ and $\partial^2 v/(\partial x_\xi\, \partial\theta)$ for each point in $R_n \times \Omega$ (except possibly for a set of probability 0) where R_n is the sample space and Ω is the parameter space (an open interval on the real line). Show that $f_n(x_1, \ldots, x_n; \theta)$ must be of form $\exp\,[K_1(\theta)g(\tilde{\theta}) + K_2(\theta) + h_n(x_1, \ldots, x_n)]$ where $K_1(\theta)$ and $K_2(\theta)$ do not depend on $(x_1, \ldots, x_n)$; $h_n(x_1, \ldots, x_n)$ does not depend on θ and $g(\tilde{\theta})$ depends on $\tilde{\theta}$ and not θ. [Koopman (1936) and Pitman (1936).]

12.28 (*Continuation*) *Functions of sufficient estimators as minimum variance estimators of their mean values* [*Rao* (1949)]. Let $u(g(\tilde{\theta}))$ be a function of $\tilde{\theta}$ whose mean value is $q(\theta)$. Let $v(g(\tilde{\theta}))$ be another function whose mean value is $q(\theta)$. Then if $w(g(\tilde{\theta})) = u(g(\tilde{\theta})) - v(g(\tilde{\theta}))$, we have

$$\int_{R_n} w(g) \exp\,[K_1(\theta)g + K_1(\theta) + h_n]\, dx_1 \cdots dx_n = 0, \text{ for all } \theta \in \Omega.$$

Show that if this expression can be repeatedly differentiated with respect to θ under the integral and if $w(g)$ can be represented by a Taylor series, then $\mathscr{E}(w(g))^2 = 0$ for all $\theta \in \Omega$, which implies that $w(g) = 0$ with probability 1 for all $\theta \in \Omega$. Hence $u(g(\tilde{\theta}))$ differs at most from $v(g(\tilde{\theta}))$ over a set of probability zero, thus showing that under the assumptions made $u(g(\tilde{\theta}))$ is a unique unbiased estimator for $g(\theta)$ and hence from **12.2.3** that $u(g(\tilde{\theta}))$ has minimum variance as an estimator of its mean value.

CHAPTER 13

Testing Parametric Statistical Hypotheses

13.1 INTRODUCTORY REMARKS AND DEFINITIONS

This chapter is devoted primarily to the basic principles of the theory of testing parametric statistical hypotheses originally set forth by Neyman and Pearson (1928, 1933), and the relations between this theory and maximum likelihood estimation theory in large samples. The ideas of statistical hypothesis-testing have been considerably extended and generalized in the last few years; however, we shall not attempt to cover these extensions in detail here. The reader interested in them will find an excellent treatment of the subject by Lehmann (1959).

A few of the basic concepts and results for finite samples are introduced in this section; later sections deal with asymptotic theory for large samples. The asymptotic theory for large samples is closely related to parametric estimation theory for large samples as has been presented in Chapter 12.

Suppose $(x_1, \ldots, x_n)$ is an n-dimensional random variable having c.d.f. $F_n(x_1, \ldots, x_n; \theta)$ where θ is an r-dimensional parameter and its space is Ω which, for the present, will be an open region in the Euclidean space R_r. The most important case of $(x_1, \ldots, x_n)$ arises if $(x_1, \ldots, x_n)$ is a random sample from a c.d.f. $F(x; \theta)$, in which case $F_n(x_1, \ldots, x_n; \theta) = \prod_{\xi=1}^{n} F(x_\xi; \theta)$. Most of this chapter is devoted to the case of random sampling.

Let ω be a subset of Ω. If θ_0 is the true value of θ in the population, we shall set up the *statistical hypothesis* $\mathscr{H}$ that $\theta_0 \in \omega$ against the alternative that $\theta_0 \in \Omega - \omega$, or more briefly, $\mathscr{H}(\omega; \Omega)$. We say that $\mathscr{H}$ is *true* if $\theta_0 \in \omega$ and *false* if $\theta_0 \in \Omega - \omega$. The assumption that $\theta_0 \in \omega$ is sometimes referred to as the *null hypothesis*. Unless ambiguity arises, we shall drop the $_0$ and write θ_0 as θ. If ω contains one point θ the hypothesis $\mathscr{H}$ is called a *simple* hypothesis; otherwise a *composite* hypothesis. We decide to accept or reject $\mathscr{H}$ on the basis of $(x_1, \ldots, x_n)$. More precisely, let W_n be a set in the sample space R_n which does not depend on θ such that if $(x_1, \ldots, x_n) \in W_n$, we reject $\mathscr{H}$, otherwise we accept $\mathscr{H}$. Two types of errors are recognized here: a *Type I error* is committed if $\mathscr{H}$ is rejected when it is true, and a *Type II error* is committed if $\mathscr{H}$ is accepted when it is false. The selection of W_n and the occurrence of $(x_1, \ldots, x_n)$ in W_n as the criterion for rejecting $\mathscr{H}$ is called a *statistical test*, or more briefly, a *test* of $\mathscr{H}$; W_n is called the *critical set* or *critical region* of the test. There will be no ambiguity if, for the sake of brevity, we refer to W_n as the *test* of $\mathscr{H}$. The quantity*

$$P(W_n \mid \theta) = \int_{W_n} dF_n \tag{13.1.1}$$

is called the *power* of the test W_n, and is a function of θ whose values lie on the interval [0, 1]. $P(\bar{W}_n \mid \theta) = 1 - P(W_n \mid \theta)$ as a function of θ is called the *operating characteristic function* of the test W_n. The probabilities of committing Type I and Type II errors are respectively,

$$P(W_n \mid \theta), \qquad \text{for } \theta \in \omega \tag{13.1.2}$$

and

$$1 - P(W_n \mid \theta), \qquad \text{for } \theta \in \Omega - \omega. \tag{13.1.3}$$

These two probabilities are frequently called the *risks of Type I and Type II* errors, respectively.

To be satisfactory, a test must, of course, satisfy certain criteria. A test W_n would be ideal if the probabilities for Type I and Type II errors, as given in (13.1.2) and (13.1.3), were both zero. This requirement, however, is obviously too much to hope for in general; we must be satisfied with less. A desirable property of a test W_n is that of being *unbiased*, that is, of satisfying the conditions

$$P(W_n \mid \theta) \leqslant \alpha, \qquad \text{if } \theta \in \omega \tag{13.1.4}$$

and

$$P(W_n \mid \theta) > \alpha, \qquad \text{if } \theta \in \Omega - \omega \tag{13.1.5}$$

for a given *level of significance* α, $0 < \alpha < 1$. This means that the Type I error is controlled by requiring its probability not to exceed α and at the same time we are assured that the probability of rejecting $\mathscr{H}$ if it is not true (that is, if $\theta \in \Omega - \omega$) exceeds α. If test W_n is such that $P(W_n \mid \theta') = \alpha$

* Some authors prefer to introduce the *characteristic function* φ_{W_n} *of the set* W_n where $\varphi_{W_n} = 1$ at all points in W_n and 0 otherwise, in which case $P(W_n \mid \theta) = \mathscr{E}(\varphi_{W_n} \mid \theta)$.

for some $\theta \in \omega$, W_n is said to be a test of *size* α for θ and is denoted by $W_{n,\alpha}$. If there is no ambiguity we shall henceforth use W_α to denote a test for a given α and sample size n. A test not unbiased is said to be *biased.*

If the test W_α is such that equality in (13.1.4) holds for all $\theta \in \omega$, then W_n is said to be *similar to the sample space*, or briefly W_α is a *similar critical region.*

If

$$\lim_{n \to \infty} P(W_\alpha \mid \theta) = 1, \qquad \text{for } \theta \in \Omega - \omega,$$

then W_α is said to be a *consistent test* of size α for the hypothesis $\mathscr{H}(\omega; \Omega)$, a term introduced by Wald and Wolfowitz (1940).

If W_α and W_α^* are two tests for $\mathscr{H}(\omega; \Omega)$ for a given α, such that

$$\text{(13.1.6)} \qquad P(W_\alpha^* \mid \theta) \leqslant P(W_\alpha \mid \theta), \qquad \text{if } \theta \in \omega$$

and

$$\text{(13.1.7)} \qquad P(W_\alpha^* \mid \theta) > P(W_\alpha \mid \theta), \qquad \text{if } \theta \in \Omega - \omega,$$

then the test W_α^* is uniformly more powerful than the test W_α; W_α^* would be preferred to W_α. If every (Borel) set W_α in R_n satisfies (13.1.6) and (13.1.7), the test W_α^* is a *uniformly most powerful test* for $\mathscr{H}$.

Example. To illustrate some of the preceding ideas with an important example, let $(x_1, \ldots, x_n)$ be a sample from a population having the normal distribution $N(\mu, \sigma^2)$. Consider the composite hypothesis $\mathscr{H}(\omega; \Omega)$ where Ω is the set of points (μ, σ^2) for which $\sigma^2 > 0$ (that is, a Euclidean half-plane) and ω is the subset of Ω for which $\mu = \mu_0$ (that is, the points of the line $\mu = \mu_0$ in Ω).

We know from **8.4.3** that if $\mathscr{H}$ is true $\sqrt{n}(\bar{x} - \mu)/s$ has the Student distribution $S(n - 1)$. Let us take as W_α the set of points in R_n for which

$$\left| \frac{(\bar{x} - \mu_0)\sqrt{n}}{s} \right| > t_\alpha$$

where

$$1 - \int_{-t_\alpha}^{+t_\alpha} f_{n-1}(t)\, dt = \alpha$$

and $f_{n-1}(t)$ is the p.d.f. of the Student distribution $S(n - 1)$ given by (7.8.4). Then we have

$$P(W_\alpha \mid (\mu, \sigma^2) \in \omega) = P(|t| > t_\alpha) = \alpha,$$

and hence W_α is a test of size α for $\mathscr{H}(\omega; \Omega)$ which has the property of being similar to the sample space. Furthermore, for any point (μ_1, σ_1^2) in $\Omega - \omega$, we have

$$P(W_\alpha \mid (\mu_1, \sigma_1^2)) = P\left(\left| \frac{(\bar{x} - \mu_0)\sqrt{n}}{s} \right| > t_\alpha \mid (\mu_1, \sigma_1^2) \right)$$

$$= P\left(\left| t + \frac{\delta}{\sqrt{u}} \right| > t_\alpha \right)$$

where $\delta = (\mu_1 - \mu_0)\sqrt{n(n-1)}/\sigma_1$ and t and u are random variables having the following p.e. over the half-plane $u \geqslant 0$:

$$\frac{\left(\frac{u}{2}\right)^{\frac{1}{2}(n-3)}}{2\sqrt{\pi(n-1)}\,\Gamma\left(\frac{n-1}{2}\right)} e^{-\frac{1}{2}(1+t^2/(n-1))u}\,dt\,du.$$

Denoting this p.e. by $f(t, u)\,dt\,du$ and noting that for all values of u in $(0, \infty)$

$$\int_{-t_\alpha - \frac{\delta}{\sqrt{u}}}^{+t_\alpha - \frac{\delta}{\sqrt{u}}} f(t, u)\,dt < \int_{-t_\alpha}^{+t_\alpha} f(t, u)\,dt$$

it follows that

$$P\left(\left|t + \frac{\delta}{\sqrt{u}}\right| > t_\alpha\right) > P(|t| > t_\alpha) = \alpha, \qquad \delta \neq 0$$

which, of course, means that W_α is unbiased for testing $\mathscr{H}$.

Unfortunately, unbiased and uniformly most powerful tests exist only in various special situations. In the case of a simple hypothesis, there is a fairly general solution for the case where $\Omega - \omega$ contains one point, and a solution under certain conditions if $\Omega - \omega$ is a set of more than one point. The case of a simple hypothesis will be considered in Section 13.2. In case of composite hypotheses, solutions have been found only in special situations, particularly where W_α is similar to the sample space. But uniformly most powerful unbiased tests exist in an asymptotic sense for some composite hypotheses for large samples in case of population distributions for which maximum likelihood estimators exist and are asymptotically normally distributed. We shall deal with this problem in Sections 13.3 and 13.4.

Historical remark. In developing the theory of acceptance sampling for accepting or rejecting lots of mass-produced articles, Dodge and Romig of the Bell Telephone Laboratories introduced the concepts of *producer's risk* and *consumer's risk* about 1925. Their first published results appeared four years later [See Dodge and Romig (1929, 1959)]. These concepts were, in fact, the forerunners respectively, of risks of Type I and Type II errors given by (13.1.2) and (13.1.3). More precisely, suppose θ is the fraction of defective articles in a *lot* of N articles, and t is the fraction of defective articles in a sample of n articles from the lot. Then for a given value of θ in the lot, t is a random variable with a c.d.f. $F_n(t; \theta)$, the sample space of t being $0, \frac{1}{n}, \ldots, \frac{r}{n}, r \leqslant n$ in the interval $[0, 1]$ and the parameter space of θ being the points $0, 1/N, \ldots, \frac{N}{N}$ on the interval $[0, 1]$. Given a value θ_0 of θ an *unacceptable* fraction defective for the lot would be one for which $\theta > \theta_0$, in which case $\omega = (\theta_0, 1)$, whereas an *acceptable* fraction defective (*zone of preference*) would be one for which $\theta \leqslant \theta_0$, in which case $\Omega - \omega = (0, \theta_0)$. Now we choose t_α and the critical set $W_\alpha = [t_\alpha, 1]$, so that if $t \in W_\alpha$ we reject the lot and so that $P(t \in W_\alpha \mid \theta \in \omega) \leqslant \alpha$. The quantity $P(t \in W_\alpha \mid \theta \in \omega)$ is the *producer's risk*, that is, the probability the inspector will reject the lot if it has an acceptable fraction defective θ. The producer's risk is controlled so as not to exceed α, being (approximately) α if $\theta = \theta_0$. In practice α is usually taken as 0.05 or 0.10. The quantity $P(t \in \overline{W}_\alpha \mid \theta \in \Omega - \omega) = \beta$ is the *consumer's risk*, that is, the probability the inspector of the lot will accept it if it has an unacceptable fraction defective θ. In practice, if N is sufficiently large, one can choose n large enough to make the consumer's risk arbitrarily small if the fraction defective in the lot has any specific value exceeding θ_0. Ordinarily, the sample size n is chosen so that the consumer's risk β has a given value (usually 0.05 or 0.10) for some given value of θ, say θ_1, slightly larger than θ_0. The interval $(\theta_0, 1)$ of unacceptable values of θ is broken into two sets: (θ_0, θ_1) a *zone of indifference*, and $(\theta_1, 1)$ a *zone of rejection*. Note that we can calculate $P(t \in \overline{W}_\alpha \mid \theta)$ for a sample of size n from a lot with any given fraction defective θ, and it is the probability of accepting the lot if the fraction of defectives is θ; it is a function of θ which may be denoted by $L(\theta)$. The graph of $L(\theta)$ is called the *operating characteristic curve* of the sampling plan specified by the sample size n and critical fraction defective t_α. It satisfies the conditions $L(\theta_0) = 1 - \alpha$ and $L(\theta_1) = \beta$.

A second important concept introduced by Dodge and Romig (1929) was that of *average outgoing quality limit*, which is defined as the maximum (over all values of θ) of the mean value of the fraction of defectives in a lot assuming that rejected lots are screened of defective items and then accepted. This assumes, of course, nondestructive testing, that is, that a defective item can be determined without destroying the item upon testing it.

The basic concepts of Type I and Type II errors have long played a fundamental role in the administration of criminal law, the counterparts of risks of Type I and Type II errors being respectively, the risks of convicting an innocent person and of acquitting a guilty one.

13.2 TEST OF A SIMPLE HYPOTHESIS

(a) Case of Two Alternatives

In case $\mathscr{H}$ is a simple hypothesis which we may conveniently describe for the moment by taking one point θ_0 in ω and one point θ_1 in $\Omega - \omega$, and if $(x_1, \ldots, x_n)$ has p.d.f. $f_n(x_1, \ldots, x_n; \theta)$ for θ_0 and θ_1, there exists, under conditions to be stated below, an unbiased and a most powerful test for $\mathscr{H}(\theta_0; \theta_0 \cup \theta_1)$. This problem was originally investigated by Neyman and Pearson (1933). The following result, due to them, provides a most powerful unbiased test for $\mathscr{H}$:

13.2.1 *Suppose* $(x_1, \ldots, x_n)$ *is a random variable with* p.d.f. $f_n(x_1, \ldots, x_n; \theta)$ *and the parameter space* Ω *consists of two distinct points which, for convenience, we label* θ_0 *and* θ_1. *For a given* $c_\alpha > 0$, *let* W_α *be the set in the sample space* R_n *for which*

$$f_n(x_1, \ldots, x_n; \theta_1) \geqslant c_\alpha f_n(x_1, \ldots, x_n; \theta_0) \tag{13.2.1}$$

and where

$$P(W_\alpha \mid \theta_0) = \alpha.$$

If W_α^* is any other set in R_n such that

$$P(W_\alpha^* \mid \theta_0) = \alpha,$$

then

$$P(W_\alpha \mid \theta_1) \geqslant P(W_\alpha^* \mid \theta_1), \tag{13.2.2}$$

that is, W_α unbiased and is a more powerful test for testing $\mathscr{H}(\theta_0; \theta_0 \cup \theta_1)$ than any other test W_α^* of size α.

To establish **13.2.1**, we first show that W_α is a more powerful test than W_α^*. We note that for a given $c_\alpha > 0$, W_α is the event in the space of $(x_1, \ldots, x_n)$ for which (13.2.1) holds, the amount of probability on W_α for $\theta = \theta_1$ being α. That is, W_α is a test of size α for $\mathscr{H}(\theta_0; \theta_0 \cup \theta_1)$.

Now suppose W_α^* is any other test for $\mathscr{H}$ of size α, that is

$$P(W_\alpha^* \mid \theta_0) = \alpha. \tag{13.2.3}$$

We then have

$$P(W_\alpha - (W_\alpha \cap W_\alpha^*) \mid \theta_0) = P(W_\alpha^* - (W_\alpha \cap W_\alpha^*) \mid \theta_0). \tag{13.2.4}$$

Now for any point in the sample space R_n contained in W_α, we have $f_n(x_1, \ldots, x_n; \theta_1) \geqslant c_\alpha f_n(x_1, \ldots, x_n; \theta_0)$, and hence

$$P(W_\alpha - (W_\alpha \cap W_\alpha^*) \mid \theta_1) \geqslant c_\alpha P(W_\alpha - (W_\alpha \cap W_\alpha^*) \mid \theta_0). \tag{13.2.5}$$

For any point not contained in W_α, we have
$f_n(x_1, \ldots, x_n; \theta_1) < c_\alpha f_n(x_1, \ldots, x_n; \theta_0)$ and therefore

$$c_\alpha P(W_\alpha^* - (W_\alpha \cap W_\alpha^*) \mid \theta_0) \geqslant P(W_\alpha^* - (W_\alpha \cap W_\alpha^*) \mid \theta_1). \tag{13.2.6}$$

Making use of (13.2.4) in (13.2.5) and (13.2.6), we obtain

$$P(W_\alpha - (W_\alpha \cap W_\alpha^*) \mid \theta_1) \geqslant P(W_\alpha^* - (W_\alpha \cap W_\alpha^*) \mid \theta_1). \tag{13.2.7}$$

Now $W_\alpha - (W_\alpha \cap W_\alpha^*)$ and $W_\alpha \cap W_\alpha^*$ are disjoint sets, and so are $W_\alpha^* - (W_\alpha \cap W_\alpha^*)$ and $W_\alpha \cap W_\alpha^*$. Hence we may add $P(W_\alpha \cap W_\alpha^* \mid \theta_1)$ to both sides of (13.2.7) and obtain (13.2.2), that is,

$$P(W_\alpha \mid \theta_1) \geqslant P(W_\alpha^* \mid \theta_1)$$

which is equivalent to the statement that W_α is a more powerful test than W_α^* for $\mathscr{H}$.

We must now show that W_α is unbiased.

Now W_α is obviously unbiased if $c_\alpha \geqslant 1$ in (13.2.1). If $c_\alpha < 1$, there exists a subset W'_α of W_α for which

(13.2.8) $$f_n(x_1, \ldots, x_n; \theta_1) \geqslant f_n(x_1, \ldots, x_n; \theta_0), \quad \text{if } (x_1, \ldots, x_n) \in W'_\alpha.$$

Otherwise it follows that throughout R_n

(13.2.9) $$0 \leqslant f_n(x_1, \ldots, x_n; \theta_1) \leqslant f_n(x_1, \ldots, x_n; \theta_0)$$

which contradicts the fact that

$$\int_{R_n} f_n(x_1, \ldots, x_n; \theta_0)\, dx_1 \cdots dx_n = \int_{R_n} f_n(x_1, \ldots, x_n; \theta_1)\, dx_1 \cdots dx_n = 1.$$

It is evident that if R_n is the entire sample space

(13.2.10) $$\begin{aligned} P(W'_\alpha \mid \theta_1) - P(W'_\alpha \mid \theta_0) &= P(R_n - W'_\alpha \mid \theta_0) - P(R_n - W'_\alpha \mid \theta_1) \\ &\geqslant P(W_\alpha - W'_\alpha \mid \theta_0) - P(W_\alpha - W'_\alpha \mid \theta_1), \end{aligned}$$

that is,

(13.2.11) $$P(W_\alpha \mid \theta_1) \geqslant P(W_\alpha \mid \theta_0) = \alpha$$

and hence W_α is unbiased.

A theorem similar to **13.2.1** can be stated in the case where $(x_1, \ldots, x_n)$ is a discrete n-dimensional random variable with p.f. $p_n(x_1, \ldots, x_n; \theta)$ where the distributions having p.f.'s $p_n(x_1, \ldots, x_n; \theta_0)$ and $p_n(x_1, \ldots, x_n; \theta_1)$ are the alternative distributions involved in the statistical hypothesis $\mathscr{H}(\theta_0; \theta_0 \cup \theta_1)$.

Finally, we remark that where $(x_1, \ldots, x_n)$ is a random sample from a distribution having p.d.f. $f(x; \theta)$ we would merely replace $f_n(x_1, \ldots, x_n; \theta)$ in **13.2.1** by $\prod_{\xi=1}^{n} f(x_\xi; \theta)$. A similar remark holds if $(x_1, \ldots, x_n)$ is a sample from a discrete distribution having p.f. $p(x; \theta)$.

(b) Case of Similar Critical Regions

For certain special p.d.f.'s $f_n(x_1, \ldots, x_n; \theta)$ the family of sets defined by (13.2.1) for all positive values of c_α and for the alternative $\theta = \theta_1$ is the same family of sets for all values of θ_1 in Ω such that $\theta_1 \neq \theta_0$ in (13.2.2). In this case, W_α is the uniformly most powerful unbiased test of size α for the hypothesis $\mathscr{H}(\theta_0; \Omega)$.

In particular, if there exists a sufficient estimator $\tilde{\theta}$ for θ having p.d.f. $v(\tilde{\theta}; \theta)$ such that the family of sets in R_n defined by $v(\tilde{\theta}; \theta_1) \geqslant c_\alpha v(\tilde{\theta}; \theta_0)$ for all positive values of c_α is the same family as that obtained by any other value of $\theta_1 \neq \theta_0$ in Ω, then for the given α one could find a uniformly most powerful unbiased test W_α from this family for $\mathscr{H}(\theta_0; \Omega)$, which would clearly be determined by the sufficient estimator $\tilde{\theta}$. But even this approach holds little hope except for the case of a one-dimensional parameter θ.

It is therefore sufficient to give an example to illustrate the existence of such a test.

Example. Suppose $(x_1, \ldots, x_n)$ is a sample from $N(\mu, 1)$. Let the parameter space of μ be the real line R_1. Now consider the hypothesis

$$\mathscr{H}(\mu_0; \mu_0 \cup \mu_1).$$

Applying (13.2.1), we find after some reduction that W_α is the set in the sample space R_n for which

$$e^{n(\mu_1 - \mu_0)\bar{x}} \geqslant c'_\alpha$$

where $c'_\alpha > 0$. Thus, W_α is determined by the sufficient estimator $\bar{x}$ for μ. When we recall that if $(x_1, \ldots, x_n)$ is a sample from $N(\mu, 1)$, then $\bar{x}$ has the distribution $N(\mu, 1/n)$, it is evident that:

(i) If $\mu_1 > \mu_0$, W_α is the set in R_n for which $\bar{x} > z_\alpha$ where z_α is defined by the equation

$$1 - \Phi(\sqrt{n}(z_\alpha - \mu_0)) = \alpha$$

$\Phi(x)$ being the c.d.f. of $N(0, 1)$.

(ii) If $\mu_1 < \mu_0$, W_α is the set for which $\bar{x} < z'_\alpha$ where

$$\Phi(\sqrt{n}(\mu_0 - z'_\alpha)) = \alpha.$$

In case (i), the test with critical set defined by $\bar{x} > z_\alpha$ is uniformly most powerful for testing the hypothesis $\mathscr{H}(\mu_0; \mu \geqslant \mu_0)$, that is, that $\mu = \mu_0$ against any alternative $\mu = \mu_1$ where $\mu_1 > \mu_0$. Since

$$1 - \Phi(\sqrt{n}(z_\alpha - \mu_1) > 1 - \Phi(\sqrt{n}(z_\alpha - \mu_0)), \qquad \text{if } \mu_1 > \mu_0$$

it is evident that this test is unbiased if Ω consists of all values of $\mu \geqslant \mu_0$.

Similar remarks hold for case (ii).

(c) Walds' Reduction of a Composite Hypothesis with one point in $\Omega - \omega$ to a Simple Hypothesis

In a composite hypothesis $\mathscr{H}(\omega; \omega \cup \theta_1)$, that is, having one point θ_1 in $\Omega - \omega$, Wald (1939) has approached the problem of constructing a test for $\mathscr{H}$ essentially by replacing the family of p.d.f's $f_n(x_1, \ldots, x_n; \theta)$ for $\theta \in \omega$ by a p.d.f. $f_G(x_1, \ldots, x_n)$ obtained by averaging $f_n(x_1, \ldots, x_n; \theta)$ with respect to θ by means of an *a priori* c.d.f. $G(\theta)$ defined over ω. More specifically, let

$$f_G(x_1, \ldots, x_n) = \int_\omega f_n(x_1, \ldots, x_n; \theta)\, dG(\theta) \tag{13.2.12}$$

which is a p.d.f, and consider the problem of testing the hypothesis $\mathscr{H}_G$ that the population p.d.f. is $f_G(x_1, \ldots, x_n)$ against the alternative that it is $f_n(x_1, \ldots, x_n; \theta_1)$.

Assuming that $f_G(x_1, \ldots, x_n)$ has all the properties ascribed to $f_n(x_1, \ldots, x_n; \theta)$ in **13.2.1**, then **13.2.1** can be applied to determine a most powerful test $W_{G\alpha}$ of size α for testing $\mathscr{H}_G$. Wald's (1939) theorem states that:

13.2.2 *If there exists a c.d.f. $G(\theta)$ for $\theta \in \omega$ such that the most powerful test $W_{G\alpha}$ of size α for $\mathscr{H}_G$ is also of size α for testing $\mathscr{H}(\omega; \omega \cup \theta_1)$, then*

$$\int_{\omega_0} dG(\theta) = 0 \tag{13.2.13}$$

where ω_0 is the subset of ω for which

$$\int_{W_{G\alpha}} f_n(x_1, \ldots, x_n; \theta)\, dx_1 \cdots dx_n = \alpha, \tag{13.2.14}$$

and $W_{G\alpha}$ is the most powerful test for $\mathscr{H}(\omega; \omega \cup \theta_1)$.

To prove **13.2.2**, note that if, for given α, $W_{G\alpha}$ is a test for $\mathscr{H}(\omega; \omega \cup \theta_1)$, then we must have

$$\int_{W_{G\alpha}} f_n(x_1, \ldots, x_n; \theta)\, dx_1 \ldots dx_n \leqslant \alpha, \qquad \theta \in \omega \tag{13.2.15}$$

and hence

$$\int_\omega \int_{W_{G\alpha}} f_n(x_1, \ldots, x_n; \theta)\, dx_1 \cdots dx_n\, dG(\theta) \leqslant \alpha \int_\omega dG(\theta). \tag{13.2.16}$$

If we interchange the order of integration, it is evident that (13.2.13) must hold in order for

$$\int_{W_{G\alpha}} f_G(x_1, \ldots, x_n)\, dx_1 \cdots dx_n = \alpha$$

that is, in order for $W_{G\alpha}$ to be of size α.

Now if $W_{G\alpha}$ is the most powerful test for $\mathscr{H}_G$, we must have

$$\int_{W_{G\alpha}} f_n(x_1, \ldots, x_n; \theta_1)\, dx_1 \cdots dx_n \leqslant \int_{W'_{G\alpha}} f_n(x_1, \ldots, x_n; \theta_1)\, dx_1 \cdots dx_n \tag{13.2.17}$$

where $W'_{G\alpha}$ is any other test of size α for $\mathscr{H}_G$. But (13.2.17) is precisely the condition for $W_{G\alpha}$ to be the most powerful test for $\mathscr{H}(\omega; \omega \cup \theta_1)$.

It should be noted that if $W_{G\alpha}$ were a similar set, that is, if the equality holds in (13.2.15) for $\theta \in \omega$, then (13.2.13) would be a trivially true statement. But, of course, in this instance the similarity property essentially solves the problem of determining a critical region.

13.3 THE LIKELIHOOD RATIO TEST

Tests for composite hypotheses having optimal properties for finite samples have been obtained for various special problems by an important principle due to Neyman and Pearson (1928, 1933) called the *likelihood ratio* principle, which is a natural extension of the test defined by (13.2.2) to a composite hypothesis.

In large samples these likelihood ratio tests have optimal asymptotic properties under the same conditions for which asymptotically normally distributed maximum likelihood estimates exist.

On the other hand, more general approaches have been made to the problem of testing composite hypotheses for finite samples by Lehmann (1950, 1959), Lehmann and Scheffé (1950), Lehmann and Stein (1948), and others.

The only tests of composite hypotheses which we shall consider in detail are likelihood ratio tests, particularly their asymptotic properties for large values of n. Likelihood ratio tests for large values of n are just as fundamental in the theory of statistics as the large-sample theory of parametric statistical estimation considered in Chapter 12.

Asymptotic distributions of likelihood ratio tests of composite hypotheses in large samples when the hypothesis tested is true were first discussed by Wilks (1938*a*). Wald (1941*a*, 1941*b*) later made a detailed study of these tests including asymptotic properties of bias, power, and other aspects. Many of the results of the next few sections are slightly less general versions of those contained in Wald's fundamental papers. The methods used here, however, are perhaps simpler than those used by Wald.

(a) Definition of a Likelihood Ratio Test

Suppose $(x_1, \ldots, x_n)$ is a sample from the c.d.f. $F(x; \theta)$ where θ is r-dimensional. For convenience we take x as one-dimensional although our results extend to the case where x is k-dimensional with minor notational changes. The likelihood element is

$$dF_n(x_1, \ldots, x_n; \theta) = \prod_{\xi=1}^{n} dF(x_\xi; \theta). \tag{13.3.1}$$

Now consider the hypothesis

$$\mathscr{H}(\omega; \Omega) \tag{13.3.2}$$

and let

$$(dF_n)_\omega = \sup_{\theta \in \omega} dF_n(x_1, \ldots, x_n; \theta) \tag{13.3.3}$$

$$(dF_n)_\Omega = \sup_{\theta \in \Omega} dF_n(x_1, \ldots, x_n; \theta). \tag{13.3.4}$$

In most ordinary applications, $(dF_n)_\omega$ and $(dF_n)_\Omega$ are simply maxima of dF_n for $\theta \in \omega$ and $\theta \in \Omega$ obtained by the usual differentiation procedure.

The *likelihood ratio* $\lambda_{\mathscr{H}}$ for testing $\mathscr{H}(\omega; \Omega)$, or more briefly, $\mathscr{H}$, is defined by

(13.3.5) $$\lambda_{\mathscr{H}} = \frac{(dF_n)_\omega}{(dF_n)_\Omega}.$$

The values of $\lambda_{\mathscr{H}}$ lie on the interval [0, 1]. The critical set W_α in R_n for testing $\mathscr{H}$ is the set for which $\lambda_{\mathscr{H}} < c_\alpha$, c_α being a constant, where

(13.3.6) $$\int_{W_\alpha} dF_n \leqslant \alpha, \qquad \theta \in \omega.$$

The test defined by $\lambda_{\mathscr{H}} < c$ is essentially an extension of that defined by (13.2.2) to the case where ω and $\Omega - \omega$ have more than one point each.

In addition to this fact, $\lambda_{\mathscr{H}}$ has intuitive appeal as a test. For if we are comparing the "plausibility" of one value of θ against another, given that we have a sample $(x_1, \ldots, x_n)$, we would intuitively be inclined to choose that value of θ which gives the likelihood element the larger value. Thus, if we cannot obtain an appreciably larger value of the likelihood element by searching for a value of θ through the entire parameter space Ω than we can by searching through the set ω, our intuition will assess the evidence as strongly favoring the proposition that the "most plausible" value of θ belongs to ω, that is, that $\mathscr{H}$ is true. We shall see that, in the case of large samples, this intuitive appeal can be rigorously supported under fairly general conditions.

Example. Suppose $(x_1, \ldots, x_n)$ is a sample from $N(\mu, \sigma^2)$, and that we wish to determine the likelihood ratio test for the hypothesis $\mathscr{H}$ defined as follows:

Ω : the half-plane for which $-\infty < \mu < +\infty$, $\sigma^2 > 0$

ω : the subset of Ω (half-line) for which $\mu = \mu_0$.

We have

$$dF_n = \left(\frac{1}{2\pi\sigma^2}\right)^{\frac{1}{2}n} \exp\left[-\frac{1}{2\sigma^2}\sum_{\xi=1}^{n}(x_\xi - \mu)^2\right] dx_1 \cdots dx_n .$$

Maximizing dF_n with respect to (μ, σ^2) in Ω, and again with respect to (μ, σ^2) in ω, and taking the ratio of the two maxima in accordance with (13.3.5), we obtain as the likelihood ratio for $\mathscr{H}$,

$$\lambda_{\mathscr{H}} = \left(\frac{S_\Omega}{S_\omega}\right)^{\frac{1}{2}n}$$

where S_Ω is the minimum of the sum of squares $\sum_{\xi=1}^{n}(x_\xi - \mu)^2$ with respect to μ, that is,

$$S_\Omega = \sum_{\xi=1}^{n}(x_\xi - \bar{x})^2$$

whereas

$$S_\omega = \sum_{\xi=1}^{n}(x_\xi - \mu_0)^2.$$

Note that $\lambda_{\mathscr{H}}$ can be written as follows

$$\lambda_{\mathscr{H}} = \left(1 + \frac{t^2}{n-1}\right)^{-\frac{1}{2}n}$$

where

$$t = \sqrt{n}(\bar{x} - \mu_0)/s$$

$\bar{x}$ and s^2 being the sample mean and variance defined in (8.2.1) and (8.2.6). The random variable t is, of course, the "Student" ratio defined in (8.4.16). Thus, the critical region W_α of size α of the likelihood ratio test is the portion of the sample space of $\lambda_{\mathscr{H}}$ for which

$$P(\lambda_{\mathscr{H}} < \lambda_\alpha) = \alpha, \qquad \text{for } \mu = \mu_0$$

where

$$\lambda_\alpha = \left(1 + \frac{t_\alpha^2}{n-1}\right)^{-\frac{1}{2}n}$$

and t_α is chosen so that

$$\int_{-t_\alpha}^{+t_\alpha} f_{n-1}(t)\,dt = 1 - \alpha$$

$f_{n-1}(t)$ being the p.d.f. of the "Student" distribution $S(n-1)$ defined by (7.8.4).

The likelihood ratio $\lambda_{\mathscr{H}}$ for testing the hypothesis $\mathscr{H}$ is therefore equivalent to the "Student" t-test.

(b) The Likelihood Ratio Test in Normal Regression Theory

In small samples one of the most important applications of the likelihood ratio test is in the testing of various hypotheses concerning the parameters of a normal distribution. The preceding example is one important illustration. In many of these applications the exact sampling distribution of the likelihood ratio can be found for finite samples of size n. Most of these tests are quite straightforward and will not be further considered except in the problems at the end of this chapter.

It is perhaps worthwhile, however, to consider the likelihood ratio test for testing the hypothesis that some of the regression parameters in normal regression theory have zero values. Referring to Section 10.3(c), suppose y_ξ, $\xi = 1, \ldots, n$ are independent random variables having normal distributions $N(\beta_1 x_{1\xi} + \cdots + \beta_k x_{k\xi}, \sigma^2)$, $\xi = 1, \ldots, n$, where $(x_{i\xi}, i = 1, \ldots, k)$ are linearly independent fixed vectors (that is, the matrix $\|a_{ij}\|$ in (10.3.8) is nonsingular). Let $\mathscr{H}$ be the hypothesis for which Ω and ω are defined as follows:

(13.3.7)
Ω: the half $(k+1)$-dimensional Euclidean space for which $-\infty < \beta_i < +\infty$, $i = 1, \ldots, k$, $\sigma^2 > 0$.
ω: the subset of Ω for which $\beta_{k'+1} = \cdots = \beta_k = 0$, $k' < k$.

The hypothesis $\mathscr{H}$ thus defined is sometimes called the *general linear hypothesis* of normal regression theory. Our sample consists of what may be regarded as the conditional random variables $(y_\xi \mid x_{1\xi}, \ldots, x_{k\xi}, \xi = 1, \ldots, n)$, and the likelihood of the parameters $\beta_1, \ldots, \beta_k, \sigma^2$ is given by

$$(13.3.8) \quad dF_n = \left(\frac{1}{\sqrt{2\pi}\sigma}\right)^n \cdot \exp\left[-\frac{1}{2\sigma^2}\sum_{\xi=1}^{n}(y_\xi - \beta_1 x_{1\xi} - \cdots - \beta_k x_{k\xi})^2\right] dy_1, \ldots, dy_n.$$

Determining $(dF_n)_\Omega$ and $(dF_n)_\omega$ by the usual differentiation procedure, we find

$$(13.3.9) \quad \lambda_{\mathscr{H}} = \frac{(dF_n)_\Omega}{(dF_n)_\omega} = \left(\frac{S_\Omega}{S_\omega}\right)^{\frac{1}{2}n},$$

where S_Ω is the minimum of the sum of squares

$$(13.3.10) \quad \sum_{\xi=1}^{n}(y_\xi - \beta_1 x_{1\xi} - \cdots - \beta_k x_{k\xi})^2$$

with respect to $\beta_1, \ldots, \beta_k$, and S_ω is the minimum of

$$(13.3.11) \quad \sum_{\xi=1}^{n}(y_\xi - \beta_1 x_{1\xi} - \cdots - \beta_{k'} x_{k'\xi})^2$$

with respect to $\beta_1, \ldots, \beta_{k'}$. Referring to Section 10.3(*b*) it will be seen that S_Ω is identical with S_1 as given by (10.3.27), that is,

$$(13.3.12) \quad S_\Omega = \frac{|a_{i_0 j_0}|}{|a_{ij}|}$$

where the a_{ij}, $i, j = 1, \ldots, k$, and $a_{i_0 j_0}$ are defined in (10.3.8), (10.3.11), and (10.3.28). $|a_{ij}|$ is, of course, the determinant of the matrix $\|a_{ij}\|$, whereas $|a_{i_0 j_0}|$ is defined by (10.3.28).

Similarly, we have

$$(13.3.13) \quad S_\omega = \frac{|a_{i_0' j_0'}|}{|a_{i' j'}|}$$

where $|a_{i_0' j_0'}|$ and $|a_{i' j'}|$ are similar to $|a_{i_0 j_0}|$ and $|a_{ij}|$ with $i', j' = 1, \ldots, k'$. It follows by an argument based on Cochran's theorem **8.4.4** that, if $\mathscr{H}$ is true, S_Ω/σ^2 and $(S_\omega - S_\Omega)/\sigma^2$ are independently distributed according to chi-square distributions $C(n - k)$ and $C(k - k')$, respectively. We shall not present the details since the argument is similar to that used in establishing the fact that S_1/σ^2 and S_2/σ^2 are independently distributed according to chi-square distributions $C(n - k)$ and $C(k)$, where S_1 and S_2 are given by (10.3.30).

Thus the likelihood ratio $\lambda_{\mathscr{H}}$ in (13.3.9) may be written as

$$\lambda_{\mathscr{H}} = \left(1 + \frac{(k - k')}{(n - k)}\mathscr{F}\right)^{-\frac{1}{2}n} \tag{13.3.14}$$

where

$$\mathscr{F} = \frac{(n - k)(S_\omega - S_\Omega)}{(k - k')S_\Omega} \tag{13.3.15}$$

has the Snedecor distribution $S(k - k', n - k)$ whose p.e. is defined by (7.8.9). Since there is a one-to-one correspondence between $\lambda_{\mathscr{H}}$ and $\mathscr{F}$, it therefore follows that the likelihood ratio test for $\mathscr{H}$ is equivalent to the $\mathscr{F}$ test. The critical region $\lambda_{\mathscr{H}} < \lambda_\alpha$ in the sample space of $\lambda_{\mathscr{H}}$, where λ_α is the 100α % point of the $\lambda_{\mathscr{H}}$ distribution, corresponds to the critical region $\mathscr{F} > [(n - k)/(k - k')](\lambda_\alpha^{-2/n} - 1)$ in the sample space of $\mathscr{F}$. The $\mathscr{F}$ test just described is the *Model I analysis of variance test* in its most general form. Daly (1940) has shown that this test is unbiased.

We therefore have the following basic result concerning the general linear hypothesis in normal regression theory:

13.3.1. *Suppose y_ξ, $\xi = 1, \ldots, n$, are independent random variables having the normal distributions $N(\beta_1 x_{1\xi} + \cdots + \beta_k x_{k\xi}, \sigma^2)$, $\xi = 1, \ldots, n$, where the matrix $\|a_{ij}\|$, $i, j = 1, \ldots, k$, defined by* (10.3.8) *is nonsingular. The likelihood ratio $\lambda_{\mathscr{H}}$ for testing the hypothesis $\mathscr{H}$ specified by* (13.3.7) *is given by* (13.3.14) *where $\mathscr{F}$ is given by* (13.3.15). *Furthermore, if $\mathscr{H}$ is true $[(n - k)/(k - k')](\lambda_{\mathscr{H}}^{-2/n} - 1)$ has the Snedecor distribution $S(k - k', n - k)$.*

The general linear hypothesis is more comprehensive than may appear at first glance. More generally, we may wish to test the hypothesis $\mathscr{H}'$ that $\beta_{k'+1} = \beta_{k'+1,0}, \ldots, \beta_k = \beta_{k0}$ rather than $\beta_{k'+1} = \cdots = \beta_k = 0$ as now specified by ω in (13.3.7). In this case, we may introduce the random variable $y'_\xi = (y_\xi - \beta_{k'+1,0}x_{k'+1\xi} - \cdots - \beta_{k0}x_{k\xi})$ in determining $(dF_n)_\omega$. We would thus have independent random variables y'_ξ, $\xi = 1, \ldots, n$, with normal distributions $N(\beta_1 x_{1\xi} + \cdots + \beta_{k'}x_{k'\xi}, \sigma^2)$ and the hypothesis $\mathscr{H}'$ would be specified by (13.3.7) with no change in Ω but with $\beta_{k'+1} = \beta_{k'+1,0}, \ldots, \beta_k = \beta_{k0}$ in ω. The likelihood ratio $\lambda_{\mathscr{H}'}$ is identical in structure to $\lambda_{\mathscr{H}}$ with y_ξ replaced by y'_ξ in the matrices $\|a_{i_0'j_0'}\|$ and $\|a_{i'j'}\|$. The distribution theory of $\lambda_{\mathscr{H}'}$ when $\mathscr{H}'$ is true is exactly the same as the distribution theory of $\lambda_{\mathscr{H}}$ when $\mathscr{H}$ is true.

Again, we may wish to test the hypothesis $\mathscr{H}''$ that there exist certain linearly independent linear constraints among the $\beta_1, \ldots, \beta_k$ of form

$$\sum_{i=1}^{k} c_{iu}\beta_i = \gamma_{u0}, \qquad u = k' + 1, \ldots, k$$

where the c_{iu} and γ_{u0} are specified numbers. In this case we may apply the transformation of regression coefficients defined by

$$\beta_1 = \gamma_1, \ldots, \beta_{k'} = \gamma_{k'},$$

$$\sum_{i=1}^{k} c_{iu}\beta_i = \gamma_u, \qquad u = k' + 1, \ldots, k$$

to the regression function $\beta_1 x_{1\xi} + \cdots \beta_k x_{k\xi}$ and obtain a regression function of form $\gamma_1 z_{1\xi} + \cdots + \gamma_k z_{k\xi}$ where $z_{1\xi}, \ldots, z_{k\xi}$ are linear functions of $x_{1\eta}, \ldots, x_{k\eta}$, $\eta = 1, \ldots, n$, with coefficients depending on the c_{pu}. The hypothesis $\mathscr{H}''$ is thus reduced to an hypothesis of type $\mathscr{H}'$ discussed above, with $\gamma_1, \ldots, \gamma_k$ playing the role of $\beta_1, \ldots, \beta_k$ and $z_{1\xi}, \ldots, z_{k\xi}$ playing the role of $x_{1\xi}, \ldots, x_{k\xi}$, $\xi = 1, \ldots, n$. The sampling theory of $\lambda_{\mathscr{H}''}$ when $\mathscr{H}''$ is true is exactly the same as that of $\lambda_{\mathscr{H}}$ when $\mathscr{H}$ is true.

13.4 ASYMPTOTIC DISTRIBUTION OF LIKELIHOOD RATIO IN LARGE SAMPLES

We now turn to large-sample problems. First we shall consider the case of a simple hypothesis

$$(13.4.1) \qquad \mathscr{H}(\theta_0; \Omega)$$

which will be referred to as $\mathscr{H}$ in this section, where θ is a one-dimensional parameter, the parameter space Ω being an interval on the real axis R_1 containing some (open) interval Ω_0 containing θ_0. In large-sample theory the only part of Ω which plays an essential role in the asymptotic sampling theory of the likelihood ratio test is Ω_0. We shall assume that for a sample of size n, there exists a maximum likelihood estimator $\hat{\theta}_n$ for θ_0, such that the sequence of estimators $\hat{\theta}_n$, $n = 1, 2, \ldots$ converges in probability to θ_0. The corresponding sequence of likelihood ratios, therefore, is given by

$$(13.4.2) \qquad \lambda_{\mathscr{H}} = \frac{\prod_{\xi=1}^{n} dF(x_\xi; \theta_0)}{\prod_{\xi=1}^{n} dF(x_\xi; \hat{\theta}_n)}, \qquad n = 1, 2, \ldots.$$

We shall show that under conditions to be stated below

$$-2 \log \lambda_{\mathscr{H}}, \qquad n = 1, 2, \ldots$$

converges in distribution to the chi-square distribution $C(1)$ when $\mathscr{H}$ is true.

We shall assume that $F(x; \theta)$ is regular with respect to its first and second θ-derivatives for θ in Ω_0 (see Section 12.1). Under these conditions

$$(13.4.3) \qquad H(\theta_0, \theta) = \int_{-\infty}^{\infty} \log dF(x; \theta)\, dF(x; \theta_0)$$

can be differentiated twice under the integral sign with respect to θ, thus yielding

$$H'(\theta_0, \theta_0) = 0$$

$$H''(\theta_0, \theta_0) = -B^2(\theta_0, \theta_0), \tag{13.4.4}$$

where $B^2(\theta_0, \theta_0)$ is defined in (12.1.8).

Therefore $H(\theta_0, \theta)$ has a relative maximum $H(\theta_0, \theta_0)$ at $\theta = \theta_0$. The quantity $H(\theta_0, \theta)$ is of basic importance in statistical information theory as developed by Kullback (1959). In general, if $F_1(x)$, $F_2(x)$ and $G(x)$ are three c.d.f.'s, absolutely continuous with respect to each other, Kullback's *mean information* or *information integral* for discriminating $F_1(x)$ against $F_2(x)$ per observation from $G(x)$ is defined by $I(1; 2) = \int_{-\infty}^{\infty} \log \frac{dF_1(x)}{dF_2(x)} dG(x)$. Hence

$$H(\theta_0, \theta_0) - H(\theta_0, \theta_1) = \int_{-\infty}^{\infty} \log \frac{dF(x; \theta_0)}{dF(x; \theta_1)} dF(x; \theta_0)$$

is, in Kullback's sense, the information integral for discriminating $F(x; \theta_0)$ against $F(x; \theta_1)$ per observation from $F(x; \theta_0)$.

Remark. A particular case of the function $H(\theta, \theta)$ first arose in statistical mechanics. Suppose x is a point whose components represent the coordinates and momenta of a given gas molecule. The space of x is called the *phase space* of the system. If $F(x; t)$ represents the c.d.f. of x in the aggregate of molecules in this system at time t, then the integral $H(t, t)$ which would be obtained by using $F(x; t)$ in (13.4.3). $H(t, t)$ is the H-function originally introduced by Boltzmann (1910) to approximate $-\log P$ (except for an additive constant) where

$$P = \frac{n!}{n_1!\, n_2! \cdots n_N!} \left(\frac{1}{N}\right)^n$$

the probability that for n molecules in the system, there will be n_i molecules in cell E_i, $i = 1, 2, \ldots, N$, at time t, where $E_1, E_2, \ldots, E_N$ are disjoint "cells" in the phase space (having equal probabilities $\frac{1}{N}$ which constitute the entire phase space. $H(t, t)$ essentially measures the deviation of the system from a "most probable" or "equilibrium" state.

Functions of type $H(\theta_0, \theta)$ also play an important role in communication theory as developed by Shannon (1948).

Recalling from Section 12.2(*c*) that $B^2(\theta_0, \theta_0)$ has been defined by Fisher as the *amount of information pertaining to* θ_0 *per observation from* $F(x; \theta_0)$ it will be noted from (13.4.4) that Fisher's amount of information per observation is the second derivative of Kullback's *information integral* $[H(\theta_0, \theta_0) - H(\theta_0, \theta)]$ at $\theta = \theta_0$. Stated another way, $B^2(\theta_0, \theta_0)$ essentially measures the curvature of $H(\theta_0, \theta)$ at its maximum value which occurs at $\theta = \theta_0$.

Summarizing, we may say that:

13.4.1 *If $F(x; \theta)$ is regular with respect to its second θ-derivative for θ in Ω_0, then $H(\theta_0, \theta)$ has a relative maximum of $H(\theta_0, \theta_0)$ at $\theta = \theta_0$. The second derivative of $[H(\theta_0, \theta_0) - H(\theta_0, \theta)]$ (Kullback's information integral) at $\theta = \theta_0$, is $B^2(\theta_0, \theta_0)$ [Fisher's amount of information pertaining to θ_0 per observation from $F(x; \theta_0)$].*

In a sample $(x_1, \ldots, x_n)$ from the c.d.f. $F(x; \theta_0)$, the quantity

$$\frac{1}{n}\sum_{\xi=1}^{n} \log dF(x_\xi; \hat{\theta}) \tag{13.4.5}$$

is the mean of a sample of size n from a distribution having mean $H(\theta_0, \theta_0)$, and hence by **9.1.1** converges in probability to $H(\theta_0, \theta_0)$, which exists if $F(x; \theta)$ is regular in its first θ-derivative in Ω_0. Also, under these conditions, $\hat{\theta}(x_1, \ldots, x_n)$, $n = 1, 2, \ldots$ converges in probability to θ_0 as $n \to \infty$, as we have seen in **12.3.2**. Furthermore, it follows from **4.3.8** that

$$\frac{1}{n}\sum_{\xi=1}^{n} \log F(x_\xi; \hat{\theta}) \tag{13.4.6}$$

converges in probability to $H(\theta_0, \theta_0)$ as $n \to \infty$.

Therefore

13.4.2 *If $(x_1, \ldots, x_n)$ is a sample from $F(x; \theta_0)$, where $F(x; \theta)$ is regular in its first θ-derivative in Ω_0, then both*

$$\frac{1}{n}\sum_{\xi=1}^{n} \log dF(x_\xi; \theta_0) \quad \textit{and} \quad \frac{1}{n}\sum_{\xi=1}^{n} F(x_\xi; \hat{\theta})$$

converge in probability to $H(\theta_0, \theta_0)$ as $n \to \infty$.

If $\hat{\theta}$ satisfies the stronger condition of being asymptotically normally distributed, then we have the following result concerning the asymptotic distribution of $-2 \log \lambda_{\mathscr{H}}$ when $\mathscr{H}$ is true:

13.4.3 *If $\hat{\theta}$ is asymptotically normally distributed in accordance with* **12.3.3**, *then*

$$\lim_{n\to\infty} P(-2 \log \lambda_{\mathscr{H}} < \chi^2 \mid \theta_0) = \frac{1}{\sqrt{2\pi}} \int_0^{\chi^2} u^{-\frac{1}{2}} e^{-\frac{1}{2}u}\, du \tag{13.4.7}$$

that is, if $\mathscr{H}$ is true, -2 log $\lambda_{\mathscr{H}}$, $n = 1, 2, \ldots$ converges in distribution to the chi-square distribution $C(1)$.

To establish **13.4.3**, we refer to the assumptions of **12.3.3.** Since $\hat{\theta}$ converges almost certainly to θ_0 as $n \to \infty$, it follows that for an arbitrary $\varepsilon > 0$ there exists an n_ε such that the probability exceeds $1 - \varepsilon$ that the following equality holds

$$\text{(13.4.8)} \quad \sum_{\xi=1}^{n} \log dF(x_\xi; \theta_0) = \sum_{\xi=1}^{n} \log dF(x_\xi; \hat{\theta}) + \frac{1}{2} \sum_{\xi=1}^{n} \left[\frac{\partial^2}{\partial\theta^2} \log dF(x_\xi; \theta)\right]_{\theta=\theta^*} (\theta_0 - \hat{\theta})^2$$

for all $n > n_\varepsilon$, where θ^* is a random variable such that $|\theta_0 - \theta^*| < |\theta_0 - \hat{\theta}|$. Now (13.4.8) can be rewritten as

(13.4.9)

$$-2\sum_{\xi=1}^{n} \log \left(\frac{dF(x_\xi; \theta_0)}{dF(x_\xi; \hat{\theta})}\right) = -\frac{1}{n} \sum_{\xi=1}^{n} \left[\frac{\partial^2}{\partial\theta^2} \log dF(x_\xi; \theta)\right]_{\theta=\theta^*} \cdot [\sqrt{n}(\theta_0 - \hat{\theta})]^2.$$

It should be noted that the left side of (13.4.9) is $-2 \log \lambda_{\mathscr{H}}$. The fact that the probability exceeds $1 - \varepsilon$ that (13.4.9) holds for all $n > n_\varepsilon$ implies that the left- and right-hand sides of (13.4.9) are sequences of random variables for $n = 1, 2, \ldots$ which converge together in distribution to the chi-square distribution $C(1)$. The expression

$$\text{(13.4.10)} \quad -\frac{1}{n} \sum_{\xi=1}^{n} \left[\frac{\partial^2}{\partial\theta^2} \log dF(x_\xi; \theta)\right]_{\theta=\theta^*}$$

converges in probability to $B^2(\theta_0, \theta_0)$. Furthermore, $\sqrt{n}B(\theta_0, \theta_0)(\theta_0 - \hat{\theta})$ converges in distribution to $N(0, 1)$.

Therefore, the sequence of random variables on the right of (13.4.9) converges in distribution to the chi-square distribution $C(1)$. Thus, we conclude that the expression on the left of (13.4.9), namely, $-2 \log \lambda_{\mathscr{H}}$, converges in probability to a random variable having the chi-square distribution $C(1)$, thereby completing the argument for **13.4.3**.

13.5 CONSISTENCY OF LIKELIHOOD RATIO TEST

Now suppose W_α is the set in R_n for which

$$\text{(13.5.1)} \quad -2 \log \lambda_{\mathscr{H}} > \chi^2_\alpha$$

where $\mathscr{H}$ is the hypothesis $\mathscr{H}(\theta_0; \Omega_0)$, and χ^2_α is the number for which $P(\chi^2 > \chi^2_\alpha) = \alpha$, this probability being computed from the chi-square distribution $C(1)$. If $F(x; \theta)$ is regular with respect to its second θ-derivative in Ω_0, then we know from **13.4.3** that

$$\text{(13.5.2)} \quad \lim_{n\to\infty} P(W_\alpha \mid \theta_0) = \alpha.$$

Now suppose $\theta_1 \neq \theta_0$ is any point in Ω_0, and consider

(13.5.3) $$\lim_{n\to\infty} P(W_\alpha \mid \theta_1).$$

We can write

(13.5.4) $$\lambda_{\mathscr{H}} = \lambda'_{\mathscr{H}} \cdot \nu$$

where

$$\lambda'_{\mathscr{H}} = \prod_{\xi=1}^{n} \left(\frac{dF(x_\xi; \theta_1)}{dF(x_\xi; \hat{\theta})}\right), \qquad \nu = \prod_{\xi=1}^{n} \left(\frac{dF(x_\xi; \theta_0)}{dF(x_\xi; \theta_1)}\right).$$

If we let

(13.5.5) $$-2 \log \lambda'_{\mathscr{H}} = u_n, \qquad -2 \log \nu = nv_n,$$

then W_α is the set in R_n for which

(13.5.6) $$u_n + nv_n > \chi^2_\alpha.$$

If the true value of θ is θ_1, then it follows from **13.4.3** with θ_0 replaced by θ_1, that u_n, $n = 1, 2, \ldots$, is a sequence of random variables converging in distribution to the chi-square distribution $C(1)$. We may write

(13.5.7) $$v_n = 2\left[\frac{1}{n}\sum_{\xi=1}^{n} \log dF(x_\xi; \theta_1) - \frac{1}{n}\sum_{\xi=1}^{n} \log dF(x_\xi; \theta_0)\right].$$

It follows from **9.1.1** that if θ_1 is the true value of θ, $\frac{1}{n}\sum_{\xi=1}^{n} \log dF(x_\xi; \theta_1)$ and $\frac{1}{n}\sum_{\xi=1}^{n} \log dF(x_\xi; \theta_0)$ converge in probability to $H(\theta_1, \theta_1)$ and $H(\theta_1, \theta_0)$ respectively. But we know from **13.4.1** that $H(\theta_1, \theta_1) > H(\theta_1, \theta_0)$. Therefore the sequence of random variables v_n, $n = 1, 2, \ldots$ converges in probability to the positive number $v_0 = 2[H(\theta_1, \theta_1) - H(\theta_1, \theta_0)]$. Since u_n, $n = 1, 2, \ldots$, converges in distribution to the chi-square distribution $C(1)$, it is seen that

(13.5.8) $$\lim_{n\to\infty} P(u_n + nv_n > \chi^2_\alpha) = 1$$

which is equivalent to the statement

(13.5.9) $$\lim_{n\to\infty} P(W_\alpha \mid \theta = \theta_1) = 1$$

that is, W_α is a consistent test for $\mathscr{H}(\theta_0; \Omega)$.

Therefore, for θ in Ω the sequence of power functions

(13.5.10) $$P(W_\alpha \mid \theta), \qquad n = 1, 2, \ldots$$

converges to the function

(13.5.11) $$\zeta(\theta) = \begin{cases} \alpha, & \theta = \theta_0 \\ 1, & \theta \neq \theta_0 \end{cases}$$

which means that for sufficiently large n, W_α, that is $-2 \log \lambda_{\mathscr{H}}$, is a consistent test for $\mathscr{H}$.

We summarize as follows:

13.5.1 *If $F(x; \theta)$ is regular with respect to its first and second θ-derivatives, the likelihood test defined by (13.5.1) is a consistent test for $\mathscr{H}(\theta_0; \Omega)$.*

13.6 ASYMPTOTIC POWER OF LIKELIHOOD RATIO TEST

We have seen from **13.4.3** that for large samples the likelihood ratio $\lambda_{\mathscr{H}}$ determines a test for the hypothesis $\mathscr{H}(\theta_0; \Omega)$ for which the probability of a Type I error is computed from a chi-square distribution with one degree of freedom, and from **13.5.1** that the test is consistent, that is, the power function of the test as $n \to \infty$ converges to $\zeta(\theta)$ as defined in (13.5.11). We shall now show that under certain conditions the power function of no other consistent test for $\mathscr{H}(\theta_0; \Omega)$ of size α converges to $\zeta(\theta)$ more rapidly than that for the test determined by $\lambda_{\mathscr{H}}$.

For this purpose, it is sufficient to return to the regular estimating function $g_n(x_1, \ldots, x_n; \theta)$ defined in Section 12.5, which, as we have seen, has properties sufficient to insure that if $\mathscr{H}$ is true the two sequences of random variables given in (12.5.11) and (12.5.12) converge together in distribution to the normal distribution $N(0, 1)$, as $n \to \infty$. Hence, the sequence

$$ng_n^2(x_1, \ldots, x_n; \theta_0), \qquad n = 1, 2, \ldots \tag{13.6.1}$$

converges in distribution to the chi-square distribution $C(1)$. Let W_α^* be the test defined by

$$ng_n^2(x_1, \ldots, x_n; \theta_0) > \chi_\alpha^2 \tag{13.6.2}$$

where χ_α^2 is the same number defined in (13.5.1). Therefore, we have

$$\lim_{n \to \infty} P(W_\alpha^* \mid \theta_0) = \alpha. \tag{13.6.3}$$

Now it follows from the properties of $g_n(x_1, \ldots, x_n; \theta)$ as a regular estimating function for θ_0 that, if the true value of θ is $\theta_1 \neq \theta_0$, where θ_1 is also a point in Ω_0

$$ng_n^2(x_1, \ldots, x_n; \theta_0), \qquad n = 1, 2, \ldots \tag{13.6.4}$$

and

$$w_n^* = \{g_n'(x_1, \ldots, x_n; \tilde{\theta}^*)[\sqrt{n}(\theta_1 - \tilde{\theta}) + \sqrt{n}(\theta_0 - \theta_1)]\}^2, \qquad n = 1, 2, \ldots \tag{13.6.5}$$

converge together in distribution if either does. But we may write

$$w_n^* = u_n^* + nv_n^* \tag{13.6.6}$$

where

$$u_n^* = [g_n'(x_1, \ldots, x_n; \tilde{\theta}^*)\sqrt{n}(\theta_1 - \tilde{\theta})]^2 \tag{13.6.7}$$

$$v_n^* = \frac{1}{n}[w_n^* - u_n^*].$$

Since the true value of θ is θ_1, the sequence u_n^*, $n = 1, 2, \ldots$, converges in distribution to the chi-square distribution $C(1)$, and v_n^*, $n = 1, 2, \ldots$, converges in probability to the constant v_0^* where

$$v_0^* = B^{*2}(\theta_1, \theta_1)(\theta_0 - \theta_1)^2. \tag{13.6.8}$$

Now $v_0^* > 0$ since $\theta_1 \neq \theta_0$ and $B^{*2}(\theta_1, \theta_1) > 0$. Therefore,

$$\lim_{n\to\infty} P(u_n^* + nv_n^* > \chi_\alpha^2) = 1 \tag{13.6.9}$$

which is equivalent to the statement that

$$\lim_{n\to\infty} P(W_\alpha^* \mid \theta = \theta_1) = 1. \tag{13.6.10}$$

But (13.6.3) and (13.6.10) together imply that the test provided by the inequality (13.6.2) is consistent for testing $\mathscr{H}$.

The problem of comparing the power functions of the likelihood ratio test W_α defined by the inequality (13.5.1) and the arbitrary test W_α^* defined by the inequality (13.6.1) is complicated by the fact that both tests are consistent and hence have the same limiting power function $\zeta(\theta)$ defined by (13.5.11) as $n \to \infty$. To overcome this difficulty, we shall introduce and utilize some new notions to be defined below.

If W_α is a test for $\mathscr{H}$ based on a sample $(x_1, \ldots, x_n)$ whose limiting power function $\eta(\theta)$, as $n \to \infty$, is defined as follows for θ in Ω_0

$$\eta(\theta_0) = \alpha \tag{13.6.11}$$

$$\alpha < \eta(\theta) < 1, \qquad \theta \neq \theta_0.$$

we shall say that W_α is an *asymptotically unbiased test* for $\mathscr{H}(\theta_0; \Omega)$ at significance level α. Note that asymptotic unbiasedness of a test is a weaker condition on the power of the test than consistency of the test.

Suppose $W_{1\alpha}$ and $W_{2\alpha}$ are two asymptotically unbiased tests of size α for $\mathscr{H}$ whose asymptotic power functions are $\eta_1(\theta)$ and $\eta_2(\theta)$ respectively for θ in Ω_0.

If

$$\eta_1(\theta) \geqslant \eta_2(\theta) \tag{13.6.12}$$

with $\eta_1(\theta_0) = \eta_2(\theta_0) = \alpha$ we shall say that $W_{1\alpha}$ is *asymptotically more powerful**

* Strictly speaking, we should use the phrase *asymptotically at least as powerful* unless there is at least one value of θ for which only $>$ holds in (13.6.12). But use of the shorter phraseology should cause no ambiguity.

than $W_{2\alpha}$ for testing $\mathscr{H}$. If the equality holds in (13.6.12) everywhere in Ω_0, we shall say that $W_{1\alpha}$ and $W_{2\alpha}$ are *equivalent asymptotically unbiased tests* of size α for $\mathscr{H}$.

Now let us return to the problem of comparing the power of tests W_α and W_α^* defined by (13.5.1) and (13.6.2), respectively. We shall consider tests $W_{c\alpha}$ and $W_{c\alpha}^*$ defined by the inequalities

$$u_n + cv_n > \chi_\alpha^2 \tag{13.6.13}$$
$$u_n^* + cv_n^* > \chi_\alpha^2$$

respectively, where $c > 0$ is arbitrary. Since u_n and u_n^* are both non-negative, it is clear that $W_{c\alpha} \subset W_\alpha$ and $W_{c\alpha}^* \subset W_\alpha^*$ for $n > c$.

We shall show that for *every* $c > 0$, $W_{c\alpha}$ is asymptotically more powerful than W_α^* for testing $\mathscr{H}$, in which case we shall adopt Wald's (1941*a*) terminology and say that W_α is *asymptotically more stringent* for testing $\mathscr{H}$ than W_α^*.

If $W_{c\alpha}$ and $W_{c\alpha}^*$ are asymptotically equivalent for *every* $c > 0$, we shall say that W_α and W_α^* are *asymptotically equally stringent* for testing $\mathscr{H}$.

An important case of asymptotically equally stringent tests may be stated as follows:

13.6.1 *Let W_α be the likelihood ratio test defined by the* (13.5.1) *and W_α^* that defined by replacing $g_n(x_1, \ldots, x_n; \theta_0)$ by $h_n(x_1, \ldots, x_n; \theta_0)$ in* (13.6.2), *where $h_n(x_1, \ldots, x_n; \theta_0)$ is defined by* (12.5.1). *Then if the* c.d.f. *$F(x; \theta)$ is regular with respect to its second θ-derivative in Ω_0, W_α and W_α^* are asymptotically equally stringent for testing $\mathscr{H}$.*

We have shown in Section 13.5 essentially that (u_n, v_n) in (13.6.13) is a pair of random variables whose distribution converges, as $n \to \infty$, to a degenerate distribution in the (u, v)-plane which is the chi-square distribution $C(1)$ along the half-line $v = v_0$, $u > 0$, where

$$v_0 = 2[H(\theta_1, \theta_1) - H(\theta_1, \theta_0)]. \tag{13.6.14}$$

Similarly, we have shown that (u_n^*, v_n^*) is a pair of random variables whose distribution converges to a degenerate distribution in the (u, v)-plane which is the chi-square distribution $C(1)$ along the half-line $v = v_0^*$ and $u > 0$, where

$$v_0^* = B^{*2}(\theta_1, \theta_1)(\theta_0 - \theta_1)^2. \tag{13.6.15}$$

But

$$H(\theta_1, \theta_0) = H(\theta_1, \theta_1) + H'(\theta_1, \theta_1)(\theta_0 - \theta_1) + \tfrac{1}{2}H''(\theta_1, \theta_1^*)(\theta_0 - \theta_1)^2 \tag{13.6.16}$$

where $|\theta_0 - \theta_1^*| < |\theta_0 - \theta_1|$. Since $H'(\theta_1, \theta_1) = 0$, we can write

$$2[H(\theta_1, \theta_1) - H(\theta_1, \theta_0)] = -H''(\theta_1, \theta_1^*)(\theta_0 - \theta_1)^2. \tag{13.6.17}$$

But

$$-H''(\theta_1, \theta_1) = B^2(\theta_1, \theta_1)$$

and since $F(x; \theta)$ is regular with respect to its second θ-derivative in Ω_0, we have the result at $\theta = \theta_1$, that

$$\frac{B^{*2}(\theta_1, \theta_1)}{B^2(\theta_1, \theta_1)} \leqslant 1 \tag{13.6.18}$$

which, of course, is similar to (12.5.21) at $\theta = \theta_0$. Therefore, since $H''(\theta_1, \theta)$ is continuous in θ over Ω_0, it follows that, for θ_0 and θ_1 sufficiently near each other, (13.6.18) holds with $B^2(\theta_1, \theta_1)$ replaced by $-H''(\theta_1, \theta_1^*)$, which, in turn, implies that

$$0 < v_0^* \leqslant v_0. \tag{13.6.19}$$

Now consider the power functions of $W_{c\alpha}$ and $W_{c\alpha}^*$ for a fixed value of c, that is, consider

$$P(u_n + cv_n > \chi_\alpha^2 \mid \theta_1) \quad \text{and} \quad P(u_n^* + cv_n^* > \chi_\alpha^2 \mid \theta_1) \tag{13.6.20}$$

as functions of θ_1. Denote the limits of these functions as $n \to \infty$, that is, the asymptotic power functions of $W_{c\alpha}$ and $W_{c\alpha}^*$, by

$$\eta_c(\theta_1) \quad \text{and} \quad \eta_c^*(\theta_1) \tag{13.6.21}$$

respectively.

Since u_n and u_n^* are non-negative, and since the distributions of (u_n, v_n) and (u_n^*, v_n^*) converge to degenerate distributions in the (u, v)-plane as $n \to \infty$ which are chi-square distributions $C(1)$ along the half-lines $v = v_0, u > 0$; and $v = v_0^*, u > 0$, respectively, it follows that

(13.6.22)

(*i*) $\eta_c(\theta_1) = \eta_c^*(\theta_1) = 1,$ for values of θ_1 such that $v_0^* \geqslant \dfrac{\chi_\alpha^2}{c}$

(*ii*) $1 > \eta_c(\theta_1) > \eta_c^*(\theta_1) > \alpha,$ for values of θ_1 such that $0 < v_0^* < \dfrac{\chi_\alpha^2}{c}$

(*iii*) $\eta_c(\theta_0) = \eta_c^*(\theta_0) = \alpha,$ for $\theta_1 = \theta_0$, that is, for θ_1 such that $v_0^* = 0$.

But it follows from (13.6.8) that the three sets of values of θ_1 indicated in (13.6.22(*i*), (*ii*), and (*iii*)) consists respectively, of (i) values of θ_1 in Ω_0 but outside some interval Ω_0' containing θ_0, (ii) values of $\theta_1 \neq \theta_0$ inside Ω_0', and (iii) the value $\theta_1 = \theta_0$.

Therefore, since the asymptotic power functions $\eta_c(\theta)$ and $\eta_c^*(\theta)$ satisfy (13.6.11) for every $c > 0$ and since the pair satisfies (13.6.12), it follows that $W_{c\alpha}$ is asymptotically more powerful than $W_{c\alpha}^*$ for testing $\mathscr{H}$, for every $c > 0$, and hence that W_α is *asymptotically more stringent than* W_α^* for testing $\mathscr{H}$.

If $v_0^* = v_0$, then (13.6.22) holds with (*ii*) replaced by

$$1 > \eta_c(\theta_1) = \eta_c^*(\theta_1) > \alpha$$

for values of θ_1, such that $0 < v_0^* < \chi_\alpha^2/c$, which means that $W_{c\alpha}$ and $W_{c\alpha}^*$ are equivalent asymptotically unbiased tests for *every* $c > 0$, that is, W_α and W_α^* are asymptotically equally stringent for testing $\mathscr{H}$. But $v_0^* = v_0$ if and only if equality holds in (13.6.18) which, in turn, holds if and only if the regular estimating function $g_n(x_1, \ldots, x_n; \theta)$ used in defining W_α^* is replaced by $h_n(x_1, \ldots, x_n; \theta)$. But we know from **13.6.1** that using $h_n(x_1, \ldots, x_n; \theta)$ in (13.6.2) yields a test W_α^* which is asymptotically equally as stringent as W_α, the test defined by (13.5.1).

We may summarize our results in the form of the following important theorem:

13.6.2 *Suppose* $(x_1, \ldots, x_n)$ *is a sample from the* c.d.f. $F(x; \theta)$ *where* $F(x; \theta)$ *is regular with respect to its second* θ*-derivative in* Ω_0. *Let* W_α *be the likelihood ratio test of size* α *defined in* (13.5.1). *Suppose* $g_n(x_1, \ldots, x_n; \theta)$ *is any regular estimating function for* θ, *as defined in Section* 12.5, *and let* W_α^* *be the test of size* α *based on* $g_n(x_1, \ldots, x_n; \theta)$ *as defined by* (13.6.2). *Then* W_α *is asymptotically more stringent than* W_α^* *for testing the hypothesis* $\mathscr{H}(\theta_0; \Omega)$. *However, if we replace* $g_n(x_1, \ldots, x_n; \theta)$ *in* (13.6.2) *by* $h_n(x_1, \ldots, x_n; \theta)$ *as defined in* (12.5.1), *the resulting test and test* W_α^* *are asymptotically equally stringent for testing* $\mathscr{H}(\theta_0; \Omega)$.

13.7 THE LIKELIHOOD RATIO TEST OF A SIMPLE HYPOTHESIS

In Sections 13.4, 13.5, and 13.6 we have presented the principal large-sample asymptotic properties of a likelihood ratio test for a simple hypothesis in which the parameter θ is one-dimensional—in fact, the important part of the parameter space Ω in the theory developed in those sections is an open interval Ω_0 in Ω containing θ_0. The basic argument by which these asymptotic properties were developed can be extended without essentially new difficulties to the case where θ is an r-dimensional parameter $(\theta_1, \ldots, \theta_r)$. It will therefore be sufficient to give the r-dimensional extensions of the results of Sections 13.4 through 13.6 with a minimum of detail. We shall regard the random variable x having c.d.f. $F(x, \theta)$ as one-dimensional, although our results will hold for k-dimensional random variables with only minor changes in notation.

The hypothesis to be considered here is the simple hypothesis

$$\mathscr{H}(\theta_0; \Omega_r)$$

which will be denoted by $\mathscr{H}_r$, where θ_0 is the point $(\theta_{10}, \ldots, \theta_{r0})$ and Ω_r is any set in the Euclidean space R_r which contains an r-dimensional open interval Ω_{r0} which, in turn, contains θ_0. In dealing with asymptotic theory of likelihood ratio tests for $\mathscr{H}_r$, the part of Ω_r which will be of major interest is Ω_{r0}.

First, it will be convenient to state the extension of **13.4.1** for the case of an r-dimensional parameter θ as:

13.7.1 *If θ is r-dimensional and if $F(x; \theta)$ is regular with respect to all of its second θ-derivatives in Ω_{r0}, then $H(\theta_0, \theta)$ as defined in* (12.1.17) *has a maximum of $H(\theta_0, \theta_0)$ at $\theta = \theta_0$. Furthermore, the negative of the matrix of second derivatives of $H(\theta_0, \theta)$ at $\theta = \theta_0$ is $\|B_{pq}(\theta_0, \theta_0)\|$, $p, q, = 1, \ldots, r$, as defined in* (12.1.19).

The proof of **13.7.1** is a straightforward extension of that of **13.4.1** to the case of an r-dimensional parameter and will be omitted. It should be noted in the r-dimensional case that $H(\theta_0, \theta_0) - H(\theta_0, \theta_1)$ is Kullback's information integral for discriminating $F(x; \theta_0)$ against $F(x; \theta_1)$ per observation from $F(x; \theta_0)$, whereas $\|B_{pq}(\theta_0, \theta_0)\|$, Fisher's *matrix of information pertaining to θ_0 per observation* from $F(x; \theta_0)$, is the matrix of second derivatives of Kullback's information integral at $\theta_1 = \theta_0$ [See Kullback and Leibler (1951)].

The r-dimensional version of **13.4.2** requires no change of wording, understanding, of course, that θ, θ_0 and $\hat{\theta}$ are r-dimensional. The r-dimensional version of **13.4.3** may be stated as follows:

13.7.2 *If θ is r-dimensional and if $F(x; \theta)$ is regular with respect to all of its second θ-derivatives, and if $\mathscr{H}_r$ is true, then*

$$\lim_{n\to\infty} P(-2 \log \lambda_{\mathscr{H}_r} < \chi^2) = \frac{1}{2^{\frac{1}{2}r}\Gamma(\frac{1}{2}r)} \int_0^{\chi^2} u^{\frac{1}{2}r-1} e^{-\frac{1}{2}u}\, du \tag{13.7.1}$$

that is, $-2 \log \lambda_{\mathscr{H}_r}$ converges in distribution to the chi-square distribution $C(r)$.

The proof is similar to that of **13.4.2** and is left to the reader.

Theorem **13.5.1** holds if θ is r-dimensional. The argument is a straightforward extension of that for **13.5.1** and is omitted.

In the extension of **13.6.1** to the case where θ is r-dimensional, W_α and W_α^* are tests defined in the sample space R_n by the inequalities

(13.7.2) $$-2 \log \lambda_{\mathscr{H}} > \chi_\alpha^2$$

and

(13.7.3) $$U_n(x_1, \ldots, x_n; \theta_0) > \chi_\alpha^2$$

where $U_n(x_1, \ldots, x_n; \theta_0)$ is defined by (12.9.1) and where χ_α^2 is the 100α % point of the chi-square distribution $C(r)$. The proof is a rather direct extension of that for **13.6.1** and is left to the reader.

Theorem **13.6.2** extends without any particular difficulty to the case of an r-dimensional parameter and is left as an exercise for the reader. The r-dimensional version of W_α^* referred to in **13.6.2** is the region in the sample space in which $V_n > \chi_\alpha^2$ where V_n is defined in (12.9.15). Instead of the one-dimensional estimating function $h_n(x_1, \ldots, x_n; \theta)$ we would, of course, use the vector function $h_{pn}(x_1, \ldots, x_n; \theta)$, $p = 1, \ldots, r$, defined in (12.9.1), that is,

(13.7.4) $$h_{pn}(x_1, \ldots, x_n; \theta) = \frac{1}{n} S_{pn}(x_1, \ldots, x_n; \theta).$$

13.8 THE LIKELIHOOD RATIO TEST OF A COMPOSITE HYPOTHESIS

In problems of hypothesis testing where several parameters are involved, we are often interested in the case where the parameter subspace $\omega_{r'}$ is an r'-dimensional Euclidean section of Ω_r, that is, a cylinder set consisting of points of form $(\theta_1, \ldots, \theta_{r'}, \theta_{r'+10}, \ldots, \theta_{r0})$, $\theta_{r'+10}, \ldots, \theta_{r0}$ having fixed values. Let us denote this composite hypothesis by $\mathscr{H}(\omega_{r'}; \Omega_r)$ or more briefly by $\mathscr{H}_{r-r'}$. The likelihood ratio test $\lambda_{\mathscr{H}_{r-r'}}$ for $\mathscr{H}_{r-r'}$ is defined in the usual way by (13.3.5). In large samples, the part of Ω_r of main interest is that lying within some open r-dimensional interval Ω_{r0} containing the true parameter point $(\theta_{10}, \ldots, \theta_{r0})$.

The basic theorem [Wilks (1938*a*)] about the likelihood ratio test can be stated as follows:

13.8.1 *Suppose $(x_1, \ldots, x_n)$ is a sample from the* c.d.f. $F(x; \theta)$, *where θ is r-dimensional, and $F(x; \theta)$ is regular in all of its second θ-derivatives, for $\theta \in \Omega_{r0}$. Then*

(13.8.1)
$$\lim_{n\to\infty} P(-2 \log \lambda_{\mathscr{H}_{r-r'}} < \chi^2 \mid \theta \in \omega_{r'}) = \frac{1}{2\Gamma\left(\frac{r-r'}{2}\right)} \int_0^{\chi^2} \left(\frac{u}{2}\right)^{[(r-r')/2]-1} e^{-\frac{1}{2}u}\, du$$

that is, if $\mathscr{H}_{r-r'}$ is true, $-2 \log \lambda_{\mathscr{H}_{r-r'}}$, $n = 1, 2, \ldots$ converges in probability to a random variable having the chi-square distribution $C(r - r')$.

To establish this theorem, let $\hat{\theta}_1, \ldots, \hat{\theta}_r$ be defined as in **12.7.2.** Let $\hat{\theta}'_1, \ldots, \hat{\theta}'_{r'}$, be the solutions of

$$S_{p'n}(x_1, \ldots, x_n; \theta) = 0, \qquad p' = 1, \ldots, r', \tag{13.8.2}$$

with respect to $\theta_1, \ldots, \theta_{r'}$, that is, $\hat{\theta}'_1, \ldots, \hat{\theta}'_{r'}$ are maximum likelihood estimators for $\theta_1, \ldots, \theta_{r'}$, the remaining components being fixed at $\theta_{r'+1\,0}, \ldots, \theta_{r0}$, respectively. Under the assumptions of **12.7.3** it follows that $(\hat{\theta}'_1, \ldots, \hat{\theta}'_{r'})$ is asymptotically distributed, for large n, according to

$$N(\{\theta_{p'0}\}; \|nB_{p'q'}\|^{-1}), \qquad p', q' = 1, \ldots, r'. \tag{13.8.3}$$

Now referring to (12.6.1) let us adopt the following notation, where $\theta_0 \in \omega_{r'}$,

$$\begin{aligned} \frac{1}{\sqrt{n}} S_{pn}(x_1, \ldots, x_n; \theta_0) &= \zeta_{np} \\ \sqrt{n}(\theta_{p0} - \hat{\theta}_p) &= \eta_{np} \\ \sqrt{n}(\theta_{p'0} - \hat{\theta}_{p'}) &= \eta'_{np'} \\ \frac{1}{n} S_{pqn}(x_1, \ldots, x_n, \theta) &= A^{(n)}_{pq}(\theta) \end{aligned} \tag{13.8.4}$$

$$p, q = 1, \ldots, r, \qquad p', q' = 1, \ldots, r'.$$

Because of the regularity of $F(x; \theta)$ in all of its second θ-derivatives it follows that for an arbitrary $\varepsilon > 0$, there is an n_ε so that the probability exceeds $1 - \varepsilon$ that all four of the following equations hold for all $n > n_\varepsilon$:

$$\zeta_{np} = \sum_{q=1}^{r} A^{(n)}_{pq}(\theta^*)\eta_{nq}, \qquad p = 1, \ldots, r \tag{13.8.5}$$

$$\zeta_{np'} = \sum_{q'=1}^{r'} A'^{(n)}_{p'q'}(\theta^{*\prime})\eta'_{nq'}, \qquad p' = 1, \ldots, r' \tag{13.8.6}$$

$$\sum_{\xi=1}^{n} \log dF(x_\xi, \theta_0) = \sum_{\xi=1}^{n} \log dF(x_\xi, \hat{\theta}_1, \ldots, \hat{\theta}_r) + \tfrac{1}{2} \sum_{p,q=1}^{r} A^{(n)}_{pq}(\theta^*_1)\eta_{np}\eta_{nq} \tag{13.8.7}$$

$$\sum_{\xi=1}^{n} \log dF(x_\xi, \theta_0) = \sum_{\xi=1}^{n} \log dF(x_\xi, \hat{\theta}'_1, \ldots, \hat{\theta}'_{r'}, \theta_{r'+10}, \ldots, \theta_{r0}) + \tfrac{1}{2} \sum_{p',q'=1}^{r'} A^{(n)}_{p'q'}(\theta^{*\prime}_1)\eta'_{np'}\eta'_{nq'} \tag{13.8.8}$$

where θ^* and θ_1^* are points in Ω_{r0} between θ_0 and $\hat{\theta}$, while $\theta^{*\prime}$ and $\theta_1^{*\prime}$ are points in $\omega_{r'} \cap \Omega_{r0}$ between θ_0 and $(\hat{\theta}_1', \ldots, \hat{\theta}_{r'}', \theta_{r'+10}, \ldots, \theta_{r0})$.

The η_{nq} and $\eta_{nq'}'$ can be expressed as

$$
\begin{aligned}
\eta_{nq} &= \sum_{p=1}^{r} A_{(n)}^{pq}(\theta^*)\zeta_{np} \\
\eta_{nq'}' &= \sum_{p'=1}^{r'} A_{(n)}'^{p'q'}(\theta^{*\prime})\zeta_{np'}
\end{aligned}
\tag{13.8.9}
$$

where

$$\|A_{(n)}^{pq}(\theta^*)\| = \|A_{pq}^{(n)}(\theta^*)\|^{-1}, \text{ and } \|A_{(n)}'^{p'q'}(\theta^{*\prime})\| = \|A_{p'q'}'^{(n)}(\theta^{*\prime})\|^{-1}.$$

Since the left-hand sides of (13.8.7) and (13.8.8) are the same, and since

$$
\begin{aligned}
-2\log\lambda_{\mathscr{H}_{r-r'}} = -2\Bigg\{&\sum_{\xi=1}^{n}\log dF(x_\xi, \hat{\theta}_1', \ldots, \hat{\theta}_{r'}', \theta_{r'+10}, \ldots, \theta_{r0}) \\
&- \sum_{\xi=1}^{n}\log dF(x_\xi, \hat{\theta}_1, \ldots, \hat{\theta}_r)\Bigg\}
\end{aligned}
\tag{13.8.10}
$$

it follows that for all $n > n_\varepsilon$ the probability exceeds $1 - \varepsilon$ that the following equality holds

$$-2\log\lambda_{\mathscr{H}_{r-r'}} = \sum_{p',q'=1}^{r'} A_{p'q'}^{(n)}(\theta_1^{*\prime})\eta_{np'}'\eta_{nq'}' - \sum_{p,q=1}^{r} A_{pq}^{(n)}(\theta_1^*)\eta_{np}\eta_{nq} \tag{13.8.11}$$

where the η_{np} and $\eta_{np'}'$ are to be expressed in terms of the ζ_{np} by equations (13.8.9).

If $\theta_0 \in \omega_{r'}$ is the true value of θ, we see from **12.7.1** that $(\zeta_{n1}, \ldots, \zeta_{nr})$, $n = 1, 2, \ldots$ is a sequence of random variables converging in distribution to $N(\{0\}; \|B_{pq}\|)$. Furthermore, the matrices $\|A_{pq}^{(n)}(\theta^*)\|$ and $\|A_{pq}^{(n)}(\theta_1^*)\|$, both converge in probability to the matrix $\|-B_{pq}\|$, whereas the matrices $\|A_{p'q'}^{(n)}(\theta^{*\prime})\|$ and $\|A_{p'q'}^{(n)}(\theta_1^{*\prime})\|$ both converge in probability to $\|-B_{p'q'}\|$ as $n \to \infty$, where B_{pq} is defined in **12.7.1**.

Carrying out the algebra on the right-hand side of (13.8.11), we find that as $n \to \infty$ the two sides of (13.8.11) converge together in distribution to the distribution of a random variable Q where

$$Q = \sum_{p,q=1}^{r} B^{pq}y_py_q - \sum_{p',q'=1}^{r'} B_{[r']}^{p'q'}y_{p'}y_{q'} \tag{13.8.12}$$

where $\|B_{[r']}^{p'q'}\| = \|B_{p'q'}\|^{-1}$ and where $(y_1, \ldots, y_r)$ has the distribution $N(\{0\}, \|B_{pq}\|)$. It can be shown, that for $\theta_0 \in \omega_{r'}$ the distribution of Q is the chi-square distribution $C(r - r')$. Thus, since the random variable $-2\log\lambda_{\mathscr{H}_{r-r'}}$ converges in distribution to the same distribution as Q, we therefore conclude the argument for **13.8.1**.

Finally, we remark without proof that under the assumptions of **13.8.1**, the likelihood ratio test for $\mathscr{H}_{r-r'}$ is consistent, that is,

$$\lim_{n\to\infty} P(-2\log\lambda_{\mathscr{H}_{r-r'}} > \chi^2_\alpha \mid \theta \in \Omega_r - \omega_{r'}) = 1.$$

By using the class of regular estimating functions $ng_{pn}(x_1, \ldots, x_n; \theta)$ mentioned in Section 12.9 rather than the score functions $S_{pn}(x_1, \ldots, x_n; \theta)$ in (13.8.5) through (13.8.8), we can construct a class of tests for $\mathscr{H}_{r-r'}$ which are consistent. However, none of the tests in this class is asymptotically more stringent than the likelihood ratio test when the notion of asymptotic stringency is extended in a fairly straightforward manner to composite hypotheses.

PROBLEMS

13.1 A sample $(x_1, \ldots, x_n)$ is assumed to come from a distribution having p.d.f. of form $\theta e^{-\theta x}$, $x > 0$, and $\Omega = (0, +\infty)$. Determine, by using the Neyman-Pearson theorem, the critical region W_α in the sample space of $(x_1, \ldots, x_n)$ for testing the hypothesis $\mathscr{H}(\theta_0; \theta_0 \cup \theta_1)$ where $\theta_1 > \theta_0$ and which satisfies

$$P(W_\alpha \mid \theta_0) = \alpha.$$

Determine the power function of this test, that is, $P(W_\alpha \mid \theta)$ as a function of θ.

13.2 Suppose $n_1, \bar{x}_1, s_1^2$ are the size, mean, and sample variance of a sample from the distribution $N(\mu_1, \sigma^2)$ and $n_2, \bar{x}_2, s_2^2$ are similar quantities for an independent sample from the distribution $N(\mu_2, \sigma^2)$. Consider the composite statistical hypothesis $\mathscr{H}(\omega; \Omega)$ defined as follows:

Ω is the Euclidean half R_3 space of (μ_1, μ_2, σ^2), $\sigma^2 > 0$

ω is the subset of Ω for which $\mu_1 = \mu_2$.

Show that the likelihood ratio for testing $\mathscr{H}$ is equivalent to the Student ratio

$$t = \frac{(\bar{x}_1 - \bar{x}_2)}{s\sqrt{1/n_1 + 1/n_2}}$$

where

$$s^2 = \frac{(n_1 - 1)s_1^2 + (n_2 - 1)s_2^2}{n_1 + n_2 - 2}$$

and t has the "Student" distribution $S(n_1 + n_2 - 2)$ when $\mathscr{H}$ is true. Show that this test is unbiased.

13.3 If the preceding problem is generalized to the case of k samples where $n_i, \bar{x}_i, s_i^2$, $i = 1, \ldots, k$, are the size, mean, and sample variance of k independent samples from $N(\mu_i, \sigma^2)$, $i = 1, \ldots, k$ respectively, and $\mathscr{H}(\omega; \Omega)$ is the statistical hypothesis for which

Ω is the Euclidean half R_{k+1} space of $(\mu_1, \ldots, \mu_k, \sigma^2)$, $\sigma^2 > 0$,

ω is the subset of Ω for which $\mu_1 = \cdots = \mu_k$

show that the likelihood ratio for $\mathscr{H}(\omega; \Omega)$ is equivalent to the Snedecor ratio

$$\mathscr{F} = \frac{(n-k)\sum_{i=1}^{k} n_i(\bar{x}_i - \bar{x})^2}{k\sum_{i=1}^{k}(n_i - 1)s_i^2},$$

where $n = n_1 + \cdots + n_k$ and $\bar{x} = \frac{1}{n}\sum_{i=1}^{k} n_i\bar{x}_i$ and $\mathscr{F}$ has the Snedecor distribution $S(k, n-k)$ if $\mathscr{H}(\omega; \Omega)$ is true. Show that this test is unbiased.

13.4 If s_1^2 and s_2^2 are variances in samples of sizes n_1 and n_2 from $N(\mu_1, \sigma_1^2)$ and $N(\mu_2, \sigma_2^2)$ respectively, and if $\mathscr{H}(\omega; \Omega)$ is the statistical hypothesis where Ω is the Euclidean quarter-R_4 space of $(\mu_1, \mu_2, \sigma^2, \sigma_2^2)$, $\sigma_1^2 > 0$, $\sigma_2^2 > 0$, and ω is the subset of Ω for which $\sigma_1^2 = \sigma_2^2$ show that the likelihood ratio for $\mathscr{H}(\omega; \Omega)$ is equivalent to the Snedecor ratio

$$\mathscr{F} = \frac{s_1^2}{s_2^2}$$

which has the Snedecor distribution $S(n_1 - 1, n_2 - 1)$ if $\mathscr{H}(\omega; \Omega)$ is true.

13.5 Suppose x_1 and x_2 are independent random variables having binomial distributions $Bi(n_1, p_1)$ and $Bi(n_2, p_2)$. Let $\mathscr{H}(\omega; \Omega)$ be the statistical hypothesis for which Ω is the space of all possible points (p_1, p_2) inside the unit square having vertices (0, 0), (0, 1), (1, 0), (1, 1), and ω is the subset of Ω for which $p_1 = p_2$. Show that the likelihood ratio for testing $\mathscr{H}(\omega; \Omega)$ is given by

$$\lambda = \left(\frac{x}{n}\right)^{x}\left(1 - \frac{x}{n}\right)^{n-x}\left(\frac{x_1}{n_1}\right)^{-x_1}\left(\frac{x_2}{n_2}\right)^{-x_2}\left(1 - \frac{x_1}{n_1}\right)^{-n_1+x_1}\left(1 - \frac{x_2}{n_2}\right)^{-n_2+x_2}$$

where $x = x_1 + x_2$ and $n = n_1 + n_2$ and that the limiting distribution of $-2\log\lambda$ as $n_1, n_2 \to \infty$ is the chi-square distribution $C(1)$ if $\mathscr{H}(\omega; \Omega)$ is true.

13.6 Generalize the preceding problem and its solution to the case where $x_1, \ldots, x_k$ are independent random variables from binomial distributions $Bi(n_1, p_1), \ldots, Bi(n_k, p_k)$.

13.7 *Test for independence in an* $r \times s$ *contingency table.* Suppose $(n_{ij}, i = 1, \ldots, r, j = 1, \ldots, s)$ is an $(rs - 1)$-dimensional random variable from the multinomial distribution

$$\frac{n!}{\prod_{j=1}^{s}\prod_{i=1}^{r} n_{ij}!}\prod_{j=1}^{s}\prod_{i=1}^{r} p_{ij}^{n_{ij}}$$

where $\sum_j\sum_i n_{ij} = n$ and where each $p_{ij} > 0$ with $\sum_j\sum_i p_{ij} = 1$. Let $\mathscr{H}(\omega; \Omega)$ be the statistical hypothesis for which Ω is the space of all possible values of the p_{ij} and ω is the subset of Ω for which $p_{ij} = p_i q_j$ where $\sum_i p_i = \sum_j q_j = 1$. Show that the likelihood ratio for $\mathscr{H}(\omega; \Omega)$ is given by

$$\lambda = \frac{\prod_{i=1}^{r}\left(\frac{n_{i.}}{n}\right)^{n_{i.}}\prod_{j=1}^{s}\left(\frac{n_{.j}}{n}\right)^{n_{.j}}}{\prod_{j=1}^{s}\prod_{i=1}^{r}\left(\frac{n_{ij}}{n}\right)^{n_{ij}}}$$

where

$$n_{i\cdot} = \sum_j n_{ij}, \qquad n_{\cdot j} = \sum_i n_{ij}$$

and that the limiting distribution of $-2 \log \lambda$ is chi-square $C((r-1)(s-1))$ if $\mathscr{H}(\omega;\Omega)$ is true.

13.8 (*Continuation*) Let g be defined as follows

$$g = \sum_{j=1}^{s} \sum_{i=1}^{s} \left(n_{ij} - \frac{n_{i}.n_{.j}}{n}\right)^2 \Big/ \left(\frac{n_{i}.n_{.j}}{n}\right)$$

where

$$n_{i\cdot} = \sum_j n_{ij}, \qquad n_{\cdot j} = \sum_i n_{ij}.$$

Show that if $\mathscr{H}$ is true g and $-2 \log \lambda$ converge together in distribution to the chi-square distribution $C((r-1)(s-1))$ as $n \to \infty$.

13.9 *Independence of layers in an $r \times s \times t$ contingency table.* Let $(n_{ijk}; i = 1, \ldots, r; j = 1, \ldots, s; k = 1, \ldots, t)$ be an $(rst-1)$-dimensional random variable having the multinomial distribution

$$\frac{n!}{\prod_{k=1}^{t} \prod_{j=1}^{s} \prod_{i=1}^{r} n_{ijk}!} \prod_k \prod_j \prod_i p_{ijk}^{n_{ijk}}$$

where $\sum_k \sum_j \sum_i n_{ijk} = n$ and $p_{ijk} > 0$ with

$$\sum_k \sum_j \sum_i p_{ijk} = 1.$$

Let $\mathscr{H}(\omega;\Omega)$ be the hypothesis where Ω is the space of all possible p_{ijk} and $\omega \in \Omega$ is the set for which $p_{ijk} = p_{ij}q_k$, $\sum_j \sum_i p_{ij} = \sum_k q_k = 1$. Show that the likelihood ratio λ for testing $\mathscr{H}$ is given by

$$\lambda = \frac{\prod_j \prod_i \left(\frac{n_{ij\cdot}}{n}\right)^{n_{ij\cdot}} \prod_k \left(\frac{n_{\cdot\cdot k}}{n}\right)^{n_{\cdot\cdot k}}}{\prod_k \prod_j \prod_i \left(\frac{n_{ijk}}{n}\right)^{n_{ijk}}}$$

where $n_{ij\cdot} = \sum_k n_{ijk}$ and $n_{\cdot\cdot k} = \sum_j \sum_i n_{ijk}$,

and that the limiting distribution of $-2 \log \lambda$ as $n \to \infty$ is the chi-square distribution $C((rs-1)(t-1))$ if $\mathscr{H}(\omega;\Omega)$ is true.

13.10 Suppose $(x_{11}, \ldots x_{1n_1})$ and $(x_{21}, \ldots, x_{2n_2})$ are independent samples from Poisson distributions $Po(\mu_1)$ and $Po(\mu_2)$ respectively. Let $\mathscr{H}(\omega;\Omega)$ be the statistical test in which Ω is the space of all possible points (μ_1, μ_2) in the first quadrant of the $\mu_1\mu_2$-plane, whereas ω is the line $\mu_1 = \mu_2$ in Ω. Show that the likelihood ratio λ for $\mathscr{H}(\omega;\Omega)$ is given by

$$\lambda = \frac{\bar{x}^{n\bar{x}}}{\bar{x}_1^{n_1\bar{x}_1}\, \bar{x}_2^{n_2\bar{x}_2}}$$

where $\bar{x}_1$ and $\bar{x}_2$ are the means of the two samples respectively,

$$\bar{x} = \frac{n_1\bar{x}_1 + n_2\bar{x}_2}{n_1 + n_2}$$

and $n = n_1 + n_2$. Show that the limiting distribution of $-2 \log \lambda$ as $n_1, n_2 \to \infty$ is the chi-square distribution $C(1)$ if $\mathscr{H}(\omega; \Omega)$ is true. Generalize to k samples.

13.11 The $(k - 1)$-dimensional random variable $(n_1, \ldots, n_k)$ has the multinomial distribution

$$\frac{n!}{n_1! \cdots n_k!} p_1^{n_1} \cdots p_k^{n_k}$$

where $n_1 + \cdots + n_k = n$ and $p_i > 0$ with $p_1 + \cdots + p_k = 1$. Show that the likelihood ratio λ for the hypothesis $\mathscr{H}(\omega; \Omega)$ where Ω is the space of all possible p_i and ω is the subset for which $p_1 = p_{10}, \ldots, p_k = p_{k0}$ is given by

$$\lambda = \left[\left(\frac{p_{10}}{\hat{p}_1}\right)^{\hat{p}_1} \cdots \left(\frac{p_{k0}}{\hat{p}_k}\right)^{\hat{p}_k}\right]^n$$

where $\hat{p}_1 = n_1/n, \ldots, \hat{p}_k = n_k/n$ and show that the limiting distribution of $-2 \log \lambda$ as $n \to \infty$ is the chi-square distribution $C(k - 1)$ if $\mathscr{H}(\omega; \Omega)$ is true.

13.12 (*Continuation*) Show that as $n \to \infty$, $-2 \log \lambda$ and $\sum_{i=1}^{k} \frac{(n_i - np_{i0})^2}{np_{i0}}$ converge together to the chi-square distribution $C(k - 1)$ when $\mathscr{H}(\omega; \Omega)$ is true.

13.13 If $s_1^2, \ldots, s_k^2$ are sample variances of independent samples of sizes $n_1, \ldots, n_k$ respectively, from the distributions $N(\mu_1, \sigma_1^2), \ldots, N(\mu_k, \sigma_k^2)$ respectively, and if $\mathscr{H}(\omega; \Omega)$ is the statistical hypothesis in which Ω is the 2^{-k} part of Euclidean space R_{2k} of all possible values of $(\mu_1, \ldots, \mu_k, \sigma_1^2, \ldots, \sigma_k^2)$, $\sigma_1^2 > 0, \ldots, \sigma_k^2 > 0$, with ω being the subset of Ω for which $\sigma_1^2 = \cdots = \sigma_k^2$, show that the likelihood ratio λ for $\mathscr{H}(\omega; \Omega)$ is given by

$$\lambda = \left[\frac{(n_1 - 1)s_1^2}{n_1 s_0^2}\right]^{\frac{1}{2}n_1} \cdots \left[\frac{(n_k - 1)s_k^2}{n_k s_0^2}\right]^{\frac{1}{2}n_k}$$

where

$$s_0^2 = \frac{(n_1 - 1)s_1^2 + \cdots + (n_k - 1)s_k^2}{n_1 + \cdots + n_k}.$$

13.14 If s^2 is the variance of a sample of size n from $N(\mu, \sigma^2)$, show that using the likelihood ratio for testing the hypothesis $\mathscr{H}(\omega; \Omega)$ where Ω is the space of points (μ, σ^2), $\sigma^2 \leqslant \sigma_0^2$, and ω is the subset in Ω for which $\sigma^2 = \sigma_0^2$, is equivalent to using the ratio

$$\frac{(n - 1)s^2}{\sigma_0^2}$$

which has the chi-square distribution $C(n - 1)$ if $\mathscr{H}(\omega; \Omega)$ is true, and hence the critical region W_α of size α for rejecting $\mathscr{H}(\omega; \Omega)$ is the set of values of s^2 for which

$$\frac{(n - 1)s^2}{\sigma_0^2} > \chi_\alpha^2$$

where χ_α^2 is the $100\alpha\,\%$ point of the chi-square distribution $C(n - 1)$.

13.15 *Test for hypothesis of parallel regression lines.* Suppose $y_{11}, \ldots, y_{1n_1}$ are independent random variables having the distributions $N(\beta_{10} + \beta_{11}x_{1\xi_1}, \sigma^2)$, $\xi_1 = 1, \ldots, n_1$, and $y_{21}, \ldots, y_{2n_2}$ are independent random variables having the distributions $N(\beta_{20} + \beta_{21}x_{2\xi_2}, \sigma^2)$, $\xi_2 = 1, \ldots, n_2$. Let $\mathscr{H}(\omega; \Omega)$ be the hypothesis in which Ω is the Euclidean half-R_5 space $(\beta_{10}, \beta_{11}, \beta_{20}, \beta_{21}, \sigma^2)$, $\sigma^2 > 0$, and ω is the subset of Ω for which $\beta_{11} = \beta_{21}$. Show that the likelihood ratio for $\mathscr{H}$ is equivalent to

$$\mathscr{F} = \frac{(S_\omega - S_\Omega)}{S_\omega/(n_1 + n_2 - 4)}$$

where $\mathscr{F}$ has the Snedecor distribution $S(1, n_1 + n_2 - 4)$ if $\mathscr{H}$ is true, and where

$$S_\omega = \sum_{p=1}^{2} \sum_{\xi_p=1}^{n_p} [(y_{p\xi_p} - \bar{y}_p) - \hat{\beta}_p(x_{p\xi_p} - \bar{x}_p)]^2$$

$$\bar{y}_p = \frac{1}{n_p} \sum_{\xi_p=1}^{n_p} y_{p\xi_p}, \qquad \bar{x}_p = \frac{1}{n_p} \sum_{\xi_p=1}^{n_p} x_{p\xi_p}, \qquad p = 1, 2,$$

$$\hat{\beta}_p = \frac{a_p}{b_p}, \qquad a_p = \sum_{\xi_p=1}^{n_p} (y_{p\xi_p} - \bar{y}_p)(x_{p\xi_p} - \bar{x}_p), \qquad b_p = \sum_{\xi_p=1}^{n_p} (x_{p\xi_p} - \bar{x}_p)^2,$$

and where

$$S_\Omega = \sum_{p=1}^{2} \sum_{\xi_p=1}^{n_p} [(y_{p\xi_p} - \bar{y}_p) - \hat{\beta}(x_{p\xi_p} - \bar{x}_p)]^2$$

$$\hat{\beta} = \frac{a_1 + a_2}{b_1 + b_2}.$$

13.16 If $x_{\xi\eta}$; $\xi = 1, \ldots, r$; $\eta = 1, \ldots, s$ are independent random variables having the distributions $N(\mu + \mu_{\xi\cdot} + \mu_{\cdot\eta}, \sigma^2)$ where $\sum_{\xi=1}^{r} \mu_{\cdot\xi} = 0$, $\sum_{\eta=1}^{s} \mu_{\cdot\eta} = 0$. Let $\mathscr{H}(\omega; \Omega)$ be the hypothesis with Ω defined as the set of all real vectors $(\mu, \mu_{1\cdot}, \ldots, \mu_{r\cdot}, \mu_{\cdot 1}, \ldots, \mu_{\cdot s}, \sigma^2)$ subject only to the conditions $\sum_{\xi} \mu_{\xi\cdot} = \sum_{\eta} \mu_{\cdot\eta} = 0$, and ω is the set in Ω for which $\mu_{1\cdot} = \cdots = \mu_{r\cdot} = 0$. Show that the likelihood ratio for testing $\mathscr{H}$ is equivalent to

$$\mathscr{F} = \frac{S_{\cdot 0}/(r - 1)}{S_{\cdot\cdot}/[(r - 1)(s - 1)]}$$

where $\mathscr{F}$ has the Snedecor distribution $S((r - 1), (r - 1)(s - 1))$ if $\mathscr{H}$ is true. And where

$$S_{\cdot 0} = s \sum_{\xi} (\bar{x}_{\xi\cdot} - \bar{x})^2$$

$$S_{\cdot\cdot} = \sum_{\eta} \sum_{\xi} (x_{\xi\eta} - \bar{x}_{\xi\cdot} - \bar{x}_{\cdot\eta} + \bar{x})^2$$

$$\bar{x} = \frac{1}{rs} \sum_{\eta} \sum_{\xi} x_{\xi\eta}, \qquad \bar{x}_{\xi\cdot} = \frac{1}{s} \sum_{\eta} x_{\xi\eta}, \qquad \bar{x}_{\cdot\eta} = \frac{1}{r} \sum_{\xi} x_{\xi\eta}.$$

13.17 *Test for equality of probability ratios in Luce's (1959) choice behavior model.* In this behavior model it is assumed that in choosing between two alternatives A_1 and A_2 the ratio of the probability of choosing A_2 to the probability of choosing A_1 remains constant for variations of a third choice alternative A_3. The problem here is to construct a test for this hypothesis on the basis of experimental results.

More precisely, suppose (n_{1i}, n_{2i}, n_{3i}), where $n_{1i} + n_{2i} + n_{3i} = n_i, i = 1, \ldots, k$, are k independent sets of random variables having the trinomial distributions

$$\frac{n_i!}{n_{1i}!\, n_{2i}!\, n_{3i}!} p_i^{n_{1i}} q_i^{n_{2i}} r_i^{n_{3i}}$$

$i = 1, \ldots, k$ where $p_i > 0, q_i > 0, r_i > 0, p_i + q_i + r_i = 1$. Let $\mathscr{H}(\omega; \Omega)$ be the hypothesis in which Ω consists of all possible p_i, q_i, r_i, that is, p_i, q_i, r_i are positive and satisfy $p_i + q_i + r_i = 1$, $i = 1, \ldots, k$, whereas ω is the subset in Ω for which $q_i/p_i = t$ and $p_i(1 + t) + r_i = 1$. Show that the likelihood ratio λ for testing $\mathscr{H}$ is given by

$$\lambda = \frac{\prod_{i=1}^{k} \{(\hat{p}_i)^{n_{1i}+n_{2i}} (\hat{t})^{n_{2i}} [1 - \hat{p}_i(1 + \hat{t})]^{n_{3i}}\}}{\prod_{i=1}^{k} [n_{1i}^{n_{1i}} n_{2i}^{n_{2i}} n_{3i}^{n_{3i}} n^{-n}]},$$

where

$$\hat{t} = \sum_i n_{2i} \Big/ \sum_i n_{1i}$$

$$\hat{p}_i = \frac{n_{1i} + n_{2i}}{n_i(1 + \hat{t})},$$

and that for large values of $n_1, \ldots, n_k$, $-2 \log \lambda$ has as its asymptotic distribution the chi-square distribution with $k - 1$ degrees of freedom if $\mathscr{H}$ is true. Also show that if $\mathscr{H}$ is true the variance of $\hat{t}$ in large samples is approximately

$$t(1 + t) \Big/ [\sum_i n_i p_i].$$

CHAPTER 14

Testing Nonparametric Statistical Hypotheses

In Chapter 13, we gave a brief account of the theory of tests of parametric statistical hypotheses with special reference to likelihood ratio tests in large samples. In that theory the class of admissible c.d.f.'s was of the form $\{F(x, \theta) : \theta \in \Omega\}$, that is, a class of c.d.f.'s of specified functional form, the members of the class corresponding to the values of a real parameter θ in some parameter space Ω, where, of course, either x or θ or both can be multidimensional.

In the theory of tests of nonparametric statistical hypotheses the class of admissible hypotheses is, in general, the class or a subclass of continuous c.d.f.'s, depending, of course, on the particular problem at hand. The general theory of nonparametric tests has not been as well developed as that for parametric tests. Therefore, in this chapter we discuss the theory of nonparametric statistical tests in terms of some of the more important problems in this field rather than attempt to set up a general theory of nonparametric tests. The reader interested in further literature on details of nonparametric tests is referred to books by Fraser (1957) and Kendall (1953), and also to survey articles by Kendall and Sundrum (1953), Moran, Whitfield, and Daniels (1950), Scheffé (1943), Wolfowitz (1949), and Wilks (1948, 1959*a*). A comprehensive bibliography has been published by Savage (1962).

14.1 THE QUANTILE TEST

The simplest type of nonparametric statistical test is one for testing the hypothesis that a sample $(x_1, \ldots, x_n)$ comes from a population whose c.d.f. $F(x)$ has a unique pth quantile $\underline{x}_p$, that is,

(14.1.1) $$F(\underline{x}_p) = p.$$

To specify the hypothesis precisely, let $\mathscr{C}_{0p}$ be the class of continuous c.d.f.'s having x_{0p} as their pth quantile and let the admissible class $\mathscr{C}_p$ be the set of all continuous c.d.f.'s. Then the hypothesis in which we are interested may be designated as $\mathscr{H}(\mathscr{C}_{0p}; \mathscr{C}_p)$, the hypothesis that the c.d.f. $F(x)$ of the population from which the sample is drawn belongs to the subclass $\mathscr{C}_{0p}$ of $\mathscr{C}_p$. Basically, $\mathscr{H}(\mathscr{C}_{0p}; \mathscr{C}_p)$ is a *nonparametric composite statistical hypothesis*, not to be confused with a parametric composite statistical hypothesis $\mathscr{H}(\omega; \Omega)$. In the latter case Ω is a set of elements representable as a set of points in a Euclidean space and ω is a subset of Ω, whereas in the case of $\mathscr{H}(\mathscr{C}_{0p}; \mathscr{C}_p)$ the set of admissible c.d.f.'s $\mathscr{C}_p$ cannot be placed into a bicontinuous one-to-one correspondence with the points of a set in a Euclidean space. Nor can this be done for $\mathscr{C}_{0p}$.

Now let $(x_1, \ldots, x_n)$ be a sample from a c.d.f. in $\mathscr{C}_{0p}$. Let r be the number of components of $(x_1, \ldots, x_n)$ whose values lie on the interval $(-\infty, x_{0p}]$; r has the binomial distribution $Bi(n, p)$. An intuitively reasonable test W_α for $\mathscr{H}(\mathscr{C}_{0p}; \mathscr{C}_p)$ is that for which r belongs to one of the two sets of integers

$$\{0, 1, \ldots, r_{\frac{1}{2}\alpha}\}, \{r'_{\frac{1}{2}\alpha}, \ldots, n\} \tag{14.1.2}$$

where $r_{\frac{1}{2}\alpha}$ is the largest integer for which

$$P(r \leqslant r_{\frac{1}{2}\alpha} \mid F \in \mathscr{C}_{0p}) \leqslant \tfrac{1}{2}\alpha \tag{14.1.3a}$$

and $r'_{\frac{1}{2}\alpha}$ is the smallest integer for which

$$P(r \geqslant r'_{\frac{1}{2}\alpha} \mid F \in \mathscr{C}_{0p}) \leqslant \tfrac{1}{2}\alpha. \tag{14.1.3b}$$

It follows from **9.2.1a** that, for large n, r is asymptotically distributed according to $N(np, np(1-p))$, from which it follows that

$$\lim_{n\to\infty} P(W_\alpha \mid F \in \mathscr{C}_{0p}) = \alpha. \tag{14.1.4}$$

Approximations for $r_{\frac{1}{2}\alpha}$ and $r'_{\frac{1}{2}\alpha}$ for large n are

$$\begin{aligned} r_{\frac{1}{2}\alpha} &= np - y_{\frac{1}{2}\alpha}\sqrt{npq} + 0(1) \\ r'_{\frac{1}{2}\alpha} &= np + y_{\frac{1}{2}\alpha}\sqrt{npq} + 0(1) \end{aligned} \tag{14.1.5}$$

where $q = 1 - p$, $y_{\frac{1}{2}\alpha} > 0$ and $\Phi(-y_{\frac{1}{2}\alpha}) = \frac{1}{2}\alpha$, where $\Phi(x)$ is the c.d.f of $N(0, 1)$.

Now let us examine the consistency of W_α. If $F(x)$ is any member of $\mathscr{C}_p - \mathscr{C}_{0p}$, then r will have the binomial distribution $Bi(n, p')$, where $p' \neq p$. Since, as $n \to \infty$, r/n converges in probability to the constant p if $F \in \mathscr{C}_{0p}$ (the interval $(r_{\frac{1}{2}\alpha}/n, r'_{\frac{1}{2}\alpha}/n)$ converges to the point p) and to p' $(\neq p)$ if $F \in \mathscr{C}_p - \mathscr{C}_{0p}$, it is evident that

$$\lim_{n\to\infty} P(W_\alpha \mid F \in \mathscr{C}_p - \mathscr{C}_{0p}) = 1. \tag{14.1.6}$$

Therefore,

14.1.1 *The test W_α for which r belongs to one of the two sets of integers in* (14.1.2) *is consistent for testing the hypothesis $\mathscr{H}(\mathscr{C}_{0p}; \mathscr{C}_p)$.*

It should be noted that the test W_α^L for which $r \in \{0, 1, \ldots, r_\alpha\}$ would be a *lower one-tail test* which would be consistent for testing any alternative in $\mathscr{C}_{0p}$ against any alternative in $\mathscr{C}_p$ having its pth quantile *greater* than $\underline{x}_{0p}$. Similarly W_α^U for which $r \in \{r'_\alpha, \ldots, n\}$ would be an *upper one-tail test* which would be consistent for testing any alternative in $\mathscr{C}_{0p}$ against any alternative in $\mathscr{C}_p$ having its pth quantile *less* than $\underline{x}_{0p}$.

Remark. When $p = 0.5$, the test W_α is sometimes referred to as the *sign test*, and the hypothesis $\mathscr{H}(\mathscr{C}_{0p}; \mathscr{C}_p)$ reduces to the hypothesis that $(x_1, \ldots, x_n)$ comes from a c.d.f. $F(x)$ having $\underline{x}_{0.5}$ as its median. A case of considerable practical importance arises when the sample components $(x_1, \ldots, x_n)$ are themselves differences of independent pairs of random variables $(u_1, v_1; u_2, v_2; \ldots; u_n, v_n)$, that is, $x_\xi = u_\xi - v_\xi$, $\xi = 1, \ldots, n$. The median $\underline{x}_{0.5}$ in this instance is usually 0. See Dixon and Mood (1946) for a fuller discussion of the sign test. Further tests concerning medians have been developed by Walsh (1949).

14.2 THE NONPARAMETRIC SIMPLE STATISTICAL HYPOTHESIS

(a) Preliminary Remarks

One of the basic nonparametric statistical hypotheses arises as follows. Suppose a sample $(x_1, \ldots, x_n)$ is from some continuous c.d.f. $F(x)$. Could the sample have "reasonably" come from the specified continuous c.d.f. $F_0(x)$? If we let $\mathscr{C}$ be the class of continuous c.d.f.'s, it will be convenient to denote this hypothesis by $\mathscr{H}(F_0; \mathscr{C})$ and refer to it as the *nonparametric simple statistical hypothesis*.

Ideally, we would like to devise a consistent test for $\mathscr{H}(F_0; \mathscr{C})$, that is, one which would discriminate between F_0 and any c.d.f. in $\mathscr{C}$ different from F_0, in the case of indefinitely large samples. This, however, is too much to ask for. The most we can do is to construct tests which are consistent for testing F_0 against alternatives contained in various subclasses of $\mathscr{C}$.

We shall consider three approaches to the problem of devising tests for $\mathscr{H}(F_0; \mathscr{C})$. The first consists of a nonparametric composite hypothesis based on Pearson's chi-square test; the second, which we shall call the *empty cell test*, is a simple nonparametric approach for c.d.f.'s in the class $\mathscr{A}$ of absolutely continuous c.d.f.'s; and the third is a nonparametric treatment for c.d.f.'s in the class of continuous c.d.f.'s based on the confidence contours discussed in Sections 11.6 and 11.7.

(b) A Nonparametric Composite Hypothesis Based on Pearson's Chi-Square Test

Let $\underline{x}_{i/m}$, be the i/mth quantile of the c.d.f. $F_0(x)$, that is, $F_0(\underline{x}_{i/m}) = i/m$, $i = 1, \ldots, m$, and let $\mathscr{C}_0$ be the subclass of c.d.f.'s in $\mathscr{C}$ having these same quantiles. Let I_i be the interval $(\underline{x}_{(i-1)/m}, \underline{x}_{i/m}]$, $i = 1, \ldots, m$, with $\underline{x}_0 = -\infty$, $\underline{x}_1 = +\infty$. Let r_i be the number of components of the sample falling into I_i. If the sample comes from any c.d.f. in $\mathscr{C}_0$, then the cell frequencies $(r_1, \ldots, r_m)$ is an m-dimensional random variable satisfying $r_1 + \cdots + r_m = n$, and having the $(m-1)$-dimensional multinomial distribution $M(n; 1/m, \ldots, 1/m)$ with p.f. given by

$$p(r_1, \ldots, r_m) = \frac{n!}{r_1! \cdots r_m!}\left(\frac{1}{m}\right)^n. \tag{14.2.1}$$

By considering m fixed for all n, we shall, at a certain sacrifice to be described later, replace the problem of testing $\mathscr{H}(F_0; \mathscr{C})$ by a fairly simple problem of testing a nonparametric composite hypothesis by using Pearson's chi-square test criterion. This test, as we shall see, is consistent for testing the nonparametric composite hypothesis $\mathscr{H}(\mathscr{C}_0; \mathscr{C})$, that is, for testing any $F(x) \in \mathscr{C}_0$ against any alternative $F(x) \in \mathscr{C} - \mathscr{C}_0$, and is an adequate substitute for more refined nonparametric tests for $H(F_0; \mathscr{C})$ for many practical statistical purposes.

For fixed m let

$$Q_n = \frac{m}{n}\sum_{i=1}^{m}\left(r_i - \frac{n}{m}\right)^2 \tag{14.2.2}$$

and let $W_{1\alpha}$ be the test for which $(r_1, \ldots, r_m)$ makes

$$Q_n > \chi^2_{\alpha,n} \tag{14.2.3}$$

where $\chi^2_{\alpha,n}$ is chosen for each n to make the size of $W_{1\alpha}$ as close as possible to α when the sample is from a c.d.f. in $\mathscr{C}_0$.

It follows from **9.3.2a** that

$$\lim_{n\to\infty} P(W_{1\alpha} \mid F \in \mathscr{C}_0) = \int_{\chi^2_\alpha}^{\infty} dF_{m-1}(\chi^2) = \alpha \tag{14.2.4}$$

where $dF_{m-1}(\chi^2)$ is the p.e. of the chi-square distribution with $m - 1$ degrees of freedom given by (7.8.1).

Now consider the power of $W_{1\alpha}$ for testing $\mathscr{H}(\mathscr{C}_0; \mathscr{C})$; that is, we wish to examine the value of $P(W_{1\alpha} \mid F \in \mathscr{C} - \mathscr{C}_0)$. Let p_i be the amount of probability in I_i as computed from some c.d.f. $F_1(x)$ in $\mathscr{C} - \mathscr{C}_0$. We have

$$p_i = \int_{I_i} dF_1(x), \qquad i = 1, \ldots, m \tag{14.2.5}$$

where, of course, the point $(p_1, \ldots, p_m)$ is different from $(1/m, \ldots, 1/m)$.

It follows from the multidimensional extension of **9.1.1** that if $(x_1, \ldots, x_n)$ is from the c.d.f. $F_1(x)$, the random variable $(r_1/n, \ldots, r_m/n)$ converges in probability to the point $(p_1, \ldots, p_m)$ as $n \to \infty$. Now for a given n consider the set E_n in the sample space of $(x_1, \ldots, x_n)$ for which

$$\sum_{i=1}^{m} \left(\frac{r_i}{n} - p_i\right)^2 < D^2 \tag{14.2.6}$$

where D^2 is any positive number less than the (Euclidean) squared distance between $(p_1, \ldots, p_m)$ and $(1/m, \ldots, 1/m)$. Then it will be seen from the fact that $W_{1\alpha}$ consists of sample points for which

$$\sum_{i=1}^{m} \left(\frac{r_i}{n} - \frac{1}{m}\right)^2 > \frac{\chi^2_{\alpha,n}}{mn} \tag{14.2.7}$$

where $\lim_{n\to\infty} \chi^2_{\alpha,n} = \chi^2_\alpha$, that for n greater than some n_1

$$E_n \subset W_{1\alpha}.$$

Therefore, for $n > n_1$, we have

$$P(E_n \mid F_1) \leqslant P(W_{1\alpha} \mid F_1). \tag{14.2.8}$$

But, since $(r_1/n, \ldots, r_m/n)$ converges in probability to $(p_1, \ldots, p_m)$ as $n \to \infty$, then for an arbitrary $\delta > 0$, there is an n_2 such that

$$P(E_n \mid F_1) > 1 - \delta$$

for $n > n_2$. Therefore, for $n > \max(n_1, n_2)$ we have

$$P(W_{1\alpha} \mid F_1) \geqslant 1 - \delta,$$

that is,

$$\lim_{n\to\infty} P(W_{1\alpha} \mid F_1) = 1. \tag{14.2.9}$$

Thus, $W_{1\alpha}$ is consistent for testing the hypothesis $\mathscr{H}(\mathscr{C}_0; \mathscr{C})$ that $F(x)$ belongs to $\mathscr{C}_0$ against any alternative c.d.f. belonging to $\mathscr{C} - \mathscr{C}_0$. It is not consistent, of course, for testing $F_0(x)$ against other alternatives in $\mathscr{C}_0$. Summarizing, we have the following result:

14.2.1 *Let $\mathscr{C}_0$ be the class of continuous c.d.f.'s [including $F_0(x)$] which have the quantiles $\underline{x}_{i/m}$, $i = 1, \ldots, m - 1$, and let $I_i = (\underline{x}_{(i-1)/m}, \underline{x}_{i/m}]$, with $\underline{x}_0 = -\infty$ and $\underline{x}_1 = +\infty$, be intervals determined by these quantiles. Let $r_1, \ldots, r_m$ be the numbers of components of a sample $(x_1, \ldots, x_n)$ falling into $I_1, \ldots, I_m$, respectively, and let $W_{1\alpha}$ be the test consisting of the points in the sample space of $(x_1, \ldots, x_n)$ for which*

$$\frac{m}{n} \sum_{i=1}^{m} \left(r_i - \frac{n}{m}\right)^2 > \chi^2_{\alpha,n}$$

where $\chi^2_{\alpha,n}$ *is chosen to make the test* $W_{1\alpha}$ *as nearly of size* α *as possible for any* $F \in \mathscr{C}_0$. *Then for any* $F \in \mathscr{C}_0$ (14.2.4) *holds. The test* $W_{1\alpha}$ *is consistent for testing any* $F \in \mathscr{C}_0$ *against any* $F \in \mathscr{C} - \mathscr{C}_0$. *However, for any* $F \in \mathscr{C}_0$ (14.2.4) *holds, and hence* $W_{1\alpha}$ *is not consistent for testing* F_0 *against any alternative in* $\mathscr{C}_0$.

(c) The Empty Cell Test

The approach to the problem of testing the nonparametric simple hypothesis $\mathscr{H}(F_0; \mathscr{C})$ which we considered above has as its major shortcoming the fact that $W_{1\alpha}$ is not a consistent test of F_0 against alternatives in $\mathscr{C}_0$. This class can, of course, be reduced by taking larger values of m but holding m fixed as $n \to \infty$. But, as long as m remains fixed as $n \to \infty$ the test is essentially a composite nonparametric statistical test which will not distinguish F_0 from all other members of $\mathscr{C}_0$. The question arises as to whether one can devise a test which will distinguish F_0 from "almost all" alternatives in $\mathscr{C}_0$ by allowing both m and n to increase indefinitely. We shall consider a simple test which will distinguish F_0 from alternatives in a subclass $\mathscr{A}$ of $\mathscr{C}$, the class of absolutely continuous c.d.f.'s, that is, the class of c.d.f.'s $\{F(x)\}$ which have derivatives (p.d.f.'s) $\{f(x)\}$, where F_0, of course, also belongs to $\mathscr{A}$.

Let $s_0, s_1, \ldots, s_n$ be *cell frequency counts* determined by the sample $(x_1, \ldots, x_n)$, that is, the number of intervals I_i, $i = 1, \ldots, m$ containing $0, 1, \ldots, n$ components of the sample respectively. Then, if the sample is from any c.d.f. in $\mathscr{C}_0$, the p.f. of the (degenerate) random variable $(s_0, s_1, \ldots, s_n)$ is readily obtainable from (14.2.1) and is seen to be given by the following expression

$$(14.2.10) \quad p(s_0, s_1, \ldots, s_n) = \frac{m!\, n!}{m^n (0!)^{s_0} (1!)^{s_1} \cdots (n!)^{s_n} s_0!\, s_1! \cdots s_n!}.$$

where, of course, $s_0, s_1, \ldots, s_n$ are non-negative integers satisfying the two conditions

$$(14.2.11) \quad \begin{aligned} s_0 + s_1 + \cdots + s_n &= m \\ s_1 + 2s_2 + \cdots + ns_n &= n. \end{aligned}$$

By using methods similar to those by which (6.1.4) and (6.1.7) were obtained, we find the general factorial moment of the components of the random variables $(s_0, s_1, \ldots, s_n)$ to be as follows:

(14.2.12)

$$\mathscr{E}(s_0^{[g_0]} s_1^{[g_1]} \cdots s_n^{[g_n]}) = \frac{m!\, n!\, (m - g_0 - g_1 - \cdots - g_n)^{n - g_1 - 2g_2 - \cdots - ng_n}}{m^n (2!)^{g_2} \cdots (n!)^{g_n} (m - g_0 - g_1 - \cdots - g_n)!\, (n - g_1 - 2g_2 - \cdots - ng_n)!}$$

where the conditions which the non-negative integers $g_0, \ldots, g_n$ must satisfy are $m - g_0 - g_1 - \cdots - g_n \geqslant 0$ and $n - g_1 - 2g_2 - \cdots - ng_n \geqslant 0$. From this expression we can find means, variances, covariances, and other moments of $s_0, s_1, \ldots, s_n$.

The *empty cell test*, proposed by David (1950), is based on s_0, the number of the intervals $I_1, \ldots, I_m$ which contain no components of the sample. The mean and variance of s_0 are found from (14.2.12) to be as follows:

$$\mu(s_0) = m\left(1 - \frac{1}{m}\right)^n$$

(14.2.13)

$$\sigma^2(s_0) = m(m-1)\left(1 - \frac{2}{m}\right)^n + m\left(1 - \frac{1}{m}\right)^n - m^2\left(1 - \frac{1}{m}\right)^{2n}.$$

As a matter of fact by examining the structure of (14.2.10) it will be seen that $p(s_0, s_1, \ldots, s_n)$ is the coefficient of $u_0^{s_0}u_1^{s_1}\cdots u_n^{s_n}v^n$ in the formal expansion of

$$\frac{n!}{m^n}\left(u_0 + \frac{u_1 v}{1!} + \frac{u_2 v^2}{2!} + \cdots + \frac{u_n v^n}{n!}\right)^m \tag{14.2.14}$$

from which it is seen that the p.f. of s_0, say $p(s_0)$, is obtained by putting $u_1 = \cdots = u_n = 1$ in (14.2.14) and taking the coefficient of $u_0^{s_0}v^n$ in the expansion of the resulting expression, namely,

$$\frac{n!}{m^n}\left(u_0 + \frac{v}{1!} + \frac{v^2}{2!} + \cdots + \frac{v^n}{n!}\right)^m. \tag{14.2.15}$$

But the coefficient of $u_0^{s_0}v^n$ in (14.2.15) is the same as the coefficient of $u_0^{s_0}v^n$ in

$$\frac{n!}{m^n}(u_0 + e^v - 1)^m \tag{14.2.16}$$

which is the same as the coefficient of v^n in $\varphi(v)$, where

$$\varphi(v) = \frac{n!}{m^n}\binom{m}{s_0}(e^v - 1)^{m-s_0}. \tag{14.2.17}$$

But the coefficient of v^n in $\varphi(v)$ is given by

$$\frac{1}{n!}\left[\frac{d^n\varphi(v)}{dv^n}\right]_{v=0}. \tag{14.2.18}$$

Performing this differentiation and setting $v = 0$ we obtain

$$p(s_0) = \frac{m!}{m^n s_0!}\sum_{i=0}^{m-s_0}\frac{(-1)^i(m - s_0 - i)^n}{i!(m - s_0 - i)!}, \tag{14.2.19}$$

where $s_0 = k, k+1, \ldots, m-1$; $k = \max(0, m-n)$.

In the case of large values of m and n the following result due to David (1950) can be established by straightforward analysis:

14.2.2 *If $(x_1, \ldots, x_n)$ is a sample from the continuous c.d.f. $F_0(x)$, then s_0 is asymptotically distributed, as $m, n \to \infty$ so that $n/m \to \rho > 0$, according to $N(me^{-\rho}, m[e^{-\rho} - e^{-2\rho}(1 + \rho)])$.*

Let $W_{2\alpha}$ be the test in the sample space of $(x_1, \ldots, x_n)$ for which $s_0 > s_{0\alpha}$ where $s_{0\alpha}$ is the smallest integer for which $\sum_{s_0=s_{0\alpha}}^{m-1} p(s_0) \leqslant \alpha$. It follows from **14.2.2** that for large m, n, with $n = \rho m + O(1)$,

$$s_{0\alpha} = m\left[e^{-\rho} + \frac{y_\alpha}{\sqrt{m}}(e^{-\rho} - e^{-2\rho}(1 + \rho))^{\frac{1}{2}} + o\left(\frac{1}{m}\right)\right] \tag{14.2.20}$$

and that as $m, n \to \infty$, so that $n/m \to \rho > 0$,

$$\lim P(W_{2\alpha} \mid F_0) = \frac{1}{\sqrt{2\pi}} \int_{y_\alpha}^{\infty} e^{-\frac{1}{2}t^2}\, dt = \alpha. \tag{14.2.21}$$

The choice of $W_{2\alpha}$ is plausible since it is intuitively evident that the distribution of s_0 tends to be pushed to the right on the s_0-axis if the sample comes from *any* distribution in $\mathscr{C}$ different from the one having c.d.f. $F_0(x)$.

We shall examine the plausibility of $W_{2\alpha}$ more formally by considering its consistency for testing an absolutely continuous F_0 against alternatives in $\mathscr{A}$ the class of absolutely continuous c.d.f.'s. If $f_0(x)$ is the derivative of $F_0(x)$ and $f_1(x)$ is that of any c.d.f. $F_1(x)$ in $\mathscr{A}$ different from $F_0(x)$, then $f_0(x)$ and $f_1(x)$ will differ over a set of positive probability as computed from either $f_0(x)$ or $f_1(x)$. Let $\mathscr{A}^*$ be the subclass of $\mathscr{A}$ such that all moments of the ratio $f_1(x)/f_0(x)$ exist with respect to $F_0(x)$ and also with respect to $F_1(x)$. $\mathscr{A}^*$ contains $F_0(x)$. We then have the following result:

14.2.3 *The test $W_{2\alpha}$ is consistent for testing $F_0(x)$ against any alternative $F_1(x) \neq F_0(x)$ in $\mathscr{A}^*$. That is,*

$$\lim P(W_{2\alpha} \mid F_1 \in \mathscr{A}^* - F_0) = 1$$

if $m, n \to \infty$ so that $n/m \to \rho > 0$.

To prove **14.2.3** it is sufficient to show that if the sample $(x_1, \ldots, x_n)$ comes from any $F_1(x)$ in $\mathscr{A}^* - F_0$ then s_0/m converges in probability to a constant which exceeds $e^{-\rho}$ as $m, n \to \infty$ so that $n/m \to \rho > 0$.

Let z_i be a random variable which has the value 1 if I_i contains no components of the sample and 0 otherwise, $i = 1, \ldots, m$. Then

$$s_0 = z_1 + \cdots + z_m.$$

Denoting $\mathscr{E}(s_0/m \mid F_1 \in \mathscr{A}^* - F_0)$, by $\mathscr{E}(s_0/m \mid F_1)$, we have

$$\mathscr{E}\left(\frac{s_0}{m} \middle| F_1\right) = \frac{1}{m}[\mathscr{E}(z_1) + \cdots + \mathscr{E}(z_m)].$$

But

$$\mathscr{E}\left(\frac{z_i}{m}\right) = \frac{1}{m}(1 - p_i)^n, \qquad i = 1, \ldots, m,$$

where p_i is the probability on I_i computed from $f_1(x)$. Therefore

$$\text{(14.2.22)} \quad \mathscr{E}\left(\frac{s_0}{m} \middle| F_1\right) = \frac{1}{m}\sum_{i=1}^{m}(1 - p_i)^n$$

$$= 1 - \frac{n}{m} + \frac{n(n-1)}{2!\,m}\sum_i p_i^2 - \frac{n(n-1)(n-2)}{3!\,m}$$

$$\cdot \sum_i p_i^3 + \cdots + (-1)^k \frac{n(n-1)\cdots(n-k+1)}{k!\,m}$$

$$\cdot \sum_i p_i^k + \cdots + \frac{(-1)^n}{m}\sum_i p_i^n.$$

We can write the general term in this expression as follows:

$$(-1)^k \frac{1}{k!}\left(\frac{n}{m}\right)\left(\frac{n-1}{m}\right)\cdots\left(\frac{n-k+1}{m}\right)\left\{\sum_{i=2}^{m-1}\left[\left(\frac{p_i}{\Delta_i}\right)\bigg/\left(\frac{\left(\frac{1}{m}\right)}{\Delta_i}\right)\right]^k \frac{1}{m}\right\} + \delta_{m,n},$$

where Δ_i is the length of I_i and

$$\delta_{m,n} = (-1)^k \frac{1}{k!}\left(\frac{n}{m}\right)\left(\frac{n-1}{m}\right)\cdots\left(\frac{n-k+1}{m}\right)\left[\left(\frac{p_1}{1/m}\right)^k + \left(\frac{p_m}{1/m}\right)^k\right]\cdot\frac{1}{m}.$$

Taking limits as $m, n \to \infty$ so that $n/m \to \rho > 0$, and recalling that we are assuming that the moments of $f_1(x)/f_0(x)$ with respect to either $F_0(x)$ or $F_1(x)$ are all finite, we find that $\delta_{m,n} \to 0$ and the limit of the expression (14.2.22) is

$$\sum_{k=0}^{\infty}(-1)^k \frac{\rho^k}{k!}\int_{-\infty}^{\infty}[h(x)]^k\,dF_0(x)$$

where $h(x) = f_1(x)/f_0(x)$. Hence, as $m, n \to \infty$ so that $n/m \to \rho > 0$, we have

$$\text{(14.2.23)} \qquad \lim \mathscr{E}\left(\frac{s_0}{m} \middle| F_1\right) = \int_{-\infty}^{\infty} e^{-\rho h(x)}\,dF_0(x).$$

Making use of the fact that if $g(x)$ is a non-negative random variable whose mean value is finite, we have

$$\text{(14.2.24)} \qquad \mathscr{E}(g(x)) \geqslant e^{\mathscr{E}(\log g(x))}$$

with equality holding only if $g(x)$ is a constant with probability 1, we have

$$\int_{-\infty}^{\infty} e^{-\rho h(x)}\, dF_0(x) \geqslant \exp\left[-\int_{-\infty}^{\infty} \rho h(x)\, dF_0(x)\right] = e^{-\rho}. \tag{14.2.25}$$

The equality holds if and only if $h(x) \equiv C$ with probability 1. But this implies that $f_0(x) = f_1(x)$ with probability 1, which in turn implies that $F_0(x) \equiv F_1(x)$. Thus if $f_0(x)$ and $f_1(x)$ differ over a set of positive probability, that is, if $F_1 \in \mathscr{A}^* - F_0$

$$\lim \mathscr{E}\left(\frac{s_0}{m} \,\middle|\, F_1\right) > e^{-\rho}. \tag{14.2.26}$$

To complete the argument for **14.2.3** it is sufficient to show that $\sigma^2(s_0/m \mid F_1) \to 0$ as $m \to \infty$ for any n. If we denote the covariance matrix of $(z_1, \ldots, z_m)$ by $\|\sigma_{ij}\|$, we have

$$\begin{aligned}\sigma^2\left(\frac{s_0}{m} \,\middle|\, F_1\right) &= \frac{1}{m^2} \sum_{i,j=1}^{m} \sigma_{ij} \\ &= \frac{1}{m^2} \sum_{i=1}^{m} (1 - p_i)^n[1 - (1 - p_i)^n] \\ &\quad + \frac{1}{m^2} \sum_{i \neq j=1}^{m} [(1 - p_i - p_j)^n - (1 - p_i)^n(1 - p_j)^n].\end{aligned} \tag{14.2.27}$$

Since $(1 - p_i - p_j)^n \leqslant (1 - p_i)^n(1 - p_j)^n$, it follows that

$$\sigma^2\left(\frac{s_0}{m} \,\middle|\, F_1\right) \leqslant \frac{1}{m^2} \sum_{i=1}^{m} (1 - p_i)^n[1 - (1 - p_i)^n].$$

But $(1 - p_i)^n[1 - (1 - p_i)^n] \leqslant \frac{1}{4}$, $i = 1, \ldots, m$, for any n. Therefore

$$\sigma^2\left(\frac{s_0}{m} \,\middle|\, F_1\right) \leqslant \frac{1}{4m},$$

from which it follows that $\sigma^2(s_0/m \mid F_1) \to 0$ as $m \to \infty$. The fact that $\mathscr{E}(s_0/m \mid F_1) > \mathscr{E}(s_0/m \mid F_0) = e^{-\rho}$ and that $\sigma^2(s_0/m \mid F_1) \to 0$ as $m \to \infty$ implies that

$$\lim P(W_{2\alpha} \mid F_1 \in \mathscr{A}^* - F_0) = 1$$

as $m, n \to \infty$ so that $n/m \to \rho > 0$, thus completing the argument for **14.2.3**.

As a matter of fact, it has been shown by Okamoto (1952) and by Kitabatake (1958), that if the sample $(x_1, \ldots, x_n)$ comes from any distribution having c.d.f. $F_1(x)$ in $\mathscr{A}^*$ the asymptotic distribution of s_0 for large m, n with $n = \rho m + O(1)$, $\rho > 0$, is

$$N\left(m \int_0^1 e^{-\rho h(x)}\, dF_0(x),\ mK^2\right)$$

where

(14.2.28)

$$K^2 = \int_{-\infty}^{\infty} [e^{-\rho h(x)} - e^{-2\rho h(x)}]\, dF_0(x) - \rho\left[\int_{-\infty}^{\infty} e^{-\rho h(x)} h(x)\, dF_0(x)\right]^2.$$

(d) Confidence Contours as Nonparametric Tests

In Sections 11.6 and 11.7, we discussed the problem of estimating the c.d.f. $F(x)$ of a distribution from a sample $(x_1, \ldots, x_n)$ from the distribution. The empirical c.d.f. $F_n(x)$ defined in (11.6.2), is the basic statistical function in those estimation problems. In dealing with the problem of devising a test for $\mathscr{H}(F_0; \mathscr{C}_1)$, where $\mathscr{C}_1$ is some subclass of the class of continuous c.d.f.'s $\mathscr{C}, F_n(x)$ suggests itself for a major role in such a test.

Anderson and Darling (1952), Cramér (1928), Kimball (1947), Kolmogorov (1933*b*), Malmquist (1954), Sherman (1950), Smirnov (1939*b*), von Mises (1931), and others have considered tests based on $F_n(x)$. We refer the reader to a comparative discussion of these tests by Birnbaum (1953*b*), and by Darling (1957). Of these various tests we shall discuss only the confidence contours of Sections 11.6 and 11.7.

Referring to Section 11.6 let $W_{3\alpha}$ be the event in the sample space of $(x_1, \ldots, x_n)$ for which

$$F_0(x) \leqslant F_n(x) + d \tag{14.2.29}$$

fails to hold for all x. If d is chosen so that $P(W_{3\alpha} \mid F_0) = \alpha$, then $W_{3\alpha}$ is a test of size α. We shall examine the power of this test. More precisely, making use of results due to Birnbaum (1953*a*), we shall determine a lower bound for $P(W_{3\alpha} \mid F_1)$ where $F_1(x)$ is a member of $\mathscr{C}$ different from $F_0(x)$ and satisfying a mild condition to be stated presently. Furthermore, under this condition it will be shown that $W_{3\alpha}$ is a consistent test for F_0 against alternatives in a broad subclass of $\mathscr{C}$.

We know from **11.6.1** that if $(x_1, \ldots, x_n)$ is a sample from $F_0(x)$ then we can write

$$P(F_0(x) \leqslant F_n(x) + d, \text{for all } x \mid F_0) = P_n(d), \tag{14.2.30}$$

that is,

$$P(W_{3\alpha} \mid F_0) = 1 - P_n(d) \tag{14.2.31}$$

where $P_n(d)$ is given by formula (11.6.5). For a given level of significance α, and sample size n, we can find a value of d, say $d'(n, \alpha)$, or briefly, d', from (11.6.5) so that $P_n(d') = 1 - \alpha$, thus making $W_{3\alpha}$ of size α, that is,

$$P(W_{3\alpha} \mid F_0) = \alpha. \tag{14.2.32}$$

Now suppose $F_1(x)$ is a c.d.f. in $\mathscr{C}$ different from $F_0(x)$. We wish to examine the power $P(W_{3\alpha} \mid F_1)$ of the test $W_{3\alpha}$ against the alternative $F_1(x)$ where

$$(14.2.33) \quad 1 - P(W_{3\alpha} \mid F_1) = P(F_0(x) \leqslant F_n(x) + d', \text{ for all } x \mid F_1).$$

But $F_0(x) \leqslant F_n(x) + d'$ holds for all x if and only if this inequality holds at the jumps of $F_n(x)$, that is, if and only if

$$(14.2.34) \qquad F_0(x_{(\xi)}) \leqslant \frac{\xi - 1}{n} + d', \qquad \xi = 1, \ldots, n$$

where $(x_{(1)}, \ldots, x_{(n)})$ are the order statistics of the sample. But (14.2.34) hold if and only if

$$(14.2.35) \qquad x_{(\xi)} \leqslant F_0^{-1}\left(\frac{\xi - 1}{n} + d'\right), \qquad \xi = 1, \ldots, n$$

where $F_0^{-1}(y)$ is the inverse of $F_0(x)$. But (14.2.35) holds if and only if

$$(14.2.36) \quad F_1(x_{(\xi)}) \leqslant F_1\left[F_0^{-1}\left(\frac{\xi - 1}{n} + d'\right)\right], \qquad \xi = 1, \ldots, n$$

holds.

Let

$$(14.2.37) \qquad F_1(F_0^{-1}(z)) = G(z)$$

and make the change of variables $y_{(\xi)} = F_1(x_{(\xi)})$, $\xi = 1, \ldots, n$. But we know from (8.7.2) that the p.e. of $(y_{(1)}, \ldots, y_{(n)})$ is

$$(14.2.38) \qquad n!\, dy_{(1)} \cdots dy_{(n)}.$$

Therefore the probability that (14.2.36) holds, that is, the value of the right-hand side of (14.2.33), is given by

$$(14.2.39) \quad n! \int_0^{G(d')} \int_{y_{(1)}}^{G(1/n+d')} \cdots \int_{y_{(n-1)}}^{G((n-1)/n+d')} dy_{(n)}\, dy_{(n-1)} \cdots dy_{(1)}.$$

Now let us assume that

$$(14.2.40) \qquad \underset{-\infty<x<+\infty}{\text{l.u.b.}} \quad (F_0(x) - F_1(x)) = \delta$$

and let x' be a value of x for which

$$(14.2.41) \qquad F_0(x') - F_1(x') = \delta.$$

If k is the largest integer contained in $n(F_0(x') - d')$, then

$$(14.2.42) \quad G\left(\frac{\xi - 1}{n} + d'\right) \leqslant F_0(x') - \delta, \qquad \xi \leqslant k + 1.$$

Furthermore,

$$G\left(\frac{\xi - 1}{n} + d'\right) \leqslant 1, \qquad \xi = k + 2, \ldots, n. \tag{14.2.43}$$

Replacing $G[(\xi - 1)/n + d']$, $\xi = 1, \ldots, n$ in the limits of the integral in (14.2.39) by the quantities on the right-hand sides of (14.2.42) and (14.2.43) does not decrease the value of the integral. The resulting integral with the new limits, which can be evaluated by induction, has the value

$$1 - \sum_{\xi=0}^{k} \binom{n}{\xi} Q^{\xi}(1 - Q)^{n-\xi} \tag{14.2.44}$$

where

$$Q = F_0(x') - \delta. \tag{14.2.45}$$

Since k is the largest integer in $n(F_0(x') - d')$, that is, in $n(Q + \delta - d')$, we have the following lower bound for the power of the test:

$$P(W_{3\alpha} \mid F_1) \geqslant \sum_{\xi=0}^{k} \binom{n}{\xi} Q^{\xi}(1 - Q)^{n-\xi}. \tag{14.2.46}$$

We summarize as follows:

14.2.4 *For a sample of size n let $W_{3\alpha}$ be the set of points in the sample space for which* (14.2.29) *fails to hold. Let $F_1(x)$ be a member of $\mathscr{C}$ such that* (14.2.40) *holds. Then $P(W_{3\alpha} \mid F_1)$, the power of the test W_α, has the lower bound indicated in* (14.2.46).

In the case of large samples it follows from **9.2.1a** that the expression on the right-hand side of (14.2.46) is approximately

$$\frac{1}{\sqrt{2\pi}} \int_{-\infty}^{y} e^{-\frac{1}{2}t^2}\, dt \tag{14.2.47}$$

where

$$y = \frac{\sqrt{n}(\delta - d')}{\sqrt{Q(1 - Q)}}. \tag{14.2.48}$$

Now d' is a function of n and α, and $\lim_{n\to\infty} d' = 0$. Therefore, if $\delta > 0$, the limit of the integral (14.2.47) as $n \to \infty$ is unity, and hence if $\delta > 0$

$$\lim_{n\to\infty} P(W_{3\alpha} \mid F_1) = 1. \tag{14.2.49}$$

Therefore, we have the following result:

14.2.5 *Let $\mathscr{C}_1$ be the subclass of $\mathscr{C}$ for which $\delta > 0$ in* (14.2.40). *Then $W_{3\alpha}$ is a consistent test for the nonparametric simple hypothesis $H(F_0; \mathscr{C}_1 \cup F_0)$.*

This result means roughly that if the graph of $F_0(x)$ lies above that of $F_1(x)$ somewhere, then the probability ultimately becomes 1, as $n \to \infty$, that the graph of $F_n(x) + d'$ will lie below that of $F_0(x)$ somewhere if $(x_1, \ldots, x_n)$ actually comes from $F_1(x)$.

The reader will note from (11.6.1) that if $W'_{3\alpha}$ is the test defined as the set of points in the sample space for which

$$F_0(x) \geqslant F_n(x) - d' \tag{14.2.29a}$$

fails to hold for all x, then $W'_{3\alpha}$ is of size α. Furthermore, if $\mathscr{C}'_1$ is the subclass of $\mathscr{C}$ for which $\delta' < 0$ in

$$\underset{-\infty<x<+\infty}{\text{g.l.b.}} (F_0(x) - F_1(x)) = \delta' \tag{14.2.40a}$$

then by argument similar to that used in establishing **14.2.5** we find that $W'_{3\alpha}$ is a consistent test for $H(F_0; \mathscr{C}'_1 \cup F_0)$. Formulations of companion theorems to **14.2.4** and **14.2.5** concerning test $W'_{3\alpha}$ are left to the reader.

It now follows from **14.2.5** and from the fact that $W'_{3\alpha}$ is consistent for testing $F_0(x)$ against alternatives in $\mathscr{C}'_1$ that if $W''_{3\alpha}$ is a test whose critical region consists of the set of points for which

$$|F_n(x) - F_0(x)| \leqslant d'' \tag{14.2.29b}$$

fails to hold for all x, d'' being chosen so that $W''_{3\alpha}$ is of size α, then $W''_{3\alpha}$ is a consistent test of size α for the nonparametric simple hypothesis $H(F_0; \mathscr{C}_1 \cup \mathscr{C}'_1 \cup F_0)$, that is, for testing $F_0(x)$ against any alternative in $\mathscr{C}_1 \cup \mathscr{C}'_1$.

14.3 THE PROBLEM OF TWO SAMPLES FROM CONTINUOUS DISTRIBUTIONS

(a) Introductory Remarks

Suppose $(x_1, \ldots, x_{n_1})$ and $(x'_1, \ldots, x'_{n_2})$ are independent samples from continuous c.d.f.'s $F_1(x)$ and $F_2(x)$, respectively, and let us denote these samples by O_{n_1} and O_{n_2} respectively. A basic problem in nonparametric statistical inference is to devise a test or tests of the statistical hypothesis that $F_1(x) \equiv F_2(x)$ which is consistent against alternatives in various classes of pairs $(F_1(x), F_2(x))$ in which $F_1(x) \not\equiv F_2(x)$, on the basis of information provided by O_{n_1} and O_{n_2}.

Tests of this type have been devised by Dixon (1940), Mathisen (1943), Smirnov (1939a), Wald and Wolfowitz (1940), Wilks (1961) and others. We shall consider several of these tests in this section.

In discussing tests of the hypothesis that $F_1(x) \equiv F_2(x)$, we must consider classes of alternatives to the assumption that$F_1(x) \equiv F_2(x)$. It will be convenient to let $\mathscr{J}$ denote the class of all pairs of continuous c.d.f.'s $(F_1(x), F_2(x))$, and let $\mathscr{J}_0$ be the subclass of $\mathscr{J}$ in which $F_1(x)$ and $F_2(x)$ are identical, that is, $F_1(x) \equiv F_2(x)$; we shall denote the common c.d.f. by $F(x)$. The most general form of a test for the hypothesis that $F_1(x) \equiv F_2(x)$ would be a test that $(F_1(x), F_2(x)) \in \mathscr{J}_0$ against any alternative $(F_1(x), F_2(x)) \in \mathscr{J} - \mathscr{J}_0$. The Smirnov (1939a) test to be considered in Section 14.3(*e*) is an example of such a general test which is consistent. The tests to be considered here are consistent for testing the hypothesis that $(F_1(x), F_2(x)) \in \mathscr{J}_0$ against alternatives only in certain subclasses of $\mathscr{J} - \mathscr{J}_0$. If $\mathscr{J}^*$ is any such subclass we shall refer to a test for testing $(F_1(x), F_2(x)) \in \mathscr{J}_0$ against alternatives $(F_1(x), F_2(x)) \in \mathscr{J}^*$ as $\mathscr{H}(\mathscr{J}^0; \mathscr{J}^0 \cup \mathscr{J}^*)$.

Before considering specific tests it will be convenient to discuss some distribution theory concerning two independent samples from identical and continuous c.d.f.'s.

(b) Distribution Theory of Basic Random Variables Generated by the Order Statistics of Two Samples

A simple but fundamental theorem on the order statistics of two samples from populations having identical and continuous c.d.f.'s can be stated as follows:

14.3.1 *If O_{n_1} and O_{n_2} are independent samples from identical continuous c.d.f.'s whose order statistics are $(x_{(1)}, \ldots, x_{(n_1)})$ and $(x'_{(1)}, \ldots, x'_{(n_2)})$ respectively, and if $z_1, \ldots, z_{n_1+n_2}$ is the ordered set of the two sets of order statistics, then*

$$P(z_1 < \cdots < z_{n_1+n_2}) = 1 \Big/ \binom{n_1 + n_2}{n_1}. \tag{14.3.1}$$

To verify this statement let $F(x)$ be the common c.d.f. from which O_{n_1} and O_{n_2} are drawn. For simplicity it is sufficient to consider the probability of a specified "meshing" of order statistics of the two samples, say

$$P(x_{(1)} < \cdots < x_{(n_1)} < x'_{(1)} < \cdots < x'_{(n_2)}). \tag{14.3.2}$$

Let us relabel the random variables in (14.3.2) as $z_1, \ldots, z_{n_1+n_2}$. If we let $y_1 = F(z_1), \ldots, y_{n_1+n_2} = F(z_{n_1+n_2})$, then the p.e. of the y's is, according to (8.7.2),

$$n_1!\, n_2!\, dy_1 \cdots dy_{n_1+n_2}. \tag{14.3.3}$$

The event $z_1 < \cdots < z_{n_1+n_2}$ occurs if and only if $y_1 < \cdots < y_{n_1+n_2}$. The probability of this latter event is

$$n_1!\, n_2! \int_0^1 \cdots \int_0^{y_3} \int_0^{y_2} dy_1\, dy_2 \cdots dy_{n_1+n_2} = 1 \Big/ \binom{n_1 + n_2}{n_1}. \tag{14.3.4}$$

It is now evident that (14.3.4) holds for *any* arrangement of the two sets of order statistics in (14.3.2), thus concluding the argument for **14.3.1**.

As before, let O_{n_1} and O_{n_2} be independent samples from $F_1(x)$ and $F_2(x)$ respectively, where $(F_1(x), F_2(x)) \in \mathscr{I}_0$, the common c.d.f. being $F(x)$. The order statistics of O_{n_1} determine sample blocks $B_1^{(1)}, \ldots, B_1^{(n_1+1)}$, as defined in Section 8.7. Let the coverages (see Section 8.7) for these blocks as determined by $F(x)$ be $u_1, \ldots, u_{n_1+1}$, respectively.

Let $r_1, \ldots, r_{n_1+1}$ be the number of elements of O_{n_2} that lie in $B_1^{(1)}, \ldots, B_1^{(n_1+1)}$, respectively, where $r_{n_1+1} = n_2 - r_1 - \cdots - r_{n_1}$. It will be convenient to refer to the random variables $(r_1, \ldots, r_{n_1+1})$ as *block frequencies* which O_{n_1} determines from O_{n_2}. We shall consider the distribution of these block frequencies.

Now $(u_1, \ldots, u_{n_1}, r_1, \ldots, r_{n_1})$ is a mixed $(n_1 + n_2)$-dimensional random variable, the u's being continuous and the r's discrete such that the conditional p.f. of the r's for fixed values of the u's is

$$\frac{n_2!}{r_1! \cdots r_{n_1+1}!} u_1^{r_1} \cdots u_{n_1+1}^{r_{n_1+1}}. \tag{14.3.5}$$

But we know from **8.7.4** that the p.e. of the u's is

$$n_1!\, du_1 \cdots du_{n_1}. \tag{14.3.6}$$

Thus, the joint p.f.-p.e. of the u's and r's is the product of the expressions in (14.3.5) and (14.3.6). To obtain the (unconditional) p.f. of the r's we integrate this product over the simplex S_{n_1}: $\{(u_1, \ldots, u_n)\colon u_1 \geqslant 0, \ldots, u_{n_1} \geqslant 0, u_1 + \cdots + u_{n_1} \leqslant 1\}$. If we use the properties of the Dirichlet distribution in Section 7.7, this integration yields

$$1\Big/\binom{n_1 + n_2}{n_1} \tag{14.3.7}$$

as the p.f. of the n_1-dimensional random variable $(r_1, \ldots, r_{n_1})$ where the r's are all 0 or positive integers such that $r_1 + \cdots + r_{n_1} \leqslant n_2$.

By following a line of argument similar to that which yielded (14.3.7) it can be shown that *any t* of the random variables $(r_1, \ldots, r_{n_1})$, which we may take as $(r_1^*, \ldots, r_t^*)$, have the p.f.

$$\binom{n_1 + n_2 - r_1^* - \cdots - r_t^* - t}{n_1 - t}\Big/\binom{n_1 + n_2}{n_1}. \tag{14.3.8}$$

Summarizing, we have the following result:

14.3.2 *Let O_{n_1} and O_{n_2} be independent samples from identical continuous c.d.f.'s. Let $(r_1, \ldots, r_{n_1+1})$ be the block frequencies which O_{n_1} determines from O_{n_2}. Then the p.f. of $(r_1, \ldots, r_{n_1+1})$, subject to* *the condition* $r_1 + \cdots + r_{n_1+1} = n_2$, *has the constant value* $1 \Big/ \binom{n_1 + n_2}{n_1}$ *over all possible sample points of* $(r_1, \ldots, r_{n_1+1})$. *Furthermore, the p.f. of any t of these random variables, say* $(r_1^*, \ldots, r_t^*)$, *is given by* (14.3.8).

If we let $(r_1', \ldots, r_{n_2+1}')$ be random variables denoting the block frequencies which O_{n_2} determines from O_{n_1}, then $(r_1', \ldots, r_{n_2+1}')$ has

$$1 \Big/ \binom{n_1 + n_2}{n_2} \tag{14.3.9}$$

as its p.f. over all sample points of $(r_1', \ldots, r_{n_2+1}')$ where, of course, $r_1' + \cdots + r_{n_2+1}' = n_1$. Since the expressions in (14.3.7) and (14.3.9) have the same value, the p.f.'s of $(r_1', \ldots, r_{n_2+1}')$ and $(r_1, \ldots, r_{n_1+1})$ therefore have the same constant value over the two sample spaces.

It should be noted that the p.f. of the block frequency counts $(r_1, \ldots, r_{n_1})$ as given by (14.3.7) is identically the same as that for the different possible arrangements of the combined order statistics of O_{n_1} and O_{n_2} as given by (14.3.1), which, of course, is not surprising since there is a one-to-one correspondence between the possible values of the vector random variable $(r_1, \ldots, r_{n_1+1})$ and the possible arrangements of the combined order statistics in O_{n_1} and O_{n_2}.

Remark. The continuous c.d.f. from which O_{n_1} and O_{n_2} are drawn can be k-dimensional. In this case we can adapt any method of cutting k-dimensional sample blocks described in Section 8.7(*c*) and let $B_k^{(1)}, \ldots, B_k^{(n_1+1)}$ be the resulting blocks formed by the n_1 points in R_k representing O_{n_2}. Then let $(r_1, \ldots, r_{n_1+1})$ be the numbers of the n_2 points in R_k representing O_{n_2} falling into these blocks respectively. Then the distribution of the random variable $(r_1, \ldots, r_{n_1+1})$ is exactly the same as that given in **14.3.2**.

Now let s_0 be the number of the blocks $B_1^{(1)}, \ldots, B_1^{(n_1+1)}$ which contain 0 elements from O_{n_2}, s_1 the number which contain 1 element from O_{n_2}, and so on; $s_0, s_1, \ldots, s_{n_2}$ will be called *block frequency counts* which O_{n_1} determines from O_{n_2}. Now $(s_0, s_1, \ldots, s_{n_2})$ is a multidimensional random variable which must satisfy the conditions

$$\begin{aligned} s_0 + s_1 + \cdots + s_{n_2} &= n_1 + 1 \\ s_1 + 2s_2 + \cdots + n_2 s_{n_2} &= n_2. \end{aligned} \tag{14.3.10}$$

Since all points in the sample space of the random variable $(r_1, \ldots, r_{n_1+1})$ are equally probable, the problem of determining the p.f. of $(s_0, \ldots, s_{n_2})$ is merely one of enumerating points in the sample space of the r's satisfying certain conditions. After a little reflection the reader will see that the number of sample points of the random variable $(r_1, \ldots, r_{n_1+1})$ which map into a given sample point $(s_0', s_1', \ldots, s_{n_2}')$ of the random variable $(s_0, s_1, \ldots, s_{n_2})$ is the coefficient of $u_0^{s_0'} u_1^{s_1'} \cdots u_{n_2}^{s_{n_2}'} v^{n_2}$ in the formal expansion of

$$(u_0 + u_1 v + u_2 v^2 + \cdots + u_{n_2} v^{n_2})^{n_1+1}. \tag{14.3.11}$$

Extracting this coefficient, dividing by $\binom{n_1 + n_2}{n_1}$, and dropping dashes, we obtain the following p.f. of $(s_0, s_1, \ldots, s_{n_2})$:

$$\frac{(n_1 + 1)!}{s_0!\, s_1! \cdots s_{n_2}!} \bigg/ \binom{n_1 + n_2}{n_1} \tag{14.3.12}$$

where $s_0, s_1, \ldots, s_{n_2}$, of course, satisfy the conditions (14.3.10). Note that the number of s's involved in (14.3.12) depends on n_2. If one desires the p.f. of only a fixed number, say $(s_0, s_1, \ldots, s_t)$, one would set $u_{t+1} = \cdots = u_{n_2} = u$ in (14.3.11) and select the coefficient of $u_0^{s_0} u_1^{s_1} \cdots u_t^{s_t} u^s v^{n_2}$ where $s_0 + s_1 + \cdots + s_t + s = n_1 + 1$, and $s = s_{t+1} + \cdots + s_{n_2}$. For $t = 0$, this procedure yields for the p.f. of s_0

$$p(s_0) = \frac{\binom{n_1 + 1}{s_0}\binom{n_2 - 1}{n_1 - s_0}}{\binom{n_1 + n_2}{n_1}} \tag{14.3.13}$$

the sample points of s_0 being $k, k + 1, \ldots, n_1$, where $k = \max(0, n_1 - n_2 + 1)$. For larger values of t the p.f. is complicated and is omitted.

Factorial moments of one or more of the random variables $(s_0, s_1, \ldots, s_{n_2})$ can be obtained from the p.f. (14.3.12) by methods similar to those used for deriving (6.1.7). In general, we find

$$\mathscr{E}(s_0^{[g_0]} s_1^{[g_1]} \cdots s_t^{[g_t]}) = (n_1 + 1)^{[g_0 + \cdots + g_t]} \tag{14.3.14}$$
$$\cdot \binom{n_1 + n_2 - g_0 - 2g_1 - \cdots - (t + 1)g_t}{n_1 - g_0 - \cdots - g_t} \bigg/ \binom{n_1 + n_2}{n_1},$$

where, of course, $g_0, \ldots, g_t$ are non-negative integers such that $g_0 + \cdots + g_t \leqslant n_1$ and $g_1 + 2g_2 + \cdots + tg_t \leqslant n_2$. From (14.3.14) one can find means, variances, covariances, and other moments of one or more of the s's.

Summarizing, we have the result:

14.3.3 *If O_{n_1} and O_{n_2} are independent samples from identical continuous c.d.f.'s the p.f. of the block frequency counts $(s_0, s_1, \ldots, s_{n_2})$, which O_{n_1} determines from O_{n_2} is given by* (14.3.12) *where the components of $(s_0, s_1, \ldots, s_{n_2})$ satisfy* (14.3.10). *Furthermore, the general factorial moment of $(s_0, s_1, \ldots, s_t)$ is given by* (14.3.14).

In particular, for s_0, we have

$$\mathscr{E}(s_0) = \frac{n_1(n_1+1)}{(n_1+n_2)}$$

(14.3.15) $$\sigma^2(s_0) = \frac{n_1^2(n_1^2-1)}{(n_1+n_2)(n_1+n_2-1)} + \frac{n_1(n_1+1)}{(n_1+n_2)} - \frac{n_1^2(n_1+1)^2}{(n_1+n_2)^2}.$$

If we let $n_2 = \rho n_1 + O(1)$, $\rho > 0$, these reduce to

(14.3.16) $$\mathscr{E}(s_0) = n_1\left(\frac{1}{1+\rho} + o\left(\frac{1}{n_1}\right)\right)$$
$$\sigma^2(s_0) = n_1\left(\frac{\rho^2}{(1+\rho)^3} + o\left(\frac{1}{n_1}\right)\right).$$

Furthermore, the following statement summarizes the situation concerning the asymptotic distribution of s_0 in large samples:

14.3.4 *If O_{n_1} and O_{n_2} are independent samples from identical continuous c.d.f.'s, and if n_1, n_2 are large so that $n_2 = \rho n_1 + O(1)$ then s_0 is asymptotically distributed according to* $N\left(\frac{n_1}{1+\rho}, \frac{n_1\rho^2}{(1+\rho)^3}\right)$.

The proof of **14.3.4** can be accomplished by approximating all factorials in the expression for $p(s_0)$ in (14.3.13) by Sterling's formula, with

(14.3.17) $$s_0 = \frac{n_1}{1+\rho} + y\sqrt{\frac{n_1\rho^2}{(1+\rho)^3}},$$

from which it will be found that

(14.3.18) $$\lim_{n_1\to\infty} \sum_{s_0=s_0'(n_1)}^{s_0''(n_1)} p(s_0) = \frac{1}{\sqrt{2\pi}}\int_{y'}^{y''} e^{-\frac{1}{2}t^2}\,dt,$$

where $s_0'(x_1)$ and $s_0''(n_1)$ are the largest integers contained in the numbers obtained by respectively substituting y' and y'' for y in the right-hand side of (14.3.17). We omit the details.

(c) An Empty Block Test

The first nonparametric two-sample test we shall consider is a simple *empty block test* suggested by the David one-sample empty cell test discussed in Section 14.2(*c*). The test has as its critical region $W_{a\alpha}$, the set of values of s_0 for which $s_0 \geqslant s_0(\alpha, n_1, n_2)$ where, $s_0(\alpha, n_1, n_2)$ is the smallest integer for which

(14.3.19) $$P(W_{a\alpha} \mid (F_1, F_2) \in \mathscr{J}_0) \leqslant \alpha.$$

Since the p.f. of s_0, if $(F_1, F_2) \in \mathscr{J}_0$, that is, $F_1(x) \equiv F_2(x)$, is given by (14.3.13), values of $s_0(\alpha, n_1, n_2)$ for small values of n_1 and n_2 can be determined from this p.f.

For large values of n_1, for which $n_2 = \rho n_1 + O(1)$, it follows from **14.3.4** that we can approximate the critical value $s_0(\alpha, n_1, n_2)$ as follows:

(14.3.20) $$s_0(\alpha, n_1, n_2) = \frac{n_1}{1+\rho} + y_\alpha \sqrt{\frac{n_1 \rho^2}{(1+\rho)^3}} + O(1)$$

where

(14.3.21) $$\frac{1}{\sqrt{2\pi}} \int_{y_\alpha}^{\infty} e^{-\frac{1}{2}t^2}\, dt = \alpha.$$

Thus, it follows from **14.3.4** that if $n_1, n_2 \to \infty$ so that $n_2/n_1 \to \rho > 0$, then

(14.3.22) $$\lim P(W_{a\alpha} \mid (F_1, F_2) \in \mathscr{J}_0) = \alpha.$$

The test $W_{a\alpha}$ is plausible since it is intuitively apparent that the distribution of s_0 tends to be pushed to the right on the s_0-axis if (F_1, F_2) does not belong to $\mathscr{J}_0$.

We shall examine this plausibility more formally by examining the consistency of $W_{a\alpha}$. Let $F_1^{-1}(u)$ be the inverse of the c.d.f. $F_1(x)$ and let $\mathscr{J}^*$ be the class of pairs of continuous c.d.f.'s $(F_1(x), F_2(x))$ such that:

(i) $F_2(F_1^{-1}(u))$ has a derivative $g(u)$ for all u on (0, 1) except possibly for a set of 0 probability,
(ii) The derivatives of $F_2(F_1^{-1}(u))$ and $F_1(F_1^{-1}(u))$ ($\equiv u$) with respect to u on (0, 1) differ over a set of positive probability.

We shall prove the following result:

14.3.5 *The test $W_{a\alpha}$ is consistent for testing any $(F_1, F_2) \in \mathscr{J}_0$ against any $(F_1, F_2) \in \mathscr{J}^*$ as $n_1, n_2 \to \infty$ so that $n_2 = \rho n_1 + O(1)$, where $\rho > 0$.*

To prove **14.3.5** it is sufficient to show that if $(F_1, F_2) \in \mathscr{J}^*$, $s_0/(n_1 + 1)$ converges in probability to a number greater than $1/(1 + \rho)$ as $n_1, n_2 \to \infty$ so that $n_2/n_1 \to \rho > 0$. It will be recalled that $1/(1 + \rho)$ is the quantity to which $s_0/(n_1 + 1)$ converges in probability if $(F_1, F_2) \in \mathscr{J}_0$.

Let z_ξ be a random variable whose value is one if sample block $B_1^{(\xi)}$ determined by O_{n_1} contains no components of O_{n_2} and 0 otherwise, $\xi = 1, \ldots, n_1 + 1$. Then

(14.3.23) $$s_0 = z_1 + \cdots + z_{n_1+1}.$$

We now compute $\mathscr{E}(s_0)$ assuming O_{n_1} and O_{n_2} are independent samples respectively, from a pair of c.d.f.'s (F_1, F_2) belonging to $\mathscr{J}^*$. Since

$$\mathscr{E}(z_\xi) = P(z_\xi = 1)$$

we have, denoting $\mathscr{E}[s_0/(n_1+1) \mid (F_1, F_2) \in \mathscr{J}^*]$ by $\mathscr{E}[s_0/(n_1+1) \mid \mathscr{J}^*]$

$$\mathscr{E}\left(\frac{s_0}{n_1+1} \middle| J^*\right) = \frac{1}{n_1+1}[P(z_1 = 1) + \cdots + P(z_{n_1+1} = 1)]. \tag{14.3.24}$$

Now for $\xi = 2, \ldots, n_1$ we have

$$P(z_\xi = 1) = \frac{n_1!}{(\xi-2)!\,(n_1-\xi)!} \cdot \int_{S_2} [1 - F_2(x_{(\xi)}) + F_2(x_{(\xi-1)})]^{n_2}[F_1(x_{(\xi-1)})]^{\xi-2} \cdot [1 - F_1(x_{(\xi)})]^{n_1-\xi}\, dF_1(x_{(\xi-1)})\, dF_1(x_{(\xi)}), \tag{14.3.25}$$

where S_2 is the region in Euclidean space R_2 for which $-\infty < x_{(\xi-1)} < x_{(\xi)} < +\infty$.

Expressions for $P(z_1 = 1)$ and $P(z_{n_1+1} = 1)$ are slightly simpler than (14.3.25) but are such that each divided by $n_1 + 1$ has a limit of 0 as $n_1, n_2 \to \infty$ with $n_2/n_1 \to \rho > 0$.

Let us change the variables as follows

$$u_1 = F_1(x_{(\xi-1)})$$
$$u_2 = F_1(x_{(\xi)}).$$

Then (14.3.25) becomes

$$P(z_\xi = 1) = \frac{n_1!}{(\xi-2)!\,(n_1-\xi)!} \times \int_{T_2} [1 - G(u_1) + G(u_2)]^{n_2} u_1^{\xi-2}(1-u_2)^{n_1-\xi}\, du_1\, du_2 \tag{14.3.26}$$

where T_2 is the triangle $0 < u_1 < u_2 < 1$, and $G(u) = F_2(F_1^{-1}(u))$. We assume that $G(u)$ has a derivative $g(u)$. Then $g(u)$ is a probability density function on (0, 1).

Thus, we have

$$\mathscr{E}\left(\frac{s_0}{n_1+1} \middle| J^*\right) = \frac{1}{n_1+1} \sum_{\xi=2}^{n_1} P(z_\xi = 1) + \delta_{n_1,n_2} \tag{14.3.27}$$

where $\delta_{n_1,n_2} = (n_1+1)^{-1}[P(z_1 = 1) + P(z_{n_1+1} = 1)]$.
Inserting the expression for $P(z_\xi = 1)$ from (14.3.26) into (14.3.27) and performing the summation we obtain

$$\mathscr{E}\left(\frac{s_0}{n_1+1} \middle| \mathscr{J}^*\right) = \frac{n_1(n_1-1)}{n_1+1} \times \int_{T_2} [1 - u_2 + u_1]^{n_1-2} [1 - G(u_2) + G(u_1)]^{n_2}\, du_1\, du_2 + \delta_{n_1,n_2}. \tag{14.3.28}$$

We now make the transformation

$$u_1 = v$$
$$u_2 = v + \frac{y}{n_1}.$$

Then (14.3.28) can be written

$$\text{(14.3.29)} \quad \mathscr{E}\left(\frac{s_0}{n_1 + 1} \,\middle|\, \mathscr{J}^*\right) = \frac{n_1(n_1 - 1)}{n_1(n_1 + 1)} \int_0^1 \int_0^{(1-v)n_1} \left[1 - \frac{y}{n_1}\right]^{n_1 - 2_1}$$
$$\cdot \left[1 - \left\{\frac{G\left(v + \dfrac{y}{n_1}\right) - G(v)}{y/n_1}\right\} \frac{y}{n_1}\right] \qquad y\, dv + \delta_{n_1, n_2}.$$

Taking limits as $n_1, n_2 \to \infty$ so that $n_2/n_1 \to \rho > 0$, we obtain

$$\text{(14.3.30)} \qquad \lim \mathscr{E}\left(\frac{s_0}{n_1 + 1} \,\middle|\, \mathscr{J}^*\right) = \int_0^1 \int_0^\infty e^{-y[1 + \rho g(v)]}\, dy\, dv.$$

Performing the integration with respect to y we finally obtain

$$\text{(14.3.31)} \qquad \lim \mathscr{E}\left(\frac{s_0}{n_1 + 1} \,\middle|\, \mathscr{J}^*\right) = \int_0^1 \frac{dv}{1 + \rho g(v)}.$$

We must now show that the integral on the right-hand side of (14.3.31) exceeds $1/(1 + \rho)$ if $g(v)$ differs from unity over a set of positive probability, which will be the case if $(F_1, F_2) \in \mathscr{J}^*$, since the derivatives of $F_2(F_1^{-1}(u))$ and u are assumed to differ over a set of positive probability on (0, 1).

We note that under this condition the following strict Schwarz inequality holds:

$$\text{(14.3.32)} \quad \int_0^1 \frac{dv}{1 + \rho g(v)} \cdot \int_0^1 [1 + \rho g(v)]\, dv$$
$$> \left[\int_0^1 \sqrt{\frac{1}{1 + \rho g(v)}} \sqrt{1 + \rho g(v)}\, dv\right]^2.$$

But the right-hand side of the inequality is 1 and the second integral on the left has the value $1 + \rho$. Therefore

$$\text{(14.3.33)} \qquad \int_0^1 \frac{dv}{1 + \rho g(v)} > \frac{1}{1 + \rho},$$

that is to say, as $n_1, n_2 \to \infty$ with $n_2 = \rho n_1 + O(1)$, $\rho > 0$,

$$\lim \mathscr{E}\left(\frac{s_0}{n_1 + 1} \,\middle|\, \mathscr{J}^*\right) > \lim \mathscr{E}\left(\frac{s_0}{n_1 + 1} \,\middle|\, \mathscr{J}_0\right).$$

To complete the proof of the consistency of the empty cell test it is sufficient to show that the variance of $s_0(n_1 + 1)$ for $(F_1, F_2) \in \mathscr{J}^*$ for large n_1, n_2 with $n_2 = n_1\rho + O(1)$, is of order $1/n_1$. We have

$$(14.3.34)\qquad \sigma^2\left(\frac{s_0}{n_1+1}\,\Big|\,\mathscr{J}^*\right) = \frac{1}{(n_1+1)^2}\left[\sum_{\xi,\eta=1}^{n_1+1}\sigma_{\xi\eta}\right],$$

where $\|\sigma_{\xi\eta}\|$ is the covariance matrix of $(z_1, \ldots, z_{n_1+1})$. The covariance $\sigma_{\xi\eta}$ between z_ξ and z_η can be expressed as follows:

$$(14.3.35)\qquad \sigma_{\xi\eta} = P(z_\xi = 1, z_\eta = 1) - P(z_\xi = 1)P(z_\eta = 1).$$

We shall now show that $\sigma_{\xi\eta} < 0$ for every $\xi \neq \eta$. We know that the elementary coverages $u_1, \ldots, u_{n_1}$ generated by the first sample O_{n_1} have probability element given by $n_1!\, du_1 \cdots du_{n_1}$, in (14.3.6). Denote this differential by $dH(u)$.

Let $p_\xi = F_2(x_{(\xi)}) - F_2(x_{(\xi-1)})$, with a similar expression for p_η. Since $u_1 + \cdots + u_\xi = F_1(x_{(\xi)})$, $\xi = 1, \ldots, n_1$ is a one-to-one transformation between the $x_{(\xi)}$ and u_ξ, it is evident that the random variable $p_1, \ldots, p_{n_1}$ are functions of $u_1, \ldots, u_{n_1}$. For a given first sample O_{n_1} we have the following conditional probability:

$$(14.3.36)\qquad P(z_\xi = 1, z_\eta = 1 \mid u_1, \ldots, u_{n_1}) = (1 - p_\xi - p_\eta)^{n_2}.$$

For the unconditional probability we have

$$(14.3.37)\qquad P(z_\xi = 1, z_\eta = 1) = \mathscr{E}(1 - p_\xi - p_\eta)^{n_2}$$

where $\mathscr{E}$ is taken with respect to the distribution of $(u_1, \ldots, u_{n_1})$. Similarly,

$$(14.3.38)\qquad \begin{aligned} P(z_\xi = 1 \mid u_1, \ldots, u_{n_1}) &= (1 - p_\xi)^{n_2} \\ P(z_\eta = 1 \mid u_1, \ldots, u_{n_1}) &= (1 - p_\eta)^{n_2} \end{aligned}$$

and for the unconditional probabilities $P(z_\xi = 1)$ and $P(z_\eta = 1)$ we have

$$(14.3.39)\qquad P(z_\xi = 1)\, P(z_\eta = 1) = \mathscr{E}(1 - p_\xi)^{n_2} \cdot \mathscr{E}(1 - p_\eta)^{n_2}.$$

By methods similar to those used in setting up the expression in (14.3.26) for $P(z_\xi = 1)$, we can set up a four-dimensional integral for

$$P(z_\xi = 1, z_\eta = 1), \qquad \xi \neq \eta,$$

that is, for

$$\mathscr{E}[(1 - z_\xi)^{n_2}(1 - z_\eta)^{n_2}],$$

and it can be shown by a considerable amount of analysis (see Blum and Weiss (1957)), which we shall omit, that

$$(14.3.40)\qquad \mathscr{E}(1 - p_\xi - p_\eta)^{n_2} \leqslant \mathscr{E}[(1 - p_\xi)^{n_2}] \cdot \mathscr{E}[(1 - p_\eta)^{n_2}].$$

But (14.3.40) is equivalent to the statement that

$$(14.3.41)\qquad P(z_\xi = 1, z_\eta = 1) - P(z_\xi = 1) \cdot P(z_\eta = 1) \leqslant 0.$$

Referring to (14.3.35) we therefore obtain

$$\sigma_{\xi\eta} \leqslant 0, \qquad \xi \neq \eta = 1, \ldots, n_1, \tag{14.3.42}$$

which implies the following inequality from (14.3.34)

$$\sigma^2\left(\frac{s_0}{n_1+1} \,\middle|\, \mathscr{J}^*\right) \leqslant \frac{1}{(n_1+1)^2}\sum_{\xi=1}^{n_1+1}(\sigma_{\xi\xi}). \tag{14.3.43}$$

But

$$\begin{aligned}\sigma_{\xi\xi} &= P(z_\xi = 1) - [P(z_\xi = 1)]^2 \\ &= P(z_\xi = 1)[1 - P(z_\xi = 1)] \leqslant \tfrac{1}{4}\end{aligned} \tag{14.3.44}$$

for $\xi = 1, \ldots, n_1 + 1$.
Thus, we finally obtain

$$\sigma^2\left(\frac{s_0}{n_1+1} \,\middle|\, \mathscr{J}^*\right) \leqslant \frac{1}{4(n_1+1)} \tag{14.3.45}$$

$$\sum_{i=0}^{k}\frac{(s_i - na_i)^2}{na_i} + \frac{u^2 + v^2}{n\rho^2(1+\rho)a_k} > \chi^2_\alpha \tag{14.3.46}$$

where $a_i = \rho^i/(1+\rho)^{i+1}$ and

$$\begin{aligned}u &= \sum_{i=0}^{k}(s_i - na_i)(i - \rho - k - 1) \\ v &= \sqrt{\rho(1+\rho)}\sum_{i=0}^{k}(s_i - na_i)\end{aligned} \tag{14.3.47}$$

which implies that $\sigma^2\left(\frac{s_0}{n_1+1} \,\middle|\, \mathscr{J}^*\right) \to 0$ as $n_1 \to \infty$ which completes the argument for **14.3.5.**

Blum and Weiss (1957) have considered subclasses of $\mathscr{J}^*$ against which arbitrary block frequency counts s_i are consistent for testing the hypothesis that $(F_1, F_2) \in \mathscr{J}_0$. Wilks (1961) has considered a chi-square-like test based on $s_0, s_1, \ldots, s_k$ for fixed k and large samples, the critical region $W^*_{a\alpha}$ being the event in the sample space of the two samples for which

$$\sum_{i=0}^{k}\frac{(s_i - na_i)^2}{na_i} + \frac{u^2 + v^2}{n\rho^2(1+\rho)a_k} > \chi^2_\alpha$$

where $a_i = \rho^i/(1+\rho)^{i+1}$ and

$$u = \sum_{i=0}^{k}(s_i - na_i)(i - \rho - k - 1)$$

$$v = \sqrt{\rho(1+\rho)}\sum_{i=0}^{k}(s_i - na_i)$$

and χ^2_α is the 100α % point of the chi-square distribution $C(k+1)$. This test, as one might expect, is somewhat more powerful than the test $W_{a\alpha}$ based only on s_0.

Remark. The test $W_{a\alpha}$ (or $W_{a\alpha}^*$) can be extended to the case of k-dimensional distributions having continuous c.d.f.'s. In this case we consider O_{n_1} as a generator of k-dimensional sample blocks $B_k^{(1)}, \ldots, B_k^{(n_1+1)}$ as described in Section 8.7(*c*). The blocks are generated in a specific order depending on the system of ordering functions used, whereas O_{n_2} generates points in these blocks. The entire configuration of numbers of points in the sequence of blocks are the components of the random variable $(r_1, \ldots, r_{n_1+1})$ where $r_1 + \cdots + r_{n_1+1} = n_2$. The number of empty blocks is the random variable s_0 and its distribution is given by (14.3.13) if the c.d.f.'s of the samples are identical. The test can be shown to be consistent in the k-dimensional case as in the one-dimensional case under certain conditions on the c.d.f.'s.

(d) The Run Test

If we denote each order statistic of O_{n_1} by C and each order statistic of O_{n_2} by $\bar{C}$, then any arrangement of all the combined order statistics of O_{n_1} and O_{n_2} corresponds to a permutation of n_1 C's, and n_2 $\bar{C}$'s. Now under the conditions of **14.3.1** all the $\binom{n_1 + n_2}{n_1}$ possible arrangements of n_1 C's and n_2 $\bar{C}$'s are equally probable. Thus, all of the theory of runs of Section 6.6 apply to the runs generated by the $\binom{n_1 + n_2}{n_1}$ possible sequences of C's and $\bar{C}$'s. (Note that in Section 6.6 $n_1 + n_2$ is denoted by n.)

The statistical function proposed by Wald and Wolfowitz (1940) as the two-sample test is *total number of runs u* in a sequence of C's and $\bar{C}$'s (that is, in a combined sequence of the order statistics of O_{n_1} and O_{n_2}), as defined in Section 6.6(*a*). The p.f. of u if $(F_1(x), F_2(x)) \in \mathscr{J}_0$ is given by (6.6.20), whereas $\mathscr{E}(u)$ and $\sigma^2(u)$ are given by (6.6.19). Note that the random variables u and s_0 are strongly negatively correlated.

Since small values of u would tend to cast doubt on the truth of the hypothesis that $(F_1(x), F_2(x)) \in \mathscr{J}_0$, we define the critical set of values of u as the set for which

$$P(u \leqslant u(\alpha, n_1, n_2)) \leqslant \alpha, \tag{14.3.48}$$

where $u(\alpha, n_1, n_2)$ is the largest integer for which (14.3.48) holds. Let us call this test $W_{b\alpha}$, its size being α approximately.

For larger values of n_1 and n_2, u is approximately normally distributed with the mean and variance given by (6.6.19). More precisely, Wald and Wolfowitz have established the following theorem:

14.3.6 *If, when $(F_1(x), F_2(x)) \in \mathscr{J}_0$, n_1, n_2 are allowed to increase indefinitely so that $\dfrac{n_2}{n_1} \to \rho > 0$ then u is asymptotically distributed according to*

$$N\left(\frac{2n_1\rho}{1 + \rho}, \frac{4\rho^2 n_1}{(1 + \rho)^3}\right).$$

The proof of **14.3.6** is similar to that given for **9.6.3**, and we shall only give a brief outline here, omitting the details.

Let

$$y = \frac{\left(u - \dfrac{2n_1\rho}{1 + \rho}\right)(1 + \rho)^{\frac{3}{2}}}{2\rho\sqrt{n_1}}$$

from which it is seen that

(14.3.49)
$$u = \frac{2\rho\sqrt{n_1}(y + \sqrt{n_1(1 + \rho)})}{\sqrt{(1 + \rho)^3}}.$$

Now denote by $p_1(u)$ and $p_2(u)$ the expressions for $p(u)$ in (6.6.20) for even and odd values of u, respectively.

Then for a given n_1 and given numbers $t' < t''$, we have

(14.3.50) $P(y' < y < y'') = \sum_1 p_1\left(\dfrac{2\rho\sqrt{n_1}(y + \sqrt{n_1(1 + \rho)})}{\sqrt{(1 + \rho)^3}}\right)$

$$+ \sum_2 p_2\left(\frac{2\rho\sqrt{n_1}(y + \sqrt{n_1(1 + \rho)})}{\sqrt{(1 + \rho)^3}}\right),$$

where $\sum_1$ denotes summation over the discrete values of y between y' and y'' yielded by (14.3.48) for even values of u, whereas $\sum_2$ is a similar sum for odd values of u.

Now consider the first sum, which may be written

(14.3.51) $\sum_1\left[p_1\left(\dfrac{2\rho n_1(y + \sqrt{n_1(1 + \rho)})}{\sqrt{(1 + \rho)^3}}\right)\sqrt{n_1\rho^2/(1 + \rho)^3}\right]\Delta y$

where $\Delta y = \sqrt{(1 + \rho)^3/(n_1\rho^2)}$. By using Stirling's formula (7.6.29) for large factorials on all factorials in the expression for $p_1(u)$ (that is, $p(n)$ for u even), it will be found that as $n_1 \to \infty$ the expression (14.3.51) converges to

(14.3.52)
$$\frac{1}{2\sqrt{2\pi}}\int_{y'}^{y''} e^{-\frac{1}{2}t^2}\, dt$$

Similarly, it will be found that the second sum in (14.3.50) also converges to the integral (14.3.52) as $n_1 \to \infty$. Hence, by adding the two limits we obtain

(14.3.53)
$$\lim_{n_1 \to \infty} P(y' < y < y'') = \frac{1}{\sqrt{2\pi}}\int_{y'}^{y''} e^{-\frac{1}{2}t^2}\, dt.$$

for any two fixed numbers $y' < y''$, which is the assertion of **14.3.6**.

If we let $n_1, n_2 \to \infty$ so that $\lim n_2/n_1 = \rho > 0$ and if $(F_1(x), F_2(x)) \in \mathcal{J}_0$, it follows from **14.3.6** that

(14.3.54)
$$\lim_{n_1 \to \infty} P(v \leqslant u(\alpha, n_1, \rho n_1)) = \alpha.$$

Wald and Wolfowitz have shown that the $W_{b\alpha}$ test is consistent for testing $(F_1(x), F_2(x)) \in \mathcal{J}_0$ against alternatives in the class $\mathcal{J}^*$ referred to in **14.3.5**.

(e) The Smirnov Test

Let $F_{1n_1}(x)$ and $F_{2n_2}(x)$ be the empirical c.d.f.'s determined by O_{n_1} and O_{n_2} as defined in (11.6.2). The Smirnov (1939*a*) statistic

(14.3.55)
$$D_{n_1,n_2} = \sup_x |F_{1n_1}(x) - F_{2n_2}(x)|$$

suggests itself as a reasonable criterion for testing the hypothesis that $(F_1(x), F_2(x)) \in \mathcal{J}_0$.

More precisely let $W_{c\alpha}$ be the critical region in the sample space of O_{n_1} and O_{n_2} for which

(14.3.56)
$$P(D_{n_1,n_2} > \delta(\alpha, n_1, n_2)) \leqslant \alpha$$

where $\delta(\alpha, n_1, n_2)$ is chosen as the smallest number for which (14.3.56) holds, if $(F_1(x), F_2(x)) \in \mathcal{J}_0$.

Smirnov has shown that if $(F_1(x), F_2(x)) \in \mathcal{J}_0$, and if $n_2/n_1 \to \rho > 0$, as $n_1, n_2 \to \infty$, then

(14.3.57)
$$\lim_{n_1,n_2\to\infty} P(D_{n_1,n_2} > \lambda\sqrt{1/n_1 + 1/n_2}) = 1 - \sum_{i=-\infty}^{\infty} (-1)^i e^{-2i^2\lambda^2}.$$

Therefore, by taking $\delta(\alpha, n_1, n_2) = \lambda_\alpha\sqrt{1/n_1 + 1/n_2}$, where λ_α satisfies

(14.3.58)
$$\sum_{i=-\infty}^{\infty} (-1)^i e^{-2i^2\lambda_\alpha^2} = 1 - \alpha$$

we see that as $n_1, n_2 \to \infty$ so that $n_2/n_1 \to \rho > 0$

(14.3.59)
$$\lim_{n_1,n_2\to\infty} P\left(D_{n_1,n_2} > \lambda_\alpha\sqrt{\frac{1}{n_1} + \frac{1}{n_2}}\right) = \alpha.$$

Now $W_{c\alpha}$ is consistent for testing $(F_1(x), F_2(x)) \in \mathcal{J}_0$ against alternatives $(F_1(x), F_2(x)) \in \mathcal{J} - \mathcal{J}_0$, that is, if $F_1(x) \not\equiv F_2(x)$, it can be shown that as $n_1, n_2 \to \infty$, with $n_2/n_1 \to \rho > 0$,

(14.3.60)
$$\lim_{n_1,n_2\to\infty} P\left(D_{n_1,n_2} > \lambda_\alpha\sqrt{\frac{1}{n_1} + \frac{1}{n_2}}\right) = 1.$$

The proof is omitted.

Note that the random variable D_{n_1,n_2} is a function of the block frequencies $(r_1, \ldots, r_{n_1+1})$, and hence for any fixed d, $P(D_{n_1,n_2} \leqslant d)$ can be computed for small samples by counting sample points in the sample space of $(r_1, \ldots, r_{n_1+1})$ for which $D_{n_1,n_2} \leqslant d$ and dividing by $\binom{n_1 + n_2}{n_1}$. This is rather involved if $n_1 \neq n_2$. However, if $n_1 = n_2 = n$, $D_{n,n}$ is a multiple of $1/n$. In this case Massey (1951) has shown by argument similar to that used in establishing **11.7.1** that

$$P\left(D_{n_1,n} \leqslant \frac{k}{n}\right) = \sum_{i=0}^{k} U(i, n) \Big/ \binom{2n}{n} \tag{14.3.61}$$

where $U(\xi, \eta)$, $\xi = 0, 1, \ldots, 2k - 1$, $\eta = 1, \ldots, n$ is the number of sample points in the sample space of $(r_1, \ldots, r_{n_1+1})$ for which $F_{1n}(x_{(\eta)}) = (\xi + \eta - k)/n$, and $|F_{1n}(x) - F_{2n}(x)| \leqslant k/n$ for $x < x_{(\eta)}$ ($x_{(\eta)}$ being the ηth order statistic of O_{n_1}). $U(\xi, \eta)$ satisfies the difference equations

$$\begin{aligned} U(0, \eta + 1) &= U(0, \eta) + U(1, \eta) \\ U(1, \eta + 1) &= U(0, \eta) + U(1, \eta) + U(2, \eta) \\ &\;\;\vdots \\ U(2k - 2, \eta + 1) &= U(0, \eta) + \cdots + U(2k - 1, \eta) \\ U(2k - 1, \eta + 1) &= U(0, \eta) + \cdots + U(2k - 1, \eta) \end{aligned} \tag{14.3.62}$$

subject to the boundary conditions

$$\begin{aligned} U(i, 0) &= 0 \qquad i = 1, \ldots, k - 1 \\ U(k, 0) &= 1. \end{aligned} \tag{14.3.63}$$

Similar formulas can be developed for $n_1 \neq n_2$, in which case D_{n_1,n_2} is a multiple of $1/n_1n_2$. Massey (1951), using such recurrence relations, has computed tables of

$$P\left(D_{n_1,n_2} \leqslant \frac{k}{n_1n_2}\right), \qquad k = 1, 2, \ldots, n_1n_2 \quad \text{for} \quad n_1 \leqslant n_2 \leqslant 10.$$

A fairly simple explicit formula for $P(D_{n_1,n_2} \leqslant k/n_1n_2)$ for the case of $n_1 = n_2 = n$ has been found by Gnedenko and Koroliuk (1951) which deserves mention here.

To obtain the Gnedenko-Koroliuk result let D_n^+ and D_n^- be random variables defined by

$$D_n^+ = \sup_x (F_{1n}(x) - F_{2n}(x)) \tag{14.3.64}$$

$$D_n^- = \inf_x (F_{1n}(x) - F_{2n}(x)). \tag{14.3.64a}$$

Let the order statistics of the two samples combined be $z_{(1)} < z_{(2)} < \cdots < z_{(2n)}$, and let ζ_t be a random variable having value $1/n$ if $z_{(t)}$ is an x

(that is, belongs to the first sample) and $-1/n$ if $z_{(t)}$ is an x' (that is, belongs to the second sample) $t = 1, \ldots, 2n$. Let

$$s_t = \zeta_1 + \cdots + \zeta_t. \tag{14.3.65}$$

Then it is seen that

$$D_n^+ = \sup_t s_t, \qquad D_n^- = \inf_t s_t. \tag{14.3.66}$$

If we assign $s_0 = 0$ and consider the graph of the points (t, s_t), $t = 0, 1, \ldots, 2n$, in the (t, s)-plane, connecting the sequence of points by line segments, we have a path which begins at $(0, 0)$ and ends at $(2n, 0)$. Note that between any two successive integral values of t the path either rises or falls. There are $\binom{2n}{n}$ possible paths, one path corresponding to each possible sequence of x's and x''s among the order statistics $z_{(1)} < \cdots < z_{(2n)}$ of the two samples. Under the assumption that $(F_1(x), F_2(x)) \in \mathscr{J}_0$ all of these paths are equally probable. The problem of determining the value of $P(D_n^+ \leqslant k/n)$ therefore reduces to that of determining the number to paths that lie entirely below the line $s = (k + 1)/n$, and then dividing this number by $\binom{2n}{n}$. Similarly, the problem of finding the value of $P(D_{n,n} \leqslant k/n)$ is equivalent to determining the number of paths lying entirely between the lines $s = \pm(k + 1)/n$ and dividing this number by $\binom{2n}{n}$. Denote these lines by L^+ and L^-, respectively. Consider $P(D_n^+ \leqslant k/n)$ first and let us examine any path which does not lie entirely below the line L^+. There is a first point A where the path meets line L^+. Now, corresponding to the portion of the path from A to $(2n, 0)$ there is a mirror image of this path with respect to L^+ passing from A to $(2n, 2(k + 1)/n)$. Therefore the number of paths from $(0, 0)$ meeting the line L^+ and passing on to $(2n, 0)$ is equal to the number from $(0, 0)$ to the line L^+ and passing along the image path to $(2n, 2(k + 1)/n)$, which is merely $\binom{2n}{n+k+1}$, that is, the number of ways $n + k + 1$ rises and $n - k - 1$ falls can be permuted. The number of paths which do not meet line L^+ is therefore $\binom{2n}{n} - \binom{2n}{n+k+1}$, and hence we have

$$P\left(D_n^+ \leqslant \frac{k}{n}\right) = 1 - \frac{\binom{2n}{n+k+1}}{\binom{2n}{n}}. \tag{14.3.67}$$

By considerations of symmetry it will be seen that

$$P\left(D_n^- \geqslant -\frac{k}{n}\right) = P\left(D_n^+ \leqslant \frac{k}{n}\right). \tag{14.3.68a}$$

If we let $\lambda = k/\sqrt{2n}$, then we have

$$P(\sqrt{n/2}\,D_n^+ \leqslant \lambda) = 1 - \frac{\binom{2n}{n + \lambda\sqrt{2n} + 1}}{\binom{2n}{n}}. \tag{14.3.69}$$

Making use of Stirling's formula (7.6.29) for large factorials we find that

$$\lim_{n\to\infty} \frac{\binom{2n}{n + \lambda\sqrt{2n} + 1}}{\binom{2n}{n}} = e^{-2\lambda^2}, \tag{14.3.70}$$

and hence

$$\lim_{n\to\infty} P(\sqrt{n/2}\,D_n^+ < \lambda) = 1 - e^{-2\lambda^2}. \tag{14.3.71}$$

Summarizing,

14.3.7 *If $F_{1n}(x)$ and $F_{2n}(x)$ are empirical c.d.f.'s in two independent samples each of size n from identical continuous c.d.f.'s and if D_n^+ and D_n^- are defined as in* (14.3.64) *and* (14.3.64*a*), *then $P(D_n^+ \leqslant k/n)$ and $P(D_n^- \geqslant -k/n)$ have the same value, say $P_n^*(k/\sqrt{2n})$, where*

$$P_n^*\left(\frac{k}{\sqrt{2n}}\right) = 1 - \frac{\binom{2n}{n + k + 1}}{\binom{2n}{n}}, \tag{14.3.72}$$

and

$$\lim_{n\to\infty} P_n^*\left(\frac{\lambda}{\sqrt{n}}\right) = 1 - e^{-2\lambda^2}. \tag{14.3.73}$$

The reader should note that $P_n(\lambda/\sqrt{n})$, where $P_n(d)$ is defined in (11.6.5) for one sample of size n, has the same limit as $P_n^*(\lambda/\sqrt{n})$ as $n \to \infty$.

Now we consider the problem of determining the value of $P(D_{n,n} \leqslant k/n)$. Let S_0^+ be the set of paths which meet L^+, and S_0^- the set which meet L^-. Then $S_0^+ \cup S_0^-$ consists of all paths which do not lie entirely between L^+ and L^-. Let $N(E)$ be the number of paths in any set E of paths. Then we have

$$N(S_0^+ \cup S_0^-) = N(S_0^+) + N(S_0^-) - N(S_0^+ \cap S_0^-) \tag{14.3.74}$$

where

$$N(S_0^+) = N(S_0^-) = \binom{2n}{n+k+1}. \tag{14.3.75}$$

Now $S_0^+ \cap S_0^-$ consists of the union of two sets S_1^+ and S_1^- where S_1^+ consists of all paths which contain at least one segment joining L^+ to L^- and similarly S_1^- consists of all paths which contain at least one segment joining L^- to L^+. Then we have

$$\begin{aligned} N(S_0^+ \cap S_0^-) &= N(S_1^+ \cup S_1^-) \\ &= N(S_1^+) + N(S_1^-) - N(S_1^+ \cap S_1^-). \end{aligned} \tag{14.3.76}$$

By reasoning similar to that by which $N(S_0^+)$ and $N(S_0^-)$ were determined we find that

$$N(S_1^+) = N(S_1^-) = \binom{2n}{n+2(k+1)}. \tag{14.3.77}$$

Defining S_i^+ as the set of paths which contain at least i segments joining L^+ and L^- the first beginning from L^+, with a similar definition for S_i^-, we therefore find that

$$\begin{aligned} N(S_0^+ \cup S_0^-) = 2\bigg[&\binom{2n}{n+k+1} - \binom{2n}{n+2(k+1)} \\ &+ \cdots + (-1)^{r-1}\binom{2n}{n+r(k+1)}\bigg] \end{aligned} \tag{14.3.78}$$

where $r = [n/(k+1)]$, the largest integer in $n/(k+1)$.

Since $N(D_{n,n} \leqslant k/n) = \binom{2n}{n} - N(S_0^+ \cup S_0^-)$, we find by dividing by $\binom{2n}{n}$, that

$$\begin{aligned} P\left(D_{n,n} \leqslant \frac{k}{n}\right) &= 1 + 2\sum_{i=1}^{r}(-1)^i \frac{\binom{2n}{n+i(k+1)}}{\binom{2n}{n}} \\ &= \sum_{i=-r}^{+r}(-1)^i \frac{\binom{2n}{n+i(k+1)}}{\binom{2n}{n}}. \end{aligned} \tag{14.3.79}$$

Now let us examine $P(D_{n,n} \leqslant k/n)$ as $n \to \infty$. It will be convenient to denote $\binom{2n}{n+i(k+1)} \Big/ \binom{2n}{n}$ by $g_n(i, k)$, and let $k = \lambda\sqrt{2n}$. Then for an arbitrary $\varepsilon > 0$, we can choose integers r' and $n_1(r', \varepsilon)$ so that

$$\left| 2\sum_{i=r'+1}^{\infty}(-1)^i e^{-2i^2\lambda^2} \right| < \frac{\varepsilon}{3}, \tag{14.3.80}$$

and

$$\left| P\left(D_{n,n} < \sqrt{\frac{2}{n}}\,\lambda \right) - \sum_{i=-r'}^{+r'} (-1)^i g_n(i, \lambda\sqrt{2n}) \right| < \frac{\varepsilon}{3} \tag{14.3.81}$$

for $n > n_1(r', \varepsilon)$. Applying Stirling's formula for large factorials to $g_n(i, \lambda\sqrt{2n})$, $i = 0, \pm 1, \ldots, \pm r'$ as we did in (14.3.70), it can be verified that there is an $n_2(r', \varepsilon)$ so that for $n > n_2(r', \varepsilon)$, we have

$$\left| \sum_{i=-r'}^{+r'} (-1)^i g_n(i, \lambda\sqrt{2n}) - \sum_{i=-r'}^{+r'} (-1)^i e^{-2i^2\lambda^2} \right| < \frac{\varepsilon}{3}. \tag{14.3.82}$$

Combining (14.3.80), (14.3.81), and (14.3.82) we find that

$$\left| P(\sqrt{n/2}\, D_{n,n} < \lambda) - \sum_{i=-\infty}^{\infty} (-1)^i e^{-2i^2\lambda^2} \right| < \varepsilon \tag{14.3.83}$$

for $n > \max(n_1, n_2)$.

Summarizing we obtain the following result:

14.3.8 *If $F_{1n}(x)$ and $F_{2n}(x)$ are empirical c.d.f.'s in independent samples of size n from identical continuous c.d.f.'s, then*

$$P\left(D_{n,n} \leqslant \frac{k}{n} \right) = \sum_{i=-r}^{+r} (-1)^i \binom{2n}{n + i(k+1)} \Big/ \binom{2n}{n}, \tag{14.3.84}$$

where r is the largest integer in $n/(k+1)$, and

$$\lim_{n\to\infty} P(\sqrt{n/2}\, D_{n,n} < \lambda) = \sum_{i=-\infty}^{\infty} (-1)^i e^{-2i^2\lambda^2}, \tag{14.3.85}$$

where $D_{n,n}$ is defined in (14.3.55).

The reader will remember, of course, that (14.3.85) is essentially a special case of (14.3.57) for samples of equal size n. It should be noted that $P(\sqrt{n}\,D_n < \lambda)$, where D_n is defined for one sample of size n in (11.7.1), has the same limit as (14.3.85).

(f) The Mann-Whitney (Wilcoxon) Test

We shall now discuss an important two-sample test originally proposed by Wilcoxon (1945) for samples of equal size, but extended to samples of unequal size by Mann and Whitney (1947), who also obtained certain asymptotic results for large samples.

Suppose $O_{n_1} : (x_1, \ldots, x_{n_1})$ and $O_{n_2} : (x'_1, \ldots, x'_{n_2})$ are samples from continuous c.d.f.'s $F_1(x)$ and $F_2(x)$, respectively. Let $z_{\xi\eta}$, $\xi = 1, \ldots, n_1$, $\eta = 1, \ldots, n_2$ be a set of random variables defined as follows

$$z_{\xi\eta} = \begin{cases} 1 & \text{if} \quad x_\xi < x'_\eta \\ 0 & \text{if} \quad x_\xi > x'_\eta \end{cases} \tag{14.3.86}$$

and let

$$U = \sum_{\eta=1}^{n_2} \sum_{\xi=1}^{n_1} z_{\xi\eta}. \tag{14.3.87}$$

Thus, if $(x_1, \ldots, x_{n_1})$ and $(x'_1, \ldots, x'_{n_2})$ are arranged in a sequence of increasing order of magnitude, U denotes the total number of times an x precedes an x'. If all of the x's and x''s in this sequence are assigned ranks from 1 to $n_1 + n_2$, the sum T of the ranks of the x's is the *Wilcoxon statistic*. It can be verified that

$$U + T = n_1 n_2 + \frac{n_1(n_1 + 1)}{2}. \tag{14.3.88}$$

Actually, Wilcoxon considered T only for the case of samples of equal size.

The Mann-Whitney test is a one-sided test based on the U statistic, and is constructed so that the critical region W_α of size α consists of the event in the sample space of the two samples for which $U = 0, 1, \ldots, U_\alpha$ where U_α is the largest integer for which

$$P(U \leqslant U_\alpha \mid (F_1, F_2) \in \mathscr{J}_0) \leqslant \alpha. \tag{14.3.89}$$

If we let $p_{n_1 n_2}(U)$ be the p.f. of U, assuming $(F_1, F_2) \in \mathscr{J}_0$, then $p_{n_1 n_2}(U)$ satisfies the difference equation

(14.3.90)

$$p_{n_1 n_2}(U) = \frac{n_1}{n_1 + n_2} p_{n_1 - 1 n_2}(U) + \frac{n_2}{n_1 + n_2} p_{n_1 n_2 - 1}(U - n_1),$$

where $P_{oi}(U) = p_{io}(U) = 1$ if $U = 0$, and 0 if $U \neq 0$. Using this difference equation Mann and Whitney (1947) have tabulated values of $p_{n_1 n_2}(U)$ for $1 \leqslant n_1 \leqslant n_2 \leqslant 8$.

Now let us consider the mean and variance of U. In order to examine the consistency of the Mann-Whitney test it will be convenient to determine the mean and variance of U for any two continuous c.d.f.'s $F_1(x)$, $F_2(x)$, that is, for $(F_1, F_2) \in \mathscr{J}$.

We have,

$$\mathscr{E}(U) = \mathscr{E}\left(\sum_{\xi,\eta} z_{\xi\eta}\right) = \sum_{\xi,\eta} \mathscr{E}(z_{\xi\eta}) = n_1 n_2 a \tag{14.3.91}$$

where

$$a = \int_{-\infty}^{\infty} F_1(x)\, dF_2(x). \tag{14.3.92}$$

Also,

$$(14.3.93)\qquad \mathscr{E}(U^2) = \mathscr{E}\left(\sum_{\xi,\eta} z_{\xi\eta}\right)^2$$

$$= \sum_{\xi,\eta} \mathscr{E}(z_{\xi\eta}^2) + \sum_{\substack{\xi,\eta,\zeta \\ \xi\neq\eta}} \mathscr{E}(z_{\xi\zeta}z_{\eta\zeta}) + \sum_{\substack{\xi,\eta,\zeta \\ \eta\neq\zeta}} \mathscr{E}(z_{\xi\eta}z_{\xi\zeta}) + \sum_{\substack{\xi\neq\zeta \\ \eta\neq\omega}} \mathscr{E}(z_{\xi\eta}z_{\zeta\omega})$$

$$= n_1n_2a + n_1n_2(n_1 - 1)b + n_1n_2(n_2 - 1)c + n_1n_2(n_1 - 1)(n_2 - 1)a^2$$

where

$$(14.3.94)\qquad b = \int_{-\infty}^{\infty} (F_1(x))^2\, dF_2(x), \qquad c = \int_{-\infty}^{\infty} (1 - F_2(x))^2\, dF_1(x).$$

Therefore

$$(14.3.95)\qquad \sigma^2(U) = n_1n_2[a + (n_1 - 1)b + (n_2 - 1)c - (n_1 + n_2 - 1)a^2].$$

If $(F_1, F_2) \in \mathscr{J}_0$, then $a = \frac{1}{2}$, $b = c = \frac{1}{3}$ and

$$(14.3.96)\qquad \mathscr{E}(U) = \frac{n_1n_2}{2}, \qquad \sigma^2(U) = \frac{n_1n_2(n_1 + n_2 + 1)}{12}.$$

Mann and Whitney have shown that if $(F_1, F_2) \in \mathscr{J}_0$, U is asymptotically distributed according to $N(n_1n_2/2, n_1n_2(n_1 + n_2 + 1)/12)$ for large n_1 and n_2.

If we take the ratio U/n_1n_2, then

$$(14.3.97)\qquad \mathscr{E}\left(\frac{U}{n_1n_2}\,\middle|\, \mathscr{J}_0\right) = \frac{1}{2}, \qquad \sigma^2\left(\frac{U}{n_1n_2}\,\middle|\, \mathscr{J}_0\right) = \frac{n_1 + n_2 + 1}{12n_1n_2},$$

from which it can be verified by Chebychev's inequality (3.3.5) that as $n_1, n_2 \to \infty$, U/n_1n_2 converges in probability to the constant $\frac{1}{2}$; also U_α/n_1n_2 converges in probability to $\frac{1}{2}$.

Now let us examine the consistency of the U test. Let $\mathscr{J}^-$ be the class of pairs $(F_1(x), F_2(x))$ such that

$$(14.3.98)\qquad a < \tfrac{1}{2}.$$

Then it can be verified that for any $(F_1(x), F_2(x)) \in \mathscr{J}^-$ we have $b < \frac{1}{3}$ and $c < \frac{1}{3}$. The mean and variance of U/n_1n_2 for any pair of c.d.f.'s in $\mathscr{J}^-$ are given by

$$\mathscr{E}\left(\frac{U}{n_1n_2}\right) = a < \tfrac{1}{2}$$

$$(14.3.99)\qquad \sigma^2\left(\frac{U}{n_1n_2}\right) = \frac{a + (n_1 - 1)b + (n_2 - 1)c - (n_1 + n_2 - 1)a^2}{n_1n_2}$$

from which it is evident that as $n_1, n_2 \to \infty$, U/n_1n_2 converges in probability to a constant $a < \frac{1}{2}$, which implies that

$$(14.3.100)\qquad \lim_{n_1,n_2\to\infty} P\left(\frac{U}{n_1n_2} < \frac{U_\alpha}{n_1n_2}\,\middle|\, (F_1, F_2) \in \mathscr{J}^-\right) = 1.$$

That is,

14.3.9 *The Mann-Whitney test is consistent for testing the hypothesis* $\mathscr{H}(\mathscr{I}_0; \mathscr{I}_0 \cup \mathscr{I}^-)$, *that is, for testing* $(F_1, F_2) \in \mathscr{I}_0$ *against any alternative* $(F_1, F_2) \in \mathscr{I}^-$.

If $\mathscr{I}^+$ denotes the class of pairs (F_1, F_2) such that $a > \frac{1}{2}$, we can interchange F_1 and F_2, thus reducing this case to the one already considered.

If the Mann-Whitney test had been defined as a two-tail test with critical region W'_α consisting of the integers $0, 1, \ldots, U_{\frac{1}{2}\alpha}$ and $U'_{\frac{1}{2}\alpha}, \ldots, n_1 n_2$ where $U_{\frac{1}{2}\alpha}$ is the largest integer for which $P(U \leqslant U_{\frac{1}{2}\alpha} \mid (F_1, F_2) \in \mathscr{I}_0) \leqslant \frac{1}{2}\alpha$ and $U'_{\frac{1}{2}\alpha}$ is the smallest integer for which $P(U \geqslant U_{\frac{1}{2}\alpha} \mid (F_1, F_2) \in \mathscr{I}_0) \leqslant \frac{1}{2}\alpha$, then W'_α would be consistent for testing $(F_1, F_2) \in \mathscr{I}_M$ against any alternative $(F_1, F_2) \in \mathscr{I} - \mathscr{I}_M$, where $\mathscr{I}_M$ is the class of all pairs of continuous c.d.f.'s for which $a = \frac{1}{2}$. Note that $\mathscr{I}_M$ is *not* the class of pairs of identical continuous c.d.f.'s but that $\mathscr{I}_0 \subset \mathscr{I}_M \subset \mathscr{I}$.

14.4 THE METHOD OF RANDOMIZATION

In this section we shall consider a procedure known as the method of randomization for constructing tests of certain nonparametric statistical hypotheses. There are two types of randomization tests: *component randomization tests*, and *rank randomization tests*. We shall consider them only briefly. The reader interested in further details is referred to Fraser (1957), Hoeffding (1948*a*), Lehmann (1959), and Lehmann and Stein (1949).

(a) Component Randomization

The method of component randomization was originally proposed by Fisher (1926*b*). This method is simple in principle but technically difficult to apply to particular problems, except for very small samples. The method in its simplest form for one sample is to consider as the sample space the $n!$ points one obtains by permuting the components of the sample $(x_1, \ldots, x_n)$. More generally, suppose $(x_1, \ldots, x_n)$ has a p.d.f. $f(x_1, \ldots, x_n)$. Except for points in a set of probability 0, there are $n!$ distinct points in the sample space R_n of $(x_1, \ldots, x_n)$ whose coordinates are permutations of the components of a given sample. Let this set of $n!$ points be denoted by S. Then if $f_0(x_1, \ldots, x_n)$ is a p.d.f. which is symmetric in $x_1, \ldots, x_n$, we have the following conditional probability attached to the point $(x_{\xi_1}, \ldots, x_{\xi_n})$, where $\xi_1, \ldots, \xi_n$ is any permutation of $1, \ldots, n$.

$$(14.4.1) \qquad p_0(x_{\xi_1}, \ldots, x_{\xi_n} \mid S) = \frac{f_0(x_{\xi_1}, \ldots, x_{\xi_n})}{\sum^* f_0(x_{\xi_1}, \ldots, x_{\xi_n})} = \frac{1}{n!}$$

where Σ^* denotes summation over all permutations $\xi_1, \ldots, \xi_n$ of $1, \ldots, n$.

If $f_1(x_1, \ldots, x_n)$ is an alternative p.d.f. the probability attached to $x_{\xi_1}, \ldots, x_{\xi_n}$ is

$$p_1(x_{\xi_1}, \ldots, x_{\xi_n} \mid S) = \frac{f_1(x_{\xi_1}, \ldots, x_{\xi_n})}{\sum^* f_1(x_{\xi_1}, \ldots, x_{\xi_n})}. \tag{14.4.2}$$

If a test W_α of size α exists which is most powerful for testing $f_0(x_1, \ldots, x_n)$ against $f_1(x_1, \ldots, x_n)$ and if for simplicity we choose α to be a multiple of $1/n!$, then W_α must be the set of $n!\,\alpha$ points in S for which $P(W_\alpha \mid f_1)$ is a maximum. Such a test exists and is unique if the values of p_1 at the $n!$ points of S are all different, or more generally, if there is a single point in S such that the value of p_1 at that point is not greater than the values of $n!\,\alpha - 1$ points in S and is greater than the values of p_1 at all other points in S. In order for such a test W_α to be effective it must have power for discriminating f_0 against f_1 whatever sample point $(x_1, \ldots, x_n)$ may occur, where the sample components are all different.

It should be noted that W_α has no power for testing $f_0(x_1, \ldots, x_n)$ against $f_1(x_1, \ldots, x_n)$ if $f_1(x_1, \ldots, x_n)$ is also symmetric in $x_1, \ldots, x_n$.

In the preceding remarks we have, for the sake of simplicity, discussed the ideas of component randomization for the case of a single sample $(x_1, \ldots, x_n)$. The most interesting and important extensions of these ideas are for problems of two or more samples, experimental designs, and multidimensional samples. In all these problems, however, there is considerable technical difficulty in the strict application of the method of component randomization, except for very small sample sizes. The difficulty is essentially that of evaluating the probability function p_1 over the permutations of the sample components under the alternative f_1 to the null hypothesis, and of selecting the required set of $\alpha n!$ points in S for the test W_α. In order to make progress those who have utilized the method of component randomization in constructing nonparametric tests have relaxed from a strict application of component randomization principles and have borrowed test functions $g(x_1, \ldots, x_n)$ from parametric testing theory to use for determining a critical region W_α. They have then investigated moments and large sample properties of the distribution functions of these test functions over the space of permutations S of the sample components under the null hypothesis and in some cases under the alternative hypothesis. A critical region W_α in the space of permutations S is then taken as the set of permutation points which yield values of g in some interval or set.

It is not feasible to consider all the useful component randomization tests that have been devised. It will be sufficient to give two examples.

(b) Examples of Component Randomization Tests

The earliest component randomization test, known as the *Fisher-Pitman test*, is a two-sample test designed to test the hypothesis that two samples are from distributions having identical means against alternatives in which the means are unequal. The test was originally treated in a specific problem by Fisher (1926*b*); it was investigated more generally by Pitman (1937*b*).

Suppose $O_{n_1} : (x_1, \ldots, x_{n_1})$ and $O_{n_2} : (x'_1, \ldots, x'_{n_2})$ are two independent samples from continuous c.d.f.'s. Let the two samples be pooled together and let an arbitrary permutation of the $n_1 + n_2$ components of the two samples be denoted by $y_1, \ldots, y_{n_1+n_2}$. Let $\bar{y}$, s^2 be the sample mean and variance of $(y_1, \ldots, y_{n_1})$ and $\bar{y}'$, s'^2 the sample mean and variance of $(y_{n_1+1}, \ldots, y_{n_1+n_2})$. The test function proposed is

(14.4.3)

$$g(y_1, \ldots, y_{n_1+n_2}) = \frac{n_1 n_2(\bar{y} - \bar{y}')^2}{(n_1 + n_2)[(n_1 - 1)s^2 + (n_2 - 1)s'^2] + n_1 n_2(\bar{y} - \bar{y}')^2}.$$

Note that $g(y_1, \ldots, y_{n_1+n_2})$ is of form $t^2/(n_1 + n_2 - 2 + t^2)$ where t is the Student t ratio for the hypothesis that the two samples are from identical normal distributions, given that they came from two normal distributions with equal variances.

Pitman has determined the first four moments of the conditional distribution of $g(y_1, \ldots, y_{n_1+n_2})$ given $y_1, \ldots, y_{n_1+n_2}$, over the set of all permutations of the $n_1 + n_2$ y's. Since the denominator of g is constant over all permutations of the y's, the only part of g which has to be considered is the numerator which has at most $\binom{n_1 + n_2}{n_1}$ different values corresponding to the number of different ways the $n_1 + n_2$ sample components can be separated into two sets, one consisting of n_1 components and the other consisting of n_2 components. Pitman has shown that the distribution of g over these permutations has approximately the beta distribution $Be(\frac{1}{2}, \frac{1}{2}(n_1 + n_2 - 2))$. In case the two samples come from identical normal distributions, the unconditional distribution of g is exactly $Be(\frac{1}{2}, \frac{1}{2}(n_1 + n_2 - 2))$. Finally, we remark that the critical region W_α consists of all permutations which yield values of g for which $P(g > g_\alpha) = \alpha$ where g_α would be approximated as the upper 100α % point of the beta distribution $Be(\frac{1}{2}, \frac{1}{2}(n_1 + n_2 - 2))$. The difference between the means of O_{n_1} and O_{n_2} would be judged significant at the 100α % level of significance if $g(x_1, \ldots, x_{n_1}, x'_1, \ldots, x'_{n_2}) > g_\alpha$. Wald and Wolfowitz (1944) have considered the asymptotic distribution of g for large n_1 and n_2 and have given a proof which is equivalent to showing that g_α as given by the beta distribution $Be(\frac{1}{2}, \frac{1}{2}(n_1 + n_2 - 2))$ is asymptotically correct for large values of $n_1 + n_2$.

As a second example of a component randomization test we consider the standard Model I analysis of variance test for testing the hypothesis that column effects are 0 in an experimental layout of r rows and s columns (see Section 10.6) under the assumption that the observations in each row constitute a sample of size s from a population having a continuous c.d.f., the c.d.f.'s corresponding to rows being identical except for means. If $(x_{\xi\eta},\ \xi = 1, \ldots, r,\ \eta = 1, \ldots, s)$ is the sample the test function considered by Welch (1937) and by Pitman (1938) is

$$g(\{x_{\xi\eta}\}) = \frac{\sum_{\xi,\eta} (\bar{x}_{\cdot\eta} - \bar{x})^2}{\sum_{\xi,\eta} (x_{\xi\eta} - \bar{x}_{\cdot\eta} - \bar{x}_{\xi\cdot} + \bar{x})^2 + \sum_{\xi,\eta} (\bar{x}_{\cdot\eta} - \bar{x})^2} \tag{14.4.4}$$

where $\sum_{\xi,\eta}$ denotes summation for $\xi = 1, \ldots, r$, $\eta = 1, \ldots, s$; $\bar{x}_{\xi\cdot}, \bar{x}_{\cdot\eta}$ being the means of the ξth row and ηth column respectively, in $\|x_{\xi\eta}\|$, whereas $\bar{x}$ is the mean of all rs x's in the sample. Pitman, using the method of component randomization, determined the first four moments of g under all permutations of the x's within rows and showed that its distribution could be satisfactorily approximated by the beta distribution $Be(\frac{1}{2}(s - 1), \frac{1}{2}(r - 1)(s - 1))$, which is the exact distribution of g in normal theory under the assumption that column effects are 0 as set forth in Section 10.6(*a*). The critical region W_α of points in the space of permutations on which this test is based consists of all permutation points for which $P(g > g_\alpha) = \alpha$, where g_α is the upper $100\alpha\%$ tail of the beta distribution given above.

Arnold (1958) has extended this randomization test to the case where $x_{\xi\eta}$ is a vector and $s = 2$. In this case the test is equivalent to Hotelling's T^2 for two samples as defined and discussed in Chapter 18. Welch (1937) has also used the component randomization method for investigating the Model I analysis of variance test for Latin Squares.

Hoeffding (1952) has made a study of the power of component randomization tests for large samples, which includes the two examples given above, and he has shown that they are asymptotically as powerful for large samples as the corresponding parametric tests.

(c) Rank Randomization Tests

The method of component randomization is not strictly a distribution-free procedure, since test functions and their distribution theory under component randomization, depend on the values of the components in an observed or realized sample. However, the property of being distribution-free can be achieved by using rank randomization tests. In rank randomization tests, the ranks of sample components are used rather than the components themselves. Thus, if $x_1, \ldots, x_n$ are components of an n-dimensional random variable from a continuous c.d.f. $F(x_1, \ldots, x_n)$ the *ranks* $a_1, \ldots, a_n$ of $x_1, \ldots, x_n$, respectively, are the subscripts of the x's when placed in increasing order. That is, a_ξ is the subscript of the ξth smallest x in $(x_1, \ldots, x_n)$. Thus, $a_1, \ldots, a_n$ is a permutation of $1, \ldots, n$. Let the space of all $n!$ permutations of $1, \ldots, n$ be S'. Every point in the sample space R_n of $(x_1, \ldots, x_n)$ for which $x_{\xi_1} < \cdots < x_{\xi_n}$ maps into the point $(\xi_1, \ldots, \xi_n)$ of the space S'. If $(x_1, \ldots, x_n)$ has a c.d.f. $F_0(x_1, \ldots, x_n)$ which is symmetric in the x's then we have

$$p_0(\xi_1, \ldots, \xi_n) = \frac{1}{n!} \tag{14.4.5.}$$

at each point in S'. As the reader will recall from (14.4.1) $p_0(x_{\xi_1}, \ldots, x_{\xi_n} \mid S)$ is a conditional probability distribution for a fixed $(x_1, \ldots, x_n)$. The corresponding distribution $p_0(\xi_1, \ldots, \xi_n)$ of ranks, however, is not a conditional distribution. If $F_1(x_1, \ldots, x_n)$ is any n-dimensional continuous c.d.f. the probability distribution $p_1(\xi_1, \ldots, \xi_n)$ is given by

$$p_1(\xi_1, \ldots, \xi_n) = \int_{E_n} dF_1(x_1, \ldots, x_n) \tag{14.4.6}$$

where E_n is the region for which $x_{\xi_1} < \cdots < x_{\xi_n}$.

The problem of constructing a critical region W'_α for testing the c.d.f. F_0 against the c.d.f. F_1 from points in S' is entirely similar to that of constructing a critical region W_α for testing the p.d.f. f_0 against the p.d.f. f_1 from points in S which has been discussed earlier. In the case of W'_α, however, the test does not depend on the observed point $(x_1, \ldots, x_n)$ as in the case of W_α. In this sense W'_α is a more truly nonparametric test than W_α. It should be noted that a rank randomization test has no power for discriminating F_0 from F_1 if both c.d.f.'s are symmetric in $x_1, \ldots, x_n$.

As in component randomization tests there is considerable technical difficulty in the strict application of the principle of rank randomization to the determination of W'_α. In order to make progress in rank randomization test theory it has been found necessary to use test functions $h(a_1, \ldots, a_n)$ defined at all points in S' suggested by analogous parametric testing problems.

Although a considerable number of rank randomization tests have been developed, we shall consider only three examples.

(d) Examples of Rank Randomization Tests

The Wilcoxon test T referred to in (14.3.88) is a rank randomization test for testing the hypothesis $\mathscr{H}(\mathscr{A}; \mathscr{B})$ where $\mathscr{A}$ (the null hypothesis) is the class of continuous and symmetric $(n_1 + n_2)$-dimensional c.d.f.'s of form $F(x_1) \cdots F(x_{n_1})F(x_1') \cdots F(x_{n_2}')$, whereas $\mathscr{B}$ is the class of $(n_1 + n_2)$-dimensional continuous c.d.f.'s of form $F_1(x_1) \cdots F_1(x_{n_1})F_2(x_1') \cdots F_2(x_{n_2}')$, with $a < \frac{1}{2}$, a being defined in (14.3.92). Denoting $x_1, \ldots, x_{n_1}, x_1', \ldots, x_{n_2}'$ by $y_1, \ldots, y_{n_1+n_2}$, with $a_1, \ldots, a_{n_1+n_2}$ as the ranks of the y's, the Wilcoxon statistic T is simply $h(a_1, \ldots, a_{n_1+n_2})$ where

$$(14.4.7) \qquad h(a_1, \ldots, a_{n_1+n_2}) = \sum_1^n a_i.$$

Another example of a randomization test based on a rank statistic is the rank correlation coefficient. Let $(x_1, y_1; x_2, y_2; \ldots; x_n, y_n)$ be a sample of size n from a population having a continuous c.d.f. $F(x, y)$. Let $a_1, \ldots, a_n$ be the ranks of $x_1, \ldots, x_n$ and $b_1, \ldots, b_n$ the ranks of $y_1, \ldots, y_n$. The rank statistic $g(a_1, \ldots, a_n, b_1, \ldots, b_n)$ used here is the ordinary correlation coefficient R between the two sets of ranks. The hypothesis to be tested by R is $\mathscr{H}(\mathscr{A}; \mathscr{B})$, where $\mathscr{B}$ is the class of continuous $2n$-dimensional c.d.f.'s of form $F(x_1, y_1) \cdots F(x_n, y_n)$ whereas $\mathscr{A}$ (the null hypothesis) is the subclass of $\mathscr{B}$ for which $F(x_\xi, y_\xi) = F_1(x_\xi)F_2(y_\xi)$, $\xi = 1, \ldots, n$, $F_1(x)$ and $F_2(y)$ being continuous c.d.f.'s. Hotelling and Pabst (1936) investigated the distribution of R when the null hypothesis is true, that is, when the distribution of $(x_1, y_1; \ldots; x_n, y_n)$ belongs to $\mathscr{A}$, and showed that $\sqrt{n}R$ is asymptotically distributed according to $N(0, 1)$ for large n. Olds (1938) tabulated the exact distribution of a quantity S related to R by the equation

$$(14.4.8) \qquad R = 1 - 6S/(n^3 - n).$$

If the critical region W_α of the R test is taken as the set in sample space for which $R^2 > R_\alpha^2$ where R_α^2 is the smallest number for which $P(R^2 \geqslant R_\alpha^2) \leqslant \alpha$, Hoeffding (1948*b*) showed that the test is not consistent for testing any alternative in $\mathscr{A}$ against any alternative in $\mathscr{B} - \mathscr{A}$.

Hoeffding (1948*b*) has devised a randomization test based on ranks for testing $\mathscr{H}(\mathscr{A}; \mathscr{B})$ as defined above. His statistic is

$$(14.4.9) \qquad D = \frac{(n-5)!}{n!}[A - 2(n-2)B + (n-2)(n-3)C]$$

where

$$A = \sum_{\xi=1}^{n} a_\xi b_\xi (a_\xi - 1)(b_\xi - 1)$$

(14.4.10)
$$B = \sum_{\xi=1}^{n} (a_\xi - 1)(b_\xi - 1)c_\xi$$

$$C = \sum_{\xi=1}^{n} c_\xi (c_\xi - 1)$$

where c_ξ is the number of sample pairs (x_ξ, y_ξ), (x_η, y_η) for which $x_\eta < x_\xi$ and $y_\eta < y_\xi$.

D is an unbiased estimator for the quantity

(14.4.11) $$\Delta = \int_{-\infty}^{\infty} \int_{-\infty}^{\infty} [F(x, y) - F(x, \infty)F(\infty, y)]^2 \, dF(x, y)$$

with variance, under the null hypothesis, given by

(14.4.12) $$\sigma^2(D) = \frac{2(n^2 + 5n - 32)}{8100n(n-1)(n-2)(n-3)} .$$

Hoeffding has determined $\sigma^2(D)$ under the hypothesis that the $2n$-dimensional distribution is in $\mathscr{B} - \mathscr{A}$.

Important rank randomization tests for testing various hypotheses in the analysis of variance have been devised by Friedman (1937), Kendall and Smith (1939), Kruskal (1952), Wallis (1939), and Wormleighton (1959). Andrews (1954) has studied the power of some of these tests in large samples.

PROBLEMS

14.1 Suppose $\mathscr{C}_0$ is the class of all continuous c.d.f.'s having quantiles $p_1, p_2, \ldots, p_k$ at $x_{01}, x_{02}, \ldots, x_{0k}$, where $0 = p_0 < p_1 < \cdots < p_k < p_{k+1} = 1$ respectively. Let $\mathscr{C}$ be the class of all continuous c.d.f.'s. Let $I_1, \ldots, I_{k+1}$ be the intervals $(-\infty, x_{01}]$, $(x_{01}, x_{02}]$, $\ldots$, $(x_{0k}, +\infty)$ respectively. In a sample of size n let $n_1, \ldots, n_{k+1}$, $(n_1 + \cdots + n_{k+1} = n)$ be the numbers of sample components falling into $I_1, \ldots, I_{k+1}$ respectively. Show that for large n the set of points W_α in the sample space of $(n_1, \ldots, n_{k+1})$ for which

$$\sum_{i=1}^{k+1} \frac{[n_i - n(p_i - p_{i-1})]^2}{n(p_i - p_{i-1})} > \chi^2_{k,\alpha}$$

is a consistent test of size α for testing the hypothesis $\mathscr{H}(\mathscr{C}_0; \mathscr{C})$ and $P(\chi^2 > \chi^2_{k,\alpha}) = \alpha$, where χ^2 is a random variable having the chi-square distribution $C(k)$.

14.2 Let $\mathscr{C}_{0p}$ be the class of all n-dimensional c.d.f.'s of form $F_1(x_1)F_2(x_2) \cdots F_n(x_n)$ where $F_1(x_{0p}) = \cdots = F_n(x_{0p}) = p$. Let $\mathscr{C}_p$ be the class of all n-dimensional c.d.f.'s of form $F_1(x_1)F_2(x_2) \cdots F_n(x_n)$ where $F_\xi(x_{0p}) = p_\xi$, $\xi = 1, \ldots, n$ and let $(p_1 + \cdots + p_n)/n \to p' \neq p$ as $n \to \infty$. Let r be the number of components of the n-dimensional random variable $(x_1, \ldots x_n)$ which lie on

$(-\infty, x_{0p}]$ and let W_α be defined as in (14.1.2). Show that as $n \to \infty$, W_α is a consistent test of size α for testing $\mathscr{H}(\mathscr{C}_{0p}; \mathscr{C}_p)$.

14.3 Verify formula (14.2.12).

14.4 Show from (14.2.12) that

$$\lim \mathscr{E}\left(\frac{s_i}{m}\right) = \frac{\rho^i e^{-\rho}}{i!}$$

as $m, n \to \infty$ so that $n/m \to \rho > 0$.

14.5 Show that the random variable s_1 defined in Section 14.2(*c*) has

$$N(m\rho e^{-\rho}, m\rho e^{-\rho}(1 - e^{-\rho} - \rho e^{-\rho}))$$

as its asymptotic distribution for large m, n when $n = \rho m + O(1)$, $\rho > 0$.

14.6 Show that the coefficient of $u_0^s v^n$ in (14.2.15) is identical with the coefficient of $u_0^s v^n$ in (14.2.16).

14.7 Prove (14.2.24).

14.8 Show that if $(r_1^*, \ldots, r_t^*)$ are any t of the random variables $(r_1, \ldots, r_{n_1})$ having the p.f. (14.3.7), the p.f. of $(r_1^*, \ldots, r_t^*)$ is given by (14.3.8).

14.9 Verify (14.3.13).

14.10 Verify (14.3.14).

14.11 From (14.3.14) show that

$$\lim \mathscr{E}\left(\frac{s_i}{n_1 + 1}\right) = \frac{\rho^i}{(1 + \rho)^{i+1}}$$

as $n_1, n_2 \to \infty$ so that $n_1/n_2 \to \rho > 0$.

14.12 (*Continuation*). Under the same limiting conditions show that

$$\lim \mathscr{E}\left(\frac{s_i}{n_1 + 1}\,\middle|\, \mathscr{I}^*\right) = \int_0^1 \frac{[\rho g(u)]^i}{[1 + \rho g(u)]^{i+1}}\, du.$$

14.13 In **14.3.1** let m_1 be the number of x's which exceed z_k, and m_2 the number of x''s which exceed z_k. Show that the p.f. of (m_1, m_2) is given by

$$p(m_1, m_2) = \frac{\binom{n_1}{m_1}\binom{n_2}{m_2}}{\binom{n_1 + n_2}{k}}$$

where m_1 and m_2 are non-negative integers such that $m_1 + m_2 = n_1 + n_2 - k$. [Mood (1950)].

14.14 Suppose $\mathscr{I}_1$ is the class of all pairs of continuous c.d.f.'s of form $(F(x), F(x - \delta))$ where δ is a constant, and let $\mathscr{I}_0$ be the class of all pairs of identical continuous c.d.f.'s. Show that the Mann-Whitney test is consistent for testing any element of $\mathscr{I}_0$ against any element of $\mathscr{I}_1$ for which $\delta > 0$.

14.15 (*Continuation*) If $(x_{(1)}, \ldots, x_{(n_1)})$ and $(x'_{(1)}, \ldots, x'_{(n_2)})$ are the order statistics of independent samples of sizes n_1 and n_2 from $F(x)$ and $F(x - \delta)$ respectively, show that

$$P(x_{(r)} - x'_{(s)} < \delta) = \sum_{i=1}^{r-1} \left(\frac{s}{s+i}\right) \binom{n_1}{i} \binom{n_2}{s} \Big/ \binom{n_1+n_2}{i+s},$$

and hence that if $r < s$ and $r' > s'$, $(x'_{(s')} - x_{(r')}, x'_{(s)} - x_{(r)})$ is a 100γ % confidence interval for δ where

$$\gamma = 1 - \sum_{i=1}^{r-1} \left(\frac{s}{s+i}\right) \binom{n_1}{i} \binom{n_2}{s} \Big/ \binom{n_1+n_2}{i+s} - \sum_{i=r'}^{n_1} \left(\frac{s'}{s'+i}\right) \binom{n_1}{i} \binom{n_2}{s'} \Big/ \binom{n_1+n_2}{i+s'}.$$

[Mood (1950)].

14.16 Verify (14.3.78).

14.17 Verify (14.3.88).

14.18 Establish the difference equation (14.3.90).

14.19 Verify (14.3.93).

14.20 Show that $\mathscr{E}(D) = \Delta$ where D is defined by (14.4.9) and Δ by (14.4.11).

14.21 *Bertrand's* (1887) *ballot problem.* Suppose there are N_1 ballots in a box marked for A, and N_2 marked for B, where $N_1 > N_2$. Suppose the ballots are drawn out one at a time, and let y_i be a random variable denoting the number of votes for A minus the number for B when i ballots have been drawn and counted. If the $\binom{N_1+N_2}{N_1}$ different possible (distinct) orders of drawing out all ballots are assigned equal probabilities, show (by methods similar to those used in Section 14.3(*e*)) that the probability is $(N_1 - N_2)/(N_1 + N_2)$ that A leads B (that is, that $y_1, \ldots, y_{N_1+N_2}$ are all > 0) throughout the counting.

14.22 (*Continuation*). Suppose $N_1 = N_2 = N$, in which case $y_{2N} = 0$, and consider the $\binom{2N}{N}$ polygonal paths reaching from (0, 0) to (2N, 0) obtained by consecutively connecting the $2N + 1$ points $(0, 0), (1, y_1), \ldots, (2N, 0)$ in the xy-plane. Note that there can be only an even number of segments of any path lying above (or below) the x-axis. Show that the number of paths having $2n$ segments lying above the x-axis (and $2N - 2n$ below) is

$$\binom{2N}{N} \Big/ (N+1),$$

and hence that the probability that a fraction n/N of a path will lie above the x-axis is given by

$$P_N(n) = \frac{1}{N+1}, \quad n = 0, 1, \ldots, N$$

if all paths are assigned equal probabilities.

14.23 *Long leads in coin-tossing.* [Chung and Feller (1949)]. If a coin is tossed $2N$ times, let y_i be a random variable denoting the number of heads minus the number of tails in the first i tosses, and consider the 2^{2N} possible polygonal paths connecting the points $(0, 0)$, $(1, y_1), \ldots (2N, y_{2N})$. Show that the number of paths having $2n$ segments above the x-axis (and, of course, $2N - 2n$ below) is

$$\binom{2n}{n}\binom{2N-2n}{N-n}.$$

And hence the probability that the fraction n/N of a path lies above the x-axis is

$$\binom{2n}{n}\binom{2N-2n}{N-n}\Big/ 2^{2N}, \quad n = 0, 1, \ldots, N,$$

all paths being assigned equal probabilities (that is, assuming $2N$ independent throws of a true coin).

14.24 (*Continuation*). *The (first) arc sine law* [Chung and Feller (1949)]. By using Stirling's formula for large factorials show that the limiting c.d.f. $G(u)$ of the random variable $\frac{n}{N}$ as $N \to \infty$ is given by

$$G(u) = \frac{2}{\pi}\sin^{-1}\sqrt{u}.$$

Note that the p.d.f. of this limiting c.d.f. is given by $g(u) = 1/[\pi\sqrt{u(1-u)}]$ which is a U-shaped density function, with infinite density at the ends of the interval $(0, 1)$, thus indicating greater likelihood that either relatively large fractions or relatively small fractions of a path lie above the x-axis rather than fractions near $\frac{1}{2}$ as intuition tends to suggest.

CHAPTER 15

Sequential Statistical Analysis

15.1 INTRODUCTORY REMARKS

In all of the theory of sampling, estimation, and testing of statistical hypotheses which has been discussed in Chapters 8 to 14, the size of the sample has been held fixed. All of the theory has involved the determination of sampling distributions of certain statistics, that is, functions of the sample components for a given sample size. The limiting forms of some of these distributions as the sample size increases indefinitely have been determined. In virtually none of this theory have we considered situations in which the sample size is itself a random variable. However, in Chapters 6 and 7 we discussed several *waiting-time distributions* in which the number of trials required to achieve some objective is a random variable. The situations leading to these waiting-time distributions are examples of simple sequential-like statistical procedures. This chapter is devoted mainly to tests of statistical hypotheses, although some results concerning sequential estimation are given.

The results we shall consider are the basic ones pertaining to the testing of simple hypotheses. Some results concerning the testing of composite hypotheses will be found in Wald's book (1947*a*), and more recent results will be found in papers by Barnard (1952), and Cox (1952). Results on sequential t-tests have been obtained by David and Kruskal (1956), and Rushton (1950, 1952).

The basic idea of a sequential test originated with Dodge and Romig (1929) who devised a *double sampling* scheme for deciding whether to accept or reject a lot of N articles containing an unknown number of defectives $N\theta$, where θ is a multiple of $1/N$ on [0, 1]. Under the Dodge-Romig scheme a sample of size n_1 is drawn from the lot and one of three alternative decisions is made, depending as follows on the number m_1 of defectives found in the sample:

(i) Accept the lot if $m_1 \leqslant c_1$.
(ii) Reject the lot if $m_1 > c_2$.
(iii) Draw a second sample of size n_2 if $c_1 < m_1 \leqslant c_2$.

If decision (iii) is taken, that is, if a second sample of size n_2 is drawn, then one of two decisions is made depending on the number m_2 of defective articles in the second sample:

(i) Accept lot if $m_1 + m_2 \leqslant c_2$.
(ii) Reject lot is $m_1 + m_2 > c_2$.

A double sampling plan is completely determined by specifying the four numbers n_1, n_2, c_1, c_2. The total sample size is clearly a random variable n which can take on two values, namely, n_1 and $n_1 + n_2$. A thorough treatment of both single and double sampling plans is given in the book by Dodge and Romig (1959).

To see that a double sampling plan together with the decisions associated with its operation essentially constitute a statistical test we merely have to note that rejection of the lot occurs if and only if the two-dimensional random variable (m_1, m_2), falls in a critical region W defined as $E_1 \cup E_2$ where E_1 is the set of points in the sample space of (m_1, m_2) for which $m_1 > c_2$ and E_2 is the set for which $c_1 < m_1 \leqslant c_2$ and $m_1 + m_2 > c_2$. $P(W \mid \theta)$ the probability of rejection of the lot if its fraction of defectives is θ, can be written with no particular difficulty in terms of probabilities given by hypergeometric distributions. $P(W \mid \theta)$, as a function of θ is, of course, the power function of the "test" W. If we let

$$L(\theta) = P(\bar{W} \mid \theta),$$

then $L(\theta)$ is the probability of accepting a lot having fraction defective θ. Using the notation and terminology in *Historical Remark* of Section 13.1, we say that the lot is in the zone of preference if $\theta \leqslant \theta_0$, in the zone of indifference if $\theta_0 < \theta < \theta_1$ and in the zone of rejection if $\theta \geqslant \theta_1$. For a given producer's risk α (risk of Type I error) at $\theta = \theta_0$ and consumer's risk β (risk of Type II error) at $\theta = \theta_1$, we then have

$$L(\theta_0) = 1 - \alpha, \qquad L(\theta_1) = \beta$$

where, in practical applications, values of α and β are usually taken in the interval [0.01, 0.10]. The operating characteristic curve, that is, the graph of $L(\theta)$ as a function of θ, passes through the points (0, 1), $(\theta_0, 1 - \alpha)$, (θ_1, β), and (1, 0).

The idea of a more truly sequential procedure is due to Bartky (1943) who extended the notion of double sampling to *multiple sampling* for the case of an infinitely large lot having an unknown fraction of defectives θ. Other early examples of mathematical studies of multistage procedures include Mahalanobis' (1940) series of sample censuses of jute acreage in Bengal, and Hotelling's (1941) notion of a series of experiments to determine the maximum of a regression function, and stochastic approximation methods by Dixon and Mood (1948). It remained largely for Wald (1945) however, to take the major step toward a general theory of sequential analysis as we know it today. In this chapter we shall present only a brief account of some of the basic results in sequential analysis due mainly to Wald. The reader interested in further details about the more classical results should consult Wald's (1947*a*) book. For more recent mathematical investigations of sequential-like procedures the reader is referred to papers by Dvoretzky, Kiefer and Wolfowitz (1953*a*, 1953*b*), Kiefer (1948, 1953, 1957), Kiefer and Wolfowitz (1952), Robbins (1952), Robbins and Monro (1951), and others.

15.2 THE BASIC STRUCTURE OF A SEQUENTIAL TEST

(a) Description of Events in a Sequential Test

The Dodge-Romig double sampling scheme described above suggests the essential structure of a general sequential test. Suppose we have a stochastic process $(x_1, x_2, \ldots)$ which depends on a real parameter θ. This means that every finite set of the components $x_1, x_2, \ldots$ has a c.d.f. depending on θ whose parameter space is Ω. Thus, $(x_1, \ldots, x_n)$ has a c.d.f. $F(x_1, \ldots, x_n; \theta)$, for $n = 1, 2, \ldots$. We shall first consider the simple hypothesis $\mathscr{H}(\theta_0; \Omega)$ (more briefly $\mathscr{H}_0$) that $\theta = \theta_0$. Let $R^{(1)}, \ldots, R_1^{(n)}$, be sample spaces of $x_1, \ldots, x_n$ respectively, and let $R_n = R^{(1)} \times \cdots \times R_1^{(n)}$ be the sample space of $(x_1, \ldots, x_n)$, $n = 1, 2, \ldots$. It will also be convenient to use R_∞ where $R_\infty = R_1^{(1)} \times R_1^{(2)} \times \cdots$.

For each positive integer n let G_n°, G_n', G_n be disjoint events in R_n such that

$$G_n^\circ \cup G_n' \cup G_n = G_{n-1} \times R_1^{(n)}. \tag{15.2.1}$$

We begin the sequential experiment by drawing x_1 and adopt the following rules for stopping or continuing the experiment:

If

$$\begin{aligned} &x_1 \in G_1^\circ && \text{accept } \mathscr{H}_0 \text{ without further drawings} \\ &x_1 \in G_1' && \text{reject } \mathscr{H}_0 \text{ without further drawings} \\ &x_1 \in G_1 && \text{draw } x_2; \end{aligned} \tag{15.2.2}$$

if

$$\begin{aligned} &(x_1, x_2) \in G_2^\circ && \text{accept } \mathscr{H}_0 \text{ without further drawings} \\ &(x_1, x_2) \in G_2' && \text{reject } \mathscr{H}_0 \text{ without further drawings} \\ &(x_1, x_2) \in G_2 && \text{draw } x_3 \end{aligned} \tag{15.2.3}$$

and, in general, if

$$\begin{aligned} &(x_1, \ldots, x_n) \in G_n^\circ \quad \text{accept } \mathscr{H}_0 \text{ without further drawings} \\ &(x_1, \ldots, x_n) \in G_n' \quad \text{reject } \mathscr{H}_0 \text{ without further drawings} \\ &(x_1, \ldots, x_n) \in G_n \quad \text{draw } x_{n+1} \end{aligned} \tag{15.2.4}$$

$n = 1, 2, \ldots.$

It is convenient to call $G_1, G_2, \ldots$ the sequence of *experiment continuation events*.

Note that $G_1^\circ, \ldots, G_{n-1}^\circ, G_1', \ldots, G_{n-1}'$ are cylinder events in R_n, and G_n°, G_n', G_n are events in R_n such that the $2n + 1$ events are disjoint and

$$G_1^\circ \cup \cdots \cup G_n^\circ \cup G_1' \cup \cdots \cup G_n' \cup G_n = R_n. \tag{15.2.5}$$

However, more conveniently, these same events, for any positive integer n, are disjoint cylinder events in R_∞, whose union is R_∞ itself.

The events $G_1^\circ, \ldots, G_n^\circ, G_n', \ldots, G_n', G_n$, $n = 1, 2, \ldots,$ together with the decision procedure defined in (15.2.2), (15.2.3), and (15.2.4) define a sequential test S for $\mathscr{H}_0$. The critical region W_α for rejecting $\mathscr{H}_0$ is $G_1' \cup G_2' \cup \cdots$ where the G's are chosen so that $P(W_\alpha \mid \theta_0) = \alpha$. The procedure for carrying out a sequential test step-by-step will be referred to as the *sequential process* for S.

It should be observed that any test for $\mathscr{H}_0$ based on a sample of fixed size n and having rejection region W in the sample space can be viewed as what might be called a *degenerate sequential test* in which $G_1^\circ, \ldots, G_{n-1}^\circ, G_1', \ldots, G_{n-1}', G_n$ are null sets while $G_n' = W$ and $G_n^\circ = \bar{W}$.

(b) Probabilities in a Sequential Test

The probabilities of accepting $\mathscr{H}_0$, rejecting $\mathscr{H}_0$, and drawing x_{n+1} upon the drawing of x_n, are

$$P(G_n^\circ \mid \theta), \quad P(G_n' \mid \theta), \quad P(G_n \mid \theta) \tag{15.2.6}$$

respectively. As we have already stated the events $G_1^\circ, \ldots, G_n^\circ, G_1', \ldots, G_n', G_n$ are disjoint and their probabilities are functions of the parameter θ. Let us set

$$\begin{aligned} L_n(\theta) &= P(G_1^\circ \mid \theta) + \cdots + P(G_n^\circ \mid \theta) \\ M_n(\theta) &= P(G_1' \mid \theta) + \cdots + P(G_n' \mid \theta) \\ N_n(\theta) &= P(G_n \mid \theta) \end{aligned} \tag{15.2.7}$$

and

$$\begin{aligned} L(\theta) &= \lim_{n\to\infty} L_n(\theta) = P(G^\circ \mid \theta) \\ M(\theta) &= \lim_{n\to\infty} M_n(\theta) = P(G' \mid \theta) \\ N(\theta) &= \lim_{n\to\infty} N_n(\theta) \end{aligned} \tag{15.2.8}$$

where $G^\circ = G_1^\circ \cup G_2^\circ \cup \ldots$, and $G' = G_1' \cup G_2' \cup \ldots$. These limits exist since $L_n(\theta)$, $M_n(\theta)$, $N_n(\theta)$ are non-negative with sum 1 and since $L_n(\theta)$ and $M_n(\theta)$ are nondecreasing as $n = 1, 2, \ldots$. $L(\theta')$ and $M(\theta')$ are the probabilities of accepting $\mathscr{H}_0$ and of rejecting $\mathscr{H}_0$ if the true value of θ in the population is θ'. $L(\theta)$, as a function of θ, is the *operating characteristic function* of the sequential test S. $N(\theta)$ is the probability that the sequential process for S continues indefinitely. $N(\theta) = 0$ means that the sequential process terminates with probability 1. It is evident that

$$L(\theta) + M(\theta) + N(\theta) = 1. \tag{15.2.9}$$

If we let $G_n^\circ \cup G_n' = G_n^*$, and put

$$p(n \mid \theta) = P(G_n^* \mid \theta) \tag{15.2.10}$$

then $p(n \mid \theta)$ is the probability that the sequential process for S terminates (with a decision to accept $\mathscr{H}_0$ or reject $\mathscr{H}_0$) with n trials, if θ is the true value of the parameter.

If the sequential process for S terminates with probability 1, that is, if $N(\theta) = 0$, then the average number of trials required to terminate the process is

$$\mathscr{E}(n \mid \theta) = \sum_{t=1}^{\infty} t p(t \mid \theta) \tag{15.2.11}$$

which, of course, is a function of θ. $\mathscr{E}(n \mid \theta)$ is commonly referred to in sequential analysis as the *average sample number*. In the case of a degenerate sequential test, that is, a test based on a sample of fixed size n, we have, of course

$$\mathscr{E}(n \mid \theta) = n.$$

It will be noted that since

$$\mathscr{E}(n \mid \theta) \geqslant \sum_{t=1}^{m} t p(t \mid \theta) + mP(G_m \mid \theta)$$

$m = 1, 2, \ldots$, it is necessary for $\lim_{m \to \infty} mP(G_m \mid \theta) = 0$, that is, for $N(\theta) = 0$, in order for $\mathscr{E}(n \mid \theta)$ to be finite.

For the case of a simple random sampling process from a distribution having a finite mean $\mathscr{E}(x)$, the following theorem concerning $\mathscr{E}(n)$, due to Wald (1945) and Blackwell (1946), will be useful in later sections:

15.2.1 *Suppose $(x_1, x_2, \ldots)$ is a sequence of independent random variables from a distribution with finite mean $\mathscr{E}(x)$, and let S be a sequential test for which $\mathscr{E}(n)$ is finite. If $x_1 + x_2 + \cdots + x_n$ denotes the sum of x's drawn until S terminates, then*

$$\mathscr{E}(x_1 + x_2 + \cdots + x_n) = \mathscr{E}(x) \cdot \mathscr{E}(n). \tag{15.2.12}$$

To prove **15.2.1** we note that

$$(15.2.13)\quad \mathscr{E}(x_1 + x_2 + \cdots + x_n) = \mathscr{E}(x_1) + \mathscr{E}(x_2 \mid x_1 \in G_1) \cdot P(G_1) + \mathscr{E}(x_3 \mid (x_1, x_2) \in G_2) \cdot P(G_2) + \cdots$$

where $G_1, G_2, \ldots$ are defined in (15.2.4). Since $x_1, x_2, \ldots$ are mutually independent, and have mean values all equal to $\mathscr{E}(x)$ it is evident that (15.2.13) reduces to

$$(15.2.14)\quad \mathscr{E}(x_1 + x_2 + \cdots + x_n) = \mathscr{E}(x)[1 + P(G_1) + P(G_2) + \cdots].$$

As before let $G_n^* = G_n^\circ \cup G_n'$, $n = 1, 2, \ldots$. Then since $G_1^*, \ldots, G_n^*, G_n$ are disjoint events in R_∞ for each n whose probabilities sum to 1, and since finiteness of $\mathscr{E}(n)$ implies termination of S with probability 1, we have

$$(15.2.15)\quad \begin{aligned} 1 &= \sum_{t=1}^{\infty} P(G_t^*) \\ P(G_t) &= \sum_{t'=t+1}^{\infty} P(G_{t'}^*), \qquad t = 1, 2, \ldots. \end{aligned}$$

Substituting these in (15.2.14), we find

$$(15.2.16)\quad \mathscr{E}(x_1 + x_2 + \cdots + x_n) = \mathscr{E}(x)\left[\sum_{t=1}^{\infty} tP(G_t^*)\right],$$

which is (15.2.12), thus concluding the argument for **15.2.1**.

(c) Criteria for Choosing a Sequential Test

It will be noted that in the formal definition of a sequential test no restrictions have been placed on the sets G_n°, G_n', G_n in our sequential tests, except that for each n they be disjoint sets satisfying (15.2.1). One of the basic problems in the construction of a sequential test S is that of choosing G_n°, G_n', and G_n. One obviously desirable requirement, and one which will always be imposed, is that the sequential process for S terminate with probability 1, that is,

$$(15.2.17)\quad N(\theta) = 0.$$

Another desirable requirement, as in the case of all statistical tests, is to fix the risk of a Type I error, that is, to require

$$(15.2.18)\quad L(\theta_0) = 1 - \alpha.$$

It will also be useful in comparing two sequential tests (as in the case of statistical tests based on samples of fixed size) to fix the risk β of a Type II error for some value of θ, say $\theta = \theta_1$, that is, to require

$$(15.2.19)\quad L(\theta_1) = \beta.$$

In considering θ_0 against the alternative θ_1, we shall let $\mathscr{H}_0$ denote the hypothesis that θ_0 is the true value of θ, and $\mathscr{H}_1$ the hypothesis that θ_1 is the true value.

The imposition of conditions (15.2.17), (15.2.18), and (15.2.19) implies, of course, that

$$M(\theta_0) = \alpha, \qquad M(\theta_1) = 1 - \beta. \tag{15.2.20}$$

Taking $\theta_0 < \theta_1$, the sets of values θ for which $\theta \leqslant \theta_0$, $\theta_0 < \theta < \theta_1$, $\theta \geqslant \theta_1$ are referred to respectively, as zones of preference for *acceptance*,

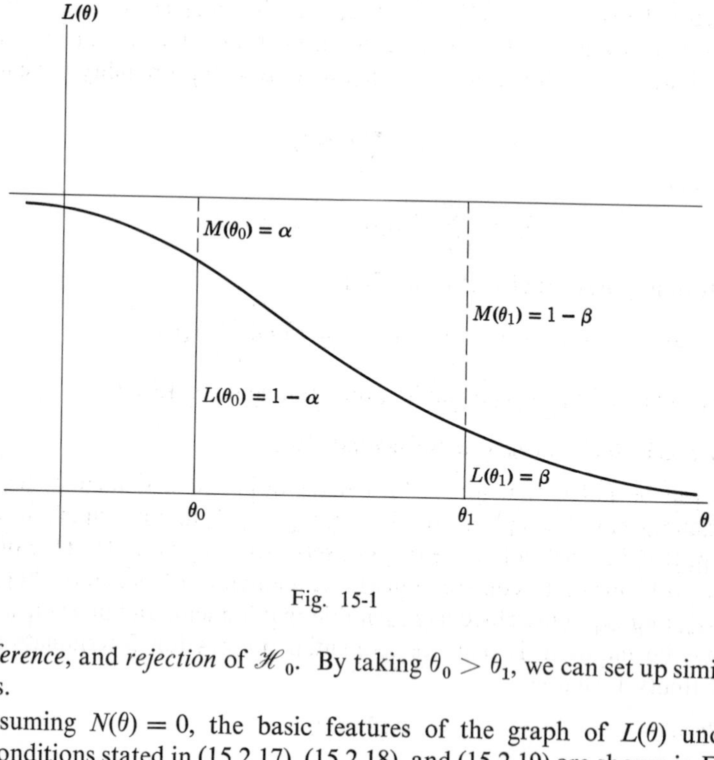

Fig. 15-1

indifference, and *rejection* of $\mathscr{H}_0$. By taking $\theta_0 > \theta_1$, we can set up similar zones.

Assuming $N(\theta) = 0$, the basic features of the graph of $L(\theta)$ under the conditions stated in (15.2.17), (15.2.18), and (15.2.19) are shown in Fig. 15.1.

A sequential test for which $L(\theta)$ satisfies (15.2.18) and (15.2.19) will be called a sequential test of *strength* $(\alpha, \theta_0; \beta, \theta_1)$.

If S_1 and S_2 are two sequential tests with operating characteristic functions $L_1(\theta)$ and $L_2(\theta)$ so that $L_1(\theta_0) = L_2(\theta_0) = 1 - \alpha$ while $L_1(\theta_1) < L_2(\theta_1)$, we shall say that S_1 is *stronger* than S_2 for testing $\mathscr{H}_0$ against $\mathscr{H}_1$. If S_1 is *stronger* than S_2 for all possible sequential tests S_2, then S_1, which we may denote by S^*, has *maximum strength*.

If the average sample numbers of two tests S_1 and S_2 of equal strength are $\mathscr{E}_1(n \mid \theta)$ and $\mathscr{E}_2(n \mid \theta)$ and if

$$\mathscr{E}_1(n \mid \theta) \leqslant \mathscr{E}_2(n \mid \theta), \qquad \theta = \theta_0, \theta_1, \tag{15.2.21}$$

with the equality not holding for both θ_0 and θ_1, S_1 will be considered as preferable to S_2 for testing $\mathscr{H}_0$ against $\mathscr{H}_1$.

If the relation (15.2.21) holds for both θ_0 and θ_1 and for all possible choices of S_2 different from S_1, then S_1 which we may denote by S^* is the sequential test of *maximum efficiency*. We shall show in Section 15.4 that such a sequential test S^* of maximum efficiency can be constructed by a method based on the probability ratio.

15.3 CARTESIAN SEQUENTIAL TESTS

(a) Description and Properties of the Test

Before dealing with the best possible sequential test that can be devised, it may be useful in fixing ideas to discuss a simple, even if not very efficient, sequential test based on the binomial waiting-time distribution (6.5.15). Suppose $(x_1, x_2, \ldots)$ is a sequence of independent random variables all having the same c.d.f. $F(x; \theta)$.

Consider the following sequential test S for testing $\mathscr{H}_0$ against $\mathscr{H}_1$ Let G_0°, G_0', G_0 be disjoint *initial sets* whose union is the entire real axis, and let $G_{(n)}^\circ$, $G_{(n)}'$, $G_{(n)}$ be the sets G_0°, G_0', G_0 in $R_1^{(n)}$, $n = 1, 2, \ldots$, where $R_1^{(n)}$ is the sample space of x_n, $n = 1, 2, \ldots$. Then it will be seen that the disjoint events $G_1^\circ, \ldots, G_n^\circ, G_1', \ldots, G_n', G_n$ defined in (15.2.4) can be constructed as follows:

$$\begin{aligned}
G_1^i &= \{x_1 \in G_{(1)}^i\}, \qquad i = {}^\circ, {}' \\
G_2^i &= \{x_1, x_2) \in G_{(1)} \times G_{(2)}^i\}, \qquad i = {}^\circ, {}' \\
&\;\;\vdots \\
G_n^i &= \{(x_1, \ldots, x_n) \in G_{(1)} \times \cdots \times G_{(n-1)} \times G_{(n)}^i\}, \quad i = {}^\circ, {}' \\
G_n &= \{(x_1, \ldots, x_n) \in G_{(1)} \times \cdots \times G_{(n)}\}.
\end{aligned} \tag{15.3.1}$$

Thus, the sequential process for S is simply to continue drawing from $(x_1, x_2, \ldots)$ one x at a time as long as an x falls in G_0 on the real axis. As soon as an x falls in G_0° we accept $\mathscr{H}_0$, or alternatively as soon as an x falls in G_0' we accept $\mathscr{H}_1$. It will be convenient to refer to such a test as a *Cartesian sequential test*.

Now let

$$(15.3.2)\qquad \begin{aligned} a(\theta) &= \int_{G_0^\circ} dF(x;\,\theta) \\ b(\theta) &= \int_{G_0'} dF(x;\,\theta) \\ c(\theta) &= \int_{G_0} dF(x;\,\theta) = 1 - a(\theta) - b(\theta). \end{aligned}$$

We shall denote $a(\theta)$, $b(\theta)$, and $c(\theta)$ by a, b, and c, and let $a(\theta_i) = a_i$, $b(\theta_i) = b_i$, and $c(\theta_i) = c_i$, $i = 0, 1$. Then

$$(15.3.3)\qquad P(G_t^\circ \mid \theta) = ac^{t-1}, \quad P(G_t' \mid \theta) = bc^{t-1}, \quad P(G_n \mid \theta) = c^n$$

$t = 1, \ldots, n$

$$(15.3.4)\qquad \begin{aligned} L_n(\theta) &= \frac{a(1 - c^n)}{1 - c} \\ M_n(\theta) &= \frac{b(1 - c^n)}{1 - c} \\ N_n(\theta) &= c^n. \end{aligned}$$

If G_0 is chosen so that $c(\theta) < 1$, we have $\lim_{n\to\infty} N_n(\theta) = 0$, and

$$(15.3.5)\qquad \begin{aligned} L(\theta) &= \frac{a}{a + b} \\ M(\theta) &= \frac{b}{a + b}. \end{aligned}$$

The probability that the sequential process terminates upon drawing x_n is given by

$$(15.3.6)\qquad p(n \mid \theta) = (a + b)c^{n-1}$$

and the average sample number is

$$(15.3.7)\qquad \mathscr{E}(n \mid \theta) = (a + b) \sum_{n=1}^{\infty} nc^{n-1} = \frac{1}{a + b}.$$

It will be seen that $p(n \mid \theta)$ is merely a special case of the binomial waiting-time distribution (6.5.15) for $k = 1$.

If the risks of Type I and Type II errors are to be α and β at $\theta = \theta_0$ and $\theta = \theta_1$, respectively, then G_0° and G_0' must satisfy the equations

$$(15.3.8)\qquad \frac{a_0}{a_0 + b_0} = 1 - \alpha, \qquad \frac{a_1}{a_1 + b_1} = \beta$$

that is, we must have

$$\frac{a_0}{b_0} = \frac{1 - \alpha}{\alpha}, \qquad \frac{a_1}{b_1} = \frac{\beta}{1 - \beta}. \tag{15.3.9}$$

If, for specified values of α and β, and $\mathscr{E}(n \mid \theta_0)$ $(= 1/(a_0 + b_0))$, a collection of choices exists for G_0° and G_0' the choice that yields the strongest Cartesian sequential test in case $F(x; \theta)$ has a p.d.f. (or p.f.) $f(x; \theta)$ is suggested by the Neyman-Pearson theorem **13.2.1**. That is, we take G_0° as the set of values of x for which the probability ratio satisfies

$$\frac{f(x; \theta_1)}{f(x; \theta_0)} \leqslant k_0, \tag{15.3.10}$$

and G_0' as the set for which

$$\frac{f(x; \theta_1)}{f(x; \theta_0)} \geqslant k_1, \tag{15.3.11}$$

where $k_1 > k_0 > 0$ are chosen so that (15.3.8) is satisfied for G_0° and G_0' for specified values of α, β and $\mathscr{E}(n \mid \theta_0)$. In other words, any other choice of sets G_0° and G_0', say $G_0^{\circ *}$ and $G_0'^{*}$, which leaves $a_0/(a_0 + b_0)$ fixed at α and $\mathscr{E}(n \mid \theta_0)$ fixed at n_0, say (and hence the probability of a Type I error and average sample number at θ_0 fixed), gives a Cartesian sequential test S for which $a_1/(a_1 + b_1) > \beta$, that is, for which the probability of a Type II error exceeds that for the sequential test based on G_0° and G_0' as provided by (15.3.10) and (15.3.11).

This statement holds under essentially the same conditions as those stated in **13.2.1**. The proof is similar to that of **13.2.1**, and is left as an exercise for the reader (Problem 15.6).

Remark. By drawing r x's at a time from the population, one can devise an *r-fold Cartesian sequential test S*. It will be sufficient to consider only the case where the acceptance and rejection sets in the sample space R_r of $(x_1, \ldots, x_r)$ are determined by the probability ratio. Thus, let G_0° be the set of points $(x_1, \ldots, x_r)$ satisfying

$$\prod_{i=1}^{r} \frac{f(x_i; \theta_1)}{f(x_i; \theta_0)} \leqslant k_0 \tag{15.3.12}$$

and G_0' the set satisfying

$$\prod_{i=1}^{r} \frac{f(x_i; \theta_1)}{f(x_i; \theta_0)} \geqslant k_1 \tag{15.3.13}$$

while G_0 is the set satisfying neither inequality, where $k_1 > k_0 > 0$ are chosen so as to satisfy (15.3.8), $a(\theta)$, $b(\theta)$, and $c(\theta)$ being defined as $P(G_0^\circ \mid \theta)$, $P(G_0' \mid \theta)$, and $P(G_0 \mid \theta)$. It can be shown by argument similar to that for **13.2.1** that this choice of G_0° and G_0' yields a stronger Cartesian sequential test than that for any other choice of G_0° and G_0' in the sample space of $(x_1, \ldots, x_r)$ which leaves $a_0/(a_0 + b_0)$ fixed at α and $\mathscr{E}(n \mid \theta_0)$ fixed.

It will be evident to the reader that if there is difficulty in fixing Type I and Type II errors at α and β for a one-fold Cartesian sequential test, one can fix these errors as close to α and β as one pleases by using an r-fold Cartesian sequential test for a suitably chosen r.

(b) Application to Nonparametric Testing

As mentioned earlier, the Cartesian sequential test may not be the best sequential test that can be devised. However, if nothing is known about $F(x; \theta)$ except that it is continuous, then a Cartesian sequential test provides a nonparametric sequential test. To see this let $F(x; \theta_0)$, $F(x; \theta_1)$ be continuous, and let them be denoted by $F_0(x)$ and $F_1(x)$ respectively. Let us choose G_0° and G_0' as the intervals $(-\infty, a)$ and $(b, +\infty)$, $a < b$, then (15.3.8) can be written as

$$\text{(15.3.14)} \qquad \frac{F_0(a)}{F_0(a) + 1 - F_0(b)} = 1 - \alpha, \qquad \frac{F_1(a)}{F_1(a) + 1 - F_1(b)} = \beta$$

and the Cartesian sequential test based on this choice of G_0° and G_0' becomes a test for accepting a continuous c.d.f. $F_0(x)$ with values $F_0(a)$ and $1 - \alpha F_0(a)/(1 - \alpha)$ at a and b respectively against the alternative of accepting a continuous c.d.f. with values $F_1(a)$ and $1 - (1 - \beta)F_1(a)/\beta$ at a and b respectively. The average sample size is $[F_0(a) + 1 - F_0(b)]^{-1}$ if $F_0(x)$ is the distribution actually sampled, and $[F_1(a) + 1 - F_1(b)]^{-1}$ if $F_1(x)$ is the distribution actually sampled.

15.4 THE PROBABILITY RATIO SEQUENTIAL TEST

(a) Definition of the Test

The discussion in the preceding section of the choices of G_0°, G_0', and G_0 which lead to strongest Cartesian sequential tests for testing $\mathscr{H}_0$ against $\mathscr{H}_1$, suggests what, indeed, turns out to be the best possible choice of the sets $G_1^\circ, \ldots, G_n^\circ$, $G_1', \ldots, G_n'$, G_n, $n = 1, 2, \ldots$, for a sequential test. Namely, to choose G_n°, G_n', and G_n, as disjoint events in R_n satisfying (15.2.1) for $n = 1, 2, \ldots$, defined for each n by the respective inequalities:

$$\text{(15.4.1)} \qquad \frac{Q_{1n}}{Q_{0n}} \leqslant k_0, \qquad \frac{Q_{1n}}{Q_{0n}} \geqslant k_1, \qquad k_0 < \frac{Q_{1n}}{Q_{0n}} < k_1$$

where

$$\text{(15.4.2)} \qquad Q_{in} = \prod_{t=1}^{n} f(x_t; \theta_i), \qquad i = 0, 1$$

and where $f(x; \theta_0)$ and $f(x; \theta_1)$ are p.f.'s or p.d.f.'s absolutely continuous with respect to each other. It is sufficient to consider only the case of p.d.f.'s throughout this section. The results obtained for p.d.f.'s hold with trivial changes for p.f.'s.

This test was originally proposed by Wald (1945), and is called the *probability ratio sequential test.* The test therefore consists of accepting $\mathscr{H}_0$, accepting $\mathscr{H}_1$, or drawing x_{n+1}, after drawing x_n according to whether $(x_1, \ldots, x_n)$ satisfies the first, second, or third inequality in (15.4.1), $n = 1, 2, \ldots$, where k_0 and k_1 are fixed so that

(15.4.3) $$L(\theta_0) = 1 - \alpha, \qquad L(\theta_1) = \beta$$

that is, so that the probabilities of Type I and Type II errors are α and β.

(b) Properties of the Distribution Function of Number of Trials Required to Terminate the Probability Ratio Sequential Process

First let us consider the question of whether the probability ratio sequential process terminates with probability 1. Let

(15.4.4) $$z = \log\left(\frac{f(x; \theta_1)}{f(x; \theta_0)}\right)$$

and let $H(z, \theta)$ be the c.d.f. of z for a fixed value of θ in the parameter space Ω. When $\theta_0 \neq \theta_1$, z is assumed to be a nondegenerate random variable. Let

(15.4.5) $$z_t = \log\left(\frac{f(x_t; \theta_1)}{f(x_t; \theta_0)}\right), \qquad t = 1, 2, \ldots.$$

Then $(z_1, z_2, \ldots)$ is a stochastic process generated by simple random sampling from a c.d.f. $H(z; \theta)$ and the sequential process terminates for the smallest integer n for which the inequality

(15.4.6) $$\log k_0 < z_1 + \cdots + z_n < \log k_1$$

fails to hold. Let us keep in mind that G_n is the event for which the inequality (15.4.6) holds, and hence is the event resulting in the drawing of more than n x's before the sequential process terminates.

Let $\zeta_1 = z_1 + \cdots + z_r$, $\zeta_2 = z_{r+1} + \cdots + z_{2r}, \ldots$, $\zeta_i = z_{(i-1)r+1} + \cdots + z_{ir}, \ldots$. If the sequential process never terminates then we must have

(15.4.7) $$\zeta_i^2 < D^2, \qquad i = 1, 2, \ldots,$$

where $D = |\log k_0| + |\log k_1|$. Since $\zeta_1, \zeta_2, \ldots$ is a sequence of independent random variables having the same c.d.f., say $J(\zeta; \theta)$, the probability that the inequalities (15.4.7) for $i = 1, 2, \ldots, m$ hold is p^m where

(15.4.8) $$p = P(\zeta_1^2 < D^2) = \cdots = P(\zeta_m^2 < D^2).$$

Since z is nondegenerate, it has a positive variance σ^2. The variance of ζ is $r\sigma^2$ and can be made to exceed D^2 by choosing $r > D^2/\sigma^2$, in which case we have $\mathscr{E}(\zeta^2) > D^2$, and hence

$$P(\zeta^2 < D^2) = p < 1.$$

Therefore, taking $r > D^2/\sigma^2$, and letting $m \to \infty$, we find the probability of failure of termination of the sequential process to be $\lim_{m\to\infty} p^m = 0$.

Summarizing,

15.4.1 *If the random variable z defined by* (15.4.4) *has a finite mean, and finite positive variance, the probability ratio sequential process terminates with probability* 1.

As a matter of fact, a stronger theorem due to Stein (1946) holds for the distribution function of the number of trials required to terminate the sequential process. A modified form of Stein's theorem is as follows:

15.4.2 *If the random variable z is nondegenerate all moments of n exist.*

To establish **15.4.2** it is sufficient to consider the moment-generating function $\psi(u)$ of n

$$\psi(u) = \mathscr{E}(e^{un}) = \sum_{t=1}^{\infty} e^{ut} P(G_t^* \mid \theta) \tag{15.4.9}$$

where $G_t^* = G_t^\circ \cup G_t'$ and is the event resulting in the termination of the sequential process at the tth drawing. By breaking up the series on the right of (15.4.9) into blocks of r terms each, it will be seen that for $u \geqslant 0$

$$\begin{aligned}\psi(u) &\leqslant e^{ru}P(0 < n \leqslant r) + e^{2ru}P(r < n \leqslant 2r) + \cdots \\ &\leqslant e^{ru}P(n > 0) + e^{2ru}P(n > r) + \cdots.\end{aligned}$$

But we know from (15.4.8) that if z has positive variance

$$P(n \geqslant mr) \leqslant p^m.$$

Hence

$$\psi(u) \leqslant e^{ru} + e^{2ru}p + e^{3ru}p^2 + \cdots = e^{ru}(1 - e^{ru}p)^{-1}. \tag{15.4.10}$$

If $u \leqslant 0$, we similarly find

$$\psi(u) \leqslant e^{u}(1 - e^{ru}p)^{-1}. \tag{15.4.10a}$$

Therefore, if u is any (real) number which satisfies $e^{ru}p < 1, \psi(u)$ converges. The u interval for which this inequality holds contains $u = 0$ as an interior point and it follows that the kth derivative of $\psi(u)$ exists at $u = 0$, which yields the kth moment of u, for $k = 1, 2, \ldots$.

Finally, the following result (which can be restated as a corollary to **15.2.1**) is important in obtaining an expression for $\mathscr{E}(n \mid \theta)$, and in examining the efficiency of a probability ratio sequential test:

15.4.3 *If z is a nondegenerate random variable, then*

$$\mathscr{E}(z_1 + z_2 + \cdots + z_n \mid \theta) = \mathscr{E}(z \mid \theta) \cdot \mathscr{E}(n \mid \theta). \tag{15.4.11}$$

The finiteness of $\mathscr{E}(n \mid \theta)$ follows at once from **15.4.2** and the remaining argument follows from **15.2.1**.

(c) Determination of the Boundary Constants k_0 and k_1 for the Probability Ratio Sequential Tests

Now consider the problem of determining the *boundary* constants k_0 and k_1 for a probability ratio sequential test of strength $(\alpha, \theta_0; \beta, \theta_1)$. The exact determination of k_0 and k_1 is a difficult problem, although close inequalities and good approximations are fairly easy to find.

It follows from the definition of the probability ratio sequential test, that $\mathscr{H}_0$ will be accepted upon the drawing of x_n if the first inequality in (15.4.1) is satisfied. Since we have denoted this event by G_n°, we have at all points in G_n°

$$Q_{1n} \leqslant k_0 Q_{0n}, \qquad n = 1, 2, \ldots. \tag{15.4.12}$$

Referring to (15.2.8), we have

$$L(\theta) = P(G_1^\circ \mid \theta) + P(G_2^\circ \mid \theta) + \cdots. \tag{15.4.13}$$

If $L(\theta)$ is to be of strength $(\alpha, \theta_0; \beta, \theta_1)$, then we must have

$$L(\theta_0) = 1 - \alpha, \qquad L(\theta_1) = \beta. \tag{15.4.14}$$

It follows from (15.4.12) that

$$P(G_n^\circ \mid \theta_1) \leqslant k_0 P(G_n^\circ \mid \theta_0), \qquad n = 1, 2, \ldots. \tag{15.4.15}$$

Therefore

$$\sum_{n=1}^{\infty} P(G_n^\circ \mid \theta_1) \leqslant k_0 \sum_{n=1}^{\infty} P(G_n^\circ \mid \theta_0)$$

that is,

$$L(\theta_1) \leqslant k_0 L(\theta_0). \tag{15.4.16}$$

Substituting from (15.4.14) we find that k_0 must satisfy the inequality

$$k_0 \geqslant \frac{\beta}{1 - \alpha}.$$

In a similar manner by considering the event G'_n, resulting in the acceptance of $\mathscr{H}_1$ upon drawing x_n, $n = 1, 2, \ldots$, in which the second inequality of (15.4.1) holds, we find that

$$M(\theta_1) \geqslant k_1 M(\theta_0). \tag{15.4.17}$$

Since $N(\theta) = 0$, we have $M(\theta) = 1 - L(\theta)$. Making the substitution in (15.4.17) and using (15.4.14), we find that k_1 must satisfy

$$k_1 \leqslant \frac{1 - \beta}{\alpha}.$$

Therefore,

15.4.4 *The constants k_0 and k_1 for the probability ratio sequential test of strength $(\alpha, \theta_0; \beta, \theta_1)$ satisfy the inequalities*

$$k_0 \geqslant \frac{\beta}{1 - \alpha}, \qquad k_1 \leqslant \frac{1 - \beta}{\alpha}. \tag{15.4.18}$$

Let us now examine the actual strength of a probability ratio sequential test if, for a given α and β, we choose

$$k_0 = \frac{\beta}{1 - \alpha}, \qquad k_1 = \frac{1 - \beta}{\alpha}. \tag{15.4.19}$$

It is convenient to think of $(\alpha, \theta_0; \beta, \theta_1)$ as the *intended strength* of the test, and $(\alpha', \theta_0; \beta', \theta_1)$ as the *actual strength* of the test. Then it follows from (15.4.18) that

$$\frac{\beta}{1 - \alpha} \geqslant \frac{\beta'}{1 - \alpha'} \quad \text{and} \quad \frac{1 - \beta}{\alpha} \leqslant \frac{1 - \beta'}{\alpha'}. \tag{15.4.20}$$

It follows from these two inequalities that

$$\beta' \leqslant \frac{\beta}{1 - \alpha}, \quad \text{and} \quad \alpha' \leqslant \frac{\alpha}{1 - \beta} \tag{15.4.21}$$

and also that

$$\alpha' + \beta' \leqslant \alpha + \beta. \tag{15.4.22}$$

The last inequality simply states that the sum of the probabilities of *actual* Type I and Type II errors for the choice of k_0 and k_1 given by (15.4.19) cannot exceed the sum of the probabilities of the *intended* Type I and Type II errors. Therefore, we have

15.4.5 *If the constants k_0 and k_1 for a probability ratio sequential test of intended strength $(\alpha, \theta_0; \beta, \theta_1)$ are actually chosen as $\beta/(1 - \alpha)$ and $\alpha/(1 - \beta)$, respectively, the actual strength of the resulting test is $(\alpha', \theta_0; \beta', \theta_1)$ where α' and β' satisfy inequalities* (15.4.21) *and* (15.4.22).

In practical applications the intended strength $(\alpha, \theta_0; \beta, \theta_1)$ has values of α and β rarely exceeding 0.10, and the inequalities (15.4.21) and (15.4.22) indicate that the difference between the actual strength and intended strength from a practical point of view is not very important.

(d) The Operating Characteristic Function of the Probability Ratio Sequential Test

Thus far we have considered the operating characteristic function $L(\theta)$ for only two values of θ, namely, θ_0 and θ_1. The exact determination of $L(\theta)$ for an arbitrary θ in the parameter space Ω is a difficult problem. However, we can determine an approximate expression for $L(\theta)$ without undue difficulty.

It will be convenient at the outset to establish the following lemma due to Wald (1945):

15.4.6 *Let* $z = \log f(x; \theta_1) - \log f(x; \theta_0)$, *where* x *is a random variable having* p.d.f. $f(x, \theta)$, θ *being a point in a parameter space* Ω *containing* θ_0 *and* θ_1. *If*

(i) $\mathscr{E}(e^{zh})$ *exists for every real* h,

(ii) $\mathscr{E}(z) \neq 0$,

(iii) *for some* $\delta_1 > 0$ *and* $0 < \delta_2 < 1$, $P(e^z > 1 + \delta_1) > 0$, *and* $P(e^z < 1 - \delta_2) > 0$,

then, for each θ *there is an* h, *say* $h(\theta)$, $\neq 0$ *such that*

$$\int_{-\infty}^{\infty} \left(\frac{f(x; \theta_1)}{f(x; \theta_0)}\right)^h f(x; \theta)\, dx = 1 \tag{15.4.23}$$

that is,

$$\mathscr{E}(e^{zh(\theta)}) = 1.$$

To establish **15.4.6**, let E_1 be the set of values of x for which $e^z > 1 + \delta_1$ and E_2 be the set for which $e^z < 1 - \delta_2$. Then we have for $h > 0$,

$$\mathscr{E}(e^{zh}) \geqslant \int_{E_1} e^{zh} f(x; \theta)\, dx > (1 + \delta_1)^h P(E_1) \tag{15.4.24}$$

and for $h < 0$

$$\mathscr{E}(e^{zh}) \geqslant \int_{E_2} e^{zh} f(x; \theta)\, dx > (1 - \delta_2)^h P(E_2). \tag{15.4.25}$$

Since $P(E_1)$ and $P(E_2)$ are $\neq 0$, it is evident from (15.4.24) and (15.4.25) that

$$\lim_{h \to +\infty} \mathscr{E}(e^{zh}) = \lim_{h \to -\infty} \mathscr{E}(e^{zh}) = +\infty. \tag{15.4.26}$$

Denoting $\mathscr{E}(e^{zh})$ by $\psi(h)$ and the first and second derivatives of $\psi(h)$ by $\psi'(h)$ and $\psi''(h)$, we have

$$\psi'(0) = \mathscr{E}(z) \neq 0 \tag{15.4.27}$$

and

$$\psi''(h) = \mathscr{E}(z^2 e^{zh}) > 0. \tag{15.4.28}$$

Since $\psi(0) = 1$, $\psi'(0) \neq 0$, and $\psi''(h) > 0$ it is evident that $\psi(h)$, which depends on θ as well as h, has a minimum less than 1 and furthermore that $\psi(h) - 1 = 0$ has two roots, $h = 0$, and $h = h(\theta)$, where $h(\theta) \neq 0$, thus concluding the argument for **15.4.6.**

Now let us consider the problem of approximating $L(\theta)$. The event G_n° for which $\mathscr{H}_0$ is accepted upon drawing x_n is determined by the first inequality in (15.4.1). The same event G_n° is determined by the inequality

$$Q_{*n} \leqslant k_0^h Q_n \tag{15.4.29}$$

where

$$Q_{*n} = \left(\frac{Q_{1n}}{Q_{0n}}\right)^h Q_n \quad \text{and} \quad Q_n = \prod_{t=1}^{n} f(x_t; \theta) \tag{15.4.30}$$

and h is any real number. Now Q_{*n} can be written as

$$Q_{*n} = \prod_{t=1}^{n} \left[\left(\frac{f(x_t; \theta_1)}{f(x_t; \theta_0)}\right)^h f(x_t; \theta)\right]. \tag{15.4.31}$$

Under the conditions of **15.4.6**, there is for each θ, an h, say $h(\theta)$, $\neq 0$ such that

$$\int \left(\frac{f(x; \theta_1)}{f(x; \theta_0)}\right)^{h(\theta)} f(x, \theta)\, dx = 1. \tag{15.4.32}$$

Therefore the function $f^*(x; \theta)$ defined by

$$f^*(x; \theta) = \left(\frac{f(x; \theta_1)}{f(x; \theta_0)}\right)^{h(\theta)} f(x; \theta) \tag{15.4.33}$$

has the properties of a p.d.f.

Let $P^*(G_n^\circ \mid \theta)$ be the probability of the event G_n° evaluated from $f^*(x; \theta)$ in exactly the same way that $P(G_n^\circ \mid \theta)$ is the probability of G_n° evaluated from $f(x; \theta)$. Defining $L^*(\theta)$ similar to $L(\theta)$ in (15.2.8), let

$$L^*(\theta) = \sum_{n=1}^{\infty} P^*(G_n^\circ \mid \theta) = P^*(G^\circ \mid \theta).$$

Then by argument similar to that used in deriving (15.4.16) we find

$$L^*(\theta) \leqslant k_0^{h(\theta)} L(\theta). \tag{15.4.34}$$

The event G_n' which results in accepting $\mathscr{H}_1$ and which is defined by the second inequality of (15.4.1) is equivalently defined by the inequality

$$Q_{*n} \geqslant k_1^h Q_n. \tag{15.4.35}$$

Following a line of argument similar to that by which (15.4.34) was obtained we find, corresponding to (15.4.17),

$$1 - L^*(\theta) \geqslant k_1^{h(\theta)}(1 - L(\theta)). \tag{15.4.36}$$

Reasoning similar to that underlying the replacement of the inequalities in (15.4.18) by the equalities (15.4.19) for approximating values of k_0 and k_1 suggests replacement of the inequalities (15.4.34) and (15.4.36) by equalities in order to approximate $L(\theta)$. The resulting approximation is found to be

$$L(\theta) \cong \frac{1 - k_1^{h(\theta)}}{k_0^{h(\theta)} - k_1^{h(\theta)}}. \tag{15.4.37}$$

Since $h(\theta_0) = 1$ and $h(\theta_1) = -1$, it will be noted that if the values of k_0 and k_1 are approximated by (15.4.19) the approximation (15.4.37) gives the correct values for $L(\theta_0)$ and $L(\theta_1)$ for the probability ratio sequential test of strength $(\alpha, \theta_0; \beta, \theta_1)$, namely, those in (15.4.14).

The approximation (15.4.37) for $L(\theta)$ is satisfactory for practical purposes. To examine precisely how accurate the approximation is for values of θ different from θ_0 and θ_1 is a rather tedious piece of analysis which is omitted. But the details of such an analysis will be found in Wald's book (1947*a*).

We may summarize as follows:

15.4.7 *Under the conditions of* **15.4.6**, *an approximation for the operating characteristic function* $L(\theta)$ *for a probability ratio sequential test of strength* $(\alpha, \theta_0; \beta, \theta_1)$ *is given by* (15.4.37), *where* $h(\theta)$ *satisfies* (15.4.32), *and where* k_0 *and* k_1 *are approximated by* (15.4.19).

(e) The Average Sample Number of the Probability Ratio Sequential Test

To obtain an approximate expression for $\mathscr{E}(n \mid \theta)$, we may proceed as follows: If $\mathscr{E}(z \mid \theta) \neq 0$, we find from (15.4.11) that

$$\mathscr{E}(n \mid \theta) = \frac{\mathscr{E}(z_1 + z_2 + \cdots + z_n \mid \theta)}{\mathscr{E}(z \mid \theta)}. \tag{15.4.38}$$

As before let $G^\circ = G_1^\circ \cup G_2^\circ \cup \cdots$ and $G' = G_1' \cup G_2' \cup \cdots$. Then $P(G^\circ \mid \theta) = L(\theta)$, and $P(G' \mid \theta) = 1 - L(\theta)$ and we can write

$$\begin{aligned} \mathscr{E}(z_1 + z_2 + \cdots + z_n \mid \theta) &= \mathscr{E}(z_1 + z_2 + \cdots + z_n \mid G^\circ; \theta)L(\theta) \\ &\quad + \mathscr{E}(z_1 + z_2 + \cdots + z_n \mid G'; \theta)(1 - L(\theta)). \end{aligned} \tag{15.4.39}$$

But $\mathscr{E}(z_1 + z_2 + \cdots + z_n \mid G^\circ; \theta) \leqslant \log k_0$ and $\mathscr{E}(z_1 + z_2 + \cdots + z_n \mid G'; \theta) \geqslant \log k_1$. Replacing k_0 and k_1 in these inequalities by the values indicated in (15.4.19) and replacing the inequalities by equalities we find that (15.4.38) yields the following approximation for $\mathscr{E}(n \mid \theta)$:

$$(15.4.40)\qquad \mathscr{E}(n \mid \theta) \cong \frac{L(\theta) \log \left(\frac{\beta}{1-\alpha}\right) + (1 - L(\theta)) \log \left(\frac{1-\beta}{\alpha}\right)}{\mathscr{E}(z \mid \theta)}.$$

At θ_0 and θ_1 the approximate values of $\mathscr{E}(n \mid \theta)$ are

$$(15.4.41)\qquad \begin{aligned} \mathscr{E}(n \mid \theta_0) &\cong \frac{(1-\alpha) \log \left(\frac{\beta}{1-\alpha}\right) + \alpha \log \left(\frac{1-\beta}{\alpha}\right)}{\mathscr{E}(z \mid \theta_0)} \\ \mathscr{E}(n \mid \theta_1) &\cong \frac{\beta \log \left(\frac{\beta}{1-\alpha}\right) + (1-\beta) \log \left(\frac{1-\beta}{\alpha}\right)}{\mathscr{E}(z \mid \theta_1)}. \end{aligned}$$

These approximations are satisfactory for practical purposes. The problem of placing bounds on the error of the approximation in (15.4.40) requires considerable analysis which we omit. The reader interested in the details is referred to Wald (1947*a*).

(f) The Efficiency of the Probability Ratio Sequential Test

The following theorem due to Wald (1945) provides lower bounds for $\mathscr{E}(n \mid \theta)$ at θ_0 and θ_1 for *any* sequential test:

15.4.8 *Let $(x_1, x_2, \ldots)$ be a sequence of independent random variables having the p.d.f. $f(x; \theta)$, such that $\mathscr{E}(z \mid \theta) \neq 0$ where z is defined in (15.4.4). Let S be any sequential test of strength $(\alpha, \theta_0; \beta, \theta_1)$ which terminates with probability* 1. *Then*

$$(15.4.42)\qquad \begin{aligned} \mathscr{E}(n \mid \theta_0) &\geqslant \frac{(1-\alpha) \log \left(\frac{\beta}{1-\alpha}\right) + \alpha \log \left(\frac{1-\beta}{\alpha}\right)}{\mathscr{E}(z \mid \theta_0)} \\ \mathscr{E}(n \mid \theta_1) &\geqslant \frac{\beta \log \left(\frac{\beta}{1-\alpha}\right) + (1-\beta) \log \left(\frac{1-\beta}{\alpha}\right)}{\mathscr{E}(z \mid \theta_1)}. \end{aligned}$$

There will be no ambiguity if we denote by G_n°, G_n', G_n, the basic events in the sample space of $(x_1, \ldots, x_n)$, $n = 1, 2, \ldots$, satisfying (15.2.1) which define an arbitrary sequential test S. [In case S is a probability ratio sequential test, G_n°, G_n', G_n, $n = 1, 2, \ldots$, are defined by (15.2.1) and the three inequalities in (15.4.1).]

As usual let $z_1, z_2, \ldots$ be defined as in (15.4.5). It follows from (15.4.11) that for S we will have

$$\mathscr{E}(n \mid \theta) = \frac{\mathscr{E}(z_1 + z_2 + \cdots + z_n \mid \theta)}{\mathscr{E}(z \mid \theta)}. \tag{15.4.43}$$

But

$$\mathscr{E}(z_1 + z_2 + \cdots + z_n \mid \theta) = \mathscr{E}(z_1 + z_2 + \cdots + z_n \mid G^\circ; \theta)P(G^\circ \mid \theta) + \mathscr{E}(z_1 + z_2 + \cdots + z_n \mid G'; \theta)P(G' \mid \theta). \tag{15.4.44}$$

First, let us consider $\mathscr{E}(n \mid \theta_0)$. Since S is of strength $(\alpha, \theta_0; \beta, \theta_1)$, we have

$$P(G^\circ \mid \theta_0) = 1 - \alpha, \qquad P(G' \mid \theta_0) = \alpha. \tag{15.4.45}$$

Making use of the fact that for any random variable y

$$\mathscr{E}(y) \leqslant \log \mathscr{E}(e^y), \tag{15.4.46}$$

we have

(15.4.47)

$$\mathscr{E}(z_1 + z_2 + \cdots + z_n \mid G^\circ; \theta_0) \leqslant \log \mathscr{E}(e^{z_1 + z_2 + \cdots + z_n} \mid G^\circ; \theta_0) = \log \mathscr{E}\left(\frac{Q_{1n}}{Q_{0n}} \,\middle|\, (G^\circ; \theta_0)\right).$$

where $\mathscr{E}\left(\frac{Q_{1n}}{Q_{0n}} \,\middle|\, G^\circ; \theta_0\right)$ is the conditional mean value, over G° $(= G_1^\circ \cup G_2^\circ \cup \cdots)$ when $\theta = \theta_0$, of the product $\left(\frac{f(x_1; \theta_1)}{f(x_1; \theta_0)}\right)\left(\frac{f(x_2; \theta_1)}{f(x_2; \theta_0)}\right) \cdots$ taken until the sequential procedure terminates. But

$$\mathscr{E}\left(\frac{Q_{1n}}{Q_{0n}} \,\middle|\, G^\circ; \theta_0\right) = P(G^\circ \mid \theta_1)/P(G^\circ \mid \theta_0) = \frac{\beta}{1 - \alpha}. \tag{15.4.48}$$

Therefore

$$\mathscr{E}(z_1 + z_2 + \cdots + z_n \mid G^\circ; \theta_0) \leqslant \log \frac{\beta}{1 - \alpha}. \tag{15.4.49}$$

Similarly,

$$\mathscr{E}(z_1 + z_2 + \cdots + z_n \mid G'; \theta_0) \leqslant \log \frac{1 - \beta}{\alpha}. \tag{15.4.50}$$

Substituting from (15.4.45), (15.4.49), and (15.4.50) into (15.4.44), we obtain the first inequality in (15.4.42). The second inequality is obtained in an entirely similar manner, thus concluding the argument for **15.4.8.**

Further studies of lower bounds for $\mathscr{E}(n \mid \theta)$ have been made by Hoeffding (1960) and Kiefer and Weiss (1957). Anderson (1960) and

Donnelly (1957) have made some modifications of the probability test in the case of sampling from a normal distribution with known variance but unknown mean, so as to reduce $\mathscr{E}(n \mid \theta)$.

In the case of a probability ratio sequential test, that is, where G_n°, G_n', G_n are determined by (15.2.1) and by the inequalities in (15.4.1), we have seen that

$$\begin{aligned} &\mathscr{E}(z_1 + z_2 + \cdots + z_n \mid G^\circ; \theta) \leqslant \log k_0 \\ &\mathscr{E}(z_1 + z_2 + \cdots + z_n \mid G'; \theta) \geqslant \log k_1. \end{aligned} \tag{15.4.51}$$

If, when a sample point $(x_1, \ldots, x_n)$ falls in G_n°, (that is, when $z_1 + \cdots + z_n \leqslant \log k_0$) upon drawing x_n, we arbitrarily assign $z_1 + \cdots + z_n$ the value $\log k_0$, and if when $(x_1, \ldots, x_n)$ falls in G_n' (that is, when $z_1 + \cdots + z_n \geqslant \log k_1$) upon drawing x_n, we arbitrarily assign $z_1 + \cdots + z_n$ the value $\log k_1$, then the inequalities in (15.4.49) and (15.4.50) become equalities and the expressions (15.4.41) become equalities for $\mathscr{E}(n \mid \theta_0)$ and $\mathscr{E}(n \mid \theta_1)$. Thus, it is seen that the equalities (15.4.42) are actually realized in the case of a probability ratio sequential test modified by the approximation under which $z_1 + \cdots + z_n$, $n = 1, 2, \ldots$, is assigned the value $\log k_0$ or $\log k_1$ in accordance with the rule mentioned above. Under these conditions, therefore, no sequential test exists which would be more efficient for testing $\mathscr{H}_0$ against $\mathscr{H}_1$ than the modified probability ratio test. This heuristic argument concerning the optimum character of the probability ratio sequential test was originally put forth by Wald (1945). A more complete and rigorous argument was given later by Wald and Wolfowitz (1948).

(g) Truncation of Probability Ratio Sequential Test

If the probability ratio sequential test does not terminate for $n = 1, \ldots, N - 1$, suppose the following rule is adopted for terminating the test upon drawing x_N:

Accept $\mathscr{H}_0$ if

$$\log k_0 < z_1 + \cdots + z_N \leqslant 0; \tag{15.4.52}$$

accept $\mathscr{H}_1$ if

$$0 < z_1 + \cdots + z_N < \log k_1. \tag{15.4.53}$$

Let this *truncated sequential test* be denoted by S_N, and let α_N and β_N be Type I and Type II errors associated with S_N. We shall consider the problem of determining upper bounds for α_N and β_N. Let $G_{(N)}^\circ$ and $G_{(N)}'$ be the events in R_N in which $\mathscr{H}_0$ and $\mathscr{H}_1$ are accepted respectively by the test S_N. Then $G_{(N)}^\circ$ and $G_{(N)}'$ are disjoint events whose union is R_N.

It will be more convenient to consider $G^{\circ}_{(N)}$ and $G'_{(N)}$ as cylinder sets in R_∞; They are disjoint, of course, and their union is R_∞.

We have

(15.4.54) $$P(G'_{(N)} \mid \theta_0) = \alpha_N.$$

As before, we let G° and G' denote the events in R_∞ in which $\mathscr{H}_0$ and $\mathscr{H}_1$ are accepted respectively, in the (nontruncated) probability ratio sequential test S. We recall that

(15.4.55) $$P(G' \mid \theta_0) = \alpha.$$

Let G'^* be the event in R_∞ in which $\mathscr{H}_1$ is accepted in the truncated case and rejected ($\mathscr{H}_0$ is accepted) in the nontruncated case.

Then

(15.4.56) $$G'_{(N)} \subset (G' \cup G'^*).$$

Now if J denotes the event for R_∞ in which (15.4.53) holds we have

$$G'^* \subset J,$$

and hence

(15.4.57) $$G'_{(N)} \subset (G' \cup J).$$

Now if we choose N large enough to make $z_1 + \cdots + z_N$ approximately normal, we have

(15.4.58) $$P(0 < z_1 + \cdots + z_N < \log k_1 \mid \theta_0) = \Phi(y_0') - \Phi(y_0) + O\left(\frac{1}{\sqrt{N}}\right)$$

where

(15.4.59) $$y_0 = -\sqrt{N}\mathscr{E}(z \mid \theta_0)/\sigma(z \mid \theta_0)$$
$$y_0' = \sqrt{N}\left[\frac{1}{N}\log k_1 - \mathscr{E}(z \mid \theta_0)\right]\bigg/\sigma(z \mid \theta_0)$$

and $\Phi(y)$ is the c.d.f. of $N(0, 1)$, whereas $\mathscr{E}(z \mid \theta_0)$ and $\sigma^2(z \mid \theta_0)$ are the mean and variance of the random variable $z = \log\left(\frac{f(x; \theta_1)}{f(x; \theta_0)}\right)$ determined from the p.d.f. $f(x; \theta_0)$.

Therefore, except for terms of order $\frac{1}{\sqrt{N}}$, $\alpha + \Phi(y_0') - \Phi(y_0)$ is an upper bound for α_N. In a similar manner, we find that except for terms of order $\frac{1}{\sqrt{N}}$, $\beta + \Phi(y_1') - \Phi(y_1)$ is an upper bound for β_N where

(15.4.59a) $$y_1 = \sqrt{N}\left[\frac{1}{N}\log k_0 - \mathscr{E}(z \mid \theta_1)\right]\bigg/\sigma(z \mid \theta_1)$$
$$y_1' = -\sqrt{N}\mathscr{E}(z \mid \theta_1)/\sigma(z \mid \theta_1),$$

whereas $\mathscr{E}(z \mid \theta_1)$ and $\sigma^2(z \mid \theta_1)$ are the mean and variance of z determined from $f(x; \theta_1)$.

For small values of α and β, the unknowns k_0 (in y_1) and k_1 (in y_0') can be closely approximated by $\beta/(1 - \alpha)$ and $(1 - \beta)/\alpha$ respectively, as indicated in Section 15.4(*c*).

Summarizing, we have the following result due to Wald (1945):

15.4.9 *If the random variable*

$$z = \log\left(\frac{f(x; \theta_1)}{f(x; \theta_0)}\right)$$

has finite mean and variance whether computed from $f(x; \theta_0)$ or $f(x; \theta_1)$, an upper bound for the Type I error α_N of the truncated probability ratio sequential test S_N is

$$\alpha + \Phi(y_0') - \Phi(y_0) + O\left(\frac{1}{\sqrt{N}}\right)$$

where α is the Type I error by the nontruncated probability ratio test S, and where y_0 and y_0' are given by (15.4.59). *Similarly, an upper bound for the Type II error β_N is*

$$\beta + \Phi(y_1') - \Phi(y_1) + O\left(\frac{1}{\sqrt{N}}\right)$$

where β is the Type II error of S, whereas y_1 and y_1' are given by (15.4.59*a*).

15.5 APPLICATION OF PROBABILITY RATIO SEQUENTIAL TEST TO BINOMIAL DISTRIBUTION

To illustrate the results of Section 15.4 we consider the case of sampling from the binomial distribution $Bi(1, \theta)$.

In this case $(x_1, x_2, \ldots)$ is a sequence of independent random variables whose p.f. is

$$f(x; \theta) = \theta^x(1 - \theta)^{1-x}, \qquad x = 0, 1. \tag{15.5.1}$$

Denoting $\sum_{t=1}^{n} x_t$ by n_1, we have

$$\frac{Q_{1n}}{Q_{0n}} = \left(\frac{\theta_1}{\theta_0}\right)^{n_1}\left(\frac{1 - \theta_1}{1 - \theta_0}\right)^{n-n_1}. \tag{15.5.2}$$

Making use of the approximation (15.4.19) for k_0 and k_1 let a_n and b_n be defined as follows:

$$
(15.5.3) \qquad a_n = \frac{\log\left(\dfrac{\beta}{1-\alpha}\right) - n\log\left(\dfrac{1-\theta_1}{1-\theta_0}\right)}{\log\left[\dfrac{\theta_1(1-\theta_0)}{\theta_0(1-\theta_1)}\right]}
$$

$$
b_n = \frac{\log\left(\dfrac{1-\beta}{\alpha}\right) - n\log\left(\dfrac{1-\theta_1}{1-\theta_0}\right)}{\log\left[\dfrac{\theta_1(1-\theta_0)}{\theta_0(1-\theta_1)}\right]}.
$$

The three inequalities in (15.4.1) then reduce, respectively, to

$$
(15.5.4) \qquad n_1 \leqslant a_n, \qquad n_1 \geqslant b_n, \qquad a_n < n_1 < b_n.
$$

Thus, the sequential process continues as long as $a_n < n_1 < b_n$; it terminates upon drawing x_n with acceptance of $\mathscr{H}_0$ if $n_1 \leqslant a_n$; and it terminates upon drawing x_n with acceptance of $\mathscr{H}_1$ (rejection of $\mathscr{H}_0$) if $n_1 \geqslant b_n$.

Making use of the approximations (15.4.19) for k_0 and k_1 in (15.4.37) we obtain the following approximation for the operating characteristic function of our sequential test:

$$
(15.5.5) \qquad L(\theta) \cong \frac{1 - \left(\dfrac{1-\beta}{\alpha}\right)^{h(\theta)}}{\left(\dfrac{\beta}{1-\alpha}\right)^{h(\theta)} - \left(\dfrac{1-\beta}{\alpha}\right)^{h(\theta)}}
$$

where $h(\theta)$ is a function of θ defined by applying (15.4.32), that is,

$$
(15.5.6) \qquad \sum_{x=0}^{1}\left[\frac{\theta_1^x(1-\theta_1)^{1-x}}{\theta_0^x(1-\theta_0)^{1-x}}\right]^h \theta^x(1-\theta)^{1-x} = 1.
$$

Simplifying (15.5.6) we obtain the following functional relationship between θ and h

$$
(15.5.7) \qquad \theta = \frac{1 - \left(\dfrac{1-\theta_1}{1-\theta_0}\right)^h}{\left(\dfrac{\theta_1}{\theta_0}\right)^h - \left(\dfrac{1-\theta_1}{1-\theta_0}\right)^h}.
$$

It will be noted that $h(\theta_0) = 1$, and $h(\theta_1) = -1$. Furthermore, it will be seen from (15.5.5) that the approximation formula for $L(\theta)$ gives correct values for $L(\theta_0)$ and $L(\theta_1)$.

Referring to (15.4.40) and computing the value of $\mathscr{E}(z \mid \theta)$, we find that the average sample number $E(n \mid \theta)$ is given approximately by

$$\mathscr{E}(n \mid \theta) \cong \frac{L(\theta) \log \left(\dfrac{\beta}{1-\alpha}\right) + (1 - L(\theta)) \log \left(\dfrac{1-\beta}{\alpha}\right)}{\theta \log \dfrac{\theta_1}{\theta_0} + (1-\theta) \log \left(\dfrac{1-\theta_1}{1-\theta_0}\right)}. \tag{15.5.8}$$

Thus, if the (approximate) expression for $L(\theta)$ in (15.5.5) and the expression for θ given by (15.5.7) are substituted into (15.5.8) we obtain an expression for $\mathscr{E}(n \mid \theta)$ as a function of the parameter h. $\mathscr{E}(n \mid \theta)$ can be quite readily determined for $\theta = 0, \theta_1, \theta_2$.

Applications of the Wald probability ratio sequential test for sampling from various specific distributions have been published by Wald (1947*a*) and by the Statistical Research Group, Columbia University (1945, 1947).

15.6 SEQUENTIAL ESTIMATION

(a) General Comments

A general theory of sequential estimation of parameters has not been developed. Wald (1947*a*) gave one formulation of the general problem of sequential estimation by intervals but did not solve it. He and Stein (1947) however, did consider a sequential procedure for determining a confidence interval of fixed length and confidence coefficient for the mean of a normal distribution with known variance. Other specific estimation problems have also been considered, having aims akin to those of sequential estimation. An account of a number of results on these problems has been given by Anscombe (1953).

The basic idea of the sequential estimation of a parameter is in general to do just enough sampling to be able to obtain an estimate which has a predetermined degree of precision, in some sense which does not depend on the unknown population parameter being estimated. Expression of degree of precision can be set forth in various ways. One simple way to express such precision would be to provide an estimator for the parameter whose variance does not depend on the parameter. This can sometimes be done by using samples whose size is fixed in advance. In the case of large samples an approximate solution of this problem under certain regularity conditions was presented in Section 12.3(*e*).

Another way to express such precision is for the parameter to have a confidence interval of length specified in advance of the sampling. This can be achieved in special cases by a fixed-size sample. For instance, suppose μ is unknown but σ^2 is known in $N(\mu, \sigma^2)$, and that $\bar{x}$ is the mean of a sample of size n from this distribution. Then $\bar{x} \pm \frac{1}{2}\delta$ is a confidence interval of length δ having coefficient $\geqslant \gamma$ estimating μ, provided $n - 1$ is chosen as the largest integer in $y_\gamma^2\sigma^2/\delta^2$ (unless this quantity is an integer, in which case n is chosen as this integer) where y_γ satisfies $\Phi(y_\gamma) - \Phi(-y_\gamma) = \gamma$, $\Phi(y)$ being the c.d.f. of $N(0, 1)$.

(b) Stein's Fixed Interval Estimator for Mean of a Normal Distribution

In the case where both σ^2 and μ are unknown, Stein (1945) showed how a double-sampling procedure could be used for establishing a confidence interval of fixed length δ for estimating μ and having confidence coefficient $\geqslant \gamma$. His result can be stated as follows:

15.6.1 *Let $(x_1, \ldots, x_n)$ be a first sample and $(x_{n+1}, \ldots, x_{n+m})$ a second independent sample from $N(\mu, \sigma^2)$. Let s^2 be the sample variance of the first sample and $\bar{x}$ the mean of both samples combined. Let $k = [2st_{n-1,\gamma}/\delta]^2$ where $t_{n-1,\gamma}$ is the upper $100(1 - \frac{1}{2}\gamma)\%$ point of the Student distribution $S(n - 1)$. Let m be chosen as 0 if $k - n \leqslant 0$ and as the smallest positive integer $\geqslant k - n$ if $k - n > 0$. Then $\bar{x} \pm \frac{1}{2}\delta$ is a confidence interval of length δ for μ having confidence coefficient $\geqslant \gamma$.*

To establish **15.6.1,** it can be readily seen that, even though $\bar{x}$ is the mean of a sample of size m from $N(\mu, \sigma^2)$, where m is a random variable, $(\bar{x} - \mu)/[\sigma/\sqrt{m + n}]$ has the distribution $N(0, 1)$, and is independent of $(n - 1)s^2/\sigma^2$, which has the chi-square distribution $C(n - 1)$. Thus, $(\bar{x} - \mu)\sqrt{m + n}/s$ has the Student distribution $S(n - 1)$, and hence we have

$$P\left(-t_{n-1,\gamma} < \frac{(\bar{x} - \mu)}{s}\sqrt{m + n} < +t_{n-1,\gamma}\right) = \gamma \tag{15.6.1}$$

or equivalently

$$P\left(\bar{x} - \frac{st_{n-,1\gamma}}{\sqrt{m + n}} < \mu < \bar{x} + \frac{st_{n-1,\gamma}}{\sqrt{m + n}}\right) = \gamma. \tag{15.6.2}$$

If $st_{n-1,\gamma}/\sqrt{n} \leqslant \frac{1}{2}\delta$, then $\bar{x} \pm \delta$ is a confidence interval of length δ for μ provided by the mean of the first sample only, and having confidence coefficient $\geqslant \gamma$. If, however, $st_{n-1,\gamma}/\sqrt{n} > \frac{1}{2}\delta$, we draw a second sample of size m such that m is the smallest positive integer for which

$$\frac{st_{n-1,\gamma}}{\sqrt{m + n}} \leqslant \tfrac{1}{2}\delta$$

that is, m is the smallest positive integer such that

$$m \geqslant \left[\frac{2st_{n-1,\gamma}}{\delta}\right]^2 - n.$$

In this case $\bar{x} \pm \frac{1}{2}\delta$ is also a confidence interval of length $\leqslant \delta$ for μ given by the mean of both samples combined and having confidence coefficient $\geqslant \gamma$, thus establishing **15.6.1**.

PROBLEMS

15.1 Show that **15.2.1** holds if $(x_1, x_2, \ldots)$ is a sequence of independent random variables defined on a finite interval (a, b) and having equal means $\mathscr{E}(x)$ and if $\mathscr{E}(n)$ is finite.

15.2 Suppose $(x_1, x_2, \ldots)$ is a sequence of independent random variables having c.d.f.'s $F_1(x; \theta)$, $F_2(x; \theta)$, ... Let $E^{(i)}$ be a set such that

$$\int_{E^{(i)}} dF_i(x; \theta) = p_i(\theta) \leqslant p < 1.$$

If S is any sequential process having the sequence of experiment continuation sets $G_1, G_2, \ldots$ defined as follows

$$G_n = E^{(1)} \times E^{(2)} \times \cdots \times E^{(n)},$$

show that the sequential process terminates with probability 1.

15.3 A sequence of independent random variables $(x_1, x_2, \ldots)$ is assumed to be drawn from a population having p.d.f. $\theta e^{-\theta x}$, $x, \theta > 0$. It is desired to set up a Cartesian sequential process for testing the hypothesis $\mathscr{H}_0$ that $\theta = \theta_0$ against the hypothesis $\mathscr{H}_1$ that $\theta = \theta_1$ where $\theta_1 > \theta_0$, and where Type I and Type II errors are α and β respectively. Show that G_0°, G_0', and G_0 yielded by the probability ratio criterion are the intervals (x_2, ∞), $(0, x_1)$, (x_1, x_2) where $x_1 < x_2$ satisfy the conditions

$$\frac{e^{-\theta_0 x_2}}{1 - e^{-\theta_0 x_1}} = \frac{1 - \alpha}{\alpha}$$

$$\frac{e^{-\theta_1 x_2}}{1 - e^{-\theta_1 x_1}} = \frac{\beta}{1 - \beta}.$$

Show that the average sample number $\mathscr{E}(n \mid \theta)$ for this sequential process attains its largest value for $\theta = \dfrac{\log x_2 - \log x_1}{x_2 - x_1}$.

15.4 (*Continuation*) Suppose an r-fold Cartesian sequential process is used for testing $\mathscr{H}_0$ against $\mathscr{H}_1$. Show that G_0°, G_0', and G_0 (in Euclidean r-space) determined by the probability ratio criterion are defined as follows:

$$G_0^\circ : \left\{(x_1, \ldots, x_r) : \sum_1^r x_i \geqslant y_2\right\}$$

$$G_0' : \left\{(x_1, \ldots, x_r) : \sum_4^r x_i \leqslant y_1\right\}$$

$$G_0 : \left\{(x_1, \ldots, x_r) : y_1 < \sum_1^r x_i < y_2\right\},$$

where (y_1, y_2) satisfy

$$\int_{y_2}^{\infty} f(y; \theta_0)\, dy = \frac{1 - \alpha}{\alpha} \int_0^{y_1} f(y; \theta_0)\, dy$$

$$\int_{y_2}^{\infty} f(y; \theta_1)\, dy = \frac{\beta}{1 - \beta} \int_0^{y_1} f(y; \theta_1)\, dy$$

and

$$f(y; \theta) = \frac{\theta^r y^{r-1} e^{-\theta y}}{\Gamma(r)}, y > 0.$$

15.5 In the Cartesian sequential test discussed in Section 15.3(*a*), if the process has not terminated upon the nth trial suppose it is truncated (arbitrarily terminated) by choosing $\mathscr{H}_0$ or $\mathscr{H}_1$ with probabilities $a/(a + b)$ and $b/(a + b)$ respectively. Show that in this case

$$L_n(\theta) = \frac{a}{a + b}$$

$$M_n(\theta) = \frac{b}{a + b}$$

$$\mathscr{E}(n \mid \theta) = \frac{1 - c^n}{a + b},$$

and that Type I and Type II errors α and β satisfy (15.3.8).

15.6 Prove that the Cartesian sequential test for which G_0° is defined by (15.3.10) and G_0' by (15.3.11) is stronger than any other Cartesian sequential test which leaves $a_0/(a_0 + b_0)$ fixed at α and $\mathscr{E}(n \mid \theta_0)$ (that is, $1/(a_0 + b_0)$) fixed at n_0.

15.7 Suppose $(x_1, x_2, \ldots)$ is a sequence of independent random variables all having the normal distribution $N(\theta, 1)$. Let $\mathscr{H}_0$ be the hypothesis that $\theta = \theta_0$ and $\mathscr{H}_1$ the hypothesis that $\theta = \theta_1 > \theta_0$. For Type I and Type II errors α and β, and taking approximations for k_0 and k_1 as given by (15.4.19), show that in a sequential probability ratio test for testing $\mathscr{H}_0$ against $\mathscr{H}_1$, that $\mathscr{H}_0$ is accepted on the nth drawing if, for the first time,

$$x_1 + \cdots + x_n \leqslant \frac{1}{(\theta_1 - \theta_0)} \log\left(\frac{\beta}{1 - \alpha}\right) + n\frac{\theta_1 + \theta_2}{2},$$

that $\mathscr{H}_1$ is accepted on the nth drawing if, for the first time,

$$x_1 + \cdots + x_n \geqslant \frac{1}{(\theta_1 - \theta_0)} \log \frac{1-\beta}{\alpha} + n\frac{\theta_1 + \theta_2}{2}$$

and an $(n + 1)$th drawing is made if $x_1 + \cdots + x_2$ lies between the two values indicated above.

Also show that

$$L(\theta) \simeq \frac{\left(\dfrac{1-\beta}{\alpha}\right)^h - 1}{\left(\dfrac{1-\beta}{\alpha}\right)^h - \left(\dfrac{\beta}{1-\alpha}\right)^h}$$

where

$$h = \frac{\theta_1 + \theta_0 - 2\theta}{\theta_1 - \theta_0}$$

and that this approximation for $L(\theta)$ passes through the points $(\theta_0, 1 - \alpha)$ and (θ_1, β), as specified by the Type I and Type II errors. Furthermore, show that the average sample number is

$$\mathscr{E}(n \mid \theta) \simeq -2\,\frac{L(\theta)\log\left(\dfrac{\beta}{1-\alpha}\right) + (1 - L(\theta))\log\left(\dfrac{1-\beta}{\alpha}\right)}{h(\theta_1 - \theta_0)^2}.$$

Extend these results to the case of sampling from $N(\theta, \sigma^2)$, where σ^2 is known.

15.8 Suppose $(x_1, x_2, \ldots)$ is a sequence of independent random variables having the normal distribution $N(0, \sigma^2)$. Let $\mathscr{H}_0$ be the hypothesis that $\sigma^2 = \sigma_0^2$, and $\mathscr{H}_1$ the hypothesis that $\sigma^2 = \sigma_1^2 > \sigma_0^2$. For Type I and Type II errors α and β and using the approximations for k_0 and k_1 given by (15.4.19) show that in a sequential probability ratio test for testing $\mathscr{H}_0$ against $\mathscr{H}_1$, that $\mathscr{H}_0$ is accepted on the nth trial if, for the first time,

$$x_1^2 + \cdots + x_n^2 \leqslant \frac{\sigma_0^2\sigma_1^2\left[2\log\left(\dfrac{\beta}{1-\alpha}\right) + n\log\left(\dfrac{\sigma_1^2}{\sigma_0^2}\right)\right]}{\sigma_1^2 - \sigma_0^2}$$

that $\mathscr{H}_1$ is accepted on the nth trial if, for the first time,

$$x_1^2 + \cdots + x_n^2 \geqslant \frac{\sigma_0^2\sigma_1^2\left[2\log\left(\dfrac{1-\beta}{\alpha}\right) + n\log\left(\dfrac{\sigma_1^2}{\sigma_0^2}\right)\right]}{\sigma_1^2 - \sigma_0^2}$$

and the $(n + 1)$th observation is taken if $x_1^2 + \cdots + x_n^2$ lies between the two values given above.

15.9 (*Continuation*) Show that the operating characteristic function $L(\sigma^2)$ is given approximately by

$$L(\sigma^2) \simeq \frac{\left(\dfrac{1-\beta}{\alpha}\right)^h - 1}{\left(\dfrac{1-\beta}{\alpha}\right)^h - \left(\dfrac{\beta}{1-\alpha}\right)^h}$$

where the functional relationship between σ^2 and h is as follows

$$\sigma^2 = \frac{\sigma_0^2\sigma_1^2\left[1 - \left(\frac{\sigma_0}{\sigma_1}\right)^{2h}\right]}{h(\sigma_1^2 - \sigma_0^2)}$$

and verify that the graph of this approximation for $L(\sigma^2)$ passes through the points $(\sigma_0^2, 1 - \alpha)$ and (σ_1^2, β).

Furthermore, show that the average sample number is given by

$$\mathscr{E}(n \mid \sigma^2) \cong \frac{2\sigma_0^2\sigma_1^2\left[L(\sigma^2)\log\left(\frac{\beta}{1-\alpha}\right) + (1 - L(\sigma^2))\log\left(\frac{1-\beta}{\alpha}\right)\right]}{(\sigma_1^2 - \sigma_0^2)\sigma^2 + \sigma_0^2\sigma_1^2\log(\sigma_0^2/\sigma_1^2)}.$$

15.10 Determine the probability ratio sequential test for the hypothesis $\mathscr{H}_0$ that a sample of size n comes from the Poisson distribution $P_0(\theta_0)$, the alternative $\mathscr{H}_1$ being that the sample comes from the Poisson distribution $P_0(\theta_1)$ where $\theta_1 > \theta_0$. Find approximations to the operating characteristic function $L(\theta)$ and the average sample number $\mathscr{E}(n \mid \theta)$ for the test.

15.11 In the binomial waiting time distribution

$$\binom{n-1}{k-1}p^k(1-p)^{n-k}, \qquad n = k, k+1, \ldots$$

let $\tilde{p} = \dfrac{k-1}{n-1}$. Show that $\tilde{p}$ is an unbiased estimator for p, that

$$\sigma^2(\tilde{p}) = \frac{p^2}{k-2}\left[1 - \left(\frac{k-1}{k-3}\right)p + \frac{2(k-1)p^2}{(k-3)(k-4)} - \cdots\right]$$

and hence that the coefficient of variation of $\tilde{p}$, that is, $\sigma(\tilde{p})/\mathscr{E}(\tilde{p})$ is

$$\frac{1}{\sqrt{k-2}}(1 + O(p)).$$

[See Haldane (1945)].

15.12 Let s^2 be the variance of a sample of size n from $N(\mu, \sigma^2)$. For an arbitrary $\varepsilon > 0$ let m be the smallest integer $\geqslant (n-1)s^2/[(n-3)\varepsilon]$. If $\bar{x}'$ is the mean of an independent sample of size m from $N(\mu, \sigma^2)$ show that $\sigma^2(\bar{x}') \leqslant \varepsilon$.

15.13 Let s_1^2 and s_2^2 be sample variances of independent samples of sizes n_1 and n_2 from $N(\mu_1, \sigma_1^2)$ and $N(\mu_2, \sigma_2^2)$ respectively. For arbitrary $\varepsilon > 0$ let m_1 be the smallest integer $\geqslant [2(n_1-1)s_1^2]/[(n_1-3)\varepsilon]$ and m_2 the smallest integer $\geqslant [2(n_2-1)s_2^2]/[(n_2-3)\varepsilon]$. Let further independent samples of sizes m_1 and m_2 from $N(\mu_1, \sigma_1^2)$ and $N(\mu_2, \sigma_2^2)$ respectively, and let $\bar{x}_1$ and $\bar{x}_2$ be the means of these samples respectively. Show that $\sigma^2(\bar{x}_1 - \bar{x}_2) \leqslant \varepsilon$.

CHAPTER 16

Statistical Decision Functions

16.1 GENERAL REMARKS

In the theory of testing statistical hypotheses as developed by Neyman and Pearson, the risks involved in falsely accepting or falsely rejecting an hypothesis are recognized and emphasized. In two pioneering papers followed by a book, Wald (1947*b*, 1949*a*, 1950) has extended the theory of such risks to a wider class of statistical problems, developing what he has called the theory of *statistical decision functions*. His development was partly motivated by the theory of games as formulated by von Neumann and Morgenstern (1944) and partly by a desire to construct a more general theory of risk-evaluation involved in statistical procedures. Wald's original work on statistical decision theory has been followed by many research papers and several books including a rather comprehensive one by Blackwell and Girshick (1954) dealing with statistical decision theory for discrete sample spaces. In this chapter we shall present a brief introduction to some of the main ideas and basic results of statistical decision theory in the simplest type of situation. We shall not go into the theory in its most general form. The reader interested in the general theory and further details is referred to the books by Wald and by Blackwell and Girshick. Books at a more elementary level have been written by Chernoff and Moses (1959), Luce and Raiffa (1957), Raiffa and Schlaifer (1961), Savage (1954), and Weiss (1961).

16.2 DEFINITIONS AND TERMINOLOGY

Suppose x is a discrete random variable with sample space R, and θ is a point in a parameter space Ω which contains only a finite set of points $\theta_1, \ldots, \theta_h$. Let $p(x \mid \theta)$ be the probability distribution of x for a given θ. Thus we have h probability distributions $p(x \mid \theta_1), \ldots, p(x \mid \theta_h)$, one of which is assumed to be the *true* distribution whenever an observation is made on x.

Suppose a is any one of a set A of decisions (or conclusions) that could be made about θ on the basis of an outcome of an observation on x. For instance, if $\theta_1, \ldots, \theta_h$ are numbers, and c is some critical number, A might consist of two elements, a_1 and a_2, where a_1 is the decision that $\theta \leqslant c$ and a_2 the decision that $\theta > c$. A could, of course, consist of any finite number or an infinite number of elements. A is called the *decision space* (although it might be more accurate terminology to call it the *conclusion space*). If the observation on x is to have any relevance as to which decision $a \in A$ is made, then a must depend on x. We must therefore have a *decision function* $d(x)$ to determine what decision a in A to make if the sample point x occurs. Thus, $d(x)$ is a single-valued function of x having as its value some $a \in A$ for each point x in the sample space R. Any decision function $d(x)$ considered will be a member of some *class of decision functions* D. It should be noted that for a given decision function $d(x)$ and for a sample space R having only a finite number of points, there will be only a finite number of elements in A.

Now, for each $\theta \in \Omega$ it is possible to make any decision $a \in A$. If the consequence of making decision a when θ is true is such that $L(\theta, a)$ is the *loss* or *cost*, then $L(\theta, a)$, which we shall take as a bounded single-valued real function, is defined at every point (θ, a) in the product space $\Omega \times A$ and is called the *loss function*. For any decision function $d \in D$ and parameter point $\theta \in \Omega$, the *risk function* $r(\theta, d)$ is defined as the mean value of the loss function, $L(\theta, d(x))$ over the sample space, that is,

$$r(\theta, d) = \mathscr{E}[L(\theta, d(x))] = \sum_{x \in R} L(\theta, d(x))p(x \mid \theta). \tag{16.2.1}$$

Thus, $r(\theta, d)$ is defined at every point in the product space $\Omega \times D$ and is bounded since $L(\theta, d(x))$ is bounded.

Remark. It may be convenient to think of some of the preceding concepts in terms of elementary game theory concepts. We may think of nature as player I and the statistician as player II. Nature's strategies or choices are $\theta_1, \ldots, \theta_h$, whereas the statistician's strategies or choices are the d's in D. Thus, for any θ chosen by nature and any d chosen by the statistician $L(\theta, d(x))$ is the statistician's loss (nature's pay off) if the sample point x occurs. The quantity $r(\theta, d)$ defined by (16.2.1) is the average loss to the statistician who adopts $d(x)$ as his strategy.

Now suppose we are given a certain loss function $L(\theta, a)$, a decision function d, and the associated risk function $r(\theta, d)$. The question now arises as to what criteria might be used for preferring one decision function to another in D. The risk function itself provides one reasonable criterion.

For suppose d and d^* are two decision functions in D. Then comparing d and d^* on the basis of the risk function, d^* would be *preferable* to d if

(16.2.2) $$r(\theta, d^*) \leqslant r(\theta, d)$$

for all θ in Ω and

(16.2.3) $$r(\theta, d^*) < r(\theta, d)$$

for at least one θ in Ω. If (16.2.2) and (16.2.3) hold d^* is called a *uniformly better decision* function than d. If no other decision function in D is uniformly better than d^*, then d^* is called an *admissible decision function*. A class D of decision functions is called a *complete class* if for any d not in D we can find a d^* in D which is uniformly better than d.

If a complete class D of decision functions contains no (proper) subclass which is complete, then D is called a *minimal complete class*. The concept of a complete class of decision functions is important in general statistical decision theory. The relationship between a complete class and a minimal complete class of decision functions is given in the following theorems **16.2.1** and **16.2.2**:

16.2.1 *If a minimal complete class exists, it is equal to the class of admissible decision functions.*

Let D_0 denote the class of admissible decision functions. Then if D is a minimal complete class, D_0 is a subset of D. Now suppose d' is an element of D which does not belong to D_0. Then there is a d'' which is uniformly better than d'. But d'' cannot be an element in D since D is a minimal complete class. Hence there is an element d''' in D that is uniformly better than d'' and therefore uniformly better than d'. But this is impossible since D is a minimal complete class. Hence D_0 cannot be a proper subset of D and therefore D_0 and D are identical.

If the class D_0 of all admissible decision functions is complete then D_0 is a minimal complete class. It follows from this fact and **16.2.1** that

16.2.2 *A necessary and sufficient condition for the existence of a minimal complete class of decision functions is that the class of admissible decision functions be complete.*

Wald (1950) has shown under rather general conditions that the set of admissible decision functions forms a complete class, and, in particular, that the set of all Bayes solutions (to be considered later) is complete.

16.3 MINIMAX SOLUTION OF THE DECISION PROBLEM

Let us return to the risk function $r(\theta, d)$. For any given decision function d, and loss function $L(\theta, a)$, the *risk vector* associated with the possible choices of θ, namely, $\theta_1, \ldots, \theta_h$, is $(r(\theta_1, d), \ldots, r(\theta_h, d))$. Now consider the maximum component of this vector. It is a function of d. Suppose d^* is a decision function in a class D which minimizes this maximum component. In other words, suppose d^* is a member of D such that

$$\text{l.u.b.}_{\theta \in \Omega}\, r(\theta, d^*) \leqslant \text{l.u.b.}_{\theta \in \Omega}\, r(\theta, d) \tag{16.3.1}$$

for all d in D. Then d^* or its associated risk vector $(r(\theta_1, d^*), \ldots, r(\theta_h, d^*))$ is called the *minimax solution* of our decision problem.

If D contains only a finite number of elements d it is evident that there exists at least one minimax solution d^* and the risk associated with such a solution would be

$$\min_{d \in D} \{\max_{\theta \in \Omega} r(\theta, d)\}. \tag{16.3.2}$$

The d (or d's) in D, that is, the d (or d's) which yield this *minimax risk*, could be found by exhaustively examining the values of $r(\theta, d)$ at the finite set of points on $\Omega \times D$ in the order indicated by (16.3.2).

If, however, D contains an infinite number of elements the situation is more complicated. In this case a geometric representation will help. For each d in D the h-tuple $(r(\theta_1, d), \ldots, r(\theta_h, d))$ can be represented as a point r in Euclidean space R_h. The set of all such points r corresponding to all d in D is a set E in R_h. E is bounded since r is bounded on $\Omega \times D$. Thus the problem of finding a minimax solution of the decision problem in this case is equivalent to taking the maximum coordinate of each point r in E and then minimizing these maximum coordinates. If a solution d^* exists it can be expressed in the following way. Let E_i be the subset of E for which $r(\theta_i, d) \geqslant \max_{j \neq i} \{r(\theta_j, d)\}$, $i = 1, \ldots, h$. Then $E = E_1 \cup E_2 \cup \cdots \cup E_h$, although it should be noted that $E_1, \ldots, E_h$ are not necessarily disjoint, since any point in E having two or more equal largest coordinates would belong to two or more of the sets $E_1, \ldots, E_h$. Now let M_i be the greatest lower bound of the ith coordinate of all points in E_i, $i = 1, \ldots, h$. Then if there were a d^* in D (a point in E) such that

$$\max_i r(\theta_i, d^*) = \min \{M_1, \ldots, M_h\} \tag{16.3.3}$$

d^* would be a minimax solution. If E is closed, that is, if E contains the limit points of all sequences of points in E, then there does exist at least one d^*.

We shall show that E is closed under some mild conditions. First, note that for any point in $D \times R$, $(L(\theta_1, d(x)), \ldots, L(\theta_h, d(x)))$ is a point L

in Euclidean space R_h. Let F be the set of all such points L. F is bounded since $L(\theta, d(x))$ is bounded. We shall show that

16.3.1 *If F is closed E is also closed.*

First, consider the case where R has a finite number of points. In this case A will have a finite number of points. Since $d(x)$ is single-valued, $L(\theta_i, d(x))$ will have a finite number of values and so will $r(\theta_i, d)$, since $r(\theta_i, d)$ is the average or mean value of $L(\theta_i, d(x))$ over the sample space R with respect to $p(x \mid \theta_i)$, $i = 1, \ldots, h$. Therefore, if R contains a finite number of points, E contains a finite number of points and is therefore closed, and a minimax solution to the decision problem therefore exists in this case.

Now consider the case in which R has a countably infinite number of points. Let these points be arranged in sequence x_t, $t = 1, 2, \ldots$. Then for any d in D (and hence its corresponding point r in E) we have

$$r(\theta_i, d) = \sum_{t=1}^{\infty} L(\theta_i, d(x_t))p(x_t \mid \theta_i). \tag{16.3.4}$$

Thus d is specified by the sequence of points

$$(L(\theta_1, d(x_t)), \ldots, L(\theta_h, d(x_t))), \qquad t = 1, 2, \ldots \tag{16.3.5}$$

in Euclidean R_h. All such sequences corresponding to all d in D generate the set of points F in R_h which is bounded since $L(\theta, d(x))$ is bounded. We assume F to be closed.

Consider next any convergent sequence of points r_α, $\alpha = 1, 2, \ldots$ in E where r_α has the coordinates

$$(r(\theta_1, d_\alpha), \ldots, r(\theta_h, d_\alpha)), \tag{16.3.6}$$

and let its limit point r^* be denoted by

$$(r(\theta_1, d^*), \ldots, r(\theta_h, d^*)). \tag{16.3.7}$$

We shall show that r^* also belongs to E, and hence that E is closed. The sequence of points in F corresponding to d_α is

$$(L(\theta_1, d_\alpha(x_t)), \ldots, L(\theta_h, d_\alpha(x_t))), \qquad t = 1, 2, \ldots \tag{16.3.8}$$

For $\alpha = 1, 2, \ldots$ (16.3.8) is thus a double sequence of points $L_{\alpha t}$ in F. Using the Cantor diagonal procedure for this double sequence, a subsequence $d_{\alpha\beta}$, $\beta = 1, 2, \ldots$ can be chosen from the sequence d_α, $\alpha = 1, 2, \ldots$ such that for each t the sequence of points

$$(L(\theta_1, d_{\alpha\beta}(x_t)), \ldots, L(\theta_h, d_{\alpha\beta}(x_t))) \tag{16.3.9}$$

in F converges to

$$(L(\theta_1, d^*(x_t)), \ldots, L(\theta_h, d^*(x_t))) \tag{16.3.10}$$

as $\beta \to \infty$. Since F is closed, all points (16.3.10) for $t = 1, 2, \ldots$ are in F. Now consider the risk function

$$r(\theta, d_{\alpha\beta}) = \sum_{t=1}^{\infty} L(\theta, d_{\alpha\beta}(x_t))p(x_t \mid \theta). \tag{16.3.11}$$

Taking the limit as $\beta \to \infty$ and in view of the convergence of the sequence (16.3.9) to that in (16.3.10) for every t, we have

$$\begin{aligned}\lim_{\beta\to\infty} r(\theta, d_{\alpha\beta}) &= \lim_{\beta\to\infty} \sum_{t=1}^{\infty} L(\theta, d_{\alpha\beta}(x_t))p(x_t \mid \theta)\\ &= \sum_{t=1}^{\infty} \lim_{\beta\to\infty} L(\theta, d_{\alpha\beta}(x_t))p(x_t \mid \theta)\\ &= \sum_{t=1}^{\infty} L(\theta, d^*(x_t))p(x_t \mid \theta).\end{aligned} \tag{16.3.12}$$

The interchange of lim and Σ is valid since the loss function is bounded and $p(x_t \mid \theta)$ is a probability distribution. But since the points in (16.3.10) are in F, it follows that $\sum_{t=1}^{\infty} L(\theta, d^*(x_t))p(x_t \mid \theta)$, which may be denoted by $r(\theta, d^*)$, is in E. Hence E is closed.

Thus, if the set F in R_h consisting of the points $(L(\theta_1, d(x)), \ldots, L(\theta_h, d(x)))$ for all $d \in D$ and $x \in R$ is closed, the set E in R_h consisting of the points $(r(\theta_1, d), \ldots, r(\theta_h, d))$ for all $d \in D$ is also closed, and there exists a minimax solution d^* in D for our statistical decision problem.

Application of the minimax procedure to specific problems usually requires a considerable amount of computation. We select a very simple example to illustrate the procedure.

Example. Suppose x is a discrete random variable with sample space (0, 1) which has the distribution

$$p(x \mid \theta_i) = \binom{1}{x}\theta_i^x(1-\theta_i)^{1-x} \qquad \theta_1 = \tfrac{1}{4}, \theta_2 = \tfrac{1}{2}$$

and let the decision space A consist of two elements a_1 and a_2. Let the loss function $L(\theta_i, a_j)$ be defined as follows:

	a_1	a_2
θ_1	1	4
θ_2	3	2

Now the possible decisions for each sample point x are a_1 and a_2. If we permit either decision a_1 or a_2 at each sample point then there are four possible decision functions d_α, $\alpha = 1, \ldots, 4$ on the sample space where

$$\begin{array}{ll} d_1(0) = a_1 & d_1(1) = a_1 \\ d_2(0) = a_1 & d_2(1) = a_2 \\ d_3(0) = a_2 & d_3(1) = a_1 \\ d_4(0) = a_2 & d_4(1) = a_2 \end{array}$$

The four risk vectors $(r(\theta_1, d_\alpha), r(\theta_2, d_\alpha))$, $\alpha = 1, \ldots, 4$ are found by applying (16.2.1); that is,

$$r(\theta_i, d_\alpha) = \sum_{x=0}^{1} L(\theta_i, d_\alpha(x))p(x \mid \theta_i), \qquad i = 1, 2; \ \alpha = 1, \ldots, 4.$$

Inserting numerical values of $d_\alpha(x)$, $L(\theta_i, d_\alpha)$ and $p(x \mid \theta_i)$ we find the numerical values of the 4 risk vectors to be $(1, 3)$, $(\frac{7}{4}, \frac{5}{2})$, $(\frac{13}{4}, \frac{5}{2})$, $(4, 2)$. The maximum coordinates of the four vectors are $3, \frac{5}{2}, \frac{13}{4}, 4$, respectively, and the minimum of these is $\frac{5}{2}$. The solution occurs for the risk vector $(r(\theta_1, d_2), r(\theta_2, d_2))$, and hence the minimax solution is the decision function $d_2(x)$ and the corresponding risk vector is $(\frac{7}{4}, \frac{5}{2})$.

16.4 BAYES SOLUTIONS OF THE STATISTICAL DECISION PROBLEM

(a) Solution against a Specified *a priori* Distribution

Now suppose a value of θ occurs in accordance with some *a priori* probability function $q(\theta)$, $\theta = \theta_1, \ldots, \theta_h$. Then for a given decision function d in the available class D, the risk function $r(\theta, d)$ would be averaged over θ with respect to the *a priori* distribution $q(\theta)$ to yield the *average risk* $\bar{r}(q, d)$ defined as follows:

$$\begin{aligned} \bar{r}(q, d) &= \sum_{i=1}^{h} r(\theta_i, d)q(\theta_i) \\ &= \sum_{t=1}^{\infty} \sum_{i=1}^{h} L(\theta_i, d(x_t))p(x_t \mid \theta_i)q(\theta_i). \end{aligned} \tag{16.4.1}$$

If we knew $q(\theta)$, and could find a decision function d^* in D such that

$$\bar{r}(q, d^*) \leqslant \bar{r}(q, d) \tag{16.4.2}$$

for any other d in D, then d^* would be regarded as an optimum solution of the statistical decision problem. A decision function which thus minimizes $\bar{r}(q, d)$ is called a *Bayes solution* relative to the particular *a priori* distribution $q(\theta)$.

Note that since $q(\theta_i) > 0$, $i = 1, \ldots, h$, and $q(\theta_1) + \cdots + q(\theta_h) = 1$, then $(q(\theta_1), \ldots, q(\theta_h))$ can be represented as a point on the $(h - 1)$-dimensional simplex G in Euclidean space R_h which is spanned by the h points $(1, 0, \ldots, 0), (0, 1, 0, \ldots, 0), \ldots, (0, \ldots, 0, 1)$.

Now suppose we have some *unknown* nondegenerate *a priori* distribution. How do we minimize the risk?

We can represent $\bar{r}(q, d)$ as

$$\bar{r}(q, d) = \sum_{t=1}^{\infty} \sum_{i=1}^{h} L(\theta_i, d(x_t))Q(\theta_i \mid x_t)P_q(x_t) \tag{16.4.3}$$

where

$$Q(\theta_i \mid x_t) = \frac{p(x_t \mid \theta_i)q(\theta_i)}{P_q(x_t)} \tag{16.4.4}$$

and

$$P_q(x_t) = \sum_{i=1}^{h} p(x_t \mid \theta_i)q(\theta_i). \tag{16.4.5}$$

$Q(\theta_i \mid x_t)$ is the *a posteriori* probability that $\theta = \theta_i$ given that $x = x_t$ and given the *a priori* distribution $q(\theta_i)$.

Now for a given $q(\theta)$ and t, let

$$\sum_{i=1}^{h} L(\theta_i, d^*(x_t))Q(\theta_i \mid x_t) \tag{16.4.6}$$

be the greatest lower bound of

$$\sum_{i=1}^{h} L(\theta_i, d(x_t))Q(\theta_i \mid x_t) \tag{16.4.7}$$

for all points $(L(\theta_1, d(x_t)), \ldots, L(\theta_h, d(x_t)))$ in F for a fixed t. Note that for fixed t, $(Q(\theta_1 \mid x_t), \ldots, Q(\theta_h \mid x_t))$ is a point in G. Thus the sequence of points $x_1, x_2, \ldots$ in the sample space R determines a sequence of vectors in F and a corresponding sequence in G. If F is extended to include the vectors

$$(L(\theta_1, d(x_t))Q(\theta_1 \mid x_t), \ldots, L(\theta_h, d(x_t))Q(\theta_h \mid x_t)) \tag{16.4.8}$$

for each t and for all d in D, and also the minimizing vector

$$(L(\theta_1, d^*(x_t))Q(\theta_1 \mid x_t), \ldots, L(\theta_h, d^*(x_t))Q(\theta_h \mid x_t)) \tag{16.4.9}$$

for each t, then E, and hence D, are closed and the expression in (16.4.6) will provide a minimum for the expression in (16.4.7). In other words, D will contain at least one d^* so that

$$\sum_{t=1}^{\infty} \sum_{i=1}^{h} L(\theta_i, d^*(x_t))Q(\theta_i \mid x_t)P_q(x_t) \leqslant \bar{r}(q, d) \tag{16.4.10}$$

for any d in D. Each such d^* provides a Bayes solution of the decision problem against the *a priori* distribution $q(\theta)$. The set of such d^* is nonempty and closed and hence the class of admissible decision functions d^* is complete.

Summarizing, we have

16.4.1 *If, for a given a priori distribution $q(\theta)$, F is extended to include the vectors* (16.4.8) *for each t and all $d \in D$ and* (16.4.9) *for each t, then E is closed and d^* corresponding to the sequence of vectors $(L(\theta_1, d^*(x_t)), \ldots, L(\theta_h, d^*(x_t)))$ defined in* (16.4.6) *minimizes the average risk $\bar{r}(q, d)$ defined in* (16.4.1).

(b) Geometrical Interpretation

If we denote the left-hand side of (16.4.10) by $\bar{r}(q, d^*)$, then we have

$$\bar{r}(q, d^*) = \min_{d \in D} \bar{r}(q, d) = \min_{r \in E} \sum_{i=1}^{h} r(\theta_i, d) q(\theta_i). \tag{16.4.11}$$

Thus, if the weighted average of the coordinates of each point (risk vector) in E is taken, using the *a priori* probabilities $q(\theta_1), \ldots, q(\theta_h)$ as weights, then $\bar{r}(q, d^*)$ is equal to the smallest of these weighted averages. Or stated geometrically, if, for a given $q(\theta)$, one takes the family of hyperplanes (C being the parameter)

$$\sum_{i=1}^{h} r_i q(\theta_i) = C \tag{16.4.12}$$

in R_h passing through all points in E, the point $r = (r_1, \ldots, r_h)$ yielding the smallest C provides the Bayes solution against the *a priori* distribution $q(\theta)$, and the value of C for this point is the minimum average risk $\bar{r}(q, d^*)$. All coordinates of r are finite since E is bounded.

(c) The Set of Solutions against All Possible *a priori* Distributions

Now let us examine the set of solutions corresponding to all possible *a priori* distributions. Note that

$$(q(\theta_1), \ldots, q(\theta_h)) \tag{16.4.13}$$

is a point in the simplex G in R_h spanned by the h points $(1, 0, \ldots, 0), \ldots, (0, \ldots, 0, 1)$. The coordinates of the point (16.4.13) are direction numbers of the normal to the hyperplane (16.4.12). Thus, if we find the point r in E yielding the smallest C for each possible *a priori* distribution $q(\theta)$, then the set of hyperplanes corresponding to this set of smallest C's forms a lower bounding *hull* or *envelope* for the set E. It is evident that the set E^* of all points r in E which are used in determining the bounding hyperplanes for the hull lies on the hull itself. The set of risk vectors E^* is the set of all Bayes solutions of the decision problem generated by all possible *a priori* distributions $q(\theta)$.

Example. At the end of Section 16.3 we gave a simple numerical example to illustrate the problem of obtaining a minimax solution. Let us return to that example and assume that θ has the *a priori* distribution $q(\theta)$ where $q(\theta_1) = \frac{1}{3}$, $q(\theta_2) = \frac{2}{3}$. Recalling that the risk vectors obtained in the minimax example corresponding to decision functions d_1, d_2, d_3, d_4 are $(1, 3)$, $(\frac{7}{4}, \frac{5}{2})$, $(\frac{13}{4}, \frac{5}{2})$, $(4, 2)$, respectively, we now calculate $\sum_i r(\theta_i, d)q(\theta_i)$ the inner products of each of these vectors and the vector $(q(\theta_1), q(\theta_2)) = (\frac{1}{3}, \frac{2}{3})$. The four inner products are $\frac{28}{12}$, $\frac{27}{12}$, $\frac{33}{12}$, and $\frac{32}{12}$. They are the average risks associated respectively with d_1, d_2, d_3, d_4 against $q(\theta)$. Thus the decision function yielding the minimum average risk against the particular $q(\theta)$ chosen is d_2. If the bounding hull as explained above is constructed from the four risk vectors it will be found that the hull itself contains d_1, d_2 and d_4. This means that d_1, d_2 or d_4 will provide a Bayes solution against any possible *a priori* distribution $q(\theta)$.

16.5 REMARKS ON EXTENSIONS AND GENERALIZATIONS

We have discussed statistical decision theory only for the simplest case, namely, that in which the random variable x is discrete, where the parameter space Ω contains a finite number of points, and where the random variable x is observed once. (Notice that x could be an n-dimensional random variable, however. For instance, x could be a sample of size n drawn from a Poisson distribution with parameter θ. The parameter θ, however, would be limited to a finite number of values.) We have derived general minimax and Bayes solutions in this relatively simple case.

There are various directions in which the results obtained can be extended and generalized. The first might be to extend the results to the case where x is an absolutely continuous random variable, thus possessing a density function $f(x \mid \theta)$ over the sample space, with the parameter space Ω held to a finite number of values. The next direction would be to let Ω have a countably infinite number of values, or even be a set of points in an Euclidean space with x being either discrete or continuous. Another direction for generalization is to introduce sequential sampling. To attempt the general theory for these various extensions here, however, would fall outside the scope of this book. Readers interested in these various extensions should refer to the books by Wald (1950), and by Blackwell and Girshick (1954), and to papers by Dvoretzky, Kiefer and Wolfowitz (1953*a*, 1953*b*), by Karlin and Rubin (1956), and by Lehmann (1957).

PROBLEMS

16.1 Suppose x is a random variable which has the binomial distribution

$$p(x \mid \theta) = \binom{2}{x}\theta^x(1-\theta)^{2-x}, \qquad x = 0, 1, 2$$

and that the parameter space of θ contains two points $\theta_1 = \frac{1}{100}$, $\theta_2 = \frac{1}{10}$. Let the decision space A contain two elements a_1, a_2 and let the loss function $L(\theta_i, a_j)$ be defined as follows:

	a_1	a_2
θ_1	1	4
θ_2	3	2

Let the space D of decision functions contain three points $d_1(x)$, $d_2(x)$, and $d_3(x)$ where

$$d_\alpha(x) = \begin{cases} a_1 & x = 0, 1, \ldots, \alpha - 1 \\ a_2 & \text{otherwise} \end{cases}$$

$\alpha = 1, 2, 3$. Determine the decision function that provides the minimax solution for this decision problem.

16.2 (*Continuation*) Find the decision functions which provide Bayes solutions of the decision problem against all possible *a priori* distributions of θ over the points $\theta_1 = \frac{1}{100}$, $\theta_2 = \frac{1}{10}$.

16.3 Suppose x is a random variable having p.f.

$$p(x \mid \theta) = (1 - \theta)\theta^{x-1}, \qquad x = 1, 2, \ldots$$

and the parameter space Ω contains two points $\theta_1 = \frac{1}{10}$, $\theta_2 = \frac{2}{10}$. Let the decision space A contain two elements (a_1, a_2), and let the loss function $L(\theta_i, a_j)$ be defined as follows:

	a_1	a_2
θ_1	1	3
θ_2	4	2

Let the space D of decision functions contain the following points:

$$d_\alpha(x) = \begin{cases} a_1 & x = 1, \ldots, \alpha \\ a_2 & x = \alpha + 1, \alpha + 2, \ldots \end{cases}$$

$\alpha = 1, 2$. Determine which decision function provides the minimax solution for the decision problem.

16.4 (*Continuation*) Find the decision functions which provide Bayes solutions of the decision problem against all possible *a priori* distributions of θ, assuming, of course, that the space of θ has only two points $\theta_1 = \frac{1}{10}$, $\theta_2 = \frac{2}{10}$.

16.5 Suppose x is a random variable that has one of the p.f.'s $p(x \mid \theta_1)$ or $p(x \mid \theta_2)$. Let the decision space A contain two elements (a_1, a_2) and let the loss function $L(\theta_i, a_j)$ be as follows:

	a_1	a_2
θ_1	0	c
θ_2	1	0

Show that of all possible decision functions a decision function $d(x)$ which provides a Bayes solution against the *a priori* distribution $q(\theta)$ is as follows: $d(x) = a_1$ for all values of x satisfying the likelihood ratio inequality

$$\frac{p(x \mid \theta_2)}{p(x \mid \theta_1)} \leqslant c\,\frac{q(\theta_1)}{q(\theta_2)}$$

while $d(x) = a_2$ for all other values of x.

CHAPTER 17

Time Series

17.1 INTRODUCTORY REMARKS

Many problems occur in science and technology in which a process produces what we may idealize as a family of random variables (stochastic process) such that there is a random variable x_t for each value of t (time) on some interval T, thus generating a *random function* of time. Such functions are called *time series*. Examples of time series are: voltage in a circuit over a period of seconds; noise in a factory over a period of minutes; height of sea waves over a period of hours.

Other time series show characteristics of random fluctuations superimposed over some smooth trend or time function. Examples of such time series are: temperature in a city over a 24-hour period; stock prices over a period of months; national employment over a period of years.

Various statistical and mathematical methods have been developed over a period of many years for studying and analyzing time series. Many of these methods have been developed to estimate the smooth trend function supposedly underlying the series by "averaging out," in some more or less empirical sense, the random fluctuations in the series.

In recent years, however, considerable attention has been given to the study of time series as stochastic processes. Such an approach often provides more insight into mechanisms by which time series are or can be generated than a more empirical approach.

There is already a large body of literature, including several books, which deals with time series from the point of view of stochastic processes. Here we shall give only a short introduction to the subject, including a few of the simpler basic results. Further details and results will be found in various books, particularly those by Bartlett (1955), Doob (1953), Grenander and Rosenblatt (1957) and Wold (1938).

17.2 STATIONARY TIME SERIES

It will be recalled from Section 4.1 that a stochastic process is defined as a family of random variables $\{x_t; t \in T\}$, such that for every finite set of choices of $t \in T$, say $t_1, \ldots, t_n$, the set $x_{t_1}, \ldots, x_{t_n}$ of random variables has a joint probability distribution function.

For time series we may think of t as the time index and T as the entire time axis (real axis) or any interval or set on the time axis. Thus, in some cases T might consist of only a sequence of equally spaced points along the time axis. For a given time unit, we shall be particularly interested in the case where $T = \ldots, -1, 0, +1, \ldots$, and *samples* consisting of finite blocks of T such as $1, \ldots, n$ or $1, \ldots, M$ or $n - h, n - h + 1, \ldots, n - 1$ or $1, \ldots, 2k + 1$.

An arbitrary stochastic process $\{x_t; t \in T\}$, where t represents a time index, is entirely too general to discuss usefully. We shall confine ourselves to what are called *stationary processes* or *stationary time series*. There are two important types which we shall consider.

A *strictly stationary* time series $\{x_t; t \in T\}$ has the property that any finite set

$$(x_{t_1}, \ldots, x_{t_n}) \tag{17.2.1}$$

of random variables from the family $\{x_t; t \in T\}$ has the same joint distribution function as the set

$$(x_{t_1+h}, \ldots, x_{t_n+h}) \tag{17.2.2}$$

for any h. (It should be pointed out that x_t can be complex or a k-dimensional random variable although we shall be concerned almost entirely with the case where x_t is a one-dimensional random variable.) Thus, the joint distributions of the sets of random variables (17.2.2) for different values of h are all identical and depend only on the time differences

$$t_2 - t_1, t_3 - t_2, \ldots, t_n - t_{n-1}. \tag{17.2.3}$$

The reader can readily verify that

17.2.1 *If $\{x_t; t \in T\}$ is a strictly stationary time series with $\mathscr{E}(|x_t|) < \infty$, then $\mathscr{E}(x_t) = c$, a constant for all t.*

The following is an illustrative example of a strictly stationary time series.

Example. Consider the (real) time series

$$x_t = \sum_{p=1}^{r} a_p \cos(t\omega_p + u_p), \qquad t = \ldots, -1, 0, +1, \ldots \tag{17.2.4}$$

where $a_1, \ldots, a_r$ are real constants, $\omega_1, \ldots, \omega_r$ are constants on the interval $[-\pi, +\pi]$ which we may take as ordered $\omega_1 < \cdots < \omega_r$ and $u_1, \ldots, u_r$ are independent random variables, each having a rectangular distribution on the interval $[-\pi, +\pi]$. Take any finite set of values of t, say $t_1, \ldots, t_n$. We then have

(17.2.5) $$x_{t_\xi+h} = \sum_{p=1}^{r} a_p \cos(t_\xi \omega_p + v_p), \qquad \xi = 1, \ldots, n$$

where the v_p are independent random variables uniformly distributed on the intervals $[h\omega_p \pm \pi]$, $p = 1, \ldots, r$. It is evident that the joint distribution of $(x_{t_1}, \ldots, x_{t_n})$ is identical with that of $(x_{t_1+h}, \ldots, x_{t_n+h})$.

Thus, the time series (17.2.4) is strictly stationary.

It should be noted that a natural complex version of the time series (17.2.4) is

(17.2.6) $$x_t = \sum_{p=1}^{r} a_p e^{i(t\omega_p + u_p)}, \qquad t = \ldots, -1, 0, +1, \ldots$$

which, of course, is also strictly stationary.

Some of the most useful studies of time series are those which do not require the assumption of strict stationarity but are based on the weaker assumptions that: (i) for all t, $\mathscr{E}(x_t)$ is a constant which may be taken as 0, and (ii) the distributions in (17.2.2) have the same covariance matrix for all h. A time series $\{x_t; t \in T\}$ satisfying these two conditions is said to be *weakly stationary*. This means that the covariance matrix depends only on the time differences (17.2.3) and hence the covariance of x_{t+h} and x_t (taking $\mathscr{E}(x_t) = 0$) will be a function of h only, namely,

(17.2.7) $$\mathscr{E}(x_{t+h}x_t) = \gamma_h.$$

The covariance γ_h considered as a function of h is called the *covariance function* of the time series $\{x_t; t \in T\}$; γ_h is sometimes called the *lag covariance* or *auto-covariance* with lag h. Note that the correlation coefficient between x_{t+h} and x_t is $\gamma_h/\gamma_0 = \rho_h$, say. Considered as a function of h, ρ_h is called the *autocorrelation function* or the *serial correlation function* of the time series.

If x_t is complex, its covariance function γ_h is defined by

(17.2.8) $$\gamma_h = \mathscr{E}(x_{t+h}\bar{x}_t)$$

where $\bar{x}_t$ is the complex conjugate of x_t.

The reader will readily see that:

17.2.2 *A strictly stationary time series with finite covariance matrix is also weakly stationary.*

A model for some time series $\{y_t: t \in T\}$ would be of the form

(17.2.9) $$y_t = m_t + x_t$$

where m_t is a constant for each t and x_t is such that $\{x_t: t \in T\}$ is a stationary time series with $\mathscr{E}(x_t) = 0$. Hence

(17.2.10) $$\mathscr{E}(y_t) = m_t$$

and the covariance function of $\{y_t: t \in T\}$ is identical with that of $\{x_t: t \in T\}$, that is (in the real case),

(17.2.11) $$\gamma_h = \mathscr{E}(y_{t+h} - m_{t+h})(y_t - m_t) = \mathscr{E}(x_{t+h}x_t).$$

One of the problems of time series is to estimate m_t or γ_h or to test hypotheses concerning m_t or γ_h from observations on a finite number n of random variables taken from the time series, where, of course, n may be allowed to $\rightarrow \infty$.

17.3 THE SPECTRAL FUNCTION OF A STATIONARY TIME SERIES

(a) A Special Case

As far as the first two moments of a stationary time series are concerned the time series is described by the covariance function γ_h. We shall show that the covariance function itself can be expressed in terms of what is called the *spectral distribution function* of the time series.

First let us consider the relationship between the covariance function and the spectral distribution function of the special stationary time series given by (17.2.4). It is readily verified that for all t we have

(17.3.1) $$\mathscr{E}(x_t) = \sum_{p=1}^{r} a_p \mathscr{E} \cos (t\omega_p + u_p) = 0.$$

For the covariance function $\gamma_h = \mathscr{E}(x_{t+h}x_t)$, we have

(17.3.2) $$\gamma_{t-s} = \sum_{p,q=1}^{r} a_p a_q \mathscr{E} \cos (s\omega_p + u_p) \cos (t\omega_q + u_q).$$

But it is seen that $\mathscr{E} \cos (s\omega_p + u_p) \cos (t\omega_q + u_q) = 0, q \neq p$. Using the fact that for $q = p$

$$\cos (s\omega_p + u_p) \cos (t\omega_p + u_p) = \tfrac{1}{2} \cos ((s + t)\omega_p + 2u_p) + \tfrac{1}{2} \cos (s - t)\omega_p$$

it is seen that

$$\mathscr{E} \cos (s\omega_p + u_p) \cos (t\omega_p + u_p) = \tfrac{1}{2} \cos (s - t)\omega_p.$$

Therefore, putting $t - s = h$, we have

(17.3.3) $$\gamma_h = \frac{1}{2}\sum_{p=1}^{r} a_p^2 \cos (h\omega_p) = \int_{-\pi}^{+\pi} \cos (h\omega)\, dF(\omega)$$

where $F(\omega)$ is the nondecreasing step function of ω on $[-\pi, +\pi]$ defined as

(17.3.4) $$F(\omega) = \frac{1}{2} \sum_{\omega_p \leqslant \omega} a_p^2.$$

At the end points, of $[-\pi, +\pi]$, we have $F(-\pi) = 0$ and $F(+\pi) = \gamma_0$. $F(\omega)$ is called the *spectral distribution function* associated with the covariance function γ_h, $k = \ldots, -1, 0, +1, \ldots$, in (17.3.3). The expression at the extreme right of (17.3.3) is also referred to as the *spectral representation* of γ_h.

In the case of complex time series, (17.2.6), it can be verified by fairly simple analysis that the covariance function γ_h is

(17.3.5) $$\gamma_h = \frac{1}{2} \sum_{p=1}^{r} a_p^2 e^{ih\omega_p} = \int_{-\pi}^{+\pi} e^{ih\omega}\, dF(\omega),$$

where $F(\omega)$ is identically the same spectral distribution as that in (17.3.3).

(b) The General Case

Now for a given time unit suppose we have an arbitrary real stationary time series

(17.3.6) $$x_t, \qquad t = \ldots, -1, 0, +1, \ldots$$

with $\mathscr{E}(x_t) \equiv 0$ and covariance function γ_h. In this case we have $\gamma_{-h} = \gamma_h$. For any integer M, let $F_M(\omega)$ be defined for any ω on $[-\pi, +\pi]$ as follows:

(17.3.7) $$\begin{aligned} F_M(\omega) &= \frac{1}{\pi M} \int_{-\pi}^{\omega} \mathscr{E}(x_1 \cos \omega + \cdots + x_M \cos M\omega)^2\, d\omega \\ &= \frac{1}{\pi M} \int_{-\pi}^{\omega} \sum_{p,q=1}^{M} \gamma_{p-q} \cos p\omega \cos q\omega\, d\omega \\ &= \frac{1}{\pi M} \left\{ \sum_{p\neq q=1}^{M} \gamma_{p-q} \left[\frac{\sin (p+q)\omega}{2(p+q)} + \frac{\sin (p-q)\omega}{2(p-q)} \right] \right. \\ &\quad \left. + \gamma_0 \sum_{p=1}^{M} \frac{\sin p\omega \cos p\omega}{2p} + \frac{\gamma_0 M}{2} (\omega + \pi) \right\}. \end{aligned}$$

$F_M(\omega)$ is nondecreasing on $[-\pi, +\pi]$ and furthermore $F_M(-\pi) = 0$, $F_M(+\pi) = \gamma_0$. As $M \to \infty$, $F_M(\omega)$ has as its limit a nondecreasing function $F(\omega)$ on $[-\pi, \pi]$ such that $F(-\pi) = 0$ and $F(+\pi) = \gamma_0$. (For existence and uniqueness of such a limit function $F(\omega)$ see discussion in proof of **5.4.1**.) Note that $dF_m(\omega)$ and $dF(\omega)$ are symmetric about $\omega = 0$.

Since $\gamma_{-h} = \gamma_h$, it is sufficient to consider h to be zero or a positive integer. We shall show that for $h = 0, 1, 2, \ldots$

$$\gamma_h = \int_{-\pi}^{+\pi} \cos h\omega \, dF(\omega). \tag{17.3.8}$$

Consider a sequence of integers $M_1, M_2, \ldots$. For $h = 0, 1, \ldots, M_s$, we have

$$\begin{aligned}\int_{-\pi}^{+\pi} \cos h\omega \, dF_{M_s}(\omega) &= \frac{1}{\pi M_s} \int_{-\pi}^{+\pi} \sum_{p,q=1}^{M_s} \gamma_{p-q} \cos p\omega \cos q\omega \cos h\omega \, d\omega \\ &= \frac{1}{\pi M_s} (A_h + B_h)\end{aligned}$$

where

$$A_h = \int_{-\pi}^{+\pi} \sum_{p \neq q=1}^{M_s} \gamma_{p-q} \cos p\omega \cos q\omega \cos h\omega \, d\omega$$

and

$$B_h = \gamma_0 \int_{-\pi}^{+\pi} \sum_{p=1}^{M_s} \cos^2 p\omega \cos h\omega \, d\omega.$$

Using the fact that

$$\begin{aligned}\cos p\omega \cos q\omega \cos h\omega &= \tfrac{1}{4}[\cos (p + q + h)\omega + \cos (p + q - h)\omega \\ &\quad + \cos (p - q + h)\omega + \cos (p - q - h)\omega]\end{aligned}$$

$$\cos^2 p\omega \cos h\omega = \tfrac{1}{4}[\cos (2p + h)\omega + \cos (2p - h)\omega + 2 \cos h\omega]$$

and that

$$\gamma_{p-q} = \gamma_{q-p}$$

we find

$$\begin{aligned}A_h &= \pi\gamma_h(M_s - h), & h &\neq 0 \\ &= 0 & h &= 0\end{aligned}$$

and

$$\begin{aligned}B_h &= \pi\gamma_0\delta_h, & h &\neq 0 \\ &= \pi\gamma_0 M_s, & h &= 0\end{aligned}$$

where $\delta_h = \frac{1}{4}[1 + (-1)^h]$.

Therefore, for $h = 0, 1, \ldots, M_s$ we have

$$\begin{aligned}\int_{-\pi}^{\pi} \cos h\omega \, dF_{M_s}(\omega) &= \left[\gamma_h\left(1 - \frac{h}{M_s}\right) + \frac{\gamma_0\delta_h}{M_s}\right], & h &\neq 0 \\ &= \gamma_0, & h &= 0.\end{aligned} \tag{17.3.9}$$

If we let $M_s \to \infty$, we obtain (17.3.8) for any non-negative integer h. Now every convergent subsequence of $\{F_M(\omega)\}$, and hence the sequence

$\{F_M(\omega)\}$ itself, yields the same result. Since $dF(\omega)$ and also $\cos h\omega$ are both symmetric in ω about $\omega = 0$ we can write (17.3.8) in the form

$$\gamma_h = 2\int_0^{\pi} \cos h\omega \, dF(\omega). \tag{17.3.10}$$

$F(\omega)$ is called the *spectral distribution function* of the stationary time series x_t. If we normalize $F(\omega)$ by dividing by γ_0, then the ratio $F(\omega')/\gamma_0$ is the fraction of *spectral mass* produced by all frequencies $\omega \leqslant \omega'$. If $F(\omega)$ is absolutely continuous with derivative $f(\omega)$, then $f(\omega)$ is called the *spectral density function* of the time series. Summarizing, we have the following result:

17.3.1 *If* x_t, $t = \ldots, -1, 0, +1, \ldots$ *is a real stationary time series with* $\mathscr{E}(x_t) = 0$, *and covariance function* γ_h, $h = 0, 1, 2, \ldots$ *then* γ_h *is given by* (17.3.10) *where the spectral distribution function* $F(\omega)$ *is* $\lim_{M\to\infty} F_M(\omega)$, *and where* $F_M(\omega)$ *is given by* (17.3.7).

For a complex stationary time series x_t, $t = \ldots, -1, 0, +1, \ldots$ with covariance function $\gamma_h = \mathscr{E}(x_{t+h}\bar{x}_t)$, $h = \ldots, -1, 0, +, \ldots$ the theorem corresponding to **17.3.1**, is due to Herglotz (1911), and states that

$$\gamma_h = \int_{-\pi}^{\pi} e^{ih\omega} \, dF(\omega) \tag{17.3.11}$$

where

$$F(\omega) = \lim_{M\to\infty} F_M(\omega) \tag{17.3.12}$$

and

$$F_M(\omega) = \frac{1}{2\pi M}\int_{-\pi}^{\omega} \mathscr{E}\left(\sum_{p=1}^{M} x_p e^{-ip\omega}\right)\left(\sum_{q=1}^{M} \bar{x}_q e^{iq\omega}\right) d\omega. \tag{17.3.13}$$

The quantity under the integral is seen to be real and non-negative.

The proof of the theorem for the complex case is similar to that of **17.3.1** and is left to the reader.

Results (17.3.10) and (17.3.11) for real and complex stationary time series are general enough for most stationary time series problems that occur in physical processes since the time unit can always be chosen to fit the problem. However, an extension of (17.3.11) to the case where the range of h is the real (time) axis rather than multiples of the given time unit has been established by Bochner (1932). His result states that if, for a complex stationary time series $\{x_t: -\infty < t < +\infty\}$, γ_h is continuous at $h = 0$ then

$$\gamma_h = \int_{-\infty}^{\infty} e^{ih\omega} \, dF(\omega). \tag{17.3.14}$$

In the case of a real time series the corresponding generalization of (17.3.10) is that

$$\gamma_h = 2\int_0^\infty \cos h\omega \, dF(\omega). \tag{17.3.15}$$

(c) White Noise

If a real stationary time series x_t, $t = \ldots, -1, 0, +1, \ldots$, has the constant spectral density function $f(\omega) = \gamma_0/2\pi$ on $[-\pi, +\pi]$ then it will be seen from (17.3.8) that $\gamma_h = 0$ for all $h \neq 0$. Conversely, suppose $\gamma_h = 0$ for all $h \neq 0$. Then we have

$$\begin{aligned} \int_{-\pi}^{+\pi} \cos h\omega \, f(\omega)\, d\omega &= 0, \qquad h \neq 0 \\ &= \gamma_0, \qquad h = 0. \end{aligned} \tag{17.3.16}$$

If $f(\omega)$ is symmetric around $\omega = 0$, and can be represented as a Fourier series $b_0 + b_1 \cos \omega + b_2 \cos 2\omega + \cdots$ we have

(17.3.17)

$$\int_{-\pi}^{+\pi} \cos h\omega \, f(\omega)\, d\omega = \int_{-\pi}^{+\pi} [b_0 + b_1 \cos \omega + b_2 \cos 2\omega + \cdots] \cos h\omega \, d\omega, \qquad h = 0, 1, \ldots.$$

But (17.3.17) reduces to

$$\int_{-\pi}^{+\pi} \cos h\omega \, f(\omega)\, d\omega = b_h \int_{-\pi}^{+\pi} \cos^2 h\omega \, d\omega. \tag{17.3.18}$$

The left-hand side vanishes for $h \neq 0$ as stated in (17.3.16) and hence $b_h = 0$ for $h \neq 0$. Therefore, $f(\omega) = b_0$, and it follows from the fact that

$$\int_{-\pi}^{+\pi} f(\omega)\, d\omega = \gamma_0,$$

that $b_0 = \gamma_0/2\pi$. Therefore on $[-\pi, +\pi]$

$$f(\omega) = \frac{\gamma_0}{2\pi}. \tag{17.3.19}$$

Thus,

17.3.3 *If a real stationary time series x_t, $t = \ldots, -1, 0, +1, \ldots$, has a symmetric spectral density function $f(\omega)$, representable as a Fourier series of form $b_0 + b_1 \cos \omega + b_2 \cos 2\omega + \cdots$ on $[-\pi, +\pi]$, a necessary and sufficient condition for the covariance function γ_h to vanish for all $h \neq 0$ is that $f(\omega)$ be constant on $[-\pi, +\pi]$, in which case $f(\omega) = \gamma_0/2\pi$.*

A stationary time series x_t, $t = \ldots, -1, 0, +1, \ldots$, in which every pair $x_t, x_{t'}$ $t \neq t'$ has zero correlation is called a *white noise*. **17.3.3** thus states a necessary and sufficient condition for a time series to be a white noise for a certain class of spectral density functions.

17.4 ESTIMATION OF MEAN AND COVARIANCE FUNCTION OF A STATIONARY TIME SERIES

(a) Estimation of the Mean

Suppose x_t, $t = \ldots, -1, 0, +1, \ldots$, is a real stationary time series with $\mathscr{E}(x_t) = \mu$, and that we wish to estimate μ from the finite sample $(x_1, \ldots, x_n)$. The mean $\bar{x} = (x_1 + \cdots + x_n)/n$ is an unbiased estimator for μ since

(17.4.1) $$\mathscr{E}(\bar{x}) = \mu.$$

The variance of $\bar{x}$ is given

(17.4.2) $$\sigma^2(\bar{x}) = \frac{1}{n}\left[\gamma_0 + 2\sum_{\xi=1}^{n-1}\left(1 - \frac{\xi}{n}\right)\gamma_\xi\right].$$

If $F(\omega)$ has a derivative $f(\omega)$ which is symmetric about $\omega = 0$, continuous at $\omega = 0$, and can be represented by the Fourier series

(17.4.3) $$\frac{1}{2\pi}(\gamma_0 + 2\gamma_1 \cos \omega + 2\gamma_2 \cos 2\omega + \cdots)$$

on $[-\pi, +\pi]$, it can be verified that as $n \to \infty$ the quantity in [ ] in (17.4.2) has the value $2\pi f(0)$ as its limit. Therefore, for large n we have, under these conditions,

(17.4.4) $$\sigma^2(\bar{x}) \simeq \frac{2\pi f(0)}{n}.$$

(b) Estimation of the Covariance

If we let our estimator for γ_h be

(17.4.5) $$c_h = \frac{1}{n-h}\sum_{\xi=1}^{n-h}(x_\xi - \bar{x})(x_{\xi+h} - \bar{x}),$$

we find after some algebraic manipulation that

(17.4.6) $$\mathscr{E}(c_h) = \gamma_h - \sigma^2(\bar{x}) + \frac{2}{n}R$$

where

(17.4.7)
$$R = \left[\sum_{\xi=1}^{n-1}\left(1 - \frac{\xi}{n}\right)\gamma_\xi - \sum_{\xi=1}^{n-h-1}\left(1 - \frac{\xi}{n-h}\right)(\gamma_\xi + \gamma_{\xi+h}) - \sum_{\xi=1}^{h}\gamma_\xi\right].$$

It can be shown that $R \to 0$ as $n \to \infty$; hence for large n the asymptotic bias in c_h as an estimator for γ_h is approximately $-\sigma^2(\bar{x})$. If the spectral distribution has a density function $f(\omega)$ continuous at $\omega = 0$, then we have for large n

$$\mathscr{E}(c_h) \cong \gamma_h - \frac{2\pi f(0)}{n}. \tag{17.4.8}$$

It should be noted that if μ is known and if $\bar{x}$ is replaced by μ in (17.4.5), the mean value of the resulting covariance is exactly γ_h.

The problem of determining the variance of c_h involves finiteness assumptions about fourth order moments of the time series x_t, $t = \ldots, -1, 0, +1, \ldots$ and some involved algebra. However, an approximation can be obtained if x_t, $t = \ldots, -1, 0, +1, \ldots$ is strictly stationary. Consider the time series $z_t = (x_t - \mu)(x_{t+h} - \mu)$, $t = \ldots, -1, 0, +1, \ldots$ which is strictly stationary if x_t is. The mean of this time series is γ_h. Let its covariance function exist and be denoted by $\gamma^*_{h,g}$. Then

$$\gamma^*_{h,g} = \mathscr{E}(z_t z_{t+g}). \tag{17.4.9}$$

Now it can be shown that for large n the variance of c_h is approximated by that of c^*_h, where

$$c^*_h = \frac{1}{n-h} \sum_{\xi=1}^{n-h} x_\xi x_{\xi+h}. \tag{17.4.10}$$

The variance of c^*_h has the same structure as that of $\sigma^2(\bar{x})$ and is given by

$$\sigma^2(c^*_h) = \frac{1}{n-h}\left[\gamma^*_{h,0} + 2\sum_{\xi=1}^{n-h-1}\left(1 - \frac{\xi}{n-h}\right)\gamma^*_{h,\xi}\right]. \tag{17.4.11}$$

The series converges if the time series z_t has a spectral density function, and for large n

$$\sigma^2(c^*_h) \cong \frac{\gamma^*_{h,0}\left(3 - \frac{1}{\pi}\right)}{n} \tag{17.4.12}$$

where it will be recalled that

$$\gamma^*_{h,0} = \mathscr{E}[(x_t - \mu)(x_{t+h} - \mu)]^2.$$

17.5 ESTIMATION OF SPECTRAL DISTRIBUTION

Suppose a stationary time series x_t, $t = \ldots, -1, 0, +1, \ldots$ has mean μ and spectral density function $f(\omega)$. We shall consider the problem of estimating $f(\omega)$ from a sample $(x_1, \ldots, x_n)$ of the time series, assuming μ is known.

In view of the definition of $F(\omega)$ as the limit of $F_M(\omega)$ as $M \to \infty$ where $F_M(\omega)$ is defined in (17.3.7), a quantity which suggests itself as an estimator for $f(\omega)$ based on $(x_1, \ldots, x_n)$ is $f_n(\omega)$ defined as

$$(17.5.1) \quad f_n(\omega) = \frac{1}{n\pi}(y_1 \cos \omega + y_2 \cos 2\omega + \ldots + y_n \cos n\omega)^2$$

where $y_t = x_t - \mu$. Taking mean values and summing some simple trigonometric series we find

$$(17.5.2) \quad \mathscr{E}f_n(\omega) = \frac{1}{2\pi}\left\{\gamma_0\left[1 + \frac{\sin n\omega \cos (n+1)\omega}{n \sin \omega}\right]\right.$$
$$\left. + 2\sum_{\xi=1}^{n-1} \gamma_\xi\left[\left(1 - \frac{\xi}{n}\right) \cos h\omega + \frac{\sin (n-\xi)\omega \cos (n+1)\omega}{n \sin \omega}\right]\right\}.$$

Allowing $n \to \infty$ and making use of the Fourier expansion of $f(\omega)$ given by (17.4.3) it is evident that for any value of ω in $[-\pi, +\pi]$ except for $\omega = 0$,

$$\lim_{n\to\infty} \mathscr{E}f_n(\omega) = f(\omega), \quad \text{whereas} \quad \lim_{n\to\infty} \mathscr{E}f_n(0) = 2f(0).$$

Thus, $f_n(\omega)$ is an asymptotically unbiased estimator for $f(\omega)$ for ω in $[-\pi, +\pi]$ except at $\omega = 0$. On the other hand it can be shown under mild conditions that the variance of $f_n(\omega)$ does not converge to 0 as $n \to \infty$. In particular, if $(x_1, \ldots, x_n)$ is a normal n-dimensional random variable, $f_n(\omega)$, except for a constant multiplier, has a chi-square distribution with one degree of freedom for all values of n.

Hence, the obvious estimator $f_n(\omega)$ for $f(\omega)$ is not a consistent estimator. Thus, some other estimator for $f(\omega)$ must be sought. Various consistent estimators for $f(\omega)$ have been suggested and treated by Bartlett (1950), Grenander and Rosenblatt (1953), Jenkins and Priestly (1957), Parzen (1957), Tukey (1949*a*), and others. Most of these procedures approach the problem by estimating $f(\omega)$ at a fixed point ω on its interval $[-\pi, +\pi]$. Tukey's method, however, is to consider the problem of estimating spectral masses within subintervals of $[-\pi, +\pi]$. We shall consider a method similar to Tukey's and make no attempt to discuss the other estimators. The reader interested in them should refer to the books by Bartlett (1950) and by Grenander and Rosenblatt (1957).

We shall assume that for the time series from which the sample $(x_1, \ldots, x_n)$ comes, there exists a spectral density function $f(\omega)$ which can be represented by the Fourier expansion (17.4.3). Now for a given positive integer m, we shall proceed to estimate from the sample

$(x_1, \ldots, x_n)$, the *spectral masses*

(17.5.3) $$\Delta_p = F\left(\frac{p\pi}{m} + \frac{\pi}{2m}\right) - F\left(\frac{p\pi}{m} - \frac{\pi}{2m}\right)$$

contained in the intervals

(17.5.4) $$I_p = \left(\frac{p\pi}{m} - \frac{\pi}{2m}, \quad \frac{p\pi}{m} + \frac{\pi}{2m}\right], \qquad p = 0, \pm 1, \pm 2, \ldots, \pm m.$$

where $F(\omega)$ is the spectral distribution function.

The number m is chosen so as to balance the uncertainties of attempting to estimate too many spectral masses against the need to study the spectrum with sufficient resolution. This means roughly that we select m as the number of covariances to give a "reasonable" description of the covariance function of the time series.

The spectral mass Δ_p can be expressed in terms of $f(\omega)$ as follows:

(17.5.5) $$\Delta_p = \int_{\frac{p\pi}{m} - \frac{\pi}{2m}}^{\frac{p\pi}{m} + \frac{\pi}{2m}} f(\omega)\, d\omega.$$

Using the Fourier expansion (17.4.3) for $f(\omega)$ and performing the integration, we obtain

(17.5.6) $$\Delta_p = \frac{\gamma_0}{2m} + \frac{1}{m}\sum_{h=1}^{\infty} \gamma_h \left(\frac{\sin(h\pi/2m)}{h\pi/2m}\right) \cos\frac{hp\pi}{m}.$$

Now if it is considered that the information in the time series is "reasonably" represented by the covariances $\gamma_0, \gamma_1, \ldots, \gamma_m$, then the spectral masses Δ_p, $p = 0, \pm 1, \ldots, \pm m$ may be approximated by truncating the infinite series in (17.5.6) at $h = m$, thus giving the approximate spectral masses

(17.5.7)

$$\tilde{\Delta}_p = \frac{\gamma_0}{2m} + \frac{1}{m}\sum_{h=1}^{m} \gamma_h \left(\frac{\sin(h\pi/2m)}{h\pi/2m}\right) \cos\frac{hp\pi}{m}, \qquad p = 0, \pm 1, \ldots, \pm m.$$

If μ is known for the time series, and if a sample $(x_1, \ldots, x_n)$ is taken, the estimator which suggests itself for $\tilde{\Delta}_p$ is $\tilde{\Delta}_{pn}$, where

(17.5.8) $$\tilde{\Delta}_{pn} = \frac{c_0}{2m} + \frac{1}{m}\sum_{h=1}^{m} c_h \left(\frac{\sin(h\pi/2m)}{h\pi/2m}\right) \cos\frac{hp\pi}{m},$$

where $c_h = \dfrac{1}{n-h}\sum_{\xi=1}^{n-h} y_\xi y_{\xi+h}$, $h = 0, 1, \ldots, n-1$, $m \leqslant n-1$, and $y_\xi = x_\xi - \mu$. It is evident that since $\mathscr{E}(c_h) = \gamma_h$, $h = 0, 1, \ldots$, we have

(17.5.9) $$\mathscr{E}(\tilde{\Delta}_{pn}) = \tilde{\Delta}_p,$$

that is, $\tilde{\Delta}_{pn}$ is an unbiased estimator for $\tilde{\Delta}_p$.

We remark that the spectral mass estimator $\tilde{\Delta}_{pn}$ is slightly different from the one suggested by Tukey (1949*a*). If $\dfrac{\sin(h\pi/2m)}{h\pi/2m}$ is replaced by the approximation $(0.46 \cos(h\pi/m) + 0.54)$ and if an end point correction is made by taking only half of the last term in the sum in (17.5.8), Tukey's estimate of Δ_p is obtained.

Since $\tilde{\Delta}_{pn}$ is an estimator for the spectral mass in the interval $p\pi/m \pm \pi/2m$, an unbiased estimator $\mathscr{E}^{-1}(\bar{f}_p(\omega))$ for the average spectral density $\bar{f}_p(\omega)$, over the interval I_p is obtained by dividing the spectral mass estimator $\tilde{\Delta}_{pn}$ by the interval length π/m, that is,

$$\mathscr{E}^{-1}(\bar{f}_p(\omega)) = \frac{m}{\pi}\tilde{\Delta}_{pn}. \tag{17.5.10}$$

It can be shown that the variance of the estimator $\dfrac{m}{\pi}\tilde{\Delta}_{pn}$ in (17.5.10) for large n and large m (where $m = 0(n)$) is

$$\sigma^2\left(\frac{m}{\pi}\tilde{\Delta}_{pn}\right) \cong \frac{2}{n}\bar{f}_p^2(\omega)\sum_{h=-m}^{m}\left(\frac{\sin(h\pi/2m)}{h\pi/2m}\right)^2\cos^2\frac{hp\pi}{m}. \tag{17.5.11}$$

If the mean μ is unknown for the time series we replace μ by the sample mean $\bar{x}$ in defining c_h and use these modified forms of c_h, $h = 0, 1, \ldots, m$ in (17.5.8) to obtain the estimator for the spectral mass $\tilde{\Delta}_{pn}$. It can be shown that (17.5.9) and (17.5.11) are still valid for large n for these changes in c_h.

17.6 STATISTICAL TESTS FOR PARAMETRIC TIME SERIES

A full treatment of the various statistical inference problems concerning stationary time series is beyond the scope of this book. Here we shall discuss estimation and statistical testing only in the cases of several classical parametric time series. The reader interested in the treatment of further problems is referred to the books by Bartlett (1955) and Grenander and Rosenblatt (1957).

(a) The Variate Difference Method

Suppose the time series x_t, $t = \ldots, -1, 0, +1, \ldots$ is known to be of the form

$$x_t = \sum_{p=0}^{k}\beta_p t^p + \varepsilon_t = \mu_t + \varepsilon_t \tag{17.6.1}$$

where $\beta_0, \beta_1, \ldots, \beta_k$ are unknown, and where ε_t is a white noise with variance σ^2.

If k is known, then for a sample $(x_1, \ldots, x_n)$, $n \geqslant k + 1$, minimum variance estimators for the β's may be obtained by least squares and the variances of the estimators may be obtained by the methods discussed in Section 10.3(*c*). The same procedure applies if each t^p is replaced by a polynomial $g_p(t)$ of degree p, $p = 0, 1, \ldots, k$, where $g_0(t), g_1(t), \ldots, g_k(t)$ are functionally independent.

However, if k is unknown in (17.6.1), one way of estimating it is by a semiempirical procedure known as the *variate difference method* which works as follows. Let $y_{h,t}$ be a time series defined by the hth forward difference of the time series x_t. Then

$$y_{h,t} = \Delta^h x_t = \Delta^h \mu_t + \Delta^h \varepsilon_t \tag{17.6.2}$$

where

$$\Delta^h \varepsilon_t = \varepsilon_{t+h} - \binom{h}{1}\varepsilon_{t+h-1} + \binom{h}{2}\varepsilon_{t+h-2} - \cdots + (-1)^h \varepsilon_t \tag{17.6.3}$$

with a similar expression for $\Delta^h \mu_t$. Now suppose we consider the sequence of samples $(x_1, \ldots, x_{n+h})$, $h = 1, 2, \ldots$ and form the ratios

$$Q_h = \sum_{\xi=1}^{n} y_{h,\xi}^2 \Big/ \left[n \binom{2h}{h} \right], \qquad h = 1, 2, \ldots. \tag{17.6.4}$$

If we take the mean value of Q_h we obtain

$$\mathscr{E} Q_h = \mathscr{E} \sum_{\xi=1}^{n} (\Delta^h \mu_\xi + \Delta^h \varepsilon_\xi)^2 \Big/ \left[n \binom{2h}{h} \right]. \tag{17.6.5}$$

Since ε_ξ is a white noise and μ_ξ is a nonrandom function of t, we find

$$\mathscr{E}(\Delta^h \mu_\xi + \Delta^h \varepsilon_\xi)^2 = (\Delta^h \mu_\xi)^2 + \mathscr{E}(\Delta^h \varepsilon_\xi)^2. \tag{17.6.6}$$

But

$$\mathscr{E}(\Delta^h \varepsilon_\xi)^2 = \sigma^2 \sum_{p=0}^{h} \binom{h}{p}^2 = \binom{2h}{h} \sigma^2. \tag{17.6.7}$$

Substituting from (17.6.7) into (17.6.6) and then into (17.6.5), we find

$$\mathscr{E} Q_h = R_h + \sigma^2 \qquad h = 1, 2, \ldots, \tag{17.6.8}$$

where $R_h = \dfrac{1}{n} \sum_{\xi=1}^{n} (\Delta^h \mu_\xi)^2 \Big/ \binom{2h}{h}$.

It will be noted that R_h is non-negative and vanishes for all $h \geqslant k + 1$. Thus, $\mathscr{E} Q_h = \sigma^2$, for all $h \geqslant k + 1$.

Thus, in a practical situation, one would calculate Q_h, $h = 1, 2, \ldots$ and choose as the estimate of k the value of h at which Q_h appears to become constant (except for "small" random fluctuations) and use Q_h itself as an estimate of σ^2. The reader interested in a full discussion of the variate difference method, its various modified forms, and its applications should consult Tintner (1940) and Quenouille (1948) on the subject.

(b) Trigonometric Time Series with Known Coefficients and Periods

Suppose the time series x_t, $t = \ldots, -1, 0, +1$, has the following periodic parametric form

(17.6.9)
$$x_\xi = \sum_{p=1}^{k} [a_p \cos \omega_p \xi + b_p \sin \omega_p \xi] + \varepsilon_\xi$$

where ε_ξ is a white noise with known variance σ^2, a_p, b_p, and ω_p are known (real) constants, and $0 \leqslant \omega_p \leqslant \pi$.

Now consider a sample $(x_1, \ldots, x_n)$ from the time series. If we multiply (17.6.9) by $\cos \omega\xi$, $0 \leqslant \omega \leqslant \pi$, sum over ξ and divide by n, we obtain

(17.6.10)
$$\alpha(\omega) = \sum_{p=1}^{k} [a_p u_p(\omega) + b_p v_p(\omega)] + \delta(\omega)$$

where

(7.6.11)
$$\begin{aligned} \alpha(\omega) &= \frac{1}{n} \sum_{\xi=1}^{n} x_\xi \cos \omega\xi \\ u_p(\omega) &= \frac{1}{n} \sum_{\xi=1}^{n} \cos \omega_p \xi \cos \omega\xi \\ v_p(\omega) &= \frac{1}{n} \sum_{\xi=1}^{n} \sin \omega_p \xi \cos \omega\xi \\ \delta(\omega) &= \frac{1}{n} \sum_{\xi=1}^{n} \varepsilon_\xi \cos \omega\xi. \end{aligned}$$

Similarly, if we multiply (17.6.9) by $\sin \omega\xi$, $0 \leqslant \omega \leqslant \pi$, sum over ξ and divide by n, we obtain

(17.6.12)
$$\alpha'(\omega) = \sum_{p=1}^{k} [a_p u_p'(\omega) + b_p v_p'(\omega)] + \delta'(\omega),$$

where $\alpha'(\omega)$, $u_p'(\omega)$, $v_p'(\omega)$, $\delta'(\omega)$ are obtained from $\alpha(\omega)$, $u_p(\omega)$, $v_p(\omega)$, $\delta(\omega)$ respectively, by replacing $\cos \omega\xi$ by $\sin \omega\xi$.

Note that $u_p(\omega)$, $v_p(\omega)$, $u_p'(\omega)$, $v_p'(\omega)$ can be written as follows:

(17.6.13)
$$\begin{aligned} u_p(\omega) &= \frac{1}{2n} \sum_{\xi=1}^{n} [\cos (\omega_p + \omega)\xi + \cos (\omega_p - \omega)\xi] \\ v_p(\omega) &= \frac{1}{2n} \sum_{\xi=1}^{n} [\sin (\omega_p + \omega)\xi + \sin (\omega_p - \omega)\xi] \\ u_p'(\omega) &= \frac{1}{2n} \sum_{\xi=1}^{n} [\sin (\omega_p + \omega)\xi - \sin (\omega_p - \omega)\xi] \\ v_p'(\omega) &= \frac{1}{2n} \sum_{\xi=1}^{n} [-\cos (\omega_p + \omega)\xi - \cos (\omega_p - \omega)\xi] \end{aligned}$$

from which it can be verified that as $n \to \infty$

$$\mathscr{E}[\alpha(\omega)] \to 0, \qquad \omega \neq \omega_p, \qquad p = 1, \ldots, k$$
$$\to \frac{a_p}{2}, \qquad \omega = \omega_p. \tag{17.6.14}$$

Similarly, as $n \to \infty$

$$\mathscr{E}[\alpha'(\omega)] \to 0, \qquad \omega \neq \omega_p, \qquad p = 1, \ldots, k$$
$$\to -\frac{b_p}{2}, \qquad \omega = \omega_p. \tag{17.6.15}$$

Thus, if $2\pi/\omega_p$ is a genuine period of the time series x_t, $t = \ldots, -1, 0, +1, \ldots$ then $\alpha(\omega_p)$ will tend to be near $a_p/2$ and $\alpha'(\omega_p)$ will tend to be near $-b_p/2$ in large samples.

If the time series is given and if the periods $2\pi/\omega_p$, $p = 1, \ldots, k$, are known, but the coefficients a_p, b_p, $p = 1, \ldots, k$ are unknown, then it follows from Section 10.3(*c*) that minimum variance linear estimators for these coefficients can be found by least squares.

(c) Trigonometric Time Series with Unknown Parameters—Periodogram Analysis

If the time series is of the form (17.6.9) where the a_p, b_p and ω_p, and even k, are unknown Schuster (1898) has proposed a method of searching for possible periods and Fisher (1929) has proposed a method of testing the significance of suspected periods. We shall discuss this method which is known as *periodogram analysis*.

The behavior of the mean values of the functions $\alpha(\omega)$ and $\alpha'(\omega)$ as described earlier suggests that $\alpha(\omega)$ and $\alpha'(\omega)$ considered as functions of ω should be useful in screening out true periods if any exist. If we assume there are no true periods at all, that is, if the a_p and b_p are all zero, then x_t is a white noise. If n is odd, say $n = 2k + 1$, let

$$\omega_p = \frac{2\pi p}{2k + 1}, \qquad p = 1, \ldots, k. \tag{17.6.16}$$

Now consider $\alpha(\omega_p)$, $\alpha'(\omega_p)$, $p = 1, \ldots, k$ which are all linear functions of x_t. The covariance of $\alpha(\omega_p)$ and $\alpha(\omega_q)$, $p \neq q$, is

(17.6.17)

$$\begin{aligned}\operatorname{cov}(\alpha(\omega_p), \alpha(\omega_q)) &= \frac{\sigma^2}{(2k + 1)^2} \sum_{\xi=1}^{2k+1} \cos \omega_p\xi \cos \omega_q\xi \\ &= \frac{\sigma^2}{2(2k + 1)^2} \sum_{\xi=1}^{2k+1} [\cos(\omega_p + \omega_q)\xi + \cos(\omega_p - \omega_q)\xi].\end{aligned}$$

But

$$\sum_{\xi=1}^{2k+1} \cos(\omega_p \pm \omega_q)\xi = \text{real part of} \left\{\frac{e^{i(\omega_p \pm \omega_q)}\left[1 - e^{i(\omega_p \pm \omega_q)(2k+1)}\right]}{1 - e^{i(\omega_p \pm \omega_q)}}\right\}.$$

Since $(\omega_p \pm \omega_q)(2k + 1)$ is an integral multiple of 2π it is evident that the quantity in [ ] in the preceding expression vanishes.

Hence

$$\operatorname{cov}(\alpha(\omega_p), \alpha(\omega_q)) = 0, \qquad p \neq q.$$

In a similar manner it follows that

$$\operatorname{cov}(\alpha'(\omega_p), \alpha'(\omega_q)) = 0, \qquad p \neq q$$

and

$$\operatorname{cov}(\alpha'(\omega_p), \alpha(\omega_q)) = 0, \qquad p, q = 1, \ldots, k.$$

Furthermore for $p = q$, (17.6.17) yields the variance of $\alpha(\omega_p)$,

$$\sigma^2(\alpha(\omega_p)) = \frac{\sigma^2}{2(2k + 1)}.$$

In a similar manner we find

$$\sigma^2(\alpha'(\omega_p)) = \frac{\sigma^2}{2(2k + 1)}.$$

Summarizing, we have the following result:

17.6.1 *If $x_1, \ldots, x_{2k+1}$ are independent random variables having mean 0 and variance σ^2, then $\alpha(\omega_p), \alpha'(\omega_p), p = 1, \ldots, k$ where ω_p is given by* (17.6.16) *are random variables having zero means, zero covariances, and variances $\sigma^2/(4k + 2)$.*

We are now in a position to examine the sample $(x_1, \ldots, x_{2k+1})$ for possible periods of lengths $2\pi/\omega_p$, $p = 1, \ldots, 2k + 1$. If $2\pi/\omega_p$ is a genuine period then we know from the behavior of $\alpha(\omega_p)$ and $\alpha'(\omega_p)$ that both of these will tend to have values away from 0 for large n $(= 2k + 1)$. This, in turn, means that $\alpha^2(\omega_p) + \alpha'^2(\omega_p)$ will tend to have a significantly large positive value if $2\pi/\omega_p$ is a genuine period. But, for the given sample $(x_1, \ldots, x_{2k+1})$, $\alpha^2(\omega_p) + \alpha'^2(\omega_p)$ has a value for each ω_p, $p = 1, \ldots, k$. So we need a way of testing whether the largest one (or the mth largest) of the quantities $\alpha^2(\omega_p) + \alpha'^2(\omega_p)$, $p = 1, \ldots, k$, is significantly large under the assumption that $\{x_t\}$ is a white noise. Another complication, of course, is that σ^2 is usually unknown. The quantity

$$I_{2k+1}(\omega) = \frac{(2k + 1)}{2\pi}\left[\alpha^2(\omega) + \alpha'^2(\omega)\right] \tag{17.6.18}$$

is called the *periodogram* of $(x_1, \ldots, x_{2k+1})$.

To deal with the problem of significance testing, we make the further assumption that $\{x_t\}$ is normal white noise, that is, $x_1, \ldots, x_{2k+1}$ are independent and all have the distribution $N(0, \sigma^2)$. Then

$$\frac{\sqrt{2(2k+1)}}{\sigma}\,\alpha(\omega_p), \qquad \frac{\sqrt{2(2k+1)}}{\sigma}\,\alpha'(\omega_p), \qquad p = 1, \ldots, k$$

are $2k$ independent random variables having normal distributions $N(0, 1)$. Let

$$(17.6.19)\quad u_p = \frac{2k+1}{\sigma^2}\,[\alpha^2(\omega_p) + \alpha'^2(\omega_p)] = \frac{2\pi}{\sigma^2}\,I_{2k+1}(\omega_p), \qquad p = 1, \ldots, k.$$

Then $u_1, \ldots, u_k$ are independent random variables having probability element

$$(17.6.20)\qquad e^{-(u_1+\cdots+u_k)}\,du_1 \cdots du_k$$

if $u_p \geqslant 0$, $p = 1, \ldots, k$, and 0 otherwise. The problem of whether the largest (or the mth largest) of the quantities $\alpha^2(\omega_p) + a'^2(\omega_p)$ (or $I_{2k+1}(\omega_p)$) is significantly large reduces to testing whether the largest or $u_1, \ldots, u_k$ is significantly large. Since σ^2 is unknown, the test which suggests itself is whether the largest (or the mth largest) of the ratios

$$(17.6.21)\qquad \frac{u_p}{u_1 + \cdots + u_k}, \qquad p = 1, \ldots, k$$

is significantly large. This ratio, it should be noted, does not depend on σ^2.

Let v_r be the rth smallest of $u_1, \ldots, u_k$, $r = 1, \ldots, k$. Then $v_1, \ldots, v_k$, $0 < v_1 < v_2 < \cdots < v_k < +\infty$, are the order statistics of $u_1, \ldots, u_k$ and their probability element is

$$(17.6.22)\qquad k!\, e^{-(v_1+\cdots+v_k)}\,dv_1 \cdots dv_k.$$

Let

$$(17.6.23)\qquad g = \frac{v_k}{v_1 + \cdots + v_k}.$$

Then g is identically the same random variable as the largest of the ratios in (17.6.21). We shall obtain the distribution function of g by first finding the hth moment of g using a simple but effective device described below.

It will be noted that the hth moment of g is given by

$$(17.6.24)\quad \mathscr{E}(g^h) = \int_{-\infty}^{0} \cdots \int_{-\infty}^{0} \mathscr{E}\{e^{(v_1+\cdots+v_k)\varphi} v_k^h\}\,dz_1 \cdots dz_h$$

where $\varphi = (z_1 + \cdots + z_h)$. But, we find by integrating with respect to $v_1, \ldots, v_{k-1}$ that

(17.6.25)
$$\mathscr{E}\{e^{(v_1+\cdots+v_k)\varphi}v_k^h\} = k(1-\varphi)^{-(k-1)}\int_0^\infty (1-e^{-v_k(1-\varphi)})^{k-1}e^{-v_k(1-\varphi)}v_k^h dv_k.$$

Putting this expression in (17.6.24), expanding $(1-e^{-v_k(1-\varphi)})^{k-1}$, performing the v_k integration, and then integrating with respect to $z_1, \ldots, z_h$ we obtain

(17.6.26) $$\mathscr{E}(g^h) = k(k-1)\sum_{p=0}^{k-1}(-1)^p\binom{k-1}{p}\frac{\Gamma(h+1)\Gamma(k-1)}{(p+1)^{h+1}\Gamma(h+k)}.$$

But

(17.6.27) $$\frac{\Gamma(h+1)\Gamma(k-1)}{\Gamma(h+k)} = \int_0^1 w^h(1-w)^{k-2}\,dw.$$

Thus, we can write

(17.6.28)
$$\mathscr{E}(g^h) = k(k-1)\sum_{p=0}^{k-1}(-1)^p\binom{k-1}{p}\int_0^1\left(\frac{w}{p+1}\right)^h(1-w)^{k-2}d\left(\frac{w}{p+1}\right).$$

Putting $w/(p+1) = y$, (17.6.28) reduces to

(17.6.29)
$$\mathscr{E}(g^h) = k(k-1)\sum_{p=0}^{k-1}(-1)^p\binom{k-1}{p}\int_0^{\frac{1}{p+1}} y^h[1-(p+1)y]^{k-2}\,dy.$$

So if we put

$$\theta_p(y) = [1-(p+1)y]^{k-2}, \qquad 0 < y < \frac{1}{p+1}$$
$$= 0, \qquad \text{otherwise}$$

then

(17.6.30) $$\mathscr{E}(g^h) = k(k-1)\int_0^1\sum_{p=0}^{k-1}(-1)^p\binom{k-1}{p}y^h\theta_p(y)\,dy$$

which holds for $h = 1, 2, \ldots$.

Thus, it follows from **5.5.1***a* that g and y have the same distribution. The p.d.f. of g is therefore

(17.6.31) $$f(g)\,dg = k(k-1)\sum_{p=0}^{k-1}(-1)^p\binom{k-1}{p}\theta_p(g)\,dg.$$

The probability that $g > g'$ is readily found by integrating $f(g)\,dg$ to be

(17.6.32) $$P(g > g') = \sum_{p=0}^{r}(-1)^p\binom{k}{p+1}[1-(p+1)g']^{k-1}$$

where r is the largest integer $\leqslant k - 1$ for which $1 - (r + 1)g' \geqslant 0$. This result was originally obtained by Fisher (1929) by a different method. Thus, for a given k and a given α the value of g_α for which $P(g > g_\alpha) = \alpha$ would be the critical value of g for the 100α % level of significance. Davis (1941) has tabulated $P(g > g_\alpha)$ for $g_\alpha k = 0.10(0.10)10.0$, $k = 10(10)70$; and for $g_\alpha k = 5.1(0.1)10.1$, $k = 80(10)160(20)300$.

Summarizing, we have the following result:

17.6.2 *Suppose $x_1, \ldots, x_{2k+1}$ are independent random variables having the distribution $N(0, \sigma^2)$ and $u_p, p = 1, \ldots, k$, are defined as in* (17.6.19), *where $\alpha(\omega_p)$ and $\alpha'(\omega_p)$ are defined in* (17.6.11) *and* (17.6.12). *Then, if $g = \max \{u_p\}/(u_1 + \cdots + u_k)$, the probability element of g is given by* (17.6.31) *while $P(g > g')$ for any g' on* (0, 1) *is given by* (17.6.32).

It can be shown by methods similar to these used above that if g is defined as the mth largest of $u_1, \ldots, u_k$ divided by $u_1 + \cdots + u_k$, then

$$P(g > g') = \frac{k!}{(m-1)!} \sum_{p=m}^{r} \frac{(-1)^{p-m}(1 - pg')^{k-1}}{p(k-p)!\,(p-m)!} \tag{17.6.33}$$

where r is the largest integer for which $1 - rg' \geqslant 0$.

Finally, it should be noted that in the preceding discussion it has been assumed that the mean of the x_t is 0. Now if the mean is unknown, approximations to the preceding results, for large n at least, can be obtained by replacing x_ξ by $x_\xi - \bar{x}$, $\xi = 1, \ldots, 2k + 1$.

17.7 TESTING A NORMAL NOISE FOR WHITENESS

Suppose we have a sample $(x_1, \ldots, x_n)$ from a normal time series $\{x_t\}$ and that we wish to test the hypotheses that the time series is a white noise, that is, that $x_1, \ldots, x_n$ are independent random variables having identical normal distributions $N(\mu, \sigma^2)$. A criterion which has been proposed by R. L. Anderson (1942) for testing this hypothesis is the ratio

$$R = \frac{c_1'}{c_0'} \tag{17.7.1}$$

where
$$\sigma^2 c_h' = \sum_{\xi=1}^{n} (x_\xi - \bar{x})(x_{\xi+h} - \bar{x}), \qquad h = 0, 1$$

and $x_{n+1} \equiv x_1$. Making c_1' *circular* by putting $x_{n+1} \equiv x_1$ as contrasted with running the summation from $t = 1$ to $n - 1$, even if it looks arbitrary, makes for considerable simplicity in dealing with the distribution theory of R. If the sample comes from a white noise then for large n, R tends to have values near 0; if not, R tends to have values away from 0.

Now c_1' and c_0' are quadratic forms in $x_1, \ldots, x_n$. The difference $c_1' - R'c_0'$ where R' is constant, is also a quadratic in $x_1, \ldots, x_n$. We can solve the distribution problem of R by determining the distribution of the quadratic $c_1' - R'c_0'$ since

$$P(c_1' - R'c_0' > 0) = P(R > R'). \tag{17.7.2}$$

If $(x_1, \ldots, x_n)$ is from normal white noise, the characteristic function of $c_1' - R'c_0'$ is given by

(17.7.3)

$$\varphi(t) = \left(\frac{1}{2\pi}\right)^{n/2} \int_{R_n} \exp\left[i(c_1' - R'c_0')t - \frac{1}{2\sigma^2}\sum_{\xi=1}^{n}(x_\xi - \mu)^2\right] dx_1 \cdots dx_n.$$

Using methods and results of Section 8.4 and noting that $c_1' - R'c_0'$ is also a quadratic in $(x_1 - \mu), \ldots, (x_n - \mu)$ we find

$$\varphi(t) = \prod_{\xi=1}^{n-1} [1 - 2it(d_\xi - R')]^{-\frac{1}{2}} \tag{17.7.4}$$

where $d_\xi = \cos(2\pi\xi/n)$, $\xi = 1, \ldots, n-1$. Since $d_\xi = d_{n-\xi}$ we can write

$$\varphi(t) = \prod_{\xi=1}^{(n-1)} [1 - 2it(d_\xi - R')]^{-1} \tag{17.7.5}$$

where for immediate convenience n is assumed to be odd. The probability density function of $c_1' - R'c_0' = y$, say, is provided by **5.1.2**, that is,

$$g(y) = \frac{1}{2\pi}\int_{-\infty}^{\infty} e^{-iyt}\varphi(t)\,dt. \tag{17.7.6}$$

Now $\varphi(z)$ has simple poles at the points $z_\xi = -\frac{1}{2}i/(d_\xi - R')$, $\xi = 1, \ldots, (n-1)/2$, in the complex plane. Thus, using the method of residues to evaluate the integral in (17.7.6) we obtain

$$g(y) = \tfrac{1}{2}\sum_{d_\xi > R'} A_\xi \exp\left\{-\frac{y}{2(d_\xi - R')}\right\} \tag{17.7.7}$$

where

$$A_\xi = \frac{(d_\xi - R')^{\frac{1}{2}(n-5)}}{\prod\limits_{j=1,\, j\neq\xi}^{\frac{1}{2}(n-1)} (d_\xi - d_j)}. \tag{17.7.8}$$

Since $P(R > R') = P(y > 0)$, we find, therefore, that

$$P(R > R') = \int_0^\infty g(y)\,dy = \sum_{d_\xi > R'} A_\xi(d_\xi - R') \tag{17.7.9}$$

from which critical values of R' can be obtained for given levels of significance. The method has been extended by R. L. Anderson to the case of even values of n. He has provided tables of R_α for which $P(R > R_\alpha) = \alpha$, for $\alpha = 0.99, 0.95, 0.05, 0.01$, and $n = 5(1)15(5)75$. Further studies of serial correlation problems have been made by T. W. Anderson (1948), Dixon (1944), Durbin and Watson (1951), Hannan (1955), Koopmans (1942), Moran (1948), Ogawara (1951), Quenouille (1948), Whittle (1951), and others.

17.8 LINEAR PREDICTION IN TIME SERIES

Suppose we have a time series x_t, $t = \cdots -2, -1, 0, \ldots$, with spectral density function $f(\omega)$ and wish to predict x_1 linearly by least squares from the entire past history of the series. A mathematical question which arises is this: What are conditions under which the least squares predictor for x_1 is *nondeterministic*, that is, has positive mean square error as against conditions under which the predictor is *deterministic*, that is, has 0 mean square error? The answer to this question has been given by Kolmogorov (1941) and Wiener (1949). We shall discuss the problem briefly by using elementary methods of analysis.

Suppose $\hat{x}_{1,n}$ is the predictor for x_1 based on the history $x_{-n}, \ldots, x_{-1}, x_0$. Then $\hat{x}_{1,n}$ is given by

$$\hat{x}_{1,n} = \sum_{\xi=-n}^{0} \hat{a}_{\xi,n} x_\xi \tag{17.8.1}$$

where $\hat{a}_{\xi,n}$, $\xi = -n, \ldots, -1, 0$ are the values of $a_{\xi,n}$ which minimize

$$\mathscr{E}\left(x_1 - \sum_{\xi=-n}^{0} a_{\xi,n} x_\xi\right)^2. \tag{17.8.2}$$

Using the method of least squares of Section 3.8 we find

$$\hat{a}_{\xi,n} = \sum_{\xi'=-n}^{0} \sigma^{\xi\xi'} \gamma_{\xi'-1} \tag{17.8.3}$$

where $\|\sigma^{\xi\xi'}\| = \|\sigma_{\xi\xi'}\|^{-1}$ and $\|\sigma_{\xi\xi'}\|$ is the $(n+1) \times (n+1)$ matrix

$$\left\| \begin{matrix} \gamma_0 & \gamma_1 & \cdots & \gamma_n \\ \gamma_1 & \gamma_0 & \cdots & \gamma_{n-1} \\ \cdot & \cdot & \cdot & \cdot \\ \cdot & \cdot & \cdot & \cdot \\ \cdot & \cdot & \cdot & \cdot \\ \gamma_n & \gamma_{n-1} & \cdots & \gamma_0 \end{matrix} \right\| \tag{17.8.4}$$

and where it will be recalled that

$$\gamma_h = \gamma_{-h} = \int_{-\pi}^{\pi} \cos h\omega f(\omega)\, d\omega. \tag{17.8.5}$$

If we let Δ_{n+1} denote determinant of $\|\sigma_{\xi\xi'}\|$, with Δ_{n+2} being similarly defined, then the mean square error of the predictor $\hat{x}_{1,n}$ given in (17.8.1) is

$$\mathscr{E}(x_1 - \hat{x}_{1,n})^2 = \frac{\Delta_{n+2}}{\Delta_{n+1}}. \tag{17.8.6}$$

Now we have for $\xi = 1, \ldots, n$,

$$\frac{\Delta_{n+2+\xi}}{\Delta_{n+1+\xi}} \leqslant \frac{\Delta_{n+2}}{\Delta_{n+1}} \leqslant \frac{\Delta_{n+2-\xi}}{\Delta_{n+1-\xi}}, \tag{17.8.7}$$

and hence

$$\prod_{\xi=1}^{n}\left(\frac{\Delta_{n+2+\xi}}{\Delta_{n+1+\xi}}\right) \leqslant \left(\frac{\Delta_{n+2}}{\Delta_{n+1}}\right)^n \leqslant \prod_{\xi=1}^{n}\left(\frac{\Delta_{n+2-\xi}}{\Delta_{n+1-\xi}}\right), \tag{17.8.8}$$

from which we obtain

$$\left(\frac{\Delta_{2n+2}}{\Delta_{n+2}}\right)^{1/n} \leqslant \frac{\Delta_{n+2}}{\Delta_{n+1}} \leqslant \left(\frac{\Delta_{n+1}}{\Delta_1}\right)^{1/n}. \tag{17.8.9}$$

Taking logarithms and denoting $\Delta_{n+2}/\Delta_{n+1}$ by σ_n^2 we find

$$A_n \leqslant \log \sigma_n^2 \leqslant B_n \tag{17.8.10}$$

where

$$\begin{aligned} A_n &= \frac{1}{n}\sum_{\xi=1}^{2n+2} \log \theta_{\xi,2n+2} - \frac{1}{n}\sum_{\xi=1}^{n+2} \log \theta_{\xi,n+2} \\ B_n &= \frac{1}{n}\sum_{\xi=1}^{n+1} \log \theta_{\xi,n+1} - \frac{1}{n}\log \gamma_0 \end{aligned} \tag{17.8.11}$$

where $(\theta_{1,2n+2}, \ldots, \theta_{2n+2,2n+2})$ are the latent roots of the matrix whose determinant is Δ_{2n+2}, with similar definitions for $(\theta_{1,n+2}, \ldots, \theta_{n+2,n+2})$ and $(\theta_{1,n+1}, \ldots, \theta_{n+2,n+1})$. Since all three matrices are positive definite with diagonal elements all equal to γ_0, all roots in each of the three sets lie on the interval $(0, \gamma_0)$ and hence the logarithms of the roots in each set lie on $(-\infty, \log \gamma_0)$.

If the average of the logarithms of the roots of matrix (17.8.4) has a finite limit K, as $n \to \infty$, then it is evident that

$$\lim_{n\to\infty} A_n = \lim_{n\to\infty} B_n = K.$$

Hence, if we let

$$\sigma^2 = \lim_{n\to\infty} \mathscr{E}(x_1 - \hat{x}_{1,n})^2 \tag{17.8.12}$$

we have

$$\sigma^2 = e^K. \tag{17.8.13}$$

If K is not finite, its value is $-\infty$, in which case $\sigma^2 = 0$, and $\hat{x}_{1,n}$ has a mean square error for predicting x_1 which $\to 0$ as $n \to \infty$.

It can be shown, further, by more advanced methods of analysis, which will not be given here, that if K is finite its value is given by

(17.8.14) $$K = \log 2\pi + \frac{1}{2\pi}\int_{-\pi}^{+\pi} \log f(\omega)\, d\omega.$$

The reader interested in the proof is referred to Kolmogorov (1941) and Wiener (1949). Proofs can also be found in the books by Doob (1953), and by Grenander and Rosenblatt (1957).

Another problem involving linear prediction in time series is the *autoregressive* scheme which we shall mention briefly. An autoregressive scheme of order r for a given time series $\{x_t\}$ states that for constants $a_1, \ldots, a_r$, the time series

(17.8.15) $$y_t = x_t - a_1 x_{t-1} - \cdots - a_r x_{t-r}$$

is a white noise. Least squares estimates for $a_1, \ldots, a_r$ can be found from a sample $(x_1, \ldots, x_n)$ in the usual way. Dixon (1944) has considered the problem of testing the hypothesis that $a_1, \ldots, a_r$ are all zero against the alternative that $a_m, a_{m+1}, \ldots, a_r$ are 0, using the method of likelihood ratios and assuming y_t is a normal white noise. Other studies of autoregressive problems have been considered by Mann and Wald (1943). Wold (1938) has considered the problem of testing the hypothesis that a time series $\{x_t\}$ is of form $y_t + a_1 y_{t-1} + \cdots + a_r y_{t-r}$ where $\{y_t\}$ is a white noise.

PROBLEMS

17.1 Suppose a stationary time series $x_t, t = \ldots, -1, 0, +1, \ldots$ has spectral density function

$$f(\omega) = \frac{1}{\pi^2}(\pi - |\omega|), \qquad -\pi < \omega < \pi.$$

Show that the covariance function γ_h is as follows

$$\gamma_h = \begin{cases} 1 & h = 0 \\ \left(\dfrac{2}{\pi h}\right)^2 & h = \text{odd} \\ 0 & h = \text{even}. \end{cases}$$

17.2 (*Slutzky's* (1937) *theorem*) If $x_t, t = \ldots, -1, 0, +1, \ldots$ is a real stationary time series with spectral density function $f(\omega)$, show that the *smoothed* time series

$$y_t = \sum_{p=0}^{k} a_p x_{t-k+p}, \qquad t = \cdots -1, 0, +1, \ldots$$

where the a's are real constants is a stationary process which has spectral density function

$$\left[\sum_{q=0}^{k} \sum_{p=0}^{k} a_p a_q \cos (p - q)\omega\right] f(\omega).$$

In particular, if $a_p = 1/(k + 1)$, $p = 0, 1, \ldots, k$ that is, if the smoothing process is to take a *moving average* of $k + 1$ elements of the time series at a time, show that the spectral density function of the resulting time series is

$$\frac{1 - \cos (k + 1)\omega}{(k + 1)^2(1 - \cos \omega)} f(\omega).$$

17.3 (*Continuation*) If the stationary time series x_t, $t = \ldots, -1, 0, +1, \ldots$ is repeatedly smoothed r times by the formula given in the preceding problem, show that the resulting time series is stationary with spectral density function

$$\left[\sum_{q=0}^{k} \sum_{p=0}^{k} a_p a_q \cos (p - q)\omega\right]^r f(\omega).$$

17.4 Suppose x_t, $t = \ldots, -1, 0, +1, \ldots$ is a white noise with unit variance. Show that the time series generated by taking the nth (forward) difference of this time series has covariance function

$$\begin{aligned} \gamma_h &= (-1)^h \frac{(2n)!}{(n - h)!\,(n + h)!} \qquad && -n \leqslant h \leqslant n, \\ &= 0 && \text{otherwise.} \end{aligned}$$

17.5 Suppose x_t, $t = \ldots, -1, 0, +1, \ldots$ is a *linear process*, that is, a time series such that

$$x_t = \sum_{p=-\infty}^{+\infty} a_{t-p}\varepsilon_p$$

where the a's are real numbers such that $\sum_{p=-\infty}^{\infty} a_p^2 < +\infty$, and the ε's are independent random variables having zero means and unit variances.

Show that the spectral density function $f(\omega)$ of this time series is given by

$$f(\omega) = \frac{1}{2\pi} \sum_{q=-\infty}^{+\infty} \sum_{p=-\infty}^{+\infty} a_p a_{p+q} \cos q\omega.$$

17.6 If x_t, $t = \ldots, -1, 0, +1, \ldots$ is a stationary time series with $\mathscr{E}(x_t) = 0$, show that its spectral distribution $F(\omega)$ is given by $\lim_{M\to\infty} F_M^*(\omega)$ where

$$F_M^*(\omega) = \frac{1}{\pi M} \int_{-\pi}^{\omega} \mathscr{E}(x_1 \sin \omega + \cdots + x_M \sin M\omega)^2 \, d\omega$$

and also by $\lim_{M\to\infty} F_M^{**}(\omega)$ where

$$F_M^{**}(\omega) = \int_{-\pi}^{\omega} \mathscr{E} I_M(\omega)\, d\omega$$

where $I_M(\omega)$ is the *periodogram* defined by

$$I_M(\omega) = \frac{1}{2\pi M}\left|\sum_{p=1}^{M} x_p e^{-ip\omega}\right|^2.$$

17.7 If x_t, $t = \ldots, -1, 0, +1, \ldots$ is a complex stationary time series, show that the covariance function

$$\gamma_h = \mathscr{E}(x_{t+h}\bar{x}_t), \qquad h = \ldots, -1, 0, +1, \ldots$$

is given by (17.3.11), where $F(\omega) = \lim_{M\to\infty} F_M(\omega)$, and $F_M(\omega)$ is given by (17.3.13).

17.8 Let x_t, $t = \ldots, -1, 0, +1, \ldots$ be a stationary time series whose covariance function γ_h is given by

$$\gamma_h = \gamma_0\rho^h, \qquad 0 < \rho < +1, \qquad h = 0, 1, \ldots.$$

Show that if σ_n^2 is the variance of the least squares linear estimate of x_1 from $x_{-n}, x_{-n+1}, \ldots x_{-1}, x_0$, then $\sigma_n^2 = \gamma_0(1 - \rho^2)$.

17.9 (*Continuation*). Show that the spectral density function of the time series in Problem 17.8 is given by

$$f(\omega) = \frac{\gamma_0(1 - \rho\cos\omega)}{\pi(1 + \rho^2 - 2\rho\cos\omega)}$$

17.10 Establish formula (17.6.33) by methods similar to those used in developing formula (17.6.32).

CHAPTER 18

Multivariate Statistical Theory

A branch of mathematical statistics that has been quite thoroughly developed during the last quarter of a century is linear and quadratic analysis of samples from multidimensional distributions. This branch has come to be known as *multivariate statistical analysis*. Most of the sampling theory and statistical tests that have been developed for problems in this field are for samples from multivariate normal distributions. There is a large body of results on multivariate statistical theory scattered throughout the literature of mathematical statistics. In this type of book it is not feasible to cover all the important and interesting results of this literature. We shall therefore confine ourselves to the main ideas and principal results of multivariate statistical theory. The reader interested in further mathematical details should consult the recent books by Anderson (1958) and Roy (1957) on this subject. Details concerning applications of these methods to numerical problems can be found in books by Rao (1952) and Kendall (1957). Anderson's book has a comprehensive bibliography on multivariate statistical analysis.

18.1 MULTIDIMENSIONAL STATISTICAL SCATTER

By way of introduction to multivariate statistical analysis it will be convenient to define and discuss what we shall call the *scatter* of a sample about a point.

(a) The One-Dimensional Case

First of all, suppose $(x_1, \ldots, x_n)$ is a sample from a one-dimensional c.d.f. $F(x)$. Let x_0 be an arbitrary real number. Note that for any given values of $x_1, \ldots, x_n$ and x_0 they can be represented as points along the real line. Then the *scatter* ${}_1S_{x_0,n}$ of the sample about the *pivotal point* x_0 is defined as follows:

$$(18.1.1)\qquad {}_1S_{x_0,n} = \sum_{\xi=1}^{n} (x_\xi - x_0)^2 = {}_1S_{\bar{x},n} + n(\bar{x} - x_0)^2.$$

Note that ${}_1S_{x_0,n}$ is merely the sum of squares of the n possible segments $(x_\xi - x_0)$ generated by taking x_0 and each x_ξ in the sample. Note also that the scatter of the sample about x_0 is equal to the scatter of the sample about its mean $\bar{x}$ plus the weighted scatter of $\bar{x}$ about x_0, the weight being the sample size n. Thus, *the scatter of a sample is a minimum when taken about the sample mean.* We also know from Section 8.2 that if $F(x)$ has mean μ and variance σ^2 then

$$(18.1.2)\qquad \mathscr{E}({}_1S_{x_0,n}) = n[\sigma^2 + (\mu - x_0)^2]$$

from which it is evident that *the value of* $\mathscr{E}({}_1S_{x,n})$ *is a minimum if* x_0 *is chosen as* μ.

In the language of mechanics, ${}_1S_{x_0,n}$ is the moment of inertia of the sample points $x_1, \ldots, x_n$ about the pivotal point x_0.

Now suppose we have two samples $(x_1^{(1)}, \ldots, x_{n_1}^{(1)})$ and $(x_1^{(2)}, \ldots, x_{n_2}^{(2)})$ from one-dimensional distributions $F_1(x)$ and $F_2(x)$, respectively. Let $\bar{x}^{(1)}, \bar{x}^{(2)}$ be the two sample means, $\bar{x}$ the mean of both samples pooled together, ${}_1S_{\bar{x}^{(1)},n_1}, {}_1S_{\bar{x}^{(2)},n_2}$ the scatters of the samples about their respective means as pivotal points, and ${}_1S_{\bar{x},n_1+n_2}$ the scatter of the pooled sample about $\bar{x}$ as the pivotal point. By *pooled sample* we mean the sample of size $n_1 + n_2$ obtained by regarding the two samples as one single (grand) sample. The reader can verify that

$$(18.1.3)\qquad {}_1S_{\bar{x},n_1+n_2} = {}_1S_{\bar{x}^{(1)},n_1} + {}_1S_{\bar{x}^{(2)},n_2} + n_1(\bar{x}^{(1)} - \bar{x})^2 + n_2(\bar{x}^{(2)} - \bar{x})^2.$$

Thus, the scatter of the pooled sample about $\bar{x}$ has as its minimum value under rigid translations of the elements in each sample, the sum of the scatters of the two samples about their respective means, this minimum occurring only when the two sample means coincide. The result expressed by (18.1.3) extends immediately to any number of samples.

(b) The Two-Dimensional Case

Now let us consider a sample $(x_{1\xi}, x_{2\xi}, \xi = 1, \ldots, n)$ of size n from a two-dimensional c.d.f. $F(x_1, x_2)$, and let (x_{10}, x_{20}) be an arbitrary pair of numbers. Note that for a given sample, the sample, together with (x_{10}, x_{20}), can be represented as a cluster of points (at most $n + 1$) in a plane. It will be convenient to call this a *sample cluster*. Take any two elements (pairs) in the sample, say $(x_{1\xi}, x_{2\xi})$ and $(x_{1\eta}, x_{2\eta})$, and the point (x_{10}, x_{20}). These three pairs can in general be represented by three points in an x_1x_2-plane.

Unless the three points are collinear, a fourth point can be chosen in three different ways so that the three original points and any one of the choices for the fourth point form the vertices of a parallelogram. The three possible parallelograms all have the same area $A_{\xi\eta,x_0}$ (except possibly for sign). It will be convenient to call the absolute value of this area the *two-dimensional content* determined by $(x_{1\xi}, x_{2\xi})$, $(x_{1\eta}, x_{2\eta})$, (x_{10}, x_{20}). The squared value of this content is

$$A^2_{\xi\eta,x_0} = \begin{vmatrix} x_{1\xi} - x_{10} & x_{2\xi} - x_{20} \\ x_{1\eta} - x_{10} & x_{2\eta} - x_{20} \end{vmatrix}^2. \tag{18.1.4}$$

Squaring the determinant on the right, we find that it may be written as the sum of four determinants, namely,

$$A^2_{\xi\eta,x_0} = \Delta_{\xi\xi} + \Delta_{\xi\eta} + \Delta_{\eta\xi} + \Delta_{\eta\eta} \tag{18.1.5}$$

where

$$\Delta_{\xi\eta} = \begin{vmatrix} (x_{1\xi} - x_{10})^2 & (x_{1\eta} - x_{10})(x_{2\eta} - x_{20}) \\ (x_{2\xi} - x_{20})(x_{1\xi} - x_{10}) & (x_{2\eta} - x_{20})^2 \end{vmatrix} \tag{18.1.6}$$

with similar definitions for $\Delta_{\xi\xi}$, $\Delta_{\eta\xi}$, and $\Delta_{\eta\eta}$. Note that $\Delta_{\xi\xi}$ and $\Delta_{\eta\eta}$ are both 0.

The pair (ξ, η) can be any two of the integers $1, \ldots, n$. There are $\binom{n}{2}$ such pairs, and the sum of $A^2_{\xi\eta,x_0}$ over all such choices is $\sum_{\eta=1}^{n} \sum_{\xi=1}^{n} A^2_{\xi\eta,x_0}$ since $A^2_{\xi\xi,x_0} = 0$. But this sum is equivalent to the sum of $\Delta_{\xi\eta}$ (or of $\Delta_{\eta\xi}$) for $\xi, \eta = 1, \ldots, n$, that is, $\sum_{\eta=1}^{\eta} \sum_{\xi=1}^{\eta} \Delta_{\xi\eta}$. Denoting this sum by ${}_2S_{x_0,n}$ and noting that for any set of determinants $\begin{vmatrix} a_\xi & b_\eta \\ c_\xi & d_\eta \end{vmatrix}$, $\xi, \eta = 1, \ldots, n$, the relation

$$\sum_{\eta} \sum_{\xi} \begin{vmatrix} a_\xi & b_\eta \\ c_\xi & d_\eta \end{vmatrix} = \begin{vmatrix} \sum_{\xi} a_\xi & \sum_{\eta} b_\eta \\ \sum_{\xi} c_\xi & \sum_{\eta} d_\eta \end{vmatrix} \tag{18.1.7}$$

holds; we therefore have the result

$${}_2S_{x_0,n} = \begin{vmatrix} \sum_{\xi} (x_{1\xi} - x_{10})^2 & \sum_{\eta} (x_{1\eta} - x_{10})(x_{2\eta} - x_{20}) \\ \sum_{\xi} (x_{2\xi} - x_{20})(x_{1\xi} - x_{10}) & \sum_{\eta} (x_{2\eta} - x_{20})^2 \end{vmatrix} \tag{18.1.8}$$

where $\sum_{\xi}$ and $\sum_{\eta}$ each denote summation over values $1, \ldots, n$.

The matrix whose determinant is ${}_2S_{x_0,n}$ is sometimes called a *Gramian matrix*, that is, a matrix product $A'A$ where A is the matrix

$$\left\| \begin{matrix} x_{11} - x_{10} & x_{21} - x_{20} \\ \vdots & \vdots \\ x_{1n} - x_{10} & x_{2n} - x_{20} \end{matrix} \right\| \tag{18.1.9}$$

and A' is the transpose of A.

The quantity ${}_2S_{x_0,n}$ is the *scatter* of the sample $(x_{1\xi}, x_{2\xi}, \xi = 1, \ldots, n)$ about the pivotal point (x_{10}, x_{20}); it is the sum of squares of the two-dimensional contents determined by (x_{10}, x_{20}) and all possible pairs of points in the sample $(x_{1\xi}, x_{2\xi}, \xi = 1, \ldots, n)$. The matrix whose determinant is ${}_2S_{x_0,n}$ will be called the *scatter matrix* of the sample about (x_{10}, x_{20}).

If we let $\bar{x}_i = \frac{1}{n} \sum_{\xi=1}^{n} x_{i\xi}$, $i = 1, 2$ and

$$u_{ij} = u_{ji} = \sum_{\xi=1}^{n} (x_{i\xi} - \bar{x}_i)(x_{j\xi} - \bar{x}_j), \qquad i, j = 1, 2 \tag{18.1.10}$$

we find that

$$\sum_{\xi=1}^{n} (x_{i\xi} - x_{i0})(x_{j\xi} - x_{j0}) = u_{ij} + n(\bar{x}_i - x_{i0})(\bar{x}_j - x_{j0}). \tag{18.1.11}$$

Therefore, we have

(18.1.12)

$$ {}_2S_{x_0,n} = \left| \begin{matrix} u_{11} + n(\bar{x}_1 - x_{10})^2 & u_{12} + n(\bar{x}_1 - x_{10})(\bar{x}_2 - x_{20}) \\ u_{21} + n(\bar{x}_2 - x_{20})(\bar{x}_1 - x_{10}) & u_{22} + n(\bar{x}_2 - x_{20})^2 \end{matrix} \right| $$

which is the two-dimensional analogue of (18.1.1). Writing the determinant on the right as the sum of four determinants in the usual way, we find that ${}_2S_{x_0,n}$ can be expressed as follows:

$$ {}_2S_{x_0,n} = \left| \begin{matrix} u_{11} & u_{12} \\ u_{21} & u_{22} \end{matrix} \right| \cdot \left[1 + n \sum_{i,j=1}^{2} u^{ij}(\bar{x}_i - x_{i0})(\bar{x}_j - x_{j0}) \right] \tag{18.1.13}$$

where $\|u^{ij}\| = \|u_{ij}\|^{-1}$, this inverse matrix existing, of course, only if $\|u_{ij}\|$, the scatter matrix of the sample about $(\bar{x}_1, \bar{x}_2)$, is nonsingular with probability 1. It will be convenient to call $\|u_{ij}\|$ the *internal scatter matrix* and its determinant $|u_{ij}|$ the *internal scatter* of the sample. It will be noted that $\|u_{ij}\|$ is nonsingular if and only if the n points $(x_{1\xi}, x_{2\xi}, \xi = 1, \ldots, n)$ in the sample are noncollinear. Thus, if $\|u_{ij}\|$ is nonsingular, the quadratic form inside [ ] on the right of (18.1.13) is positive definite, in which case it is evident that

18.1.1 *The choice of the pivotal point (x_{10}, x_{20}) which minimizes the scatter of the sample about (x_{10}, x_{20}) is the sample mean $(\bar{x}_1, \bar{x}_2)$, in which case the minimum scatter ${}_2S_{\bar{x},n}$ is the internal scatter of the sample given by*

(18.1.14)
$$ {}_2S_{\bar{x},n} = \begin{vmatrix} u_{11} & u_{12} \\ u_{21} & u_{22} \end{vmatrix}. $$

Next we shall show that

18.1.2 *The mean value of the two-dimensional scatter about (x_{10}, x_{20}) is given by*

(18.1.15)
$$ \mathscr{E}({}_2S_{x_0,n}) = 2\binom{n}{2}\begin{vmatrix} \sigma_{11} + (\mu_1 - x_{10})^2 & \sigma_{12} + (\mu_1 - x_{10})(\mu_2 - x_{20}) \\ \sigma_{21} + (\mu_2 - x_{20})(\mu_1 - x_{10}) & \sigma_{22} + (\mu_2 - x_{20})^2 \end{vmatrix} $$

where $\|\sigma_{ij}\|$ is the covariance matrix and (μ_1, μ_2) is the vector of means of $F(x_1, x_2)$. The minimum of $\mathscr{E}({}_2S_{x_0,n})$ occurs if and only if $(x_{10}, x_{20}) = (\mu_1, \mu_2)$, in which case $\min_{(x_{10},x_{20})} \mathscr{E}({}_2S_{x_0,n}) = \mathscr{E}({}_2S_{\mu,n}) = 2\binom{n}{2}|\sigma_{ij}|$.

Formula (18.1.15) is the two-dimensional analogue of (18.1.2). To establish this result we first write down the expression defining $\mathscr{E}(\Delta_{\xi\eta})$ namely,

(18.1.16)
$$ \mathscr{E}(\Delta_{\xi\eta}) = \int_{R_2}\int_{R_2} \Delta_{\xi\eta}\, dF(x_{1\xi}, x_{2\xi})\, dF(x_{1\eta}, x_{2\eta}). $$

Noting that (18.1.7) holds for integrals as well as finite sums, we may therefore write the quantity on the right of (18.1.16) as a determinant $|D_{ij}|$ where

$$ D_{21} = \int_{R_2} (x_{2\xi} - x_{20})(x_{1\xi} - x_{10})\, dF(x_{1\xi}, x_{2\xi}) $$

with similar expressions for D_{11}, D_{12}, and D_{22}. Thus, we have

$$ D_{ij} = \mathscr{E}(x_i - x_{i0})(x_j - x_{j0}) = \sigma_{ij} + (\mu_i - x_{i0})(\mu_j - x_{j0}). $$

Therefore, for $\xi \neq \eta = 1, \ldots, n$

$$ \mathscr{E}(\Delta_{\xi\eta}) = \begin{vmatrix} \sigma_{11} + (\mu_1 - x_{10})^2 & \sigma_{12} + (\mu_1 - x_{10})(\mu_2 - x_{20}) \\ \sigma_{21} + (\mu_2 - x_{20})(\mu_1 - x_{10}) & \sigma_{22} + (\mu_2 - x_{20})^2 \end{vmatrix} $$

But $\mathscr{E}(A^2_{\xi\eta,x_0}) = \mathscr{E}(\Delta_{\xi\xi}) + \mathscr{E}(\Delta_{\xi\eta}) + \mathscr{E}(\Delta_{\eta\xi}) + \mathscr{E}(\Delta_{\eta\eta}) = 2\mathscr{E}(\Delta_{\xi\eta})$.

Furthermore, ${}_2S_{x_0,n}$ consists of the sum of all possible $A^2_{\xi\eta,x_0}$, the number of these being $\binom{n}{2}$. Hence

$$\mathscr{E}({}_2S_{x_0,n}) = 2\binom{n}{2}\mathscr{E}(\Delta_{\xi\eta})$$

which is seen to be equal to the right-hand side of (18.1.15), thereby completing the argument for **18.1.2**.

Note that if $\|\sigma_{ij}\|$ is positive definite then $\|\sigma^{ij}\| = \|\sigma_{ij}\|^{-1}$ exists and is also positive definite and we can write

$$(18.1.17)\quad \mathscr{E}({}_2S_{x_0,n}) = 2\binom{n}{2}|\sigma_{ij}|\left[1 + \sum_{i,j=1}^{2}\sigma^{ij}(\mu_i - x_{i0})(\mu_j - x_{j0})\right].$$

Thus, it is evident from (18.1.17) that

18.1.3 *If $\|\sigma_{ij}\|$ is positive definite, $\mathscr{E}({}_2S_{x_0,n})$ attains its minimum if (x_{10}, x_{20}) is chosen as the mean (μ_1, μ_2) of $F(x_1, x_2)$ in which case this minimum is*

$$(18.1.18)\quad \mathscr{E}({}_2S_{\mu,n}) = \mathscr{E}|v_{ij}| = 2\binom{n}{2}|\sigma_{ij}|$$

where

$$(18.1.19)\quad v_{ij} = v_{ji} = \sum_{\xi=1}^{n}(x_{i\xi} - \mu_i)(x_{j\xi} - \mu_j), \qquad i, j = 1, 2.$$

Now suppose $(x^{(1)}_{1\xi_1}, x^{(1)}_{2\xi_1}, \xi_1 = 1, \ldots, n_1)$ and $(x^{(2)}_{1\xi_2}, x^{(2)}_{2\xi_2}, \xi_2 = 1, \ldots, n_2)$ are two samples from the c.d.f.'s $F_1(x_1, x_2)$, $F_2(x_1, x_2)$, respectively. Let $(\bar{x}^{(1)}_1, \bar{x}^{(1)}_2)$ and $(\bar{x}^{(2)}_1, \bar{x}^{(2)}_2)$ be the means of the samples and $\|u^{(1)}_{ij}\|$ and $\|u^{(2)}_{ij}\|$, $i, j, = 1, 2$, the internal scatter matrices of the two samples. If these samples, when pooled, have mean $(\bar{x}_1, \bar{x}_2)$ it is straightforward to verify that the scatter of the pooled samples about $(\bar{x}_1, \bar{x}_2)$ is given by

(18.1.20)

$${}_2S_{\bar{x},n_1+n_2} = \left|u^{(1)}_{ij} + u^{(2)}_{ij} + n_1(\bar{x}^{(1)}_i - \bar{x}_i)(\bar{x}^{(1)}_j - \bar{x}_j) + n_2(\bar{x}^{(2)}_i - \bar{x}_i)(\bar{x}^{(2)}_j - \bar{x}_j)\right|.$$

The second-order determinant on the right of (18.1.20) is the two-dimensional analogue of the right-hand side of (18.1.3).

(c) The *k*-Dimensional Case

One encounters no new kinds of difficulties in defining the scatter ${}_kS_{x_0,n}$ of a sample $(x_{1\xi}, \ldots, x_{k\xi}, \xi = 1, \ldots, n)$ from a k-dimensional c.d.f. $F(x_1, \ldots, x_k)$ about a pivotal point $(x_{10}, \ldots, x_{k0})$. The sample and pivotal point can be represented as $n + 1$ points in Euclidean space R_k, which will be called a *sample cluster*. Thus for each choice of k different sample points, say $(x_{1\xi_i}, \ldots, x_{k\xi_i})$, $i = 1, \ldots, k$, together with the pivotal point

$(x_{10}, \ldots, x_{k0})$, there are $k + 1$ ways to choose a further point so that this point together with the $k + 1$ points already mentioned form a k-dimensional parallelotope. These $k + 1$ different possible parallelotopes have equal k-dimensional volumes (except possibly for sign), the absolute value of which would be called the *k-dimensional content* determined by $(x_{1\xi_i}, \ldots, x_{k\xi_i})$, $i = 1, \ldots, k$, and $(x_{10}, \ldots, x_{k0})$. The *scatter* ${}_kS_{x_0,n}$ is then defined as the sum of squares of the k-dimensional contents determined by $(x_{10}, \ldots, x_{k0})$ and each of $\binom{n}{k}$ different possible choices of $(x_{1\xi_i}, \ldots, x_{k\xi_i})$, $i = 1, \ldots, k$. It is straightforward to show that ${}_kS_{x_0,n}$ is the extension of ${}_2S_{x_0,n}$ as given by (18.1.11), to k dimensions, that is,

$$\text{(18.1.21)} \qquad {}_kS_{x_0,n} = |u_{ij} + n(\bar{x}_i - x_{i0})(\bar{x}_j - x_{j0})|$$

where the definition of u_{ij}, $i, j = 1, \ldots, k$, is evident from (18.1.10). If $\|u_{ij}\|$, the internal scatter matrix of the sample, is positive definite, which will occur if and only if the n sample points $(x_{1\xi}, \ldots, x_{k\xi}, \xi = 1, \ldots, n)$ do not lie in a flat space of less than k-dimensions, then ${}_kS_{x_0,n}$ can be written in the alternative form

$$\text{(18.1.22)} \qquad {}_kS_{x_0,n} = |u_{ij}| \cdot \left[1 + n \sum_{i,j=1}^{k} u^{ij}(\bar{x}_i - x_{i0})(\bar{x}_j - x_{j0})\right],$$

where, of course, $\|u^{ij}\| = \|u_{ij}\|^{-1}$ and is positive definite if $\|u_{ij}\|$ is. We note that (18.1.22) is the k-dimensional extension of (18.1.13). Since the quadratic form inside [ ] is positive definite, the k-dimensional version of **18.1.1** readily follows, that is, ${}_kS_{x_0,n}$ has its minimum value if $(x_{10}, \ldots, x_{k0})$ is chosen as the vector of sample means $(\bar{x}_1, \ldots, \bar{x}_k)$, in which case the minimum value of ${}_kS_{x_0,n}$ is ${}_kS_{\bar{x},n}$, where

$$\text{(18.1.23)} \qquad {}_kS_{\bar{x},n} = |u_{ij}|, \qquad i, j = 1, \ldots, k$$

which, of course, is the internal scatter of the sample.

It will be noted that ${}_kS_{x_0,n}$ is the determinant of the *Gramian* matrix $A'A$ where A is the $n \times k$ matrix

$$\left\| \begin{matrix} (x_{11} - x_{10}) \cdots (x_{k1} - x_{k0}) \\ \cdot \qquad\qquad \cdot \\ \cdot \qquad\qquad \cdot \\ \cdot \qquad\qquad \cdot \\ (x_{1n} - x_{10}) \cdots (x_{kn} - x_{k0}) \end{matrix} \right\|$$

and A' is the transpose of A. Also $\|u_{ij}\|$ is the *Gramian matrix* $B'B$ where B is A with x_{i0} replaced by $\bar{x}_i$, $i = 1, \ldots, k$ and B' is the transpose of B.

If the sample $(x_{1\xi}, \ldots, x_{k\xi},\ \xi = 1, \ldots, n)$ is from a k-dimensional distribution whose vector of means is $(\mu_1, \ldots, \mu_k)$ and whose (nonsingular) covariance matrix is $\|\sigma_{ij}\|$, then following an argument similar to that by which (18.1.15) was established, it will be found that

$$\mathscr{E}({}_kS_{x_0,n}) = k!\binom{n}{k} \cdot |\sigma_{ij} + (\mu_i - x_{i0})(\mu_j - x_{j0})| \tag{18.1.24}$$

which can be written in the alternative form

$$\mathscr{E}({}_kS_{x_0,n}) = k!\binom{n}{k} \cdot |\sigma_{ij}| \cdot \left[1 + \sum_{i,j=1}^{k} \sigma^{ij}(\mu_i - x_{i0})(\mu_j - x_{j0})\right]. \tag{18.1.25}$$

This is the k-dimensional analogue of (18.1.17) from which the k-dimensional form of **18.1.2** follows, that is, the minimum of $\mathscr{E}({}_kS_{x_0,n})$ occurs if and only if $(x_{10}, \ldots, x_{k0})$ is chosen as the vector of means $(\mu_1, \ldots, \mu_k)$, this minimum being $\mathscr{E}({}_kS_{\mu,n})$ and having the value

$$\mathscr{E}({}_kS_{\mu,n}) = \mathscr{E}|v_{ij}| = k!\binom{n}{k} |\sigma_{ij}| \tag{18.1.26}$$

where

$$v_{ij} = v_{ji} = \sum_{\xi=1}^{n} (x_{i\xi} - \mu_i)(x_{j\xi} - \mu_j). \tag{18.1.27}$$

The determinant $|\sigma_{ij}|$ is sometimes called the *generalized variance* of $F(x_1, \ldots, x_k)$. It can also be regarded as the *internal scatter* of the distribution having c.d.f. $F(x_1, \ldots, x_k)$.

The reader who has followed the details of establishing (18.1.20) will have very little difficulty in seeing that the k-dimensional version of (18.1.20), that is, the scatter ${}_kS_{\bar{x},n_1+n_2}$ of two samples pooled together, about the mean of the pooled samples is given by (18.1.20) with $i, j = 1, \ldots, k$. For a detailed discussion of multidimensional statistical scatter the reader is referred to Wilks (1960*a*).

18.2 THE WISHART DISTRIBUTION

(a) Derivation of the Wishart Distribution

In the preceding section we have seen that the mean value of the scatter of a sample from a k-dimensional distribution is relatively easy to determine for samples from an arbitrary distribution having finite means and covariance matrix. However, the problem of determining higher moments of a scatter seems to be very difficult for samples from such distributions, except in the case of sampling from a k-dimensional normal distribution. As a matter of fact, we can find not only the moments of the scatter, but the distribution function of the elements of the scatter matrix for

k-dimensional samples from a k-dimensional normal distribution if the pivotal point $(x_{10}, \ldots, x_{k0})$ is chosen as the mean of the normal distribution. The distribution of the elements of the scatter matrix in this case is a remarkable result known as the *Wishart* (1928) *distribution* which is fundamental in the theory of multivariate statistical inference thus far developed. We shall give a derivation of the Wishart distribution based on the use of characteristic functions that will be seen to be valid for every point in the space of $\{v_{ij}\}$ for which $\|v_{ij}\|$ is positive definite, provided $\|\sigma_{ij}\|$ is positive definite and $k \leqslant n$.

Let $(x_{1\xi}, \ldots, x_{k\xi}, \xi = 1, \ldots, n)$ be a sample from the k-dimensional normal distribution $N(\{\mu_i\}; \|\sigma_{ij}\|)$ whose p.d.f. is given in (7.4.1). A sample of size n from a k-dimensional distribution is defined in Section 8.1. Consider the scatter ${}_kS_{\mu,n}$ of this sample about the population mean $(\mu_1, \ldots, \mu_k)$. We have

(18.2.1) $$ {}_kS_{\mu,n} = |v_{ij}| $$

where

(18.2.2) $$ v_{ij} = v_{ji} = \sum_{\xi=1}^{n} (x_{i\xi} - \mu_i)(x_{j\xi} - \mu_j), \qquad i, j = 1, \ldots, k. $$

If we denote by $\varphi(\{t_{ij}\})$, $t_{ij} \equiv t_{ji}$, the characteristic function of $\{v_{ii}, i = 1, \ldots, k;\ 2v_{ij}, i > j = 1, \ldots, k\}$ we have

(18.2.3)
$$ \varphi(\{t_{ij}\}) = \left[\frac{\sqrt{|\sigma^{ij}|}}{(2\pi)^{k/2}}\right]^n \int_{R_{nk}} \exp\left[-\frac{1}{2}\sum_{i,j=1}^{k} \sigma^{ij} v_{ij} + i \sum_{i,j=1}^{k} v_{ij} t_{ij}\right] \prod_{i,\xi} dx_{i\xi}. $$

Since $\varphi(0) = 1$ we note that the integral in (18.2.3) has the value $[(2\pi)^{\frac{1}{2}k}/\sqrt{|\sigma^{ij} - 2it_{ij}|}]^n$. Hence,

(18.2.4) $$ \varphi(\{t_{ij}\}) = |\tau_{ij}|^{\frac{1}{2}n} \cdot |\tau_{ij} - it_{ij}|^{-\frac{1}{2}n} $$

where

(18.2.5) $$ \tau_{ij} = \tfrac{1}{2}\sigma^{ij}, \qquad i, j = 1, \ldots, k. $$

Applying the multidimensional version of Lévy's theorem, namely, **5.2.1**, the probability density function of $\{v_{ii}, i = 1, \ldots, k; 2v_{ij}, i > j = 1, \ldots, k\}$ is given by

(18.2.6)
$$ f(\{v_{ii}, 2v_{ij}\}) = \left(\frac{1}{2\pi}\right)^{\frac{1}{2}k(k+1)} \int_{R_{\frac{1}{2}k(k+1)}} \exp\left[-i \sum_{i,j=1}^{k} v_{ij} t_{ij}\right] \varphi(t_{ij}) \prod_{i \geqslant j=1}^{k} dt_{ij}. $$

Let $\tau_{ij} - it_{ij} = \theta_{ij}$. Then $\theta_{ij} = \theta_{ji}$, $-it_{ij} = \theta_{ij} - \tau_{ij}$, $dt_{ij} = i\, d\theta_{ij}$, and

(18.2.7) $$ f(\{v_{ii}, 2v_{ij}\}) = A \int_{\substack{\tau_{ij} - i\infty \\ i \geqslant j = 1, \ldots, k}}^{\tau_{ij} + i\infty} \exp\left[\sum_{i,j=1}^{k} v_{ij}\theta_{ij}\right] \cdot |\theta_{ij}|^{-\frac{1}{2}n} \prod_{i \geqslant j=1}^{k} d\theta_{ij} $$

where

$$(18.2.8) \qquad A = \frac{|\tau_{ij}|^{\frac{1}{2}n} \exp\left[-\sum_{i,j=1}^{k} \tau_{ij}v_{ij}\right]}{(2\pi i)^{\frac{1}{2}k(k+1)}}.$$

We shall perform the integration iteratively as follows: First with respect to θ_{kk}, then with respect to $\theta_{1k}, \ldots, \theta_{k-1,k}$, next with respect to $\theta_{k-1,k-1}$, then with respect to $\theta_{1,k-1}, \ldots, \theta_{k-2,k-1}$ and so on through k such cycles. We may write (18.2.7) as follows:

$$(18.2.9) \qquad f(\{v_{ii}, 2v_{ij}\}) = AH_1 \int_{\substack{\tau_{ik}-i\infty \\ i=1,\cdots,k-1}}^{\tau_{ik}+i\infty} \exp\left[2\sum_{i=1}^{k-1} v_{ik}\theta_{ik}\right] H_2 \prod_{i=1}^{k-1} d\theta_{ik}$$

where

$$(18.2.10) \qquad \begin{aligned} H_1 &= \int_{\tau_{ij}-i\infty}^{\tau_{ij}+i\infty} \exp\left[\sum_{i,j=1}^{k-1} v_{ij}\theta_{ij}\right] \cdot |\theta_{ij}|_{(k-1)}^{-\frac{1}{2}n} \prod_{i\geqslant j=1}^{k-1} d\theta_{ij} \\ H_2 &= \int_{\tau_{kk}-i\infty}^{\tau_{kk}+i\infty} \exp\left(v_{kk}\theta_{kk}\right) \cdot \left[\theta_{kk} - \sum_{i,j=1}^{k-1} \theta_{(k-1)}^{ij}\theta_{ik}\theta_{jk}\right]^{-\frac{1}{2}n} d\theta_{kk} \end{aligned}$$

and where $\|\theta_{(k-1)}^{ij}\| = \|\theta_{ij}\|_{(k-1)}^{-1}$ and $\|\theta_{ij}\|_{(k-1)} = \left\|\begin{matrix} \theta_{11} & \cdots & \theta_{1,k-1} \\ \vdots & & \vdots \\ \theta_{k-1,1} & \cdots & \theta_{k-1,k-1} \end{matrix}\right\|.$

Now in H_2 let

$$(18.2.11) \qquad \theta_{kk} - \sum_{i,j=1}^{k-1} \theta_{(k-1)}^{ij}\theta_{ik}\theta_{jk} = w.$$

Then, we have

$$(18.2.12) \qquad H_2 = \exp\left[v_{kk}\sum_{i,j=1}^{k-1} \theta_{(k-1)}^{ij}\theta_{ik}\theta_{jk}\right] \cdot \int_{c-i\infty}^{c+i\infty} e^{v_{kk}w}\, w^{-\frac{1}{2}n}\, dw$$

where c is a real and positive constant whose value we need not write down. The integral in (18.2.12) is basically a Hankel integral (see the example in Section 5.1) and has a value I_1 given by

$$(18.2.13) \qquad I_1 = \frac{(v_{kk})^{\frac{1}{2}n-1}(2\pi i)}{\Gamma(\frac{1}{2}n)}$$

Inserting the value of H_2 in (18.2.9), we obtain

$$(18.2.14) \qquad f(\{v_{ii}, 2v_{ij}\}) = A \cdot H_1 \cdot I_1 \cdot G$$

where

$$(18.2.15) \qquad G = \int_{\substack{\tau_{ik}-i\infty \\ i=1,\ldots,k-1}}^{\tau_{ik}+i\infty} \exp\left[v_{kk}\sum_{i,j=1}^{k-1} \theta_{(k-1)}^{ij}\theta_{ik}\theta_{jk} + \sum_{i=1}^{k-1} v_{ik}\theta_{ik}\right] \prod_{i=1}^{k-1} d\theta_{ik}.$$

Making the transformation $\theta_{ik} = i\psi_{ik}$, $i = 1, \ldots, k-1$ in G, we find

(18.2.16)
$$G = (i)^{k-1}\int_{\substack{-\infty+i\tau_{ik} \\ i=1,\ldots,k-1}}^{+\infty+i\tau_{ik}} \exp\left[-v_{kk}\sum_{i,j=1}^{k-1}\theta_{(k-1)}^{ij}\psi_{ik}\psi_{jk} + 2i\sum_{i=1}^{k-1}v_{ik}\psi_{ik}\right]\prod_{i=1}^{k-1}d\psi_{ik}.$$

It will be noted that the integral in (18.2.16) is an integral of a function of $(k-1)$ complex variables taken over the entire $(k-1)$ dimensional space of the real parts of these complex variables for fixed values of the imaginary parts. It can be shown that the integral does not depend on what fixed (finite) values of the imaginary parts are chosen. Hence, the fixed values may be chosen as 0. Thus, the integral has the same value as if $\psi_{1k}, \ldots, \psi_{k-1,k}$ were real, with the integral taken over the $(k-1)$-dimensional space of $\psi_{1k}, \ldots, \psi_{k-1,k}$. The value of this integral can be written down by referring to (7.4.13) and (7.4.17). Thus

(18.2.17)
$$G = J_1 \cdot |\theta_{(k-1)}^{ij}|^{-\frac{1}{2}} \exp\left[-\sum_{i,j=1}^{k-1}\theta_{ij}\frac{v_{ik}v_{jk}}{v_{kk}}\right]$$

where

(18.2.18)
$$J_1 = \frac{\pi^{\frac{1}{2}(k-1)}i^{k-1}}{(v_{kk})^{\frac{1}{2}(k-1)}}.$$

Putting this value of G in (18.2.14) we may now write

(18.2.19)
$$f(\{v_{ii}, 2v_{ij}\}) = A(I_1J_1)\int_{\substack{\tau_{ij}-i\infty \\ i\geqslant j=1,\ldots,k-1}}^{\tau_{ij}+i\infty} \exp\left[\sum_{i,j=1}^{k-1}v_{ij}^{(1)}\theta_{ij}\right]\cdot|\theta_{ij}|_{(k-1)}^{\frac{1}{2}(n-1)}\prod_{i\geqslant j=1}^{k-1}d\theta_{ij}.$$

where

(18.2.20)
$$v_{ij}^{(1)} = v_{ij} - \frac{v_{ik}v_{jk}}{v_{kk}}, \qquad i, j = 1, \ldots, k-1.$$

It will now be seen that the integral appearing in (18.2.19) has the same structure as that appearing in (18.2.7) except that k, n, v_{ij} are replaced by $k-1$, $n-1$, $v_{ij}^{(1)}$ respectively.

Therefore, by successively repeating cycles of integration similar to that already performed, we obtain

(18.2.21)
$$f(v_{ii}, 2v_{ij}\}) = A(I_1J_1)(I_2J_2)\cdots(I_kJ_k)$$

where I_1 and J_1 are given by (18.2.13) and (18.2.18) and

(18.2.22)
$$I_2 = \frac{(v_{k-1,k-1}^{(1)})^{\frac{1}{2}(n-1)-1}(2\pi i)}{\Gamma\left(\dfrac{n-1}{2}\right)}, \ldots, \qquad I_k = \frac{(v_{11}^{(k-1)})^{\frac{1}{2}(n-k+1)-1}(2\pi i)}{\Gamma\left(\dfrac{n-k+1}{2}\right)}$$

$$J_2 = \frac{\pi^{\frac{1}{2}(k-2)}i^{k-2}}{(v_{k-1,k-1}^{(1)})^{\frac{1}{2}(k-2)}}, \ldots, \qquad J_k = \frac{\pi^{\frac{1}{2}(k-k)}i^{k-k}}{(v_{11}^{(k-1)})^{\frac{1}{2}(k-k)}} = 1.$$

and where $v_{ij}^{(1)}$ is given in (18.2.20) and

(18.2.23)

$$v_{ij}^{(2)} = v_{ij}^{(1)} - \frac{v_{i,k-1}^{(1)} v_{j,k-1}^{(1)}}{v_{k-1,k-1}^{(1)}}, \qquad i, j = 1, \ldots, k-2$$
$$\vdots$$
$$v_{11}^{(k)} = v_{11}^{(k-1)} - \frac{v_{12}^{(k-1)} v_{12}^{(k-1)}}{v_{22}^{(k-1)}}.$$

Substituting these values of $I_1, \ldots, I_k, J_1, \ldots, J_k$ into (18.2.21), we find

(18.2.24)

$$f(\{v_{ii}, 2v_{ij}\}) = \frac{|\tau_{ij}|^{\frac{1}{2}n}[v_{kk}v_{k-1,k-1}^{(1)} \cdots v_{11}^{(k-1)}]^{\frac{1}{2}(n-k-1)} \exp\left[-\sum_{i,j=1}^{k} \tau_{ij}v_{ij}\right]}{2^{\frac{1}{2}k(k-1)}\pi^{k(k-1)/4}\Gamma\left(\frac{n}{2}\right)\Gamma\left(\frac{n-1}{2}\right)\cdots\Gamma\left(\frac{n-k+1}{2}\right)}$$

at any point in the sample space of $\{v_{ii}, 2v_{ij}\}$ in which $\|v_{ij}\|$ is positive definite and 0 otherwise. But the structure of the elements in the sequence $v_{kk}, v_{k-1,k-1}^{(1)}, \ldots, v_{11}^{(k-1)}$ is similar to that of the sequence $\sigma^{11}, \sigma_{(1)}^{22}, \ldots, \sigma_{(k-1)}^{kk}$ defined in Section 7.4(a). It therefore follows from (7.4.6) that

(18.2.25) $$v_{kk}v_{k-1,k-1}^{(1)} \cdots v_{11}^{(k-1)} = |v_{ij}|.$$

Now the expression on the right of (18.2.24) is the p.d.f. of $\{v_{ii}, i = 1, \ldots, k;\ 2v_{ij}, i > j = 1, \ldots, k\}$. Thus, if $g(\{v_{ij}\})$ is the p.d.f. of the $\{v_{ij}, i \geqslant j = 1, \ldots, k\}$, we have

(18.2.26) $$g(\{v_{ij}\}) = 2^{\frac{1}{2}k(k-1)} f(\{v_{ii}, 2v_{ij}\}).$$

Remembering that $\tau_{ij} = \frac{1}{2}\sigma^{ij}$ we finally obtain the following basic result due to Wishart (1928):

18.2.1 *If* $(x_{1\xi}, \ldots, x_{k\xi}, \xi = 1, \ldots, n), k \leqslant n,$ *is a sample from the k-dimensional distribution* $N(\{\mu_i\}; \|\sigma_{ij}\|)$ *and if* $\|v_{ij}\|$ *is the scatter matrix of the sample about the population mean* $(\mu_1, \ldots, \mu_k)$, *as defined by* (18.2.2), *then the elements of* $\|v_{ij}\|$ *have the p.d.f.*

(18.2.27)

$$g(\{v_{ij}\}) = \frac{|\sigma^{ij}|^{\frac{1}{2}n}|v_{ij}|^{\frac{1}{2}(n-k-1)} \exp\left[-\frac{1}{2}\sum_{i,j=1}^{k} \sigma^{ij}v_{ij}\right]}{2^{\frac{1}{2}kn}\pi^{k(k-1)/4}\Gamma\left(\frac{n}{2}\right)\Gamma\left(\frac{n-1}{2}\right)\cdots\Gamma\left(\frac{n-k+1}{2}\right)}$$

in the $\frac{1}{2}k(k+1)$-dimensional region for which $\|v_{ij}\|$ is positive definite and 0 otherwise.

It is convenient to say that any matrix $\|v_{ij}\|$ of random variables whose elements have the p.d.f. (18.2.27) has the *Wishart distribution* $W(k, n, \|\sigma_{ij}\|)$ which is characterized by the parameters k, n, and $\|\sigma_{ij}\|$.

The distribution was originally established by Wishart (1928) by a method of geometric argument, whereas a slightly different version of the geometric argument was presented later by Mahalanobis, Bose, and Roy (1937). The distribution has also been derived by various other methods by Ingham (1933), Wishart and Bartlett (1932), Madow (1938), Hsu (1939a), Sverdrup (1947), and Rasch (1948). The derivation presented above is essentially that given by Ingham, Wishart, and Bartlett.

For the case $k = 1$ it should be noted that $\sigma^{11} = 1/\sigma^2$ where σ^2 is the variance of the normal distribution from which the sample comes, and the Wishart distribution $W(1, n, \sigma^2)$ reduces to a distribution in which $\sigma^{11}v_{11}$ has the chi-square distribution with n degrees of freedom.

It should be noted that the number of random variables in the sample is nk and the number in the Wishart distribution is $\frac{1}{2}k(k+1)$ which means that the Wishart distribution is a $\frac{1}{2}k(k+1)$-dimensional marginal distribution determined from the nk-dimensional distribution of the sample elements. James (1954) has shown that a transformation of the $x_{i\xi}$ exists which transforms the probability element of the sample into the probability elements of three independent distributions, one of which is the Wishart distribution, another is a distribution concerning the k-plane spanned by the k n-dimensional vectors $[(x_{i1} - \mu_i), \ldots, (x_{in} - \mu_i)]$, $i = 1, \ldots, k$, and the third one is the distribution of the orthogonal $k \times k$ matrix which determines the orientation of these k vectors in the k-plane.

(b) Moments and Distribution of the Scatter in Samples from a Normal Distribution

In (18.1.26) we have seen that the mean value of the scatter $|v_{ij}|$ of a sample $(x_{1\xi}, \ldots, x_{k\xi}, \xi = 1, \ldots, n)$ about the mean $(\mu_1, \ldots, \mu_k)$ of the distribution from which the sample is drawn is $k!\binom{n}{k} \cdot |\sigma_{ij}|$ where $\|\sigma_{ij}\|$ is the covariance matrix of the distribution. In the particular case of a k-dimensional normal distribution, the rth moment of $|v_{ij}|$ can be found as follows.

First, it will be convenient to let

$$g(\{v_{ij}\}) = K(k, n, \{\sigma_{ij}\}) \cdot h(k, n, \{\sigma_{ij}\}, \{v_{ij}\}) \tag{18.2.28}$$

where

$$K(k, n, \{\sigma_{ij}\}) = \frac{|\sigma^{ij}|^{\frac{1}{2}n}}{2^{\frac{1}{2}kn}\pi^{k(k-1)/4}\prod_{i=1}^{k}\Gamma\left(\frac{n+1-i}{2}\right)} \tag{18.2.29}$$

and, over the space of the $\{v_{ij}\}$ in which $\|v_{ij}\|$ is positive definite,

$$h(k, n, \{\sigma_{ij}\}, \{v_{ij}\}) = |v_{ij}|^{\frac{1}{2}(n-k-1)} \exp\left[-\tfrac{1}{2}\sum_{i,j=1}^{k}\sigma^{ij}v_{ij}\right] \tag{18.2.30}$$

and 0 elsewhere. Then

$$\mathscr{E}(|v_{ij}|^r) = K(k, n, \{\sigma_{ij}\})\int_{R^{\frac{1}{2}k(k+1)}} |v_{ij}|^r h(k, n, \{\sigma_{ij}\}, \{v_{ij}\}) \prod_{i\geqslant j=1}^{k} dv_{ij}. \tag{18.2.31}$$

Now since the integral of $g(\{v_{ij}\})$ over the entire sample space of the $\{v_{ij}\}$ is unity, we have

$$\int_{R^{\frac{1}{2}k(k+1)}} h(k, n, \{\sigma_{ij}\}, \{v_{ij}\}) \prod_{i\geqslant j=1}^{k} dv_{ij} = \frac{1}{K(k, n, \{\sigma_{ij}\})}. \tag{18.2.32}$$

But the integrand of (18.2.31) is $h(k, n + 2r, \{\sigma_{ij}\}, \{v_{ij}\})$ and the integral of this function over the sample space of the $\{v_{ij}\}$ is $1/K(k, n + 2r, \{\sigma_{ij}\})$. Therefore

(18.2.33)

$$\mathscr{E}(|v_{ij}|^r) = \frac{K(k, n, \{\sigma_{ij}\})}{K(k, n + 2r, \{\sigma_{ij}\})} = |2\sigma_{ij}|^r \prod_{i=1}^{k}\left[\frac{\Gamma\left(\frac{n+1-i}{2}+r\right)}{\Gamma\left(\frac{n+1-i}{2}\right)}\right].$$

Putting $r = 1$ we find for the case of $|v_{ij}|$ in a sample of size n from $N(\mu_i, \|\sigma_{ij}\|)$

$$\mathscr{E}(|v_{ij}|) = k!\binom{n}{k}|\sigma_{ij}|.$$

This result, it will be recalled from (18.1.26), holds for the more general case of $|v_{ij}|$ in a sample of size n from an arbitrary k-dimensional distribution with covariance matrix $\|\sigma_{ij}\|$.

Note that we may write $\mathscr{E}(|v_{ij}|^r)$ as follows:

$$\mathscr{E}(|v_{ij}|^r) = \int_0^\infty \cdots \int_0^\infty (|2\sigma_{ij}|z_1 \cdots z_k)^r \prod_{i=1}^{k}\left[\frac{z_i^{\frac{1}{2}(n+1-i)-1}e^{-z_i}}{\Gamma\left(\frac{n+1-i}{2}\right)}dz_i\right] \tag{18.2.34}$$

from which it is evident that the distribution of $|v_{ij}|$ is identical with the distribution of $|2\sigma_{ij}|(z_1 \cdots z_k)$ where the z_i are independent random variables having gamma distributions $G(\frac{1}{2}(n + 1 - i))$, $i = 1, \ldots, k$ respectively.

Summarizing we have the following result:

18.2.2 *If $|v_{ij}|$ is the scatter of a sample of size n from $N(\{\mu_i\}; \|\sigma_{ij}\|)$ about the population mean $(\mu_1, \ldots, \mu_k)$, the distribution of $|v_{ij}|$ is identical with that of $|2\sigma_{ij}| \prod_{i=1}^{k} z_i$ where the z_i are independent random variables having gamma distributions $G(\frac{1}{2}(n + 1 - i))$, $i = 1, \ldots, k$ respectively.*

For an explicit form of the distribution function of $|v_{ij}|$ expressed as an integral the reader is referred to Wilks (1932).

(c) Reproductivity of the Wishart Distribution

The law of reproductivity for the Wishart distribution may be stated as follows:

18.2.3 *If $\{v_{ij}^{(1)}\}$ and $\{v_{ij}^{(2)}\}$ are independent sets of random variables having Wishart distributions $W(k, n_1, \|\sigma_{ij}\|)$ and $W(k, n_2, \|\sigma_{ij}\|)$, the set of random variables $\{v_{ij}^{(1)} + v_{ij}^{(2)}\}$ has the Wishart distribution $W(k, n_1 + n_2, \|\sigma_{ij}\|)$.*

This can be established at once by characteristic functions. For it follows from (18.2.4) that the characteristic functions of $\{v_{ij}^{(1)}\}$ and $\{v_{ij}^{(2)}\}$ are

$$|\tau_{ij}|^{\frac{1}{2}n_1} \cdot |\tau_{ij} - i\varepsilon_{ij}t_{ij}|^{-\frac{1}{2}n_1} \quad \text{and} \quad |\tau_{ij}|^{\frac{1}{2}n_2} \cdot |\tau_{ij} - i\varepsilon_{ij}t_{ij}|^{-\frac{1}{2}n_2}$$

respectively, where $\tau_{ij} = \frac{1}{2}\sigma^{ij}$ and where $\varepsilon_{ii} = 1$ and $\varepsilon_{ij} = \frac{1}{2}, i \neq j$. Since $\{v_{ij}^{(1)}\}$ and $\{v_{ij}^{(2)}\}$ are independent, the characteristic function of $\{v_{ij}^{(1)} + v_{ij}^{(2)}\}$ is the product of these two characteristic functions, that is,

$$|\tau_{ij}|^{\frac{1}{2}(n_1+n_2)} \cdot |\tau_{ij} - i\varepsilon_{ij}t_{ij}|^{-\frac{1}{2}(n_1+n_2)}$$

which is the characteristic function of the Wishart distribution

$$W(k, n_1 + n_2, \|\sigma_{ij}\|).$$

A basic reproductive-like result for Wishart and normal distributions which will be useful later may be stated as follows:

18.2.4 *Let $\{a_{ij}\}$ be a set of random variables having the Wishart distribution $W(k, n, \|\sigma_{ij}\|)$ and let $(b_{1\beta}, \ldots, b_{k\beta})$, $\beta = 1, \ldots, p$, be p independent sets of random variables having identical distributions $N(\{0\}; \|\sigma_{ij}\|)$, which are also independent of the set $\{a_{ij}\}$. Then the set of random variables $\left\{a_{ij} + \sum_{\beta=1}^{p} b_{i\beta}b_{j\beta}\right\}$ has the Wishart distribution $W(k, n + p, \|\sigma_{ij}\|)$.*

The proof **18.2.4** by characteristic functions is similar to that of **18.2.3** and is left as an exercise for the reader. For $p \geqslant k$ **18.2.4** is an immediate consequence of **18.2.3**.

18.3 INDEPENDENCE OF MEANS AND INTERNAL SCATTER MATRIX IN SAMPLES FROM *k*-DIMENSIONAL NORMAL DISTRIBUTIONS

Suppose $(x_{1\xi}, \ldots, x_{k\xi}, \xi = 1, \ldots, n)$ is a sample from the k-dimensional distribution $N(\{\mu_i\}, \|\sigma_{ij}\|)$ and let $\|u_{ij}\|$ be the internal scatter matrix of the sample. We shall show that:

18.3.1 *The elements of the internal scatter matrix $\|u_{ij}\|$ and the sample means $(\bar{x}_1, \ldots, \bar{x}_k)$ are independent sets of random variables having the distributions $W(k, n-1, \|\sigma_{ij}\|)$, and $N(\{\mu_i\}; \|\sigma_{ij}/n\|)$ respectively.*

To establish this result we shall consider the characteristic function of $\{v_{ii}, i = 1, \ldots, k, 2v_{ij}, i > j = 1, \ldots, k\}$ and $((\bar{x}_1 - \mu_1), \ldots, (\bar{x}_k - \mu_k))$ defined as

$$\varphi(\{t_{ij}\}, \{t_i\}) = \mathscr{E}\left\{\exp\left[i\sum_{i,j=1}^{k} v_{ij}t_{ij} + i\sum_{i=1}^{k}(\bar{x}_i - \mu_i)t_i\right]\right\} \tag{18.3.1}$$

where $\|v_{ij}\|$ is the scatter matrix of the sample about the population mean $(\mu_1, \ldots, \mu_k)$ as defined in (18.1.27). It will be noted that the right-hand side of (18.3.1) can be expressed as

$$\left(\mathscr{E}\left\{\exp\left[i\sum_{i,j=1}^{k} y_i y_j t_{ij} + i\sum_{i=1}^{k}\frac{y_i}{n}t_i\right]\right\}\right)^n \tag{18.3.2}$$

where $y_i = x_i - \mu_i$. Evaluating (18.3.2) from results given in Section 7.4, we find that

$$\varphi(\{t_{ij}\}, \{t_i\}) = |\sigma^{ij}|^{\frac{1}{2}n} \cdot |\sigma_*^{ij}|^{-\frac{1}{2}n} \cdot \exp\left[-\tfrac{1}{2}\sum_{i,j=1}^{k}\sigma_{ij}^*\frac{t_i t_j}{n}\right] \tag{18.3.3}$$

where $\|\sigma_*^{ij}\| = \|\sigma^{ij} - 2it_{ij}\|$ and $\|\sigma_{ij}^*\| = \|\sigma_*^{ij}\|^{-1}$.

Now applying the multidimensional form of Lévy's theorem **5.2.1**, the p.d.f. of $\{v_{ii}, (\bar{x}_i - \mu_i)\, i = 1, \ldots, k; 2v_{ij}, i > j = 1, \ldots, k\}$ is given by

$$\left(\frac{1}{2\pi}\right)^{\frac{1}{2}k(k+3)}\int_{R_{\frac{1}{2}k(k+3)}}\exp\left[-i\sum_{i,j=1}^{k}v_{ij}t_{ij} - i\sum_{i=1}^{k}(\bar{x}_i - \mu_i)t_i\right] \cdot \varphi(\{t_{ij}\}, \{t_i\})\prod_{i=1}^{k}dt_i\prod_{i\geqslant j=1}^{k}dt_{ij} \tag{18.3.4}$$

Making use of results in Section 7.4 and performing the integration with respect to the t_i we find first that expression (18.3.4) reduces to

$$(18.3.5)\qquad I\cdot\frac{\sqrt{|n\sigma^{ij}|}}{(2\pi)^{\frac{1}{2}k}}\exp\left[-\frac{n}{2}\sum_{i,j=1}^{k}\sigma^{ij}(\bar{x}_i-\mu_i)(\bar{x}_j-\mu_j)\right]$$

where

$$(18.3.6)\qquad I=\left(\frac{1}{2\pi}\right)^{\frac{1}{2}k(k+1)}\int_{R_{\frac{1}{2}k(k+1)}}|\sigma^{ij}|^{\frac{1}{2}(n-1)}|\sigma^{ij}-2it_{ij}|^{-\frac{1}{2}(n-1)}\cdot\exp\left[-i\sum_{i,j=1}^{k}u_{ij}t_{ij}\right]\prod_{i\geqslant j=1}^{k}dt_{ij}$$

and

$$u_{ij}=v_{ij}-n(\bar{x}_i-\mu_i)(\bar{x}_j-\mu_j)=\sum_{\xi=1}^{n}(x_{i\xi}-\bar{x}_i)(x_{j\xi}-\bar{x}_j).$$

Thus, by comparing (18.3.6) with (18.2.6) it is seen that I is the p.d.f. of the set of random variables $(u_{ii}, i=1,\ldots,k;\ 2u_{ij}, i>j=1,\ldots,k)$ where $\{u_{ij}\}$ have the the Wishart distribution $W(k, n-1, \|\sigma_{ij}\|)$.

Thus, (18.3.5) is the product of the p.d.f. of the Wishart distribution $W(k, n-1, \|\sigma_{ij}\|)$ and the p.d.f. of the normal distribution $N\left(\{\mu_i\}; \left\|\frac{\sigma_{ij}}{n}\right\|\right)$, thereby establishing **18.3.1**.

18.4 HOTELLING'S GENERALIZED STUDENT DISTRIBUTION

(a) Case of One Sample

Suppose we have a sample from a k-dimensional normal distribution and we wish to test the hypothesis that the vector of means of the distribution has the value $(\mu_1^\circ, \ldots, \mu_k^\circ)$. A test based on scatter analysis which suggests itself is to compare the internal scatter $|u_{ij}|$ of the sample with the scatter $|v_{ij}|$ of the sample about the specified mean $(\mu_1^\circ, \ldots, \mu_k^\circ)$ of the distribution. This suggests taking the ratio

$$(18.4.1)\qquad R_1=\frac{|u_{ij}|}{|v_{ij}|}=\frac{|u_{ij}|}{|u_{ij}+n(\bar{x}_i-\mu_i^\circ)(\bar{x}_j-\mu_j^\circ)|}.$$

Since

$$(18.4.2)\qquad |v_{ij}|=|u_{ij}|\cdot\left[1+n\sum_{i,j=1}^{k}u^{ij}(\bar{x}_i-\mu_i^\circ)(\bar{x}_j-\mu_j^\circ)\right],$$

we can write R_1 in the alternative form

$$(18.4.3)\qquad R_1=\frac{1}{1+nD^2/(n-1)}$$

where

$$(18.4.4)\qquad D^2=(n-1)\sum_{i,j=1}^{k}u^{ij}(\bar{x}_i-\mu_i^\circ)(\bar{x}_j-\mu_j^\circ)$$

and $\|u^{ij}\|=\|u_{ij}\|^{-1}$.

It is evident that R_1 lies on [0, 1], and is unity if and only if the vector sample means $(\bar{x}_1, \ldots, \bar{x}_k)$ is equal to the vector of population means $(\mu_1^\circ, \ldots, \mu_k^\circ)$. We assume, of course, that $\|u_{ij}\|$ is positive definite, which occurs with probability 1 if $n > k$. The quantity D^2 is known as the *Mahalanobis* (1936) (*squared*) *distance between the sample and population* or more briefly Mahalanobis' D^2. The larger the value of D^2 the smaller the value of R_1, of course. The quantity T^2 defined by

$$T^2 = nD^2 \tag{18.4.5}$$

is called Hotelling's (1931) (*squared*) *generalized Student ratio*, or more briefly Hotelling's T^2.

It should be remarked that $R_1 = \lambda^{2/n}$ where λ is the Neyman-Pearson likelihood ratio for testing the composite statistical hypothesis (see Section 13.3) $\mathscr{H}(\omega; \Omega)$ where Ω is the admissible set of points in the parameter space consisting of all values of σ_{ij}, $i \geqslant j = 1, \ldots, k$, for which $\|\sigma_{ij}\|$ is positive definite and all real values of the μ_i, whereas ω is the subset of Ω consisting of all points of Ω for which $\mu_i = \mu_i^\circ$, $i = 1, \ldots, k$. The proof of this statement is left as an exercise for the reader.

We now consider the sampling distribution of R_1 if $\mathscr{H}(\omega; \Omega)$ is true, that is, if the normal distribution from which the sample is drawn has the vector of means $(\mu_1^\circ, \ldots, \mu_k^\circ)$. A relatively simple way to do this is by first determining the moments of R_1, then inferring the distribution from the moments. Thus we wish to find the value of

$$\mathscr{E}(R_1^r) = \mathscr{E}(|u_{ij}|^r |v_{ij}|^{-r}), \qquad r = 1, 2, \ldots.$$

Since the elements of $\|v_{ij}\|$ have the Wishart distribution $W(k, n, \|\sigma_{ij}\|)$, it follows that $\mathscr{E}(|v_{ij}|^{-r})$ is given by (18.2.33) with r replaced by $-r$. But $v_{ij} = u_{ij} + b_i b_j$, where $b_i = \sqrt{n}(\bar{x}_i - \mu_i^\circ)$, and where the $\{u_{ij}\}$ and $\{b_i\}$ are independent sets of random variables having distributions $W(k, n - 1, \|\sigma_{ij}\|)$ and $N(\{0\}; \|\sigma_{ij}\|)$ respectively, as we have seen in **18.3.1**.

Therefore, we may write

$$K(k, n-1, \{\sigma_{ij}\}) \left[\frac{|\sigma^{ij}|}{(2\pi)^k}\right]^{\frac{1}{2}} \int_{R^{\frac{1}{2}k(k+1)}} \int_{R^k} |v_{ij}|^{-r} \exp\left[-\tfrac{1}{2} \sum_{i,j=1}^{k} \sigma^{ij} b_i b_j\right] \tag{18.4.6}$$

$$\cdot\ h(k, n-1, \{\sigma_{ij}\}, \{u_{ij}\}) \prod_{i=1}^{k} db_i \prod_{i \geqslant j=1}^{k} du_{ij}$$

$$= |2\sigma_{ij}|^{-r} \prod_{i=1}^{k} \frac{\Gamma\left(\dfrac{n+1-i}{2} - r\right)}{\Gamma\left(\dfrac{n+1-i}{2}\right)}.$$

where $K(k, n-1, \{\sigma_{ij}\})$ and $h(k, n-1, \{\sigma_{ij}\}, \{u_{ij}\})$ are defined in Section 18.2(*b*).

If n is replaced by $n + 2r$ throughout (18.4.6) and if both sides of the resulting equation are multiplied by $K(k, n-1, \{\sigma_{ij}\})/K(k, n+2r-1, \{\sigma_{ij}\})$, it will be seen that the left-hand side of the equation is the integral which defines $\mathscr{E}(R_1^r)$ and the right-hand side is the value of the integral. Therefore, we have

(18.4.7)

$$\mathscr{E}(R_1^r) = \left[\frac{|2\sigma_{ij}|^{-r}K(k, n-1, \{\sigma_{ij}\})}{K(k, n+2r-1, \{\sigma_{ij}\})}\right] \prod_{i=1}^{k} \frac{\Gamma\left(\dfrac{n+1-i}{2}\right)}{\Gamma\left(\dfrac{n+1-i}{2}+r\right)}$$

Substituting the values of $K(k, n-1, \{\sigma_{ij}\})$, $K(k, n+2r-1, \{\sigma_{ij}\})$ from (18.2.29) into (18.4.7), we obtain, after simplification

$$\mathscr{E}(R_1^r) = \frac{\Gamma\left(\dfrac{n}{2}\right)\Gamma\left(\dfrac{n-k}{2}+r\right)}{\Gamma\left(\dfrac{n}{2}+r\right)\Gamma\left(\dfrac{n-k}{2}\right)} \tag{18.4.8}$$

which holds for $r = 1, 2, \ldots$. It will be observed that the right-hand side of (18.4.8) can be expressed as the rth moment of a beta distribution as follows:

$$\mathscr{E}(R_1^r) = \frac{\Gamma\left(\dfrac{n}{2}\right)}{\Gamma\left(\dfrac{k}{2}\right)\Gamma\left(\dfrac{n-k}{2}\right)} \int_0^1 R_1^{\frac{1}{2}(n-k)-1+r}(1-R_1)^{\frac{1}{2}k-1}\, dR_1$$

$$r = 0, 1, 2, \ldots.$$

We conclude from **5.5.1a** that R_1 has the beta distribution $Be(\frac{1}{2}(n-k), \frac{1}{2}k)$, that is, the probability element of R_1 is

$$\frac{\Gamma\left(\dfrac{n}{2}\right)}{\Gamma\left(\dfrac{k}{2}\right)\Gamma\left(\dfrac{n-k}{2}\right)} R_1^{\frac{1}{2}(n-k)-1}(1-R_1)^{\frac{1}{2}k-1}\, dR_1. \tag{18.4.9}$$

Summarizing, we have

18.4.1 *If* $(x_{1\xi}, \ldots, x_{k\xi}, \xi = 1, \ldots, n)$, $n > k$, *is a sample from the normal distribution* $N(\{\mu_i^\circ\}; \|\sigma_{ij}\|)$, *the ratio of scatters* $R_1 = |u_{ij}|/|v_{ij}|$ *has the beta distribution* $Be(\frac{1}{2}(n-k), \frac{1}{2}k)$.

Applying the transformation

$$R_1 = \frac{1}{1 + T^2/(n-1)}$$

to (18.4.9) we obtain, noting that the sample space of T^2 is $(0, +\infty)$,

(18.4.10) $$\frac{\Gamma\left(\dfrac{n}{2}\right)}{(n-1)^{\frac{1}{2}k}\Gamma\left(\dfrac{k}{2}\right)\Gamma\left(\dfrac{n-k}{2}\right)} (1 + T^2/(n-1))^{-\frac{1}{2}n}(T^2)^{\frac{1}{2}k-1}\, d(T^2)$$

as the probability element of Hotelling's T^2. It will be noted that for $k = 1$, (18.4.10) reduces to the probability element of the ordinary Student t^2 with $n - 1$ degrees of freedom. (See Section 7.8(b).)

(b) Case of Two Samples

Suppose we have two samples $(x_{1\xi_1}^{(1)}, \ldots, x_{k\xi_1}^{(1)}, \xi_1 = 1, \ldots, n_1)$ and $(x_{1\xi_2}^{(2)}, \ldots, x_{k\xi_2}^{(2)}, \xi_2 = 1, \ldots, n_2)$ from normal distributions having identical sets of parameters. Denote the common distribution by $N(\{\mu_i\}; \|\sigma_{ij}\|)$. Let $\|u_{ij}^{(1)}\|$ and $\|u_{ij}^{(2)}\|$ be the internal scatter matrices of the two samples and $\|u_{ij}\|$ the internal scatter matrix of the two samples pooled together as a single sample. The matrix $\|u_{ij}^{(1)} + u_{ij}^{(2)}\|$ will be called the *within-sample scatter matrix* for the two samples. Geometrically, it is the scatter matrix for the k-dimensional cluster one obtains by rigidly translating one sample cluster with respect to the other (without rotation) until the means of both samples coincide, and then pooling the two sample clusters together as a single cluster.

Consider the ratio

(18.4.11) $$R_1' = \frac{|u_{ij}^{(1)} + u_{ij}^{(2)}|}{|u_{ij}|}.$$

We shall show that R_1' has a structure and distribution similar to R_1 as defined in (18.4.3). Let us put

(18.4.12) $$u_{ij}^{w} = u_{ij}^{(1)} + u_{ij}^{(2)}.$$

Then u_{ij}, which is the typical element on the right-hand side of (18.1.20), can be expressed as follows

(18.4.13) $$u_{ij} = u_{ij}^{w} + b_i b_j$$

where

(18.4.14) $$b_i = \sqrt{\frac{n_1 n_2}{n_1 + n_2}}\, (\bar{x}_i^{(1)} - \bar{x}_i^{(2)}).$$

Then we may write

$$(18.4.15) \qquad R_1' = \frac{|u_{ij}^{\text{w}}|}{|u_{ij}^{\text{w}} + b_i b_j|}.$$

It follows from **18.2.3** that $\{u_{ij}^{\text{w}}\}$ has the Wishart distribution $W(k, n_1 + n_2 - 2, \|\sigma_{ij}\|)$. Also, since the b_i are functions of the means of the two samples it follows from **18.3.1** that the $\{b_i\}$ are independent of the $\{u_{ij}^{\text{w}}\}$. As a matter of fact the reader will readily verify that the $\{b_i\}$ have the normal distribution $N(\{0\}; \|\sigma_{ij}\|)$.

The problem of finding the distribution of R_1' is thus seen to be similar to that of finding the distribution of R_1 in (18.4.1). In fact, we note, that the distribution of R_1' is identical with that of R_1 with $n - 1$ replaced by $n_1 + n_2 - 2$.

We may write

$$(18.4.16) \qquad R_1' = \frac{1}{1 + \dfrac{n_1 n_2}{(n_1 + n_2)(n_1 + n_2 - 2)} D'^2}$$

where

$$(18.4.17) \quad D'^2 = \frac{(n_1 + n_2)(n_1 + n_2 - 2)}{n_1 n_2} \sum_{i,j=1}^{k} u_{\text{w}}^{ij} b_i b_j$$

$$= (n_1 + n_2 - 2) \sum_{i,j=1}^{k} u_{\text{w}}^{ij} (\bar{x}_i^{(1)} - \bar{x}_i^{(2)})(\bar{x}_j^{(1)} - \bar{x}_j^{(2)})$$

and where $\|u_{\text{w}}^{ij}\| = \|u_{ij}^{\text{w}}\|^{-1}$.

The quantity D'^2 is the Mahalanobis (1930, 1936) (*squared*) *generalized distance between the two samples.*

Hotelling's generalized Student ratio for the two-sample problem is defined by the relation

$$T'^2 = \frac{n_1 n_2}{n_1 + n_2} D'^2$$

and the probability element of the distribution of T'^2 is given by (18.4.10) with $n - 1$ replaced by $n_1 + n_2 - 2$, assuming, of course, that both samples are independently drawn from identical k-dimensional normal distributions.

Finally, we should remark that $R_1' = \lambda^{2/n_1 + n_2}$ where λ is the Neyman-Pearson likelihood ratio for testing the statistical hypothesis $\mathcal{H}(\omega; \Omega)$, where Ω is the parameter space for which $\|\sigma_{ij}^{(1)}\| = \|\sigma_{ij}^{(2)}\| = \|\sigma_{ij}\|$ is positive definite, and where the vectors of means $(\mu_1^{(1)}, \ldots, \mu_k^{(1)})$ and $(\mu_1^{(2)}, \ldots, \mu_k^{(2)})$ are real vectors, while ω is the subspace of Ω in which the two vectors of means are equal; $N(\{\mu_i^{(1)}\}; \|\sigma_{ij}^{(1)}\|)$ and $N(\{\mu_i^{(2)}\}; \|\sigma_{ij}^{(2)}\|)$

being used to label the normal distributions from which the two samples are drawn. Verification of this is left to the reader.

18.5 THE MULTIDIMENSIONAL MODEL I ANALYSIS OF VARIANCE TEST

The one-dimensional Model I analysis of variance test in its most general form is the test of the statistical hypothesis $\mathscr{H}$, the linear hypothesis of normal regression theory, defined in Section 13.3(*b*). The likelihood ratio λ for this hypothesis is of the form [see (13.3.9)]

$$\lambda = \left[\frac{S_\Omega}{S_\Omega + (S_\omega - S_\Omega)}\right]^{\frac{1}{2}n} \tag{18.5.1}$$

where S_Ω and $S_\omega - S_\Omega$ are independent random variables whose distributions are the one-dimensional Wishart distributions $W(1, m_1, \sigma^2)$ and $W(1, m_2, \sigma^2)$ respectively, if $\mathscr{H}$ is true, where $m_1 = n - k$ and $m_2 = k - k'$, the numbers of degrees of freedom in S_Ω and $S_\omega - S_\Omega$ respectively. As pointed out in Section 13.3(*b*), the likelihood ratio λ is equivalent to the Snedecor $\mathscr{F}$ test, where

$$\mathscr{F} = \frac{m_1(S_\omega - S_\Omega)}{m_2 S_\Omega}. \tag{18.5.2}$$

For testing $\mathscr{H}$, λ is equivalent to $\lambda^{2/n}$, of course, which is the ratio inside [ ] in (18.5.1). The numerator and denominator of this ratio are simply one-dimensional scatters with the numerator being the smaller.

The multidimensional extension of the Model I analysis of variance test in its most general form is a generalization of the ratio inside [ ] in (18.5.1). The generalization is a ratio of the form

$$R_s = \frac{|a_{ij}|}{\left|a_{ij} + \sum_{\beta=1}^{s} b_{i\beta}b_{j\beta}\right|} \tag{18.5.3}$$

$i, j = 1, \ldots, k$, $k \leqslant m$, where the $\{a_{ij}\}$ have the Wishart distribution $W(k, m, \|\sigma_{ij}\|)$ and where $(b_{1\beta}, \ldots, b_{k\beta})$, $\beta = 1, \ldots, s$, are independent vectors of random variables all having the normal distribution $N(\{0\}; \|\sigma_{ij}\|)$ and are all independent of the $\{a_{ij}\}$.

Using the same method for determining the rth moment of R_s as that used for finding the rth moment of R_1 as given by (18.4.8), we find

$$\mathscr{E}(R_s^r) = \prod_{i=1}^{k} \frac{\Gamma\left(\frac{m+1-i}{2} + r\right)\Gamma\left(\frac{m+s+1-i}{2}\right)}{\Gamma\left(\frac{m+1-i}{2}\right)\Gamma\left(\frac{m+s+1-i}{2} + r\right)}. \tag{18.5.4}$$

Note that if $s \geqslant k$, $\mathscr{E}(R_s^r)$ can be expressed as follows:

(18.5.4*a*)

$$\mathscr{E}(R_s^r) = \prod_{i=1}^{k}\left[\frac{\Gamma\left(\frac{m+s+1-i}{2}\right)}{\Gamma\left(\frac{m+1-i}{2}\right)\Gamma\left(\frac{s}{2}\right)} \cdot \frac{\Gamma\left(\frac{m+1-i}{2}+r\right)\Gamma\left(\frac{s}{2}\right)}{\Gamma\left(\frac{m+s+1-i}{2}+r\right)}\right]$$

from which it is seen that the distribution of R_s is identical with the distribution of the product $\prod_{i=1}^{k} z_i$, where the z_i are independent random variables having the beta distributions

$$Be\left(\frac{m+1-i}{2}, \frac{s}{2}\right), \qquad i = 1, \ldots, k$$

respectively.

Similarly, if $1 \leqslant s < k$ $\mathscr{E}(R_s^r)$ can be expressed in the following form:

$$(18.5.4b)\ \mathscr{E}(R_s^r) = \prod_{i=1}^{s}\left[\frac{\Gamma\left(\frac{m+i}{2}\right)}{\Gamma\left(\frac{m-k+i}{2}\right)\Gamma\left(\frac{k}{2}\right)} \cdot \frac{\Gamma\left(\frac{m-k+i}{2}+r\right)\Gamma\left(\frac{k}{2}\right)}{\Gamma\left(\frac{m+i}{2}+r\right)}\right]$$

from which it is seen that the distribution of R_s is identical with that of the product $\prod_{i=1}^{s} w_i$, where the w_i are independent random variables having the beta distributions

$$Be\left(\frac{m-k+i}{2}, \frac{k}{2}\right), \qquad i = 1, \ldots, s$$

respectively.

Summarizing, we have the following result:

18.5.1 *Suppose* $\|a_{ij}\|$ *is a symmetric matrix, positive definite with probability* 1, *whose elements are random variables having the Wishart distribution* $W(k, m, \|\sigma_{ij}\|)$ *and let*

$$(b_{1\beta}, \ldots, b_{k\beta}), \qquad \beta = 1, \ldots, s$$

be s independent k-dimensional random variables all having the normal distribution $N(\{0\}; \|\sigma_{ij}\|)$ *and which are also independent of the random variables in* $\|a_{ij}\|$. *Let*

$$R_s = \frac{|a_{ij}|}{\left|a_{ij} + \sum_{\beta=1}^{s} b_{i\beta}b_{j\beta}\right|}.$$

Then:

(i) *If* $s \geqslant k$, *the distribution of* R_s *is identical with the distribution of the product of* k *independent random variables having the beta distributions*

$$Be\left(\frac{m+1-i}{2}, \frac{s}{2}\right), \qquad i = 1, \ldots, k.$$

(ii) *If* $1 \leqslant s < k$, *the distribution of* R_s *is identical with the distribution of the product of* s *independent random variables having the beta distributions*

$$Be\left(\frac{m-k+i}{2}, \frac{k}{2}\right), \qquad i = 1, \ldots, s.$$

It will be observed that R_1 as defined in (18.4.1) is a special case of R_s with $s = 1$, $m = n - 1$, $a_{ij} = u_{ij}$, and $b_i = \sqrt{n}(\bar{x}_i - \mu_i)$, $i, j = 1, \ldots, k$. Similarly, R_1' in (18.4.11) or (18.4.15) is a special case of R_s with $s = 1$, $m = n_1 + n_2 - 2$, $a_{ij} = u_{ij}^{(1)} + u_{ij}^{(2)}$, and

$$b_i = \sqrt{\frac{n_1 n_2}{n_1 + n_2}}\,(\bar{x}_i^{(1)} - \bar{x}_i^{(2)}), \qquad i, j = 1, \ldots, k.$$

An interesting special case of R_s occurs for $s = 2$, in which case $\sqrt{R_2}$ has the beta distribution $Be(m + 1 - k, k)$, that is, the probability element of $\sqrt{R_2}$ is

$$\frac{\Gamma(m+1-k)}{\Gamma(m+1)\Gamma(k)} (\sqrt{R_2})^{m-k}(1 - \sqrt{R_2})^{k-1}\, d\sqrt{R_2}. \tag{18.5.5}$$

To verify this we start with the rth moment of R_2 which we find from (18.5.4b), to be

(18.5.6)

$$\mathscr{E}(R_2^r) = \frac{\Gamma\left(\frac{m+2}{2}\right)\Gamma\left(\frac{m+1}{2}\right)\Gamma\left(\frac{m+2-k}{2} + r\right)\Gamma\left(\frac{m+1-k}{2} + r\right)}{\Gamma\left(\frac{m+2}{2} + r\right)\Gamma\left(\frac{m+1}{2} + r\right)\Gamma\left(\frac{m+2-k}{2}\right)\Gamma\left(\frac{m+1-k}{2}\right)}.$$

Using the Legendre duplication formula (7.6.15), that is,

$$\Gamma(\nu)\Gamma(\nu + \tfrac{1}{2}) = \frac{\sqrt{\pi}\,\Gamma(2\nu)}{2^{2\nu-1}} \tag{18.5.7}$$

we can telescope four pairs of gamma functions in (18.5.6) and obtain

$$\mathscr{E}(R_2^r) = \frac{\Gamma(m+1)\Gamma(m+1-k+2r)}{\Gamma(m+1+2r)\Gamma(m+1-k)} \tag{18.5.8}$$

which may be written as

$$(18.5.9)\qquad \mathscr{E}(R_2^r) = \frac{\Gamma(m+1)}{\Gamma(m+1-k)\Gamma(k)} \int_0^1 z^{m-k+2r}(1-z)^{k-1}\, dz$$

$r = 1, 2, \ldots.$

It is seen from (18.5.9) and **5.5.1a** that the distribution of R_2 is identical with that of z^2 where z has the beta distribution $Be(m+1-k, k)$. Hence $\sqrt{R_2}$ has this distribution, that is, it has the probability element given by (18.5.5). An expression for the distribution of R_s in the form of an integral has been given by Wilks (1932).

Example. Suppose $(x_{1\xi_\gamma}^{(\gamma)}, \ldots, x_{k\xi_\gamma}^{(\gamma)}, \xi_\gamma = 1, \ldots, n_\gamma)$, $\gamma = 1, 2, 3$, are three independent samples from identical normal distributions $N(\{\mu_i\}; \|\sigma_{ij}\|)$. Let $\|u_{ij}^{(\gamma)}\|$, $\gamma = 1, 2, 3$, be the internal scatter matrices of the three samples and let $\|u_{ij}\|$ be the internal scatter matrix of the grand sample obtained by pooling the three samples. Let $\|u_{ij}^{(1)} + u_{ij}^{(2)} + u_{ij}^{(3)}\| = \|u_{ij}^W\|$ be the within-sample scatter matrix for the three samples. Then the ratio

$$(18.5.10)\qquad R_2' = \frac{|u_{ij}^W|}{|u_{ij}|}$$

is the three-sample extension of the ratio R_1' in (18.4.15). R_2' is a special case of R_s in (18.5.3) with $s = 2$ and $m = n_1 + n_2 + n_3 - 3$, and hence the probability element of $\sqrt{R_2'}$ is given by the expression (18.5.5) with m replaced by $n_1 + n_2 + n_3 - 3$.

It will be noted that the system of random variables $\{u_{ij}^W\}$ has the Wishart distribution $W(k, n_1 + n_2 + n_3 - 3, \|\sigma_{ij}\|)$. Furthermore, the *between-sample scatter matrix* $\|u_{ij}^B\|$ is defined by

$$(18.5.11)\qquad u_{ij}^B = u_{ij} - u_{ij}^W = \sum_{\gamma=1}^{3} n_\gamma(\bar{x}_i^{(\gamma)} - \bar{x}_i)(\bar{x}_j^{(\gamma)} - \bar{x}_j)$$

$i, j = 1, \ldots, k$, where $(\bar{x}_1, \ldots, \bar{x}_k)$ is the mean of the grand sample. Two independent vectors $(b_{1\beta}, \ldots, b_{k\beta})$, $\beta = 1, 2$ can be found, which are also independent of $\{u_{ij}^W\}$ where $b_{i\beta} = \sum_{\gamma=1}^{3} c_{i\beta}^{(\gamma)}\bar{x}_i^{(\gamma)}$, such that the components of each vector have the distribution $N(\{0\}; \|\sigma_{ij}\|)$ and such that the right-hand side of (18.5.11) is $\sum_{\beta=1}^{2} b_{i\beta}b_{j\beta}$. However, it is not necessary to determine these vectors in order to establish the distribution of R_2'. This distribution can be readily found from its moments by a procedure similar to that by which the distribution of R_1 was found.

It can be shown, of course, that R_2' is equivalent to the three-sample Neyman-Pearson extension of the likelihood ratio described at the end of Section 18.4(*b*).

18.6 PRINCIPAL COMPONENTS

(a) Eigenvalues and Eigenvectors of a Scatter Matrix

In this section we shall deal with the following problem: Consider a sample of size n from a k-dimensional distribution, $k \leqslant n$. This sample may be represented geometrically as a sample cluster of n points in a k-dimensional Euclidean space R_k. Suppose we wish to project this cluster orthogonally onto an s-dimensional Euclidean space R_s, $s \leqslant k$, so as to obtain the greatest possible s-dimensional scatter of the projected points. The problem is to determine (i) the direction of projection with respect to the coordinate system of R_k and (ii) the size of the scatter in R_s.

The solution of this statistical problem can be stated in the following result due to Hotelling (1933):

18.6.1 *Suppose* $(x_{1\xi}, \ldots, x_{k\xi}, \xi = 1, \ldots, n)$ *is a sample of size* $n > k$ *from a k-dimensional distribution whose covariance matrix is positive definite. Let* $\|u_{ij}\|$ *be the internal scatter matrix of this sample and let it be positive definite with probability* 1. *Let*

(18.6.1) $$(c_{1p}, \ldots, c_{kp}), \qquad p = 1, \ldots, s$$

be s k-dimensional unit vectors, that is, such that $\sum_{i=1}^{k} c_{ip}^2 = 1, p = 1, \ldots, s$, *and set*

(18.6.2) $$z_{p\xi} = \sum_{i=1}^{k} c_{ip} x_{i\xi}, \qquad p = 1, \ldots, s.$$

Let $\|\tilde{u}_{pq}\|$ *be the internal scatter matrix of the sample* $(z_{1\xi}, \ldots, z_{s\xi}, \xi = 1, \ldots, n)$.

The values of the vectors (18.6.1) *which maximize the scatter* $|\tilde{u}_{pq}|$ *are the solutions of the s sets of equations*

(18.6.3) $$\sum_{j=1}^{k} (u_{ij} - l_p \delta_{ij}) c_{jp} = 0, \qquad i = 1, \ldots, k$$

$p = 1, \ldots, s$ *where* $l_1, \ldots, l_s$, *are the s largest roots of the characteristic equation*

$$|u_{ij} - l\delta_{ij}| = 0,$$

δ_{ij} *being the Kronecker* δ, *and where* $l_s < \cdots < l_1$ *with probability* 1. *Furthermore, these vectors are orthogonal, and the maximum value of* $|\tilde{u}_{pq}|$ *is the product* $l_1 l_2 \cdots l_s$.

To prove **18.6.1** consider the following general linear function of the components of each element of the sample:

(18.6.4) $$z_\xi = \sum_{i=1}^{k} c_i x_{i\xi}, \qquad \xi = 1, \ldots, k$$

and form the sum of squares

(18.6.5) $$Q = \sum_{\xi=1}^{n} (z_\xi - \bar{z})^2 = \sum_{i,j=1}^{k} u_{ij} c_i c_j$$

where $\sum_{i=1}^{k} c_i^2 = 1$.

Now let us find the values of $(c_1, \ldots, c_k)$ for which Q is stationary. Using the method of Lagrange multipliers, we obtain vectors from the equations:

$$\frac{\partial \varphi}{\partial c_i} = 0, \qquad i = 1, \ldots, k \tag{18.6.6}$$

where

$$\varphi = Q + l\left(1 - \sum_1^k c_i^2\right).$$

The equations (18.6.6) are

$$\sum_{j=1}^{k} (u_{ij} - l\delta_{ij})c_j = 0, \qquad i = 1, \ldots, k. \tag{18.6.7}$$

To obtain solutions of (18.6.7) other than the trivial solution $c_i = 0$, $i = 1, \ldots, k$, which, of course, is ruled out by the assumption that $\sum_i c_i^2 = 1$, it is necessary for

$$|u_{ij} - l\delta_{ij}| = 0. \tag{18.6.8}$$

Since the sample of size $n > k$ is assumed to be from a k-dimensional distribution and since $\|u_{ij}\|$ is assumed to be positive definite with probability 1, (18.6.8) has k positive roots which we assume to be different, that is $0 < l_k < \cdots < l_1$ with probability 1. These roots are called the *eigenvalues* of the matrix $\|u_{ij}\|$. These roots are also called the *characteristic roots*, or *latent roots* of the matrix $\|u_{ij}\|$.

If $l_{p'}$ is any one of these roots, then $(c_{1p'}, \ldots, c_{kp'})$ is the p'th *eigenvector* and satisfies

$$\sum_{j=1}^{k} (u_{ij} - l_{p'}\delta_{ij})c_{jp'} = 0. \tag{18.6.9}$$

If we multiply (18.6.9) by $c_{ip'}$ and sum over i we get

$$\sum_{i,j=1}^{k} (u_{ij} - l_{p'}\delta_{ij})c_{ip'}c_{jp'} = 0 \tag{18.6.10}$$

from which it follows that

$$l_{p'} = \sum_{i,j=1}^{k} u_{ij}c_{ip'}c_{jp'}. \tag{18.6.11}$$

Similarly, for any two roots $l_{p'}$, $l_{q'}$, and $p' \neq q'$, we have

$$\begin{aligned} &\sum_{i,j=1}^{k} (u_{ij} - l_{p'}\delta_{ij})c_{ip'}c_{jq'} = 0 \\ &\sum_{i,j=1}^{k} (u_{ij} - l_{q'}\delta_{ij})c_{iq'}c_{jp'} = 0, \end{aligned} \tag{18.6.12}$$

from which we find (subtracting the first equation from the second)

$$(18.6.13) \qquad (l_{p'} - l_{q'})\sum_{i=1}^{k} c_{ip'}c_{iq'} = 0.$$

Therefore, for $p' \neq q'$ we have

$$(18.6.14) \qquad \sum_{i=1}^{k} c_{ip'}c_{iq'} = 0,$$

that is, the *eigenvectors* $(c_{1p}, \ldots, c_{kp})$ $p = 1, \ldots, k$ are mutually orthogonal. These vectors are also called *characteristic vectors* or *latent vectors* of $\|u_{ij}\|$.

Using the result (18.6.14) in either of the equations (18.6.12), we find, that for $p' \neq q'$,

$$(18.6.15) \qquad \sum_{i,j=1}^{k} u_{ij}c_{ip'}c_{jq'} = 0.$$

Now suppose we choose the s eigenvectors corresponding to the s $(s \leqslant k)$ largest roots $l_s < \cdots < l_1$, and form the sample $(z_{1\xi}, \ldots, z_{s\xi}, \xi = 1, \ldots, n)$ where

$$z_{p\xi} = \sum_{i=1}^{k} c_{ip}x_{i\xi}, \qquad p = 1, \ldots, s.$$

It follows from the orthogonality of the unit eigenvectors that this sample is an orthogonal projection of the original sample onto an s-dimensional Euclidean space. The internal scatter of this projected sample is therefore $|\tilde{u}_{pq}|$, $p, q = 1, \ldots, s$, where

$$(18.6.16) \qquad \tilde{u}_{pq} = \sum_{\xi=1}^{n} (z_{p\xi} - \bar{z}_p)(z_{q\xi} - \bar{z}_q) = \sum_{i,j=1}^{k} u_{ij}c_{ip}c_{jq}.$$

But making use of (18.6.11) and (18.6.15) we see that

$$(18.6.17) \qquad |\tilde{u}_{pq}| = l_1 l_2 \cdots l_s$$

which concludes the proof of **18.6.1**.

The reader should note that if we take $s = k$, then $|\tilde{u}_{pq}| = |u_{ij}| = l_1 l_2 \cdots l_k$. This is also evident from the fact that the product of the roots of (18.6.8) has the value $|u_{ij}|$.

Referring again to (18.6.8) it will be seen that

$$(18.6.18) \qquad l_1 + \cdots + l_k = \sum_{i=1}^{k} u_{ii}.$$

In other words, the sum of the eigenvalues of $\|u_{ij}\|$ is equal to the sum of the scatters of the individual components of the sample, each taken about its own mean. The eigenvalues $l_1, \ldots, l_k$ of the internal scatter matrix $\|u_{ij}\|$ of the sample, each divided by $n - 1$, are called the *principal components* of the total sample variance by Hotelling (1933), since

$$\frac{l_1}{n-1} + \cdots + \frac{l_k}{n-1} = s_1^2 + \cdots + s_k^2 \tag{18.6.19}$$

where $s_i^2 = u_{ii}/(n-1)$, as will be seen from (18.6.18).

(b) Sampling Theory of Eigenvalues of a Scatter Matrix

Since the eigenvalues $l_1, \ldots, l_k$ are the roots of the determinantal equation (18.6.8), where $\|u_{ij}\|$ is the internal scatter matrix of the sample the eigenvalues themselves are random variables. The question arises, of course, as to what one can say about the sampling theory of these roots. The problem of determining the sampling theory of the roots in exact form has been solved only in the case where the sample $(x_{1\xi}, \ldots, x_{k\xi}, \xi = 1, \ldots, n)$ essentially comes from a k-dimensional spherical normal distribution, that is, one whose covariance matrix $\|\sigma_{ij}\|$ has eigenvalues which are all equal. In this section we shall present this sampling theory. This problem was originally solved approximately simultaneously by Fisher (1939), Girshick (1939), Hsu (1939b), Mood (1951), and Roy (1939). Mood's results were not published until twelve years after they were obtained. More recently it has been treated by James (1954), Olkin (1951), Olkin and Roy (1954), and others, using different methods.

The basic result can be expressed as follows:

18.6.2 *Let $(x_{1\xi}, \ldots, x_{k\xi}, \xi = 1, \ldots, n)$ be a sample from the normal distribution $N(\{u_i\}; \|\sigma_{ij}\|)$ and let $\|u_{ij}\|$ be the internal scatter matrix of the sample. The roots $0 < l_k < \cdots < l_1$ of the characteristic equation*

$$|u_{ij} - l\sigma_{ij}| = 0 \tag{18.6.20}$$

have a distribution with probability element

$$\frac{\pi^{\frac{1}{2}k}\left(\prod_{i=1}^{k} l_i\right)^{\frac{1}{2}(n-k-2)} \prod_{i>j=1}^{k} (l_j - l_i) \exp\left(-\frac{1}{2}\sum_{i=1}^{k} l_i\right)}{2^{\frac{1}{2}k(n-1)}\prod_{i=1}^{k} \Gamma\left(\frac{k+1-i}{2}\right)\Gamma\left(\frac{n-i}{2}\right)} \, dl_1 \cdots dl_k \tag{18.6.21}$$

in the region for which $0 < l_k < l_{k-1} < \cdots < l_1 < +\infty$, and 0 *otherwise.*

To establish **18.6.2** we proceed as follows. We know that $\{u_{ij}\}$ has the Wishart distribution $W(k, n-1, \|\sigma_{ij}\|)$. Furthermore, we know from the theory of positive definite quadratic forms (see Birkhoff and MacLane (1953), for instance) that we can find a nonsingular linear transformation

$$(18.6.22)\quad (x_{i\xi} - \mu_i) = \sum_{j=1}^{k} c_{ij} y_{j\xi}, \qquad i = 1, \ldots, k,\ \xi = 1, \ldots, n$$

so that $(y_{1\xi}, \ldots, y_{k\xi}, \xi = 1, \ldots, n)$ is a sample from $N(\{0\}; \|\delta_{ij}\|)$ where $\|\delta_{ij}\|$ is the unit matrix. Denoting the matrices $\|c_{ij}\|$, $\|\sigma_{ij}\|$ and $\|\delta_{ij}\|$ by c, σ and I and the inverses of c and σ by c^{-1} and σ^{-1}, and the transpose of a matrix with a dash, this means that

$$(18.6.23)\qquad c\sigma^{-1}c' = I, \qquad (c^{-1})'\sigma c^{-1} = I.$$

Hence, if $\|u^*_{ij}\|$ is the scatter matrix of the sample of y's, then the $\{u^*_{ij}\}$ have the Wishart distribution $W(k, n - 1, \|\delta_{ij}\|)$. Denoting the matrices $\|u_{ij}\|$ and $\|u^*_{ij}\|$ by u and u^*, we can express u and u^* in terms of each other as follows:

$$(18.6.24)\qquad u = c'u^*c, \qquad u^* = (c^{-1})'uc^{-1}.$$

Now suppose $l_1 > \cdots > l_k$ are the roots of (18.6.20), that is, of

$$(18.6.25)\qquad |u - l\sigma| = 0.$$

The probability of equality of two or more roots is zero. Multiplying the matrix $\|u - l\sigma\|$ on the left by $(c^{-1})'$ and on the right by c^{-1}, we obtain

$$(18.6.26)\qquad (c^{-1})'uc^{-1} - l(c^{-1})'\sigma c^{-1},$$

which, in view of the second equations in (18.6.23) and (18.6.24), is

$$(18.6.27)\qquad u^* - lI.$$

The determinants of (18.6.26) and (18.6.27) are, of course, identical in value. But the determinant of (18.6.26) is $|(c^{-1})'| \cdot |u - l\sigma| \cdot |c^{-1}|$, in which the factor $|u - l\sigma|$ vanishes for $l = l_1, \ldots, l_k$. Therefore $|u^* - lI|$ vanishes for these same values of l.

Hence, the distribution function of the roots of $|u - l\sigma| = 0$, where $\{u_{ij}\}$ have the Wishart distribution $W(k, n - 1, \|\sigma_{ij}\|)$ is identically the same as that of the roots of $|u^* - lI| = 0$, where $\{u^*_{ij}\}$ have the Wishart distribution $W(k, n - 1), \|\delta_{ij}\|)$. The probability element of $\{u^*_{ij}\}$ is

$$(18.6.28)\qquad \frac{|u^*_{ij}|^{\frac{1}{2}(n-k-2)} \exp\left(-\frac{1}{2}\sum_{i=1}^{k} u^*_{ii}\right) \prod_{i \geqslant j=1}^{k} du^*_{ij}}{2^{\frac{1}{2}k(n-1)}\pi^{k(k-1)/4}\Gamma\left(\dfrac{n-1}{2}\right)\cdots\Gamma\left(\dfrac{n-k}{2}\right)}.$$

From the theory of positive definite quadratic forms, there exists an orthogonal matrix $\|e_{ij}\|$ of random variables and a set of positive random variables $f_k < \cdots < f_1$ such that

$$u^* = e'fe \tag{18.6.29}$$

where e is the matrix $\|e_{ij}\|$, and f is a diagonal matrix with $f_1, \ldots, f_k$ as its diagonal elements. Also, since e is an orthogonal matrix, we know that $e'e = I$. Thus, if we replace u^* and I in the matrix $\|u^* - lI\|$ by $e'fe$ and $e'e$ respectively, we obtain

$$u^* - lI = e'fe - le'e = e'(f - l)e. \tag{18.6.30}$$

Taking determinants of the matrices in (18.6.30), the roots $l_k < \cdots < l_1$ are seen to be equal to $f_k < \cdots < f_1$ respectively, since $|e| \neq 0$. Therefore, the distribution of the roots $l_1, \ldots, l_k$ is identically the same as the distribution of $f_1, \ldots, f_k$, respectively. To find the distribution of $f_1, \ldots, f_k$, we apply the transformation (18.6.29) to the probability element (18.6.28) and take the marginal distribution with respect to $f_1, \ldots, f_k$. Since e is an orthogonal matrix, the $\{e_{ij}\}$ are functions of $\frac{1}{2}k(k-1)$ parameters which we may denote by g_t, $t = 1, \ldots, \frac{1}{2}k(k-1)$. The Jacobian of the transformation is

$$J = \left| \frac{\partial(u^*_{ij})}{\partial(g_t, f_{i'})} \right| = \overbrace{\left| \begin{matrix} \dfrac{\partial(u^*_{ij})}{\partial(g_t)} \\ \cdots\cdots \\ \dfrac{\partial(u^*_{ij})}{\partial(f_{i'})} \end{matrix} \right|}^{\frac{1}{2}k(k+1) \text{ columns}} \begin{matrix} \left.\vphantom{\dfrac{\partial}{\partial}}\right\} \frac{1}{2}k(k-1) \text{ rows} \\ \\ \left.\vphantom{\dfrac{\partial}{\partial}}\right\} k \text{ rows} \end{matrix} \tag{18.6.31}$$

where $i \geqslant j = 1, \ldots, k$; $i' = 1, \ldots, k$, $t = 1, \ldots, \frac{1}{2}k(k-1)$.
Now any element in the tth row of the upper $\frac{1}{2}k(k-1)$ rows of J is of form $\sum_{j=1}^{k} q_j f_j$ where q_j depends only on the $\{e_{ij}\}$ and their first derivatives with respect to g_t. Furthermore, any element in the lower k rows depends only on the $\{e_{ij}\}$. Hence, it is evident that J is a homogeneous polynomial in $f_1, \ldots, f_k$ of degree $\frac{1}{2}k(k-1)$, whose coefficients depend on the $\{e_{ij}\}$ and their derivatives with respect to the $\{g_t\}$.

The transformation (18.6.29) is not unique if $f_i = f_j$ for any $i \neq j$. Hence, for every $i > j$ we must have $(f_j - f_i)^{a_{ij}}$ as a factor of J where a_{ij} is a positive integer $\geqslant 1$. Therefore, J must be of form $\prod_{i>j=1}^{k} (f_j - f_i)^{a_{ij}} G$. Now since each a_{ij} is a positive integer $\geqslant 1$, since there are $\frac{1}{2}k(k-1)$ factors of form $(f_j - f_i)^{a_{ij}}$ and since J is a homogeneous polynomial in

$f_1, \ldots, f_k$ of degree $\frac{1}{2}k(k-1)$ it follows that every $a_{ij} = 1$ and that G does not depend on the f_i but only on the e_{ij} and their first derivatives with respect to the g_t. Thus, G depends only on the g_t and may be written $G(\{g_t\})$. Therefore we have

$$J = \prod_{i>j=1}^{k} (f_j - f_i) G(\{g_t\}). \tag{18.6.32}$$

Now since $|e_{ij}| = 1$, we have from (18.6.29)

$$|u_{ij}^*| = |e'| \cdot |f| \cdot |e| = f_1 \cdots f_k. \tag{18.6.33}$$

Furthermore,

$$\sum_{i=1}^{k} u_{ii}^* = \sum_{i=1}^{k} f_i. \tag{18.6.34}$$

Making use of the results (18.6.32), (18.6.33), and (18.6.34) we therefore obtain the following result after applying the transformation (18.6.29) to (18.6.28):

(18.6.35)

$$KG(g_t) \prod_{t=1}^{\frac{1}{2}k(k-1)} dg_t \left(\prod_{i=1}^{k} f_i\right)^{\frac{1}{2}(n-k-2)} \prod_{i>j=1}^{k} (f_j - f_i) \exp\left(-\tfrac{1}{2}\sum_{i=1}^{k} f_i\right) \prod_{i=1}^{k} df_i$$

where K is a constant. Since the $\{g_t\}$ and $\{f_i\}$ are independent sets of random variables, the p.e. of the $\{f_i\}$ is

$$C\left\{\left(\prod_{i=1}^{k} f_i\right)^{\frac{1}{2}(n-k-2)} \prod_{i>j=1}^{k} (f_j - f_i) \exp\left(-\tfrac{1}{2}\sum_{i=1}^{k} f_i\right)\right\} \prod_{i=1}^{k} df_i, \tag{18.6.36}$$

where C is a constant to be determined. Denoting the function in $\{\ \}$ by $R(k, n, \{f_i\})$ we may write

$$\frac{1}{C} = \int_{E_k} R(k, n, \{f_i\})\, df_1 \cdots df_k, \tag{18.6.37}$$

where E_k is the region in R_k for which $0 < f_k < f_{k-1} < \cdots < f_1 < +\infty$. Denoting the integral on the right of (18.6.37) by $\varphi_k[\frac{1}{2}(n-k-2)]$ it will be observed that since $|u_{ij}^*|^r = \left(\prod_{i=1}^{k} f_i\right)^r$, we may write the rth moment of $|u_{ij}^*|$ as follows

$$\mathscr{E}(|u_{ij}^*|^r) = \frac{\varphi_k\left(\dfrac{n-k-2}{2} + r\right)}{\varphi_k\left(\dfrac{n-k-2}{2}\right)}. \tag{18.6.38}$$

But since the system $\{u_{ij}^*\}$ has the Wishart distribution $W(k, n-1, \|\delta_{ij}\|)$ whose p.e. is (18.6.28), the value of the rth moment of $|u_{ij}^*|$ is given by (18.2.33) with $\|\sigma_{ij}\| = \|\delta_{ij}\|$ and n replaced by $n-1$. Therefore,

(18.6.39) $$\frac{\varphi_k\left(\dfrac{n-k-2}{2}+r\right)}{\varphi_k\left(\dfrac{n-k-2}{2}\right)} = 2^{rk}\prod_{i=1}^{k}\frac{\Gamma\left(\dfrac{n-i}{2}+r\right)}{\Gamma\left(\dfrac{n-i}{2}\right)}.$$

If we put $r = -\frac{1}{2}(n-k-2)$, it will be seen that

(18.6.40) $$C = \frac{1}{\varphi_k\left(\dfrac{n-k-2}{2}\right)} = \frac{2^{-\frac{1}{2}k(n-k-2)}}{\varphi_k(0)}\prod_{i=1}^{k}\frac{\Gamma\left(\dfrac{k+2-i}{2}\right)}{\Gamma\left(\dfrac{n-i}{2}\right)}.$$

But

(18.6.41) $$\varphi_k(0) = \int_{E_k}\prod_{i>j=1}^{k}(f_j - f_i)\exp\left(-\tfrac{1}{2}\sum_{i=1}^{k} f_i\right) df_1\cdots df_k.$$

Making the transformation $f_k = f'_k, f_{k-1} = f'_{k-1} + f'_k, \ldots, f_1 = f'_1 + f'_k$, we find that

(18.6.42) $$\varphi_k(0) = \frac{2}{k}\,\varphi_{k-1}(1).$$

But we see from (18.6.40), by setting $n = k + 4$, that

(18.6.43) $$\frac{\varphi_k(0)}{\varphi_k(1)} = 2^{-k}\prod_{i=1}^{k}\frac{\Gamma\left(\dfrac{k+2-i}{2}\right)}{\Gamma\left(\dfrac{k+4-i}{2}\right)} = \frac{1}{(k+1)!}.$$

It follows from (18.6.42) and (18.6.43) that

$$\varphi_k(1) = (k+1)!\,\frac{2}{k}\,\varphi_{k-1}(1).$$

Replacing k successively by $k-1, k-2, \ldots, 2$ and noting that $\varphi_1(1) = 4$, we find that

(18.6.44) $$\varphi_k(1) = 2^k(k+1)!\prod_{i=1}^{k}\Gamma(k+1-i),$$

which, when substituted in (18.6.43), gives

(18.6.45) $$\varphi_k(0) = 2^k\prod_{i=1}^{k}\Gamma(k+1-i).$$

But making use of the Legendre duplication formula for gamma functions,

(18.6.46) $$\Gamma(k+1-i) = \frac{2^{k-i}}{\sqrt{\pi}}\,\Gamma\left(\frac{k+1-i}{2}\right)\Gamma\left(\frac{k+2-i}{2}\right)$$

we find upon substituting this in (18.6.45) and then substituting the resulting expression for $\varphi_k(0)$ in (18.6.40) that

$$C = \frac{\pi^{\frac{1}{2}k}}{2^{\frac{1}{2}k(n-1)} \prod_{i=1}^{k} \Gamma\left(\frac{k+1-i}{2}\right)\Gamma\left(\frac{n-i}{2}\right)}. \tag{18.6.47}$$

Putting this value of C in (18.6.36), we finally obtain the distribution of the $f_1, \ldots, f_k$. But it will be recalled that $f_1, \ldots, f_k$ are identically equal to the random variables $l_1, \ldots, l_k$, respectively. Therefore the distribution $l_1, \ldots, l_k$ is given by (18.6.36) simply by relabeling $f_1, \ldots, f_k$ by $l_1, \ldots, l_k$ which yields (18.6.21), thus completing the argument for **18.6.2**.

The distribution of $l_1, \ldots, l_k$ given by (18.6.21) is sometimes called the distribution for the *null case*. The general case would be that in which the covariance matrix in the normal distribution from which the sample comes would be different from the matrix $\|\sigma_{ij}\|$ in (18.6.20). The distribution of the eigenvalues in this more general case has been found by James (1960). It is a considerably more complicated distribution than that given by (18.6.21).

18.7 DISCRIMINANT ANALYSIS

(a) Case of Two Samples

The problem that we shall consider in this section is as follows: Suppose we have two samples from k-dimensional distributions. These can be represented geometrically as two sample clusters in Euclidean k-space. We want to project these two sample clusters orthogonally onto a line so that the variation between the two projected samples is as large as possible, relative to the variation within the two projected samples. The problem is to find the direction of projection which will accomplish this. In other words, what we want to do is to project the two sample clusters back into one dimension so that the two sample clusters after projection are as far apart as possible relative to the within-sample variability. In practical situations if we can find a direction of projecting two k-dimensional sample clusters into one dimension so that the two projected samples are reasonably well separated, whereas they would not be thus separated by projecting into the space of one or some small number of the k-components, we would have a way of discriminating between samples from two distributions by a suitable linear combination of the k-components of the vector on which the measurements are made in the two samples. This problem was originally considered by Fisher (1938) and the method of statistical analysis which was developed from the solution of the problem is called *discriminant analysis*. We shall now consider this problem more precisely.

Suppose $(x_{1\xi_\gamma}^{(\gamma)}, \ldots, x_{k\xi_\gamma}^{(\gamma)}, \xi_\gamma = 1, \ldots, n_\gamma)$, $\gamma = 1, 2$ where $n_1 > k$, $n_2 > k$ are two samples. Let $(\bar{x}_1^{(\gamma)}, \ldots, \bar{x}_k^{(\gamma)})$, $\gamma = 1, 2$, be the vectors of means and $\|u_{ij}^{(\gamma)}\|$, $\gamma = 1, 2$, the internal scatter matrices of the two samples respectively, which are assumed to be nonsingular with probability 1. Let $(\bar{x}_1, \ldots, \bar{x}_k)$ be the vector of sample means, and $\|u_{ij}\|$ the internal scatter matrix of the grand sample composed of the two samples pooled together and $\|u_{ij}^W\| = \|u_{ij}^{(1)} + u_{ij}^{(2)}\|$ the within-samples catter matrix for the two samples. The matrix $\|u_{ij}^B\| = \|u_{ij} - u_{ij}^W\|$ is the between-sample scatter matrix. For an arbitrary vector $(c_1, \ldots, c_k)$ let

$$z_{\xi_\gamma}^{(\gamma)} = \sum_{i=1}^{k} c_i x_{i\xi_\gamma}^{(\gamma)}, \; \xi_\gamma = 1, \ldots, n_\gamma, \; \gamma = 1, 2. \tag{18.7.1}$$

Then $(z_1^{(1)}, \ldots, z_{n_1}^{(1)})$ and $(z_1^{(2)}, \ldots, z_{n_2}^{(2)})$, except for scaling, are one-dimensional samples obtained, respectively, by projecting the original k-dimensional samples onto a line whose direction cosines in the original k-dimensional space are proportional to $(c_1, \ldots, c_k)$. Let $\bar{z}^{(1)}$ and $\bar{z}^{(2)}$ be the means of the two samples of z's and $\bar{z}$ the mean of the pooled samples. Let

$$S_W = \sum_{\gamma=1}^{2} \sum_{\xi_\gamma=1}^{n_\gamma} (z_{\xi_\gamma}^{(\gamma)} - \bar{z}^{(\gamma)})^2, \quad S_B = \sum_{\gamma=1}^{2} n_\gamma (\bar{z}^{(\gamma)} - \bar{z})^2. \tag{18.7.2}$$

It will be noted that if S is the scatter of the grand sample obtained by pooling the two samples of z's, then $S = S_W + S_B$. S_W is the within-sample component and S_B the between-sample component of S. Now the basic problem is to determine $(c_1, \ldots, c_k)$ so as to maximize S_B (that is, to minimize $S_W/(S_W + S_B)$) for a fixed value of S_W.

The basic results concerning this problem may be stated as follows:

18.7.1 *Let $\|u_{ij}^W\|$ and $\|u_{ij}^B\|$ be within-sample and between-sample scatters of two samples from a k-dimensional distribution where the sample sizes both exceed k and where $\|u_{ij}^W\|$ is positive definite with probability* 1. *Let S_B and S_W be defined as in* (18.7.2). *The value of $(c_1, \ldots, c_k)$, say $(c_1', \ldots, c_k')$, which minimizes the ratio*

$$Q = \frac{S_W}{S_W + S_B} \tag{18.7.3}$$

so that S_W has a fixed value $C \neq 0$, is the solution of the equation

$$\sum_{j=1}^{k} (u_{ij}^B - l_1 u_{ij}^W) c_j = 0, \qquad i = 1, \ldots, k, \tag{18.7.4}$$

where l_1 is the nonzero root of the characteristic equation

$$|u_{ij}^B - l u_{ij}^W| = 0 \tag{18.7.5}$$

given by

$$l_1 = \frac{n_1 n_2}{n_1 + n_2} \sum_{i,j=1}^{k} u_W^{ij}(\bar{x}_i^{(1)} - \bar{x}_i^{(2)})(\bar{x}_j^{(1)} - \bar{x}_j^{(2)}) = m^2, \tag{18.7.6}$$

where

$$m = \frac{1}{\sqrt{C}} \sum_{i=1}^{k} c_i' b_i, \quad \|u_W^{ij}\| = \|u_{ij}^W\|^{-1}, \tag{18.7.7}$$

and b_i is given by (18.4.14). *Furthermore, the minimum value of Q for $S_W = C \neq 0$ is $1/(1 + l_1)$.*

To prove **18.7.1** observe that

$$\begin{aligned} S_B &= \sum_{\gamma=1}^{2} n_\gamma \left[\sum_{i=1}^{k} c_i(\bar{x}_i^{(\gamma)} - \bar{x}_i) \right]^2 = \sum_{i,j=1}^{k} u_{ij}^B c_i c_j \\ S_W &= \sum_{\gamma=1}^{2} \sum_{\xi_\gamma=1}^{n} \left[\sum_{i=1}^{k} c_i(x_{i\xi_\gamma}^{(\gamma)} - \bar{x}_i^{(\gamma)}) \right]^2 = \sum_{i,j=1}^{k} u_{ij}^W c_i c_j. \end{aligned} \tag{18.7.8}$$

To minimize $S_W/(S_W + S_B)$ subject to the condition that $S_W = C \neq 0$ is equivalent to (using a Lagrange multiplier l) maximizing $S_B + (C - S_W)l$, which we denote by φ, say, with respect to $(c_1, \ldots, c_k)$ and l. The maximum of φ is given by the solution of the equations

$$\begin{aligned} \frac{\partial \varphi}{\partial c_i} &= 0, \qquad i = 1, \ldots, k \\ \frac{\partial \varphi}{\partial l} &= 0, \end{aligned} \tag{18.7.9}$$

which may be written as

$$\sum_{j=1}^{k} (u_{ij}^B - l u_{ij}^W) c_j = 0, \qquad i = 1, \ldots, k. \tag{18.7.10}$$

To have a solution $(c_1', \ldots, c_k')$ other than $(0, \ldots, 0)$ for (18.7.10) it is necessary that

$$|u_{ij}^B - l u_{ij}^W| = 0. \tag{18.7.11}$$

Recalling from Section 18.4(*b*) that

$$u_{ij}^B = b_i b_j$$

where, as we have seen in (18.4.14),

$$b_i = \sqrt{\frac{n_1 n_2}{n_1 + n_2}} (\bar{x}_i^{(1)} - \bar{x}_i^{(2)}), \tag{18.7.12}$$

we find that (18.7.11) is equivalent to

$$l^k \left| u_{ij}^W - \frac{b_i b_j}{l} \right| = 0$$

which can be written as

$$l^{k-1}|u_{ij}^W| \cdot \left[l - \sum_{i,j=1}^{k} u_W^{ij}\, b_i b_j \right] = 0 \tag{18.7.13}$$

where $\|u_W^{ij}\| = \|u_{ij}^W\|^{-1}$. The nonzero root l_1 of (18.7.11), is therefore given by

$$l_1 = \sum_{i,j=1}^{k} u_W^{ij} b_i b_j. \tag{18.7.14}$$

Therefore, the solution $(c_1', \ldots, c_k')$ of (18.7.10) satisfies

$$\sum_{j=1}^{k} (u_{ij}^B - l_1 u_{ij}^W) c_j' = 0, \qquad i = 1, \ldots, k. \tag{18.7.15}$$

If (18.7.15) is multiplied by c_i' and summed over i, recalling that

$$S_W = \sum_{i,j=1}^{k} u_{ij}^W c_i' c_j' = C,$$

we find

$$\sum_{i,j=1}^{k} u_{ij}^B c_i' c_j' - l_1 C = 0, \tag{18.7.16}$$

for which, as stated in (18.7.6),

$$l_1 = \frac{1}{C} \left(\sum_{i=1}^{k} c_i' b_i \right)^2$$

where b_i is given by (18.7.12).

Substituting c_i' for c_i in (18.7.8) and the resulting expressions for S_B and S_W into the ratio Q, and making use of (18.7.16), we note that the minimum value of Q is $1/(1 + l_1)$, thus completing the argument for **18.7.1**.

The problem of discriminant analysis for the case of two small samples from normal distributions in a large number of dimensions, that is, where n_1 and n_2 are less than k, has been treated by Dempster (1958).

(b) Case of Several Samples

In the problem of two samples, we have shown how to project two k-dimensional samples onto a line so that the means of the two one-dimensional samples of points along this line are as far apart as possible relative to the within-sample scatter of these two one-dimensional samples. In the case of three k-dimensional samples we would want to project these three samples onto a two-dimensional space (ordinary plane) so that the scatter of the pooled two-dimensional samples is as large as possible relative to the within-sample scatter of the three projected samples. In general, if we have $s + 1$ k-dimensional samples $s + 1 \leqslant k$, we want to project these samples onto an s-dimensional Euclidean space so that the scatter of the $s + 1$ pooled s-dimensional samples, resulting from the projection, is as large as possible relative to the within-sample scatter of the $s + 1$ s-dimensional samples.

More precisely, suppose $(x_{1\xi_\gamma}^{(\gamma)}, \ldots, x_{k\xi_\gamma}^{(\gamma)}; \xi_\gamma = 1, \ldots, n_\gamma), \gamma = 1, \ldots, s + 1$, are $s + 1$ k-dimensional samples, $s + 1 \leqslant k$. Let $\|u_{ij}^{(\gamma)}\|$ be the scatter matrices of these samples about their respective means $(\bar{x}_1^{(\gamma)}, \ldots, \bar{x}_k^{(\gamma)})$, $\gamma = 1, \ldots, s + 1$, and let $\|u_{ij}\|$ be the internal scatter matrix of the pooled samples. Let $\|u_{ij}^W\| = \|u_{ij}^{(1)} + \cdots + u_{ij}^{(s+1)}\|$ be the within-sample scatter matrix, assumed to be nonsingular with probability 1, and let $\|u_{ij}^B\| = \|u_{ij} - u_{ij}^W\|$ be the between-sample scatter matrix of the samples. It is to be noted that if $\|u_{ij}^W\|$ is nonsingular with probability 1 so is $\|u_{ij}\|$. Let $z_{\xi_\gamma}^{(\gamma)}$ be defined as in (18.7.1) with $\gamma = 1, \ldots, s + 1$, where $(c_{1p}, \ldots, c_{kp})$, $p = 1, \ldots, s$, are linearly independent vectors. For the γth sample, $\gamma = 1, \ldots, s + 1$, let the coordinates of the projected s-dimensional sample be

$$(18.7.17) \qquad z_{p\xi}^{(\gamma)} = \sum_{i=1}^{k} c_{ip} x_{i\xi}^{(\gamma)}, \qquad p = 1, \ldots, s, \qquad \xi = 1, \ldots, n_\gamma.$$

Let $\|\tilde{u}_{pq}^{(\gamma)}\|$ be the internal scatter matrix of the γth sample of z's, $\gamma = 1, \ldots, s + 1$, and $\|\tilde{u}_{pq}\|$ the internal scatter matrix of the $s + 1$ pooled samples of z's. The within-sample scatter matrix $\|\tilde{u}_{pq}^{(1)} + \cdots + \tilde{u}_{pq}^{(s+1)}\|$ of the $s + 1$ samples of z's will be denoted by $\|\tilde{u}_{pq}^W\|$. The between-sample scatter matrix $\|\tilde{u}_{pq} - \tilde{u}_{pq}^W\|$ of the z's will be denoted by $\|\tilde{u}_{pq}^B\|$. Now, our problem is to find s linearly independent vectors $(c_{1p}, \ldots, c_{kp}), p = 1, \ldots, s$ which will maximize the scatter $|\tilde{u}_{pq}|$ relative to the scatter $|\tilde{u}_{pq}^W|$. This is equivalent to finding a direction of projection of the $s + 1$ samples onto an s-dimensional space R_s so that the ratio of within-sample scatter to the scatter of the pooled samples in R_s is the same as the ratio of within-sample scatter to scatter of pooled samples in the original space R_k. In other words, we want to find vectors $(c_{1p}, \ldots, c_{kp})$, $p = 1, \ldots, s$, so that

$$(18.7.18) \qquad \frac{|\tilde{u}_{pq}^W|}{|\tilde{u}_{pq}|} = \frac{|u_{ij}^W|}{|u_{ij}|}.$$

To find the required vectors we proceed as we did in the two-sample case, and for an arbitrary vector $(c_1, \ldots, c_k)$, we define S_B and S_W as in (18.7.8). We then find the values of this vector for which S_B is stationary, subject to the condition that S_W has a fixed value, $C \neq 0$, say. The values of this vector are given by the equations (18.7.9) where $\varphi = S_B + (C - S_W)l$. The resulting equations are given by (18.7.10), where, of course,

$\|u_{ij}^B\|$ and $\|u_{ij}^W\|$ are the between-sample and within-sample scatter matrices for $s + 1$ samples, rather than for two samples. The condition under which (18.7.10) can be solved for the case of $s + 1$ samples is, of course, the $(s + 1)$-sample version of (18.7.11), that is, we must have

$$|u_{ij}^B - lu_{ij}^W| = 0. \tag{18.7.19}$$

Now if $n_1 + \cdots + n_{s+1} - (s + 1) \geqslant k$ and if $s + 1 \leqslant k$, it can be verified that $\|u_{ij}^B\|$ is of rank s and $\|u_{ij}\|$ is of rank k (with probability 1). Thus, $\|u_{ij}^W\|$ is positive definite, and $\|u_{ij}^B\|$ is positive semidefinite. If we denote these two matrices by u^W and u^B, we know that it follows from the theory of such matrices (see Birkhoff and McLane (1953), for instance) that there exists a real nonsingular matrix $\|e_{ij}\| = e$, and s real numbers. all different with probability 1, namely $0 < l_s < \cdots < l_1 < +\infty$, such that

$$e'u^We = I, \qquad e'u^Be = L \tag{18.7.20}$$

where L is a $k \times k$ diagonal matrix whose diagonal elements are $l_1, \ldots, l_s$, $0, \ldots, 0$. Therefore, if we multiply the matrix $(u^B - lu^W)$ on the left by e' and on the right by e' we obtain

$$(e'u^Be - le'u^We). \tag{18.7.21}$$

Taking determinants of this matrix, we find

$$|e'u^Be - le'u^We| = |L - lI| = l^{k-s}(l_1 - l) \cdots (l_s - l). \tag{18.7.22}$$

But since

$$|e'u^Be - leu^We| = |e'| \cdot |u^B - lu^W| \cdot |e| \tag{18.7.23}$$

and since $|e| = |e'| \neq 0$, it follows that the values of l for which $|u^B - lu^W|$ vanishes are identical with those for which $|e'u^Be - le'u^We|$ vanishes. But we see from (18.7.22) that the nonzero roots (eigenvalues) of the latter are $l_1, \ldots, l_s$. Now let $(c_{1p}, \ldots, c_{kp})$, $p = 1, \ldots,$ s, be the solutions (eigenvectors) of the $(s + 1)$-sample version of (18.7.10); that is, $(c_{1p}, \ldots, c_{kp})$ satisfies the following conditions,

$$\sum_{j=1}^{k}(u_{ij}^B - l_pu_{ij}^W)c_{jp} = 0, \qquad i = 1, \ldots, k \tag{18.7.24}$$

$$\sum_{i,j=1}^{k} u_{ij}^W c_{ip} c_{jp} = C. \tag{18.7.24a}$$

It should be noted that the equation in (18.7.24*a*) is merely $C - S_W = 0$, where S_W is evaluated for $(c_1, \ldots, c_k) = (c_{1p}, \ldots, c_{kp})$.

We shall show that $(c_{1p}, \ldots, c_{kp})$, $p = 1, \ldots, s$, are the required vectors.

If we multiply the equations (18.7.24) by c_{ip} and sum over i, making use of (18.7.24a), we obtain

(18.7.25) $$\sum_{i,j=1}^{k} u_{ij}^{B} c_{ip} c_{jp} = l_p C.$$

If we now multiply (18.7.24) by c_{iq}, $q \neq p$, and sum over i, we obtain

(18.7.26) $$\sum_{i,j=1}^{k} u_{ij}^{B} c_{jp} c_{iq} - l_p \sum_{i,j=1}^{k} u_{ij}^{W} c_{jp} c_{iq} = 0.$$

Similarly,

(18.7.27) $$\sum_{i,j=1}^{k} u_{ij}^{B} c_{jq} c_{ip} - l_q \sum_{i,j=1}^{k} u_{ij}^{W} c_{jq} c_{ip} = 0.$$

Since $l_q \neq l_p$, it follows by taking the difference between (18.7.26) and (18.7.27) that

(18.7.28) $$\sum_{i,j=1}^{k} u_{ij}^{W} c_{jp} c_{iq} = 0.$$

It then follows from either (18.7.26) or (18.7.27) that

(18.7.29) $$\sum_{i,j=1}^{k} u_{ij}^{B} c_{jp} c_{iq} = 0.$$

Now it can be verified that

(18.7.30) $$\tilde{u}_{pq}^{W} = \sum_{i,j=1}^{k} u_{ij}^{W} c_{ip} c_{jq} = \delta_{pq} C$$

and

(18.7.31) $$\tilde{u}_{pq} = \tilde{u}_{pq}^{W} + \tilde{u}_{pq}^{B} = \sum_{i,j=1}^{k} (u_{ij}^{W} + u_{ij}^{B}) c_{ip} c_{jq} = \delta_{pq} C(1 + l_p),$$

where $\delta_{pq} = 1$, if $p = q$ and 0 if $p \neq q$. Therefore, we have

(18.7.32) $$\frac{|\tilde{u}_{pq}^{W}|}{|\tilde{u}_{pq}|} = \frac{1}{(1 + l_1) \cdots (1 + l_s)}.$$

Now consider the ratio of scatters $|u_{ij}^{W}|/|u_{ij}|$ in the original k-dimensional space. It follows from (18.7.20) that since

$$|e'u^{W}e| = |e'| \cdot |u^{W}| \cdot |e| = 1$$

we have $|u_{ij}^{W}| = 1/|e_{ij}|^2$. Since $|u_{ij}| = |u_{ij}^{W} + u_{ij}^{B}|$, it follows from (18.7.20) similarly that $|u_{ij}| = (1 + l_1) \cdots (1 + l_s)/|e_{ij}|^2$. Therefore,

(18.7.33) $$\frac{|u_{ij}^{W}|}{|u_{ij}|} = \frac{1}{(1 + l_1) \cdots (1 + l_s)},$$

and thus the two ratios of scatters on the left-hand sides of (18.7.32) and (18.7.33) are equal, the common value being $[1/(1 + l_1) \cdots (1 + l_s)]$.

We may summarize as follows:

18.7.2 *Let* $(x_{1\xi_\gamma}^{(\gamma)}, \ldots, x_{k\xi_\gamma}^{(\gamma)}; \xi_\gamma = 1, \ldots, n_\gamma)$, $\gamma = 1, \ldots, s + 1$, $n_1 + \cdots + n_{s+1} - (s + 1) \geqslant k$, $s + 1 \leqslant k$, *be* $s + 1$ *independent* k-*dimensional samples whose ratio of within-sample scatter to pooled-sample internal scatter is* $|u_{ij}^W|/|u_{ij}|$ *where* $|u_{ij}^W| \neq 0$ *with probability* 1. *Let these sample points be linearly mapped (projected) into an* s-*dimensional space by* (18.7.17) *and let* $|\tilde{u}_{pq}^W|/|\tilde{u}_{pq}|$ *be the ratio of within-sample scatter to pooled-sample internal scatter of the mapped sample points. The eigenvectors* $(c_{1p}, \ldots, c_{kp})$, $p = 1, \ldots, s < k$ *all subject to condition* (18.7.24*a*), *for which the two scatter ratios* $|u_{ij}^W|/u_{ij}|$ *and* $|\tilde{u}_{pq}^W|/u_{pq}|$ *are equal are the solutions of* (18.7.24) *where the eigenvalues* $0 < l_s < \cdots < l_1 < +\infty$ *are the* s *nonzero roots of* (18.7.19). *The common value of these two ratios is* $1/[(1 + l_1) \cdots (1 + l_s)]$. *The eigenvalues* l_p, $p = 1, \ldots, s$ *are given by* (18.7.25), *and they are all different with probability* 1. *Any two of the eigenvectors* $(c_{1p}, \ldots, c_{kp})$ *and* $(c_{1q}, \ldots, c_{kq})$, $p \neq q$, *satisfy conditions* (18.7.28) *and* (18.7.29).

The main significance of this theorem is that if the s eigenvectors $(c_{1p}, \ldots, c_{kp})$, $p = 1, \ldots, s$, are used in (18.7.17) to map linearly the k-dimensional sample points into an s-dimensional space, the scatter ratio $|\tilde{u}_{pq}^W|/|\tilde{u}_{pq}|$ of the mapped points in R_s is exactly the same as the scatter ratio $|u_{ij}^W|/|u_{ij}|$ in the original k-dimensional space R_k. This is equivalent to projecting the k-dimensional sample points into an s-dimensional Euclidean space so that the pooled-sample scatter $|\tilde{u}_{pq}|$ is as large as possible relative to the within-sample scatter $|u_{pq}^W|$. If we should desire to project the $s + 1$ samples onto a one-dimensional space so as to obtain as large a pooled-sample scatter of these one-dimensional points as possible relative to the within-sample scatter of the points, we would use in (18.7.17) the eigenvector $(c_{11}, \ldots, c_{k1})$ corresponding to the largest eigenvalue l_1. The ratio of within-sample scatter to pooled sample scatter for sample points as mapped onto the one-dimensional line is $1/(1 + l_1)$. Similarly, if we want to project the sample points into a t-dimensional space, $t < s$, we use the eigenvectors $(c_{1p}, \ldots, c_{kp})$, $p = 1, \ldots, t$ corresponding to the eigenvalues $l_1, \ldots, l_t$. The ratio of within-sample scatter to pooled-sample scatter for the sample points in this t-dimensional space is $1/[(1 + l_1) \cdots (1 + l_t)]$.

It should be noted that if we assume that the $s + 1$ samples came from k-dimensional normal distributions all having the same covariance matrix $\|\sigma_{ij}\|$, and if we wish to test the hypothesis that those normal distributions also have identical vectors of means, then the scatter ratio $|u_{ij}^W|/|u_{ij}|$ is equivalent to the Neyman-Pearson likelihood ratio criterion for making this test. Furthermore, if the hypothesis is true this ratio has the same distribution as R_s in (18.5.3) with $m = n_1 + \cdots + n_{s+1} - (s + 1)$.

18.8 DISTRIBUTION OF EIGENVALUES IN DISCRIMINANT ANALYSIS

In the discriminant analysis problem of Section 18.7 the eigenvalues $l_s < \cdots < l_1$, and the corresponding eigenvectors $(c_{1p}, \ldots, c_{kp})$, $p = 1, \ldots, s$ play key roles. The extent to which k-dimensional sample clusters are separated in k-space depends on the magnitudes of the eigenvalues—the larger the eigenvalues the greater the separation. If the samples are random samples from identical k-dimensional normal distributions, the eigenvalues have a distribution that we can determine without great difficulty. In the particular case of two samples from identical k-dimensional normal distributions the ratio of within-sample scatter to total scatter has the value $1/(1 + l_1)$ where l_1 is the only eigenvalue involved. Furthermore, it is seen from Section 18.4(*b*) that $1/(1 + l_1)$ is exactly the same as R_1' defined in (18.4.15), and, of course, has the same distribution as R_1'. It will be further noted that $(n_1 + n_2 - 2)l_1$ is Hotelling's T^2 for the two-sample problem.

In this section we shall consider the sampling distribution of the eigenvalues $l_1, \ldots, l_s$ in the case where the $s + 1$ samples all come from identical k-dimensional normal distributions. There are two important cases to be considered: (i) the case where $s + 1 > k$, that is, where the number of eigenvalues is equal to k, and (ii) the case where $s + 1 \leqslant k$, that is, where the number of eigenvalues is s which is less than k.

(a) The Case of k Eigenvalues

In deriving the basic distribution theory of these eigenvalues, it is convenient to establish first the following general result:

18.8.1 *Suppose $\{v_{ij}\}$ and $\{v_{ij}'\}$ are two independent systems of random variables having Wishart distributions $W(k, n, \|\sigma_{ij}\|)$ and $W(k, n', \|\sigma_{ij}\|)$, respectively, where $n, n' > k_1$. Then the roots $0 < g_k < \cdots < g_1 < +\infty$ of the equation*

(18.8.1)
$$|v_{ij}' - (v_{ij} + v_{ij}')g| = 0$$

have a distribution with probability element

(18.8.2)
$$K\left[\prod_{i=1}^{k}(1 - g_i)\right]^{\frac{1}{2}(n-k-1)}\left[\prod_{i=1}^{k} g_i\right]^{\frac{1}{2}(n'-k-1)} \prod_{i>j=1}^{k}(g_j - g_i)\, dg_1 \cdots dg_k$$

in the region for which $0 < g_k < g_{k-1} < \cdots < g_1 < 1$ *and* 0 *otherwise, where*

$$\mathrm{K} = \pi^{\frac{1}{2}k} \prod_{i=1}^{k} \frac{\Gamma\left(\dfrac{n+n'+1-i}{2}\right)}{\Gamma\left(\dfrac{n+1-i}{2}\right)\Gamma\left(\dfrac{n'+1-i}{2}\right)\Gamma\left(\dfrac{k+1-i}{2}\right)}.$$

It can be shown by argument similar to that following **18.6.2** that the proof of **18.8.1** is equivalent to proving that the roots of

$$(18.8.1a) \qquad |v_{ij}'^{*} - (v_{ij}^{*} + v_{ij}'^{*})g| = 0$$

have probability element (18.8.2) where $\{v_{ij}^{*}\}$ and $\{v_{ij}'^{*}\}$ are independent sets of random variables having Wishart distributions $W(k, n, \|\delta_{ij}\|)$ and $W(k, n', \|\delta_{ij}\|)$. So we proceed as follows.

Since $\|v_{ij}^{*} + v_{ij}'^{*}\|$ and $\|v_{ij}'^{*}\|$ are both positive definite, (with probability one), there exists a real nonsingular matrix $\|e_{ij}\| = e$, say, and a diagonal matrix m with diagonal elements all different with probability 1, namely $1 > m_1 > \cdots > m_k > 0$, so that (denoting $\|v_{ij}^{*} + v_{ij}'^{*}\|$ by $v^{*} + v'^{*}$ and $\|v_{ij}'^{*}\|$ by v'^{*})

$$(18.8.3) \qquad \begin{aligned} v^{*} + v'^{*} &= e'e \\ v'^{*} &= e'me. \end{aligned}$$

But equations (18.8.3) are equivalent to

$$(18.8.3a) \qquad \begin{aligned} v'^{*} &= e'me \\ v^{*} &= e'(I - m)e. \end{aligned}$$

It follows from (18.8.3) that the equation (18.8.1a) can be written as

$$(18.8.4) \qquad |e'me - ge'e| = 0$$

or

$$(18.8.4a) \qquad |e'| \cdot |m - gI| \cdot |e| = 0.$$

Hence the roots $g_1, \ldots, g_k$ are, respectively, identically equal to $m_1, \ldots, m_k$.

Now the probability element of $\{v_{ij}^{*}\}$ and $\{v_{ij}'^{*}\}$ is

$$(18.8.5) \qquad A|v_{ij}^{*}|^{\frac{1}{2}(n-k-1)}|v_{ij}'^{*}|^{\frac{1}{2}(n'-k-1)}e^{-\frac{1}{2}\sum_{i=1}^{k}(v_{ii}^{*}+v_{ii}'^{*})} \prod_{i \geq j=1}^{k} dv_{ij}\, dv_{ij}'^{*}$$

where A is a normalizing constant.

If we apply the transformation (18.8.3a) to (18.8.5), we find the Jacobian J to be of the following form:

$$J = \left| \begin{array}{c:c} \overbrace{\dfrac{\partial(v_{ij}^*)}{\partial(e_{i'j'})}}^{k^2 \text{ columns}} & \overbrace{\dfrac{\partial(v_{ij}^*)}{\partial(m_{i'})}}^{k \text{ columns}} \\ \hdashline \dfrac{\partial(v_{ij}'^*)}{\partial(e_{i'j'})} & \dfrac{\partial(v_{ij}'^*)}{\partial(m_{i'})} \end{array} \right| \begin{array}{l} \left.\vphantom{\dfrac{a}{b}}\right\} \tfrac{1}{2}k(k+1) \text{ rows} \\ \left.\vphantom{\dfrac{a}{b}}\right\} \tfrac{1}{2}k(k+1) \text{ rows.} \end{array} \tag{18.8.6}$$

If we add the bottom $\frac{1}{2}k(k + 1)$ rows to the top $\frac{1}{2}k(k + 1)$ rows, respectively, we obtain a form for J in which the elements in the upper right-hand block are of form $\partial(v_{ij}^* + v_{ij}'^*)/\partial(m_{i'})$ which, in view of (18.8.3), are all 0. The only part of the resulting form of J which involves the m's are the elements in the lower left-hand block, and are all homogeneous linear forms in $m_1, \ldots, m_k$ with coefficients depending only on the e_{ij}. Hence, if the resulting form of J is expanded by Laplace's method (see Bôcher (1907)) with respect to the top $\frac{1}{2}k(k + 1)$ rows, it is evident that every complementary minor picked from the bottom $\frac{1}{2}k(k + 1)$ rows will have $\frac{1}{2}k(k - 1)$ columns selected from the lower left-hand block, and hence every term in the Laplace expression will be a homogeneous polynomial in $m_1, \ldots, m_k$ of degree $\frac{1}{2}k(k - 1)$. If $m_i = m_j$, $i > j$, then the transformation (18.8.3a) is indeterminate and $J = 0$, and hence $(m_j - m_i)^{a_{ij}}$ is a factor of J, where a_{ij} is a positive integer $\geqslant 1$. Therefore, J is of form $G \prod_{i>j=1}^{k} (m_j - m_i)^{a_{ij}}$, where G depends only on the e_{ij}. But the fact that J is a polynomial of degree $\frac{1}{2}k(k - 1)$ implies that each $a_{ij} = 1$, hence

$$J = G \prod_{i>j=1}^{k} (m_j - m_i). \tag{18.8.7}$$

It is seen from (18.8.3) and (18.8.3a) that

$$\begin{aligned} |v_{ij}'^*| &= |e_{ij}|^2 \prod_{i=1}^{k} m_i \\ |v_{ij}^*| &= |e_{ij}|^2 \prod_{i=1}^{k} (1 - m_i) \\ \sum_{i=1}^{k} (v_{ii}^* &+ v_{ii}'^*) = k. \end{aligned} \tag{18.8.8}$$

Therefore, applying the results of (18.8.7) and (18.8.8) to (18.8.5) we see that the $\{e_{ij}\}$ and $\{m_i\}$ are independent sets of random variables and that the probability element of the $\{m_i\}$ is

(18.8.9)

$$K\left\{\left[\prod_{i=1}^{k}(1-m_i)\right]^{\frac{1}{2}(n-k-1)}\left[\prod_{i=1}^{k}m_i\right]^{\frac{1}{2}(n'-k-1)}\prod_{i>j=1}^{k}(m_j-m_i)\right\}dm_1\cdots dm_k$$

over the region E_k for which $0<m_k<\cdots<m_1<1$ where K is a constant to be determined.

Let us denote the integral of the function in { } over E_k by $\varphi_k(\frac{1}{2}(n-k-1), \frac{1}{2}(n'-k-1))$. Then we have

$$\frac{1}{K}=\varphi_k\left(\frac{n-k-1}{2}, \frac{n'-k-1}{2}\right). \tag{18.8.10}$$

Since $\{v_{ij}^*\}$ and $\{v_{ij}'^*\}$ have independent Wishart distributions $W(k, n, \|\delta_{ij}\|)$ and $W(k, n', \|\delta_{ij}\|)$ we may write down immediately from (18.2.33) that

$$\mathscr{E}(|v_{ij}^*|^r|v_{ij}'^*|^{-r})=\prod_{i=1}^{k}\frac{\Gamma\left(\dfrac{n+1-i}{2}+r\right)\Gamma\left(\dfrac{n'+1-i}{2}-r\right)}{\Gamma\left(\dfrac{n+1-i}{2}\right)\Gamma\left(\dfrac{n'+1-i}{2}\right)}. \tag{18.8.11}$$

But it follows from (18.8.8) that

$$\begin{aligned}\mathscr{E}(|v_{ij}^*|^r|v_{ij}'^*|^{-r})&=\mathscr{E}\left(\left[\prod_{i=1}^{k}(1-m_i)\right]^r\left[\prod_{i=1}^{k}m_i\right]^{-r}\right)\\&=\frac{\varphi_k\left(\dfrac{n-k-1}{2}+r, \dfrac{n'-k-1}{2}-r\right)}{\varphi_k\left(\dfrac{n-k-1}{2}, \dfrac{n'-k-1}{2}\right)}.\end{aligned} \tag{18.8.12}$$

Putting $r=-\frac{1}{2}(n-k-1)$ in (18.8.11) and (18.8.12), we obtain

(18.8.13)

$$\frac{\varphi_k\left(0, \dfrac{n+n'-2k-2}{2}\right)}{\varphi_k\left(\dfrac{n-k-1}{2}, \dfrac{n'-k-1}{2}\right)}=\prod_{i=1}^{k}\frac{\Gamma\left(\dfrac{k+2-i}{2}\right)\Gamma\left(\dfrac{n+n'-k-i}{2}\right)}{\Gamma\left(\dfrac{n+1-i}{2}\right)\Gamma\left(\dfrac{n'+1-i}{2}\right)}.$$

Our problem of finding K is now reduced to determining $\varphi_k(0, M)$, where $M=\frac{1}{2}(n+n'-2k-2)$. We note that

$$\varphi_k(0, M)=\int_{E_k}\left(\prod_{i=1}^{k}m_i\right)^M\prod_{i>j=1}^{k}(m_j-m_i)\,dm_1\cdots dm_k. \tag{18.8.14}$$

Let $m_1 = t, m_2 = tm_2', \ldots, m_k = tm_k'$. Then we find

$$\varphi_k(0, M) = \frac{1}{k\left(M + \dfrac{k+1}{2}\right)} \varphi_{k-1}(1, M). \tag{18.8.15}$$

From (18.8.13) we have

$$\frac{\varphi_k(0, M)}{\varphi_k(1, M-1)} = \prod_{i=1}^{k} \frac{\left(M + \dfrac{k-i}{2}\right)}{\left(\dfrac{k+2-i}{2}\right)}. \tag{18.8.16}$$

By solving the system of two-parameter difference equations defined by (18.8.15) and (18.8.16) we obtain

$$\varphi_k\left(0, \frac{n+n'-2k-2}{2}\right) = \pi^{-\frac{1}{2}k} \prod_{i=1}^{k} \frac{\Gamma\left(\dfrac{k+2-i}{2}\right)\Gamma\left(\dfrac{n+n'-k-i}{2}\right)\Gamma\left(\dfrac{k+1-i}{2}\right)}{\Gamma\left(\dfrac{n+n'+1-i}{2}\right)}. \tag{18.8.17}$$

Substituting this in (18.8.13) and solving for $1/\varphi_k(\frac{1}{2}(n-k-1), \frac{1}{2}(n'-k-1))$ which we recall from (18.8.10) is the value of K, we obtain the value of K given in **18.8.1**.

In the discriminant analysis problem we are interested primarily in the eigenvalues of the equation

$$|v_{ij}' - lv_{ij}| = 0 \tag{18.8.18}$$

rather than those of (18.8.1).

But the roots of (18.8.18) are related to those of (18.8.1) as follows:

$$g_1 = \frac{l_1}{1+l_1}, \ldots, g_k = \frac{l_k}{1+l_k}. \tag{18.8.19}$$

Thus, if we apply the transformation (18.8.19) to (18.8.2) we obtain the following result.

18.8.2 *If $\{v_{ij}\}$ and $\{v_{ij}'\}$ are independent sets of random variables having Wishart distributions $W(k, n, \|\sigma_{ij}\|$ and $W(k, n', \|\sigma_{ij}\|)$ then the eigenvectors $0 < l_k < \cdots < l_1$ of $|v_{ij}' - lv_{ij}| = 0$ have the probability element*

(18.8.20)

$$K\left[\prod_{i=1}^{k}(1+l_i)\right]^{-\frac{1}{2}(n+n')}\left[\prod_{i=1}^{k} l_i\right]^{\frac{1}{2}(n'-k-1)} \prod_{i>j=1}^{k}(l_j - l_i)\, dl_1 \cdots dl_k$$

in the region for which $0 < l_k < l_{k-1} < \cdots < l_1 < +\infty$ *and* 0 *otherwise, and where* K *is given in* **18.8.1**.

Returning now to the discriminant analysis problem for $s + 1$ samples where $s + 1 > k$ we have the following corollary of **18.8.2**.

18.8.2a *If* $\|u_{ij}^B\|$ *and* $\|u_{ij}^W\|$ *are the between-sample and within-sample scatter matrices of* $s + 1$ *samples of sizes* $n_1, \ldots, n_{s+1}$, *respectively, where* $s + 1 > k$ *and* $n_1 + \cdots + n_{s+1} - (s + 1) \geqslant k$, *then the eigenvalues* $0 < l_k < \cdots < l_1$ *of*

$$|u_{ij}^B - lu_{ij}^W| = 0 \tag{18.8.21}$$

have probability element

$$K_0\left[\prod_{i=1}^{k}(1 + l_i)\right]^{-\frac{1}{2}(n+s)}\left[\prod_{i=1}^{k} l_i\right]^{\frac{1}{2}(s-k-1)}\prod_{i>j=1}^{k}(l_j - l_i)\, dl_1 \cdots dl_k, \tag{18.8.22}$$

in the region for which $0 < l_k < l_{k-1} < \cdots < l_1 < +\infty$ *and* 0 *otherwise, and where*

$$K_0 = \pi^{\frac{1}{2}k}\prod_{i=1}^{k}\frac{\Gamma\left(\dfrac{n+s+1-i}{2}\right)}{\Gamma\left(\dfrac{n+1-i}{2}\right)\Gamma\left(\dfrac{s+1-i}{2}\right)\Gamma\left(\dfrac{k+1-i}{2}\right)} \tag{18.8.23}$$

where $n = n_1 + \cdots + n_{s+1} - (s + 1)$.

This proof follows at once from **18.8.2** after noting that under the assumptions stated, $\{u_{ij}^B\}$ and $\{u_{ij}^W\}$ are independent sets of random variables having Wishart distributions $W(k, s, \|\sigma_{ij}\|)$ and $W(k, n, \|\sigma_{ij}\|)$ where $n = n_1 + \cdots + n_{s+1} - (s + 1)$.

(b) The Case of s Eigenvalues $(s < k)$

Now, in **18.8.1** suppose $n' < k$; then the elements of $\{v'_{ij}\}$ will have a degenerate Wishart distribution. In fact, $\{v'_{ij}\}$ will have the same distribution as $\left\{\sum_{\xi=1}^{n'} z_{i\xi}z_{j\xi}\right\}$ where $(z_{1\xi}, \ldots, z_{k\xi}, \xi = 1, \ldots, n')$ is a sample of size n' from the k-dimensional normal distribution $N(\{0\}, \|\sigma_{ij}\|)$. Then (18.8.1) has n' eigenvalues $0 < g_{n'} < \cdots < g_1$, and it is evident that the probability element of $g_1, \ldots, g_{n'}$ has the same form as (18.8.2) with k, n, n' replaced by $n', n + n' - k, k$ respectively.

Returning to the discriminant analysis problem for $s + 1$ samples, where $s + 1 \leqslant k$, it will be seen that, in this case, (18.8.21) has eigenvalues $0 < l_s < \cdots < l_1$, $s < k$. Therefore, the probability element of these eigenvalues is given by (18.8.22) with k, n, s replaced by $s, n + s - k$,

k, respectively, remembering, of course, that $n = n_1 + \cdots + n_{s+1} - (s + 1)$.

The results expressed in **18.8.1** and **18.8.2** are companion results to those given in **18.6.2** and were obtained by the various authors referred to in the early part of Section 18.6(b).

18.9 CANONICAL CORRELATION

(a) Determination of Canonical Correlation Coefficients

In this section we shall consider the following problem: Suppose we have a sample from a k-dimensional distribution. We wish to find a linear function of the first s-components and a linear function of the last t-components ($s + t = k$), so that these two linear functions have the highest possible correlation coefficient. In a practical situation the two linear functions may be regarded as indices constructed from the first s and last t variables, respectively, and one of these indices is to be used for predicting or estimating the other. Then it is natural to inquire how these two linear functions should be constructed so that the ordinary correlation coefficient between them is as large as possible.

More precisely, suppose $(x_{i\xi};\, i = 1, \ldots, k;\, \xi = 1, \ldots, n)$, $n > k$, is a sample of size n from a k-dimensional distribution, and let $\|u_{ij}\|$ be the internal scatter matrix of this sample. Let $(x_{p\xi};\, p = 1, \ldots, s;\, \xi = 1, \ldots, n)$ be the first s components of the sample, and let $\|u_{pq}\|$, $p, q = 1, \ldots, s$ be its internal scatter matrix. Similarly let $(x_{v\xi};\, v = s + 1, \ldots, k;\, \xi = 1, \ldots, n)$ be the last t components ($s + t = k$ and $t \geqslant s > 0$) of the sample with internal scatter matrix $\|u_{vw}\|$; $v, w = s + 1, \ldots, k$. We note that

(18.9.1) $$\|u_{ij}\| = \left\|\begin{array}{c|c} u_{pq} & u_{pw} \\ \hline u_{vq} & u_{vw} \end{array}\right\|.$$

We assume that $\|u_{ij}\|$ is nonsingular with probability 1. Now consider real vectors $(c_{1p};\, p = 1, \ldots, s)$ and $(c_{2v};\, v = s + 1, \ldots, k)$ and let

(18.9.2) $$z_{1\xi} = \sum_{p=1}^{s} c_{1p} x_{p\xi}$$
$$z_{2\xi} = \sum_{v=s+1}^{k} c_{2v} x_{v\xi}$$

$\xi = 1, \ldots, n$. Let

$$\left\|\begin{matrix} \tilde{u}_{11} & \tilde{u}_{12} \\ \tilde{u}_{21} & \tilde{u}_{22} \end{matrix}\right\|$$

be the scatter matrix of the sample $(z_{1\xi}, z_{2\xi}, \xi = 1, \ldots, n)$ about its own mean. Our problem is to determine the two vectors $(c_{1p}; p = 1, \ldots, s)$ and $(c_{2v}; v = s + 1, \ldots, k)$ subject to some normalizing condition, which we may take as $\tilde{u}_{11} = \tilde{u}_{22} = 1$ without loss of generality, so that the correlation coefficient

$$R = \frac{\tilde{u}_{12}}{\sqrt{\tilde{u}_{11}\tilde{u}_{22}}} \tag{18.9.3}$$

is a maximum.

The solution of this problem is due to Hotelling (1935) and can be summarized in the following statement:

18.9.1 *The eigenvectors* $(c_{1p}^{(1)}; p = 1, \ldots, s)$ *and* $(c_{2v}^{(1)}; v = s + 1, \ldots, k)$ *which maximize* R *subject to the conditions* $\tilde{u}_{11} = \tilde{u}_{22} = 1$ *are the solutions of the equations*

$$\begin{aligned} -\sqrt{l_1}\sum_{q=1}^{s} u_{pq}c_{1q} + \sum_{w=s+1}^{k} u_{pw}c_{2w} &= 0, \qquad p = 1, \ldots, s \\ \sum_{q=1}^{s} u_{vq}c_{1q} - \sqrt{l_1}\sum_{w=s+1}^{k} u_{vw}c_{2w} &= 0, \qquad v = s + 1, \ldots, k \end{aligned} \tag{18.9.4}$$

where l_1 *is the largest eigenvalue of*

$$\begin{vmatrix} lu_{pq} & \cdots & -u_{pw} \\ \cdot & \cdots & \cdot \\ -u_{vq} & \cdots & u_{vw} \end{vmatrix} = 0. \tag{18.9.5}$$

Furthermore, the maximum value of R *is* $\sqrt{l_1}$.

More generally, (18.9.5) *has* s *eigenvalues* $l_1, \ldots, l_s$ *on the interval* (0, 1). *We assume that the sample* $(x_{i\xi}; i = 1, \ldots, k; \xi = 1, \ldots, n)$ *comes from a distribution such that these roots are all different with probability* 1, *in which case they may be labeled as follows:* $0 < l_s < \cdots < l_1 < 1$. *The eigenvectors* $(c_{1p}^{(g)}, p = 1, \ldots, s)$ *and* $(c_{2v}^{(g)}, v = s + 1, \ldots, k)$ *are solutions of equations* (18.9.4) *with* l_1 *replaced by* $l_g, g = 1, \ldots, s$. *The value of* R *when computed from these eigenvectors is* $\sqrt{l_g}$. *The value of* R *computed by using the eigenvectors* $(c_{1p}^{(g)}; p = 1, \ldots, s)$ *and* $(c_{2v}^{(g')}; v = s + 1, \ldots, k)$, $g \neq g'$, *is zero.*

The correlation coefficients $R^{(1)}, \ldots, R^{(s)}$, which have the values $\sqrt{l_1}, \ldots, \sqrt{l_s}$, respectively, are called the *canonical correlation coefficients* between the first s components $(x_{1\xi}, \ldots, x_{s\xi}; \xi = 1, \ldots, n)$ and the last t components $(x_{s+1\xi}, \ldots, x_{k\xi}; \xi = 1, \ldots, n)$ of the original sample $(x_{1\xi}, \ldots, x_{k\xi};\ \xi = 1, \ldots, n)$. The canonical correlation coefficient of greatest practical interest is, of course, the largest one, namely $R^{(1)}$.

To prove **18.9.1** we proceed as follows. It can be verified that

$$\begin{aligned} \tilde{u}_{11} &= \sum_{p,q} u_{pq} c_{1p} c_{1q} \\ \tilde{u}_{22} &= \sum_{v,w} u_{vw} c_{2v} c_{2w} \\ \tilde{u}_{12} &= \sum_{p,w} u_{pw} c_{1p} c_{2w} \end{aligned} \tag{18.9.6}$$

where p, q range over $1, \ldots, s$ and v, w range over $s + 1, \ldots, k$. Now consider the vectors $(c_{1p}; p = 1, \ldots, s)$ and $(c_{2v}; v = s + 1, \ldots, k)$ which will make R stationary subject to the conditions $\tilde{u}_{11} = \tilde{u}_{22} = 1$. Using Lagrange multipliers λ and μ, the same vectors will also make

$$\varphi = \tilde{u}_{12} + (1 - \tilde{u}_{11})\lambda + (1 - \tilde{u}_{22})\mu \tag{18.9.7}$$

stationary. The vectors which yield extrema of φ are given by solutions of the equations

$$\begin{aligned} \frac{\partial \varphi}{\partial c_{1p}} &= 0, \qquad p = 1, \ldots, s \\ \frac{\partial \varphi}{\partial c_{2v}} &= 0, \qquad v = s + 1, \ldots, k, \end{aligned} \tag{18.9.8}$$

that is, the equations

$$\begin{aligned} -\lambda \sum_q u_{pq} c_{1q} + \sum_w u_{pw} c_{2w} &= 0, \qquad p = 1, \ldots, s \\ \sum_q u_{vq} c_{1q} - \mu \sum_w u_{vw} c_{2w} &= 0, \qquad v = s + 1, \ldots, k. \end{aligned} \tag{18.9.9}$$

Multiplying the first equation in (18.9.9) by c_{1p} and summing over p and the second by c_{2v} and summing over v it is evident that $\lambda = \mu$. Thus, replacing μ by λ in (18.9.9) we must solve for the required vectors. But to obtain such solutions we must, of course, have

$$\left| \begin{array}{c|c} -\lambda u_{pq} & u_{pw} \\ \hline u_{vq} & -\lambda u_{vw} \end{array} \right| = 0. \tag{18.9.10}$$

We can factor the determinant so that (18.9.10) reads as follows:

$$(-1)^k \lambda^{k-2s} \left| \begin{array}{c|c} l u_{pq} & -u_{pw} \\ \hline -u_{vq} & u_{vw} \end{array} \right| = 0 \tag{18.9.11}$$

where $l = \lambda^2$. It is evident that the determinant in (18.9.11), that is, the determinant (18.9.5), is a polynomial in l of degree s, and it can be shown that its roots are real and positive. Let its roots be $l_1, \ldots, l_s$. The eigenvectors $(c_{1p}^{(g)}; p = 1, \ldots, s)$ and $(c_{2v}^{(g)}; v = s+1, \ldots, k)$ corresponding to l_g are the solution of

$$-\sqrt{l_g}\sum_q u_{pq}c_{1q} + \sum_w u_{pw}c_{2w} = 0, \qquad p = 1, \ldots, s$$

(18.9.12) $$\sum_q u_{vq}c_{1q} - \sqrt{l_g}\sum_w u_{vw}c_{2w} = 0, \qquad v = s+1, \ldots, k.$$

If we insert these eigenvectors in (18.9.12), multiply the first equation by $c_{1p}^{(g)}$ and sum over p, noting that $\sum_{p,q} u_{pq}c_{1p}^{(g)}c_{1q}^{(g)} = 1$, we obtain

(18.9.13) $$\sum_{p,w} u_{pw}c_{2w}^{(g)}c_{1p}^{(g)} = \sqrt{l_g}.$$

But due to the conditions $\tilde{u}_{11} = \tilde{u}_{22} = 1$, the left-hand side of (18.9.13) is the value of the correlation coefficient R between z_1 and z_2 in (18.9.2) using the two eigenvectors given by (18.9.12). Hence, $l_g < 1, g = 1, \ldots, s$ and the roots $l_1, \ldots, l_s$, which are assumed to be all different with probability 1, may be ordered $l_s < \cdots < l_1$, lie on (0, 1). Since l_1 is the largest root, then the eigenvectors which correspond to l_1 are the solutions of (18.9.4). It can be shown that for any sample size $n > k$, $l_1 = 1$ with probability 1 if and only if the first s and last t $(s + t = k)$ random variables in the k-dimensional distribution from which the sample is drawn are linearly dependent.

To establish the fact that the correlation coefficient between z_1 and z_2 is zero for any two different eigenvectors $(c_{1p}^{(g)}; p = 1, \ldots, s)$ and $(c_{2v}^{(g')}; v = s+1, \ldots, k)$, $g \neq g'$, we consider the equations (18.9.12) with the eigenvectors $(c_{1p}^{(g)}; p = 1, \ldots, s)$, $(c_{2v}^{(g)}; v = s+1, \ldots, k)$ inserted. Multiply the first and second equations by $c_{1p}^{(g)}$ and $c_{2v}^{(g')}$, respectively, and sum over p and v. Then consider the corresponding equations and operations with g and g' interchanged. Since $l_g \neq l_{g'}$ the reader will find by suitably combining equations, that $\sum_{p,w} u_{pw}c_{1p}^{(g)}c_{2w}^{(g')} = 0$. That is, the correlation coefficient R obtained by using $(c_{1p}^{(g)}; p = 1, \ldots, s)$ and $(c_{2v}^{(g')}; v = s + 1, \ldots, k)$ vanishes if $g \neq g'$. He will similarly find that for $g \neq g'$

$$\sum_{p,q} u_{pq}c_{1p}^{(g)}c_{1q}^{(g')} = \sum_{v,w} u_{vw}c_{2v}^{(g)}c_{2w}^{(g')} = 0.$$

(b) Sampling Theory of Canonical Correlation Coefficients

The problem of determining the sampling distribution of the squared canonical correlation coefficients $l_1, \ldots, l_s$ under general conditions is very complicated. However, under certain conditions the distribution of $l_1, \ldots, l_s$ is a special case of the distribution given in (18.8.2). More precisely we have the following result:

18.9.2 *Let* $(x_{p\xi}; p = 1, \ldots, s; \xi = 1, \ldots, n)$ *and* $(x_{v\xi}; v = s + 1, \ldots, k; \xi = 1, \ldots, n)$ *be independent samples, the first being from the s-dimensional normal distribution* $N(\{\mu_p\}; \|\sigma_{pq}\|)$ *and the second from an arbitrary t-dimensional distribution* $s \leqslant t, s + t = k$ *such that its internal scatter matrix* $\|u_{vw}\|$ *is nonsingular with probability* 1. *Then the roots* $0 < l_s < \cdots < l_1 < 1$ *of* (18.9.5) *have a distribution with probability element*

(18.9.14)
$$K^*\left[\prod_{p=1}^{s}(1 - l_p)\right]^{\frac{1}{2}(n-t-s-2)}\left[\prod_{p=1}^{s} l_p\right]^{\frac{1}{2}(t-s-1)}\prod_{p>q=1}^{s}(l_q - l_p)\, dl_1 \cdots dl_s$$

in the region for which $0 < l_s < l_{s-1} < \cdots < l_1 < 1$ *and* 0 *otherwise, and where*

$$K^* = \pi^{\frac{1}{2}s}\prod_{p=1}^{s}\frac{\Gamma\left(\frac{n-p}{2}\right)}{\Gamma\left(\frac{n-t-p}{2}\right)\Gamma\left(\frac{t+1-p}{2}\right)\Gamma\left(\frac{s+1-p}{2}\right)}. \tag{18.9.15}$$

In proving **18.9.2** it is sufficient to consider the case where $(x_{v\xi}; v = s + 1, \ldots, k; \xi = 1, \ldots, n)$ are arbitrary numbers and are linearly independent (that is, $\|u_{vw}\|$ is nonsingular). We will show that in this case $l_1, \ldots, l_s$ has distribution (18.9.14). Since (18.9.14) holds for any linearly independent set of numbers, it clearly holds if $\{x_{v\xi}\}$ is a sample from an arbitrary t-dimensional distribution provided $\|u_{vw}\|$ is nonsingular with probability 1.

To proceed with the proof of **18.9.2** let us first multiply each side of the equation in (18.9.5) on the left by the determinant

$$\begin{vmatrix} \delta_{pq} & \sum_v u_{pv}u^{vw} \\ 0 & u^{vw} \end{vmatrix}.$$

Using ordinary matrix multiplication, where $\|u^{vw}\| = \|u_{vw}\|^{-1}$ and δ_{pq} is the Kronecker δ, we obtain

$$\begin{vmatrix} lu_{pq} - \sum_{v,w} u^{vw}u_{pv}u_{wq} & -u_{pw} + \sum_{v,w'} u^{vw'}u_{pv}u_{w'w} \\ -\sum_v u_{vq}u^{vw} & \sum_{w'} u^{vw'}u_{w'w} \end{vmatrix} = 0 \tag{18.9.16}$$

where p,q ranges over $1, \ldots, s$ and v, w, w' ranges over $s + 1, \ldots, k$. Since $\sum_{w'} u^{vw'}u_{w'w} = \delta_{vw}$, it is evident that each element of the upper right-hand block of (18.9.16) is 0. Therefore the roots of (18.9.5) are identical with those of

$$\left|\sum_{v,w} u^{vw}u_{pv}u_{wq} - lu_{pq}\right| = 0. \tag{18.9.17}$$

Now let us denote $(x_{p\xi} - \bar{x}_p)$ by $y_{p\xi}$ and $(x_{v\xi} - \bar{x}_v)$ by $y_{v\xi}$. Then

$$u_{pq} = \sum_{\xi=1}^{n} y_{p\xi}y_{q\xi} = u_{pq}^{(1)} + u_{pq}^{(2)} \tag{18.9.18}$$

where

$$u_{pq}^{(1)} = \sum_{\xi=1}^{n}\left(y_{p\xi} - \sum_{w} b_{pw}y_{w\xi}\right)\left(y_{q\xi} - \sum_{v} b_{qv}y_{v\xi}\right)$$

$$u_{pq}^{(2)} = \sum_{v,w=s+1}^{k} u^{vw}b_{pw}b_{qv}$$

$$b_{pw} = \sum_{v=s+1}^{k} u^{vw}u_{pw}.$$

Substituting the expression for b_{pw} into the expression for $u_{pq}^{(2)}$ we find that $u_{pq}^{(2)} = \sum_{v,w} u^{vw}u_{pv}u_{wq}$. Therefore (18.9.17) can be written as

$$|u_{pq}^{(2)} - (u_{pq}^{(1)} + u_{pq}^{(2)})l| = 0. \tag{18.9.19}$$

It can be verified that under the assumptions of **18.9.2** the two sets of random variables $\{u_{pq}^{(1)}\}$ and $\{u_{pq}^{(2)}\}$ are independent sets of random variables having Wishart distributions $W(s, n - t - 1, \|\sigma_{pq}\|)$ and $W(s, t, \|\sigma_{pq}\|)$. It follows from **18.8.1**, therefore, that the roots of (18.9.18) have the distribution given in (18.8.2) with k replaced by s, n' replaced by t, and n replaced by $n - t - 1$. This resulting distribution is (18.9.14), thus completing the proof of **18.9.2.**

PROBLEMS

18.1 If $(x_{1\xi}, \ldots, x_{k\xi}, \xi = 1, \ldots, n)$ is a sample of size $n(n > k)$ from the normal distribution $N(\{\mu_i\}; \|\sigma_{ij}\|)$, show that the maximum likelihood estimator of $(\mu_1, \ldots, \mu_k)$ is $(\bar{x}_1, \ldots, \bar{x}_k)$ and of $\|\sigma_{ij}\|$ is $\|u_{ij}/n\|$ where $(\bar{x}_1, \ldots, \bar{x}_k)$ is the vector of sample means, and $\|u_{ij}\|$ is the internal scatter matrix of the sample.

18.2 If the elements of the symmetric matrix $\|a_{ij}\|$, $i, j = 1, \ldots, k < m$ are random variables having the Wishart distribution $W(k, m, \|\sigma_{ij}\|)$ show by the use of characteristic functions that the elements of $\|a_{pq}\|$, $p, q = 1, \ldots, s < k$ have the Wishart distribution $W(s, m, \|\sigma_{pq}\|)$.

18.3 *Distribution of correlation coefficients in sample from k-dimensional normal distribution having independent components.* If a sample $(x_{1\xi}, \ldots, x_{k\xi}$, $\xi = 1, \ldots, n)$, $n > k$, comes from the normal distribution $N(\{\mu_i\}; \|\sigma_{ij}\delta_{ij}\|)$

and if $\|u_{ij}\|$ is the internal scatter matrix of the sample, show that the p.d.f. of the elements of the correlation matrix $\|r_{ij}\|$ where $r_{ij} = \dfrac{u_{ij}}{\sqrt{u_{ii}u_{jj}}}$ is given by

$$\frac{\Gamma^k\left(\dfrac{n-1}{2}\right)|r_{ij}|^{\frac{1}{2}(n-k-2)}}{\pi^{k(k-1)/4}\prod_{i=1}^{k}\Gamma\left(\dfrac{n-i}{2}\right)}$$

over the part of the space of the $\{r_{ij}\}$ for which $\|r_{ij}\|$ is positive definite, and 0 otherwise.

18.4 (*Continuation*) Show that

$$\mathscr{E}(|r_{ij}|^g) = \prod_{i=2}^{k}\left[\frac{\Gamma\left(\dfrac{n-i}{2}+g\right)\Gamma\left(\dfrac{n-1}{2}\right)}{\Gamma\left(\dfrac{n-i}{2}\right)\Gamma\left(\dfrac{n-1}{2}+g\right)}\right] \qquad g = 0, 1, 2, \ldots$$

and hence that the distribution of $|r_{ij}|$ is identical with the distribution of $\prod_{i=2}^{k} z_i$, where the z_i are independent random variables having beta distributions $Be(\frac{1}{2}(n-i), \frac{1}{2}(i-1)), i = 2, \ldots, k$ respectively.

18.5 *Distribution of sample correlation coefficient.* If $\|u_{ij}\|$ is the internal scatter matrix of a sample of size n from the two-dimensional distribution $N(\{\mu_i\}; \|\sigma_{ij}\|), i, j = 1, 2,$ and making use of the fact that the elements of $\|u_{ij}\|$ have the Wishart distribution $W(2, n-1, \|\sigma_{ij}\|)$, show that the p.d.f. of the sample correlation coefficient $r = u_{12}/\sqrt{u_{11}u_{22}}$ can be expressed in the form

$$\frac{(1-\rho)^{\frac{1}{2}(n-1)}(1-r^2)^{\frac{1}{2}(n-4)}}{\sqrt{\pi}\,\Gamma\left(\dfrac{n-1}{2}\right)\Gamma\left(\dfrac{n-2}{2}\right)}\sum_{i=0}^{\infty}\frac{(2\rho r)^i}{i!}\Gamma^2\left(\frac{n-1+i}{2}\right)$$

for $-1 \leqslant r \leqslant +1$ and 0 otherwise, where $\rho = \sigma_{12}/\sqrt{\sigma_{11}\sigma_{22}}$, the correlation coefficient in the population. [Fisher (1915).]

18.6 (*Continuation*) Applying the transformation

$$r = u_{12}/\sqrt{u_{11}u_{22}}, \quad s = \sqrt{u_{11}u_{22}}, \quad t = \tfrac{1}{2}\log(u_{22}/u_{11}),$$

to the p.e. of the Wishart distribution $W(2, n-1, \|\sigma_{ij}\|)$ show that the p.d.f. of r can also be expressed in the form (due to Hotelling (1953)):

$$\frac{(n-2)\Gamma(n-1)(1-\rho^2)^{\frac{1}{2}(n-1)}(1-r^2)^{\frac{1}{2}(n-4)}(1-\rho r)^{-\frac{1}{2}(2n-3)}}{\sqrt{2}\pi^{\frac{3}{2}}} \cdot \sum_{i=0}^{\infty}\frac{(\frac{1}{2}+\frac{1}{2}\rho r)^i\Gamma^2(\frac{1}{2}+i)}{i!\,\Gamma(n-\frac{1}{2}+i)}.$$

18.7 (*Continuation*) Show that the Hotelling form of the probability element of r can be written in the form

$$\frac{(n-2)\Gamma(n-1)(1-\rho^2)^{\frac{1}{2}(n-1)}(1-r^2)^{\frac{1}{2}(n-4)}}{\sqrt{2\pi}\,\Gamma(n-\frac{1}{2})(1-\rho r)^{\frac{1}{2}(2n-3)}} \cdot \left\{\int_0^1\int_0^1 f(x)g(y)[1-\tfrac{1}{2}(1+\rho r)xy]^{-1}\,dy\,dx\right\} dr$$

where $f(x)$ and $g(y)$ are the p.d.f.'s of the beta distributions $Be(\frac{1}{2}, n-1)$ and $Be(\frac{1}{2}, \frac{1}{2})$ respectively. Hence, by making the transformation

$$r = \rho + \frac{1-\rho^2}{\sqrt{n}}u,\ x = \frac{v}{n},\ y = w$$

show that the limiting distribution of

$$u = \frac{(r-\rho)\sqrt{n}}{1-\rho^2} \qquad \text{as } n \to \infty \text{ is } N(0, 1).$$

18.8 (*Continuation*) Show, by making use of **9.3.1**, that if

$$z = \frac{1}{2}\log\left(\frac{1+r}{1-r}\right)$$

and

$$\zeta = \frac{1}{2}\log\left(\frac{1+\rho}{1-\rho}\right)$$

the limiting distribution of $(z-\zeta)\sqrt{n}$ as $n \to \infty$ is $N(0, 1)$.

18.9 *Confidence ellipsoids for vector of means of a normal distribution.* If $(\bar{x}_1, \ldots, \bar{x}_k)$ is the vector of means and $\|u_{ij}\|$ the internal scatter matrix of a sample of size n, $n > k$, from the k-dimensional distribution $N(\{\mu_i\}; \|\sigma_{ij}\|)$, show that

$$n\sum_{i,j=1}^{k} u^{ij}(\mu_i - \bar{x}_i)(\mu_j - \bar{x}_j) < (1 - z_\gamma)/z_\gamma$$

is a $100\gamma\%$ confidence region (ellipsoid) for the vector of population means $(\mu_1, \ldots, \mu_k)$ where z_γ is the upper $100\gamma\%$ point of the beta distribution $Be(\frac{1}{2}(n-k), \frac{1}{2}k)$. [Hotelling (1931).]

18.10 *Equivalence of Hotelling's T^2 with likelihood ratio test that a normal distribution $N(\{\mu_i\}, \|\sigma_{ij}\|)$ has a given vector of means.* Let $\mathscr{H}(\omega; \Omega)$ be the statistical hypothesis in which Ω is the $\frac{1}{2}k(k+3)$ dimensional parameter space for which $\mu_1, \ldots, \mu_k$ are real and $\|\sigma_{ij}\|$ is positive definite and ω is the subset of Ω for which $\mu_1 = \mathring{\mu}_1, \ldots, \mu_k = \mathring{\mu}_k$. Given a sample of size n from $N(\{\mu_i\}; \|\sigma_{ij}\|)$ with vector of means $(\bar{x}_1, \ldots, \bar{x}_k)$ and internal scatter matrix $\|u_{ij}\|$ show that the likelihood ratio λ for testing $\mathscr{H}$ is

$$\lambda = R_1^{\frac{1}{2}n}$$

where R_1 is given by (18.4.1). Hence R_1 (as well as Hotelling's T^2) is equivalent to λ for testing $\mathscr{H}$, and its distribution when $\mathscr{H}$ is true is the beta distribution $Be(\frac{1}{2}(n-k), \frac{1}{2}k)$.

18.11 (*Continuation*). *Equivalence of Hotelling's* T^2 *and Mahalanobis'* D^2 *with likelihood ratio test that two normal distributions with equal covariance matrices also have equal vectors of means.*

Let $\|u_{ij}^{(1)}\|$ and $\|u_{ij}^{(2)}\|$ be the internal scatter matrices of samples of sizes n_1 and n_2 respectively, from $N(\{\mu_i^{(1)}\}; \|\sigma_{ij}^{(1)}\|)$ and $N(\{\mu_i^{(2)}\}; \|\sigma_{ij}^{(2)}\|)$, $n_1, n_2 > k$, $i, j = 1, \ldots, k$.

Let $\|u_{ij}\|$ be the internal scatter matrix of the two samples pooled together into a single sample. Let $\mathscr{H}(\omega; \Omega)$ be the statistical hypothesis in which Ω is the $\frac{1}{2}k(k + 5)$-dimensional space of the parameters $\{\mu_i^{(1)}\}, \{\mu_i^{(2)}\}, \|\sigma_{ij}\|$, where $\|\sigma_{ij}^{(1)}\| = \|\sigma_{ij}^{(2)}\| = \|\sigma_{ji}\|$. Let ω be the $\frac{1}{2}k(k + 3)$-dimensional subspace in Ω for which $\mu_i^{(1)} = \mu_i^{(2)}$ $i = 1, \ldots, k$. Show that the likelihood ratio λ for testing $\mathscr{H}$ is

$$\lambda = (R_1')^{\frac{1}{2}(n_1+n_2)}.$$

Hence R_1' (as well as Hotelling's T^2 and Mahalanobis' D^2 for two samples) is equivalent to λ for testing $\mathscr{H}$ and its distribution when $\mathscr{H}$ is true in the beta distribution $Be(\frac{1}{2}(n_1 + n_2 - k - 1), \frac{1}{2}k)$.

18.12 Generalize Problem 18.11 to the case of $s + 1$ samples of sizes $n_1, \ldots, n_{s+1}$ and show that the likelihood ratio λ for testing the hypothesis $\mathscr{H}$ that all samples come from normal distributions having equal vectors of means given that they come from normal distributions having equal covariance matrices is equivalent to R_s given in (18.5.3) with $\|a_{ij}\|$ as the sum of the internal scatter matrices of the $s + 1$ samples $\|a_{ij} + \sum_{\beta=1}^{s} b_{i\beta}b_{j\beta}\|$ as the internal scatter of all samples pooled together as a single sample and $m = n_1 + \cdots + n_{s+1} - s - 1$. Hence show that the distribution of R_s in this case if $\mathscr{H}$ is true is given by **18.5.1** with the value of m just given.

18.13 *Testing the hypothesis that the covariance matrices of two normal distributions are equal.*

In Problem 18.11 consider the hypothesis $\mathscr{H}(\omega; \Omega)$ where Ω remains unchanged but where ω is the subspace in Ω for which $\|\sigma_{ij}^{(1)}\| = \|\sigma_{ij}^{(2)}\|$. Show that the likelihood ratio λ for $\mathscr{H}$ is

$$\lambda = \left|\frac{u_{ij}^{(1)}}{n_1}\right|^{\frac{1}{2}n_1} \left|\frac{u_{ij}^{(2)}}{n_2}\right|^{\frac{1}{2}n_2} \Big/ \left|\frac{u_{ij}^{(1)} + u_{ij}^{(2)}}{n}\right|^{\frac{1}{2}n}$$

where $n = n_1 + n_2$, and that if $\mathscr{H}$ is true

$$\mathscr{E}(\lambda^r) = \left[\left(\frac{n}{n_1}\right)^{n_1}\left(\frac{n}{n_2}\right)^{n_2}\right]^{\frac{1}{2}kr} \cdot \prod_{i=1}^{k} \frac{\Gamma\left(\frac{n-1-i}{2}\right)\Gamma\left(\frac{n_1 - i + n_1 r}{2}\right)\Gamma\left(\frac{n_2 - i + n_2 r}{2}\right)}{\Gamma\left(\frac{n-1-i+nr}{2}\right)\Gamma\left(\frac{n_1 - i}{2}\right)\Gamma\left(\frac{n_2 - i}{2}\right)} \Bigg],$$

$r = 0, 1, 2, \cdots$.

18.14 *Model I analysis of variance for vectors in two-factor experimental design.*

In an experimental layout of r rows $R_1, \ldots, R_r$ and s columns $C_1, \ldots, C_s$ suppose we have a k-dimensional vector random variable $(x_{1\xi\eta}, \ldots, x_{k\xi\eta})$ associated with the cell formed by the ξth row and ηth column, $\xi = 1, \ldots, r$,

$\eta = 1, \ldots, s$. Suppose these rs vector random variables are independent, such that $(x_{1\xi\eta}, \ldots, x_{k\xi\eta})$ has the k-dimensional normal distribution

$$N(\{\mu_i + \mu_{i\xi\cdot} + \mu_{i\cdot\eta}\}, \|\sigma_{ij}\|)$$

where $\sum_{\xi} \mu_{i\xi\cdot} = \sum_{\eta} \mu_{i\cdot\eta} = 0$, $i = 1, \ldots, k$. Now let $\mathscr{H}(\omega; \Omega)$ be the statistical hypothesis in which Ω is the space of the $k(r + s - 1) + \frac{1}{2}k(k + 1)$ independent parameters involved in the rs distributions mentioned above, and ω is the $(ks + \frac{1}{2}k(k + 1))$-dimensional subspace of Ω for which the "row effects" are all zero, that is,

$$\mu_{i1\cdot} = \mu_{i2\cdot} = \cdots = \mu_{ir\cdot} = 0, \qquad i = 1, \ldots, k.$$

Let the quantities $\bar{x}_{\cdot\cdot}$, $\bar{x}_{\xi\cdot}$, $\bar{x}_{\cdot\eta}$, m, $m_{\xi\cdot}$, $m_{\cdot\eta}$ defined in (10.6.3), when computed from $x_{i\xi\eta}$, $\xi = 1, \ldots, r$, $\eta = 1, \ldots, s$, be designated by $\bar{x}_{i\cdot\cdot}$, $\bar{x}_{i\xi\cdot}$, $\bar{x}_{i\cdot\eta}$, m_i, $m_{i\xi\cdot}$, $m_{i\cdot\eta}$, respectively. Let

$$S_{ij\cdot\cdot} = \sum_{\xi,\eta} (x_{i\xi\eta} - m_i - m_{i\xi\cdot} - m_{i\cdot\eta})(x_{j\xi\eta} - m_j - m_{j\xi\cdot} - m_{j\cdot\eta})$$

$$S_{ij\cdot 0} = \sum_{\xi,\eta} m_{i\xi\cdot} m_{j\xi\cdot}$$

Show that the likelihood ratio λ for testing $\mathscr{H}$ is

$$\lambda = L^{\frac{1}{2}rs}$$

where

$$L = \frac{|S_{ij\cdot\cdot}|}{|S_{ij\cdot\cdot} + S_{ij\cdot 0}|}$$

and that if $\mathscr{H}$ is true (that is, if $\mu_{i1\cdot} = \cdots = \mu_{ir\cdot} = 0$, $i = 1, \ldots, k$) the distribution of L is identical with that of R_s in **18.5.1** with m and s replaced by $(r - 1)(s - 1)$ and $(r - 1)$ respectively.

18.15 *Distribution of sum of squares of least squares residuals.* Suppose $(x_{1\xi}, \ldots, x_{k\xi}, \xi = i, \ldots, n)$, where $n > k$, is a sample from the k-dimensional distribution $N(\{\mu_i\}; \|\sigma_{ij}\|)$ and let $\|u_{ij}\|$ be the internal scatter matrix of this sample. For a fixed sample show that the minimum of $\sum_{\xi=1}^{u} [x_{1\xi} - \beta_1 - \beta_2 x_{2\xi} - \cdots - \beta_k x_{k\xi}]^2$ with respect to $\beta_1, \ldots, \beta_k$ is $|u_{ij}|/|u_{vw}|$, $i, j = 1, \ldots, k$, $v, w = 2, \ldots, k$. By using methods similar to those used in Section 18.4 for finding moments, show that

$$\mathscr{E}\left(\frac{|u_{ij}|}{|u_{vw}|}\right)^r = (2/\sigma^{11})^r \frac{\Gamma\left(\frac{n-k}{2} + r\right)}{\Gamma\left(\frac{n-k}{2}\right)} \qquad r = 0, 1, 2, \ldots$$

where σ^{11} is the element in the first row and first column of $\|\sigma_{ij}\|^{-1}$. From this sequence of moments deduce that

$$\sigma^{11}|u_{ij}|/|u_{vw}|$$

has the chi-square distribution $C(n - k)$.

18.16 (*Continuation*) *Sampling distribution of multiple correlation coefficient.* If the minimizing values of $\beta_1, \ldots, \beta_k$ are $\hat{\beta}_1, \ldots, \hat{\beta}_k$ show that the correlation coefficient R in the sample between $x_{1\xi}$ and $\hat{\beta}_1 + \hat{\beta}_2 x_{2\xi} + \cdots + \hat{\beta} x_{k\xi}$, $\xi = 1, \ldots, n$ is given by

$$1 - R^2 = \frac{|u_{ij}|}{u_{11}|u_{vw}|}.$$

Let

$$\psi(r, \theta) = \mathscr{E}\left[\left(\frac{|u_{ij}|}{|u_{vw}|}\right)^r e^{-\theta u_{11}}\right].$$

Show that

$$\psi(r, \theta) = (1 + 2\sigma_{11}\theta)^{-\frac{1}{2}(n-1)}(\tfrac{1}{2}\sigma^{11} + \theta)^{-r} \frac{\Gamma\left(\dfrac{n-k}{2} + r\right)}{\Gamma\left(\dfrac{n-k}{2}\right)}$$

$$= \frac{\Gamma\left(\dfrac{n-k}{2} + r\right)}{(2\sigma_{11})^{\frac{1}{2}(n-1)}\Gamma\left(\dfrac{n-1}{2}\right)\Gamma\left(\dfrac{n-k}{2}\right)}$$

$$\cdot \sum_{i=0}^{\infty} \frac{\left[\dfrac{\sigma_{11}\sigma^{11} - 1}{2\sigma_{11}}\right]^i \Gamma\left(\dfrac{n-1}{2} + i\right)(\tfrac{1}{2}\sigma^{11} + \theta)^{-\frac{1}{2}(n-1)-r-i}}{i!}.$$

Noting that

$$\mathscr{E}(1 - R^2)^r = \int_0^{\infty} \cdots \int_0^{\infty} \psi(r, \xi_1 + \cdots + \xi_r)\, d\xi_1 \cdots d\xi_r, \qquad r = 0, 1, 2, \ldots,$$

show that

$$\mathscr{E}(1 - R^2)^r = \frac{(1 - \rho^2)^{\frac{1}{2}(n-1)}}{\Gamma\left(\dfrac{n-1}{2}\right)\Gamma\left(\dfrac{n-k}{2}\right)} \sum_{i=0}^{\infty} \frac{\rho^{2i}\,\Gamma\left(\dfrac{n-1}{2} + i\right)\Gamma\left(\dfrac{n-k}{2} + r\right)}{i!\,\Gamma\left(\dfrac{n-1}{2} + i + r\right)}$$

where ρ^2 is the squared multiple correlation coefficient between x_1 and $x_2, \ldots, x_k$ in the population given by

$$1 - \rho^2 = \frac{1}{\sigma_{11}\sigma^{11}}.$$

From this value of the rth moment of $1 - R^2$ show that the p.d.f. of R^2 is

$$\frac{(1 - \rho^2)^{\frac{1}{2}(n-1)}(1 - R^2)^{\frac{1}{2}(n-k-2)}(R^2)^{\frac{1}{2}(k-3)}}{\Gamma\left(\dfrac{n-1}{2}\right)\Gamma\left(\dfrac{n-k}{2}\right)} \sum_{i=0}^{\infty} \frac{(\rho^2 R^2)^i\,\Gamma^2\left(\dfrac{n-1}{2} + i\right)}{i!\,\Gamma\left(\dfrac{k-1}{2} + i\right)}$$

for $0 \leqslant R^2 \leqslant 1$, and 0 otherwise, [Fisher (1928b).]

18.17 (*Continuation*) *Distribution of scatter of residuals.* Consider the residuals

$$y_{p\xi} = \left(x_{p\xi} - \beta_p - \sum_{v=s+1}^{k} \beta_{pv} x_{v\xi}\right) \qquad p = 1, \ldots, s, \qquad \xi = 1, \ldots, n, \qquad s < k$$

and the scatter

$$\left|\sum_{\xi=1}^{n} y_{p\xi} y_{q\xi}\right|$$

of these residuals. Show by methods similar to those of Section 18.4 that the minimum of this scatter with respect to the β's is

$$|u_{ij}|/|u_{vw}|, \qquad i, j = 1, \ldots, k, \qquad v, w = s + 1, \ldots, k$$

and if the sample $(x_{1\xi}, \ldots, x_{k\xi}, \xi = 1, \ldots, n)$ comes from the normal distribution $N(\{\mu_i\}; \|\sigma_{ij}\|)$ that

$$\mathscr{E}\left(\frac{|u_{ij}|}{|u_{vw}|}\right)^r = \left(\frac{2^s|\sigma_{ij}|}{|\sigma_{vw}|}\right)^r \prod_{i=k-s+1}^{k} \left[\frac{\Gamma\left(\dfrac{n-i}{2} + r\right)}{\Gamma\left(\dfrac{n-i}{2}\right)}\right].$$

Hence, verify that the distribution of $|u_{ij}|/|u_{uw}|$ is identical with that of

$$\frac{2^s|\sigma_{ij}|}{|\sigma_{vw}|} \prod_{i=1}^{s} z_i$$

where $z_1, \ldots, z_s$ are independent random variables having gamma distributions $G(\frac{1}{2}(n - k), G(\frac{1}{2}(n - k + 1)), \ldots, G(\frac{1}{2}(n - k + s - 1))$ respectively.

18.18 *Principal components (eigenvalues) of a k-dimensional probability distribution.* The principal components of a k-dimensional probability distribution with covariance matrix $\|\sigma_{ij}\|$ are the eigenvalues $\lambda_1, \ldots, \lambda_k$ of $\|\sigma_{ij}\|$ (that is, the roots of the equation $|\sigma_{ij} - \lambda\delta_{ij}| = 0$) where $+\infty > \lambda_1 \geqslant \cdots \geqslant \lambda_k \geqslant 0$. Show that $\lambda_1 + \cdots + \lambda_k = \sigma_{11} + \cdots + \sigma_{kk}$ and $\lambda_1\lambda_2 \cdots \lambda_k = |\sigma_{ij}|$. The unit eigenvector $c_{1p}, \ldots, c_{kp}$ corresponding to an eigenvalue λ_p which is distinct from all other eigenvalues is the unit vector satisfying

$$\sum_{j=1}^{k} (\sigma_{ij} - \lambda_p\delta_{ij})c_{jp} = 0, \qquad i = 1, \ldots, k.$$

If the eigenvalues $\lambda_1, \ldots, \lambda_k$ are all distinct, show that the unit eigenvectors corresponding to these eigenvalues are mutually orthogonal, and hence that if the axes are rotated so that the new axes have directions given by the eigenvectors $(c_{11}, \ldots, c_{k1}), \ldots, (c_{1k}, \ldots, c_{kk})$ respectively, the probability distribution in the new k-dimensional space has covariance matrix $\|\delta_{ij}\lambda_i\|$.

18.19 (*Continuation*) Suppose a k-dimensional probability distribution has covariance matrix $\|\sigma_{ij}\|$ where $\sigma_{ii} = \sigma^2$, $i = 1, \ldots, k$ and $\sigma_{ij} = \rho\sigma^2$, $i \neq j = 1, \ldots, k$. Show that $\lambda_1 = \sigma^2[1 + (k - 1)\rho]$ and $\lambda_2 = \cdots = \lambda_k = \sigma^2(1 - \rho)$ and that the unit eigenvector $(c_{11}, \ldots, c_{k1})$ associated with λ_1 is given by

$$c_{11} = \cdots = c_{k1} = \frac{1}{\sqrt{k}}.$$

18.20 *Test of independence of two sets of variables.* Let $\|u_{ij}\|$ be the internal scatter matrix of a sample of size $n > k$ from the k-dimensional distribution

$N(\{\mu_i\}; \|\sigma_{ij}\|)$. Let $\|u_{pq}\|$ $p, q = 1, \ldots, s$ be the internal scatter matrix of the sample using only the first s variables, and $\|u_{vw}\|, v, w = s + 1, \ldots, k$, the internal scatter matrix of the sample using only the last t $(t = k - s, s \leqslant t)$ variables. Let $\mathscr{H}(\omega; \Omega)$ be the statistical hypothesis in which Ω is the $\frac{1}{2}k(k + 3)$-dimensional parameter space in which $\mu_1, \ldots, \mu_k$ are real numbers and $\|\sigma_{ij}\|$ is positive definite, whereas ω is the $(\frac{1}{2}k(k + 3) - st)$-dimensional subset of Ω for which

$$\sigma_{pw} = 0, \qquad p = 1, \ldots, s, \; w = s + 1, \ldots, k.$$

Show that the Neyman-Pearson likelihood ratio λ for testing $\mathscr{H}$ is given by

$$\lambda = L^{\frac{1}{2}n}$$

where

$$L = \frac{|u_{ij}|}{|u_{pq}| \cdot |u_{vw}|}.$$

Using methods similar to those used in Section 18.4(a) show that

$$\mathscr{E}(L^r) = \prod_{p=1}^{s} \left[\frac{\Gamma\left(\frac{n-p}{2}\right) \Gamma\left(\frac{n-t-p}{2} + r\right)}{\Gamma\left(\frac{n-p}{2} + r\right) \Gamma\left(\frac{n-t-p}{2}\right)} \right]$$

$r = 0, 1, 2, \ldots$ if $\mathscr{H}$ is true.

Verify for $s = 1, t = k - 1$ that L has the beta distribution $Be(\frac{1}{2}(n - k), \frac{1}{2}(k - 1))$ and for $s = 2, t = k - 2$ that $\sqrt{L}$ has the beta distribution $Be(n - k, k - 2)$. [Wilks (1935).]

18.21 *Test for sphericity of a normal distribution* Let $\|u_{ij}\|$ be the internal scatter matrix of a sample of size n from the k-dimensional distribution $N(\{\mu_i\}; \|\sigma_{ij}\|)$. Let $\mathscr{H}(\omega; \Omega)$ be the statistical hypothesis in which Ω is the $\frac{1}{2}k(k + 3)$-dimensional parameter space in which the $\{\mu_i\}$ are real and $\|\sigma_{ij}\|$ is positive definite, and ω is the $(k + 1)$-dimensional subset of Ω in which $\|\sigma_{ij}\| = \|\delta_{ij}\sigma^2\|$ where δ_{ij} is the Kronecker δ, and $\sigma^2 > 0$. Show that the Neyman-Pearson likelihood ratio λ for testing $\mathscr{H}$ is given by

$$\lambda = L^{\frac{1}{2}n}$$

where

$$L = \frac{|u_{ij}|}{\bar{u}^k}, \qquad \bar{u} = \frac{1}{k}(u_{11} + \cdots + u_{kk})$$

and hence that L is equivalent to λ as a test for $\mathscr{H}$. Furthermore, show that if $\mathscr{H}$ is true

$$\mathscr{E}\left(\frac{|u_{ij}|}{\bar{u}^k}\right)^r = \int_0^\infty \cdots \int_0^\infty \psi(r, \xi_1 + \cdots + \xi_{rk})\, d\xi_1 \cdots d\xi_{rk}$$

$$= \frac{k^{kr}\, \Gamma\left(\frac{(n-1)k}{2}\right)}{\Gamma\left(\frac{(n-1)k}{2} + rk\right)} \prod_{i=1}^{k} \left[\frac{\Gamma\left(\frac{n-i}{2} + r\right)}{\Gamma\left(\frac{n-i}{2}\right)} \right], \qquad r = 0, 1, 2, \ldots,$$

where

$$\psi(r, \theta) = \mathscr{E}(|u_{ij}|^r e^{-\theta \bar{u}})$$

$$= \left(\frac{2}{\sigma^2}\right)^{kr}\left(1 + \frac{2\theta}{k\sigma^2}\right)^{-\frac{1}{2}[(n-1)+2rk]} \prod_{i=1}^{k}\left[\frac{\Gamma\left(\frac{n-i}{2}+r\right)}{\Gamma\left(\frac{n-i}{2}\right)}\right].$$

Verify that for $k = 2$, $\sqrt{L}$ has the beta distribution $Be(n - 2; 1)$. [Mauchly (1940).]

18.22 (*Continuation*) *Test of the statistical hypothesis that a k-dimensional normal distribution is symmetric in the k variables.* Let $\mathscr{H}^*(\omega; \Omega)$ be the statistical hypothesis in which Ω is defined as in the preceding problem but ω is the $(k + 3)$-dimensional subset of Ω in which $\sigma_{ii} = \sigma^2$, $\sigma^2 > 0$, $i = 1, \ldots, k$ and $\sigma_{ij} = \sigma_{ji} = \rho\sigma^2$, $i \neq j$, $-\dfrac{1}{k-1} \leqslant \rho < 1$.

Show that the Neyman-Pearson likelihood ratio λ for $\mathscr{H}^*$ is given by

$$\lambda = (L^*)^{\frac{1}{2}n}$$

where

$$L^* = \frac{|u_{ij}|}{(\bar{u} - \bar{u}^*)^{k-1}(\bar{u} + (k-1)\bar{u}^*)}$$

and $\bar{u} = \dfrac{1}{k}(u_{11} + \cdots + u_{kk})$, $\bar{u}^* = \dfrac{1}{k(k-1)}\sum_{i \neq j} u_{ij}$, and hence L^* is equivalent to λ for testing $\mathscr{H}^*$.

Furthermore, show that if $\mathscr{H}^*$ is true,

$$\mathscr{E}(L^*)^r = \int_0^\infty \cdots \int_0^\infty \psi(r, \xi_1 + \cdots + \xi_{r(k-1)}, \eta_1 + \cdots + \eta_k)\, d\xi_1 \cdots d\xi_{r(k-1)} d\eta_1 \cdots d\eta_k$$

$$= (k-1)^{r(k-1)} \frac{\Gamma\left(\frac{(n-1)(k-1)}{2}\right)}{\Gamma\left(\frac{(n-1)(k-1)}{2} + r(k-1)\right)} \prod_{i=2}^{k}\left[\frac{\Gamma\left(\frac{n-i}{2}+r\right)}{\Gamma\left(\frac{n-i}{2}\right)}\right],$$

$r = 0, 1, 2, \ldots$

where

$$\psi(r, p, q) = \mathscr{E}[|u_{ij}|^r e^{-p(\bar{u}-\bar{u}^*)-q(\bar{u}+(k-1)\bar{u}^*)}]$$

$$= 2^{kr} \prod_{i=1}^{k} \frac{\Gamma\left(\frac{n-i}{2}+r\right)}{\Gamma\left(\frac{n-i}{2}\right)} (A^{k-1} - B)^{\frac{1}{2}(n-1)}$$

$$\cdot \left[A + \frac{2p}{k-1}\right]^{-\frac{1}{2}(k-1)[(n-1)+2r]} \cdot [B + 2q]^{-\frac{1}{2}(n-1+2r)}$$

and

$$A = \frac{1}{\sigma^2(1-\rho)}, \quad B = \frac{1}{\sigma^2[1 + (k-1)\rho]}.$$

Verify that for $k = 2$, L^* has the beta distribution $Be(\frac{1}{2}(n-2), \frac{1}{2})$ and for $k = 3$, $\sqrt{L^*}$ has the beta distribution $Be(n-2, 1)$. [Wilks (1946).]
[For generalization of the test L^* to testing $N(\{\mu_i\}; \|\sigma_{ij}\|)$ for symmetry in variables within blocks see Votaw (1948).]

18.23 *Noncentral T^2 distribution.* If, in section 18.4(*a*), the sample comes from $N(\{\mu_i\}; \|\sigma_{ij}\|)$ where $(\mu_1, \ldots, \mu_k) \neq (\mu_1^\circ, \ldots, \mu_k^\circ)$, show that T^2 as defined in (18.4.5) has p.d.f.

$$\frac{e^{-\frac{1}{2}\delta^2}(T^2)^{\frac{1}{2}k-1}[1 + T^2/(n-1)]^{-\frac{1}{2}n}}{(n-1)^{\frac{1}{2}k}\Gamma\left(\dfrac{n-k}{2}\right)} \cdot \sum_{i=0}^{\infty} \frac{\left[\dfrac{\delta^2 T^2}{2(n-1)}\right]^i \Gamma\left(\dfrac{n}{2} + i\right)}{i!\,\Gamma\left(\dfrac{k}{2} + i\right)[1 + T^2/(n-1)]^i}$$

where

$$\delta^2 = n \sum_{i,j=1}^{k} \sigma_{ij}\,(\mu_i - \mu_i^\circ)(\mu_j - \mu_j^\circ).$$

[Hsu (1938).]

18.24 *Generating moments of the scatter $|v_{ij}|$ of a sample from $N(\{\mu_i\}; \|\sigma_{ij}\|)$ without using Wishart distribution* [Wilks (1934)]. Suppose a sample $(x_{1\xi}, \ldots, x_{k\xi}, \xi = 1, \ldots, n)$, $n \geqslant k$ is from a k-dimensional normal distribution $N(\{\mu_i\}; \|\sigma_{ij}\|)$. Let $\|v_{ij}\|$ be the scatter matrix about the sample mean $(\mu_1, \ldots, \mu_k)$.

Show (without the use of the Wishart distribution) that if $2r < n + 1 - k$

$$\mathscr{E}(|v_{ij}|^{-r}) = \pi^{kr} \int_{-\infty}^{\infty} \cdots \int_{-\infty}^{\infty} \mathscr{E}\left[\exp\left(-\sum_{p=1}^{2r} \sum_{i,j=1}^{k} v_{ij} z_{ip} z_{jq}\right)\right] \prod_{i,p} dz_{ip}$$

$$= |2\sigma_{ij}|^{-r} \prod_{i=1}^{k} \left[\frac{\Gamma\left(\dfrac{n+1-i}{2} - r\right)}{\Gamma\left(\dfrac{n+1-i}{2}\right)}\right].$$

References and Author Index

[Pages on which authors are cited are given in brackets]

R. L. Anderson (1942), Distribution of the serial correlation coefficient, *Ann. Math. Stat.,* Vol. 13, pp. 1–13. [533]

T. W. Anderson (1948), On the theory of testing serial correlation, *Skand. Aktuar.,* Vol. 31, pp. 88–116. [535]

T. W. Anderson and D. A. Darling (1952), Asymptotic theory of certain "goodness-of-fit" criteria based on stochastic processes, *Ann. Math. Stat.,* Vol. 23, pp. 193–212. [438]

T. W. Anderson (1958), *An Introduction to Multivariate Statistical Analysis,* John Wiley, New York. [540]

T. W. Anderson (1960), A modification of the sequential probability ratio test to reduce the sample size, *Ann. Math. Stat.,* Vol. 31, pp. 165–197. [491]

F. C. Andrews (1954), Asymptotic behavior of some rank tests for analysis of variance, *Ann. Math. Stat.,* Vol. 25, pp. 724–736. [468]

F. J. Anscombe (1953), Sequential estimation, *Jour. Roy. Stat. Soc., Series B,* Vol. 15, pp. 1–29. [496]

Army Ordnance Corps (1952), *Tables of the Cumulative Binomial Probabilities,* ORDP, 20–11. [138]

H. E. Arnold (1958), *Permutation support for multivariate techniques,* Ph.D. Dissertation, Princeton University. [465]

R. R. Bahadur (1954), Sufficiency and statistical decision functions, *Ann. Math. Stat.,* Vol. 25, pp. 423–462. [356]

R. R. Bahadur (1958), Examples of inconsistency of maximum likelihood estimates, *Sankhyā,* Vol. 20, pp. 207–210. [364]

E. W. Barankin and J. Gurland (1951), On asymptotically normal efficient estimators: I, *Univ. Calif. Pub. in Stat.,* Vol. 1, No. 6, pp. 89–129. [360, 364, 381]

G. A. Barnard (1952), The frequency justification of certain sequential tests, *Biometrika,* Vol. 39, pp. 144–150. [472]

W. Bartky (1943), Multiple sampling with constant probability, *Ann. Math. Stat.,* Vol. 14, pp. 363–377. [473]

M. S. Bartlett (1950), Periodogram analysis and continuous spectra, *Biometrika,* Vol. 37, pp. 1–16. [524]

M. S. Bartlett (1953), Approximate confidence intervals, II. More than one unknown parameter, *Biometrika,* Vol. 40, pp. 306–317. [388]

M. S. Bartlett (1955), *An Introduction to Stochastic Processes,* Cambridge University Press. [514, 524, 526]

I. L. Battin (1942), On the problem of multiple matching, *Ann. Math. Stat.,* Vol. 13, pp. 294–305. [154]

E. M. L. Beale (1960), Confidence regions in non-linear estimation, *Jour. Roy. Stat. Soc., Ser. B,* Vol. 22, pp. 41–88. [388]

J. Bernoulli (1713), *Ars Conjectandi*, Paris. [137]

J. Bertrand (1889), *Calcul des Probabilités*, Gauthier-Villars, Paris. [470]

G. Birkhoff and S. MacLane (1953), *A Survey of Modern Algebra* (Revised Edition), Macmillan, New York. [213, 569, 578]

Z. W. Birnbaum and F. H. Tingey (1951), One-sided confidence contours for probability distribution functions, *Ann. Math. Stat.,* Vol. 22, pp. 592–596. [336, 337, 338]

Z. W. Birnbaum (1953*a*), On the power of a one-sided test of fit for continuous probability functions, *Ann. Math. Stat.,* Vol. 24, pp. 484–489. [438]

Z. W. Birnbaum (1953*b*), Distribution-free tests of fit for continuous distribution functions, *Ann. Math. Stat.,* Vol. 24, pp. 1–8. [438]

D. Blackwell (1946), On an equation of Wald, *Ann. Math. Stat.,* Vol. 17, pp. 84–87. [476]

D. Blackwell (1947), Conditional expectation and unbiased sequential estimation, *Ann. Math. Stat.,* Vol. 18, pp. 105–110. [357]

D. Blackwell and M. A. Girshick (1954), *Theory of Games and Statistical Decisions,* John Wiley, New York. [502, 511]

J. R. Blum and L. Weiss (1957), Consistency of certain two-sample tests, *Ann. Math. Stat.,* Vol. 28, pp. 242–246. [451]

M. Bôcher (1907), *Introduction to Higher Algebra*, Macmillan, New York. [90, 213, 287, 583]

S. Bochner (1932), *Vorlesungen über Fouriersche Integrale,* Leipzig. [520]

L. Boltzmann (1898), *Vorlesungen über Gastheorie,* Part I, J. A. Barth, Leipzig. [409]

R. C. Bose (1938), On the application of the properties of Galois fields to the problem of construction of hyper-Graeco-Latin squares, *Sankyhā,* Vol. 3, pp. 323–338. [234]

R. C. Bose (1939), On the construction of balanced incomplete block designs, *Ann. Eugen.,* Vol. 9, pp. 353–399. [232]

G. E. P. Box and K. B. Wilson (1951), On the experimental attainment of optimum conditions, *Jour. Roy. Stat. Soc., Series B,* Vol. 13, pp. 1–45. [297]

G. E. P. Box (1952), Multifactor designs of first order, *Biometrika,* Vol. 39, pp. 49–57. [297]

G. E. P. Box (1954), The exploration and exploitation of response surfaces: Some general considerations and examples, *Biometrics,* Vol. 10, pp. 16–60. [297]

G. E. P. Box and J. S. Hunter (1957), Multifactor experimental designs for exploring response surfaces, *Ann. Math. Stat.,* Vol. 28, pp. 195–241. [297]

G. E. P. Box and M. E. Muller (1958), A note on the generation of random normal deviates, *Ann. Math. Stat.,* Vol. 29, pp. 610–611. [188]

M. G. Bulmer (1957), Approximate confidence limits for components of variance, *Biometrika,* pp. 159–167. [308]

C. Carathéodory (1927), *Vorlesungen über reelle Funktionen,* Leipzig-Berlin. [15]

A. G. Carlton (1946), Estimating the parameters of a rectangular distribution, *Ann. Math. Stat.*, Vol. 17, pp. 355–358. [248]

A. L. Cauchy (1853), Sur les résultats moyens d'observations de même nature et sur les résultats les plus probables, *Comp. Rend. Acad. Sci.*, Paris, Vol. 37, pp. 198–206. [130, 255]

D. C. Chapman and H. Robbins (1951), Minimum variance estimation without regularity assumptions, *Ann. Math. Stat.*, Vol. 22, pp. 581–586. [352, 393]

P. L. Chebyshev (1867), Des valeurs moyennes, *Jour. Math. Pures et Appl.*, Vol. 12, pp. 177–184. [75]

P. L. Chebyshev (1890), Sur deux théorèmes relatifs aux probabilités, *Acta Math.*, Vol. 14, pp. 305–315. [257]

H. Chernoff (1954), On the distribution of the likelihood ratio, *Ann. Math. Stat.*, Vol. 25, pp. 573–578. [422]

H. Chernoff and L. E. Moses (1959), *Elementary Decision Theory*, John Wiley, New York. [502]

J. H. Chung and D. B. DeLury (1950), *Confidence Limits for the Hypergeometric Distribution*, University of Toronto Press. [369]

K. L. Chung (1941), On the probability of the occurrence of at least m events among n arbitrary events, *Ann. Math. Stat.*, Vol. 17, pp. 447–465. [29]

K. L. Chung (1946), The approximate distribution of Student's statistic, *Ann. Math. Stat.*, Vol. 17, pp. 447–465. [266]

K. L. Chung and W. Feller (1949), Fluctuations in coin tossing, *Proc. Nat. Acad. Sci., U.S.A.*, Vol. 35, pp. 605–608. [471]

C. J. Clopper and E. S. Pearson (1934), The use of confidence or fiducial limits illustrated in the case of the binomial, *Biometrika*, Vol. 26, pp. 404–413. [369]

W. G. Cochran (1934), The distribution of quadratic forms in a normal system, with applications to the analysis of covariance, *Proc. Camb. Phil. Soc.*, Vol. 30, pp. 178–191. [212]

W. G. Cochran (1953), *Sampling Techniques*, John Wiley, New York. [314]

W. G. Cochran and G. M. Cox (1957), *Experimental Designs*, second edition, John Wiley, New York. [232, 297, 308]

W. S. Connor, Jr. (1952), On the structure of balanced incomplete block designs, *Ann. Math. Stat.*, Vol. 23, pp. 57–71. [232]

E. A. Cornish and R. A. Fisher (1937), Moments and cumulants in the specification of distributions, *Rev. Int. Stat. Inst.*, Vol. 4, pp. 1–14. [200]

D. R. Cox (1952), Sequential tests for composite hypotheses, *Proc. Camb. Phil. Soc.*, Vol. 48, pp. 290–299. [472]

A. T. Craig (1932), On the distributions of certain statistics, *Amer. Jour. Math.*, Vol. 54, pp. 353–366. [237]

C. C. Craig (1928), An application of Thiele's semivariants to the sampling problem, *Metron*, Vol. 7, No. 4, pp. 3–74. [200]

H. Cramér (1928), On the composition of elementary errors, *Skand. Aktuar.*, Vol. 11, pp. 13–74, 141–180. [438]

H. Cramér (1936), Über eine Eigenschaft der normalen Verteilungsfunktion, *Math. Zeits.*, Vol. 41, pp. 405–414. [211]

H. Cramér (1937), *Random Variables and Probability Distributions*, Cambridge Tracts in Mathematics, No. 36, Cambridge University Press. [122, 266]

H. Cramér (1946), *Mathematical Methods of Statistics*, Princeton University Press. [92, 93, 123, 125, 131, 177, 353, 354]

J. F. Daly (1940), On the unbiased character of likelihood-ratio tests for independence in normal systems, *Ann. Math. Stat.,* Vol. 11, pp. 1–32. [407]

D. A. Darling (1957), The Kolmogorov-Smirnov, Cramér-von-Mises tests, *Ann. Math. Stat.,* Vol. 28, pp. 823–838. [341, 438]

F. N. David (1950), Two combinatorial tests of whether a sample has come from a given population, *Biometrika,* Vol. 37, pp. 97–110. [434, 435]

H. T. David and W. H. Kruskal (1956), The WAGR sequential t-test reaches a decision with probability 1, *Ann. Math. Stat.,* Vol. 27, pp. 797–805. [472]

H. T. Davis (1941), *The Analysis of Economic Time Series,* Principia Press, Bloomington, Indiana. [533]

A. DeMoivre (1718), *The Doctrine of Chances,* London. [257]

A. P. Dempster (1955), Personal communication. [339]

A. P. Dempster (1958), A high dimensional two-sample significance test, *Ann. Math. Stat.,* Vol. 29, pp. 995–1010. [576]

P. G. L. Dirichlet (1839), Sur un nouvelle methode pour la determination des integrales multiples, *Comp. Rend. Acad. Sci.,* Vol. 8, pp. 156–160. [178]

W. J. Dixon (1940), A criterion for testing the hypothesis that two samples are from the same population, *Ann. Math. Stat.,* Vol. 11, pp. 199–204. [441]

W. J. Dixon (1944), Further contributions to the problem of serial correlation, *Ann. Math. Stat.,* Vol. 15, pp. 119–144. [535, 537]

W. J. Dixon and A. M. Mood (1946), The statistical sign test, *Jour. Amer. Stat. Assn.,* Vol. 41, pp. 557–566. [430]

W. J. Dixon and A. M. Mood (1948), A method of obtaining and analyzing sensitivity data, *Jour. Amer. Stat. Assn.,* Vol. 43, pp. 109–126. [474]

E. L. Dodd (1923), The greatest and least variate under general laws of error, *Trans. Amer. Math. Soc.,* Vol. 25, pp. 525–539. [272]

H. F. Dodge and H. G. Romig (1929), A method of sampling inspection, *Bell System Tech. Jour.,* Vol. 8, pp. 613–631. [397, 398, 472]

H. F. Dodge and H. G. Romig (1959), *Sampling Inspection Tables,* second edition, John Wiley, New York. [397, 473]

T. G. Donnelly (1957), *A family of sequential tests,* Ph.D. Thesis, University of North Carolina. [492]

M. D. Donsker (1952), Justification and extension of Doob's heuristic approach to the Kolmogorov-Smirnov theorems, *Ann. Math. Stat.,* Vol. 23, pp. 277–281. [341]

J. L. Doob (1949), Heuristic approach to the Kolmogorov-Smirnov theorems, *Ann. Math. Stat.,* Vol. 20, pp. 393–403. [341]

J. L. Doob (1953), *Stochastic Processes,* John Wiley, New York. [1, 26, 98, 99, 514, 537]

R. Dorfman (1943), The detection of defective members of large populations, *Ann. Math. Stat.,* Vol. 14, pp. 436–440. [152]

D. Dugué (1937), Application des proprietés de las limité au sens du calcul des probabilitiés á l'étude des diverses questions d'estimation, *École. Poly.,* Vol. 3, No. 4, pp. 305–372. [353, 354]

D. B. Duncan (1952), On the properties of the multiple comparison test, *Virginia Jour. of Sci.,* Vol. 3, pp. 49–67. [290]

J. Durbin and G. S. Watson (1951), Exact tests of serial correlation using noncircular statistics, *Ann. Math. Stat.,* Vol. 22, pp. 446–451. [535]

A. Dvoretzky, J. Kiefer, and J. Wolfowitz (1953*a*), Sequential decision problems for processes with continuous time parameter. Testing hypotheses, *Ann. Math. Stat.*, Vol. 24, pp. 254–264. [474, 511]

A. Dvoretzky, J. Kiefer, and J. Wolfowitz (1953*b*), Sequential decision problems for processes with continuous time parameter. Problems of estimation, *Ann. Math. Stat.*, Vol. 24, pp. 403–415. [474, 511]

M. Dwass (1959), Multiple confidence procedures, *Ann. Inst. Stat. Math.*, Vol. 10, pp. 277–282. [290, 297]

P. S. Dwyer (1938), Combined expansions of products of symmetric power sums and of sums of symmetric power products with applications to sampling, *Ann. Math. Stat.*, Vol. 9, pp. 1–47, 97–132. [200]

F. Y. Edgeworth (1905), The law of error, *Trans. Camb. Phil. Soc.*, Vol. 20, pp. 36–65. [265]

C. Eisenhart (1947), The assumptions underlying the analysis of variance, *Biometrics*, Vol. 3, pp. 1–21. [308]

G. Elfving (1947), The asymptotical distribution of range in samples from a normal population, *Biometrika*, Vol. 34, pp. 111–119. [249]

B. Epstein (1954), Tables for the distribution of the number of exceedances, *Ann. Math. Stat.*, Vol. 25, pp. 762–768. [251]

C. E. Esseen (1944), Fourier analysis of distribution functions: A mathematical study of the Laplace-Gaussian law, *Acta. Math.*, Vol. 77, pp. 1–125. [266]

W. Feller (1935), Über den zentralen Grenzwertsatz der Wahrscheinlichkeitsrechnung, *Math. Zeits.*, Vol. 40, pp. 521–559. [257, 258]

W. Feller (1948), On the Kolmogorov-Smirnov limit theorems for empirical distributions, *Ann. Math. Stat.*, Vol. 19, pp. 177–189. [341]

W. Feller (1957), *An Introduction to Probability Theory and its Applications*, second edition, John Wiley, New York. [1, 99, 109, 193]

T. S. Ferguson (1958), A method of generating best asymptotically normal estimates with application to the estimation of bacterial densities, *Ann. Math. Stat.*, Vol. 29, pp. 1046–1062. [360, 364]

E. C. Fieller (1932), The distribution of the index in a normal bivariate population, *Biometrika*, Vol. 24, pp. 428–440. [188]

R. A. Fisher (1915), Frequency distribution of the values of the correlation coefficient in samples from an indefinitely large population, *Biometrika*, Vol. 10, pp. 507–521. [195, 593]

R. A. Fisher (1922), On the mathematical foundations of theoretical statistics, *Phil. Trans. Roy. Soc. London, Series A*, Vol. 222, pp. 309–368. [351, 353, 356, 360, 362, 381]

R. A. Fisher (1924), On a distribution yielding the error functions of several well-known statistics, *Proc. Int. Math. Congress*, Vol. II, Toronto, pp. 805–813. [187]

R. A. Fisher (1925*a*), *Statistical Methods for Research Workers*, first edition, twelfth edition (1954), Oliver and Boyd, Edinburgh. [183, 185, 187, 276, 301]

R. A. Fisher (1925*b*), Theory of statistical estimation, *Proc. Camb. Phil. Soc.*, Vol. 22, pp. 700–725. [351]

R. A. Fisher (1926*a*), Applications of "Student's" distribution, *Metron*, Vol. 5, No. 4, pp. 90–104. [211]

R. A. Fisher (1926*b*), On the random sequence, *Quart. Jour. Roy. Meteor. Soc.*, Vol. 52, pp. 250–258. [462, 464]

R. A. Fisher (1928*a*), Moments and product moments of sampling distributions, *Proc. London Math. Soc.*, Vol. 30, pp. 199–238. [200]

R. A. Fisher (1928*b*), The general sampling distribution of the multiple correlation coefficient, *Proc. Roy. Soc. London, Series A*, Vol. 121, pp. 654–673. [247, 597]

R. A. Fisher and L. H. C. Tippett (1928), Limiting forms of the frequency distribution of the largest or smallest member of a sample, *Proc. Camb. Phil. Soc.*, Vol. 24, pp. 180–190. [272]

R. A. Fisher (1929), Tests of significance in harmonic analysis, *Proc. Roy. Soc. London, Series A*, Vol. 125, pp. 54–59. [529, 533]

R. A. Fisher (1930), Inverse probability, *Proc. Camb. Phil. Soc.*, Vol. 26, pp. 528–535. [369]

R. A. Fisher (1934), Two new properties of mathematical likelihood, *Proc. Roy. Soc. Series A*, Vol. 144, pp. 285–307. [393]

R. A. Fisher (1935*a*), *The Design of Experiments*, Oliver and Boyd, Edinburgh. [297]

R. A. Fisher (1935*b*), The fiducial argument in statistical inference, *Ann. Eugen.*, Vol. 6, pp. 391–398. [369, 371]

R. A. Fisher (1938), The statistical utilization of multiple measurements, *Ann. Eugen.*, Vol. 8, pp. 376–386. [573]

R. A. Fisher (1939), The sampling distribution of some statistics obtained from non-linear equations, *Ann. Eugen.*, Vol. 9, pp. 238–249. [568]

R. A. Fisher and F. Yates (1938), *Statistical Tables for Biological, Agricultural and Medical Research* (fifth edition, 1957), Oliver and Boyd, Edinburgh. [297]

D. A. S. Fraser (1951), Sequentially determined statistically equivalent blocks, *Ann. Math. Stat.*, Vol. 22, pp. 372–381. [243]

D. A. S. Fraser (1953), Nonparametric tolerance regions, *Ann. Math. Stat.*, Vol. 24, pp. 44–55. [243]

D. A. S. Fraser and I. Guttman (1956), Tolerance regions, *Ann. Math. Stat.*, Vol. 27, pp. 162–179. [243]

D. A. S. Fraser (1957), *Nonparametric Methods in Statistics*, John Wiley, New York. [235, 428, 462]

M. Fréchet (1927), Sur la loi de probabilitié de l'écart maximum, *Ann. Soc. Polonaise Math.*, Vol. 6, pp. 92–116. [272]

M. Friedman (1937), The use of ranks to avoid the assumption of normality implicit in the analysis of variance, *Jour. Amer. Stat. Assn.*, Vol. 32, pp. 675–701. [468]

F. Garwood (1936), Fiducial limits for the Poisson distribution, *Biometrika*, Vol. 28, pp. 437–442. [369]

K. F. Gauss (1809*a*), *Theoria motus corporum coelestium in sectimibus conicis solem aurbientium*, Perthes and Besser, Hamburg. [285]

K. F. Gauss (1809*b*), *Werke*, Vol. 4, Göttingen, pp. 1–93. [257]

R. C. Geary (1936), Distribution of Student's distribution for non-normal samples, *Jour. Roy. Stat. Soc., Series B*, Vol. 3, pp. 178–184. [211]

M. A. Girshick (1939), On the sampling theory of roots of determinantal equations, *Ann. Math. Stat.*, Vol. 10, pp. 203–224. [568]

M. A. Girshick, F. Mosteller, and L. J. Savage (1946), Unbiased estimates for certain binomial sampling problems with applications, *Ann. Math. Stat.*, Vol. 17, pp. 13–23. [144]

B. V. Gnedenko and V. S. Korolink (1951), On the maximum discrepancy between two empirical distributions, *Dok. Akad. Nauk.* SSR 80, Vol. 4, pp. 525–528. [455]

B. V. Gnedenko and A. N. Kolmogorov (1954), *Limit Distributions for Sums of Independent Random Variables* (translated from the 1949 Russian edition by K. L. Chung; with an appendix by J. L. Doob), Addison-Wesley Publishing Company, Reading, Mass. [1, 189, 257]

W. S. Gosset (1908), "Student," The probable error of a mean, *Biometrika,* Vol. 6, pp. 1–25. [211]

F. A. Graybill (1961), *An Introduction to Linear Statistical Models,* McGraw-Hill, New York. [297, 308]

J. A. Greenwood and H. O. Hartley (1961), *Guide to Tables in Mathematical Statistics,* Princeton University Press. [157, 187, 297]

U. Grenander and M. Rosenblatt (1953), Statistical spectral analysis of time series arising from stationary stochastic processes, *Ann. Math. Stat.,* Vol. 24, pp. 537–558. [524]

U. Grenander and M. Rosenblatt (1957), *Statistical Analysis of Stationary Time Series,* John Wiley, New York; Almqvist and Wiksell, Stockholm. [514, 524, 526, 537]

E. J. Gumbel (1935), Les valeurs extrêmes des distributions statistiques, *Ann. de l'Institut Henri Poincaré,* Vol. 4, pp. 115–158. [272]

E. J. Gumbel (1958), *Statistics of Extremes,* Columbia University Press. [235, 272]

I. Guttman (1960), Personal communication. [249]

J. B. S. Haldane (1945), On a method of estimating frequencies, *Biometrika,* Vol. 33, pp. 222–225. [144, 501]

P. R. Halmos (1946), The theory of unbiased estimation, *Ann. Math. Stat.,* Vol. 17, pp. 34–43. [280]

P. R. Halmos and L. J. Savage (1949), Application of the Radon-Nikodym theorem to the theory of sufficient statistics, *Ann. Math. Stat.,* Vol. 20, pp. 225–241. [356]

P. R. Halmos (1950), *Measure Theory,* D. Van Nostrand, Princeton, N. J. [1, 15, 16, 21, 25]

E. J. Hannan (1955), Exact tests for serial correlation, *Biometrika,* Vol. 42, pp. 133–142. [535]

M. H. Hansen, W. N. Hurwitz, and W. G. Madow (1953), *Sample Survey Methods and Theory,* Vols. I and II, John Wiley, New York. [314]

T. E. Harris (1948), Branching processes, *Ann. Math. Stat.,* Vol. 19, pp. 474–494. [132]

H. O. Hartley and H. A. David (1954), Universal bounds for mean range and extreme observation, *Ann. Math. Stat.,* Vol. 25, pp. 85–99. [250]

Harvard Computation Laboratory (1955), *Tables of the Cumulative Binomial Probability Distribution,* Harvard University. [138]

F. R. Helmert (1876*a*), Ueber die Wahrscheinlichkeit der Potensummen der Beobachtungsfehler und über einige damit im Zusammenhange stehende Fragen, *Zeits. für Math. und Phys.,* Vol. 21, pp. 192–218. [208]

F. R. Helmert (1876*b*), Die Genauigkeit der Formel von Peters zur Berechnung des Wahrscheinlichen Beobachtungsfehlers directer Beobachtungen glicher Genauigkeit, *Astron. Nachr.*, Vol. 88, pp. 112–131. [211]

G. Herglotz (1911), Über Potenzreihen mit positivem rellen Teil im Einheits kreis, *Berichte Verh. König. Sächs. Ges. Wiss., Leipzig, Math.-Phys. Klasse,* Vol. 63, pp. 501–511. [520]

W. Hoeffding (1948*a*), A class of statistics with asymptotically normal distribution, *Ann. Math. Stat.*, Vol. 19, pp. 293–325. [202, 462]

W. Hoeffding (1948*b*), A non-parametric test of independence, *Ann. Math. Stat.*, Vol. 19, pp. 546–557. [467]

W. Hoeffding (1952), The large-sample power of tests based on permutations of observations, *Ann. Math. Stat.*, Vol. 23, pp. 169–192. [465]

W. Hoeffding (1960), Lower bounds for the expected sample size and the average risk of a sequential procedure, *Ann. Math. Stat.*, Vol. 31, pp. 352–368. [491]

R. Hooke (1956*a*), Symmetric functions of a two-way array, *Ann. Math. Stat.*, Vol. 27, pp. 55–79. [226]

R. Hooke (1956*b*), Some applications of bipolykays to the estimation of variance components and their moments, *Ann. Math. Stat.*, Vol. 27, pp. 80–98. [308]

H. Hotelling (1931), The generalization of Student's ratio, *Ann. Math. Stat.*, Vol. 2, pp. 360–378. [290, 557, 594]

H. Hotelling (1933), Analysis of a complex of statistical variables into principal components, *Jour. Educ. Psych.*, Vol. 24, pp. 417–441, 498–520. [565]

H. Hotelling (1935), The most predictable criterion, *Jour. Educ. Psych.*, Vol. 26, pp. 139–142. [588]

H. Hotelling and M. R. Pabst (1936), Rank correlation and tests of significance involving no assumption of normality, *Ann. Math. Stat.*, Vol. 7, pp. 29–43. [467]

H. Hotelling (1941), Experimental determination of the maximum of a function, *Ann. Math. Stat.*, Vol. 12, pp. 20–45. [474]

H. Hotelling (1944), Some improvements in weighing and other experimental techniques, *Ann. Math. Stat.*, Vol. 15, pp. 297–306. [286]

H. Hotelling (1953), New light on the correlation coefficient and its transforms, *Jour. Roy. Stat. Soc., Series B,* Vol. 15, pp. 193–232. [593]

P. L. Hsu (1938), Notes on Hotelling's generalized *T*, *Ann. Math. Stat.*, Vol. 9, pp. 231–243. [601]

P. L. Hsu (1939*a*), A new proof of the joint product moment distribution, *Proc. Camb. Phil. Soc.*, Vol. 35, pp. 336–338. [552]

P. L. Hsu (1939*b*), On the distribution of roots of certain determinantal equations, *Ann. Eugen.*, Vol. 9, pp. 250–258. [568]

P. L. Hsu (1945*a*), The approximate distributions of the mean and variance of a sample of independent variables, *Ann. Math. Stat.*, Vol. 16, pp. 1–29. [266]

P. L. Hsu (1945*b*), The asymptotic distribution of ratios, *Ann. Math. Stat.*, Vol. 16, pp. 204–210. [266]

V. S. Huzurbazar (1948), The likelihood equation, consistency and the maxima of the likelihood function, *Ann. Eugen.*, Vol. 14, pp. 185–200. [360]

A. E. Ingham (1933), An integral that occurs in statistics, *Proc. Camb. Phil. Soc.*, Vol. 29, pp. 271–276. [552]

J. O. Irwin (1930), On the frequency distribution of the means of samples from populations of certain of Pearson's types, *Metron,* Vol. 8, No. 4, pp. 51–105. [205]

A. T. James (1954), Normal multivariate analysis and the orthogonal group, *Ann. Math. Stat.*, Vol. 25, pp. 40–75. [552, 568]

A. T. James (1960), The distribution of the latent roots of the covariance matrix, *Ann. Math. Stat.*, Vol. 31, pp. 151–158. [573]

G. M. Jenkins and M. B. Priestly (1957), The spectral analysis of time series, *Jour. Roy. Stat. Soc., Series B,* Vol. 19, pp. 1–12. [524]

I. Kaplansky and J. Riordan (1945), Multiple matching and runs by the symbolic method, *Ann. Math. Stat.*, Vol. 16, pp. 272–277. [154]

S. Karlin and H. Rubin (1956), The theory of decision procedures for distributions with monotone likelihood ratio, *Ann. Math. Stat.*, Vol. 27, pp. 272–299. [511]

T. Kawata and H. Sakamoto (1949), On the characterization of the normal population by the independence of the sample mean and sample variance, *Jour. Math. Soc. Japan,* Vol. 1, pp. 111–115. [211]

J. H. B. Kemperman (1956), Generalized tolerance limits, *Ann. Math. Stat.*, Vol. 27, pp. 180–186. [243]

O. Kempthorne (1952), *The Design and Analysis of Experiments,* John Wiley, New York. [232, 297, 308]

D. G. Kendall (1951), Some problems in the theory of queues, *Jour. Roy. Stat. Soc., Series B,* Vol. 13, pp. 151–185. [193]

D. G. Kendall (1953), Stochastic processes occurring in the theory of queues and their analysis by the method of the Markov chain, *Ann. Math. Stat.*, Vol. 24, pp. 338–354. [193]

D. G. Kendall and K. S. Rao (1950), On the generalized second limit-theorem in the calculus of probabilities, *Biometrika,* Vol. 37, pp. 224–230. [127, 128]

M. G. Kendall and B. B. Smith (1939), The problem of *m* rankings, *Ann. Math. Stat.*, Vol. 10, pp. 275–287. [468]

M. G. Kendall (1943 and 1946), *The Advanced Theory of Statistics,* Vol. I (1943), Vol. II (1946) (Vol. I of new three-volume edition appeared in 1958), Charles Griffin, London. [200]

M. G. Kendall and R. M. Sundrum (1953), Distribution-free methods and order properties, *Rev. Int. Stat. Inst.*, Vol. 23, pp. 124–134. [428]

M. G. Kendall (1953), *Rank Correlation Methods,* second edition (first edition, 1948), Charles Griffin, London. [235, 428]

M. G. Kendall (1957), *A Course in Multivariate Analysis,* Charles Griffin, London. [540]

A. Khintchine (1929), Sur la loi des grands nombres, *Comp. Rend. Acad. Sci.*, Vol. 188, pp. 477–479. [110, 254]

J. Kiefer (1948), *Sequential Determination of the Maximum of a Function,* Ph.D. Thesis, Massachusetts Institute of Technology. [474]

J. Kiefer (1952), On minimum variance estimators, *Ann. Math. Stat.*, Vol. 23, pp. 627–629. [352]

J. Kiefer (1953), Sequential minimax search for a maximum, *Proc. Amer. Math. Soc.*, Vol. 4, pp. 502–506. [474]

J. Kiefer (1957), Optimum sequential search and approximation methods under minimum regularity conditions, *Jour. Ind. Appl. Math.*, Vol. 5, pp. 105–136. [474]

J. Kiefer and L. Weiss (1957), Some properties of generalized sequential probability ratio test, *Ann. Math. Stat.*, Vol. 28, pp. 57–74. [491]

J. Kiefer and J. Wolfowitz (1952), Stochastic estimation of the maximum of a regression function, *Ann. Math. Stat.*, Vol. 23, pp. 462–466. [474]

B. F. Kimball (1947), Some basic theorems for developing tests of fit for the case of the non-parametric probability distribution function, I, *Ann. Math. Stat.*, Vol. 18, pp. 540–548. [438]

K. Kishen (1945), On the design of experiments for weighing and making other types of measurements, *Ann. Math. Stat.*, Vol. 16, pp. 294–300. [286]

S. Kitabatake (1958), A remark on a non-parametric test, *Math. Japan,* Vol. 4, pp. 45–49. [437]

T. Kitagawa and M. Mitome (1953), *Tables for the Design of Factorial Experiments,* Baifukan, Tokyo. [297]

A. Kolmogorov (1928), Über die Summen durch den Zufall bestimmer unabhängiger Grössen, *Math. Ann.*, Vol. 99, pp. 309–319. [107]

A. Kolmogorov (1933*a*), *Grundbegriffe der Wahrscheinlichkeitsrechnung,* Ergeb. Math. No. 3, Berlin. (English translation by N. Morrison (1950), Chelsea, New York). [1, 11, 26, 98]

A. Kolmogorov (1933*b*), Sulla determinazione empirica di una legge di distribuzione, *Giorn. dell' Inst. Ital. Attuari,* Vol. 4, pp. 83–91. [339, 341, 438]

A. Kolmogorov (1941), Stationary sequences in Hilbert space (in Russian), *Bull. Math. Univ. Moscow,* Vol. 2, No. 6. [535, 537]

B. O. Koopman (1936), On distributions admitting sufficient statistics, *Trans. Amer. Math. Soc.*, Vol. 39, pp. 399–409. [393]

T. Koopmans (1942), Serial correlation and quadratic forms in normal variables, *Ann. Math. Stat.*, Vol. 13, pp. 14–33. [535]

C. Kraft and L. LeCam (1956), A remark on the roots of the maximum likelihood equation, *Ann. Math. Stat.*, Vol. 27, pp. 1174–1177. [364]

W. H. Kruskal (1952), A nonparametric test for the several-sample problem, *Ann. Math. Stat.*, Vol. 23, pp. 525–540. [468]

S. Kullback and R. A. Leibler (1951), On information and sufficiency, *Ann. Math. Stat.*, Vol. 22, pp. 79–86. [418]

S. Kullback (1959), *Information Theory and Statistics,* John Wiley, New York. [409]

R. G. Laha (1954), On a characterization of the gamma distribution, *Ann. Math. Stat.*, Vol. 25, pp. 784–787. [249]

P. S. Laplace (1814), *Théorie Analytique des Probabilités,* second edition, Paris. [144, 204, 257, 366]

L. LeCam (1956), On the asymptotic theory of estimation and testing hypotheses, *Proc. Third Berkeley Symp. on Math. Stat. and Prob.*, Vol. 1, pp. 129–156, University of California Press. [360, 364]

E. L. Lehmann and C. Stein (1948), Most powerful tests of composite hypotheses. I. Normal distributions, *Ann. Math. Stat.*, Vol. 19, pp. 495–516. [403]

E. L. Lehmann and C. Stein (1949), On the theory of some non-parametric hypotheses, *Ann. Math. Stat.*, Vol. 20, pp. 28–45. [462]

E. L. Lehmann (1950), Some principles of the theory of testing hypotheses, *Ann. Math. Stat.*, Vol. 21, pp. 1–26. [403]

E. L. Lehmann and H. Scheffé (1950), Completeness, similar regions and unbiased estimates, *Sankhyā,* Vol. 10, pp. 305–340. [403]

E. L. Lehmann (1957), A theory of some multiple decision problems, *Ann. Math. Stat.*, Vol. 28, pp. 1–25. [511]

E. L. Lehmann (1959), *Testing Statistical Hypotheses,* John Wiley, New York. [394, 403, 462]

P. Lévy (1925), *Calcul des Probabilités*, Gauthier-Villars, Paris. [1, 116, 119]

P. Lévy (1935), Propriétes asymptotiques des sommes des variables aléatoires indépendantes ou enchainees, *Journ. Math. Pures et Appl.*, Vol. 14, p. 347. [257]

P. Lévy (1937), *Théorie de l'addition des variables aléatoires* (third edition, 1954), Gauthier-Villars, Paris. [1, 122]

G. J. Lieberman and D. B. Owen (1961), *Tables of the Hypergeometric Probability Distribution*, Stanford University Press. [135]

J. W. Lindeberg (1922), Eine neue Herleitung des Exponentialgesetzes in der Wahrscheinlichkeitsrechnung, *Math. Zeits.*, Vol. 15, pp. 211–225. [256–258]

M. Loève (1955), *Probability Theory*, D. Van Nostrand, Princeton, N. J. (second edition, 1960, third edition, 1963). [1, 21, 99, 123]

F. M. Lord (1955), Sampling fluctuations resulting from the sampling of test items, *Psychometrika*, Vol. 20, pp. 1–22. [224]

A. J. Lotka (1939), A contribution to the theory of self-renewing aggregates, with special reference to industrial replacement, *Ann. Math. Stat.*, Vol. 10, pp. 1–25. [191]

R. D. Luce and H. Raiffa (1957), *Games and Decisions*, John Wiley, New York. [502]

R. D. Luce (1959), *Individual Choice Behavior*, John Wiley, New York. [426]

E. Lukacs (1942), A characterization of the normal distribution, *Ann. Math. Stat.*, Vol. 13, pp. 91–93. [211, 250]

A. Lyapunov (1900), Sur une proposition de la théorie des probabilités, *Bull. de l'Acad. Imp. des Sciences, de St. Petersbourg*, Vol. 13, pp. 359–386. [257]

A. Lyapunov (1901), Nouvelle forme da théoreme sur la limite de probabilité, *Mem. Acad. Science of St. Petersbourg*, Vol. 12, pp. 1–24. [257]

W. G. Madow (1938), Contributions to the theory of multivariate statistical analysis, *Trans. Amer. Math. Soc.*, Vol. 44, pp. 454–495. [552]

P. C. Mahalanobis (1930), On tests and measures of group divergence, *Jour. Asiatic Soc. of Bengal*, Vol. 26, pp. 541–588. [560]

P. C. Mahalanobis (1936), On the generalized distance in statistics, *Proc. Nat. Inst. Sci. Calcutta*, Vol. 12, pp. 49–55. [557, 560]

P. C. Mahalanobis, R. C. Bose, and S. N. Roy (1937), Normalization of statistical variates and the use of rectangular coordinates in the theory of sampling distributions, *Sankhyā*, Vol. 3, pp. 1–40. [552]

P. C. Mahalanobis (1940), A sample survey of the acreage under jute in Bengal, with discussion on planning of experiments, *Proc. Second Indian Stat. Conference*, Statistical Publishing Society, Calcutta. [474]

S. Malmquist (1951), On a property of order statistics from a rectangular distribution, *Skand. Aktuar.*, Vol. 33, pp. 214–222. [250]

S. Malmquist (1954), On certain confidence contours for distribution functions, *Ann. Math. Stat.*, Vol. 25, pp. 523–542. [438]

H. B. Mann (1943), On the construction of sets of orthogonal Latin squares, *Ann. Math. Stat.*, Vol. 14, pp. 401–414. [234]

H. B. Mann and A. Wald (1943), On the statistical treatment of linear stochastic difference equations, *Econometrica*, Vol. 11, pp. 173–220. [537]

H. B. Mann and D. R. Whitney (1947), On a test of whether one of two random variables is stochastically larger than the other, *Ann. Math. Stat.*, Vol. 18, pp. 50–60. [459, 460, 461]

H. B. Mann (1949), *Analysis and Design of Experiments*, Dover Publications, New York. [297]

A. A. Markov (1900), *Wahrscheinlichkeitsrechumung,* Teubner, Leipzig. (German translation of Russian second edition (1908) appeared in 1912; original Russian edition appeared in 1900.) [257, 284, 285]

F. J. Massey (1950), A note on the estimation of a distribution function by confidence limits, *Ann. Math. Stat.,* Vol. 21, pp. 116–119. [339, 340, 341]

F. J. Massey (1951), The distribution of the maximum deviation between two sample cumulative step functions, *Ann. Math. Stat.,* Vol. 22, pp. 125–128. [455]

H. C. Mathisen (1943), A method of testing the hypothesis that two samples are from the same population, *Ann. Math. Stat.,* Vol. 14, pp. 188–194. [441]

K. Matusita (1957), Decision rule based on the distance for the classification problem, *Ann. Inst. Stat. Math.,* Vol. 8, pp. 67–77. [252]

J. W. Mauchly (1940), Significance test for sphericity of a normal n-variate distribution, *Ann. Math. Stat.,* Vol. 11, pp. 204–209. [600]

P. J. McCarthy (1947), Approximate solutions for means and variances in a certain class of box problems, *Ann. Math. Stat.,* Vol. 18, pp. 349–383. [144]

E. J. McShane (1944), *Integration,* Princeton University Press. [348]

E. J. McShane and T. Botts (1959), *Real Analysis,* D. Van Nostrand, Princeton, N. J. [21, 348]

R. von Mises (1931), *Wahrscheinlichskeitsrechnung, und ihre Anwendung in der Statistik und Theoretischen Physik,* Deuticke, Leipzig. [10, 438]

E. C. Molina (1942), *Poisson's Exponential Binomial Limit,* D. Van Nostrand, Princeton, N. J. [141]

A. M. Mood (1940), The distribution theory of runs, *Ann. Math. Stat.,* Vol. 11, pp. 367–392. [145, 150]

A. M. Mood (1946), On Hotelling's weighing problem, *Ann. Math. Stat.,* Vol. 17, pp. 432–446. [286]

A. M. Mood (1950), *Introduction to the Theory of Statistics,* McGraw-Hill, New York. [470]

A. M. Mood (1951), On the distribution of the characteristic roots of normal second-moment matrices, *Ann. Math. Stat.,* Vol. 22, pp. 266–273. [568]

A. M. Mood (1954), On the asymptotic efficiency of certain nonparametric two-sample tests, *Ann. Math. Stat.,* Vol. 25, pp. 514–522. [470]

P. A. P. Moran (1948), Some theorems on time series, II. The significance of the serial correlation coefficient, *Biometrika,* Vol. 35, pp. 253–260. [535]

P. A. P. Moran, J. W. Whitfield, and H. E. Daniels (1950), Symposium on ranking methods, *Jour. Roy. Stat. Soc., Series B,* Vol. 12, pp. 153–191. [428]

P. M. Morse (1958), *Queues, Inventories and Maintenance,* John Wiley, New York. [193]

S. Moriguti (1954), Confidence limits for a variance component, *Reports of Statistical Applications in Research, Japanese Union of Scientists and Engineers,* Vol. 3, No. 2, pp. 29–41. [308]

F. C. Mosteller (1946), On some useful "inefficient" statistics, *Ann. Math. Stat.,* Vol. 17, pp. 377–408. [274]

M. E. Munroe (1953), *Introduction to Measure and Integration,* Addison-Wesley, Cambridge, Mass. [1, 26]

R. B. Murphy (1948), Non-parametric tolerance limits, *Ann. Math. Stat.,* Vol. 19, pp. 581–589. [335]

K. R. Nair (1940), Table of confidence intervals for the median in samples from any continuous population. *Sankhyā,* Vol. 4, pp. 551–558. [331]

National Bureau of Standards (1942), *Tables of Probability Functions,* Vol. 2, New York. [157]

National Bureau of Standards (1949), *Tables of the Binomial Probability Distribution, Appl. Math. Series,* Vol. 6. [138]

J. von Neumann (1950), *Functional Operators,* Vol. 1, Princeton University Press. [49]

J. von Neumann and O. Morgenstern (1944), *Theory of Games and Economic Behavior,* third edition 1953, Princeton University Press. [502]

J. Neyman and E. S. Pearson (1928), On the use and interpretation of certain test criteria for purposes of statistical inference, *Biometrika,* Vol. 20 A, Part I, pp. 175–240, Part II, pp. 263–294. [394, 403]

J. Neyman and E. S. Pearson (1933), On the problem of the most efficient tests of statistical hypotheses, *Trans. Roy. Soc. London, Series A,* Vol. 231, pp. 289–337. [394, 398, 403]

J. Neyman (1934), On the two different aspects of the representative method: The method of stratified sampling and the method of purposive selection, *Jour. Roy. Stat. Soc.,* Vol. 97, pp. 558–625. [317]

J. Neyman (1935), Su un teorema concernente le cosiddetti statistiche sufficienti, *Giorn. Inst. Ital. Attuari,* Vol. 6, pp. 320–334. [356]

J. Neyman (1937), Outline of a theory of statistical estimation based on the classical theory of probability, *Philos. Trans. Roy. Soc. London, Series A,* Vol. 236, pp. 333–380. [366]

J. Neyman (1939), On a new class of "contagious" distributions, applicable in entomology and bacteriology, *Ann. Math. Stat.,* Vol. 10, pp. 35–57. [152]

M. Ogawara (1951), A note on the test of serial correlation coefficients, *Ann. Math. Stat.,* Vol. 22, pp. 115–118. [535]

M. Okamoto (1952), On a non-parametric test, *Osaka Math. Jour.,* Vol. 4, pp. 77–85. [437]

E. G. Olds (1938), Distribution of sums of squares of rank differences for small numbers of individuals, *Ann. Math. Stat.,* Vol. 9, pp. 133–148. [467]

I. Olkin (1951), *On Distribution Problems in Multivariate Analysis,* Institute of Statistics Mimeograph Series, Report No. 8, University of North Carolina. [568]

I. Olkin and S. N. Roy (1954), On multivariate distribution theory, *Ann. Math. Stat.,* Vol. 25, pp. 329–339. [568]

I. Olkin and J. W. Pratt (1958), A multivariate Chebyshev inequality, *Ann. Math. Stat.,* Vol. 29, pp. 226–234. [112]

E. Parzen (1957), On consistent estimates of the spectrum of a stationary time series, *Ann. Math. Stat.,* Vol. 28, pp. 329–348. [524]

E. Parzen (1960), *Modern Probability Theory and its Applications,* John Wiley, New York. [250]

E. S. Pearson and H. O. Hartley (1954), *Biometrika Tables for Statisticians,* Vol. 1, Cambridge University Press. [183, 185]

K. Pearson (1900), On a criterion that a system of deviations from the probable in the case of a correlated system of variables is such that it can be reasonably supposed to have arisen in random sampling, *Phil. Mag.,* Vol. 50, pp.157–175. [184, 262, 389, 431]

K. Pearson (1906), On the curves which are most suitable for describing the frequency of random samples of a population, *Biometrika,* Vol. 5, pp. 172–175. [171]

K. Pearson (1920), On the probable errors of frequency constants, Part III, *Biometrika,* Vol. 13, pp. 113–132. [274]

K. Pearson, editor (1922), *Tables of the Incomplete Gamma Function,* Cambridge University Press. [171]

K. Pearson, editor (1934), *Tables of the Incomplete Beta Function,* Cambridge University Press. [174, 187]

E. J. G. Pitman (1936), Sufficient statistics and intrinsic accuracy, *Proc. Camb. Phil. Soc.,* Vol. 32, pp. 567–579. [393]

E. J. G. Pitman (1937*a*), The "closest" estimates of statistical parameters, *Proc. Camb. Phil. Soc.,* Vol. 33, pp. 212–222. [249]

E. J. G. Pitman (1937*b*), Significance tests which may be applied to samples from any populations, *Jour. Roy. Stat. Soc., Series B,* Vol. 4, pp. 119–130. [464]

E. J. G. Pitman (1938), Significance tests which may be applied to samples from any populations, III. The analysis of variance test, *Biometrika,* Vol. 29, pp. 322–335. [465]

S. D. Poisson (1837), *Recherches sur la probabilité des jugements en matiére criminelle et en matiére civile, precédées des régles générales du calcul des probabilités,* Paris. [140]

M. H. Quenouille (1948), Some results in the testing of serial correlation coefficients, *Biometrika,* Vol. 35, pp. 261–267. [528, 535]

H. Raiffa and R. Schlaifer (1961), *Applied Statistical Decision Theory,* Graduate School of Business Administration, Harvard University. [502]

C. R. Rao (1945), Information and accuracy attainable in the estimation of statistical parameters, *Bull. Calcutta Math. Soc.,* Vol. 37, pp. 81–91. [353, 354, 357]

C. R. Rao (1949), Sufficient statistics and minimum variance estimates, *Proc. Camb. Phil. Soc.,* Vol. 45, pp. 213–218. [364, 393]

C. R. Rao (1952), *Advanced Statistical Methods in Biometric Research,* John Wiley, New York. [540]

G. Rasch (1948), A functional equation for Wishart's distribution, *Ann. Math. Stat.,* Vol. 19, pp. 262–266. [552]

A. Renyi (1953), On the theory of order statistics, *Act. Math. Acad. Sci. Hungary,* Vol. 4, pp. 191–231. [250]

W. E. Ricker (1937), The concept of confidence or fiducial limits applied to the Poisson frequency distribution, *Jour. Amer. Stat. Assn.,* Vol. 32, pp. 349–356. [369]

P. R. Rider (1955), The distribution of the product of maximum values in samples from a rectangular distribution, *Jour. Amer. Stat. Assn.,* Vol. 50, pp. 1142–1143. [249]

H. Robbins (1944*a*), On the measure of a random set, *Ann. Math. Stat.,* Vol. 15, pp. 70–74. [250]

H. Robbins (1944*b*), On distribution-free tolerance limits in random sampling, *Ann. Math. Stat.,* Vol. 15, pp. 214–216. [334]

H. Robbins (1948), Convergence of distributions, *Ann. Math. Stat.,* Vol. 19, pp. 72–76. [112]

H. Robbins (1952), Some aspects of the sequential design of experiments, *Bull. Amer. Math. Soc.,* Vol. 58, pp. 527–535. [474]

H. Robbins (1954), A remark on the joint distribution of cumulative sums, *Ann. Math. Stat.,* Vol. 25, pp. 614–616. [112]

H. Robbins and S. Monro (1951), A stochastic approximation method, *Am. Math. Stat.,* Vol. 22, pp. 400–407. [474]

H. G. Romig (1953), 50–100 *Binomial Tables,* John Wiley, New York. [138]

S. N. Roy (1939), *p*-statistics or some generalizations in analysis of variance appropriate to multivariate problems, *Sanhkyā,* Vol. 4, pp. 381–396. [568]

S. N. Roy and R. C. Bose (1953), Simultaneous confidence interval estimation, *Ann. Math. Stat.,* Vol. 24, pp. 513–536. [290]

S. N. Roy (1954), Some further results in simultaneous confidence interval estimation, *Ann. Math. Stat.,* Vol. 25, pp. 752–761. [290]

S. N. Roy (1957), *Some Aspects of Multivariate Analysis,* John Wiley, New York; Indian Statistical Institute, Calcutta. [540]

S. Rushton (1950), On a sequential t-test, *Biometrika,* Vol. 37, pp. 326–333. [472]

S. Rushton (1952), On a two-sided sequential t-test, *Biometrika,* Vol. 39, pp. 302–308. [472]

S. Saks (1937), *Theory of the Integral,* Second Edition (English translation by L. C. Young), Stechert, New York. [21, 348]

I. R. Savage (1962), *Bibliography of nonparametric statistics,* Harvard University Press, Cambridge. [235, 428]

L. J. Savage (1954), *The Foundations of Statistics,* John Wiley, New York. [502]

S. R. Savur (1937), The use of the median in tests of significance, *Proc. Indian Acad. Sci.,* Section A, Vol. 5, pp. 564–576. [331]

H. Scheffé (1943), Statistical inference in the non-parametric case, *Ann. Math. Stat.,* Vol. 14, pp. 305–332. [428]

H. Scheffé (1947), A useful convergence theorem for probability distributions, *Ann. Math. Stat.,* Vol. 18, pp. 434–438. [112]

H. Scheffé (1953), A method for judging all contrasts in the analysis of variance, *Biometrika,* Vol. 40, pp. 87–104. [290, 291]

H. Scheffé (1959), *The Analysis of Variance,* John Wiley, New York. [297, 308]

A. Schuster (1898), On the investigation of hidden periodicities with application to a supposed 26-day period of meteorological phenomena, *Terrestrial Magnetism,* Vol. 3, pp. 13–41. [529]

W. F. Sheppard (1898), On the calculation of the most probable values of frequency-constants for data arranged according to equidistant divisions of a scale, *Proc. London Math. Soc.,* Vol. 29, pp. 353–380. [328]

C. E. Shannon (1948), A mathematical theory of communication, *Bell System Tech. Jour.,* Vol. 27, pp. 379–423 and 623–656. [409]

B. Sherman (1950), A random variable related to the spacing of sample values, *Ann. Math. Stat.,* Vol. 21, pp. 339–361. [438]

W. A. Shewhart (1931), *Economic Control of Quality of Manufactured Product,* D. Van Nostrand, Princeton, N. J. [334]

S. S. Shrikhande (1952), On the dual of some balanced incomplete block designs, *Biometrics,* Vol. 8, pp. 66–72. [232]

E. Slutzky (1937), The summation of random causes as the source of cyclic processes, *Econometrica,* Vol. 5, pp. 105–146. [537]

N. Smirnov (1935), Über die Verteiling des allgemeinen Gliedes in der Variationsreihe, *Metron,* Vol. 12, No. 2, pp. 59–81. [272, 274]

N. Smirnov (1939*a*), On the estimation of the discrepancy between empirical curves of distribution for two independent samples, *Bull. Math. Univ. Moscow,* Vol. 2, No. 2, pp. 3–16. [336, 339, 441, 454]

N. Smirnov (1939*b*), Sur les écarts de la courbe de distribution empirique (In Russian, with French summary), *Rec. Math. N. S.*, Vol. 6, pp. 3–26. [438]

N. Smirnov (1944), Approximation of a distribution function from a sample (in Russian), *Uspehi Matemat. Nauk*, Vol. 10, pp. 179–206. [336]

W. L. Smith (1958), Renewal theory and its ramifications, *Jour. Roy. Stat. Soc., Series B*, Vol. 20, pp. 243–302. [191]

G. W. Snedecor (1937), *Statistical Methods*, Iowa State College Press, Ames, Iowa. [187]

P. N. Somerville (1958), Tables for obtaining non-parametric tolerance limits, *Am. Math. Stat.*, Vol. 29, pp. 599–601. [335]

Statistical Research Group, Columbia University (1945), *Sequential Analysis of Statistical Data: Applications*, Columbia University Press. [496]

Statistical Research Group, Columbia University (1947), *Techniques of Statistical Analysis*, McGraw-Hill, New York. [496]

C. Stein (1945), A two-sample test for a linear hypothesis whose power is independent of the variance, *Ann. Math. Stat.*, Vol. 16, pp. 243–258. [497]

C. Stein (1946), A note on cumulative sums, *Ann. Math. Stat.*, Vol. 17, pp. 498–499. [484]

C. Stein and A. Wald (1947), Sequential confidence intervals for the mean of a normal distribution with known variance, *Ann. Math. Stat.*, Vol. 18, pp. 427–433. [496]

F. F. Stephan and P. J. McCarthy (1958), *Sampling Opinion*, John Wiley, New York. [314]

W. L. Stevens (1939), Distributions of groups in a sequence of alternatives, *Ann. Eugen.*, Vol. 9, pp. 10–17. [149]

P. V. Sukhatme (1954), *Sampling Theory of Surveys with Applications*, Iowa State College Press, Ames, Iowa. [314]

E. Sverdrup (1947), Derivation of the Wishart distribution of the second order sample moments by straightforward integration of a multiple intergral, *Skand. Aktuar*, Vol. 30, pp. 151–166. [552]

F. S. Swed and C. Eisenhart (1943), Tables for testing randomness of grouping in a sequence of alternatives, *Ann. Math. Stat.*, Vol. 14, pp. 66–87. [149]

P. C. Tang (1938), The power function of the analysis of variance tests with tables and illustrations of their use, *Stat. Res. Mem.*, University College, London, Vol. 2, pp. 128–149. [247]

T. N. Thiele (1903), *Theory of Observations*, Layton, London (Reprinted in *Ann. Math. Stat.*, Vol. 2, 1931, pp. 165–307). [115]

W. R. Thompson (1936), On confidence ranges for the median and other expectation distributions for populations of unknown distribution form, *Ann. Math. Stat.*, Vol. 7, pp. 122–128. [331]

G. Tintner (1940), *The Variate Difference Method*, Principia Press, Bloomington, Indiana. [528]

L. H. C. Tippett (1925), On the extreme individuals and the range of samples taken from a normal population, *Biometrika*, Vol. 17, pp. 364–387. [248]

J. W. Tukey (1946), An inequality for deviations from medians, *Ann. Math. Stat.*, Vol. 17, pp. 75–78. [112]

J. W. Tukey (1947), Nonparametric estimation, II. Statistically equivalent blocks and tolerance regions—the continuous case, *Ann. Math. Stat.*, Vol. 18, pp. 529–539. [240]

J. W. Tukey (1948), Nonparametric estimation, III. Statistically equivalent blocks and multivariate tolerance regions—the discontinuous case, *Ann. Math. Stat.*, Vol. 19, pp. 30–39. [243]

J. W. Tukey (1949*a*), The sampling theory of power spectrum estimates, *Proceedings on Applications of Autocorrelation Analysis to Physical Problems* (NAVEXOS-P-735). [524, 526]

J. W. Tukey (1949*b*), Moments of random group size distributions, *Ann. Math. Stat.*, Vol. 20, pp. 523–539. [153, 154]

J. W. Tukey (1950), Some sampling simplified, *Jour. Amer. Stat. Assn.*, Vol. 45, pp. 501–519. [220, 226]

J. W. Tukey (1953), *The Problem of Multiple Comparisons*, Unpublished Manuscript, Princeton University. [290, 291, 294]

J. W. Tukey (1956*a*), Keeping moment-like sampling computations simple, *Ann. Math. Stat.*, Vol. 27, pp. 37–54. [220]

J. W. Tukey (1956*b*), Variances of variance components: I. Balanced designs, *Ann. Math. Stat.*, Vol. 27, pp. 722–736. [308]

J. W. Tukey (1957*a*), Variances of variance components: II. The unbalanced single classification, *Ann. Math. Stat.*, Vol. 28, pp. 43–56. [308]

J. W. Tukey (1957*b*), Variances of variance components: III. Third moments in a balanced single classification, *Ann. Math. Stat.*, Vol. 28, pp. 378–384. [308]

J. W. Tukey (1957*c*), Some examples with fiducial relevance, *Ann. Math. Stat.*, Vol. 28, pp. 687–695. [371]

D. F. Votaw, Jr. (1948), Testing compound symmetry in a normal multivariate distribution, *Ann. Math. Stat.*, Vol. 19, pp. 447–473. [601]

A. Wald and J. Wolfowitz (1939), Confidence limits for continuous distribution functions, *Ann. Math. Stat.*, Vol. 10, pp. 105–118. [339]

A. Wald (1939), Contributions to the theory of statistical estimation and testing hypotheses, *Ann. Math. Stat.*, Vol. 10, pp. 299–326. [401, 402]

A. Wald and J. Wolfowitz (1940), On a test whether two samples are from the same population, *Ann. Math. Stat.*, Vol. 11, pp. 147–162. [441, 452]

A. Wald (1941*a*), Asymptotically most powerful tests of statistical hypotheses, *Ann. Math. Stat.*, Vol. 12, pp. 1–19. [403, 415]

A. Wald (1941*b*), Some examples of asymptotically most powerful tests, *Ann. Math. Stat.*, Vol. 12, pp. 396–408. [403]

A. Wald (1942), Asymptotically shortest confidence intervals, *Ann. Math. Stat.*, Vol. 13, pp. 127–137. [376]

A. Wald (1943), An extension of Wilks' method for setting tolerance limits, *Ann. Math. Stat.*, Vol. 14, pp. 45–55. [240]

A. Wald and J. Wolfowitz (1944), Statistical tests based on permutations of the observations, *Ann. Math. Stat.*, Vol. 15, pp. 358–372. [266, 267, 464]

A. Wald (1945), Sequential tests of statistical hypotheses, *Ann. Math. Stat.*, Vol. 16, pp. 117–186. [474, 476, 483, 487, 490, 492, 494]

A. Wald and J. Wolfowitz (1946), Tolerance limits for a normal distribution, *Ann. Math. Stat.*, Vol. 17, pp. 208–215. [343]

A. Wald (1947*a*), *Sequential Analysis*, John Wiley, New York. [472, 474, 489, 490, 496]

A. Wald (1947*b*), Foundations of a general theory of statistical decision functions, *Econometrica*, Vol. 15, pp. 279–313. [502]

A. Wald (1948), Asymptotic properties of the maximum likelihood estimate of an unknown parameter of a discrete stochastic process, *Ann. Math. Stat.*, Vol. 19, pp. 40–46. [364]

A. Wald and J. Wolfowitz (1948), Optimum character of the sequential probability ratio test, *Ann. Math. Stat.*, Vol. 19, pp. 326–339. [492]

A. Wald (1949*a*), Statistical decision functions, *Ann. Math. Stat.*, Vol. 20, pp. 165–205. [502]

A. Wald (1949*b*), Note on the consistency of the maximum likelihood estimate, *Ann. Math. Stat.*, Vol. 20, pp. 595–601. [360]

A. Wald (1950), *Statistical Decision Functions*, John Wiley, New York. [502, 504, 511]

D. L. Wallace (1958), Asymptotic approximations to distributions, *Ann. Math. Stat.*, Vol. 29, pp. 635–654. [266]

W. A. Wallis (1939), The correlation ratio for ranked data, *Jour. Amer. Stat. Assn.*, Vol. 34, pp. 533–538. [468]

J. E. Walsh (1946), Some order statistic distributions for samples of size 4, *Ann. Math. Stat.*, Vol. 17, pp. 246–248. [253]

J. E. Walsh (1949), Some significance tests for the median which are valid under very general conditions, *Ann. Math. Stat.*, Vol. 20, pp. 64–81. [430]

L. Weiss (1961), *Stastical Decision Theory*, McGraw-Hill, New York. [502]

B. L. Welch (1937), On the z-test in randomized blocks and Latin squares, *Biometrika*, Vol. 29, pp. 21–52. [465]

E. T. Whittaker and G. N. Watson (1927), *A Course in Modern Analysis* (fourth edition), Cambridge University Press. [118, 177]

P. Whittle (1951), *Hypothesis Testing in Time Series*, University of Uppsala. [535]

D. V. Widder (1947), *Advanced Calculus*, Prentice-Hall, New York. [57]

N. Wiener (1949), *Extrapolation, Interpolation and Smoothing of Stationary Time Series*, John Wiley, New York. [535, 537]

F. Wilcoxon (1945), Individual comparisons by ranking methods, *Biometrics*, Vol. 1, pp. 80–83. [459]

S. S. Wilks (1932), Certain generalizations in the analysis of variance, *Biometrika*, Vol. 24, pp. 471–494. [554, 564]

S. S. Wilks (1934), Moment-generating operators for determinants of product moments in samples from a normal system, *Ann. Math.*, Vol. 35, pp. 312–340. [601]

S. S. Wilks (1935), On the independence of k sets of normally distributed statistical variables, *Econometrica*, Vol. 3, pp. 309–326. [599]

S. S. Wilks (1938*a*), The large-sample distribution of the likelihood ratio for testing composite hypotheses, *Ann. Math. Stat.*, Vol. 9, pp. 60–62. [403, 419]

S. S. Wilks (1938*b*), Shortest average confidence intervals from large samples, *Ann. Math. Stat.*, Vol. 9, pp. 166–175. [376]

S. S. Wilks and J. F. Daly (1939), An optimum property of confidence regions associated with the likelihood function, *Ann. Math. Stat.*, Vol. 10, pp. 225–235. [388]

S. S. Wilks (1941), On the determination of sample sizes for setting tolerance limits, *Ann. Math. Stat.*, Vol. 12, pp. 91–96. [334]

S. S. Wilks (1942), Statistical prediction with special reference to the problem of tolerance limits, *Ann. Math. Stat.*, Vol. 13, pp. 400–409. [334]

S. S. Wilks (1946), Sample criteria for testing equality of means, equality of variances, and equality of covariances in a normal multivariate distribution, *Ann. Math. Stat.*, Vol. 17, pp. 257–281. [601]

S. S. Wilks (1948), Order statistics, *Bull. Am. Math. Soc.*, Vol. 54, pp. 5–50. [428]

S. S. Wilks (1959*a*), Nonparametric statistical inference, *Probability and Statistics, The Harald Cramér Volume*, Almqvist and Wiksell, Stockholm. [428]

S. S. Wilks (1959*b*), Recurrence of extreme observations, *Jour. Austral. Math. Soc.*, Vol. 1, pp. 106–112. [251]

S. S. Wilks (1960*a*), Multidimensional statistical scatter, *Contributions to Probability and Statistics in Honor of Harold Hotelling*, Stanford University Press. [547]

S. S. Wilks (1960*b*), A two-stage scheme for sampling without replacement, *Bull. de l'Institut Intern.*, Vol. 37, No. 2, pp. 241–248. [319]

S. S. Wilks (1961), A combinatorial test for the problem of two samples from continuous distributions, *Proc. Fourth Berkeley Symp. on Math. Stat. and Prob.*, University of California Press. [441, 451]

J. D. Williams (1946), An approximation to the probability integral, *Ann. Math. Stat.*, Vol. 17, pp. 363–365. [188]

E. B. Wilson (1927), Probable inference, the law of succession, and statistical inference, *Jour. Amer. Stat. Assn.*, Vol. 27, pp. 209–212. [366]

J. Wishart (1928), The generalized product moment distribution in samples from a normal multivariate population, *Biometrika*, Vol. 20A, pp. 32–52. [548, 551, 552]

J. Wishart and M. S. Bartlett (1932), The distribution of the second order moment statistics in a normal system, *Proc. Camb. Phil. Soc.*, Vol. 28, pp. 455–459. [552]

H. Wold (1938), *A Study in the Analysis of Stationary Times Series*, University of Uppsala. [514, 537]

J. Wolfowitz (1949), Non-parametric statistical inference, *Proc. (First) Berkeley Symposium on Math. Stat. and Prob.*, pp. 93–113, University of California Press. [428]

H. Working and H. Hotelling (1929), Application of the theory of error to the interpretation of trends, *Jour. Amer. Stat. Assn.*, March Supplement, pp. 73–85. [297]

R. Wormleighton (1959), Some tests of permutation symmetry, *Ann. Math. Stat.*, Vol. 30, pp. 1005–1017. [468]

F. Yates (1949), *Sampling Methods for Censuses and Surveys*, Charles Griffin, London. [314]

G. U. Yule (1924), A mathematical theory of evolution based on the conclusions of Dr. J. C. Willis, *Phil. Trans. Roy. Soc., London, Series B*, Vol. 213, pp. 21–87. [192]

Subject Index

Admissible statistical decision functions, 504
Almost certain convergence, 106
Amount of information, 353
Analysis of variance, Model I, 297–305
examples, *see* Experimental design
Model II, 308–313
most general form of, for Model I, 407
multidimensional, 561–564
table, 301, 310
A posteriori probability, 509
A priori distribution, 401, 508
Arc sine law, 471
Asymptotic efficiency of an estimator, 363, 380, 381
Asymptotic normality of a distribution, 256
Asymptotically shortest confidence intervals, 374–376
Asymptotically unbiased tests, definition of, 414
equivalent, 415
Autocorrelation function, 516
Autocovariance, *see* Covariance function of time series
Autoregressive time series, 537
Average outgoing quality limit of a sampling plan, 398
Average sample number, of a Cartesian sequential test, 480
of a general sequential test, 476
of a probability ratio sequential test, 489–490
Bayes solutions of statistical decision problem, 508–511
Behrens-Fisher problem, 371
Bertrand's ballot problem, 470
Beta distribution, definition of, 173
mean of, 174
moments of, 174
relation between gamma distribution and, 174
relation between Snedecor distribution and, 187
relation between Student distribution and, 186
variance of, 174
Beta function, 174
Between-strata component of variance, 314
Bias, of estimators, 354
of a statistical test, 395
Binomial distribution, asymptotic normality of, for large number of trials, 257
asymptotically shortest confidence interval for parameter in, 391
characteristic function of, 137
confidence interval for parameter in, 369
definition of, 137
distribution of sum (and mean) of sample from, 206
efficiency of sample mean for estimating parameter in, 389
inverse sine transformation for large samples from, 274, 391

Binomial distribution, maximum likelihood estimator for parameter in, 391
mean of, 137
probability ratio sequential test for, 494–496
reproductivity of, 137
sufficiency of sample mean for estimating parameter in, 389
variance of, 137
Binomial distributions, k-sample problem for, 423
likelihood ratio test for equality of parameters in several, 423
two-sample problem for, 423
Binomial waiting-time distribution, characteristic function of, 144
definition of, 144
logarithmic transformation for large samples from, 274
mean of, 144
variance of, 144
Bivariate, cumulative distribution function, 41
probability density function, 46
probability function, 44
Bivariate normal distribution, characteristic function of, 161
conditional distribution from, 163
correlation coefficient between marginal c.d.f. transforms of variables in, 188
correlation coefficient between one variable and the marginal c.d.f. transform of the other in, 246
covariance matrix of, 160
definition of, 158
distribution of correlation coefficient in samples from, 593–594
distribution of ratio of random variables having, 188
marginal distributions from, 162
means of, 160
probability density function of, 161
regression functions from, 163
Blackwell-Rao theorem for obtaining improved estimators from sufficient statistics, 357
Block frequencies, definition of, 443
distribution of, 443
Block frequency counts, definition of, 444
distribution of, 445
moments of, 445
Bloodtesting problem, 152
Boltzmann's H-function, 409
Boolean field, definition of, 8
generation of, 8
Borel cylinder set, 97
Borel field, definition of, 8
generation of, 9
in k-space, 10
in R_∞, 98
minimal, 11
on the real line, 9
Branching process, 131–132

Card-matching problem, 154
Cartesian product, of probability spaces, 19
of sample spaces, 16
of sets, 16
Cartesian sequential test, as a nonparametric sequential test, 482
average sample number of, 480
definition of, 479
for exponential distribution, 498
optimum construction for, 481, 499
r-fold, 481–482
r-fold for exponential distribution, 499
Canonical correlation, 587–592
Canonical correlation coefficients, definition of, 588
distribution of, 591
Cauchy distribution, definition of, 130
distribution of mean of samples from, 130
efficiency of sample median for estimation of location parameter of, 391
failure of convergence in probability of mean of sample from, 256
Cell frequencies, definition of, 431
distribution of, 431
Cell frequency counts, definition of, 433
distribution of, 433
moments of, 433–434

Central limit theorem, 257–258
Change of variable in a probability element, 55
Characteristic function, of binomial distribution, 137
of bivariate normal distribution, 161
of chi-square distribution, 183
of gamma distribution, 171
of independent random variables, 120
of linear function of random variables, 121
of multinomial distribution, 139
of multivariate normal distribution, 168
of normal distribution, 157
of a random variable, 113
of a set, 395
of a vector random variable, 119
Chebychev inequality, 75, 255
multidimensional, 92, 112, 274
Chi-square distribution, approximate normality of, for large number of degrees of freedom, 189
characteristic function of, 183
definition of, 183
degrees of freedom of, 183
mean of, 183
moments of, 183
noncentral, 247
relation between gamma distribution and, 183
reproductivity of, 183
variance of, 183
Chi-square test, for independence in a contingency table, 424
for a multinomial distribution, 425
Choice behavior model, likelihood ratio test for, 426–427
Classes of sets, Boolean, 8
Borel, 8
completely additive, 8
finitely additive, 8
sigma-algebra, 8
Class-size distributions, case of block frequency counts, 445
case of cell frequency counts, 433
the hypergeometric case, 153–154
the multinomial case, 153
Cochran's theorem for a sample from a normal distribution, 212–214
Coefficient of correlation, *see* Correlation coefficient
Coefficient of variation, 74
Coin-tossing, long leads in, 471
Column effects, in Latin square experimental designs, 304
in three-way experimental designs, 302
in two-way experimental designs, 297–298
Component of variance, between-strata, 314
within-strata, 314
Components of variance, *see* Variance components
Composite statistical hypothesis, definition of, 395
nonparametric, 429
Wald's reduction of, to simple statistical hypothesis, 401–402
Conditional cumulative distribution function, 60, 61, 65
Conditional probability, 24
Conditional random variable, 61–66
mean of, 84
probability density function of, 64, 66, 68
probability function of, 62, 66, 68
variance of, 84
Conditional random variables, continuous, 63, 66, 68
definition of, 61
discrete, 62, 66, 68
Confidence band for a continuous c.d.f., 339–341
Confidence coefficient, 282, 366
Confidence contours, as nonparametric tests, 438–441
for a continuous c.d.f., 336–339
Confidence interval, asymptotic, from large samples, 372
definition of, general case, 365–366
for difference of means of two normal distributions with equal variances, 324
for difference of location parameters of two continuous distributions, 469–470

Confidence interval, for mean of a normal distribution, 282
for mean of a Poisson distribution, 369
for median of a continuous c.d.f., 330, 342
for median of a finite population, 342–343
for median of a second sample from order statistics of a first sample, 342
for parameter of a binomial distribution, 369
for parameter of a hypergeometric distribution, 369
for the $(n+1)$st observation from a normal distribution, 328
for range of a rectangular distribution, 390
for variance of a normal distribution, 282–283
of fixed length for mean of a normal distribution, 497–498
Confidence intervals, asymptotically shortest, 374–376
construction of, from samples from continuous c.d.f.'s, 366–368
construction of, from samples from discrete distributions, 368–369
for main effects in experimental designs, 300
for main effects in one-factor experimental design, 325–326
for quantiles, 329–332
for quantiles in finite populations, 333
for quantile intervals, 332
for regression coefficients in normal regression theory, 289
simultaneous, *see* Simultaneous confidence intervals
Confidence limits, *see* Confidence intervals
Confidence regions, asymptotically equivalent, 385
asymptotically smallest, 384–388
definition of, general case, 381–382
for mean and variance of a normal distribution, 383
for parameters of a multinomial distribution, 388–389
Confidence regions, for regression coefficients in normal regression theory, 290, 324
for vectors of main effects in experimental designs, 300, 325–326
for vector of means of a multivariate normal distribution, 594
Consistency of a statistical test, 396
Consistent estimator, definition of, 351
multidimensional, 380
Contagious distribution, 152
Contingency table, chi-square test of independence in, 424
likelihood ratio test for independence in two-way, 423
likelihood ratio test for independence in three-way, 424
Continuous cumulative distribution function, confidence contour for, 336–339
confidence band for, 339–341
definition of, 36, 45, 52
Continuous random variable, definition of, 36, 45
probability density function of, 37, 47, 52
probability element of, 37, 46
Continuous waiting-time distribution, 172–173
Consumer's risk, 397
Convergence, almost certain, 106
in distribution, 100, 103
in probability, 99, 103, 105
in the mean, 100
of functions of components in stochastic processes, 102–105
of maximum likelihood estimators, 359–360, 379–380
of sample mean in probability, 254
of vector of sample means, 274
set, 106
stochastic, 99
with probability one, 106
Convolution of distribution functions, 204
Correlation coefficient, definition of, 78
distribution of, in samples from a normal distribution, 593–594
Fisher's transformation for, 276
multiple, 91, 95
partial, 94

Correlation coefficient, sampling distribution of multiple, 596–597
Correlation coefficients, canonical, 588
  distribution of canonical, 591
  distribution of matrix of, 592–593
Correlation ratio, definition of, 86
  linear, 88, 91
  multiple, 86
Covariance, between two linear functions of random variables, 83
  definition of, 78
Covariance function of time series, definition of, 516
  estimator for, 522–523
  examples, 537
  spectral representation of, 517, 520
Covariance matrices, test for equality of, in two multivariate normal distributions, 595
Covariance matrix, definition of, 80
  inverse of, 80
  of finite population, 222
  of hypergeometric distribution, 136
  of linear functions of random variables, 83
  of likelihood estimators of parameters, 380–381
  of multinomial distribution, 139
  of normal distribution, 160, 166
  of sample mean and median in large samples from a normal distribution, 275
Coverages, covariance matrix of, 248
  distribution of sums of, 237–238
  distribution of subsets of, 238
  for normal distribution, 343
  large-sample distribution of sums of, 269–271
  multidimensional, 239–243
  one-dimensional, 235
Covering theorem for probabilities, 14
Characteristic roots of a scatter matrix, 566
Characteristic vectors of a scatter matrix, 564–567
Craps, game of, 27
Critical region, definition, 395
  similar, 396
Critical set, 395
Cumulants, 115
Cumulative distribution function, absolutely continuous case, 36
  conditional, 60, 61, 65
  confidence band for continuous, 339–341
  definition of, the multidimensional case, 50
  definition of, the one-dimensional case, 33
  definition of, the two-dimensional case, 41
  determination of, from characteristic functions, 116–120
  empirical, 336
  marginal, 42, 50–51
  of a continuous random variable, 36, 45, 52
  of a degenerate random variable, 36, 44, 52
  of a discrete random variable, 34, 43–44, 51–52
  of independent random variables, 42–43, 51
  of mixed random variables, 47, 53
  of vector (multidimensional) random variables, 41, 50
  reproductivity of, 121
Cylinder set, Borel, 17, 97
  definition of, 97

Decision functions, *see* Statistical decision functions
Decision space, 503
Degenerate, cumulative distribution function, 36, 44, 52
  random variable, 36, 44, 52
Degrees of freedom, of chi-square distribution, 183
  of the Snedecor distribution, 186
  of the Student distribution, 184
DeMoivre-Laplace theorem, on convergence of binomial distribution to normal distribution, 257
  multidimensional version, 259
Difference of means, confidence intervals for, in normal distributions with equal variances, 324
Differentiation of parametric distribution functions, 345–346, 348–349
Dirichlet distribution, conditional random variables in, 180

Dirichlet distribution, covariance matrix of, 179
definition of, 177–178
distribution of sums of random variables having, 180–181
examples of, 191, 237–238
marginal distributions from, 179, 180–181
moments of, 179
ordered, 182
probability density function of, 177
sums of random variables having, 180–181
vector of means of, 179
Discrete random variable, definition of, 34, 44, 52
mass points of, 34, 44, 52
probability function of, 35, 44, 50
Discriminant analysis, distribution of eigenvalues in, 581–587
for case of several samples, 576–581
for case of two samples, 573–576
Distance between samples, Mahalanobis', 560
Matusita's, 252
Distribution, *see* Cumulative distribution function, Random variable
Double sampling, 472–473

Edgeworth's theorem on approximation to distribution of sample sum (or mean), 262–266
Efficiency of an estimator, 351, 363, 378
Efficient estimator, definition of, 351, 352
for multidimensional case, 378
Eigenvalues, distribution of, associated with one scatter matrix, 568–573
distribution of, associated with two scatter matrices, 581–587
distribution of, in discriminant analysis, 581–587
of a multivariate probability distribution, 598
of a scatter matrix, 566
Eigenvectors, associated with a pair of scatter matrices, 581–587
in discriminant analysis, 574, 580
Eigenvectors, of a multivariate probability distribution, 598
of a scatter matrix, 567
Ellipsoidal estimator, *see* Confidence region
Empirical cumulative distribution function, 337
Empty block test for two-sample problem, 446–452
Empty cell test for one-sample problem, 433–438
Empty set, 3
Equality of variances, test for, in samples from normal distributions, 423, 425
Equivalent random variables, 57
Estimating function, definition of, 372–374
likelihood, 371
multidimensional, 385
regular, 373, 385
Estimator, asymptotic efficiency of, 363, 380
bias of, 354
Blackwell-Rao theorem for improving, 357
consistent, 351, 376
efficient, 351, 378
for covariance function in time series, 516
for mean of normal distribution having preassigned variance, 501
for parameter of Poisson distribution, 354
for residual variance in linear regression, 286–287
interval, *see* Confidence interval
linear, 277
linear unbiased, 277
lower bound for variance of, of a parameter, 353
minimum variance linear, 278
minimum variance linear estimator for population mean, 279–280
minimum variance quadratic, for population variance, 280–281
multidimensional, 376
point, 344, 350, 376
quadratic, 278
region, *see* Confidence region

Estimator, sufficient, 351, 356
unbiased in the mean, 350, 376
unbiased in the median, 350
variance, 199, 218
Expectation, *see* Mean value
Events, *see also* Sets
definition of, 2
disjoint, 3
independent, 24
intersection of (product of), 3
joint occurrence of, 5
mutually disjoint, 4
mutually independent, 24
probability of, 11
sequence of, 3
union of (sum of), 4
Event point, 1
Experiment continuation events in sequential analysis, 475
Experimental design, complete two-factor, in Model I analysis of variance, 297–301
complete two-factor, in Model II analysis of variance, 308–310
complete three-way, in Model I analysis of variance, 301–304
complete three-way, in Model II analysis of variance, 311–312
the balanced incomplete two-way, in Model II analysis of variance, 310–311
Latin square, in Model I analysis of variance, 304–305
Latin square, in Model II analysis of variance, 312–313
likelihood ratio test for zero main effects in two-way, 426
replicated one-factor, 325–326
replicated two-factor, 326
Exponential distribution, Cartesian sequential test for, 498
distribution of sample median in large samples from, 275
maximum likelihood estimator for parameter in, 390
order statistics from an, 249
sufficiency of mean as estimator for parameter in, 390
testing simple hypothesis for, 422
Factorials, Stirling's formula for large, 175–177
Fiducial interval, 370
Fiducial probability, 370
Finite population, confidence interval for median of, 343
confidence intervals for quantiles in, 333
covariance matrix of, 222
covariance matrix of a sample from, 217
covariance matrix of vector of sample means in samples from, 222
distribution of a pair of order statistics in sample from, 252
distribution of largest element in a sample from, 251
distribution of median of samples from, 251
limiting distribution of sample means in large samples from a large, 268
mean of, 217
mean of sample covariance matrix in sample from, 222
mean of sample mean in samples from, 218
mean of sample variance in samples from, 218
mean of symmetric functions of samples from, 219–221
minimum variance linear estimator for mean of, 280
minimum variance quadratic estimator for variance of, 280–281
order statistics from, 243–245
probability function of a sample from, 216
random sampling from, 214–222
variance of, 217
variance of difference of means of two samples from, 246
variance of linear function of sample elements in sample from, 246
variance of sample mean in samples from, 218
variance of sample sum in samples from, 219
Fisher's k-statistics, 200–201
Fractiles of a random variable, 37

Gamma distribution, characteristic function of, 171
definition of, 171
distribution of sample sum (mean) in samples from, 207
independence of sample mean and certain other functions of samples from, 249
logarithmic transformation for large samples from, 391
maximum likelihood estimator for parameter of, 391
mean of, 171
moments of, 171
relation between beta distribution and, 174
renewal process based on, 190
reproductivity of, 171
variance of, 171
Gamma function, 170
incomplete, 171
Gaussian distribution, *see* Normal distribution
Gauss-Markov theorem, and weighing problems, 286
on estimators for regression coefficients, 285
Generalized distance, between a sample and a population, 557
between two samples, 560
Generalized Student distribution, *see* Hotelling's T^2
Generalized variance of a multidimensional distribution, 547
Generating function, card-matching, 154
factorial-moment, 114, 119
moment, 114, 119
probability, 114, 130–132
Goodness-of-fit criterion, Pearson's, 262, 388–389

Hankel's integral, 118
Hotelling's T^2, distribution for one-sample case, 556–559
distribution for two-sample case, 559–560
equivalence with a likelihood ratio test, 594
equivalence with Mahalanobis' D^2, 557, 560
Hotelling's T^2, noncentral, 601
Hypergeometric distribution, confidence interval for parameter in, 369
covariance matrix of, in k-variate case, 136
definition of, the one-variate case, 134
definition of, the k-variate case, 136
factorial moments of, 135, 136, 150
mean of, 135, 136
variance of, 135, 136
Hypergeometric waiting-time distribution, definition, 141
examples of, 28, 153, 480
factorial moments of, 143
mean of, 143
variance of, 143

Interval estimator, *see* Confidence interval
Incomplete beta function, 174
Incomplete gamma function, 171
Independence, nonparametric test for, 467–468
of events, 24
of random variables, 42, 51
of two sets of random variables, test for, 598–599
test for, by rank correlation, 467
test for, in contingency table, 423, 424
Infinitely divisible random variable, 189
Information integral, 409
Information matrix, 418
Internal scatter matrix of a sample, 543
Internal scatter of a sample, 543, 546
Integral, evaluating, by random sampling, 328
Lebesgue-Stieltjes, 22
Inverse of a covariance matrix, 80
Inverse sine transformation for large samples from a binomial distribution, 274, 391

Jacobian of a transformation, 57, 59

k–dimensional content, 546
k-sample problem, for binomial distributions, 423
for normal distributions, 425
for Poisson distributions, 425
k-statistics, 200–201

Khintchine's theorem on convergence in probability of sample mean, 254
Kolmogorov inequality, 107, 111
Koopman-Pitman theorem on distributions admitting sufficient statistics, 393
Kurtosis of a distribution, 265

Large numbers, strong law of, 108
 weak law of, 99, 255
Large samples, asymptotic distribution of order statistics in, 268–274
 asymptotic distribution of sample median in, 273
 asymptotic distribution of means in, from large finite populations, 268
 asymptotic distribution of sums of coverages in, 269–271
 asymptotic expansion of distribution of sample sum (or mean) in, 262–266
 asymptotic joint distribution of several order statistics in, 274
 asymptotic normality of distribution of maximum likelihood estimators in, 360–362, 380
 asymptotic normality of distribution of sample sums (or means) in, 256
 asymptotic normality of distribution of score in, 358, 379
 asymptotic normality of distribution of vector of sample means in, 258–259
 asymptotic normality of Student distribution in, 189, 275
 distribution of quadratic form in sample means in, 261
 inverse sine transformation for, from binomial distribution, 274, 391
 limiting distribution of coverage on sample range in, 275
 limiting distributions of functions of sample means in, 259–261
 limiting distribution of sample mean and median in, from a normal distribution, 275
 limiting form of multinomial distribution in, 262
 logarithmic transformation for, from a rectangular distribution, 275
Large samples, logarithmic transformation for, from waiting-time distribution, 274
 square root transformation for, from Poisson distribution, 274, 365
Largest element in a sample, distribution of, in samples from a finite population, 251
 distribution of, in samples from a rectangular distribution, 248
 inequality for mean value of, 250
 probability element of, 237
Largest segment, distribution of, generated by n points on an interval, 253
Latent roots of a scatter matrix, *see* Eigenvalues
Latent vectors of a scatter matrix, *see* Eigenvectors
Latin square, definition of, 233
 in experimental designs, 304–305, 312–313
Layer (treatment) effects, in Latin square experimental designs, 304
 in three-way experimental designs, 302
Least squares, *see also* Normal regression theory
 linear regression function, 87
 residual variance, 88, 91
Lebesgue-Stieltjes integral, 22
Legendre's duplication formula for gamma functions, 175
Level of significance, 395
Lévy-Cramér theorem, 122
Lévy's theorem, for one random variable, 116
 for a vector random variable, 119
Likelihood estimating function, 371
Likelihood, definition of, 351
 element, 351
Likelihood ratio, definition of, 403–404
 large-sample distribution of, for composite hypotheses, 419–420
 large-sample distribution of, for simple hypotheses, 410–411
Likelihood ratio test, *see also* Binomial distribution, Contingency table, Exponential distribution, Multinomial

distribution, Normal distribution, Poisson distribution
Likelihood ratio test, asymptotic power of, 413–417
consistency of, 411–413
definition of, 403–404
equivalence of Hotelling's T^2 with a, 594
for choice behavior model, 427
for the general linear statistical hypothesis, 405–408
for two-way experimental design, 426
in normal regression theory, 405–408
of a composite hypothesis, 419–422
of a simple hypothesis, 417–419
Linear dependence of random variables, 56, 58
Linear independence of random variables, 56, 58
Linear estimator, *see also* Minimum variance linear estimator
definition of, 277
unbiased, 278–279
Linear function of random variables, 191
distribution of, in case of normality, 158, 168–169
mean of, 82
variance of, 82
Linear functions of random variables, asymptotic distribution of, in large samples from large finite populations, 266–267
correlation between, 93, 94
covariance matrix of, 83
Linear prediction in time series, 535–537
Linear process in time series, 538
Linear regression estimators, for coefficients in, 283–286
for residual variance in, 286–287
Linear regression function, definition, 83, 85
in bivariate normal distribution, 163
in multivariate normal distribution, 170
least squares, 87
Linear statistical hypothesis, general, 405
Logarithmic transformation for large samples, from a gamma distribution, 391
from a rectangular distribution, 275
from a waiting-time distribution, 274
Loss function, 503
Lower bound of variance of estimator, of parameter, 351–353, 392
of function of parameter, 390

Mahalanobis' D^2, definition of, for one sample, 557
definition of, for two samples, 560
relation between, and Hotelling's T^2, 557, 560
Mann-Whitney test, definition of, 460
consistency of, for testing difference of location parameters, 469
Marginal cumulative distribution function, 42, 51
Marginal probability density function, 46, 52
Marginal probability function, 44, 52
Marginal sample spaces, 17
Markov chains, 98
Markov's theorem on estimators for regression coefficients, 285
Matrix, covariance, 80
Gramian, 343, 546
positive definite, 81
scatter, 543
Matrix sample, balanced incomplete, 232–234
complete second-order, 222–225
complete third-order, 229–232
for finite populations, 226–228
Latin square, 233–234
Matrix sampling, components of sum of squares in, 225, 230, 233, 234
from finite populations, 226–228
from infinite populations, 223, 230, 232
mean values of components of sum of squares in, 225, 230, 233, 234
Maximum likelihood estimators, asymptotic efficiency of, 363, 380

Maximum likelihood estimators, asymptotic normality of, in large samples, 360–362, 380
convergence of, 359–360
definition of, 360
for parameter in exponential distribution, 390
for parameters in normal regression theory, 392
for parameter of gamma distribution, 391
for parameters of multinomial distribution, 392
functions of, having large sample variance independent of parameter, 364–365
large sample distribution of, for multidimensional case, 380
multidimensional case, 379–380
Mean deviation, 187
of sample, 250
efficiency of, as estimator for σ in a normal distribution, 391
Mean value of a random variable, 73–74
Mean of a sample, *see* Sample mean
Measure, *see* Probability measure
Median, asymptotic efficiency of sample, as estimator of mean of normal distribution, 364
confidence interval for, of a distribution, 331, 342
confidence interval for, of finite population, 343
confidence interval for, of a second sample from order statistics of a first sample, 342
distribution of sample, in samples from a finite population, 251
distribution of sample, in samples from a continuous c.d.f., 237
efficiency of sample, as estimator for location parameter of Cauchy distribution, 391
Minimal Borel field, 11
Minimal complete class of statistical decision functions, 504
Minimax risk, 505
Minimax solution of statistical decision problem, 505
Minimum variance estimators, for end points of rectangular distribution, 390
for mean values of functions of sufficient estimators, 393
for range of rectangular distribution, 390
Minimum variance linear estimator, definition of, 278
of mean of several random variables having equal means, variances and covariances, 323
for difference of two population means, 323
for population mean, 280
from several unbiased estimators of a parameter, 323, 327
Minimum variance quadratic estimator for population variance, 280–281
Moment-generating function, definition of, 114, 119
factorial, 114, 119
of a vector random variable, 119
Moment-sequence, determination of distributions from a, examples, 181, 245, 532, 553, 562–564, 596–601
uniqueness of a distribution with a given, 125–129
Moments, 75, 79
absolute, 76, 79
absolute central, 76, 79
central, 76, 79
factorial, 76, 79
of a sample, 245
of a scatter, 545, 547, 552–554
Sheppard's corrections for, 328
Monotone sequence of sets, definition of, 8
probability law for, 13
Multinomial distribution, asymptotic normality of, for large samples, 262
characteristic function of, 139
confidence region for parameters of a, 388–389

Multinomial distribution, covariance matrix of estimators for parameters in, 139
definition of, 139
distribution of vector of sample sums (or means) in samples from, 206
likelihood ratio test for a simple hypothesis concerning a, 425
marginal distribution from, 151
Matusita's inequality for two samples from a, 252
maximum likelihood estimators for parameters of, 392
Pearson's goodness-of-fit criterion for samples from, 262, 425
reproductivity of, 139
test of simple hypothesis for, 425
Multidimensional analysis of variance, general Model I, 561–564
Model I, for three samples, 564
Model I, for two-factor experimental design, 596
Multidimensional distribution, *see* Cumulative distribution function, Random variable
Multidimensional estimator, *see* Estimator
Multiple comparisons, *see* Simultaneous confidence intervals
Multiple correlation coefficient, definition of, 91
sampling distribution of, 596–597
Multiple correlation ratio, 86
Multiple sampling, 473–474
Multivariate cumulative distribution function, 50
Multivariate normal distribution, characteristic function of, 168
conditional distributions from a, 169–170, 191
covariance matrix of a, 164
definition of, 164
distribution of exponent in, 184
distribution of Hotelling's T^2 in samples from a, 556–561
distribution of scatter in samples from a, 554
distribution of vector of sample sums (or means) in samples from, 207
Multivariate normal distribution, independence of means and scatter matrix in samples from a, 555–556
Lagrange's transformation for, 164–165
marginal distributions from, 168
maximum likelihood estimators for parameters in, 392
moments of a scatter of a sample from a, 552–554
probability density function of, 164
regression functions in, 169–170
reproductivity of, 190
spherical, 190
sphericity test for, 599
test for symmetry of, in the variables, 600
vector of means of, 164
Wishart distribution of variances and covariances in a sample from a, 551
Multivariate statistical analysis, 540
Mutually disjoint events, 4

Negative binomial distribution, *see* Binomial waiting-time distribution
Neyman-Pearson theorem for most powerful test, 398–399
Noncentral chi-square distribution, 247
Noncentral Student distribution, 247
Noncentral Hotelling T^2 distribution, 601
Nonparametric composite statistical hypothesis, 429
Nonparametric sequential test, Cartesian sequential test as a, 482
Nonparametric simple statistical hypothesis, 430
Nonparametric statistical tests, chi-square test, 431
confidence contours as, 438–441
one-sample empty cell test, 433–435
the Mann-Whitney test, 459–462
the Smirnov test, 454–459
two-sample empty block test, 446–452
two-sample run test, 452–454
Nonparametric test for independence, 467–468

Normal distribution, *see also* Bivariate normal distribution, Multivariate normal distribution
a problem of order statistics from a, 253
asymptotic distribution of median of large samples from a, 273
asymptotic efficiency of median of large sample from, 364
asymptotic normality of mean and median of large samples from a, 275
bivariate, 158–163
characteristic function of, 157, 161, 168
correlation coefficient between a random variable having a, and its c.d.f. transform, 246
Cochran's theorem for a sample from a, 212–214
confidence interval for mean of, 282
confidence region of parameters in, 383
confidence interval for variance of, 282–283
definition of, 156
distribution of Student *t* in samples from, 211
distribution of sum (or mean) of samples from, 206
distribution of variance of samples from, 208
distribution of sum of squares in samples from, 208
estimator for mean of, with preassigned variance, 501
independence of mean and variance in samples from, 208–11
inequality for integral of, 188
likelihood ratio test that mean of, has specified value, 404–405
likelihood ratio test that variance of, has specified value, 425
Lukacs' condition for a sample to be from a, 250
mean of, 156
multivariate, 163–170
probability density function of, 156, 161, 164
probability ratio sequential test for mean of, 499–500
Normal distribution, probability ratio sequential test for variance of, 500–501
reproductivity of, 158, 161, 190
standardized form of, 156
Stein's confidence interval of fixed length for mean of, 497–498
sufficient statistics for estimating parameters in, 390
uniformly most powerful test for mean of, 401
variance of, 156
Normal distributions, confidence interval for difference of means of, 324
distribution of difference of means of samples from two, 207
estimator for difference of means of, with preassigned variance, 501
likelihood ratio test for equality of means of several, with equal variances, 422–423
likelihood ratio test for equality of means of two, with equal variances, 422
likelihood ratio test for equality of variances of, 423, 425
problem of, *k*-samples from, 425
Normal noise, definition of, 533
test for whiteness of, 533–535
Normal regression theory, confidence intervals for regression coefficients in, 289
confidence regions for regression coefficients in, 290, 324
definition of, 288
estimators for regression coefficients in, 247, 288
in experimental designs, 297–305
likelihood ratio test in, 405–408
maximum likelihood estimators for parameters in, 392
test for parallel regression lines in, 426
Null hypothesis, 394

Observable random variable, 278
Operating characteristic function, of probability ratio sequential test, 487–489
of a sampling plan, 398
of a sequential test, 476

Operating characteristic function, of a statistical test, 395
Optimum stratified sample, 317
Ordered Dirichlet distribution, 182
Order statistics, asymptotic distribution of, in large samples, 268–274
  coverages determined by, 235, 239–240
  definition of, 234
  in samples from finite populations, 243–245
  in two samples, 442–446
  for a probability density function symmetric in the variables, 70
  multidimensional coverages generated by, 238–243
  one-dimensional coverages generated by, 237–238
  ordering functions in theory of, 238
  probability element of, 236
  sample blocks generated by, 235
  two-sample problems concerning, 469, 470

Parallel regression lines, test for, in normal regression theory, 426
Parameter, *see* Population parameter
Parameter space, 344
Parametric cumulative distribution function, differentiation of, 345, 348
  regular, 348–350
Partial correlation coefficient, 94
Pascal distribution, *see* Binomial waiting-time distribution
Pearson's goodness-of-fit criterion, 262, 388–389
Pearson Type III distribution, *see* Gamma distribution
Percentile, 37
Periodogram, analysis, 529–533
  definition of, 530
  of stationary time series, 538–539
Pivotal point of the scatter of a sample, 541
Pivotal function for determining fiducial distributions, 370
Point estimator, 344, 350, 376
Poisson distribution, asymptotically shortest confidence interval for parameter in, 391
Poisson distribution, characteristic function of, 140
  confidence interval for mean of, 369
  cumulative distribution function of, as a definite integral, 152
  definition of, 140
  distribution of sum (or mean) of samples from, 206
  efficiency of estimator for parameter of, 354
  estimator for parameter of, 354
  mean of, 140
  probability ratio sequential test for mean of, 501
  problem of k samples from, 425
  reproductivity of, 140
  square root transformation for large samples from, 274, 365
  sufficiency of estimator of parameter in, 356–357
  variance of, 140
Poisson distributions, likelihood ratio test for equality of parameters in two, 424–425
Poisson process, 192
Population parameter, definition of, 344
  lower bound of estimator for, 353
  score of a, 353
Positive definite matrix, 81
Power of a statistical test, 395
Principal components of a scatter matrix, 568
Probability density function, of a conditional random variable, 64, 66, 68
  of a random variable, 37
  of a vector random variable, 46, 52
  marginal, 46, 52
Probability element, of a random variable, 37
  of a vector random variable, 46
  transformation (change of variable) of, 55, 57, 59
Probability function, of a conditional random variable, 62, 66, 68
  marginal, 44, 52
  of a random variable, 35
  of a vector random variable, 43, 52

Probability-generating function, 114
examples of, 130–132
Probability measure, covering theorem for, 14
definition of, 11
extension of, 15
for stochastic process, 96–99
Probability of occurrence, of all n events (of intersection of events), 28
of at least one of n events (of union of events), 12, 28
of m or more of n events, 28, 29
of exactly m of n events, 28
Probability ratio sequential test, average sample number of, 489–490
boundary constants for, 485
definition of, 482
efficiency of, 490–492
for binomial distribution, 494–496
for mean of a normal distribution, 499–500
for mean of Poisson distribution, 501
for variance of a normal distribution, 500–501
operating characteristic function of, 487–489
termination of, with probability one, 483–484
truncation of, 492–494
Probability spaces, component, 19
definition of, 11
statistical independence of, 19
Problem of k samples, from binomial distributions, 423
from normal distributions, 425
from Poisson distributions, 425
Producer's risk, 397
Product space, 16
Propagation of errors, 190
Proper linear dependence of random variables, 56, 58
Proportional stratified sample, 316

Quadratic estimator, 278
Quantile, definition, 37
test, 428–430
Quantile intervals, confidence intervals for, 332–333
Quantiles, confidence intervals for, 329–332
confidence intervals for, in finite populations, 333
Quartile, lower, 38
upper, 38
Queuing, 193

Radon-Nikodym theorem, 25
Random function, 514
Random intervals, mean of union of, 250
Randomization tests, Fisher-Pitman test, 464–465
for sample components, 462–465
for sample ranks, 465–468
for two-way experimental design, 465
Hoeffding's test of independence, 467–468
Mann-Whitney test, 459–462
rank correlation test, 467
Random sampling, from finite population, 214–222
from infinite population, 195–198
unbiased estimator of definite integral by, 328
Random set, 367
Random variable, absolute central moments of, 76, 79
absolute moments of, 76, 79
bounded, 20
central moments of, 76, 79
coefficient of variation of, 74
conditional, 25, 61, 66
continuous, 36, 45, 52
cumulants of, 115
definition of, 19, 20
degenerate, 36, 44, 52
discrete, 34, 43, 51
factorial moments of, 76, 79
integration of, 21–24
mean of, 73–74
measurable function of, 53
moments of, 75, 79
multidimensional (vector), 19, 20
observable, 278
sample space of, 19
semi-invariants of, 115
simple, 21
standard deviation of, 74

Random variable, variance of, 74
vector (multidimensional), 19
Random variables, conditional, 61–66
correlation coefficient between two, 78
covariance between two, 78
distribution of product of two, 69
distribution of ratio of two, 69
distribution of sum of two, 69
equivalent, 57
measurable functions of, 53
independent, 42, 51
linearly independent, 56, 58
linearly dependent, 56, 58
mixed, 47, 53
mutually independent, 51
properties of means of, 73-74
uncorrelated, 78
Range, of a random variable, 34
of a rectangular distribution, 155
of a sample, 237
Rank correlation test for independence, 467
Rao-Kendall theorem on moment-sequences, 128–129
Rectangular distribution, confidence interval for range of, 390
definition of, 155
distribution of geometric mean of a sample from a, 249
distribution of largest element of a sample from a, 248
distribution of largest gap in a sample from, 253
distribution of mean (or sum) of samples from, 204
distribution of median of samples from a, 248
distribution of order statistics in sample from a, 235
distribution of product of elements of sample from, 189
distribution of range of samples from, 248
efficient estimator for range of, 390
logarithmic transformation for large samples from a, 275
mean of product of largest elements in several samples from a, distribution of, 249
Rectangular distribution, minimum variance estimators of range and midpoint of, 390
order statistics of a sample from a, 235
range of, 155
sufficient estimator for endpoints of, 390
sufficient estimator for range of, 390
Regression coefficients, confidence intervals for, in normal regression theory, 289
confidence region for, in normal regression theory, 290
definition of, 83, 85
estimators for, 246–247, 283–286
Gauss-Markov theorem on estimators for, 285
Markov theorem on estimators for, 284
Regression function, definition of, 83, 85
in bivariate normal distribution, 163
in multivariate normal distribution, 170
least squares, 87
linear, 83, 85
Regular parametric distribution functions, 348–350
Renewal process, 190–191
Reproductivity of a distribution, 121
Residual variance, definition of, 84
estimator for, in linear regression, 286–287
least squares, 88, 91
Response surface analysis, 297
Risk function, 503
Risk vectors, 504
Row effects, in Latin square experimental designs, 304
in two-way experimental designs, 298
in three-way experimental designs, 302
Run test, for two-sample problem, 452–454
Runs, definition of, 145
distribution functions of, 146–147
factorial moments of, 148
mean of, 148

Runs, of successes, 153
variance, 148

Sample, covariance matrix, 197
definition of, from finite population, 215
definition of, from infinite population, 195
Fisher's *k*-statistics of a, 200–201
matrix, *see* Matrix sample
mean deviation of, 250
mean of, 196
moments of a, 245
semi-invariants of a, 200–201
symmetric functions of a, 201–203
variance of a, 199
Sample blocks, 235, 238–239
Sample cluster, 541, 545
Sample mean, as minimum variance linear unbiased estimation for population mean, 279–280
asymptotic expansion of distribution of, in large samples, 262–266
asymptotic normality of, in large samples, 256
asymptotic normality of functions of, in large samples, 259–260
characteristic function of, 205
convergence in probability of, 254–255
definition of, 196
distribution of, in large samples from large finite populations, 266–268
distribution of, in samples from a Cauchy distribution, 130
distribution of, in samples from a normal distribution, 206
efficiency of an unbiased linear estimator of a population mean relative to the, 327
geometric, distribution of, in samples from a rectangular distribution, 249
mean of, in samples from an infinite population, 198
mean of, in samples from a finite population, 218
variance of, in samples from an infinite population, 198
variance of, in samples from a finite population, 218
Sample means, asymptotic distribution of Studentized difference of, in large samples, 275
asymptotic normality of vector of, in large samples, 258–259
distribution of difference of, in samples from two normal distributions, 207
distribution of vector of, in samples from multivariate normal distributions, 207
limiting distribution of quadratic form in, for large samples, 261
probability inequalities for vector of, 274
variance of difference of two, from finite population, 246
vector of, 197
Sample median, asymptotic distribution of c.d.f. transform of, 275
asymptotic distribution of, in large samples, 273
asymptotic efficiency of, in large samples from normal distribution, 364
definition of, 237
distribution of, in large samples from exponential distribution, 275
distribution of, in samples from a rectangular distribution, 248
efficiency of, for estimating center of Cauchy distribution, 391
probability element of, 237
Sample moments, variance of, 245
Sample point, 1
Sample range, cumulative distribution function of, in samples from a continuous c.d.f., 248
distribution of, in samples from rectangular distribution, 248
limiting distribution of coverage on, in large samples, 275
mean value of, in samples from a continuous c.d.f., 248
probability element of, 237
Sample space, Cartesian product, 17
definition of, 1
marginal, 17
of a random variable, 19

Sample sum, asymptotic expansion of distribution of, in large samples, 262–266
asymptotic normality of, in large samples, 256
characteristic function of, 205
definition of, 196
distribution of, in samples from a binomial distribution, 206
distribution of, in samples from a gamma distribution, 207
distribution of, in samples from a normal distribution, 206
distribution of, in samples from a Poisson distribution, 206
distribution of, in samples from a rectangular distribution, 204
general distribution of, 203, 205
Sample sums, distribution of vector of, in samples from a multinomial distribution, 206
distribution of vector of, in samples from a multivariate normal distribution, 207
vector of, 197
Sample variance, definition of, 196
distribution of, in samples from a normal distribution, 208
mean of, in samples from an infinite population, 199
mean of, in samples from a finite population, 218
variance of, 200
Sampling without replacement, *see* Sampling from a finite population
Scatter of a multidimensional distribution, 547
Scatter of a sample, distribution of, in samples from a multivariate normal distribution, 554
internal, 543
mean value of a, 545, 547
moments of, in samples from a multivariate normal distribution, 552–554
one-dimensional, 541
two-dimensional, 543
multidimensional, 546
pivotal point of, 541, 543, 545
Scatter matrices, distribution of eigenvalues associated with a pair of, 581–587
Scatter matrix of a sample, between-sample, 564, 574
characteristic roots of, 566
characteristic vectors of, 567
definition of, 543
distribution of eigenvectors of, 567
eigenvalues of, 566
eigenvectors of, 567
internal, of a sample, 543, 546
latent roots of a, 566
latent vectors of a, 567
principal components of a, 568
within-sample, 559, 564, 574
Scatter of residuals, sampling distribution of, 597–598
Score of a parameter, asymptotic normality of, in large samples, 358, 379–380
definition of, 353
Semi-invariants, of a random variable, 115
of a sample, 200–201
Sequence of events, *see* Sequence of sets
Sequence of sets (events), 3
contracting (decreasing), 7
countably infinite, 3
expanding (increasing), 7
inferior limit of, 7
limit of, 7
monotone, 8
superior limit of, 7
Sequential analysis, *see* Sequential estimation, Sequential process, Sequential test
Sequential estimation, 496–497
Sequential process, Cartesian, 479, 498
definition of, 475
experiment continuation events in, 475
probability ratio, 482
termination of probability ratio, with probability one, 483–484
Sequential test, *see also* Cartesian sequential test, Probability ratio sequential test
average sample number of, 476
average sample number in probability ratio, 489–490

Sequential test, boundary constants for probability ratio, 485
Cartesian, 479–481, 498
Cartesian, as a nonparametric sequential test, 482
criteria for choosing a, 477–479
degenerate, 475
efficiency of probability ratio, 490–492
operating characteristic function of, 476
operating characteristic function of probability ratio, 487–489
optimum Cartesian, 481, 499
probability ratio, for binomial distribution, 494–496
probability ratio, for mean of normal distribution, 499–500
probability ratio, for mean of Poisson distribution, 501
probability ratio, for variance of a normal distribution, 500–501
probability ratio, general, 482
r-fold Cartesian, 481–482
strength of a, 478
structure of a, 474–479
truncation of probability ratio, 492–494
zones of acceptance, indifference and rejection for a, 478
Serial correlation coefficient, 533–535
Serial correlation function of time series, 516
Set function, completely additive, 11
definition of, 11
Sets, associative law for, 6
Boolean field (Boolean algebra) of, 8
Borel field (sigma-algebra) of, 8
bounded, 31
Cartesian product of, 16
commutative law for, 6
complement of, 4
completely additive class of, 8
difference of, 4
disjoint, 3
distributive law for, 6
empty (null), 3
equal, 3
fields of, 8
finitely additive class of, 8
Sets, intersection of (product of), 3
mutually disjoint, 4
outer measure of, 16
random, 367
union of (sum of), 4
Sheppard's corrections for moments, 328
Sign test, 430
Significance level, 395
Simple random sampling, from a finite population, 215
from an infinite population, 195
Simple statistical hypothesis, definition, 395
nonparametric, 430
Simultaneous confidence intervals, a probability inequality for, 291
definition of, 290
for differences of interactions in three-way experimental design, 327
for differences of main effects in two-way experimental design, 326
for differences of means of several normal distributions, 326
for linear combinations of regression coefficients, 325
for linear combinations of several normal variables, 325
Scheffé's method, 291–294
Tukey's method, 294–297
Skewness of a distribution, 265
Slutzky's theorem in time series, 537–538
Smallest element, probability element of, in sample, 237
Smirnov test for two samples, 454–459
Snedecor (variance-ratio) distribution, chi-square distribution as a limiting form of, 191
definition of, 186
degrees of freedom of, 186
in Model I analysis of variance, 407
mean of, 187
moments of, 187
probability density function of, 186
relation between beta distribution and, 187
variance of, 187
Spectral density function, definition of, 520

Spectral density function, of a geometrically decreasing covariance function, 539
of a linear process, 538
of smoothed time series, 537–538
Spectral distribution function, definition of, 520
estimation of, 523
Spectral mass, 520
Spectral representation of covariance function, 517, 520
Square root transformation, for large samples from gamma distribution, 274
for large samples from Poisson distribution, 274, 365
Standard deviation of a random variable, 74
Statistic, *see also* Estimator
consistency of a, 351, 376
definition of, 196
efficiency of a, 351, 378
sufficiency of a, 351
Statistical decision functions, admissible, 504
complete class of, 504
definition of, 502
minimal class of, 504
Statistical decision problem, Bayes' solution of, 508–511
minimax solution of, 505, 504–508
Statistical hypothesis, composite, 394–395
definition of, 394–395
general linear, 405
simple, 394–395
Statistical independence, of probability spaces, 19
of random variables, 42, 51
Statistical test, *see also* Nonparametric statistical tests, Probability ratio tests, and Sequential tests
consistent, 396
critical region of, 395
definition of, 395
for whiteness of normal noise, 533–535
operating characteristic function of, 395
power of, 395
Statistical test, size of a, 396
unbiased, 395
uniformly most powerful, 396
Stirling's formula for large factorials, 175–177
Stochastic convergence, 99
Stochastic process, birth, 192
branching, 131–132
definition of, 96
finite, 68
linear process, 538
Markov, 98
Poisson, 192
queuing, 193
random sampling, 195
sequential process, 475
stationary, 515
strictly stationary, 515
weakly stationary, 516
Stratified population, definition of, 313
estimator for mean of, from general stratified sample, 315
estimator for mean of, from optimum stratified sample, 317
estimator for mean of, from proportional stratified sample, 316
estimator for mean of, by two-stage sampling, 320
estimator from stratified sample for mean of, with minimum variance at fixed cost, 318
estimator from two-stage sample for mean of, with minimum variance at fixed cost, 322–323
two-stage sampling of, with unknown strata sizes, means and variances, 327
two-stage sampling of, having many strata, 318–321
Stratified sample, definition of, 314
estimator of population mean from general, 315
estimator of population mean from optimum, 317
estimator of population mean from proportional, 316
Strong law of large numbers, 108
Student distribution, asymptotic normality of, in large samples, 189, 275

Student distribution, definition of, 184
  degrees of freedom in, 184
  for a sample from a normal distribution, 211
  for two samples, 245
  mean of, 185
  moments of, 185
  noncentral, 247
  probability density function of, 184
  variance of, 185
Studentized range, 294
Student ratio, as a likelihood ratio test, 404–405
  definition of, 184, 211
  example of unbiased test, 396–397
Sufficient statistics, Blackwell-Rao theorem on, 357
  definition of, 351
  factorability criterion for, 354–356
  form of distribution admitting, 393
  multidimensional, 356
  Rao's theorem on, 393
Sum of squares, components of, in matrix sampling, 225, 230, 233, 234
  distribution of, in samples from a normal distribution, 208
Symmetric functions, mean of, in samples from a finite population, 219–221
  minimum variance property of, for samples, 280–281
  of a sample, mean of, 201–203
  variance of, 202

Test, *see* Statistical test
Time series, auto-covariance, 516
  autoregressive, 537
  covariance function of, 516
  definition of, 514
  estimator for covariance function of, 522
  estimator for mean of, 522
  estimator for spectral distribution of, 523
  estimators for coefficients in trigonometric, 528–529
  Fisher's test for periods in, 529–533
  lag covariance, 516
  linear prediction in, 535–537
  linear process in, 538
Time series, periodogram analysis of, 529–533
  serial correlation function of, 516
  Slutzky's theorem, 537–538
  smoothed, 537–538
  spectral density function of, 520, 538
  spectral distribution function of, 517, 520
  stationary, 515
  strictly stationary, 515
  variate difference method in, 526–527
  weakly stationary, 516
  white noise, 521, 522
Tolerance intervals, distribution-free, 334, 342
  for the problem of two samples, 343
  for normal distribution, 343
Tolerance limits, distribution-free, 334
  for finite populations, 335
  for normal distribution, 343
Tolerance regions, 335, 343
Two-sample problem, distribution-free, 441–442
  empty block test for, 446–452
  for binomial distributions, 423
  for continuous distributions, 441–442
  for normal distributions with identical variances, 422
  for Poisson distributions, 425
  Hotelling's T^2 for, 559–560
  Mann-Whitney test for, 459–462
  Matusita's inequality for the, 252
  run test for, 452–454
  Smirnov test for, 454–459
  Student's ratio for the, 245
Two-stage sampling, confidence interval of fixed length of mean or normal distribution by, 497–498
  of finite populations with many strata, 318–323
  of populations with unknown strata sizes, means and variances, 327
  to minimize variance of estimator of finite population mean at fixed total cost, 318
  variance of fixed size of difference of means of two normal distributions by, 501

Two-stage sampling, variance of fixed size of mean of normal distribution by, 501
Type I error of a statistical test, 395
Type II error of a statistical test, 395

Unbiased estimator, 277
Unbiased linear estimator, 277–278
Unbiased quadratic estimator, 278–279
Uncorrelated random variables, 78
Uniformly most powerful statistical test, definition of, 396
for mean of normal distribution, 401

Variance, definition of, 74
of an estimator of a parameter, lower bound of, 353
of linear function of random variables, 82
of a sample, 196
unbiased estimator for, 199, 218
Variance components, for balanced incomplete two-way layout, 233
estimators for these components, 311
for complete two-way layout design, 223
estimators for these components, 309
for complete three-factor layout, 230
estimators for these components, 312
for Latin square layout design, 234
estimators for these components, 313
Variance-ratio distribution, *see* Snedecor distribution
Variate difference method for time series, 526–527
Venn diagram, 4, 5

Waiting-time distributions, binomial, 143–144
continuous, 172–173
hypergeometric, 142–143
logarithmic transformation for large samples from, 274
Wald-Blackwell theorem, generalization of, 498
on sum of random variables in sequential process, 476
Weak law of large numbers, 99, 255
Weighing problems, 286
White noise, 521
covariance function of a, 538
test for, 533–535
Wilcoxon two-sample test, *see* Mann-Whitney test
Wishart distribution, characteristic function of, 554
definition of, 551
derivation of, 547–552
reproductivity of, 554
Within-sample scatter matrix, 559, 564, 574
Within-strata component of variance, 313

Yule's birth process, 192